Advances in Intelligent and Soft Computing 137

Editor-in-Chief: J. Kacprzyk

Advances in Intelligent and Soft Computing

Editor-in-Chief

Prof. Janusz Kacprzyk
Systems Research Institute
Polish Academy of Sciences
ul. Newelska 6
01-447 Warsaw
Poland
E-mail: kacprzyk@ibspan.waw.pl

Further volumes of this series can be found on our homepage: springer.com

Vol. 121. Tomasz Traczyk and Mariusz Kaleta (Eds.)
Modeling Multi-commodity Trade: Information Exchange Methods, 2011
ISBN 978-3-642-25648-6

Vol. 122. Yinglin Wang and Tianrui Li (Eds.)
Foundations of Intelligent Systems, 2011
ISBN 978-3-642-25663-9

Vol. 123. Yinglin Wang and Tianrui Li (Eds.)
Knowledge Engineering and Management, 2011
ISBN 978-3-642-25660-8

Vol. 124. Yinglin Wang and Tianrui Li (Eds.)
Practical Applications of Intelligent Systems, 2011
ISBN 978-3-642-25657-8

Vol. 125. Tianbiao Zhang (Ed.)
Mechanical Engineering and Technology, 2012
ISBN 978-3-642-27328-5

Vol. 126. Khine Soe Thaung (Ed.)
Advanced Information Technology in Education, 2012
ISBN 978-3-642-25907-4

Vol. 127. Tianbiao Zhang (Ed.)
Instrumentation, Measurement, Circuits and Systems, 2012
ISBN 978-3-642-27333-9

Vol. 128. David Jin and Sally Lin (Eds.)
Advances in Multimedia, Software Engineering and Computing Vol.1, 2011
ISBN 978-3-642-25988-3

Vol. 129. David Jin and Sally Lin (Eds.)
Advances in Multimedia, Software Engineering and Computing Vol.2, 2011
ISBN 978-3-642-25985-2

Vol. 130. Kusum Deep, Atulya Nagar, Millie Pant, and Jagdish Chand Bansal (Eds.)
Proceedings of the International Conference on Soft Computing for Problem Solving (SOCPROS 2011) December 20–22, 2011, 2012
ISBN 978-81-322-0486-2

Vol. 131. Kusum Deep, Atulya Nagar, Millie Pant, and Jagdish Chand Bansal (Eds.)
Proceedings of the International Conference on Soft Computing for Problem Solving (SocProS 2011) December 20–22, 2011, 2012
ISBN 978-81-322-0490-9

Vol. 132. Suresh Chandra Satapathy, P.S. Avadhani, and Ajith Abraham (Eds.)
Proceedings of the International Conference on Information Systems Design and Intelligent Applications 2012 (India 2012) held in Visakhapatnam, India, January 2012, 2012
ISBN 978-3-642-27442-8

Vol. 133. Sabo Sambath and Egui Zhu (Eds.)
Frontiers in Computer Education, 2012
ISBN 978-3-642-27551-7

Vol. 134. Egui Zhu and Sabo Sambath (Eds.)
Information Technology and Agricultural Engineering, 2012
ISBN 978-3-642-27536-4

Vol. 135. Honghua Tan (Ed.)
Knowledge Discovery and Data Mining, 2012
ISBN 978-3-642-27707-8

Vol. 136. Honghua Tan (Ed.)
Technology for Education and Learning, 2012
ISBN 978-3-642-27710-8

Vol. 137. Jia Luo (Ed.)
Affective Computing and Intelligent Interaction, 2012
ISBN 978-3-642-27865-5

Jia Luo (Ed.)

Affective Computing and Intelligent Interaction

Springer

Editor
Jia Luo
National Kinmen Institute of Technology
Jinning Township
Kinmen
Taiwan

ISSN 1867-5662 e-ISSN 1867-5670
ISBN 978-3-642-27865-5 e-ISBN 978-3-642-27866-2
DOI 10.1007/978-3-642-27866-2
Springer Heidelberg New York Dordrecht London

Library of Congress Control Number: 2011946229

Printed on acid-free paper

Springer is part of Springer Science+Business Media (www.springer.com)

Preface

2012 International Conference on Affective Computing and Intelligent Interaction (ICACII 2012) was the most comprehensive conference focused on the various aspects of advances in Affective Computing and Intelligent Interaction. The conference provided a rare opportunity to bring together worldwide academic researchers and practitioners for exchanging the latest developments and applications in this field such as Intelligent Computing, Affective Computing, Machine Learning, Business Intelligence and HCI.

This volume is a collection of 119 papers selected from 410 submissions from universities and industries all over the world, based on their quality and relevancy to the conference. All of the papers have been peer-reviewed by selected experts.

Taiwan has been chosen as the hosting place of ICACII 2012, which lies like a spindle in the southeast of Asia. Geographically separated from Mainland China, this precious island is known as Typhoon Island from whence its name. It is a rare tropical mountain-island in the world, two thirds of which is covered with high mountain ranges. The most famous of the mountains, Ali Mountain, is the symbol of charming Taiwan. Located in the subtropical ocean, Taiwan offers tourists comfortable weather and has no winter all the year round. With the pleasant climate, Taiwan can sustain vast biology resources. It is worth to visit Taiwan for its natural sights, such as the high and steep mountains and the dense atmosphere of seashore holidays.

The island has another Portuguese name, Formosa, which means 'beautiful island'. In addition, Taiwan has many wonderful man-made tourist recourses, such as the famed Matsu Temple and Taipei National Palace Museum. During the past decades, Taiwan's economy emerged as a miracle. As of today, it is the primary source of foreign capital for South-East Asian countries, thus allowing them to provide complete and high-quality tourist services and facilities. These made our conference more impressive and memorable.

Finally, we would like to appreciate the organizers for providing the generous financial support, and are extremely grateful to editorial broad for their hard work which contributed towards the success of the proceedings. Last but not least, special thanks are given to all those authors who attend the conference and give important contributions to complete the proceedings.

Best wishes,

Jia Luo
Publication Chair

Contents

Comparison and Evaluation on Performance of Customer Collaboration Innovation Implementation

Zhang Xue[1,2] and Zhang Qing-pu[2]

[1] School of Management, Harbin University of Commerce Harbin, Heilongjiang Province, China
[2] School of Management, Harbin Institute of Technology Harbin, Heilongjiang Province, China
dou112@sina.com.cn, Lwtong2003@yahoo.com.cn

Abstract. Customer Collaboration Innovation has become an effective approach for enterprises' commodity innovation. According to the mechanism of extension evaluation theory, this study analyzed the characters of customer collaboration innovation as well as the critical impacting factors of its implementation and established the matter element model for performance evaluation. Through case study, the application of model was illustrated by qualitative and quantitative evaluating on performance of customer collaboration innovation so as to provide important references to the decision making of customer collaboration innovation.

Keywords: Customer collaboration innovation, Performance of collaboration innovation implementation, Matter element model, Extension evaluation.

1 Introduction

Since client-oriented strategy becomes the key for enterprise to obtain the market competitive strength, the communication between enterprise and customer presents to be rapid, efficient and collaborative. In 2002, professor Hippel (Massachusetts Institute of Technology) and his team came up with concepts of leader user and customer innovation by large-scale investigating on the enterprise innovation in the U.S. Afterwards, both academic and industry field has conducted a large amount of attempts. According to above situation, some scholars come up with an innovation model which could highly utilize the customer potential: enterprise-customer collaboration innovation (Customer Collaboration Innovation, CCI). Unlike customer innovation, the main sponsor of CCI is enterprise, rather than the customer itself. Under the requirement and instruction of enterprise, customer joins in the innovation system and the entire innovation process. Recently, studies on CCI mainly focus on system design techniques[1-3], simulation[4], customer collaboration mechanism in the commodity customization[5], commodity innovation model of CCI[6], partner choosing and staff distribution of CCI[7-9], fulfillment of customer collaboration platform, collaboration optimization between enterprise and customer[10], etc. Although above studies have effectively supplemented the theory of customer

J. Luo (Ed.): Affective Computing and Intelligent Interaction, AISC 137, pp. 1–9.
springerlink.com

collaboration innovation, few have focused on analyzing and evaluating the performance of customer collaboration innovation implementation. Until now, only the researcher at Chongqing University came up with evaluation methods by adopting BP neural network, whose operation, however, is complex. Based on existed studies, this study pointed out principal impacting factors for performance of customer collaboration innovation implementation, established the matter element model for evaluating innovation performance, and demonstrated the application and operation of model by case study. With matter element model, performance of customer collaboration innovation was analyzed both qualitatively and quantitatively, which provided new viewpoint and approach for analyzing and evaluating on the performance of customer collaboration innovation implementation.

2 Impacting Factor of Customer Collaboration Innovating Implementation

Presently, some enterprises, both in domestic and abroad, have started to implement customer collaboration innovation, such as printing industry, textile machinery, ceramics making, motor bicycle design, automobile manufacturing, etc. Therefore, what is the performance of customer collaboration innovation implementation and what model and approach should be adopted for its evaluation? This question is critical for enterprises to answer.

2.1 Characters of Customer Collaboration Innovation

1) Collaboration and Conflict

Collaboration is the most basic character of customer collaboration innovation. During the process of collaboration innovation between customer and enterprise, different participators, with different innovation task, collaborate with each other and make endeavor to achieve the same objectives. Meanwhile, since customers and participators from various department of enterprise are different in expertise knowledge, skills, experience and background, conflicts are inevitable on collaboration innovation.

2) Agility and Dynamic

With the advance of technology and the improvement of people's living standard, the customers' demands are changing continually and rapidly, which result in a corresponding acceleration of enterprises' commodity innovation design. Collaboration innovation with customers could reduce the times of revisions of commodity design, and come up with a satisfying product at one time. Meanwhile, the schedule, personnel distribution, resources allocation and conflicts occur dynamically. Thus, the collaboration innovation has the character of prompt and dynamic.

3) Coexistence of Synchronicity and Asynchronicity

In the process of collaboration innovation, commodity design at different level is completed by both customers and expert team, where on one hand each participator could thoroughly express his opinions or suggestions, on the other hand, opinions of others could be received on time. If the task need completing synchronically, a face-to-face collaboration and cooperation is presented, otherwise, asynchronic approaches, such as email and correspondence, could be adopted.

4) Openness and Restriction
Collaboration innovation allows customers to participate in the innovation design of commodities, and customers, with ample degrees of freedom, could adjust and check their own commodities at all times. However, it does not mean that customer could design commodity solely at their will, and restrictions exist in different level of design, such as principle restriction, economic restriction and timeliness restriction.

5) Stage and Repeatability
The work of collaboration innovation is gradually completed in many stages, thus has the character of stage. Meanwhile, collaboration innovation is hard to be completed at one time, which requires multi-round revisions to achieve optimum design, thus has the character of repeatability.

2.2 Impacting Factor Analysis of Customer Collaboration Innovation

In the light of characters of customer collaboration innovation and based on large amount of case analysis, this study comes up with six impacting factors that affect the performance of customer collaboration innovation implementation.

1) Economic Benefit
Economic benefit is the core for any economic activities, which is the comparison of costs and outcomes of economic activities. Serving as the important index to judge if the economic activity is feasible or not, economic benefit, in general, is expressed by subtracting costs from benefits. Enterprise, serving as the cell of any society, aims at gaining benefits. Without economic benefits, enterprise could not expand reproduction and achieve farther development. Whatever styles that innovation activities adopt, the expected economic benefits must be clearly calculated and considered. Specific economic benefits that should be considered in customer collaboration innovation include: sales income of new commodity, market shares, sales rate, the motivating effects on other commodities, etc.

2) Social Benefit
In general, social benefit is focusing on the degree of satisfying social demands. The more social demands are satisfied, the higher social benefits that economic activity has. Otherwise, the social benefit of economic activities is low or even null. With the improving of social civilization, people become to emphasize social benefit and make effort to achieve its maximum value. To check if the optimum social benefit could be achieved has become the starting point for any activities. Social benefit is the prerequisite of economic benefit, without which economic benefit loses its practical foundation. Being the cell of society, enterprise, in addition to economic benefits, must respect social benefit. Specific social benefits that should be considered in customer collaboration innovation include: the quality and quantity of new commodities that better satisfy social demands, impacts on social employment, pollution loads exerted on environmental system, etc.

3) Technology Benefit

Technology benefit means the spirit right, economic right and other right obtained by technology, which in practice mainly refers to the rights of translation and use of non-patent technology and the economic benefit afterword gained. In customer collaboration innovation, technology benefits mainly include: the motivation impacts of innovation project on technology development, new technology competiveness gained in the process of innovation, and the amount of patents generated in innovation.

4) Working Efficiency

Working efficiency, to be simplelized, is the ratio of working performance to the time consumed. It is a comprehensive index for evaluating the state of working activities, in that it not only involve individual performance, achievement and career development, but entire social productivity and economic growth. Thus, working efficiency is the objective commonly emphasized and pursued by individuality, manager, organization and enterprise. In customer collaboration innovation, working efficiency refers to the ration of effective innovation generation to the time consumed, and the average amount of innovations generated by each customer.

5) Synergistic Effect

Synergistic effect refers to the entirety effects achieved by different sections, stages and aspects of enterprises' production, sales, and management that take advantage of the same kind of resources. In customer collaboration innovation, the performance of synergistic effect is mainly demonstrated by degree of effective communication between enterprises and customers, degree of mutual respects, degree of satisfaction to each other's work, and the degree of coordination of conflict and problem solution.

6) Process Evaluation

A complete innovation process must be comprised of a series of ordered activities, by which the collaboration innovation system could operate more effectively and each participator could work more efficiently. In general, the evaluations on the process of customer collaboration innovation mainly include: fluency of working flow, reasonability of working section setup, and quickness of conflict solution.

3 Evaluation Model of Performance of Customer Collaboration Innovation Implementation

3.1 Extension Assessment Method

Extension Method is a frequency or method that solves problem qualitatively and quantitatively. Fundamental logical cells of Extension Method included: subject, character and subject's character value, that is R=（subject, character, value）=（N, C, C（N））. Change of three matter elements and subjects' internal structure lead to change of subject, and matter element become a basic tool to describe subject's changeability.

1) Determination of Classic Field

Assume that the amount of assessment index are "n", C1, C2……Cn; the amount of assessment level are "m", N1, N2……Nm; then matter-element subject with the same characteristics can be demonstrated as:

$$R=\begin{bmatrix} N & N_1 & N_2\cdots N_m \\ C_1 & X_{11} & X_{12}\cdots X_{1m} \\ C_2 & X_{21} & X_{22}\cdots X_{2m} \\ \vdots & \vdots & \vdots \quad \vdots \\ C_{n1} & X_{n1} & X_{n2}\cdots X_{nm} \end{bmatrix}=\begin{bmatrix} N & N_1 & N_2 & \cdots & N_m \\ C_1 & <a_{11},b_{11}> & <a_{12},b_{12}> & \cdots & <a_{1m},b_{1m}> \\ C_2 & <a_{21},b_{21}> & <a_{22},b_{22}> & \cdots & <a_{2m},b_{2m}> \\ \vdots & \vdots & \vdots & \vdots & \vdots \\ C_n & <a_{n1},b_{n1}> & <a_{n2},b_{n2}> & \cdots & <a_{nm},b_{nm}> \end{bmatrix}$$

where $(X_{ij})_{n*m}$ is matter-element matrix; $<a_{ij}, b_{ij}>$ is the value range of index N_j relative to index C_i. Determinations of segment field matter elements and assessment-waiting matter elements.

2) Determinations of segment field matter elements and assessment-waiting matter elements.

$$\text{Let } R_P=(P,C,X_P)=\begin{bmatrix} P & C_1 & X_{P1} \\ & C_2 & X_{P2} \\ & \vdots & \vdots \\ & C_n & X_{Pn} \end{bmatrix}=\begin{bmatrix} P & C_1 & <a_{p1},b_{p1}> \\ & C_2 & <a_{p2},b_{p2}> \\ & \vdots & \vdots \\ & C_n & <a_{pn},b_{pn}> \end{bmatrix}$$

Where P is the set of type; X_{pi} is value rang of P relative to C_i, that is segment field $<a_{pi}, b_{pi}>$.Indicate investigated data by matter elements, then assessment-waiting matter elements P_0 can be illustrated as:

$$R_0=(P_0,C_i,x_i)=\begin{bmatrix} P_0 & C_1 & x_1 \\ & C_2 & x_2 \\ & \vdots & \vdots \\ & C_n & x_n \end{bmatrix}$$

Where x_i is value of P_0 relative to index C_i, that is practical data of assessment-waiting matter.

3) Determinations of weighted parameter and correlation function of assessment category.

a) Let w_i represent weighted parameter of index C_i, $\sum_{i=1}^{n} w_i = 1$。

b) Calculate position value

$$D(x_i,X_{ij},X_{pi})=\begin{cases} \rho(x_i,X_{pi})-\rho(x_i,X_{ij}) & x_i \bar{\in} X_{ij} \\ -1 & x_i \in X_{ij} \end{cases} \tag{1}$$

Where $\rho(x_i, X_{pi})$ represents distance between dot and interval X_{pi}, represents distance between dot and interval X_{ij}.

c) Calculation of Correlation Function

$$K_j(x_i)=\frac{\rho(x_i,X_{ij})}{D(x_i,X_{ij},X_{pi})}=\begin{cases}\dfrac{\rho(x_i,X_{ij})}{\rho(x_i,X_{pi})-\rho(x_i,X_{ij})} & D(x_i,X_{ij},X_{pi})\neq 0\\ \rho(x_i,X_{ij})-1 & D(x_i,X_{ij},X_{pi})=0\end{cases} \tag{2}$$

Correlation function $K_j(x_i)$ represents attribution of assessment-waiting matter's each index relative to category "j".

d)Calculate correlation degree of assessment-waiting matter relative to level "j"

$$K_j(P_0)=\sum_{i=1}^{n} w_i K_j(x_i) \tag{3}$$

If $K_{j0}(P_0)=\max\limits_{j\in\{1,2\cdots m\}} K_j(P_0)$, then assessment-waiting matter belongs to category j_0, Let

$$j^* = \frac{\sum_{j=1}^{m} j\times \overline{K_j(P_0)}}{\sum_{j=1}^{m}\overline{K_j(P_0)}} \tag{4}$$

Where j^* is called level variable value of P_0.

3.2 Extension Model for Performance Evaluation of Customer Collaboration Innovation Implementation

According to the main impacting factors of performance of collaboration innovation implementation, select evaluation index C1 as economic benefit of collaboration innovation, C2 as social benefit, C3 as technology benefit, C4 as working efficiency, C5 as synergistic effect, and C6 as process evaluation. Based on Delphi Method, establish matter-element entirety with the same characteristics (Classical fields).

$$R=\begin{bmatrix} N & N_1 & N_2 & N_3 \\ C_1 & <0,0.5> & <0.5,0.8> & <0.8,1> \\ C_2 & <0.1,0.3> & <0.3,0.7> & <0.7,1> \\ C_3 & <0,0.3> & <0.3,0.5> & <0.5,1> \\ C_4 & <0,0.4> & <0.4,0.7> & <0.7,0.9> \\ C_5 & <0,0.2> & <0.3,0.8> & <0.8,0.9> \\ C_6 & <0.1,0.4> & <0.4,0.7> & <0.7,0.9> \end{bmatrix}$$

Then segment field is:

$$R_P = \begin{bmatrix} P & C_1 & <0,1> \\ & C_2 & <0.1,1> \\ & C_3 & <0,1> \\ & C_4 & <0,0.9> \\ & C_5 & <0,0.9> \\ & C_6 & <0.1,0.9> \end{bmatrix}$$

Where P is the entirety of levels for performance evaluation of customer collaboration innovation implementation, which in this study is categorized as three levels: low, medium and high. This study selects five groups of customer collaboration innovation projects, and then establishes the matter element to be evaluated:

$$R_0 = \begin{bmatrix} P & P_1 & P_2 & P_3 & P_4 & P_5 \\ C_1 & 0.35 & 0.83 & 0.95 & 0.77 & 0.92 \\ C_2 & 0.32 & 0.90 & 0.94 & 0.78 & 0.95 \\ C_3 & 0.65 & 0.35 & 0.06 & 0.45 & 0.28 \\ C_4 & 0.85 & 0.42 & 0.08 & 0.49 & 0.36 \\ C_5 & 0.55 & 0.25 & 0.16 & 0.44 & 0.38 \\ C_6 & 0.75 & 0.41 & 0.28 & 0.59 & 0.46 \end{bmatrix}$$

Weight coefficient plays an important role in superiority assessment. Since man-made weight coefficients are usually of subjective cognitions which affect objectivity and reliability of assessment process, this paper adopts SHAPLEY to determine the weight coefficient of risk assessment index in collaboration innovation (see Table 1).

Put data and weight coefficient of R0 into formula (1),(2),(3),(4); Figure out correlation degree of assessment-waiting project relative to each assessment level; Figure out value of level variable (j*) from formula (5) and (6). Assessment results are shown in Table 2.

According to evaluation results in Figure 2, the group 3 of customer collaboration innovation belongs to the bad implementation performance level 1; group 2 and group 5 belong to the medium implementation performance level 2; group 1 and group 4 belong to the good implementation performance level 3. Besides, within level 2, group 2 performs better than group 5; within level 3, group 1 performs better than group 4.

It can be seen that the performance of customer collaboration innovation implementation could be directly and quantitatively reflected by extension evaluation method, which undoubtedly facilitates the decision making of enterprises' collaboration innovation.

Table 1. Weight factor of assessment index

assessment index	C_1	C_2	C_3	C_4	C_5	C_6
weight factor	0.18	0.15	0.20	0.12	0.13	0.12

Table 2. Correlation and assessment result

Project	N_1 (低)	N_2 (中)	N_3 (高)	j	j^*	Assessment result
1	-0.465	-0.443	0.078	3	3.01	High
2	-0.032	-0.165	-0.387	2	2.11	medium
3	0.058	-0.791	-0.912	1	1.19	low
4	-0.123	0.089	-0.411	3	2.67	high
5	0.047	-0.350	-0.639	2	1.85	medium

4 Conclusions

Matter element is an effective formal tool with wide-range of application which is capable of directly reflecting the quality and quantity of matter and vividly depicting its changing process. Summarizing the characters of customer collaboration innovation as well as the impacting factors of performance of customer collaboration innovation implementation, this study establishes multi-indices matter element model for performance evaluation of customer collaboration innovation implementation by applying extension method and provides quantitative evaluation results. In generic, this study deepens current research on qualitative analysis on performances of customer collaboration innovation and illustrates a quantitative evaluation approach, which undoubtedly provides significant references for the decision making of customer collaboration innovation.

References

1. Shen, L., Zhao, W., Li, M., Zheng, J.J.: A New Method for Customer-Oriented Virtual Collaborative Design with VRML Product Model. In: Shen, W., Luo, J., Lin, Z., Barthès, J.-P.A., Hao, Q. (eds.) CSCWD. LNCS, vol. 4402, pp. 197–206. Springer, Heidelberg (2007)
2. Kobayashi, M., Shinozaki, M.: Collaborative customer services using synchronous web browser sharing. In: Proceedings of the ACM Conference on Computer Supported Cooperative Work, pp. 99–108 (1998)
3. Sun, T.-L.: Supporting customer-company interaction in product customization using a web-based collaborative VR environment. Institute of Industrial Engineers 22(3), 35–42 (2005)
4. Elofson, G., Robinson: Collective customer collaboration impacts on supply- chain performance. International Journal of Production Research 45, 2567–2594 (2007)

5. Zheng, H.-L., Liu, Q.-Y., Wan, X.-Y., Zhang, J.-W., Xiao, X.-H.: Research on the Development Mode of Customizing Product with Customer Collaborating. Modular Machine Tool & Automatic Marnufacturing Technique, 109–112 (June 2006)
6. Yang, Y., Guo, B., Yin, S., Wang, W.-L., Zhang, X.-D.: Connotation, theory framework and application of CCI. Computer Integrated Manufacturing Systems 14(5), 449–459 (2008)
7. Wang, W.-L., Yang, Y., Wang, M.-K., Song, L.-J.: RS and SVM-based partner selection research for CCI. Computer Engineering and Applications 43(29), 245–248 (2007)
8. Tan, H.-Y., Yang, Y., Song, L.-J., Yang, J.: Identification and Fuzzy Evaluation of Innovative User in Collaborative Product Innovation. Journal of Chongqing University 31, 130–135 (2008)
9. Zhao, X.-H., Wu, F.-Z., Tuo, X.-B.: Method of collaboration customer selection based on fuzzy clustering and incidence matrix. Basic Sciences Journal of Textile Universities 20, 310–317 (2007)
10. Li, G.-M.: Research on theMould-making order management for customer collaborative platform. Guangdong University of Technology, 25–29 (2008)

A Hybrid Design and Implementation of Socket Directly and Web Browser Online Email Address Verifier

Sun Weizhen, Fu Yutong, and Luo Liming

Information Engineering College Capital Normal University No. 56,
North road of Western 3rd-Ring, 100089
sunweizhen@263.net, superfyt@163.com, luolm@mail.cnu.edu.cn

Abstract. Email address verification is a feasible approach to lessen and avoid trash emails, especially when people send a large quantity of emails. However, verifying the addresses manually takes not only much manpower, but also large amount of time. Another method which utilizes the SMTP protocol by Socket in the verification leads to a low accuracy. The present article described a hybrid verifier design integrating the Socket directly verification and the Web Browser online verification, which combined the advantages of speed and validity brought by Socket directly verification and Web Browser online verification respectively. In the test, the present design obtained a 20 percent to 50 percent increase in speed to verify the Email address, which provided hard evidence that the hybrid method was a practical verifier to accelerate the speed of email address verification.

Keywords: Email verification, Socket, Proxy pattern, Web browser.

1 Introduction

Email address verification is required before sending emails in special occasions such as informing a conference, in which it is hoped that the notice is delivered to specific readers whose addresses are valid. Besides, email address verification is also an approach to minimize the company's manpower consumption as well as trash emails.

In the SMTP Protocol, the VRFY command can be provided to SMTP client users, who can send to the SMTP servers to verify whether a particular user's name is in the server's request [1]. However, because it is easy to be used by hackers, many mail servers turn off this instruction for security considerations, which makes it really complicated when querying the e-mail address is needed.

2 Background

The defects of VRFY command makes it difficult to verify email address; however, there are some other ways that can be applied in the verification, which will be introduced in this section.

J. Luo (Ed.): Affective Computing and Intelligent Interaction, AISC 137, pp. 11–19.
springerlink.com

2.1 Use Websites Which Provide Verification Services

In order to verify the validity of email address, we can use email address verification websites.

Ipaddresslocation(http://www.ipaddresslocation.org/)is such a verification website which provides Check Email service. Users can verify the validity of email simply by inputting the address into the textbox and clicking on the Email Lookup button, and the verification information can be obtained, which is shown in Figure 1.

However, the disadvantage of this method can not be neglected: manual intervention is needed in every check, which takes about five to ten seconds. For large assignment with thousands or more emails to be verified, such a speed is unacceptable.

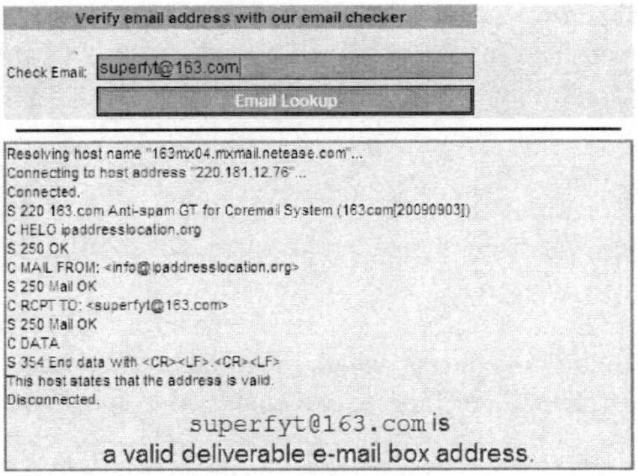

Fig. 1. Email Ckeck page of the Ipaddresslocation website

2.2 Use Socket and Multithreading to Verify Directly

Since the key problem lies in the increasing of verification speed and decreasing the amount of manual intervention, then the best choice might be the use of software with multithreading.

Normally, verification websites adopt SMTP protocol to verify the validation of email address, but the VRFY command is altered into RCPT TO. In such case, when the email address is valid, the server will return 250 response; on the other hand, when the email address is invalid, the server will return 550 response [2].

A new email address verifier can be designed based on this feature.

1. look up MX record from the email server;
2. set up TCP connection by Socket between client and the email server;
3. send SMTP protocol command;
4. judge whether the email address is valid according to the response from the server;

Such method has been proved to be practical in the experiments. Many servers can verify the validation of email address quickly through this method. However, there are also some servers which require the domain contained in the HELO command sent by the verifier matches with its IP address [3]. Take Microsoft.com as an example, when the server finds that the IP address of the verifier is dynamic or doesn't match with the domain, the service will be rejected [4].

2.3 Use Web Browser Control and Multithreading to Verify the Address Automatically

To solve the conflict between domain and IP address, we can refer to the service provided by email address verification websites, which, of course, needs to be adjusted to realize multithreading verification without manual intervention.

Provided by Microsoft's Visual Studio, WebBrowser class enables the interaction between windows form application and the Internet explorer. We can design a verifier based on this property.

1. define a WebBrowser object in the application;
2. define the Url property of WebBrowser object as the address of the Check Email page;
3. fill the e-mail address to be verified in the email address textbox and click the OK button automatically through the application;
4. save the page with verification results in the Document Completed event of the Web Browser Object;
5. use the IHTMLTxtRange object to search the Dom Document properties of the Web Browser Object, and judge whether the email address is valid.

This method introduced above meets the requirement of decreasing manual intervention, but since the Web Browser is actually the kernel of an IE browser [5], the use of each thread is like opening an IE browser, which takes much of the system resources. Besides, the opening of the web pages occupies a large amount of the network bandwidth, which means the number of threads to be opened can not be too big. These limitations restrict the speed of the verification to some extent.

In this chapter three different approaches were analyzed which can realize the verification of email address, but they all have their own limitations: the use of websites which provide verification service takes too much manpower; Socket directly verification can not ensure the validation of the results; Web Browser online verification needs a large amount of time. With the purpose to solve these problems, a hybrid design of verifier was proposed in the present study.

3 The Design of Hybrid Email Address Verifier

With the attempt to verify more email address within shorter time, a hybrid design was adopted, which tried to package the Socket directly verification and the Web Browser online verification into the same application.

3.1 Verifier Design

First of all, a high-speed Socket directly verifier is used to verify all of the email addresses and the results will be stored in the memory. Then an invalid address extractor is set up to find out automatically the email addresses which can not be verified by Socket directly verification. These addresses will then be sent to WebBrowser online verifier which will provide online verification service with high reliability. Next, the results of this verification will be returned by the invalid address extractor to the email verification result list. The time performance of this process is as below:

$$t_i = \begin{cases} t_{Socket}, & \text{can verify by socket} \\ t_{WebBrowser}, & \text{can verify by website} \end{cases}$$

$$\begin{aligned} t_{All} &= \sum_{i=1}^{n} t_i \\ &= n_{Socket} \times t_{Socket} + n_{WebBrowser} \times t_{WebBrowser} \end{aligned}$$

Because $t_{Socket} << t_{WebBrowser}$, the Socket directly verification should be used more often to save the time.

When $n_{Socket} \to n_{All}$:

$$t_{All} = n_{All} \times t_{Socket} + (n_{All} - n_{Socket}) \times t_{WebBrowser}$$

$$\begin{aligned} t_{\min} &= \lim_{n_{Socket} \to n_{All}} t_{All} \\ &= \lim_{n_{Socket} \to n_{All}} n_{All} \times t_{Socket} + (n_{All} - n_{Socket}) \times t_{WebBrowser} \\ &= n_{All} \times t_{Socket} \end{aligned}$$

$$\begin{aligned} t_{\max} &= \lim_{n_{Socket} \to 0} t_{All} \\ &= \lim_{n_{Socket} \to 0} n_{All} \times t_{Socket} + (n_{All} - n_{Socket}) \times t_{WebBrowser} \\ &= n_{All} \times t_{Socket} + n_{All} \times t_{WebBrowser} \end{aligned}$$

Although theoretically $t_{\max}$ might be longer than $n_{All} \times t_{WebBrowser}$, which means the time the hybrid verifier takes might be longer than that of the WebBrowser online verification, in the real testing process, however, nearly half of the addresses can be verified by Socket directly verification, for example, Gmail, Hotmail, 163, etc.. Therefore, $t_{\max}$ will be far shorter than $n_{All} \times t_{WebBrowser}$ in reality.

The invalid address extractor can be realized by the proxy design patterns[6]. The WebBrowser online verifier needs only to send "pick up" command to the proxy and the extractor will complete the smart reference of the addresses which can not be verified by the Socket directly verification. After that, the results of the WebBrowser online verification will be returned to the proxy, which will match the verification results with each item in the email address list.

The design of the proxy can solve the problems such as picking up the address and thread synchronization. In this way, the WebBrowser can be more focused on the verification of the email address.

The design of the verifier is as below:

1. read the email address to be verified into the application from the database;
2. use the Socket directly verifier to verify all the address one by one;
3. adopt the proxy pattern to set up an invalid address extractor as the smart reference;
4. filtrate the data by the extractor and pick out the email addresses which can not be verified by the Socket directly verifier;
5. utilize the WebBrowser online verification to verify the invalid address for the second time;
6. return the verification results to the database through the proxy;

The design of the verifier's function pattern is shown in Figure 2.

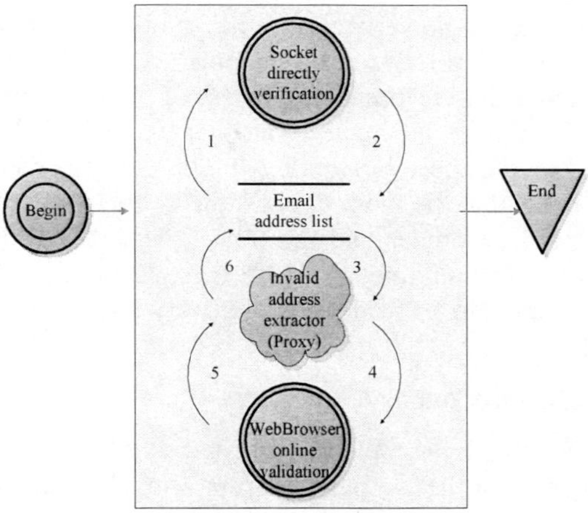

Fig. 2. Hybrid verifier's function pattern

3.2 Experiment

To attest the effect of the hybrid verifier, three lists which contains 358, 1117 and 17324 email addresses respectively were tested. The testing platform was Intel Core2 Duo P8400 with 3G EMS memory, 250G 5400rpm hard disk and Windows XP SP3 operating system. The network was 2M ADSL by China Unicom.

1. Test of the Socket directly verification

The number of threads used in the test of Socket directly verification is 100, 200, 400 respectively, which is shown in Table 1.

Table 1. The results of Three e-mail address list Socket directly verification

		100 Threads Time(S)	200 Threads Time(S)	400 Threads Time(S)
	90%	23.5	16.9	19.2
354	95%	28.3	26.5	26.1
	100%	58.9	59.6	61.8
	90%	63.3	26.3	25.2
1117	95%	67.3	29.4	30.1
	100%	134.5	76.4	70.0
	90%	1135.3	595.0	392.9
17324	95%	1193.5	625.3	416.2
	100%	2138.4	1398.7	1167.6

From Table 1 it can be seen that in the test of Socket directly verification, the number of email addresses to be verified and the number of threads impacted on the speed of verification together: in the test of 358 email addresses list, the speed was the quickest when 200 threads were used; in the test of 1117 addresses list, the speed under 400 threads was a little bit higher than that under 200 threads; when 17324 addresses list was tested, the speed brought by 400 threads was the best. Consequently, we can infer it is not true that the more threads, the better, because the use of threads also takes up the system resources and the time. Therefore, 100 threads should be used when dealing with a small amount of data while 200 threads are proper to deal with more data and, if the data to be dealt with is very large, then 400 threads should be used to bring its best effect.

2. Test of the WebBrowser online verification

Since the IE kernel adopted by the WebBrowser takes a large capacity of the system resources and the opening of the webpage consumes much of the network bandwidth, the number of threads used in the test of WebBrowser online verification is 4, 8 and 12 respectively, which is shown in Table 2.

Table 2. The results of Three e-mail address list WebBrowser online verification

		4 Threads Time(S)	8 Threads Time(S)	12 Threads Time(S)
	90%	661.1	701.6	726.1
354	95%	696.1	758.8	753.2
	100%	727.4	770.1	779.1
	90%	2080.8	2091.3	2599.9
1117	95%	2294.6	2176.0	2721.5
	100%	2682.5	2676.9	2858.2
	90%	35416.2	31974.1	29177.8
17324	95%	37216.4	33627.9	31013.3
	100%	39357.7	35526.5	32492.5

The result of the test showed that the same problem rose with that in the test of Socket directly verification: it was a wrong belief that the more of WebBrowser objects brought better effect. 4, 8 and 12 threads were recommended in the test according to the number of the email addresses.

3. Test of the hybrid verifier

Based on the results of the above two tests, in the hybrid email address verification test, 200 Socket threads and four WebBrowser threads were used in the test of 358 email list; 200 Socket threads and eight WebBrowser threads were used in the test of 1117 email list; for the test of 17324 email list, 400 Socket threads and 12 WbeBrowser threads were applied. Specifically, there were 134 addresses among the 358 email list, 236 addresses among the 1117 email list and 6582 addresses among the 17324 email list that could be verified by the Socket directly verification. The results of the hybrid verification test are shown in Figure 3.

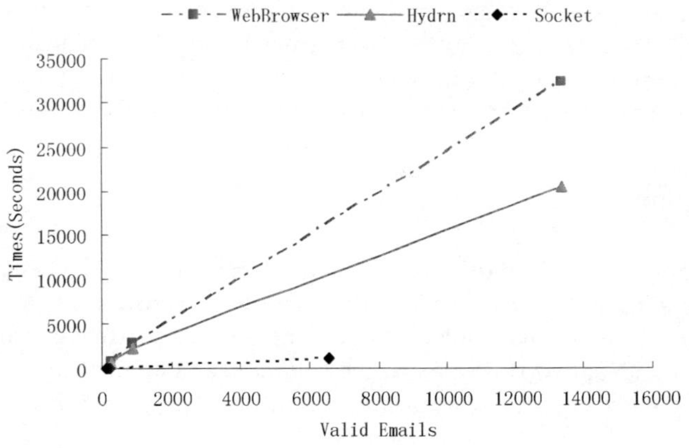

Fig. 3. The results test comparing of three e-mail address list verification

As can be seen in Chart 3, the slope of polyline from the hybrid address verifier was smaller than that from the WebBrowser online verifier, which meant when the valid addresses verified increased, the time taken by the hybrid address verifier was smaller thant that by the WebBrowser. On the other hand, although the slope of line from the Socket directly verifier was much smaller than the other two methods, the general efficiency was lower than the hybrid address verifier because its number of valid verification results was limited. Moreover, there was no need of manual intervention in the hybrid address verification, and the speed of verification was increased dramatically, which can be seen from the test results.

3.3 Analysis of the Experiment

In the process of the Socket directly verification, there were several addresses that took a very long time to verify, hence affected the speed of the whole verification. The reason accounted for this phenomenon was that the parallelizable use of many threads lead to the huddle of the network resources or the connection speed of the MX server was slow. Sometimes these addresses just can not be verified. In the experiment, it only took 50% of the gross time to verify 95% of all the addresses, while the other 5% of addresses took large amount of time. Therefore, it is necessary to adjust the verifier by adding a Counter which will set up a specific time and number and, if the number of addresses having been verified is smaller than the given number within the time, the WebBrowser online verification is used automatically in order to minimize the time.

The result of the testing also showed that the speed of the WebBroswer online verification with 20 threads is not as ideal as 20 times the speed of manual web page verification. This was because the DocumentCompleted event in the WebBrowser was defined to search the key words only after all the web pages were loaded. Because there were many advertisements in the Check Email web page on the Ipaddresslocation, and these advertisements were distributed in different servers with various connection speed, the gross speed was affected. In manual verification, however, judgment can be made as long as the verification information is shown without all the advertisements on the web page having been loaded. In the future studies and experiments, a real time judgment mechanism can be added to the threads in order to increase the speed of obtaining valid information.

4 Conclusion

In the present design, the use of Socket directly verification can increase the speed of verification to some extent, and the WebBrowser online verification can guarantee the reliability and validity of the results. Through the process the time consumed was less than that of the manual verification and the software, and no manual intervention was involved, which realized the original purpose of designing the hybrid email address verifier.

References

1. RFC 821, Simple Mail Transfer Protocol, pp. 8–9 (August 1982)
2. RFC 821, pp. 4
3. Luo, W.-B., Gao, X.-Y., Ou, X.-C., Liu, Y., Li, D.: The data security and secure content of e-mail. Computer Applications 22(3), 22–24 (2002)
4. Microsoft.com now blocks incoming SMTP mail from servers that use dynamic IP addresses (July 25, 2007), `http://support.microsoft.com/kb/886692/en-us/`
5. Edwards, M., Roberts, S.: Microsoft Corporation, Reusing Internet Explorer and the WebBrowser Control: An Array of Options (July 30, 1998), `http://msdn.microsoft.com/en-us/library/bb250464(VS.85).aspx`
6. Gamma, E., Helm, R., Johnson, R., Vlissides, J.M.: Design Patterns: Elements of Reusable Object-Oriented Software, p. 233 (November 10, 1994)

7. Ao, T.-H., Li, D.-Y.: A Verification System Based on SMTP Server. Journal of InnerMongolia University 38(4), 473–475 (2007)
8. Wang, G.-L., Qing, S.-H., Zhou, Z.-F.: Some New Attacks upon Authentication Protocols. Journal of Software 12(6), 907–913 (2001)
9. Paul Hoffman, R.: Unsolicited Bulk Email : Definitions and Problems, Internet Mail Consortium Report: UBE - DEF, IMCR - 004 (October 1997)
10. Abadi, M., Burrows, M., Lampson, B., Plotkin, G.: A Calculus for Access Control in Distributed Systems. ACM Transactions on Programming Languages and Systems 15(4), 706–734 (1993)

Artificial Emotion Elicitor Based on Sequential Check Theory*

Yun-feng Huang[1], Ling Kang[2], and Shi-long Wang[3]

[1] College of Automation, Chongqing University, Chongqing 400044, China
[2] College of Mechanical Engineering, Chongqing University, Chongqing 400044, China
[3] The State Key Lab. of Mechanical Transmission,
Chongqing University, Chongqing 400044, China
hyf525@126.com, erica-cq@tom.com, slwang@cqu.edu.cn

Abstract. Emotion elicitor is used to generate emotions in real time for agents, and is undoubtedly the basis of affective computing. To afford support from emotion sciences for the design of artificial emotion elicitor, Sequential Check Theory (SCT) was selected. Under the guidance of endogenesis strategy, relevance strategy, and operation strategy, the general emotion elicitor was built up through the construction and reduction of the decision tree. And in the process, information ratio was used for attributes selection. With the flexibility of each attribute's evaluation logic, and the close tie SCT has with emotion experiment evidence, the emotion elicitor is of universality and worth further study.

Keywords: Emotion elicitor, Sequential Check Theory, Decision tree, Agent.

1 Introduction

Just as Marvin Minsky wrote, the question is not whether intelligent machines can have any emotions, but whether machines can be intelligent without any emotions[1].With the need to deal with emotion-related affairs such as HCI (Human-Computer Interaction), and the need to migrate emotional functions to improve the system performance in decision making, action selection, behavior control and etc, an upsurge in the study of emotional agents is taking shape.

The mechanisms of how emotions are elicited are basic problems not only in original emotion sciences, but in the implementation of affective computing. Researchers in either domain have to face different and even controversial theories. If it is believed that the two domains should be closely connected with mutual benefits, just as this paper has advocated, then it is worth getting support from the former domain while considering the design of artificial emotion elicitor in the latter one.

Appraisal theories based approach has become a very popular source of inspiration for computational models, in particular the more 'cognitive-oriented' ones[2]. A kind

* This work is supported by National Nature Science Foundation of China under Grant No. 51005260.

J. Luo (Ed.): Affective Computing and Intelligent Interaction, AISC 137, pp. 21–27.
springerlink.com

of appraisal theory named Sequential Check Theory (SCT)[3] was introduced, and based on it, the emotion elicitor that is easy to implement in affective computing has been built up through the construction of an emotion decision tree.

2 Three Principles in Design of Emotion Elicitor

Three principles, that is, the endogenesis, relevance and operation strategy of emotion modeling in affective computing has been introduced by us in [4]. They are also basic principles for the design of emotion elicitor. For the sake of convenience, they are listed in the following.

1) Endogenesis strategy: Concerned about the *endogenesis* of emotion means the construction of emotion model should get clues from and be inspired by emotion-related subjects such as emotion physiology, emotion psychology, emotion sociology. This strategy provides additional testimonies and verification, and gives sustainable power to the further development of the mode.

2) Relevance strategy: Only emotions that are related to the objects being studied is worthy of more attention. This strategy puts a practical eye on emotion modeling.

3) Operation strategy: This strategy usually means the elicitation of emotion should be quantified, bear clear mechanisms and meet the requirements of business logic. Operation strategy tends to lay a good foundation for further processing.

3 SCT–Based Emotion Elicitor

3.1 SCT: A Simple Introduction

Since the early 1980s, Sequential Check Theory (SCT) was committed by Scherer as part of his Component Process Model (CPM) to differentiate emotions. The theory has been developed over 20 years with support and evidence from psychological experiments.

The core of SCT is to use a series of Stimulus Evaluation Checks (SECs) to appraise the importance of events. The objects of the appraisal come from four kinds of information, that is, *relevance*, *implication*, *coping potential* and *normative significance*. It was pointed out by Scherer that the fundamental tenet of appraisal theory is that people evaluate events in terms of the perceived relevance for their current needs and goals, including considerations of their ability to cope with consequences and the compatibility of the underlying actions with social norms and self-ideals[5].

In SCT, besides the overall four dimensions, every dimension has its own sub-dimensions. It can be seen from table 1 that the number of sub-dimensions totaled 16 in SCT. The existence of the sub-dimensions entitles extensiveness of the theory. Thus, if started from Scherer's theory, there is less chance of missing some important dimension than started with a simpler, possibly less complete theory[6].In this paper, the sub-dimensions are selected concerning implementation matters based on SCT.

Table 1. Appraisal dimensions of SCT

Relevance	Implication	Coping	Normative Significance
Novelty (Suddenness Familiarity, Predictability)	Causality: agent	Control	Internal standards compatibility
	Causality: motive	Power	External standards compatibility
Intrinsic Pleasantness	Outcome probability	Adjustment	
	Discrepancy from expectation		
Goal/need relevance	Conduciveness		
	Urgency		

3.2 Emotion Selection

Though Scherer did not think there are only restricted categories of emotion, he did admit there are some main types of emotion adaption mode. Moreover Scherer brought up with predicted appraisal patterns for some major modal emotions[3].These emotions include enjoyment/happiness (ENJ/HAP), elation/joy(ELA/JOY), displeasure/disgust (DISP/DISG), contempt/scorn (CON/SCO), sadness/dejection (SAD/DEJ), irritation/cold anger (IRR/COA), rage/hot anger (RAGE/HOA), boredom/indifference (BOR/IND), shame, guilt (GLT), and pride (PRD).

In consideration of relevance strategy, and to simplify emotion categories fairly, the following emotion adaptation modes are included for further consideration. That is, the emotion set EM is defined as:

EM={ HAP, ELA, DISG, DEJ, DESP, ANX, FEAR, IRR, BOR , GLT, PRD } . (1)

3.3 Decision Tree of Emotions

It can be observed from the literature that emotion type can be uniquely identified by the sub-dimension value defined by Scherer. In accordance with the characteristics of the check order SCT, these sub-dimensions, in theory, constitute an "emotion decision tree".

One of the basic principles of constructing a tree is to be simple, which is a version of Occam's Razor Theory, and means fewer nodes and compact tree with the full ability to explain the data.

It is noticeable that many parts of sub-dimensions of emotional value are "uncertain", and some emotion evaluation differs only between a small number of sub-dimensions, and this offers the optimization possibility of the emotion decision tree and fast recognition of emotions.

In summary, for the convenience of practice, the attributes of decision tree should be carefully selected based on the SCT appraisal dimensions. The selection of the decision tree attributes will consider the following factors:

1) Factor 1: Attributes inspection sequence should be the same as the SCT's sub-dimension check sequence in most cases.

2) Factor 2: Decision tree should be simple to reduce the decision dimensions and the cost of computation.

3) Factor 3: The corresponding decision value is better to be distinct, and easy to determine.

The first point is the constraint of SCT, and reflects the endogenesis strategy of the emotion elicitor design principle. The second and the third principles reflect concerns of operation strategy.

Let E denote the emotion set, num(E) denote the number of emotion types. The steps to construct the emotion decision tree are as following.

Step 1: Evaluate the information measure of all the attributes in Set(or sub set) E. Select the attribute xi with the largest information measure as the node of the decision tree. Emotion sets are divided into a number of sub sets Sj, 2≤j≤num(E), here S= $\bigcup_2^{num(E)} S_j$.

Step 2: If $\exists S_k \in S$ and the number of emotion types in S_k >1, let the attributes following x_i according to the sequence in SCT form the new set of attributes, and go to step1.

Step 3: If in some decision path, all the dimensions have been considered but failed to differentiate any emotions, then the attribute with information measure inferior to the former selected attribute backward the decision path, and again extend the tree branch. In the process, the sequence along the branch should be in accordance with the sequence of SCT. Once the leaf is reached along the path, the attributes selection work along the branch is temporarily done.

Step 4: If all the branches of S_k have been checked, delete S_k from S.

Step 5: Repeat step 2 to 5 until all the leaf nodes in decision tree are entitled with a certain emotion type.

Step 6: According to *Factor 3,* optimize the selection of attributes, repeat step 1 to 5 if necessary.

There are many approaches to measure the information of the decision attributes. In C4.5, Quinlan used information ratio (see formula 2 to 4). Generally speaking, the larger the value, the higher useful information ratio there will be. Information ratio is also used in this paper.

$$GainRatio(S,A) \equiv \frac{Gain(S,A)}{SplitInfo(S,A)} \quad (2)$$

Here,

$$Gain(S,A) \equiv Entropy(S) - \sum_{v \in Values(A)} \frac{|S_v|}{|S|} Entropy(S_v) \quad (3)$$

$$SplitInfo(S,A) \equiv -\sum_{i=1}^{c} \frac{|S_i|}{|S|} \log_2 \frac{|S_i|}{|S|} \quad (4)$$

Take the root node as an example, the information ratio of *novelty* can be calculated like the following.

To facilitate the attributes selection, suddenness is used to stand for novelty. Denote the set of emotion types with uncertain novelty as *OpenNovelty*. In emotion set EM. according to [3], *OpenNovelty* can be defined as following,

$$OpenNovelty=\{DISG,\ GLT,\ PRD\}. \tag{5}$$

Emotion in OpenNovelty can be allocated over all the branches with equal probability.

$$\left.\begin{aligned} Values(Novelty) &= high\ ,\ medium\ ,\ low\ ,\ verylow \\ S_{high} &= \{\ ELA, DESP, FEAR, DISG, GLT, PRD\ \} \\ S_{medium} &= \{\ ELA, DISG, GLT, PRD\ \} \\ S_{low} &= \{\ HAP, DEJ, ANX, IRR, DISG, GLT, PRD\ \} \\ S_{verylow} &= \{\ BOR, DISG, GLT, PRD\ \} \end{aligned}\right\} \tag{6}$$

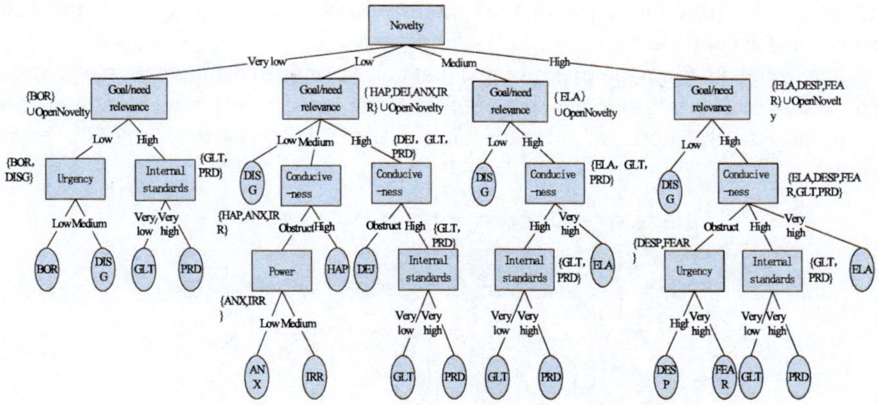

Fig. 1. The emotion decision tree

Take formula (6) into (3) and (4), and consider formula (2), there comes the gain ratio of novelty in EM in the root node,

$$GainRatio(EM,Novelty)=\ 1.184943/1.821307=0.650600\ .$$

With the same process, information gain of other attributes in root node can be calculated.

$$GainRatio(EM,Pleasantness)=\ 0.272728$$

$$GainRatio(EM,\ GoalRelevence)=\ 1.000000$$

$$GainRatio(EM,\ AgentCause)=\ 0.281141$$

$$GainRatio(EM,\ MotiveCause)=\ 0.379338$$

…

Among these, attributes with larger information gain will be the focus to study. For example, the calculation above indicate that the information gain of intrinsic pleasantness, that is *GainRatio(EM,Pleasantness)*, is relatively low, thus not fit for the role of root node attribute. After all, information gain is only a reference in attribute selection, since factor 1 and factor 3 should also be considered.

According to above approach, the decision tree was constructed based on emotion set EM (see figure 1).

It can be seen from the emotion decision tree constructed, although the dimensions in different branches are not necessarily the same, the maximum appraisal dimension number of any emotion is no greater than 4. This significantly lowered the theoretical dimension number of SCT (which is 11), and greatly accelerated the process of emotion determination. On the other hand, it brings about a consequence that there may be misjudgments when facing some unknown type of emotions. But, it is for certain that there exists great similarity between the real emotion type and the category it has been judged (at least in dimensions tested). In view of the facts that the set EM has covered main categories of emotion, and emotions in EM can be differentiated thoroughly through the emotion decision tree, the cost of dimension deduction is acceptable.

Since each path from the tree root to a leaf corresponds to a conjunction of attribute tests, and the tree itself to a disjunction of these conjunctions[7], the emotion decision tree is easy to transform to a set of rules and to direct practical emotion elicitor design.

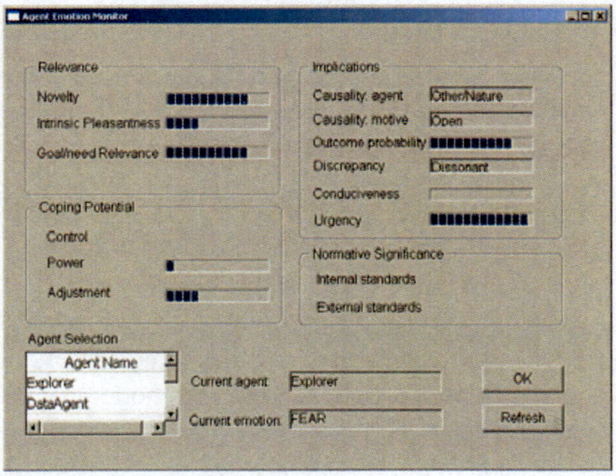

Fig. 2. Agent emotion monitor

The value of attributes in the decision tree can be calculated by the function that may be defined with domain or problem specific knowledge. And this is an embodiment of flexibility in application of emotion mechanism. Figure 2 is the interface of an agent emotion monitor in a scene of scheduling based on the SCT emotion elicitor.

4 The Summary

There seems to be a misleading bias that affective computing could thrive with little knowledge of original emotion sciences. Some researchers hold different view, e.g., Aaron Sloman has discussed the importance of deep emotion model[8]. Driven by the three principles in emotion elicitor design, the emotion elicitor based on Scherer's Sequence Check Theory (SCT) was discussed in this paper. The emotion decision tree was built up and simplified in accordance with three design factors. Information ratio was afforded as a clue to choose attribute in each node during the process. With the flexibility of each attribute's evaluation logic, and the close tie SCT has with emotion experiment evidence, the artificial emotion elicitor introduced in this paper may be expected to play a wider role in different application fields related with affective computing.

References

1. Minsky, M.L.: The society of mind. Heinemann, London (1987)
2. Cañamero, L.: Emotion understanding from the perspective of autonomous robots research. Neural Networks 18, 445–455 (2005)
3. Scherer, K.R.: Appraisal considered as a process of multi-level sequential checking. In: Scherer, K.R., Schorr, A., Johnstone, T. (eds.) Appraisal Processes in Emotion: Theory, Methods, Research, pp. 92–120. Oxford University Press, New York (2001)
4. Huang, Y.F., Wang, S.L.: Affective computing and engineering rationality. Application Research of Computers 26, 3227–3231 (2009)
5. Scherer, K.R.: Introduction: Cognitive Components of Emotion. In: Davidson, R.J., Scherer, K.R., Goldsmith, H.H. (eds.) Handbook of Affective Sciences, pp. 563–571. Oxford University Press, New York (2003)
6. Marinier III, R.P., Laird, J.E., Lewis, R.L.: A computational unification of cognitive behavior and emotion. Cognitive Systems Research 10, 48–69 (2009)
7. Mitchell, T.M.: Machine Learning. McGraw-Hill, New York (1997)
8. Sloman, A.: Beyond Shallow Models of Emotion. Cognitive Processing 2, 177–198 (2001)

Causality and Reasons of Agricultural Production and Agricultural Logistics Practitioners in China*

Wang Daoping, Li Feng, and Cheng Lei

School of Economics and Management, University of Science and Technology, Beijing
No. 30, Xueyuan Road, Haidian District, Beijing, China, 100083
dpwang@ustb.edu.cn, Lifeng20511@sina.com, Chenglei1980@126.com

Abstract. This paper employs the tool of root test in Co-integration method to analyze the causal relationship and long-run equilibrium between the number of employees and productions of grain, oil crops, and fruit. The findings indicate there is no causal relationship between the agricultural production growth and the number of employees. Oil crops stimulate an increased number of logistic workers due to the lower entry barrier; the low proportion of third-party logistics can be used to explain why there is no casual relation between the production amount and employees in the relative logistic sector.

Keywords: Agricultural logistics practitioners, Co-integration, Granger causality test, Third-party logistics.

1 Introduction

As a leading agricultural country, China ranks Top 10 in the world in terms of main agricultural products, and ranks first in vegetable and fruit productions. Meanwhile, due to its large population, China witnesses a huge consumption in various agricultural products. Therefore, the logistics of agricultural products, as a bridge between production and consumption, weighs significantly important. According to statistics, the decay rate of vegetable and fruit in logistics accounts for 25%-30% in the whole chain. In addition, due to the huge base of agricultural output, the amount of waste in logistics chain is shocking. Therefore, a study on logistics of agricultural products is significant to build "resource-conserving society" and avoid waste in land, human resources and resource (this sentence is not clear).

Since "the issues of agriculture, farmer and rural area" attract attentions both at home and abroad, logistics of agricultural product, as one of the hot-spot issues, is of high on the government's agenda at different levels. Since August 2007, seven ministries of China have jointly set up a green pathway so-called "five vertical lines and two horizontal lines", and some provincial level green paths have also occurred, which covered all key production bases and markets of agricultural products in China. Taking into account this background, by focusing on the relationship between agricultural

* This work is partially supported by NSFC Grant #70872010 to Wang Daoping and BPOPSS Grant # 07BfJG185 to Wang Daoping.

J. Luo (Ed.): Affective Computing and Intelligent Interaction, AISC 137, pp. 29–34.
springerlink.com

output and employment in logistics of agricultural products, this paper intends to explore the crucial question whether employees in this field are directly related to the output of agricultural products or not.

2 A Review of Granger Causality Test

Invested by Clive W. J. Granger, the winner of 2003 Nobel Prize in Economics, Granger causality test aims to analyze the causality of economic variables. The principle of Granger causality test is as follow: If X causes Y, then, changes of X happened first then followed by changes of Y. If X causes Y, two conditions must be established: first, Y could by predicted according to X. In other words, while doing the regression of Y on other variables (including its past value), the predictability of Y can be significantly enhanced if lagged value of X could be included. Second, X could not be predicted according to Y. If X and Y are two-way predictable, chances are that both of them are caused by other variables.

To be specific, there are three steps in the test: unit root test in time series, cointegration test among variables, and Granger Causality Test.

3 An Analysis on the Causality between Agricultural Output and Employees in Logistics of Agricultural Products

According to qualitative research, the growth of agricultural production should be one of the reasons for the increasing employment of logistics of agricultural products. Meanwhile, the variety of agricultural products and increase of agricultural output have different impacts on the employment increase of logistics of agricultural products. To this end, the agricultural products in the thesis will be divided into three types: food crops, oil crops and fruit.

Data in this part comes from *China Statistical Yearbook* and *China Rural Staistical Yearbook*. In this part, the author studies the causality of the output of three types of agricultural products and the number of employees in logistics chain. Table 1 shows the results:

Table 1. The Causality coefficient of Logistic Employment and Three Types of Agricultural Products

	Food crops	Oil crops	Fruit
Logistics employment	0.6881	0.9568	0.9041

Table 1 shows a strong causality between logistics employment and three types of agricultural products. Therefore, in this part, the author will use Granger causality test to testify the causality of the two.

Original hypothesis H_1: the change of food crop F_t does not cause the change of the employment in logistics S_t; original hypothesis H_2: the change of the employment in logistics S_t does not cause the change of food crop F_t. The model is set up as follow:

$$S_t=a_0+a_1S_{t-1}+a_2F_t+a_3F_{t-1}+u_t \quad (1)$$

$$F_t=a_0+a_1F_{t-1}+a_2S_t+a_3S_{t-1}+v_t \quad (2)$$

In Granger causality test, the economics variable should be stationary series. ADF test shows that: both of the leg length of S_t and F_t are both 3, stationary series with trend. In the formula above, S_t stands for employment in logistics of agricultural products, F_t stands for the output of food crops. Put the data from 1985-2005 into the formula, the results are showed in Table2.

Table 2. Output of food crops and employees in logistics: causality test

Original hypothesis	observation	statistics	level of significance
H1 the employment in logistics is not the cause of the output of food crops	18	0.96813	0.4423
H2: the output of food crops is not the cause of the employment in logistics		0.38569	0.7655

While doing unit root test for oil crops Z_t, the author finds out that Z_t is integrated of order. Therefore, it could be stationary through first order difference. The model of correlation between oil crops Z_t and employees of agricultural logistics S_t is as follow:

$$\Delta Z_t=a_0+a_1\triangle Z_{t-1}+a_2\triangle S_t+a_3\triangle S_{t-1}+u_t \quad (3)$$

$$\Delta S_t=a_0+a_1\triangle S_{t-1}+a_2\triangle Z_t+a_3\triangle Z_{t-1}+v_t \quad (4)$$

Put related data from 1985-2005 into the model, the results are listed in Table 3.

Table 3. Output of oil crops and employees in logistics: causality test

Original hypothesis	observation	statistics	level of significance
H1the increase of logistics employment is not the cause of the rise in oil crops output	18	0.10232	0.9035
H2 the increase of oil crops output is not the cause of the rise in logistics employment		7.16432	0.0080

While doing the unit root test for fruit output G_t, the author finds out that G_t is integrated of order., Therefore, it could be stationary through first order difference. The model of correlation between fruit output G_t and employees of agricultural logistics S_t is as follow :

$$\Delta G_t=a_0+a_1\triangle G_{t-1}+a_2\triangle S_t+a_3\triangle S_{t-1}+u_t \quad (5)$$

$$\Delta S_t=a_0+a_1\Delta S_{t-1}+a_2\Delta G_t+a_3\Delta G_{t-1}+v_t \quad (6)$$

Put related data from 1985-2005 into the model, the results are listed in Table4.

Table 4. Fruit output and logistics employment: causality test

Original hypothesis	observation	statistics	level of significance
H1 : the increase of logistics employment is not the cause of the rise in of fruit output	18	0.97333	0.4037
H2 : the increase of fruit output is not the cause of the rise in logistics employment		0.18566	0.8327

The test results show that despite the increase in the output of food crops, oil crops and fruit is closely related to the rise in logistics employment, there is no causality between the output of food crops and logistics employment. The rise in the output of oil crops causes the increase in logistics employment. There is no causality between the output of fruit and the rise in logistics employment.

3 An Analysis on the Cause of Causality between the Output of Agricultural Products and Logistics Employment in Agricultural Sector

3.1 There Is No Causality between the Output of Food Crops and Logistics Employment

Thanks to China's heavy density of population, tradition and culture, food crops have been considered to be one of the strategic stockpiles. From the founding of P.R.C. to the beginning of the 21st century, food crops had considered to be a subject in planned economy in China's agricultural market. Therefore, its production, purchase and circulation have been strictly controlled by the vertical management system of state grain reserve, which helps to stabilize the food price. Since 1985, after the "dual tracks" being applied to grain, more freedom and room have been given to the production and circulation.

3.2 The Increase of Oil Crops Is the Cause of Rise in Logistics Employment

The five major oil crops in China are soy, peanut, rape, sesame and cotton. The demand of oil crops, compared with other agricultural products, is more concentrated, since it is an important raw material of cereal and oil processing industry. This helps to reduce the logistics cost. Besides, the nonperishable oil crops are less demanding in packaging, transport and storage, leading to a lower logistics threshold. Therefore, from the practical point of view, it sounds reasonable that an increase in oil crops output may lead to the rise in the number of logistics employees.

3.3 There Is Non-casualty between the Increase of Fruit Output and a Rise in the Number of Logistics Employees

Fruits are perishable agricultural products. The seasonal and regional plantation determines that the packaging, storage and transport of fruit are quite demanding. So people usually conclude that the increase in fruit output will cause a rise in logistics employment. But according to Granger Test, it is not true. The reasons are listed as follow:

I. Fruit logistics is run by supermarket and big dealer. Third-party logistics takes a small share in the pie. With the further development of "turning farm product market into supermarket", the sales of fresh agricultural products in the supermarket play a positive role in food safety and standard packaging. However, after comparing self-owned logistics and third-party logistics, regional supermarket is likely to choose self-owned logistics for the sake of easier control of the whole logistic chain. They only prefer third-party logistics in cross-region transport. But the most of the perishable fruit are sold locally. Therefore, fruits in supermarket are transported by self-owned logistics. Fruit dealer, as another branch of fruit logistics, also prefer self-owned logistics.

II. The lack of cold chain logistics has a big impact on fruit logistics. Till now, the mainstream method of fruit logistics in China is still normal temperature logistics. Apart from cross-provincial transport from south to north, cold chain cannot be guaranteed on the whole way. In addition, refrigerated vehicles in China are short in supply. This will sharply shorten the radius of logistics. As we all know, the development of logistics employees in agricultural sector means a lot to the development of logistics industry, therefore, it is likely that has no causality has been shown between the increase in fruit output and logistics employees.

III. Fruit consumption weighs a lot in daily life of rural and urban residents.Objectively speaking, for the agricultural products that are scattered in production and demand, the logistics cost is relatively high. While for those products that are more concentrated in production and demand, the logistics cost keeps at a lower level, which is a better choice for the third-party logistics to come.

4 Conclusion and Significance

Based on the analysis above, the author arrives at such conclusions:

1. The logistics of food crops is a special type of logistics of agricultural products. The production, circulation and sales of food crops matters to state strategic reserve. Therefore, the logistics of agricultural products should be considered as a special type of logistics.
2. The demand of oil crops, a type of raw material of grain and oil processing industry, is concentrated. Its imperishable nature calls for a less demanding condition in package, transport and stock. As a result, the threshold of logistics is relatively low. Oil crops provide convincing evidence for the implementation of third-party logistics in rural area. A higher proportion of third-party logistics will make the increase of oil product output as a cause of the rise in logistics employment.
3 Self-owned logistics takes a big share in fruit logistics, while the proportion of third-party logistics remains low. Fruit consumption weighs a lot in daily life of residents. The lack of cold chain equipments shortens the radius of fruit logistics. Therefore, no causality has been proved between fruit logistics and logistics employment.

To sum up, apart from food crops, oil crops have causality with logistics employment and no causality has been found between fruit output and logistics employment. It is possible that both oil crops and fruit have something related with third-party logistics: the third-party takes a bigger share in oil crops logistics and smaller share in that of fruit.

In the follow-up research, the author will try to figure out to what extent the proportion of third-party logistics affects the correlation between output of agricultural product and logistics employment. The author will have a closer look on the effect of third-party logistics to the logistics of agricultural products.

References

1. Ding, J.: Develop Logistics of Agricultural Products with Great Effort. Supply and Marketing Cooperative Economy in China (June 2002)
2. Zhou, J., Li, Z.: Applicability of Granger Causality Test. Journal of Tsinghua University (Science and Technology) (March 2004)
3. Huang, Z., Liu, D.: China's Logistics System for Agricultural Products: Their Development and Institutional Aspects. Issues in Agricultural Economy (April 2005)
4. Zhang, Q.: Status Quo of Farm Produce Logistics in the Foreign Countries. World Agriculture (November 2004)
5. Ding, H.: On the Application of Supply Chain Theory in Enterprises Distributing Farm Produce. China Business and Market (January 2004)
6. Wang, N., Huang, L.: A Study on Agricultural Product Logistics Supply Chain Management Mode Based on Information Network. Research of Agricultural Modernization (March 2005)
7. Lu, W., Shen, P.: The Implementation of Granger Cause and Effect Test in the Study of Economic Cycle of China. Statistical Research (February 2002)
8. Zhang, X.: A Guide to Using EVIEWS. China Machine Press, Beijing (2007)

The Research on Modes and Application of Inventory Financing

Zhou Li[1], Jiang Jinsheng[2], and Liu Bingwu[1]

[1] School of Information, Beijing Wuzi University, Beijing, 101149, China
[2] Graduate Department, Beijing Wuzi University, Beijing, 101149, China
zhoulibit@126.com, i.melon@163.com, Liubingwu@bwu.edu.cn

Abstract. Inventory financing, as a new financial derivative, has boarded the stage of history for many years to solve the problem of "difficult financing "of minor enterprises. Although there are many deficiencies about inventory financing in normalization and multiformity, it solves, to a certain degree, the problem. This article illustrates inventory impawn financing emphatically from two aspects of warehouse receipt financing and confirming storehouse financing and gives two specific cases to further explanation.

Keywords: Inventory financing, the third party logistics, Minor enterprises.

1 Introduction

In recent years, minor enterprises have become fresh troops of Chinese economic development. According to statistics, minor enterprises, created for national output 60%, accounted for almost 90% of the national enterprise. But the mainly reason that restricting the development of minor enterprises is difficulty of financing which is more prominent after finance storm. The state has also issued relevant policies to support the development of minor enterprises, but due to their poor information transparency and management standardization and higher unsystematic risk, in addition to commercial banks, whose management principle is security, profitability, liquidity, lack of professional knowledge of product evaluation and warehouse management, so the condition of commercial banks would borrow no money nor loan to minor enterprises still exists. In 1999, China National Materials Storage and Transportation Co. (CMST), as the first logistics enterprises, joined with commercial banks launched inventory financing. Later, particularly in recent years, Huaxia Bank, China Merchants Bank launched logistics financial services. As a result, inventory financing in China has achieved great development and become, to a certain extent, the link of logistics enterprises, banks, and minor enterprises and provided them with a win-win platform. Subsequently, the generation of supply chain finance became inventory financing new stage. Presently, including Shenzhen development bank, Shanghai pudding bank, and a number of commercial banks have also launched a supply chain finance business. Then, inventory financing based on supply chain finance also further upgrade which provides more opportunities for minor enterprises [1,2].

J. Luo (Ed.): Affective Computing and Intelligent Interaction, AISC 137, pp. 35–42.
springerlink.com

2 Logistics Finance and Supply Chain Finance

Logistics finance and supply chain finance are both financial innovation of financial products. Although there are a slightly differences in interpretation of the concept between them ,it`s not hard to see supply chain finance is the development and deepening of logistics finance, in another word ,logistics finance is part of supply chain finance. In practical application, supply chain finance also solved some problems which logistics finance unable to solve at the beginning of generation. Firstly, let`s get an insight into the concrete concept of logistics finance and supply chain finance.

2.1 Logistics Finance

Logistics finance, logistics services which is included in financial services function, refers to the loan enterprises which are in production and logistics business in order to reduce transaction cost and risks and through logistics enterprises to obtain financial support from financial institutions. Meanwhile, logistics enterprises will provide logistics supervision and corresponding service, for example financing and financial settlement, so logistics finance is a value-added service created by logistics. Understandably, third party logistics companies, the link of finance and minor enterprises, exist as main operation part in logistics finance. When banks on inventory financing for minor enterprises , the information and value of inventory of minor enterprises is not very understand by them and the cost of monitoring mortgage stock is also high. Therefore, only relying on banks themselves against credit risk, legal risk and operation risk occurred during inventory financing is very difficult; on the contrary, it will easier if let the third party logistics enterprises involved as main operation party. In addition, inventory financing of logistics financing is only facing one of enterprises in supply chain or a loan enterprises out of supply chain and the link involved in is more one fold . Therefore, the process of this financing way is more compact and not involves affiliate guarantee, relatively, risk is more minor.

2.2 Supply Chain Finance

Supply chain finance is a portfolio investment derived from receivables , prepayments and inventory finance under commodities trading items of enterprises and base on core enterprises as entry point in order to control information flow , logistics ,capital flow effectively or bind related party by duty and provide financing services for the suppliers and retailers cooperate with the enterprises in the long-term. In supply chain finance ,the third party logistics enterprise ,for banks , is agent and provide services for suppliers and retailers located in the upstream and downstream of supply chain. Correspondingly, financial institution is the operation part. The standard paradigm of supply chain is "1+N" put forward by Shenzhen Development Bank firstly which means the core enterprises, symbolized by "1", drive the minor enterprises, symbolized by "1", in upstream and downstream of supply chain. In addition, "+" represents the connection of interests risks between the core enterprises and minor ones. From this mode of operation can be seen, supply chain finance services for one or more enterprises in supply chain, so the process of this financing mode is more complicated relatively. Meanwhile, it also refers to association guarantee and has much higher risk.

3 Inventory Financing

Inventory financing, also called financing warehouse, produced with logistics financing, refers to enterprises gain financing support by their own operating or producing inventory as pledge and is a loan way according to the total inventory to obtain revolving credit. The two most common practices are warehouse receipt pledge financing and confirming storehouse financing.

3.1 Warehouse Receipt Pledge Financing

warehouse receipt pledge financing, a capital circulation pattern that financial institutions involved in directly, just as the name implies , is that getting financing from bank by warehouse warrant as pledge and can be divided into impawn assures financing and credit guarantee financing[3,4].

a. Impawn Assures Financing

In impawn assures financing ,loan enterprises can bring some productions which already produced but cannot be converted into cash immediately or some raw materials which already purchased but cannot be used in a short time into the warehouse that the third party logistics enterprises allocated . Alternatively, they can also use their own warehouse, but pledges must under the supervision of the third party logistics enterprises. The third party logistics enterprises evaluate and review these good and draw up warehouse warrant. And minor enterprises can rely on it to apply for loan from bank (the specific process see fig. 1). In this pattern, the goods have been supervised by the third party logistics enterprises trusted by both bank and loan enterprises. The goods as pledge realize realization, thus speed up the circulation of capital of loan enterprises and guarantee production. Meanwhile, if loan enterprises want to delivery some goods, they must hand in margin, then bank inform the third party logistics enterprises to issue corresponding quantity of goods for loan enterprises.

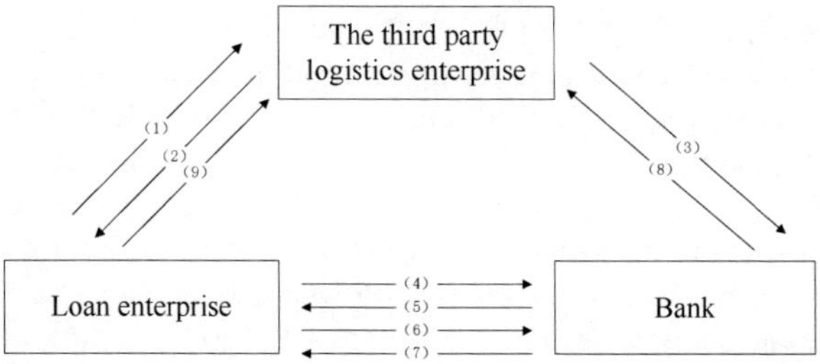

Fig. 1. Inventory financing mode I

(1) pledge (2) open warehouse receipt (3) evaluation certificate (4) apply for loan rely on warehouse receipt (5) offer loan (6) hand in margin (7) bill of landing (8) dispatch notification (9) dispatch.

b. Credit guarantee financing

In credit guarantee financing mode, bank would assess the third party logistics enterprises and provide certain independent credit quota for the third party logistics enterprises according to the scale, operating performance, balance sheet rating and credit rating of it. The third party logistics enterprises can choose a minor enterprises as cooperation partner from long- term cooperative enterprises and provide credit guarantee for it . Loan enterprises put can put goods into warehouse of the third party logistics enterprises and supervised by the third party logistics enterprises within 24 hours. When goods of loan unit need been outbound or put in storage, loan enterprises only need confirmation of the third party logistics enterprises and doesn`t need to margin to bank and confirmation of bank. This kind is more convenient and less cost than impawn assures financing (the specific process see fig.2).

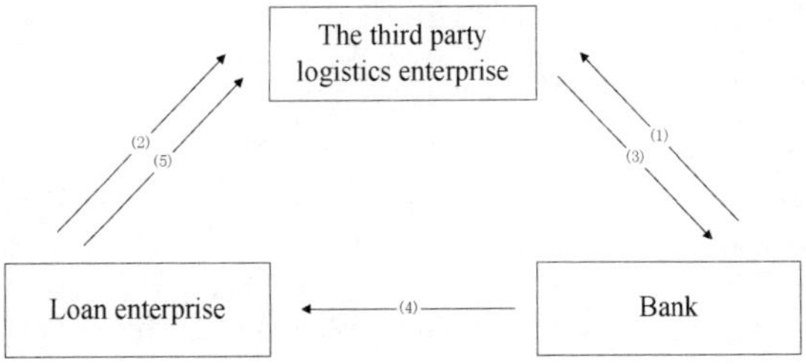

Fig. 2. Inventory financing mode II

(1) Independent credit quota (2) pledge (3) credit guarantee (4)offer loan (5) apply for ex-warehouse or putting in warehouse

3.2 Confirming Storehouse Financing

Confirming storehouse financing is produced with the generation of supply chain finance, needs upstream suppliers, downstream vendors in the supply chain, the third party logistics enterprises and banks participated in together, and sign business cooperation agreement. Vendor pays certain margin to bank who issue acceptance bill and purchases goods from supplier relying on the acceptance bill. Then the goods are delivered to and supervised by the designated third party logistics enterprises who issue warehouse receipt of the pledges. When the bill expires, bank will settle account with supplier. During this time, if the loan enterprise wants to put forward some goods, it needs to pay a certain number of margins to bank again. The bank send out bill of lading and meanwhile notice warehouse to sent out corresponding quantity of goods. Accordingly, the third party logistics enterprise checks out the bill and sends out goods (the specific process see fig. 3). Confirming storehouse financing adopt a opposite pattern of warehouse warrant pledge loans. the loan enterprise acquire acceptance bill from bank firstly and depend on it purchase goods from suppliers and then these goods are supervised by the third party logistics enterprise. Obviously, confirming storehouse financing is the deepening and upgrade of warehouse receipt finance. In most cases, non-core minor enterprises in supply chain to obtain loans are still not well-off under warehouse receipt finance. But with a view that the core enterprises in supply chain are generally more powerful and higher credit rate, as a result, confirming storehouse finance is solve the problem of financing difficulty of minor enterprise in upstream and downstream of supply chain. Accordingly, this pattern, in which the core enterprise in supply chain is primary or is jointly and severally liable for minor enterprise in supply chain, is chain model of financing. So it has a "one-wing all-wing, one loss" effect. Any tache goes wrong will affect the whole chain capital withdrawn successfully. Therefore, the supervision of confirming storehouse financing would be more complicated[5,6].

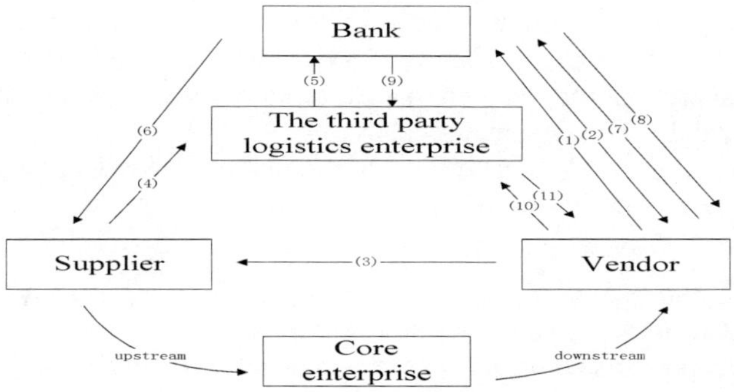

Fig. 3. Inventory financing mode III

(1) hand in margin (2) acceptance bill (3) purchase rely on bill (4) dispatch (5) warehouse receipt of pledge (6) expire settlement (7) hand in margin again (8) bill of lading (9) dispatch notification (10) delivery rely on BL (11) dispatch

Table 1. Specific comparison of two kinds of inventory

Kinds of inventory	**warehouse receipt pledge financing**	**confirming storehouse financing**
The third party logistics enterprise	Main operation party	Agent of bank , provide service for minor enterprise in supply chain
Loan enterprise	Individual enterprise out of supply chain or core enterprise in supply chain	Minor enterprise in upstream or downstream of supply chain
Association guarantee	No	Yes
Bank	Offer loan	Main operation party

4 The Specific Application of Warehouse Receipt Financing

4.1 Simple Warehouse Receipt Financing

Background introduction: Yantai Tom Aquatic Products Ltd. located in the Jincheng town of Lanzhou city (the following short for Tom Aquatic Products), is a small-scale export processing corporation of aquatic products, parts of whose raw materials depend on import. Because of the impact of financial crises and the delay of payment by clients, circulating capital fails to circulate. However, according to the regulation of large commercial bank, the load needs to be mortgaged with the form of immovable property. But taking the small evaluation of real estate of Tom Aquatic Products into account, it is hard to get the loan. Meanwhile, the agricultural and commercial bank of Lanzhou (the following short for ACB) launched a loan form of warehouse receipt finance the clients of which are those business entities, other economic organizations and privately owned business that possess certain amount of capital and could do checking independently.

Case introduction: In XX of 2009, Tom Aquatic Products apply for using the standard warehouse receipt finance of raw material. Conditions of ACB are as followed:

(a). As the third-party logistics enterprise of financing warehouse which is admitted by ACB, it should have well-deserved reputation and economic power.

(b). The mortgage should be not easy to wear down and depreciate. They should be raw materials and semi-finished products with great capacity of liquidity.

(c). 20% of the amount of deposit and loan should be paid as margin. The amount of loan should be measured at the level of 67% of the value of goods certificated by the ACB.

Tom Aquatic Products first raise an application to ACB with the copies of business license and certificate of enterprise code, loan card, financial statements for the last year and recent days and the warehouse receipt given by the third party, after the ratification and registration of stock guaranty in the administrative department of industry and commerce, Tom Aquatic Products pays the margin and sign the tripartite agreement with the third party determined by ACB. After all these procedures, ACB will provide a loan of 8,000,000 to Tom Aquatic Products for helping her walking through the difficulties.

4.2 Case of confirming Warehouse Financing

Background Introduction: Group Corporation A is one of the famous professional coal trading companies. Four sub companies: Qinhuangdao subcompany, Guangzhou sub company, Yangyuan Company and Datong Company make up the buying and selling system of the corporation. After an offer is set to the Guangzhou Company, based on the ability of three other related companies, the process of transporting coal will be soon accomplished. For the rising in coal price recent years, Group A has branched out and raise the stockpiling to reach goal of monitoring costs and raising the ability of making profits, but has taken a large portion of the group's floating capital. Shenzhen Development Bank (SDB) had launched "whole course logistic" supply chain financial transaction model (to be mentioned as "whole course logistic" model in the following passages), under the canonical form of "N+1" and the throughout supervise of the mortgage goods by the third-party logistics enterprise confirmed by the bank to ensure the total vale to be higher than the value requested by the bank for offering certification.

Case Introduction: Group A apply a stockpiling mortgage financing and SDB will investigate into the situation of the subcompanies, and set the Guangzhou subcompany in charging of selling as the borrowing totality. The loan company and three other subcompanies acting as the third parties pledging with coal together, with the supervise of stock pilings, warehouse receipt and mortgage debit note shown by the Logistics Center of China Ocean Shipping Company, SDB will offer financing to Group A. The procedure in detail is like this:

(a). SDB, the Logistics Center of China Ocean Shipping Company, Guangzhou sub company, Qinhuangdao sub company, Yangyuan sub company, Datong sub company sign a floating mortgage supervision agreement of six participants. Guangzhou sub company is mortgagor; Qinhuangdao sub company, Yangyuan sub company and Datong sub company are all Third-party mortgagor along with joint security.

(b). While Guangzhou sub company is pledging, SDB send a mortgage notice to the Logistics Center of China Ocean Shipping Company consulting about the mortgage goods. The Logistics Center of China Ocean Shipping Company will show Review Confirmation and mortgage list to SDB and sealed them for identification.

(c). The corresponding coal of terms in mortgage list are mortgaged to SDB by Guangzhou sub company. And Guangzhou sub company should sign an agreement and related legal documents with SDB. SDB offers loan to Guangzhou sub company.

(d). When Guangzhou sub company applies for picking up goods, if it is higher than the lowest reserve mortgaged to SDB, it can pick freely, otherwise, it needs to complement the rest or apply for pay backing loan.

(e). SDB will deduct the corresponding margin or sign the notice of picking up goods.

(f). The Logistics Center of China Ocean Shipping Company should deal with formalities to let related mortgage pass according to the notice of picking up goods, and at the same time sign the notice and send it to the bank. This business help the goods stocked up in a corporation change into cash in time and solve the problem of turnover of capital.

5 Summary

Using flexible operating pattern, inventory financing join banks, minor enterprises and logistics companies together, to a certain extent, provide a feasible way to solve the problem of financing difficulty of minor enterprises. Not only help minor enterprises obtain capital, but also achieving the effective combination of logistics, commercial flow and cash flow. Although inventory financing, in China only developed for more than ten years, is an emerging industry, as inventory financing is more diversified and various institution is more normalized, it will play a role better and help minor enterprises pull through capital difficulty and realize their own progress.

Acknowledgment. This paper is sponsored by 2009 Youthful Funding Project for Supply Chain Risk Management Based on Financial Econometrics Technology of Beijing Wuzi University, and Funding Project for Academic Human Resources Development in Institutions of Higher Learning Under the Jurisdiction of Beijing Municipality (PHR200906210, PHR201007145, PHR200907134), Beijing Philosophy and Social Science Planning Project (09BaJG258), Funding Project for Base Construction of Scientific Research of Beijing Municipal Commission of Education (WYJD200902).

References

1. Zhang, J., Zhu, J.: Comparative study between logistics finance and supply chain finance. Financial Theory and Practice (10), 35–38 (2009)
2. Shenzhen development bank: China Europe international business school. Supply chain finance——the new finance under the new economy. Shanghai Yuandong Press, Shanghai (2009)
3. Zhu, X.: Management school, inventory finance under supply chain finance. Shanxi Agriculture Science (2), 155–158 (2010)
4. Zheng, H.: The reason of minor financing difficulties and countermeasures. Copperative Economy and Science 380, 32–33 (2009)
5. Xie, S.: College of Logistics, overview of researches on logistics finance abroad. Logistics Sci-Tech (4), 63–67 (2010)
6. Coleman, A., Valeri, L.M.: Storage and warehouse receipts as financing instruments [EB/OL] (May 21, 2007) (December 25, 2009), http://www.eea-esem.com/files/parpers/EEA-ESEM/2006/2046/WR-malaguzzivaleri.pdf

Technology Road Mapping Method Combined Requirements-Pull with Technology-Push Analysis*

Hongqi Han[1,2], Xiaomi An[1,2], Donghua Zhu[3], and Xuefeng Wang[3]

[1] Key Laboratory of Data Engineering and Knowledge Engineering (Renmin University of China), Ministry of Education;
[2] School of Information Resource Management, Beijing 100872, China
[3] School of Management and Economics, Beijing Institute of Technology, Beijing 100081, China
bithhq@163.com, xiaomia@yahoo.com, zhudh111@bit.du.cn, wxf5122@gmail.com

Abstract. Most existing technology roadmaps were built based on requirements-pull analyses which were driven by future market needs. The major problem of the roadmapping method is the released technology roadmaps have to be revised frequently with rapid changing technology. The paper presented a new roadmapping process which consists of three stages aiming at resolving the problem. The three stages are preliminary activities, development activities and follow-up activities. Two workgroups of requirements-pull and technology-push adopt different work process for consistent objectives in the second stage. The new roadmapping method can make use of present technology assets like patents and integrate ideas of business experts and technology experts.

Keywords: Technology roadmap, Requirements-pull, Technology-phsh.

1 Introduction

Technology Roadmap (TRM) is an emerging technology strategy management technique. TRM can be used to tighten links between company plan and technology plan. Compared with traditional management methods, TRM is a more effective tools of connecting business and technology [1,2], and is a more powerful tools in supporting technology management and planning [3]. To be a strategic decision technology, TRM is an important means to connect technology and strategy planning in science, technology and product strategic management [4].TRM in country or industry level can be used to forecast the technology development trajectory to help achieving competitive advantages by allocating investments or funds into prospective research fields; TRM in company level can reveal changes of market, product and technology with time[5], so it provides a way to improve competitiveness by investigating competitive environment and identifying technology opportunity and risk[6].

* This work is supported by the Fundamental Research Funds for the Central Universities, and the Research Funds of Renmin University of China under contract 20100301682.

J. Luo (Ed.): Affective Computing and Intelligent Interaction, AISC 137, pp. 43–50.
springerlink.com

TRM was initially developed by Motorola in 1970s[7], and the technique spread to other technology-intensive firms like Phillips and Corning. A significant milestone of TRM is the success development and application of international technology roadmap for semiconductors [8]. Now TRM has developed through two generations, and the third generation is now emerging. First generation began in 1970s and end in mid-1980s, the roadmapping methodologies aimed at clear and accurate technology forecasting; second generation began in mid-1980s and continued to end-1990s, the roadmapping methodologies aimed at improving strategic technology planning decisions; third generation began in end-1990s and continued to today, the roadmapping methodologies aimed at producing integrated technology management activities.[9]

Along with wide spread and diffusion, TRM was used in country, industry company and product level. Owing to TRM performed better than original conceives, researchers now focus on finding a better roadmapping process to build build TRM. Most of the developed TRM are requirements-pull which were driven by analyzing future market needs. Researchers found that forecasts by such kind of TRM often deviate from practical situation quickly because technology often changes rapidly and beyond original anticipation, so they had to revise TRM frequently to adapt changes. Apparently, repetitive adjustment will influence using value and increase maintenance cost of TRM. Therefore, a roadmapping process combining requirements-pull with technology-push was presented here trying to resolve the problem and build a more adaptive TRM.

2 Requirements-Pull and Technology-Push TRM

There are two major types of TRM analyzed from a temporal perspective: retrospective analyses and prospective analyses [2,10]. The European Industrial Research Management Association simply refers to these two approaches as "backward" and "forward". Typically, retrospective analyses cover time frames from decades past to the present, while prospective analyses cover time frames from present to a decade or more into the future. Since retrospective analyses concern more about history than future technology evolution, however people pay close attention on future technology and market, therefore most existing TRM is prospective. Kostoff(2001) [10] mentioned there are also combination retrospective-prospective roadmaps, however it is hard to find such TRM in literature survey.

There have been two types of prospective analysis by roadmapping drivers: requirements-pull and technology-push. Requirements-pull analysis starts with a satisfied technology or products and works backward to identify the crucial researches and developments which support to reach future destination; Technology-push analysis starts with present developed or funded science and technology projects, and traces evolution forward to identify their potential influences. Roadmapping process driven by technology-push is just opposite to requirements-pull, which starts from R&D (Research and Development) planning, then goes through technology planning and product planning, and ends at market planning [10]. Fig.1 shows the basic structures of technology-push and requirements-pull TRM.

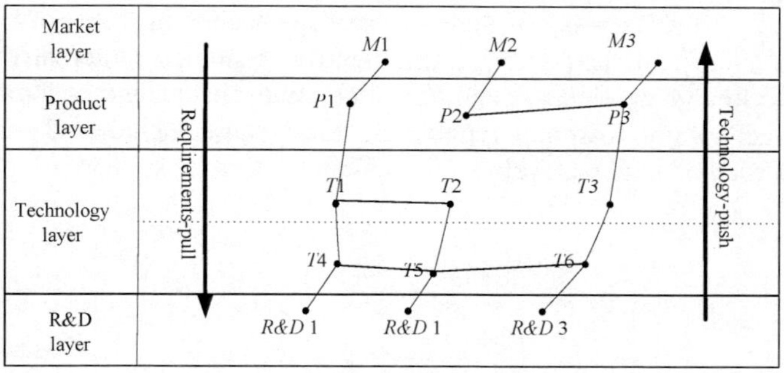

Fig. 1. Requirements-pull and technology-push roadmapping process

Most existing TRM has focused on requirements-pull roadmapping, aiming to identify necessary technologies from analysis of future market or technology needs. Few TRM were drawn up using technology-push method to find promising business opportunities from the analysis of existing technological assets, e.g. patents. The main reason lies in the technology of analyzing technology assets is hard to control, but emerging data mining technology throw lights on resolving such problems. Although most future-market-oriented TRM have been regarded as very useful and effective, it may ignore the possible business opportunities of various valuable technology applications, particularly for high-tech industries like 6T [10], that are IT(information technology), BT (bio-technology), NT (nano-technology), ST (space technology), ET (environmental technology), and CT (culture technology). Generally, there often exists great uncertainties at each stage of new technology development, application and diffusion of high-tech industry, technology-driven roadmapping approach will help reduce such uncertainties.

3 General Technology Roadmapping Process

The roadmapping process differs from one organization to the other because organizations may have different backgrounds, markets and cultures; the roadmapping process also differs from different developed roadmap types, e.g. a product technology roadmap and an industry technology roadmap may be made a great difference. It is hard to define a total same process for all kinds of TRM. Groenveld(1997) [1] presented an outline about how the overall roadmap-building process can be structured(Fig.2), and addressed that essential elements of the roadmap-building process are establishing a multidisciplinary project team that guides the process, and workshops that are organized to ensure integral involvement of, and input by, the organization.

Garcia& Bray (1997) [11] presented a well accepted technology roadmapping process which consists of three phases, namely preliminary activity, development of the technology roadmap, and follow-up activity. The first stage includes (1) Satisfy essential conditions, (2) Provide leadership/sponsorship, and (3) Define the scope and boundaries for the technology roadmap; the second stage includes (1) Identify the "product" that will be the focus of the roadmap, (2) Identify the critical system requirements and their targets, (3) Specify the major technology areas, (4) Specify the technology drivers and their targets, (5) Identify technology alternatives and their time lines, (6) Recommend the technology alternatives that should be pursued, and (7) Create the technology roadmap report; the third stage includes (1) Critique and validate the roadmap, (2) Develop an implementation plan, and (3) Review and update.

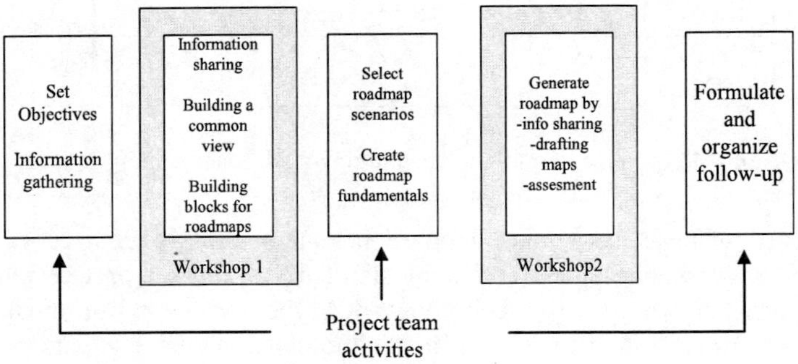

Fig. 2. General technology roadmapping process

4 Roadmapping Process Combined Requirements-Pull with Technology-Push Analysis

Apparently, TRM driver modes determine technology roadmapping process and building means to a great extent. No doubt, requirements-pull driver and technology-push driver have merits and application areas respectively. When experts can forecast future market and technology needs better, requirements-pull analysis may be a good way, while when there are many uncertainties in future markets and experts can understand present technologies better, technology-push analysis may be an effective way. Generally, future market and technology environment is very complicated and difficult to forecast, on the other hand, present developing crude technologies are also very difficult to understand, and it is painful to find business opportunities from technology assets analysis, so only use one of the two analysis methods may be not a good choice, and combination the two analyses may be a better roadmapping method.

Based on previous research achievements, a roadmapping process combining requirements-pull with technology-push method was presented (Fig 3). The proposed roadmapping process consists of three stages like common TRM process: preliminary activities, development activities and follow-up activities. The major difference from existing roadmapping process is that two groups are to be organized in the development activities stage. One group named requirements-pull workgroup constituted mainly by market experts, focusing on finding important technology and R&D project list from analysis of future market and technology needs; another group named technology-push workgroup constituted mainly by technology experts, focusing on finding important technology and R&D project list from analysis of current technology assets, such as patents, papers. Two R&D project lists is merged into one and prominent R&D projects are selected to be drawn in the developing

TRM after value evaluation, risk analysis and resource analysis. Each workgroup may include several workshops, and each workshop has a specific and detailed task. In order to coordinate works of two workgroups, the leading committee should make rules and standards for all team members to comply with, and provide effective institutional arrangement to share information.

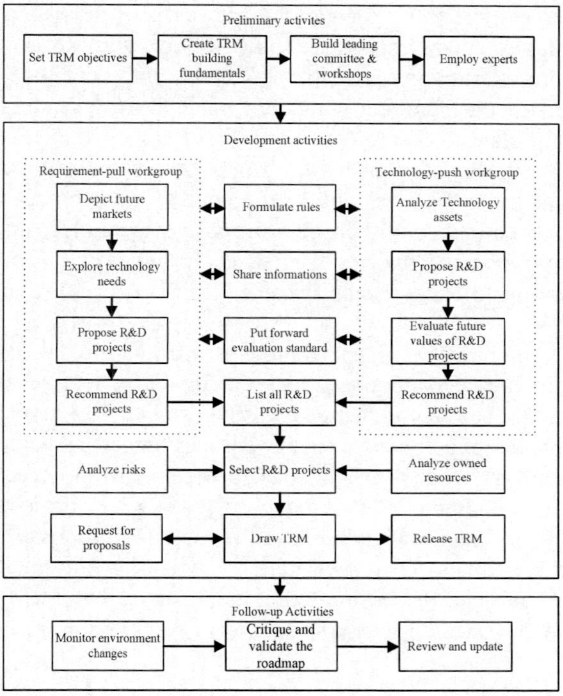

Fig. 3. Roadmapping process combining requirements-pull and technology-push

Following is the introduction of activities in three stages:

4.1 Priliminary Activities

The first stage prepares for developing TRM project and includes four activities mainly: set TRM objectives, create TRM building fundamentals, build leading committee and workshops and employ experts. In the stage, TRM development sponsors must first set development objectives based on their essential conditions and what problems they hope the TRM will help them, and then they should create TRM building fundamentals and organize a committee to make a plan to implement it. The TRM committee then selects and employs experts to build workshops for detailed tasks. Workshops are classified into two groups of requirements-pull workgroup and technology-push workgroup in the roadmapping process.

4.2 Development Activities

The second stage achieves developing technology roadmap by means of a series of activities by leading committee, requirements-pull workgroup and technology-push workgroup.

For workshops in requirements-pull workgroup, they first depict future markets scenario in order to explore future technology and market needs, then they analyze the supported technologies, and propose technologies list and R&D projects so as to arrive at the future destination smoothly. Finally, they evaluate proposed R&D projects according to the common standards made by the leading committee and recommend the important R&D projects.

For workshops in technology-push workgroup, they start from current technology assets analysis and try to find valuable technologies and projects which will bring great business opportunities, and then they propose a technology and R&D projects list. Similarly, they evaluate the proposed R&D projects with common standards. Finally, they recommend R&D projects that will influence future markets greatly to be drawn in TRM.

The Leading committee is responsible for formulating consistent rules and standards for team members to obey, and making an effective system for teams sharing information. The recommended R&D projects by two workgroups are evaluated by employed experts considering risk and owned resource conditions, and only those important and indispensable R&D projects will be selected and included into the developing roadmap. After having selected R&D projects which will be drawn, technology roadmap draft is built and request for proposals. TRM draft may be revised for several times. Then revised TRM will be released. During the whole period of the TRM project, the leading committee carries out TRM project plan and ensure project to be finished on schedule.

4.3 Follow-Up Activities

The third stage accomplishes subsequent works and includes three main activities: monitor environment changes, critique and validate the roadmap, and review and update roadmap. Generally speaking, only few people were involved in developing the roadmap, it must be critiqued, validated so as to be accepted by a larger group that will be involved in future implementation, on the other hand, the business and technology environment is changing rapidly and dynamically, so roadmap must be reviewed and updated to meet new conditions.

5 An Example of Technology Assets Analysis Used in Roadmapping Process

There are many papers explaining Roadmapping process based on requirements-pull analysis, so here just gave an example to illustrate how technology assets analysis to be used in the presented roadmapping process.

Fig.4 is a fuel-cell technology association diagram in one of our patents analysis projects. Each node in the diagram represents a kind of fuel-cell technology. The digit beside each technology node is its number. Node size indicates corresponding patent quantities. Line between two nodes indicates the two technologies have some close association.

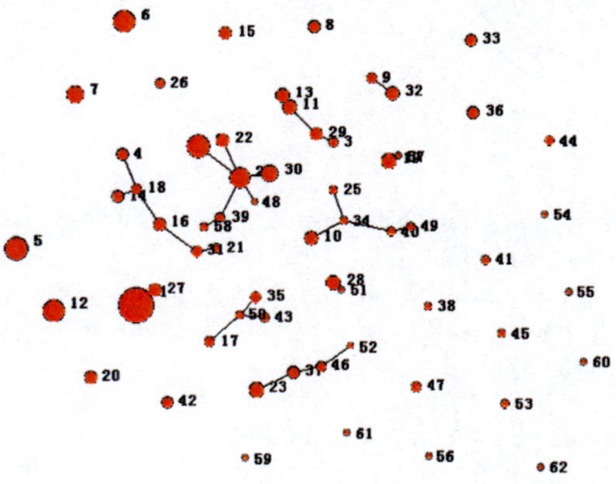

Fig. 4. A technology association diagram in fuel-cell patents analysis

When the experts of technology-push workgroup had an association diagram after analysis of the fuel-cell patents, they could found some more important technologies or R&D projects. For examples, node 1 has the max size, indicated the technology was attracting many researchers or technology experts to study; technology of node 2 may be important because it links between five other nodes. After finding and evaluating the interesting technologies, technology experts and business experts could co-work to list important R&D projects which will bring great future business opportunities. We can also consider each sub-diagram of Fig.4 as a R&D project choice and evaluate them in the second stage of presented roadmapping process. Having done these works, experts could recommend the important R&D projects to be included into developing TRM.

6 Summary

TRM is an effective strategic planning tool for technology management. TRM has two major types analyzed by time frame: retrospective analyses and prospective analyses. Prospective analysis is widely used owing to focusing on future technology evolution. There have been two types of prospective analysis by roadmapping drivers: requirements-pull and technology-push. Most existing TRM were requirements-pull driven by markets. However, TRM only with Requirements-pull analysis may bring troubles in frequent revision after released. The paper presented a roadmapping process combined requirements-pull method and technology-push method to solve the problem. The process consists of three stages: preliminary activities, development activities and follow-up activities. The first stage includes several activities preparing for the TRM project; the second stage focuses on select important R&D projects for drawing in roadmap based on works of requirements-pull workgroup and technology-push workgroup; the third stage emphases on subsequent review and update of released roadmap. The presented roadmapping process can make use of current technology assets and integrate ideas from business experts and technology specialists; thereby it may create a better roadmap and avoid revising released TRM frequently.

References

1. Groenveld, P.: Roadmapping integrates business and technology. Research Technology Management 40(5), 48–55 (1997)
2. Lee, S., Yoon, B., Lee, C., et al.: Business planning based on technological capabilities: Patent analysis for technology-driven roadmapping. Technological Forecasting and Social Change 76(6), 769–786 (2009)
3. Dhillon, I.S., Modha, D.S.: Concept decompositions for large sparse text data using clustering. Machine Learning 42(1), 143–175 (2001)
4. Li, X., Tong, Y., Tan, Y.: Technology roadmap and technology roadmap thinking. Science of Science and Management of S.T. (8), 26–28, 59 (2005)
5. Yoon, B., Phaal, R., Probert, D.: Structuring technological information for technology roadmapping: data mining approach. In: 7th WSEAS Int. Conf. on Artificial Intelligence, Knowledge Engineering and Data Bases (AIKED 2008). University of Cambridge, UK (2008)
6. Walsh, S.T.: Roadmapping a disruptive technology: A case study The emerging microsystems and top-down nanosystems industry. Technological Forecasting and Social Change 71(1-2), 161–185 (2004)
7. Willyard, C.H., McClees, C.W.: Motorola's Technology Roadmap Process. Research Management, 13–19 (September/October 1987)
8. Ding, Y., Tan, C.: Technology roadmap used as a tool of technology foresight and its future in application. Journal of Public Management 3(4), 40–45 (2006)
9. Gindy, N.N.Z., Cerit, B., Hodgson, A.: Technology roadmapping for the next generation manufacturing enterprise. Journal of Manufacturing Technology Management 17(4), 404–416 (2006)
10. Kostoff, R.N., Schaller, R.R.: Science and technology roadmaps. IEEE Transactions on Engineering Management 48(2), 132–143 (2001)
11. Garcia, M.L., Bray, O.H.: Fundamentals of Technology roadmapping, Strategic Business Development Department, Sandia National Laboratories (1997)

Multi-sensor Augmented Reality Tracking Based on Robot Hand-Eye Calibration*

Yang Zhang[1], Lin Li[1], Ya Zhou[1,*], Tao Su[1], Jin-Tao Ma[1], and Xiao-Ming Hu[2]

[1] School of Optoelectronics, Beijing Institute of Technology, Beijing 100081, China
[2] School of Life Science, Beijing Institute of Technology, Beijing 100081, China
zhouya@bit.edu.cn, bithxm@bit.edu.cn

Abstract. Calibration of coordinate systems between cameras and world coordinate system is of key importance in augmented reality. In this paper, a multi-sensor mechanical tracking method is proposed based on robot hand-eye calibration. The original camera pose estimation is obtained using computer vision algorithm by improved orthogonal iteration. In tracking process, the camera extrinsic parameters, i.e. rotation and translation matrixes are got by rotary angles of sensors mounted on the multi-arm mechanical system. Users can observe augmented scene on the video see-through display set on the end of the arms and move in six degree of freedom by linkage motion. The tracking accuracy and test application are presented.

Keywords: Augmented reality, Hand-eye calibration, Tracking.

1 Introduction

Sensor-based augmented reality has been an active area of research. Xinyu Li et al. presented a ubiquitous augmented reality prototype system by using wireless sensor networks to gather needed monitoring data throughout the real object or physical environment [1]. Bleser, G. et al. presented a visual-inertial tracking device which relies on a 3D model of the scene that enables to predict the appearances of the features by rendering the model using the prediction data of the sensor fusion filter [2]. Dongpyo Hong et al. extended the Composar augmented reality authoring tool to add support for sensor-based interaction [3]. Schall, G. et al. presented a system that uses Kalman filtering for fusion of Differential GPS [4].

In this paper, a multi-sensor mechanical tracking method based on robot hand-eye calibration is used in augmented reality.

Many research based on hand-eye calibration has been presented. Martin Hülse et al. introduced a scheme of fast learning and readaptation of robotic sensorimotor mappings [5]. Guilaume Walck et al. presented a complete 3D-reconstruction method optimized for online object modeling in context of object grasping by a robot hand

* This research work was supported by the National Natural Science Foundation of China (30900385).

* Corresponding author.

J. Luo (Ed.): Affective Computing and Intelligent Interaction, AISC 137, pp. 51–57.
springerlink.com

[6]. Mitchell Dinham et al. presented a low cost hand-eye calibration method for arc welding robots [7]. R. C. Luo et al. presented a multi-sensor based object grasping with eye-in-hand laser ranger [8].

The paper is organized as four sections. Section II (A) presents the system architecture including hardware and software, Section II (B) presents the hand-eye tracking algorithm in our system and Section II (C) presents the error alignment algorithm. Section III presents the test application in augmented reality.

2 Hand-Eye Calibration Algorithm

2.1 System Architecture

The augmented reality system named Video See-through Tabletop Augmented Reality (VSTAR) consists of the handheld display and the linkage mechanism based on the following hardware: a Intel Pentium 4 CPU 2.80 GHz PC with the camera capture card, the USB encoder data acquisition system, an 8-inch 800*600-pixel LCD and an color CCD 512*582-pixel camera mounted on the bottom, 6-DOF linkage mechanism and rotary encoders mounted on the joints (shown in Fig. 1). The software running on the system is based on Microsoft Visual Studio, Matlab, OpenGL library and 3DS models.

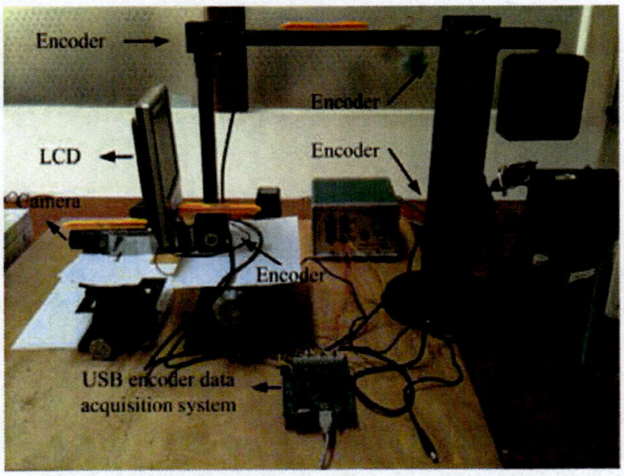

Fig. 1. System hardware

2.2 Hand-Eye Coordination

Robot hand-eye calibration is the process of determining the geometrical relationship between a camera system and a robotic manipulator. The robot motions are guided by feedback information acquired by vision sensors. In our system, the geometrical relationship between the camera and the linkage mechanism is obtained through the rotary angles of the encoders mounted on the joints.

Augmented reality registration is to get the coordinate systems transformation between the camera coordinate system and the world coordinate system, i.e. camera extrinsic parameters, the rotation and translation matrixes R and T which can be represented in (1).

$$\begin{bmatrix} u \\ v \\ 1 \end{bmatrix} = \begin{bmatrix} K & 0 \end{bmatrix} \begin{bmatrix} R & T \\ 0 & 1 \end{bmatrix} \begin{bmatrix} p \\ 1 \end{bmatrix} \frac{1}{z_C} . \tag{1}$$

In computer vision registration, R and T can be calculated by iteration in (1) using different world coordinate p and the corresponding image coordinate u and v in several continuous 2D images. The computer vision registration method has obvious disadvantages that the virtual images drift in tracking process and iteration operation is time-consuming.

In our system, using hand-eye calibration, the transformation between the bottom of linkage mechanism coordinate system and the camera coordinate system, named as ΔR and ΔT, can be obtained real-time in tracking process, which is different from previous computer vision tracking method and overcomes the image drift and time-consuming problems.

Because the rotary angles output from encoders are relative values, an original gesture should be established. The spirit levels and lift tables are used to assist for establishing of the horizontal and vertical gestures. In original gesture, the encoder data are set to zero. Using computer vision algorithm, just one picture in the original gesture is captured so that R and T can be calculated. The encoder data are zero and ΔR, ΔT are known values. So the transformation between the world coordinate system and the bottom of linkage mechanism coordinate system, named as R_0 and T_0, can be obtained. The position of the world coordinate system and the bottom of linkage mechanism coordinate system can not be changed in tracking process so they can be used as fixed values and ΔR, ΔT can be obtained real-time through encoder angles' data. By matrix multiplication, R and T in tracking can be calculated as (2) (shown in Fig. 2 (a)).

$$\begin{bmatrix} u \\ v \\ 1 \end{bmatrix} = \begin{bmatrix} K & 0 \end{bmatrix} \begin{bmatrix} \Delta R & \Delta T \\ 0 & 1 \end{bmatrix} \begin{bmatrix} R_0 & T_0 \\ 0 & 1 \end{bmatrix} \begin{bmatrix} p \\ 1 \end{bmatrix} \frac{1}{z_C} . \tag{2}$$

In our mechanical tracking system, the coordinate systems are established on every encoder and on the camera (shown in Fig. 2 (a) (b)). The sine and cosine values of the rotary angles of the encoders compose rotation and translation matrixes. By matrix multiplication, ΔR and ΔT can be calculated.

In matrixes operation, M_0, ΔM, M are defined as (3).

$$M_0 = \begin{bmatrix} R_0 & T_0 \\ 0 & 1 \end{bmatrix}, \ \Delta M = \begin{bmatrix} \Delta R & \Delta T \\ 0 & 1 \end{bmatrix}, \ M = \begin{bmatrix} R & T \\ 0 & 1 \end{bmatrix} . \tag{3}$$

In Fig. 2 (b), coordinate 6 is the camera coordinate; the matrix between the camera coordinate 6 and the bottom of the mechanical system coordinate 0 could be represented as (4). In Equation (4), M_{012}, M_{23}, M_{34}, M_{45}, M_{56} are the rotation and translation matrixes between two adjacent encoders.

$$\Delta M = M_{56} \cdot M_{45} \cdot M_{34} \cdot M_{23} \cdot M_{012} = \begin{bmatrix} \Delta R & \Delta T \\ 0 & 1 \end{bmatrix}. \tag{4}$$

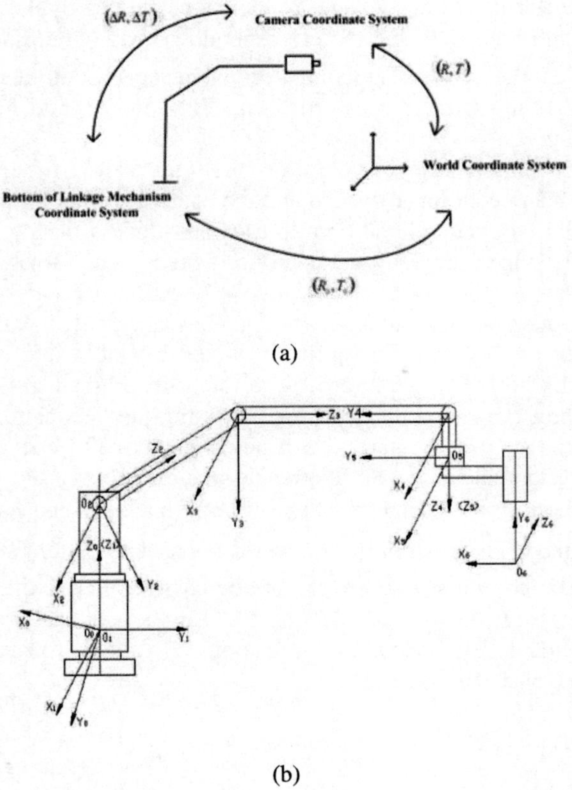

Fig. 2. Calibration of coordinate systems (a) The transformation of three coordinate systems (b) The transformation of every joint coordinate system

2.3 Error Alignment Algorithm

In practical situation, the arms of the linkage mechanism are not in a plane and the length measurement error exists, in addition, the optical center of the camera is not in the origin of the camera coordinate system we have established, so extra rotation and translation matrixes should be added in (4) and can be represented as (5). For example, M_{13}, M_{14}, M_{15} are the extra matrixes in x, y and z directions of the first arm between coordinate 1 and coordinate 2.

$$\Delta M = (M_1 \cdot M_2 \cdot M_3) \cdot M_{56} \cdot (M_4 \cdot M_5 \cdot M_6) \cdot M_{45} \cdot (M_7 \cdot M_8 \cdot M_9) \cdot M_{34} \cdot (M_{10} \cdot M_{11} \cdot M_{12}) \cdot M_{23} \cdot (M_{13} \cdot M_{14} \cdot M_{15}) \cdot M_{012} . \quad (5)$$

Through "fmincon" function in Matlab, the errors can be aligned by using nonlinear least square minimization method as (6).

$$f = f + \sum (u_1 - u_2)^2 + \sum (v_1 - v_2)^2 . \quad (6)$$

3 Test Applications in Augmented Reality

Because sensor-based hand-eye calibration is used as the tracking method, markers can be removed in tracking process, which gives users a more intuitive view field. In continuous video frames, a virtual teapot is superimposed on the real scene at a fixed position not rely on the marker (shown in Fig. 3). To verify the tracking accuracy, a virtual square is overlaid on one side of the marker cube and the marker can be sheltered by a book in tracking process (shown in Fig. 4). As the experiment result show, the image error of the four vertexes of the square is 0~4 pixels (shown in Fig. 5).

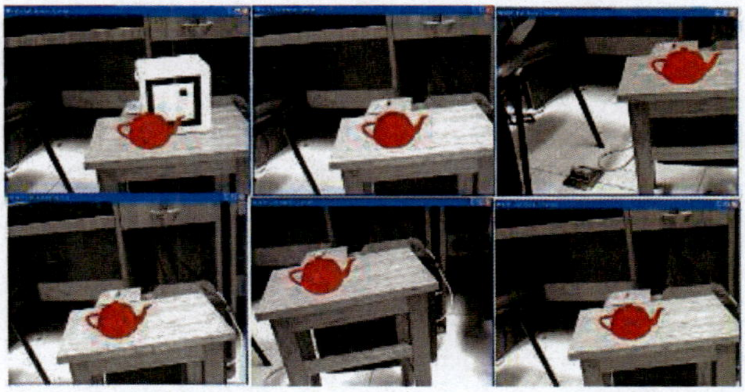

Fig. 3. The virtual teapot superimposed on the real scene not rely on markers

Fig. 4. The virtual square superimposed on one side of the cube not rely on markers with high accuracy

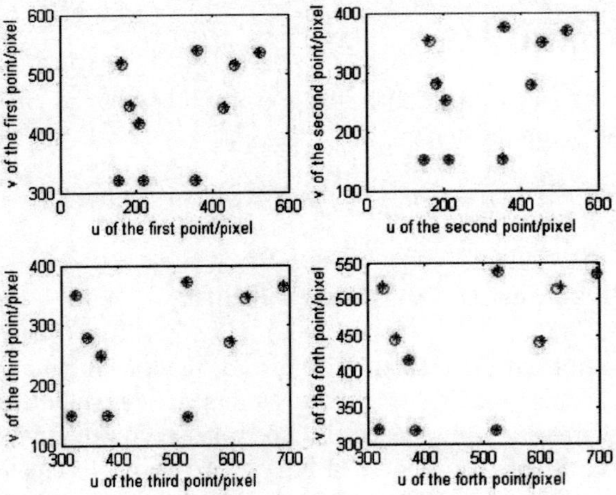

Fig. 5. The image error of the four vertexes of the square

4 Conclusion and Future Work

In this paper, multi-sensor augmented reality tracking method based on robot hand-eye calibration is presented. The advantages of our research are that tracking based on hand-eye calibration does not rely on markers which gives users a more intuitive scene and users can change the view field by holding and controlling the display driven by mechanical transmission. The tracking accuracy is tested by superimposing virtual objects and virtual wire frames. As the experiment result show, the image error is less than 4 pixels and our system offers possibility of developing novel applications in augmented reality area which demand high accuracy.

References

1. Li, X., Chen, D., Xiahou, S.: Ubiquitous Augmented Reality System. In: Second International Symposium on Knowledge Acquisition and Modeling, vol. 3, pp. 91–94 (2009)
2. Bleser, G., Stricker, D.: Advanced tracking through efficient image processing and visual-inertial sensor fusion. In: Virtual Reality Conference, pp. 137–144 (March 2008)
3. Hong, D., Looser, J., Seichter, H., Billinghurst, M., Woo, W.: A Sensor-Based Interaction for Ubiquitous Virtual Reality Systems. In: International Symposium on Ubiquitous Virtual Reality, pp. 75–78 (July 2008)
4. Schall, G., Wagner, D., Reitmayr, G., Taichmann, E., Wieser, M., Schmalstieg, D., Hofmann-Wellenhof, B.: Global pose estimation using multi-sensor fusion for outdoor Augmented Reality. In: 8th IEEE International Symposium on Mixed and Augmented Reality, pp. 153–162 (October 2009)

5. Hülse, M., McBride, S., Lee, M.: Fast Learning Mapping Schemes for Robotic Hand-Eye Coordination. Cognitive Computation 2(1), 1–16 (2009)
6. Walck, G., Drouin, M.: Progressive 3D reconstruction of unknown objects using one eye-in-hand camera. In: International Conference on Robotics and Biomimetics, pp. 971–976 (2009)
7. Dinham, M., Fang, G.: A Low Cost Hand-Eye Calibration Method for Are Welding Robots. In: International Conference on Robotics and Biomimetics, pp. 1889–1993 (2009)
8. Luo, R.C., Li, M.H., Jhu, H.L., Chen, J.W.: Multi-Sensor Based Object Grasping With Eye-in Hand Laser Ranger. In: 35th Annual Conference of the IEEE Industrial Electronics Society, pp. 2289–2294 (2009)

Measuring Software Product Quality with ISO Standards Base on Fuzzy Logic Technique

Haijun Yang

School of Information Science and Engineering, Lanzhou University of Finance and Economics Lanzhou, Gansu Province, China
yanghj@lzcc.edu.cn

Abstract. ISO25000 (SQuaRE) provides new series of international standards on software product quality measurement. This paper, based on ISO 25000 (SQuaRE) Software Quality Measurement Model, describes how to establish software quality assessment system, using Fuzzy measures to quantize Fuzzy characteristic (or sub-characteristic), and then apply Choquet integral for synthetically evaluation. An example is also provided to demonstrate how to use this approach.

Keywords: Software product quality, λ-Measure, Choquet integral, ISO 25000-SQuaRE.

1 Introduction

Software quality is one of the important issues in Software Development Life Cycle (SDLC). Software quality evaluation is the key factor for software quality assurance. In order to quantitative evaluate the quality of software; we have to make assessment to software characteristics, and evaluate how well the software meets requirements in quantity.

Nowadays, there are two software quality models: 1). Quality model that evaluates specific quality characteristics, i.e. Albrecht function point model [1] ,Halstead model [2] etc.; 2). Integrated hierarchy model, such as McCall model [3], Boehm model [4], and quality models in SQuaRE ISO 25000 standard serials [5, 6, 7]. The former model only focuses on evaluation of specific quality characteristics. The model and algorithm are relatively simple: data process is less heavy; data sampling is not hard to do. It is not difficult to create model and implement algorithm. However, it cannot evaluate overall quality, and it is hard to assess system quality as a whole.

Integrated hierarchy model represents all the factors and relationships that affect software quality. We can use this model to present relationship among these factors. The lower level can be divided into more detailed targets, then evaluate software overall quality. This model needs more data, and has more difficulties to sample data. It is more complex to represent algorithm and implement. So researchers are focusing on target selection in quality characteristics and comprehensive evaluation method. Most definitions of measurement target and judgments are largely based on subjective factors from end users and QA group, which comes with significant deviations. The

J. Luo (Ed.): Affective Computing and Intelligent Interaction, AISC 137, pp. 59–67.
springerlink.com

measurement target has a kind of fuzziness and uncertainty, which is better to be measured by fuzzy measure. On the other hand, due to relationship and constraint among measurement targets, we can reach more realistic result using fuzzy integrals for integrated evaluation.

The λ-fuzzy measure is one of fuzzy measures used widely, and has plenty applications recently, including pattern recognition, speaker verification, and public attitude analyzing because it reduces the difficulty of identification effectively [8, 9, 10, 11].

Using fuzzy measuring For maintainability, reliability, and related costs of lines of code (LOC) and Function Point ,which is a single characteristics of software quality, has plenty papers for discussion. References [12,13,14] using fuzzy measure measure characteristics and sub-characteristics of software quality, then incorporate AHP (Analytic Hierarchy Process)approach or GA (Genetic Algorithm) or fuzzy neural computing approach to the software quality evaluation.

2 Software Quality Models of ISO 25000 (SQuaRE)

The general goal of creating the ISO 25000 , which will be referred to as Software Product Quality Requirements and Evaluation (SQuaRE), is to move to a logically organized, enriched and unified series covering two main processes: software quality requirements specification and software quality evaluation, supported by a software quality measurement process. The SQuaRE series consist of five divisions: ISO 2500n-Quality Management Division,ISO 2501n-Quality Model Division,ISO 2502n-Quality Measurement Division,ISO 2503n-Quality Requirements Division, andISO 2504n-Quality Evaluation Division. SQuaRE series of standard replace ISO 9126 and ISO 14598 series of standards.

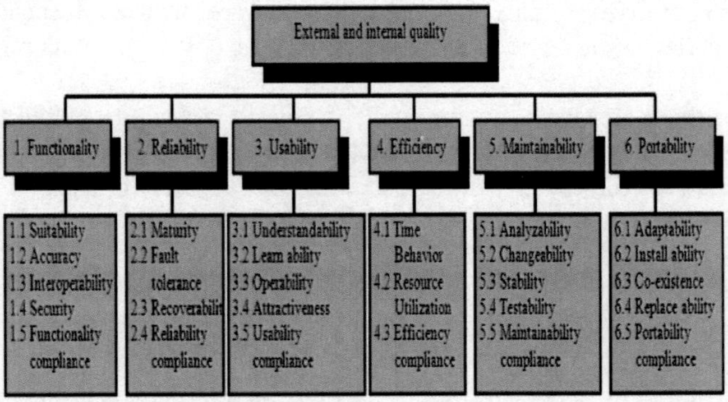

Fig. 1. ISO 2501n quality model for external and internal quality (characteristics and sub-characteristics) [7]

ISO 2501n - quality model is results of concatenation, unification and revision of more than one document from ISO 9126-1 series. SO 2501n describes a two-part quality model of software product quality, which are Internal and external quality model, Quality in-use model. The first part of the two-part quality model determines six characteristics in which they are subdivided into twenty-seven subcharacteristics for internal and external quality, as in Fig. 1. These subcharacteristics are a result of internal software attributes and are noticeable externally when the software is used as a part of a computer system. The second part of the two-part model indicates four quality in-use characteristics. All the quality characteristics and their corresponding sub-characteristics are defined in ISO 2501n [7].

Part of quality assessment can be figured out by classical mathematics directly, and get detailed value. However, lots of characteristics, sub-characteristics, and properties have uncertainties. We can only get fuzzy data by subjective assessment. When thinking of impact of multiple factors to software quality, we need to think about dimensional value to get comparability of result. On the other hand, we need to take care of fuzzification and defuzzification.

3 Background on Fuzzy Measures and Integrals

3.1 Key Concepts

Definition 1. A fuzzy measure on X is a set function $\mu : P(X) \to [0,1]$ (P(X) is the power set of X),satisfying the following conditions:

$$1.g_\lambda(\varnothing)=0, g_\lambda(X)=1$$

$$2.\forall A, \forall B \in P(X), if\ A \subseteq B, then\ \mu(A) \le \mu(B);$$

$$3.if\ \forall A_i \in P(X), A1 \subseteq A2 \subseteq A3,\cdots, or\ A1 \supseteq A2 \supseteq A3,\cdots, then$$

$$\lim_{i\to\infty} \mu(A_1) = \mu(\lim_{i\to\infty} A_i)$$

$\mu(A)$ can be viewed as the weight of importance of the set of elements A.

λ fuzzy measure is one of fuzzy measure the most widely used , which describe additive using parameters λ, and is a measure of the limit value of λ.

Definition 2. let $\lambda \in (-1,\infty)$ and let $X = \{x_1, x_2, \cdots, x_n\}$ be a finite set. if $(X, P(X))$ is a measurable space and if set function $g_\lambda : P(X) \to [0,1]$ satisfies the following conditions, then g_λ is denoted by a Sugeno λ measure and λ denoted by measure coefficient of determination [16].

$$1.g_\lambda(\Phi)=0, g_\lambda(X)=1$$

$$2.\forall A, \forall B \in P(X), A \cap B = \Phi, A \cup B \ne X$$

$$\Rightarrow g_\lambda(A \cup B) = g_\lambda(A) + g_\lambda(B) + \lambda g_\lambda(A) g_\lambda(B) \quad (1)$$

$$3.\prod_{i=1}^{n} (1+\lambda g_\lambda(x_i)) = \lambda + 1 > 0 \quad (2)$$

As a convention, $g_\lambda(x_i)$ is called a fuzzy measure, also fuzzy density.

Definition 3. Let set function $g : P(X) \to [0,1]$ be a fuzzy measure on measureable space $(X, P(X))$, and $h : X \to [0,1]$ be a measureable function on X, if $h(x_1) \le h(x_2) \le \cdots \le h(x_n), A_i = \{x_i, x_{i+1}, \cdots, x_n\}$,then

$$E \overset{def}{=} \int h dg \overset{def}{=} h(x_1) g(A_1) + \sum_{i=2}^{n} (h(x_i) - h(x_{i-1})) g(A_i) \quad (3)$$

Let us denote $E \overset{def}{=} \int hdg$ by the Choquet integral of measureable function h with respect to fuzzy measure g [17] .we denote E the overall evaluation. $h(x_i)$ is viewed as the performance of an alternative on characteristic x_i . The fuzzy measure, $g(A_i)$, can be used to express the grade of subjective importance for subset A_i . The fuzzy integral of $h(x_i)$ with respect to g gives the overall evaluation of an alternative.

Using Choquet integral is approach that is based on all kinds of important index in software assessment. If we use fuzzy measures to evaluate impacts between different assessment indexes, we definitely can understand better how the impact come from system states that caused by state characteristic.

3.2 Application Procedures of Choquet Fuzzy Integral

Fuzzy integrals are useful tool for global evaluation models. However, the number of parameter of fuzzy measure is large. λ fuzzy measure is one class of fuzzy measures which can be identified by fuzzy densities of individual evaluation items[15].

Application procedures for: (1) evaluation fuzzy densities for indexes of each level by expert (2) computing value of the parameter λ of each level (3) ascending arrangement $X_k^{i1}, X_k^{i_2}, \cdots, X_k^{i_{nk}}$ $(i_j = 1, 2, \cdots n_k)$ (4) computing $g(A_i)$ values of each level (5) computing choquet integral value for each level (6)capture overall evaluation.

Since $g_\lambda(X) = 1$ $(|X| = n)$ from the boundary conditions, value of the parameter λ can be uniquely determined by equation (2).

4 Mothed for Software Quality Synthetic Assessment

Based on the actual situation, we can use some characteristics, sub-characteristics and properties from 2501n ISO internal and external quality model as evaluation index, then create quality evaluation model using fuzzy measure and the fuzzy integral approach.

4.1 Setting Up Index System

When evaluating a software product or software project, we have to set up evaluation index. The evaluation index could be single layer, or multi-layer, which is based on the actual situation.

Taking Fig. 1 as example, we have characteristics and sub-characteristics as evaluation index. I will denote the main characteristics collection: functionality, reliability, maintainability, usability, efficiency, portability by $a=(a_1,a_2,\cdots;a_6)$ .in similar way, sub-characteristics collections is denoted by $a_i=(a_{i1},a_{i2},\cdots,a_{ik}),i=(1,2,\cdots 6)$. So we have two levels.

For some evaluation index that can be measured directly by regular statistic method, we can evaluate using detailed data. But for other evaluation index with fuzzy characteristics that are hard to represent, we need use this new model.

For some original data index with different dimension, since there is no any comparability, we cannot compare them directly.

We can use Dimensionless method for quantitative index with standard transformation that is widely used in Statistic.

4.2 Determine Values of Importance Index and Parameter λ

To evaluate the quality of a software product or related product, we need to set up the weigh for each characteristics or sub-characteristics. Usually, it takes a long time experience to get appropriate weigh, or we can get appropriate weigh through expert panel.

According to the assessment of the software products, we have to identity importance index (fuzzy densities) of elements in the a and a_i respectively, and parameter λ. Based on the statistical methods or expert evaluation method, we could get value for each index in the layers, then the parameter λ can be uniquely determined by equation(2).

Assume that there are m experts to evaluate index importance of n_k indices in n layers:

$$g_r=\left\{g_r\left(X_k^i\right)\mid k=1,2,\ldots,n;i=1,2,\ldots,n_k;r=1,2,\ldots,m\right\}$$

The $g_r\left(X_k^i\right)$ stand for value of importance index of ith index X_k^i in X_k th level estimated by rth expert. Calculating the mean value, we could acquire value of the importance index, as follow:

$$g=\left\{\frac{1}{m}g\left(X_k^i\right)\mid k=1,2,\ldots,n;i=1,2,\ldots,n_k;r=1,2,\ldots,m\right\}$$

Using the fuzzy integral integrated computation

First of all, we have to capture evaluation values of each level using Choquet integral, in similar way, then calculate evaluation values of higher level on the basis of results acquired.

Calculating fuzzy integral of X_k level, process is as follows:

(1) Renumber X_k^i 's according to the $h(X_k^i)$ value such that

$$h(X_k^{i_1}) \le h(X_k^{i_2}) \le \cdots \le h(X_k^{i_{nk}}), i_j = 1, 2, \cdots n_k$$

(2) Calculate importance index for $g(A_i)$ according to equation (1), where $A_i = \{x_i, x_{i+1}, \cdots, x_n\}, i = 1, 2, \cdots, n$

(3) Calculate importance index for $E \overset{def}{=} \int h dg$ according to equation (3), where $A_i = \{x_i, x_{i+1}, \cdots, x_n\}, i = 1, 2, \cdots, n$.

(4) Repeat steps (1) to (3), until the evaluation of the level of all.

5 Empirical Application

Due to the needs of purchase, various financial software system are evaluated by proposed software quality evaluation model.

Now let us apply proposal model to evaluate two kind of financial software A,B.Considering the characteristics of financial software system, we have to investigate emphatically three quality characteristics': functional, usability and reliability. Table I lists importance index value of each index, which estimated by common staffs,leadershipand expert. Consequently according to equation (2, 3), we could capture values of a_1, a_2, a_3 and parameter λ.

Table 1. Evaluation index system and importance index

Level	charateristics	importance index	λ
1 level a	Functionality a1 Usability a2 Reliability a3	0.75 0.618 0.675	-0.9599
2 level a1	Suitability a11 Accuracy a12 Security a13	0.6 0.5 0.5	-0.8264
2 level a2	Understandability a21 Learnability a22 Operability a23	0.6 0.6 0.5	-0.8716
2 level a3	Maturity a31 Fault tolerance a32 Recoverability a33	0.7 0.4 0.4	-0.7920

From table 2, we know human-provided values and results.

Table 2. Sample and results

Product	a11	a12	a13	a21	a22	a23	a31	a32	a33
A	72.3	65	76	84	80	78	75	71	75
B	50	67	74	85	82	75	80	68	70

Table 3. Overall evaluation results

Product	a1	a2	a3	Overall evaluation
A	71.6093	81.7724	76.5128	79.2366
B	66.9878	83.0034	74.7564	79.0187

Fuzzy integral evaluation procedure for A is as follows:

1) *Renumber a_{1i} 's according to the value such that*

$$h(a_{12}) \leq h(a_{11}) \leq h(a_{13});$$

2) *Calculate following fuzzy measure respectively according to equation(1):*

$$g(\{a_{13}\}) = 0.5$$

$$g(\{a_{11}, a_{13}\}) = g(\{a_{11}\}) + g(\{a_{13}\}) + (-0.8264) \times g(\{a_{11}\}) \times g(\{a_{13}\}) = 1.1 - 0.2479 = 0.8521$$

$$g(\{a_{12}, a_{11}, a_{13}\}) = g(\{a_{12}\}) + g(\{a_{11}, a_{13}\}) + (-0.8264) \times g(\{a_{12}\}) \times g(\{a_{11}, a_{13}\}) = 1$$

3) *Calculate evaluation value for level a1 according to equation (3).*

$$E \overset{def}{=} \int h dg \overset{def}{=} h(x_1) \times g(\{a_{12}, a_{11}, a_{13}\}) + (h(x_2) - h(x_1)) g(\{a_{11}, a_{13}\}) + (h(x_3) - h(x_2)) g(\{a_{13}\})$$
$$= 72.5 \times 1 + (65 - 72.5) \times 0.8521 + (76 - 65) \times 0.5 = 71.6093$$

Similarly, we could capture values of a_2, a_3 *and* E.

6 Conclusion

Lot of software measurement target has some sorts of fuzziness and uncertainties, and some kind of relationships or constrains among them. So it is appropriate to measure using fuzzy measure method. The Choquet integral based software quality synthetic assessment model provides a scientific basis for software quality assurance and evaluation. During the implementation phase, we need to adjust assessment index and corresponding importance index to make the valuation more scientific, accurate, and realistic.

References

1. Albrecht, A.J.: Measuring application development productivity. In: Proc. Joint SHARE/GUIDE/IBM Application Development Symp., pp. 83–92 (October 1979)
2. Halstead, M.H.: Elements of Software Science. Operating and programming systems series. Elsevier Science Inc., New York (1977)
3. McCall, J., Richards, P., Walters, G.: Factors in Software Quality, vol 1. Technical Report CDRL A003, US Rome Air Development Centre (1977)
4. Boehm, B.W., Brown, J., Kaspar, H., et al.: Characteristics of Software Quality. TRW Serious of Software Technology, vol. 1. North-Holland, New York (1978)
5. Bilsel, R.U., Buyukozkan, G., Ruan, D.: A fuzzy preference-ranking model for a quality evaluation of hospital web sites. Int. J. Intell. Syst. 21(3), 1181–1197 (2006)
6. ISO 9126, Information Technology–Software Product Evaluation–Quality Characteristics and Guidelines for Their Use. International Organisation for Standardization (September 1992)
7. ISO, ISO/IEC FCD 25000, Software Engineering – Software Product Quality Requirements and Evaluation (SQuaRE) - Guide to SQuaRE. International Organization for Standardization, Geneva (2004)
8. Yuan, B., Klir, G.J.: Constructing fuzzy measures: a new method and its application to cluster analysis, pp. 567–571. IEEE (1996)
9. Grabisch, M.: The application of fuzzy integrals in multicriteria decision making. European Journal of Operational Research 89, 445–456 (1995)
10. Lee, K.-M., Leekwang, H.: Identification of fuzzy measure by genetic algorithms. Fuzzy Sets and Systems 75, 301–309 (1995)
11. Grabisch, M.: A new algorithm for identifying fuzzy measures and its application to pattern recognition. In: Int. Joint Conf. of the 4th IEEE Int. Conf. on Fuzzy Systems and the 2nd Int. Fuzzy Engineering Symposium, Yokohama, Japan, pp. 145–150 (March 1995)
12. Chang, C.-W., Wu, C.-R., Lin, H.-L.: Integrating fuzzy theory and hierarchy concepts to evaluate software quality. Software Quality Journal 16(2), 263–276 (2008)
13. Chen, C.-B., Lin, C.-T., Wang, C.-H., Chang, C.-W.: Model for measuring quality of software in DVRS using the gap concept and fuzzy schemes with GA. Information and Software Technology 48(3), 187–203 (2006)

14. Pedrycz, W., Han, L., Peters, J.F., Ramanna, S., Zhai, R.: Calibration of software quality: fuzzy neural and rough neural computing approaches. Neurocomputing 36(1-4), 149–170
15. Takahagi, E.: On Identification methods of λ-fuzzy measures using weights and λ. Japanese Journal of Fuzzy Sets and Systems 12(5), 665–676 (2000)
16. Sugeno, M.: Theory of fuzzy integrals and its applications [PhD Disseration]. Tokyo Institute of Technology (1974)
17. Choquet, G.: Theory of capacities. Annales de l'Institut Fourier 5, 131–295 (1953)

A Framework for Moving Target Detection, Recognition and Tracking in UAV Videos

Jiabao Wang, Yafei Zhang, Jianjiang Lu, and Weiguang Xu

Institute of Command Automation, PLA University of Science and Technology,
Nanjing 210007, China
boa.plaust@gmail.com, miipl606@gmail.com

Abstract. In this paper, we present a compound framework for moving target detection, recognition and tracking based on different altitude UAV-captured videos. The novel idea of "Divide and Merge" included in our framework is expressed as follows. Firstly, we detect the small and slow moving targets using forward-backward MHI. Secondly, two distinct tracking algorithms, Particle Filter and Mean Shift, are applied to track moving targets in different altitude UAV-captured videos. Then, recognition module divides into two classes: instance recognition and category recognition. The former identifies the target, which is occluded by trees or buildings and reappears later, and the latter classifies the detected target into one category by HoG-based SVM classifier. Besides, recognition-based abnormal target detection and clustering-based abnormal trajectory detection are added to our framework. Armed with this framework, the moving targets can be tracked in real-time and the recognized target or abnormal trajectory gives the alarm in seconds.

Keywords: Target detection, Recognition, Tracking, UAV videos.

1 Introduction

In the last few years, Unmanned Aerial Vehicles (UAVs) have often played an important role in surveillance and reconnaissance, and an effective method is needed to analyse the laborious and boring aerial videos captured by UAVs. In many military and civil applications, such as COCOA system[1] and disaster management[2], the most important research topic is about detection, recognition and tracking of moving targets in UAV-captured videos.

The COCOA system developed by Ali and Shah is composed of three modules: motion compensation, moving object detection and object tracking[1]. Reilly discussed detection and tracking problems of large number of moving targets in wide area surveillance, overcoming many challenging problems, such as strong parallax, small targets[3]. Yu proposed a new framework of moving target detection and tracking for UAV video application, in which motion compensation, motion detection and object tracking were processed in parallel[4]. Xiao presented two tracking approaches for different demands between tracking precision and speed in UAV videos, and classified vehicle and person by a HoG-based classifier[5]. Oreifej

J. Luo (Ed.): Affective Computing and Intelligent Interaction, AISC 137, pp. 69–76.
springerlink.com

addressed the problem of human identity recognition over the low quality aerial images, and solved it by a weighted voter-candidate formulation[6]. In addition, several famous software packages for UAV video exploitation are referred in [7]. From above, moving targets detection, recognition and tracking is a popular and challenge problem in the state of the art.

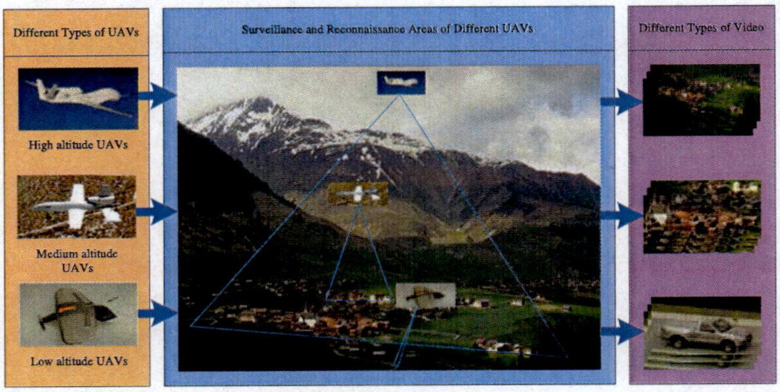

Fig. 1. Three types of UAVs and their corresponding videos

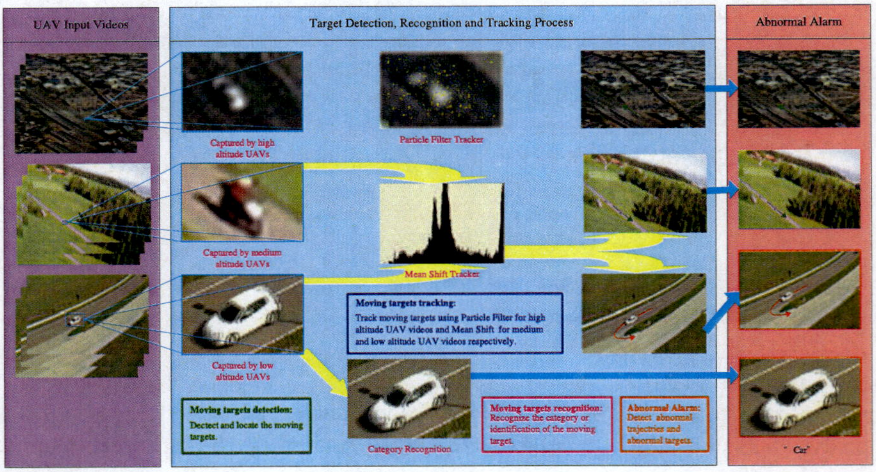

Fig. 2. The framework flow chart for target detection, recognition and tracking in several kinds of UAV-captured videos

As we known, UAVs are divided into three broad categories, high altitude UAVs (Global Hawk), medium altitude UAVs (Predator) and low altitude UAVs (include mini-UAVs and micro-UAVs). Videos captured by these UAVs are usually quite different in coverage area, the number of moving target, target's appearance and category. However, researchers only have devoted their attentions to one special kind of UAV-captured videos. In reality, different videos may have distinct characteristics, and multiple processing methods should be applied to them. High altitude UAV videos have a wide surveillance area, hundreds or thousands of targets, few pixels on one target, whereas low altitude UAVs videos may only cover one or two targets, which have clear color, shape and details. Fig. 1 gives a comparison of three types of UAVs and their corresponding videos. As a result, we believe that different methods should be assembled together to process the diverse videos.

Xiao proposed a joint probabilistic relation graph approach to detect and track a large number of vehicles simultaneously, with low frame rate and low spatial resolution[8]. But the target detection method, three-frame subtraction algorithm for fast moving vehicles and background subtraction algorithm for slow moving vehicles or the stopped vehicles, is not feasible in high frame rate videos and not practical for small and tiny moving targets in high altitude UAV videos. However, Yin and Collins could accurately detect and locate the small moving objects in airborne thermal video based on forward-backward motion history image (MHI), which inspired us to detect the location and the shape of small moving objects for tracker initialization[9].

Generally, target recognition is divided into two classes: category recognition and instance recognition. The former always involves recognizing a special target class such as "car", "bicycle", or "human"[10]. The most two famous classes to be recognized in aerial are "vehicle" and "human", because they are the major moving targets of interest for UAV videos. The latter involves re-recognizing a known target, potentially being viewed from a novel viewpoint, moving into a new background, and with partial occlusions[10]. Instance recognition can also be used for re-acquiring the target and persistent tracking after occlusion. Besides, it can identify the VIP (Very Important Person), that is why it becomes popular by the alias of identification recognition, but the framework does not contain it at present.

In high altitude UAV videos, moving targets of interest are often very small relative to their distance frame the camera, such as a car being viewed from thousands meters away[11]. Due to the lack of color, shape and texture for fast moving small targets, tracking becomes an extremely hard challenge[5]. In low altitude UAV video, parallax, which is produced by camera motion, and occlusion are the two major problems in target tracking[11,12]. Here, we use robust Particle Filter[13] to track small and tiny targets in high altitude UAV videos and Mean Shift[14] for medium and low altitude UAV videos.

After the processes of target detection, recognition and tracking, abnormal detection is performed by the recognized targets and the generated trajectories. It can give the alarm to the analyzer in seconds when an unusual target appears or a target moves in an unusual mode, i.e. moving in inverse direction or "U" turning. Therefore, we add the abnormal alarm module to our framework.

At this stage, we present the novel ideas of this paper as follows:

- Using two distinct tracking algorithms for different altitude UAV videos.
- Associating the broken trajectory by instance recognition, and recognizing car and person by a HOG-based classifier.
- Giving the alarm when abnormal target is detected based on recognition or abnormal trajectory is detected based on clustering.

The remainder of this paper is organized as follows. Section 2 describes the framework we proposed. Section 3, 4 and 5 discuss individual parts of moving target detection, recognition and tracking respectively. Abnormal detection alarm is presented in section 6. Section 7 concludes the paper and highlights future works.

2 Framework

Fig. 2 shows our framework that includes four major modules: (1) Moving target detection; (2) Target recognition; (3) Target tracking; (4) Abnormal alarm. At the very start, the input videos, which are captured by different types of UAVs, need to eliminate the camera ego motion before detecting the moving targets. After that, Moving targets detection and localization can be executed. We use forward-backward MHI to locate the target with rectangle bounding box for all kinds of UAV videos. Then, the detected target is classified by HoG-based classifier[15] and identified with weighted region matching. The next step is to track the detected targets. Two distinct tracking algorithms are applied to tracking targets in different altitude UAV videos. The generated trajectories and the recognized targets can be used for detecting abnormal actions and abnormal objects in the post-processing.

3 Moving Target Detection

The most famous approach to detection problem of moving target is frame-differencing, whereby the pixels of two video frames are subtracted from each other, yielding so-called "frame-differenced images". However, it can be done only in a static camera scenario directly. The scenario of VAU-captured video data does not conform to the static background description, since the background of the scene continuously changes as the UAV flies over land. Hence, to overcome this problem, the video needs to do compensation for the frame-to-frame camera ego motion.

To solve the camera ego motion problem, we have to investigate the camera motion model, which is often depicted by a homography. Feature-based method can compute the homography through a set of matched feature points extracted from two adjacent frames to be compensated. Given a set of matched feature points $\{(P_i,Q_i)\}$, where P_i or Q_i is the homogeneous coordinate of the matched points in the frame I_P and I_Q, and a planar parametric transformation of the form

$$Q_i = HP_i$$

where H is the 3×3 homography, and its best value can be obtained by minimize the sum of squared residuals

$$E_{LS} = \sum \| HP_i - Q_i \|^2.$$

Due to the existence of inaccurate matched points on the moving targets and few other matched points in error, the precision of H is influenced and reduced. To overcome this problem, the RANSAC algorithm[16] is introduced. It can improve the precision by kick out the outliers (inaccurate matched points on the moving targets and few other matched points in error). We make use of the improved RANSAC algorithm[17] to speed up this process.

One of the major problems about moving target detection by frame-differencing concerns the intervals among the two subtracted frames. If the interval is too small, the targets may have only moved a very small amount, resulting in very small highlighted regions in the frame-differenced images may not be classified from the noise floor[18]. Because a variety of targets may move at different speeds through the scenes, no single interval value could be suited to a given sequence[18]. Besides, the small highlighted areas of moving targets obtained by frame-differencing may not be distinguished from the noise floor. Hence, we adopt forward-backward MHI to solve above problems. The detail is referred in [9].

4 Target Recognition

To classify the detected moving targets into different categories, we design a HoG-based classifier for this task. The HoG descriptor, which is a good choice for aerial scenes[6], outperforms existing feature sets for human detection[15], but it is a time-consuming detector. To improve the speed, we use the cascade techniques to detect target only in the detected area, a smaller area relative to the whole frame. The classifiers are trained with multiple views and multiple poses of the targets (here only person and car). With the trained classifiers, we can recognize the detected moving target category.

Meanwhile, Instance recognition for persistent tracking is performed when the target is occluded by trees or buildings and reappears later. Firstly, we represent the target by two steps: (1) Computing the main gradients orientation in the detected areas like the SIFT process, and aligning them in the same direction; (2) Splitting the aligned area into several parts, and representing them by histogram. After that, we define a similarity measure between the two targets to be matched. Here, we use a weighted region matching approach with the form

$$L = \sum w_i f(F_i, G_i).$$

where, w_i is the weight value of the i-th part, f is the similarity measure function, and F_i, G_i are histogram vectors of the i-th part for detected areas F and G.

5 Target Tracking

Particle Filter is a modelling approach of nonlinear and non-gaussian system, belonging to sequential Monte Carlo methods based on point mass representations of probability densities[13]. It consists of two major steps: prediction and updating.

Prediction: Given all observations $y_{1:t-1}=\{y_1,y_2,\ldots,y_{t-1}\}$ up to time t-1, the prediction stage compute the posterior at time t as

$$p(x_t|y_{t-1}) = \int p(x_t|x_{t-1})p(x_{t-1}|y_{1:t-1})dx_{t-1}$$

The state can be updated using Bayes law when the observation y_t is observed at time t. The updated result

$$p(x_t|y_{1:t}) = \frac{p(y_t|x_t)p(x_t|y_{1:t-1})}{p(y_t|y_{1:t-1})}$$

where $p(y_t|x_t)$ is given by the observation equation.

Updating: The posterior $p(x_t|y_{1:t})$ is estimated by a finite set of N samples $\{x_t^i\}_{i=1...N}$ with importance weights w_t^i. The candidate sample x_t^i is drawn from an importance distribution $q(x_t|x_{1:t-1},y_{1:t})$ and the weight of the sample is

$$w_t^i = w_{t-1}^i \frac{p(y_t|\overline{x}_t^i)p(\overline{x}_t^i|x_{t-1}^i)}{q(\overline{x}_t|x_{1:t-1},y_{1:t})}.$$

The samples are resampled to generate an unweighted particle set according to their importance weights to avoid degeneracy. Generally, $q(x_t|x_{1:t-1},y_{1:t})=p(x_t|y_{t-1})$ and the weights become the observation likelihood $p(y_t|x_t)$.

Meanwhile, Mean Shift is used for the low altitude UAV-captured videos, in which the targets have clear appearances. Let $\{x_i^*\}_{i=1...n}$ and $\{x_i\}_{i=1...n_h}$ be the normalized pixel locations of the target and the candidate, centered at 0 in model region and y in the current frame, and $k(x)$ to be a convex and monotonic decreasing kernel profile. Then, the target model and the candidate model, which both can be represented as m-bin histograms, are computed as

$$\text{target model:} \begin{cases} \hat{q} = \{\hat{q}_u\}_{u=1...m} \\ \hat{q}_u = C\sum_{i=1}^{n} k\left(\| x_i^* \|^2\right)\delta[b(x_i^*)-u] \\ \sum_{u=1}^{m}\hat{q}_u = 1 \end{cases}$$

$$\text{candidate model:} \begin{cases} \hat{p}(y) = \{\hat{p}_u(y)\}_{u=1...m} \\ \hat{p}_u = C_h\sum_{i=1}^{n_h} k\left(\left\|\frac{y-x_i}{h}\right\|^2\right)\delta[b(x_i)-u] \\ \sum_{u=1}^{m}\hat{p}_u = 1 \end{cases}$$

where δ is the Kronecker delta function, and C and C_h both are normalization constants and derived by imposing the conditions $\sum_{u=1}^{m}\hat{q}_u = 1$ and $\sum_{u=1}^{m}\hat{p}_u = 1$ respectively.

The similarity between target model and candidate models is defined as the Bhattacharyya coefficient

$$\hat{\rho}(y) \equiv \rho[\hat{p}(y),\hat{q}] = \sum_{u=1}^{m}\sqrt{\hat{p}_u(y)\hat{q}_u}.$$

Using Taylor expansion around the values $\hat{p}_u(\hat{y}_0)$, where $\hat{y}_0$ is the initial location in the current frame, the linear approximation of the similarity is obtained as

$$\rho[\hat{p}(y),\hat{q}] \approx \frac{1}{2}\sum_{u=1}^{m}\sqrt{\hat{p}_u(\hat{y}_0)\hat{q}_u} + \frac{C_h}{2}\sum_{i=1}^{n_h} w_i k(\left\|\frac{y-x_i}{h}\right\|^2)$$

where $w_i = \sum_{i=1}^{m} \sqrt{\frac{\hat{q}_u}{\hat{p}_u(\hat{y}_0)}} \delta[b(x_i) - u]$.

To get the maximum similarity value, we only to maximize the second term, whose maximum can be found by employing the mean shift procedure[19].

6 Abnormal Alarm

This module includes two major aspects: abnormal targets detection and abnormal trajectories detection. They can run into two modes: detection with or without a priori knowledge base. We provide an interface through which analysers can give an abnormal definition by their purposes, e.g. a VIP person image or an "S" trajectory, and the system can also learn the abnormal trajectories in an unsupervised way. In the former case, users can add or cut off the rules (the provided images) defined previously. The abnormal target detection depends on the identification recognition in part 4, whilst the abnormal trajectory is based on clustering method. In the latter case, Clustering-based methods try to identify unusualness of trajectories by clustering all trajectories into groups. The dominant clusters correspond to normal trajectories, while the few outliers correspond to abnormal ones[20].

7 Conclusions

In this paper, we present the novel idea of "Divide and Merge". Two distinct tracking algorithms are applied to track moving targets in different altitude UAV-captured videos. Instance recognition and category recognition are assembled together to identifies and classifies the detected targets of interest. Besides, our framework includes recognition-based abnormal target detection and clustering-based abnormal trajectory detection. In the future, armed with the "Divide and Merge" idea, we will focus on (i) improving the precision of moving target detection to eliminate the parallax and the noise, and (ii) investigating new algorithm for target merging and splitting (e.g. one target is splitted by another static object or two targets are merged due to too close).

References

1. Ali, S., Shab, M.: COCOA - Tracking in aerial imagery. In: Proceedings of SPIE Airborne Intelligence, Surveillance, Reconnaissance(ISR) Systems and Applications, Orlando (2006)
2. Quaritsch, M., et al.: Networked UAVs as aerial sensor network for disaster management applications. Elektrotechnik & Informationstechnik(e&i) 127(3), 56–63 (2010)
3. Reilly, V., Idrees, H., Shah, M.: Detection and tracking of large number of targets in wide area surveillance. In: European Conference on Computer Vision (2010)
4. Yu, W., Yu, X., Zhang, P., Zhou, J.: A new framework of moving target detection and tracking for UAV video application. In: International Archives of the Photogrammetry, Remote Sensing and Spatial Information Sciences, vol. XXXVII, Part B3b, Beijing (2008)

5. Xiao, J., Yang, C., Han, F., Cheng, H.: Vehicle and person tracking in UAV video. In: IEEE Conference on Computer Vision and Pattern Recognition (2007)
6. Oreifej, O., Mehran, R., Shah, M.: Human identity recognition in aerial images. In: IEEE Conference on Computer Vision and Pattern Recognition (2010)
7. Se, S., et al.: Automated UAV-based mapping for airborne reconnaissance and video exploitation. In: Proceedings of SPIE Airborne Intelligence, Surveillance, Reconnaissance (ISR) Systems and Applications, Orlando (2009)
8. Xiao, J., Cheng, H., Sawhney, H., Han, F.: Vehicle detection and tracking in wide field-of-view aerial video. In: IEEE Conference on Computer Vision and Pattern Recognition (2010)
9. Yin, Z., Collins, R.: Moving object localization in thermal imagery by forward-backward MHI. In: IEEE Conference on Computer Vision and Pattern Recognition (2005)
10. Szeliski, R.: Computer vision: algorithms and applications, Electronic draft (2010) (unpublished) `http://szeliski.org/Book/`
11. Burns, J.B.: Detecting independently moving objects and their interactions in georeferenced airborne video. In: IEEE Workshop on Dectection and Recognitin of Events in Video (2001)
12. Ali, S., Reilly, V., Shah, M.: Motion and appearance contexts for tracking and re-acquiring targets in aerial videos. In: IEEE Conference on Computer Vision and Pattern Recognition (2007)
13. Arulampalam, M.S., Maskell, S., Gordon, N., Clapp, T.: A tutorial on particle filters for online nonlinear non-gaussian bayesian tracking. IEEE Transactions on Signal Processing 50(2), 174–188 (2002)
14. Comaniciu, D., Ramesh, V., Meer, P.: Kernel-based object tracking. IEEE Transactions on Pattern Analysis and Machine Intelligence 25(5), 564–577 (2003)
15. Dalal, N., Triggs, B.: Histograms of oriented gradients for human detection. In: IEEE Conference on Computer Vision and Pattern Recognition (2005)
16. Fischler, M.A., Bolles, R.C.: Random sample consensus: a paradigm for model fitting with applications to image analysis and automated cartography. Communications of the ACM 24(6), 381–395 (1981)
17. Chum, O., Matas, J.: Matching with PROSAC-progressive sample consensus. In: IEEE Conference on Computer Vision and Pattern Recognition (2005)
18. Sadlier, D., O'Connor, N.: Vehicle tracking in UAV-captured video data incorporating adaptive weighting of features, Final Report, Dublin City Univ. (Ireland) (2009)
19. Comaniciu, D., Meer, P.: Mean shift: a robust approach toward feature space analysis. IEEE Transactions on Pattern Analysis and Machine Intelligence 24(5), 603–619 (2002)
20. Piciarelli, C., Micheloni, C., Foresti, G.L.: Trajectory-based anomalous event detection. IEEE Transactions on Circuits and Systems for Video Technology 18(11), 1544–1553 (2008)

A Grammar Based Method for Video Event Indexing and Retrieval

Weiguang Xu, Yafei Zhang, Jianjiang Lu, and Jiabao Wang

Institute of Command Automation, PLA University of Science and Technology
Nanjing 210007, China
vigorxu@gmail.com, miipl606@gmail.com

Abstract. In this paper, we present a novel framework for indexing and retrieving video events. The framework has a hierarchical structural and based upon a grammatical model. We firstly define and detect some event primitives, and induce the parameters of the SCFG to describe the event sequences, by using Liang's nonparametric model of HDP-SCFG and variational inference algorithm. The MES parser and viterbi algorithm is employed to acquire the parse tree, which is index and matched next. We mainly contribute in four aspects. 1) It is the first time that transplant nonparametric grammar, ISCFG, to the domain of video event retrieval. 2) The cross data link table structure of indexing is efficient enough for searching in real time. The computational complexity of locating a non-terminal at any time point is O(log). 3) The novel matching scheme allows people to search content of interest by input a sample of video segment. The measure of similarity works well because both the structure and content of the parse trees are took into account.

Keywords: Statistical grammar, Video event, Index, Retrieval, Nonparametric.

1 Introduction

Video event retrieval has gained much attention because of the wide usage of video systems in TV broadcasting, video library, and especially the surveillance cameras located in kinds of scenes. However, most traditional video retrieval systems work based upon the literal information manually annotated by workers, thus the labor intensive work manner limits the efficiency of the systems.

To overcome the limitation, we present a novel framework (see Fig. 1) for automatically analyzing, indexing and retrieving video event without any prior knowledge or manual annotation. Actually, the framework is successor work of our previous work [1] that analyzes video events in an unsupervised manner. In the framework, we firstly define and detect a list of event primitives in the most common scenes like parking pot, street, etc. After this, we adopt the methods of Ivanov et al. [2] and Moore et al. [3], to represent events by using *SCFG* (Stochastic Context Free Grammar) [4] with the predefined primitives as terminals. Then, the non-terminals, and production rules are all learned in an unsupervised manner from the training set of detected sequences, with the power of *HDP-SCFG* (Hierarchical Drichlet

J. Luo (Ed.): Affective Computing and Intelligent Interaction, AISC 137, pp. 77–84.
springerlink.com

Process-*SCFG*) [5, 8] and variational inference [6]. The learned grammar reflects the different behaviors patterns of the objects. To robustly parse detected sequences given the learned grammar, we make use of the *MES* parser presented in our previous work [1] by extending the standard Earley-Stolcke parser [4]. Then, a data structure of cross data link table is constructed to index events happened in video, a matching scheme based on parse trees is used for retrieve relative ones given a sample of video segment.

The rest of the paper is organized as follows. In section 2, we discuss and clarify the meaning of many different terms that related to "event" in domain of video analysis, as well as some previous works related to this paper. Next, we introduce how the *ISCFG* based hierarchical model is used to represent and recognize complex event in section III.A. The method of indexing and retrieval is illustrated in III.B and III.C respectively. At last, a conclusion is drawn in section IV. Some future work is also discussed.

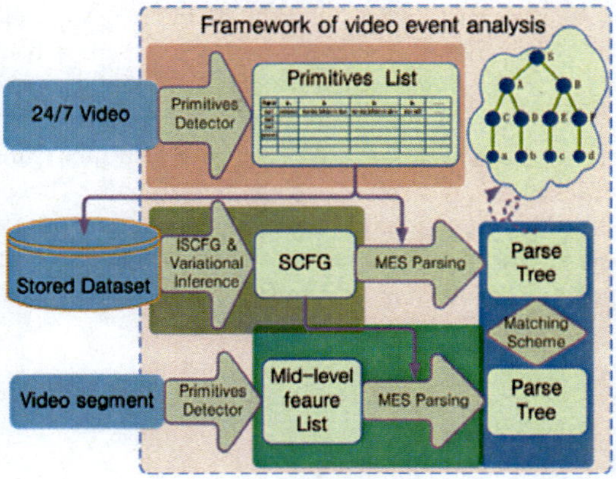

Fig. 1. The framework of video event indexing and retrieval

2 Background

2.1 Terms Related to Event

The term of "event" can be seen frequently in previous works of event analysis in videos, and it was sometimes used as an alternative of "activity". However, the exact meaning of "event" and "activity" is usually ambiguous, even misunderstood. Generally speaking, events in videos can be categorized into simple event and complex event. To make the meaning and boundary clear, simple events are defined as the ones that are detected by directly using digital features of video clips, and cannot be divided anymore. They are always called "simple events", "sub-events", "atomic events" or "event primitives". In contrast, complex events are defined as the ones that are made up of series sub-events, these sub-events could be simple or complex events, and in the case recursion, the sub-event could be the base event itself. They are always called "complex events", "activities". When human are involved as agent, simple events are always called "human actions" or "primitive actions", and complex events are called "human activities" or "human interactions". We call "event primitives" for simple events and "events" of complex events respectively here.

2.2 Related Work

Video event analysis is the foundation stone of event retrieval. In order to search event of interest accurately and fast, events happened need to be represented by some kind of formal model and recognized by some kind of inferential algorithm. A number of literatures surveyed the related works in last decades [10~14]. Xie et al. [14] classified event analysis models into knowledge/data-driven, generative/discriminant and supervised/unsupervised models respectively from different perspectives. These kinds of models have their respective pros and cros, but it is noteworthy that grammar based models have come forth to the domain of computer vision recently due to its success in speech recognition. The grammar based model is more capable of describing the sequential features than that describes events with bag-of-features. It is also more capable of modeling the hierarchical structure of events than the complanate graphical ones like *HMM*. At last it is rich of semantics and easier for human understanding than other knowledge-driven models, such as *VERL* and *VEML* [15, 16] based on ontology. All the advantages are the reason of attempting to indexing and retrieving video event based on grammatical models.

Ryoo et al. [17] used a deterministic grammar model *CFG* [7] to detect human activities and interactions. A straightforward extension with probabilities to *CFG*, i.e. *SCFG*, was used to analysis complex events in various domains [2, 3, 9, 18~21]. However all of them had to define grammar production rules manually beforehand, which limits their capability of generalization. Liang et al. [5] made use of a nonparametric model, *HDP-SCFG*, and variational inference to automatically induce grammar from a set of sentences. Their work gives us the inspiration that transplant the method to the domain of video event analysis, so as to overcome the limitation of previous grammar based event models.

3 Video Event Indexing and Retrieval

3.1 Grammar Based Event Analysis

Event analysis, representation, and recognition are foundation stone of event indexing and retrieval. We analyze events hierarchically in three steps.

Firstly, we define and detect the most common event primitives based on trajectories of moving objects, which are generated from the low-level processing, such as foreground detection, object recognition, tracking etc. The work of this step differs from previous works in three aspects: 1) They construct one sequence totally for a video segment with the detected primitives, even though the atomics are performed by different agents. We construct different sequence respectively for each single agent and agent pair. 2) The primitives they defined are all single-agent, even though the complex events to detect are multi-agent. Multi-agent primitives is also defined in our framework, they are recorded when the distance between different agents is smaller than an adaptive threshold. 3) The primitives they defined usually needs some rule-based or stochastic model that costs quite a few complexity of computing. We make the primitives as simple as possible, and they can all be detected in real-time. Specific definition can be referred from our previous work [1], and are omitted here.

Secondly, we adopt the *HDP-PCFG* model and variational inference algorithm, which was originally used for processing natural language, to learn a SCFG describing video events. Because the *HDP-PCFG* is nonparametric with infinite states, we call it *ISCFG* (infinite state *SCFG*) for short. In the *ISCFG*, probabilities of the top-level symbols z=1,2,... are drawn by using $\beta \sim GEM(\gamma)$. Non-terminals are partitioned into two types: pre-terminal symbols that emit only terminals (primitives defined in previous section) and constituent symbols. Thus, a type parameter is drawn from $T_z \sim Dir(0,0)$ for each z. The zero-prior forces T_z to put mass on only one type. For each constituent symbol, a pair of binary production parameters is drawn by using $B_z^{L-child} \sim DP(\alpha_B, \beta)$ and $B_z^{R-child} \sim DP(\alpha_B, \beta)$; for each pre-terminal symbol, an emission parameter is drawn by using $E_z \sim Dir(\alpha_E)$, where α_E is the prior distribution over primitives. All the parameters can be induced by using variational inference algorithm [6], and the SCFG is acquired. An induced *SCFG* includes the number of non-terminal symbols, type of the each symbol, emission parameters of each pre-terminal symbol and binary production parameters of each constituent symbol.

Thirdly, the learned SCFG is used to parse sequences stored in database and newly detected sequences. To enhance the tolerance to unexpectable errors caused by the inaccuracy of low-level process, we extended the standard Earley-Stolcke parser to the *MES* parser [1]. The *MES* focus on improving the step of scanning. Besides the prefix probability α and the inner probability γ, we add a new variable ε to accumulate the penalty caused by the errors. ε of the state generated by the standard parser is initialized to 1, and it decreases drastically for the state generated by the modified parser. When ε is smaller than a threshold, the corresponding state will be deleted. During each scanning step in the procedure, we do three more operations to add three additional states to corresponding state sets:

$$i: X_k \to \lambda.a\mu\,[\alpha,\gamma,\varepsilon] \Rightarrow i+1: X_k \to \lambda.a\mu\,[\alpha,\gamma,\varepsilon \cdot p_{insertion}] \tag{1}$$

$$i: X_k \to \lambda.a\mu\,[\alpha,\gamma,\varepsilon] \Rightarrow i: X_k \to \lambda a.\mu\,[\alpha,\gamma,\varepsilon \cdot p_{deletion}] \tag{2}$$

$$i: X_k \to \lambda.a\mu\,[\alpha,\gamma,\varepsilon] \Rightarrow i+1: X_k \to \lambda a.\mu\,[\alpha,\gamma,\varepsilon \cdot p_{substitution}] \tag{3}$$

The three operations are done in order to deal with the error of insertion, deletion, substitution respectively. After the parse chart acquired, the *Viterbi* algorithm is employed in a manner similar to *HMMs* to discover the derivation path or parse tree with the biggest probability. The best parse tree will be used for indexing and searching stored video.

3.2 Indexing

Parse trees are said to be an extremely rich class of objects from a statistical perspective [5]. We use parse tree to index the stored video so as to offer a semantic search facility. In specification, the objective of index is to make it fast to search any given non-terminal symbol in video. The information need to achieve include its starting time, ending time, and the whole parse tree expanded from the symbols should be reconstructed quickly. We use a data structure (Fig. 2) of cross data link table to accomplish this.

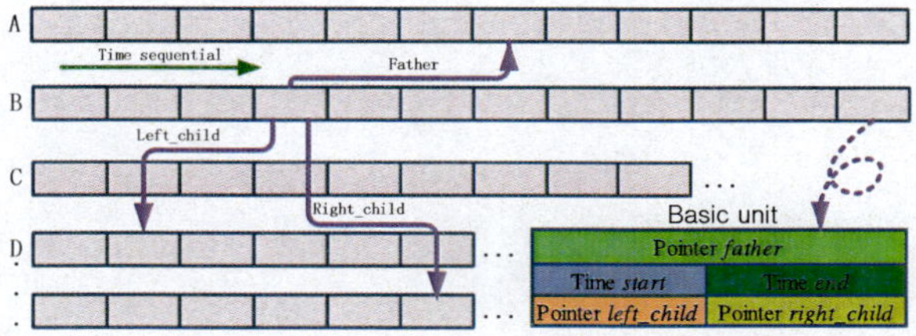

Fig. 2. Cross data link table of the index

The basic structure unit is made up of three links pointing to its father node and two child nodes. The starting time and ending time describes its time interval. Units of the same symbol are stored in an array and sorted by their starting time. Thus, given a non-terminal symbol, all the occurrences can be achieved easily. Then, through the links in the units, parse tree can also be reconstructed. If a non-terminal at certain time is needed, binary search algorithm can be employed to locate it in time of $O(\log)$.

3.3 Matching Scheme

In practice, people may search some video segment of interest, but he/she isn't an expert and doesn't have the ability to search by composing a non-terminal symbol. In this case, he/she can search by inputting a sample video segment. The segment will be processed by the primitive detector as well, and generate one or more sequences of primitives. The sequences are parsed to its best parse trees. The left is to match the trees with the ones stored in database. The problem is how to measure the similarity between two parse trees. In the framework, we present a novel algorithm to compute the similarity. A tree is preprocessed to a vector as its descriptor, whose elements are weights of symbols, reflecting their level and scale. Suppose there are total N non-terminal symbols in the grammar: $S:\{s_1, s_2, \cdots, s_N\}$, and the vector of a parse tree T is denoted by $W^T = (w_1^T, w_2^T, \cdots, w_N^T)$. The similarity between two parse tree T_1 and T_2 is measured by their description vectors' include angle cosine:

$$Similarity(T_1, T_2) = \frac{W^{T_1} \cdot W^{T_2}}{\|W^{T_1}\| \|W^{T_2}\|} . \quad (4)$$

Obviously, the key point of the measurement is how to compute the weights in the vector. Algorithm 1 in Fig. 3 shows the procedure of generating a descriptor of a parse tree. The symbols locating at higher level and expanding a bigger sub tree will get more weight. Thus symbols with similar weights can reflect the similarity between the trees they appears. Before the similarity is computed, normalization should be done by dividing *W* by *sum*(*W*), so as to eliminate the affection of trees' scale.

Algorithm 1. Generate a descriptor of a parse tree *T* by using *weight*(*T*)

Input: The parse tree *T*;

Output: The descriptor *W*;

Vector $W \leftarrow (0,0,\cdots,0)$;

Int *weight*(*Tree T*)

{

Symbol $s \leftarrow T$.root;

if *is_pre_terminal*(*s*)

$w_s \leftarrow 1$;

else $w_s \leftarrow weight(left\_subTree(s)) + weight(right\_subTree(s))$;

$W \leftarrow W + (0,\cdots,w_s,\cdots,0)$;

return w_s ;

}

Fig. 3. Algorithm of generating descriptors for parse trees

4 Conclusion and Future Work

In the paper, we present a novel framework for unsupervised analyzing events in videos. The framework can discover behavior patterns of object in videos automatically. Compared to previous works, our contribution includes four main aspects.

1) It is the first time that transplant nonparametric grammar, *ISCFG*, to the domain of video event retrieval. *ISCFG* and its inference algorithm enable the framework to analyze and index video events automatically without human intervention.

2) The cross data link table structure of indexing is efficient enough for searching in real time. The computational complexity of locating a non-terminal at any time point is *O*(log), a parse tree can be reconstructed easily and quickly given any node in the tree. The storage space is economized without any redundancy.

3) The novel matching scheme allows people to search content of interest by input a sample of video segment. The measure of similarity between two event takes into account both the structure and content of their parse trees, thus it measures the similarity well.

In future, some experiments will be done on benchmark video datasets to test effectiveness and efficiency of the framework. Some details of the models, parameters, and algorithm may be adjusted to fit the practical scene.

References

1. Xu, W.G., Lu, J.J., et al.: A unsupervised framework of video event analysis. To appear in Proceedings of the 2010 Second Pacific-Asica Conference on Knowledge Engineering and Software Engineering, Chongqing, China (December 2010)
2. Ivanov, Y., Bobick, A.: Recognition of visual activities and interactions by stochastic parsing. IEEE Transaction on Pattern Analysis and Machine Intelligence 22(8), 852–872 (2000)
3. Moore, D., Essa, I.: Recognizing multitasked activities from video using stochastic context-free grammar. In: National Conference on Artificial Intelligence, pp. 770–776 (2002)
4. Stolcke: An efficient probabilistic context-free Parsing Algorithm that computes prefix probabilities. Computational Linguistics 21(2), 164–201 (1995)
5. Liang, P., Jordan, M.I., Klein, D.: Probabilistic Grammars and Hierarchical Dirichlet Process. To appear in the Handbook of Applied Bayesian Analysis (2009)
6. Lawrence, N.D.: Variational inference in probabilistic models, PhD Thesis (2000)
7. Earley, J.: An efficient context-free parsing algorithm. Communication of the ACM 13(2), 94–102 (1970)
8. Teh, Y.W., Jordan, M.I., Beal, M., Blei, D.: Hierarchical dirichlet processes. Journal of the American Statistical Association 101, 1566–1581 (2006)
9. Lin, L., Gong, H., Wang, L.: Semantic event representation and recognition using syntactic attribute graph grammar. Pattern Recognition Letters 30, 180–186 (2009)
10. Aggarwal, J.K., Cai, Q.: Human motion analysis: a review. Computer Vision and Image Understanding 73(3), 429–440 (1999)
11. Turaga, P., Chellappa, R., Subrahmanian, V.S., Udrea, O.: Machine recognition of human activities: a survey. IEEE Transactions on Circuits, Systems and Video Technology 18(11), 1473–1488 (2008)
12. Hu, W., Tan, T., Wang, L., Maybank, S.: A survey on visual surveillance of object motion and behaviours. IEEE Transaction on Systems, Man and Cybernetic 34(3), 334–352 (2004)
13. Morris, T., Trivedi, M.M.: A survey of vision-based trajectory learning and analysis for surveillance. IEEE Transaction on Circuits and Systems For Video Technology 18(8), 1114–1127 (2008)
14. Xie, L., SunDaram, H., Cambell, M.: Event mining in multimedia streams. Proceeding of the IEEE 96(4), 623–647 (2008)
15. Francois, A.R.J., Nevatia, R., Hobbs, J., Bolles, R.C.: Verl: An ontology framework for representing and annotating video events. IEEE Multimedia 12(4), 76–86 (2005)
16. Nevatia, R., Zhao, T., Hongeng, S.: Hierarchical language-based representation of events in video streaming. In: Computer Vision and Pattern Recognition Workshop (2003)

17. Ryoo, M.S., Aggarwal, J.K.: Semantic representation and recognition of continued and recursive human activities. International Journal of Computer Vision 82, 1–24 (2009)
18. Yamamoto, M., Mitomi, H., Fujiwara, F., Tlikeato: Bayesian classification of task oriental actions based on stochastic context-free grammar. In: Proceedings of the 7th International Conference on Automatic Face and Gesture Recognition, USA, pp. 317–323 (2006)
19. Minnen, D., Essa, I., Starner, T.: Expectation grammars: leveraging high-level expectation for activity recognition. In: Computer Vision and Pattern Recognition, vol. 2, pp. 626–632 (2003)
20. Zhang, Z., Huang, K., Tan, T.: Multi-thread Parsing for Recognizing Complex Events in Videos. In: Forsyth, D., Torr, P., Zisserman, A. (eds.) ECCV 2008, Part III. LNCS, vol. 5304, pp. 738–751. Springer, Heidelberg (2008)
21. Ogale, A., Karapurkar, A., Aloimonos, Y.: View-invariant modeling and recognition of human actions using grammars. In: IEEE Workshop on Dynamical Vision (2005)

Recovery Model for Task Allocation Using Meta-level Information*

He Luo, Xiaoxuan Hu, and Fei Cheng

School of Management, Hefei University of Technology, Hefei, Anhui, China
luohe2008@gmail.com, huxiaoxuan@vip.sina.com,
chengfei2009@gmail.com

Abstract. The multi agent system is allocating tasks to different resources so as to complete those tasks. The limitation of the resources will reduce successful allocation ratio of the system, and de-commitments maybe happen at the same time when the resources cannot meet the requirement of the coming tasks, or there are some conflicts among different tasks. Under this situation, agents need to make a decision of how to re-allocate those de-committed tasks again to increase the performance of the system. In this paper, we present a failure recovery mechanism (FRM) for task allocation in the multi agent system to solve this problem. Traditional ways of re-allocation or scheduling are centralized with full or plenty information. We build a mechanism from a de-centralized view with limited information between tasks and the resources. We also show how FRM works in an open environment to recover the failure of task allocation. Experiments show the performance of the failure recovery mechanism.

Keywords: Multi agent system, Task allocation, Failure recovery, Negotiation.

1 Introduction

Task allocation problem is one of the important problems in both multi agent system researches and its applications. Agents negotiate with each other to allocate tasks. In the priority based multi agent system, the higher priority of the task it is, the higher probability of successful allocation the task it is. The basic rule for scheduling in this kind of multi agent system is that, tasks which have higher priority will always grab the resources where lower priority tasks are scheduled. Therefore, agents need to preempt the task with lower priority. In a dynamic, open environment with limited resources and unpredictable coming tasks, agents in the multi agent system need to make their own decisions to allocate these tasks in order to the coordinate resources under certain constrains, such as priority, start time, utility and so on.

* This Work is partially supported by the National Natural Science Foundation of China (Grant No. 70971035, 70631003, 70801024), the Ph.D. Programs Foundation of Ministry of Education of China (Grant No. 200803590007), and the Ph.D. Foundation of Hefei University of Technology (Grant No. 2010HGBZ0259).

J. Luo (Ed.): Affective Computing and Intelligent Interaction, AISC 137, pp. 85–92.
springerlink.com

In the centralized decision process, there is a super agent which can be viewed as the centralized scheduler. This super agent has all the status of the agents in the multi agent system, and also has the information of all the coming tasks. Therefore, by using this knowledge, this super agent can make an optimized decision or near-optimized decision to maximize the utility of the multi agent system, or other criteria. The centralized decision process can also be modeled as the integer programming (IP), or constrain satisfaction problem (CSP). The core idea of these models is to optimize the object function which is proposed on the basis of the system requirement.

In a distributed system, there is no super agent exists, and each single agent will make its own decision for multiple tasks based on its own status. There are two different strategies for agents in a distributed system. The one is selfish agent which is always trying to maximize its own utility, and the other one is cooperative agent which is to maximize the system utility when making its decision. Abdallah [1] uses termination schemes to help agents make their own decisions for the concurrent tasks. Luo [2] is focusing on the high level information in the task allocation, and suggests a concurrent negotiation framework based on the Semi Markov Decision Process by using the concurrent action model. Raja [3] presented a framework of meta-level control for agents to operate in the resource-bounded multi agent environment, and help agents to decide between dynamic deliberative actions and domain action. Zhang [4] proposed a commitment model to manage concurrent negotiations under constrains.

In this paper, we are mainly focus on the failure recovery for the de-committed tasks which are refused by the agents. We model the failure recovery problem and proposed the failure recovery mechanism (FRM) for task allocation in the multi agent system. The traditional methods for re-allocation or re-scheduling tasks are centralized with complete or plenty information. The mechanism from a de-centralized view with limited information between tasks and resources has been suggested. The following parts are arranged as follows. In section 2, we describe the background knowledge for tasks allocation. In section 3, we model the failure recovery problem and proposed the failure recovery mechanism. The experiments are given in section 4, and the conclusion and future works are in section 5.

2 Backgrounds

There are two agent roles in the system, resource manager (RM) and the process manager (PM). Resources in our scheduling system are inherently distributed, and controlled by the resource manager. Each resource manager is only responsible for one resource, where only one task can be active within the same time window. It has some privacy and limited information disclosure. Each process manager can only handle one process, but may have several tasks in it. Process manager will create a negotiator instance to negotiate different tasks with the appropriate RMs based on a partial order. The scheduling process is on the basis of priority and allows preemption, which means that a RM will de-commit from the lower task's reservation when a task with higher priority is requesting the same time window.

Task allocation problem can be viewed as the problem of assigning a set of tasks to different resources with limited constrains. Each task has a set of optional resource sets to choose from so that agents need to make decision of which resources should take the task. The goal is to allocate all the requiring tasks successfully with maximization of the system utility or minimization of the penalty of the de-commitment. Task allocation problem in the MAS can be formally defined as follows.

- A set of tasks, $T = \{T_1, T_2, \ldots , T_n\}$
- The priority of the task T_i, P_i
- The earliest start time of T_i, est_i
- The deadline of T_i, dl_i
- The duration of T_i, dur_i
- A set of resources, $R = \{R_1, R_2, \ldots , R_m\}$
- Resource set of task T_i, $RS_i = \{R_{i1}, R_{i2}, \ldots, R_{ik}\}$, where $R_{ij} \in R, j = 1, 2, \ldots, k$
- The number of available time windows for task T_i on the resource R_{ij}, N_{ip}
- The available time windows for task T_i on the resource R_{ij}, $TW_{ij} = \{tw_{ij}^1, tw_{ij}^2, \ldots, tw_{ij}^{Nip}\}$
- Start time of the time window tw_{ij}^q, st_{ij}^q
- End time of the time window tw_{ij}^q, et_{ij}^q
- Utility gained if task T_i is carried out on the resource R_{ij}, u_{ij}, $0 \le u_{ij} \le 1$
- Penalty gained if task T_i is de-committed from the resource R_{ij}, pen_{ij}
- A variable x_{ij}^q, the value of x_{ij}^q is 1 if task T_i is carried out on the qth time window tw_{ij}^q on the resource R_{ij}, otherwise the value is 0.

The goal of the task allocation problem is to get an allocation to meet:

$$Max(U) = Max \sum_{i \in n} \sum_{j \in R} u_{ij} \left(\sum_{q=1}^{N_{ip}} x_{ij}^q \right) \tag{1}$$

where $\sum_{j \in R} \sum_{q=1}^{N_{ip}} x_{ij}^q = 1, \quad \forall i \in T$, making sure every task has, and only has executed once.

And we also need to make sure that higher priority tasks will be scheduled as many as possible before lower priority tasks in the priority-based system. When task T_i and resource R_{ij} make a commitment with each other, the system will get the utility U_{ij} from this commitment. Therefore, the goal is to maximize the utility of the whole system. In the priority-based system, another basic rule is to have all higher priority tasks allocated. Therefore, pre-emption may happen when a lower priority task has already taken the time window. The penalty needs to be paid due to the de-commitment.

3 Failure Recovery

3.1 Problem Description

When a new task T_H is coming from the process manager, but there are no available resources to allocate this task. All resources in the new task's (T_H) resources set have already taken by other tasks, which priority may be or not be higher than task T_H. If the time window requested by T_H is taken by a higher priority task, then T_H has no authority to de-commit this kind of task in the priority-based system. If the time window is occupied by a task which priority is lower than task T_H, then T_H can choose this kind of task to be de-committed. Therefore, agents need to make a decision which tasks should be de-committed among all lower priority tasks. When pre-emption happens, we also need to answer two questions:

- Which lower priority task show we choose to de-commit?
- How to re-schedule the de-committed task?

In order to solve this problem, we propose three criteria which have effects on the decision of de-commitment. The first one is the penalty fees paid for de-commitment, the second one is the effects of the sub tasks, and the last one is the resources for re-scheduling the de-committed tasks. In the following part, the detailed descriptions will be introduced.

3.2 Three Criteria

In order to make a decision, we firstly defined three criteria to help agent decide. We defined "*penalty*" to describe the penalty to be paid for the de-commitment of task T_i.

$$penalty(T_i) = \left[utility(T_i)\right]^{priority(T_i)} \tag{2}$$

Here, *utility*(*T*) represents the utility u_{ij} gained when the task T_i (with lower priority) makes commitment with resource *j*, and *priority*(T_i) is the priority of the task T_i.

This formulation shows the relationship between utility and penalty. The utility is what a resource gets when it gets the commitment with a task. So, if we want to de-commit an existed commitment, corresponding penalty need to be paid. That is to say, the higher priority of task T_i it is, the higher penalty it is to de-commit this task.

For example, task T_1 with priority 5 has a commitment C_1 with resource *A*, and receives utility 0.5. While task T_2 with priority 3 has a commitment C_2 with resource *B*, and receives utility 0.8. When agent want do de-commit the commitment C_1, the penalty needed to pay is:

$$penalty(T_1) = (0.5)^5 = 0.03125$$

When agent want do de-commit the commitment C_2, the penalty needed to pay is:

$$penalty(T_2) = (0.8)^3 = 0.512$$

So, we can see that, de-committing C_2 will pay more penalty fees, and agent will de-commit the lower priority task T_1.

Another example is when task T_3 with priority 5 has a commitment C_3 with resource *D*, and receives utility 0.6. When agent wants to de-commit the commitment C_3, the penalty needed to pay is:

$$penalty(T_3) = (0.4)^5 = 0.01024$$

So, considering the commitment C_3, agent will choose to de-commit C_3 because of the lowest penalty fee. This shows that, agent should de-commit the task with lower utility gained, if optional tasks are the same priority.

We also defined "*sub_task_priority*", which is used to show the importance of task besides the priority view.

$$sub\_task\_priority(T_i) = \left[priority(T_i) \right]^{|sub\_task(T_i)|} \quad (3)$$

Here, *priority(task)* represents the priority p_i of a certain task T_i, and $|sub\_task(T_i)|$ is the number of the sub tasks of a certain task T_i.

For example, if agent has two options. The one is to de-commit the task T_1 which has three sub tasks with priority 4, and the other one is to de-commit the task T_2 which has 2 sub tasks with priority 2.

$$sub\_task\_priority(T_1) = 4^3 = 64$$

$$sub\_task\_priority(T_2) = 2^2 = 4$$

Therefore, the decision of the agent is to de-commit the task which has the lower *sub_task_priority*. In this formulation, we show a second criterion of a set of task with relations, considering the priority of sub tasks.

We then defined "*resource_availability*" for the task which is supposed to be de-committed.

$$resource\_availablity(T) = \frac{|sub\_task\_has\_resource(T)|}{|sub\_task(T)|} \quad (4)$$

$$sub\_task\_has\_resource(T) = \sum_{i \in sub\_task(T)} \sum_{j \in R} \sum_{q=1}^{N_{ip}} x_{ij}^{q} \quad (5)$$

Here, $|sub\_task\_has\_resource(task)|$ is the number of sub tasks of a certain task T which has the available resource to do the tasks, and the $|sub\_task\ (task)|$ is the number of the sub tasks of this task T. This formulation is to show the degree of availability of resources for de-commitment.

For example, task T has two sub tasks T_1 and T_2, then the $|sub\_task(T)|$ is 2. If one of the resources for sub task T_1 is not available, and all resources for sub task T_2 are available, then the $|sub\_task\_has\_resource(T)|$ is 1. This means that, there is only one sub task can have the resources to re-schedule. And the *resource_availablity* is 0.5 showing half of the resources for re-scheduling the task T is availability.

Finally, we defined the function of de-commitment fees.

$$\begin{aligned} Decommitment(T_i) &= \alpha \cdot penalty(T_i) \\ &+ \beta \cdot sub\_task\_priority(T_i) \\ &+ \gamma \cdot resource\_availablity(T) \end{aligned} \quad (6)$$

Where α, β, γ is the weight for three criteria showing the importance of them. This will be a generally object function for choosing the task to be de-committed considering penalty, sub tasks, and resources. Initially, these values will be given by experience, and then learned from the environment. Therefore, when an agent needs to make a decision of the de-commitment, it should the task with lowest de-commitment fees.

$$task = \arg\min_{T_i} \left[Decommitment(T_i) \right] \quad (7)$$

4 Experiments

4.1 Task Generator

In this experiment, there are two resources (Resource *RA* and Resource *RB*) serving three different type tasks shown in table 1.

Table 1. Task type Information

	Resource RA	Resource RB
Task A	√	
Task B		√
Task C	√	√

The tasks which need resource *RA* are defined as *Type A*, and tasks which need resource I are defined as *Type B*. If the tasks need both resource *RA* and *RB*, they are defined as *Type C*, and can be divided into two sub-tasks Task C_1 and Task C_2.

We also build a task generator to generate different tasks among these three types. This generator can generate four tasks (including sub-tasks) every 20 time steps, and each task need 10 time steps on the coordinated resource. The priority of the task is generated from 1 to 10. These four tasks will come into the system under the batch mode.

For example, if the generator generates a *Type C* task T_1, it will be automatically decomposed into two sub-tasks T_{1-1} and T_{1-2} with the same features except the resource requirement. This task generation process can be carried out under different policies, such as randomly generation policy, distribution generation policy, and so on. Table 2 shows an example of the tasks generated by the task generator under randomly generation policy. These tasks are arrived to the system under the batch mode. In this situation, there are four tasks arrived every 20 time steps.

Table 2. Task generator example

Task	Task Arriving Time	Earliest Start Time (est)	Deadline (dl)	Duration (dur)	Priority [1,10]	Resource Set
T1	0	5	25	10	8	*RA*
T2	0	5	25	10	2	*RB*
T3	0	5	25	10	3	*RA*
T4	0	5	25	10	2	*RA*
T5-1	20	25	45	10	7	*RA*
T5-2	20	25	45	10	7	*RB*
T7	20	25	45	10	5	*RA*
T8	20	25	45	10	8	*RB*
...						

Every task, which has a commitment with the coordinated resource, is regarded as a successful allocated task. Especially, tasks of *Type C* cannot be regarded as successful allocated tasks unless all of the two sub-tasks are regarded as successful allocated tasks. Otherwise, both of the sub-tasks should be rejected.

4.2 Task Allocation Strategy

Here, we are comparing four different strategies:

(a) First Come First Service (FCFS): The agent will accept the task when it has available time windows without considering other aspects, and there are no de-commitments allowed under this strategy.

(b) Sophisticated Strategy with De-commitment (SSD): The agent will accept the task when it has available time windows, and can de-commit the previous commitment when a task with higher priority requiring.

(c) Deterministic Strategy with De-commitment (DSD): The agent will decide whether to accept the coming tasks under some pre-defined decision rules, and de-commitments are also allowed.

(d) Deterministic Strategy with De-commitment and Learning (DSDL): The agent makes its own decision based on the updated policy which is learned from three criteria.

4.3 Experiment

We have generated 10000 tasks using task generator, and then allocate these tasks to resource A and B. The experiment has been done 100 times, and the result is the average number over all the experiments.

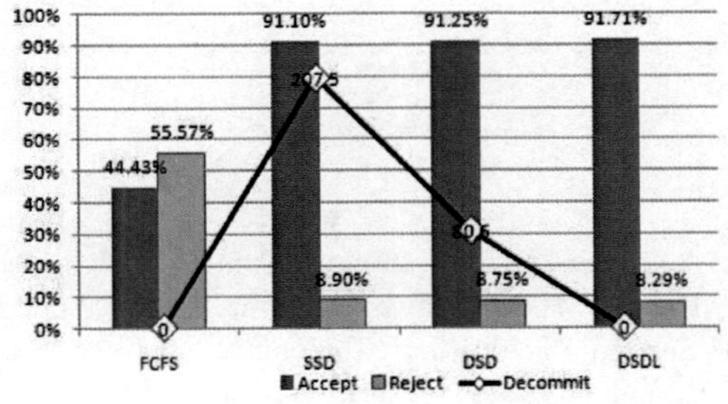

Fig. 1. Experiment results

In Figure 1, FCFS strategy rejects more than half of the coming tasks because of not using any information. When a new task comes into the system, the resource will take it when it is available. This will be some high priority tasks which comes later cannot be allocated. So, the task arrival order will change the result of allocation dramatically. On the other hand, de-commitment is not allowed, there is no de-commitment happening under this strategy.

The SSD (Sophisticated Strategy with De-commitment) greatly increases the rate of task allocation. More than 90% of the tasks are allocated, when de-commitment is allowed. When a new task with higher priority requires the resource, this resource can de-commit with the previous task and made a new commitment with the new task, and the de-committed task will also be re-allocated. This strategy has improved the allocation rate, but also increases the de-commitments times.

The DSD (Deterministic Strategy with De-commitment) has the similar performance as SSD on task allocation. When a new task comes into the system, the resource will make a decision whether to accept this task under the previous some defined decision rules. These decision rules will help the resource manager to make a smarter decision, and reduce the de-commitment times. The de-commitment time of DSD is almost one third of that of SSD.

The DSDL (Deterministic Strategy with De-commitment and Learning) gets the best performance on task allocation among these strategies. This strategy uses Q-Learning method to update the action of resource at different states based on the three criteria. The updated policy of resource manager helps it to reduce the de-commitment. In this experiment, the de-commitment has been reduced to 0, this is because the resource manager will make a commitment with highest task in the batch of tasks.

5 Conclusion and Future Work

In this paper, we present a failure recovery mechanism (FRM) for task allocation in the multi agent system to solve this problem, and build a mechanism from a de-centralized view with limited information between tasks and the resources. We use three criteria to describe the feature of the task allocation, and also show how FRM works in an open environment to recover the failure of task allocation. Experiments show the performance of the failure recovery mechanism. In the future work, we will introduce more different learn methods into the failure recovery mechanism to improve its performance.

References

1. Raja, A., Lesser, V.: A Framework for Meta-level Control in Multi-Agent Systems
2. Luo, H., Yang, S., Hu, X.: The framework for multi agent concurrent negotiation. In: Proceedings of the Second Pacific-Asia Conference on Web Mining and Web-Based Application, Wuhan, China, pp. 54–57 (2009)
3. Abdallah, S., Lesser, V.: Modeling Task Allocation Using a Decision Theoretic Model. In: Proceedings of Fourth International Joint Conference on Autonomous Agents and Multiagent Systems, Honolulu, Hawaii, pp. 719–726 (May 2007)
4. Zhang, X., Lesser, V.: Meta-Level Coordination for Solving Negotiation Chains in Semi-Cooperative Multi-Agent Systems. In: Proceedings of the Sixth International Joint Conference on Autonomous Agents and Multi-Agent Systems, Honolulu, Hawaii, pp. 50–57 (May 2007)

Intensity Weighted Stereo Matching Algorithm*

Qinyou Yang, Xianwei Rong**, Xiaoyan Yu, and Chengxiang Guan

Heilongjiang Key Laboratory for Advanced Functional Materials and Excited State Processes,
School of Physics and Electrical Engineering, Harbin Normal University,
Heilongjiang Province, China
qinyouyang@gmail.com, r7xw@163.com, xiaoyanyu2006@gmail.com,
chengxian-guan@163.com

Abstract. Stereo matching is one of the most active research topics in binocular vision. Numerous stereo matching algorithms are widely applied to three-dimensional (3-D) reconstruction and robotics. This paper presents an efficient intensity weighted matching algorithm for stereo vision applications. The proposed algorithm makes use of intensity weighted values instead of grayscale value for each pixel in the support window in order to improve the quality of disparity map. The idea is derived from the bionics fundamentals, namely retinal center detector commonly permits high resolution in the interested regions, whereas others do not require the identical resolution. Therefore, this algorithm enables the intensity weight of the kernel pixel to be the highest in the window. With the proposed algorithm, the disparity map is improved significantly with little incremental computation cost compared with the fixed-window (FW) SAD algorithm. Experimental evaluation is performed using benchmark stereo pairs. The simulation results demonstrate that the proposed algorithm can achieve bad pixel percentage of disparity map lower than the SAD algorithm.

Keywords: Stereo matching, Intensity weighted method, Binocular vision.

1 Introduction

Disparity map image is very important for reconstructing stereo objects and computer vision. Many stereo images can be recovered by means of depth map, which can be used for exploration and modeling of unstructured environments, detection of production in the industry. In order to produce high-quality real-time depth map, accurate stereo matching is necessary by using the fast and reliable algorithm. Stereo matching is an exciting and challenging field, so a large number of stereo matching schemes specific to concrete applications have been presented in the decades [1-3]. The existing algorithms are generally grouped into two different strategies, namely

* The work is partially supported by the National Natural Science Foundation of China under Grant No.60777006 and Provincial Natural Science Foundation under Grant No.F201039, as well as project-sponsored by Provincial Educational Commission under Grant No.11551126.

** Corresponding author.

J. Luo (Ed.): Affective Computing and Intelligent Interaction, AISC 137, pp. 93–100.
springerlink.com

local algorithms and global algorithms. Compared with pixel-based matching approach, local algorithms use the simple winner takes all (WTA) disparity selection strategy but increase the signal to noise ratio by aggregating matching costs over a support window. Global algorithms search for disparity assignments that minimize an energy function over the whole stereo pair with a pixel-based matching cost. The latter one commonly generates more accurate disparity map image at the expense of sophisticated matching calculation compared with the former one, so that it is hard to satisfy real-time application requirements. In practice, local algorithms, especially fixed-window approach, are applied in real-time stereo vision system thanks to easy implementation and limited memory requirements as well as less computational load. Nevertheless, fixed-window approaches generally endure the unsatisfied disparity map image, especially in the less textured and depth border regions.

According to the matching contents, the existing stereo matching techniques are broadly classified into intensity-based stereo matching [4-5] and feature-based stereo matching [6-7]. The former can directly establish dense disparity map for the modern applications of stereo such as view synthesis and image-based rendering, which requires the depth estimation in all image regions in spite of some occluded or less textured regions [8]. Nevertheless, the latter can not generate dense disparity map directly so that they require extracting the interested features for matching and interpolation techniques for obtaining disparity map [9]. Therefore, the feature-based matching approaches usually endure more sophisticated matching computation than the intensity-based ones.

To improve the depth image quality of fixed window matching, an intensity weighted stereo matching algorithm is proposed based upon the bionics fundamentals, i.e. retinal center detector commonly permits high resolution in the interested regions and low resolution in other regions. Therefore, the kernel pixel owns the highest weight value in the support window. Moreover, the proposed algorithm employs intensity-based matching other than feature-based matching in order to produce the dense disparity map image.

Based upon stereo geometry of image capturing process and selective matching algorithm, there are various matching constraints used for obtaining correct matches. In our research, due to selection of canonical stereo configuration and window-based matching, there are stereo constraints, such as epipolar constraint, uniqueness constraint, ordering constraint, and photometric compatibility constraint, used for finding the homologous points in the stereo pair.

Accordingly, our stereo matching algorithm contains two novelties. First, matching cost computation is performed by means of sum of absolute difference between weighted pixel intensities (WSAD) instead of sum of absolution difference between pixel intensities (SAD), which aims at enhancing the kernel pixel in the support window. Second, the proposed algorithm using WSAD has been demonstrated experimentally that it can produce dense disparity map by implementing on personal computer by C. Especially, the proposed algorithm can obtain lower match errors than SAD.

This contribution is organized as follows. Section 2 introduces the intensity weighted stereo matching algorithm proposed in this paper, including relevant backgrounds, WSAD formulation, and summary of WSAD algorithm for depth map.

Section 3 expounds the evaluation of WSAD algorithm. Some experimental results are given and discussed. Additionally, a comparative study between WSAD and SAD is performed by varying window size, and corresponding results are illustrated. The conclusion remarks are summarized at the end of this paper.

2 Intensity Weighted Stereo Matching Algorithm

The idea of the intensity weighted stereo matching algorithm was derived from the bionics fundamentals, namely attention and fovea sensing. The human vision usually focuses on attention of some interested regions even though boundary regions are also detected, so that the bionic systems extensively adopt multi-resolution vision system. Such theory of spatial structure to change mainly lies in two aspects: for one thing, given one point in the visual space, the existing regions are represented as eight levels of varied size; the other thing, the area is varied at a fixed angle within the visual range. The central detector of retina enables high resolution of interested regions and low resolution of around interested regions.

According to the bionics principle, the conventional fixed window based stereo matching algorithm is modified to improve the matching precision especially in the depth borders. To further decrease the computational load, the area-based matching cost employs the SAD other than sum of squared differences (SSD) and normalized cross correlation (NCC), even though SSD and NCC may produce more accurate disparity map image than SAD [1]. Accordingly, the proposed algorithm is adapted to the lower-precision vision, such as the interested object detection and traffic navigation. Thus, the compromise of matching cost computation simplicity and reasonable matching precision is considered when designing the WSAD algorithm.

2.1 Sum of Absolute Difference between Weighted Pixel Intensities (WSAD)

In order to simplify the matching cost computation, most of the existing stereo algorithms are presented on the basis of the standard stereo vision system, in which two optical axes of stereo cameras used for acquiring the stereo pairs are parallel to each other. Given stereo camera in a standard form, the searching domains for homologous points in the stereo pair can be narrowed from two dimensions (2D) to one dimension, namely the corresponding points are constrained on the identical scanline [1]. The proposed algorithm also rely on the assumption that stereo rig is in stand form so that the epipolar line and scanline coincide with each other.

The simplest area-based matching cost is sum of absolution difference (SAD) represented as equation (1).

$$SAD(x,y,d) = \sum_{i,j=-n}^{n} | R(x+j, y+i) - T(x+d+j, y+i) | \tag{1}$$

where x, y is the horizontal and vertical coordinates of the central pixel in the reference (R) and target (T) windows, respectively; d denotes the deference between x-coordinate of two central pixels; i and j denotes the row and column indices of the support window; the size of the support window is $(2n+1)$ $(2n+1)$.

Each SAD represents the cost of the correspondence between the reference window and target window. The cost aggregation strategy adopts the disparity selection winner takes all (WTA), namely in the situation of the minimum SAD within the searching range the difference between the x coordinates of two central points in the reference window and the target window is the disparity (*d*) of corresponding points in the stereo pair, which is denoted as the equation (2).

$$d = x_R - x_T \tag{2}$$

SAD-based fixed-window stereo algorithms commonly endure the unsatisfactory depth image quality. Moreover, the size of window significantly impact on the matching precision especially in the special domains such as less texture and depth discontinuities. The large size of the support window will result in the typical inaccurate matching in the depth discontinuity domains. On the other hand, too small window will make the less textured domains poor matching results. In general, the optimal size of the support window is acquired empirically. However, in practice, in spite of its limitation, the SAD-based stereo algorithms are most frequently used for real-time applications due to very low computation cost.

To improve the matching accuracy of the conventional fixed-window stereo algorithm, the sum of absolute difference between the weighted pixel intensities, named as WSAD, is presented based upon the SAD described in equation (2). WSAD is formulated as equation (3). Compared to equation (2), only the weight is additional in equation (3).

$$WSAD = \sum_{i,j=-n}^{n} w_{i,j} \mid R(x+j, y+i) - T(x+d+j, y+i) \mid \tag{3}$$

where $w_{i,j}$ is the weight of the pixel in the support window. The weight for each pixel is relevant to the distance from the central pixel in the support window, i.e. weight is higher for the pixels close to the central pixel, otherwise small weight. The large support window includes too more information to make incorrect decision for corresponding points in the stereo pair, while small support window may not contain enough information. Therefore, the proposed algorithm partially reduces the impact of window size on the depth image quality by means of weighted pixel intensity.

2.2 Procedure for the Whole Image

As mentioned above, stereo correspondence for a standard stereo vision system can take full advantage of epipolar constraint to find the homologous points in the stereo pair. The concrete steps of the proposed algorithm can be summarized as:

0. Input of the algorithm

The appropriate stereo pair, such as rectified images is first selected, and then the suitable size of block is decided based upon the input image magnitude and disparity range. The size of support window obviously affects the quality of depth image since either over large or too small window cannot acquire the smooth disparity map. Additionally, the choice of the optimal weight map of the support window is done empirically. Then the optimal weight map is another input of the proposed algorithm.

1. Stereo correspondence computation with WSAD

The corresponding points can be found by means of WSAD. The reference window can be considered as a template. i is the indices of pixels in the support window. By shifting the reference window along the epipolar line in the target image pixel by pixel, the similarity between two windows can be calculated by WSAD, namely less value is WSAD; the higher similarity is these two windows. Thus, two support windows with the minimum WSAD are the corresponding window. After that, the disparity of horizontal direction for two blocks can be obtained from the difference of-axis coordinates.

2. Depth computation

Given the disparity map, the baseline and the focal length, triangulation computes the depth of a 3-D scene point. Then the depth is transformed into the grayscale image. Closer points are brighter in the depth map image.

3. Outputting the grayscale disparity map image.

3 Experimental Evaluation

The experimental evaluation of the proposed algorithm was performed using benchmark stereo pair with different features, that is, "Tsukuba" with dimensions of 384 pixels by 288 lines and "Sawtooth" with dimensions of 434 pixels by 380 lines as well as "Map" with dimensions of 284 pixels by 216 lines. The "Tsukuba" contains objects at different depth, some occluded regions and less textured regions. The "Sawtooth" includes three planes, in which frontal plane is slanted and form a Sawtooth shape. The "Map" includes the occluded and depth discontinuity as well as smooth domains. The experiments focus on testing disparity map quality along with the variable size of support window as well as the performance comparison between the conventional SAD fixed-window stereo algorithm and the proposed algorithm.

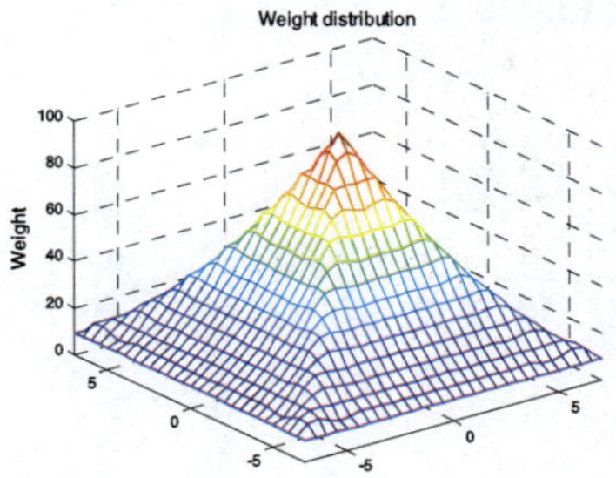

Fig. 1. The weight distribution

The weight distribution of the support window used is illustrated in Fig. 1, in which Z-axis represents the weight (w); x- and y-axis denotes the size of the support window. It can be seen that the highest weight is allocated for central pixel, while the neighbour pixel possess the weight that is relevant to the distance from the central pixel. That is to say, the neighbour pixels are farther from central pixel and they are distributed lower weight. Therefore, pixels in the support window have diverse effect on the cost aggregation by means of the weighted intensities.

Fig. 2 shows the raw disparity map of "Tsukuba" with the proposed algorithm for the window size of 11×11. We can observe that the rough 3D structure has been recovered: a trestle of the camera on the background has been reconstructed and the statue as well as lamp.

The performance comparison focuses on bad pixel percentage. Fig. 3, 4, 5 depict the matching error percentage using SAD and WSAD for "Map", "Sawtooth" and "Tsukuba", respectively. It can be observed that the proposed algorithm using "Map" and "Sawtooth" can achieve bad pixel percentage lower than the existing SAD stereo algorithm with little increased computation cost. However, for the "Tsukuba" the proposed algorithm produces matching error percentage little higher than SAD algorithm for the window size less than 11×11 due to its complex background. It can be derived that the bad pixel percentage produced by proposed algorithm will be decreased significantly along with the larger size of the support window.

Fig. 2. "Tsukuba" disparity map

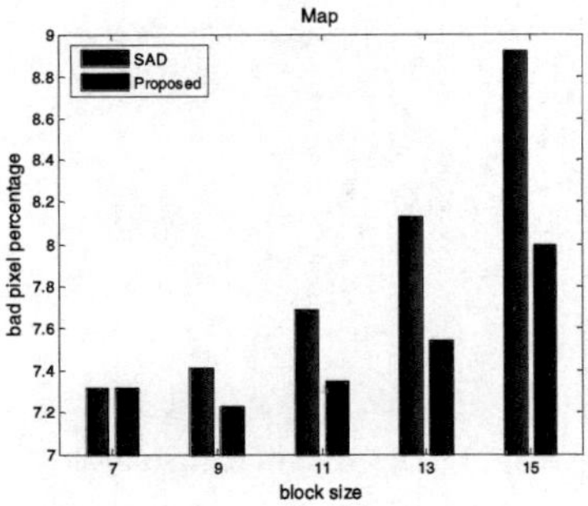

Fig. 3. The comparison of bad pixel percentage between SAD and the proposed algorithm using Map

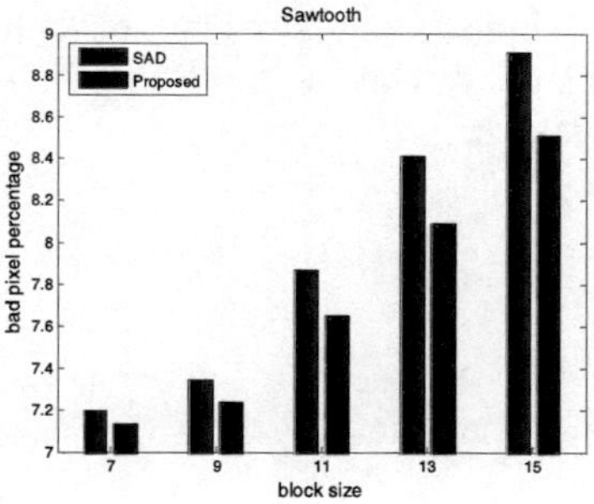

Fig. 4. The comparison of bad pixel percentage between SAD and the proposed algorithm using Sawtooth

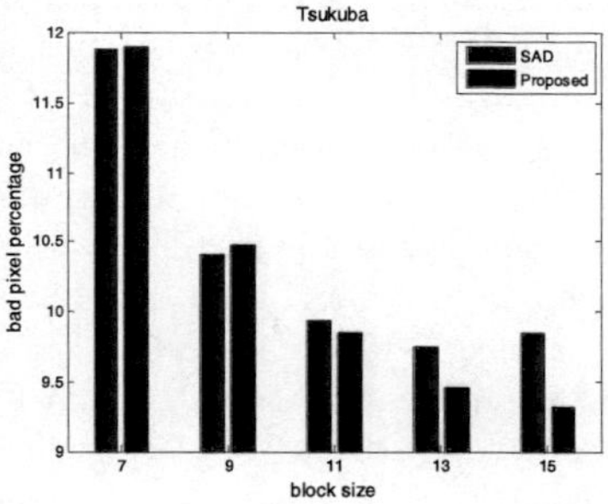

Fig. 5. The comparison of bad pixel percentage between SAD and the proposed algorithm using Tsukuba

4 Conclusions

In this paper, an intensity weighted stereo matching algorithm has been proposed for generating dense disparity map. The experimental results demonstrate that the weight distribution heavily affects matching accuracy along with varied window and the proposed algorithm is more suitable for the large support window. Moreover, the proposed algorithm can produce the bad pixel percentage of the disparity map lower than the conventional fixed-window SAD algorithm. Furthermore, the proposed algorithm has potentially numerous applications in real-time computer vision due to its low computational load. WSAD as a matching cost computation coupled with other refinement algorithms is capable of high-level matching applications such as teleconference.

In our future work, the optimal weight distribution will be examined by numerous experiments. The parallelism inherent in the proposed algorithm is going to be exploited, and then a parallel stereo vision system will be developed for various real-time applications. Additionally, several image filtering techniques such as median filtering and bilateral filtering will be used to refine the raw disparity map in order to further improve the performance of the proposed algorithm.

Acknowledgment. The benchmark stereo pairs are obtained from the web site http://www.middlebury.edu/stereo/. The authors thank the Prof. Makoto Iwata for his helpful comments on our research.

References

1. Scharstein, D., Szeliski, R.: A taxonomy and Evaluation of Two-Frame Stereo Correspondence Algorithms. International Journal of Computer Vision 47(1/2/3), 7–42 (2002)
2. Mattoccia, S., Tombari, F., Di Stefano, L.: Stereo Vision Enabling Precise Border Localization within a Scanline Optimization Framework. In: Yagi, Y., Kang, S.B., Kweon, I.S., Zha, H. (eds.) ACCV 2007, Part II. LNCS, vol. 4844, pp. 517–527. Springer, Heidelberg (2007)
3. Mattoccia, S., Tombari, F., Di Stefano, L.: Fast full-search equivalent template matching by Enhanced Bounded Correlation. IEEE Transactions on Image Processing 17(4), 528–538 (2008)
4. Wang, Z., Zheng, Z.: A region based stereo matching algorithm using cooperative optimization. In: IEEE Computer Vision and Pattern Recognition, pp. 1–8 (June 2008)
5. Sun, C.M.: Rectangular subregioning and 3-D maximum-surface techniques for fast stereo matching. In: IEEE Workshop on Stereo and Multi-Baseline Vision, in Conjunction with IEEE CVPR 2001, Kauai, December 9-10, pp. 44–53 (2001)
6. Zabih, R., Miller, J., Mai, K.: A feature-based algorithm for detecting and classifying production effects, `http://ww.cs.cornell.edu/rdz/Papers/Archive`
7. Mckeown, D., Hsieh, Y.: Hierarchical Waveform Matching: A New Feature-Based Stereo Technique. In: Computer Vision and Pattern Recognition, June 15-18, pp. 513–519 (1992)
8. Zhang, L.: Hierarchical block-based disparity estimation using mean absolute difference and dynamic programming. In: International Workshop on Very Low Bitrate Video Coding (October 2002)
9. Tombari, F., Mattoccia, S., Di Stefano, L.: Full search-equivalent pattern matching with Incremental Dissimilarity Approximations. IEEE Transactions on Pattern Analysis and Machine Intelligence 3(1), 129–141 (2009)

Joint LBP and DCT Model for Visual Speech

Zheng MeiXia[1] and Jia XiBin[1,2]

[1] Beijing University of Technology, Beijing, China
[2] Multimedia and Intelligent Software Technology, Beijing Municipal Key Laboratory
maria5201314@126.com, jiaxibin@bjut.edu.cn

Abstract. The paper aims to establish a effective feature form of visual speech to realize the Chinese viseme recognition. We propose and discuss a representation model of the visual speech which bases on the local binary pattern (LBP) and the discrete cosine transform (DCT) of mouth images. The joint model combines the advantages of the local and global texture information together, which shows better performance than using the global feature only. By computing LBP and DCT of each mouth frame capturing during the subject speaking, the Hidden Markov Model (HMM) is trained based on the training dataset and is employed to recognize the new visual speech. The experiments show this visual speech feature model exhibits good performance in classifying the difference speaking states.

Keywords: LBP, DCT, HMM, Visual speech feature.

1 Introduction

To let the computer understand and interact with the user is a promising area in human and machine interactive area. For speaking is the basis ability of people, speech recognition research has been widely done in recent decades. Except the audio speech based speech recognition, the research has been done on audio/visual bimodal based speech recognition and lip reading, which also called visual speech recognition. Visual speech recognition has broad application prospects in auxiliary speech recognition, auxiliary sign language recognition, lip-motion synthesis, the lip biometric identification of the speaker based on the dynamic characteristics. Recently, in visual speech recognition area, one of main job is the visual feature extraction of visual speech images and the analysis of visual speech characteristics to identify speech phonemes. The visual speech feature extraction methods are mainly divided into the following three ways[1]: 1) lip geometry(shape); and 2) lip motion features; 3) lip texture. The first, lip geometry, usually requires tracking of the lip contour and fitting contour model parameters and/or computing geometric features such as horizontal/vertical openings, contour perimeter, lip area, etc. However, lip tracking and contour fitting are very challenging tasks, since contour tracking algorithms are in general very sensitive to lighting conditions and image quality. The second is the use of explicit lip motion features, which have requirement that the pixel mobile of adjoining moving object frames cannot too big. The last option is the use of texture

J. Luo (Ed.): Affective Computing and Intelligent Interaction, AISC 137, pp. 101–107.
springerlink.com

information itself might sometimes carry useful discrimination information. Research shows that the method based on holistic texture feature has high recognition rate[2].

DCT plays good roles in identifying the visual speech according to the global appearance, it is not good at revealing the local information such as the texture inside the lip area reflecting the states of teeth and tongue. Actually the local textures represent the different speaking states and play an important part in classifying the visual speech. LBP(Local binary pattern) is a kind of effective local texture feature describing operator. Therefore, considering both global and local information together, this paper puts forward a visual speech recognition algorithm based on joint LBP (Local binary pattern) and discrete cosine transform (DCT)feature, and Hidden Markov Model(HMM) is employed in training and recognizing the visual speech at viseme level.

The remainder of this paper is organized as follows. In Section II, the overview of the system is given. Section III illustrates the joint feature the paper proposed and Section IV briefly introduces the viseme HMM model we used. Some experimentation results are exhibited and analyzed. Finally, the conclusion and future job is given.

2 System Overview

Our system aims to realize the Chinese viseme recognition based on analyzing the mouth images captured during speaking. One of important parts is how to establish an effective feature for visual speech to improve the ability of the system to identify viseme. In the paper, we make an analysis of the feature based on the global and local texture information. Considering the good performance of DCT and LBP in face recognition[3], we propose a joint LBP and DCT feature to represent the visual speech. To obtain the visual speeches, which are the mouth area images captured during speaking, we use the face detection algorithm that Paul Viola etc. proposed based on Harr classifier[4]. After the face is detection, the mouth area is segmented from the face area based on the geometric knowledge of the face. One of the processing results is illustrated in the mouth area segmentation part shown in Fig 1. Then HMM is employed in viseme recognition. The Chinese Viseme HMMs are trained based on the dataset recorded during the subject is speaking the Mandarin with the front face. The block diagram of our system is illustrated in Fig. 1.

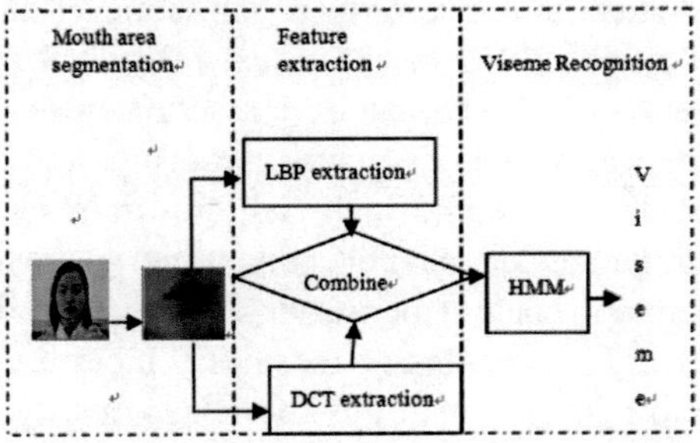

Fig. 1. Block diagram of our system

3 The Visual Speech Feature

3.1 Discrete Cosine Transform

Although the appearance mouth image has the complete information of the visual speech, it is computation consuming with high dimension without decreasing the information redundancy. The discrete cosine transform is a predominant method first introduced by Ahmed 1974 to deal with this problem[5]. Since then, it was widely used as feature extraction and compression in various applications on signal and image processing and analysis due to its fine properties[6].

Some important properties of DCT is: the more changes of the frequency domain factor u, v, the less changes of the DCT factor C(u, v). The research shows the larger factor C (u, v) is mainly distributed in upper-left corner area, where the u, v is smaller and contains the most useful information. Fig. 2 is one of results for a visual speech image's DCT coding. The figure shows the most information is concentrated on lower dimension area and the higher dimension area with less information can be neglected to reduce the redundancy of feature. According to the usual approach[7], our system extracts DCT coefficients via zigzag scanning and preserves 54 Act coefficients as eigenvector.

Fig. 2. The resulting graph of DCT coding

3.2 Local Binary Pattern

LBP[8] is an operator for computing the local texture feature of the image which bases on comparing the neighboring pixels with the target pixels. It has good ability in describing the local texture and be widely used in the area of unsupervised face recognition[9]. It has advantage of computing efficiently and high robustness of light. So the paper employs it in representing the visual speech to get the structure of lip contoure and texture of inner mouth like visibility of teeth. The initial LBP operator is using for processing the image of 3 * 3 areas. The LBP encoding process is illustrated in Fig.3. If neighborhood pixel values is greater than center pixel value, the corresponding binary is coding for 1. Conversely, it's for 0. Then we obtain the binary number by clockwise as the pixel value of the new center. Finally, we convert binary number to decimal number, thus getting the local binary mode of center pixel.

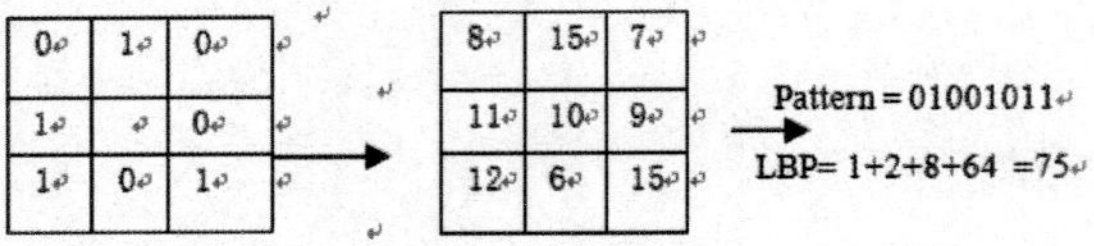

Fig. 3. Illustration of the basic LBP operator

In order to overcome the original disadvantage of high dimension of LBP operator, this system adopted 3 * 3 areas of uniform model of LBP operator[10]. If the binary coding string changes between "0" and "1" less than twice, called "Uniform" form and the rests are defined as non-uniform mode. For instance, 00000000, 11100011, 01111100 are called "Uniform" form. Compared with basic LBP form, "Uniform" form has P(P - 1) + 3 kinds of coding pattern(Here, P is the pixel number of the neighboring pixels around targeting pixel), whereas the basic LBP has 2^P kinds of coding pattern. Using "Uniform" form for expressing labial ministry characteristics can greatly save storage capacity. In addition, "Uniform" form exhibits important texture, such as the dot, line, edges and Angle, etc.

Moving the LBP operator on the training image and calculating the LBP value of each point. The statistic histogram of the LBP value expresses image texture as feature. Some results of LBP coding is shown in Fig.3. From the Fig. 3(a)-(c), the structure of the lip such as lip contour is clearly shown. From the Fig. 3(b) and (c), we can clearly tell the difference of teeth visibility states, which actually plays much important roles in identifying the different visemes.

(a) (b) (c)

Fig. 3. The resulting graph of LBP coding

3.3 Feature Level Fusion Strategy of LBP and DCT

LBP and DCT shows good performance in representing the mouth images, but it still has some shortcomings. LBP emphasis on the local texture and loses global detail of appearance. DCT plays well in describing the global appearance in low dimension space but can't emphasis the local information which plays more important roles in identifying the visual speech. To overcome the above shortage, the paper proposes to combine the two features together and different feature vectors complement each other to improve the recognition rate.

It is broadly divided into two categories of fusion theory. One kind is realized the fusion at decision level and the other is at the feature level. We select the later strategy to establish new vector for visual speech images. It is feasible for we concentrate on research of establish effective image feature rather than decision strategy. At the fusion stage the LBP and DCT coefficient are being assigned the same weight. Just like ai={ai1,ai2, ... , ai59} is the vector of LBP and bi={bi1, bi2, ... ,bi54} represents the vector of DCT, the joint feature vector is vi = {ai1, ai2, ..., ai59, bi1, bi2, ..., bi54 }.

4 HMM-Based Chinese Viseme Recognition

Hidden Markov Models(HMM)[11] has good ability in describing the dynamic attribution. In view of the visual speech is continuous procedure, the dynamic information plays important roles in identifying the visime. Therefore the paper employs the HMM in recognition visemes and recovering the dynamic property of visual speech. State transition process is not observed and random process of the observable events is covert state transition process of random functions. The diagram of HMM model is illustrated in Fig.4.

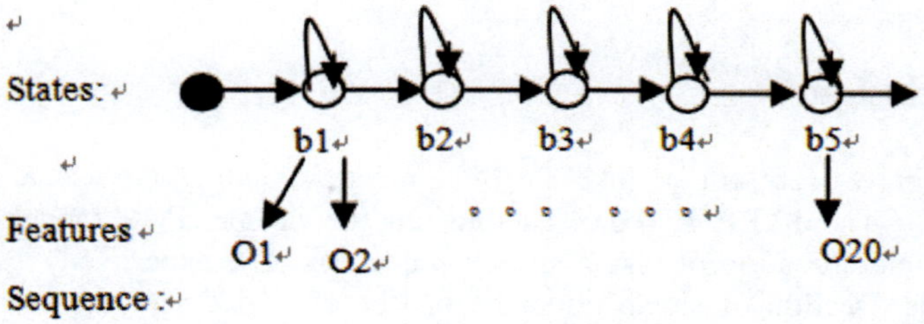

Fig. 4. The graph of HMM model

4.1 Definitions of Basic Visemes for Chinese Mandarin

This experiment is mainly directed against the Chinese identification. Generally speaking, a Chinese syllables composed by the initial and final. There are 21 initials and 38 finals, some of which share the same viseme[12]. In Table 1, initials/finals sharing the same viseme are grouped together. There is a viseme for "silence" (SIL), i.e. the natural state of he face without speaking. Selected text contains all the phonemes and the data set is embouchure uniform distributed.

Table 1. Definitions of basic visemes for chinese mandarin

Initial		Final		
b, p, m	g, k, h	a, ang	ou	i
f	j, q, x	ai, an	e, eng	u
d, t, n	zh,ch, sh, r	ao	ei, en	v (/yu/)
l	z, c, s	o	er	SIL

4.2 Classification of Basic Visemes for Chinese Mandarin

In the experiment, we extracted the LBP and DCT features of the visual speech image respectively and combined them as the feature vector. Then the k-means approach is used to classify the training dataset represented in the joint feature into different visemes classes and the centric vector is computed as the codebook. According to the note book, the class number label is obtained for each visual speech frame. At last, observer sequences is converted into a numerical sequence just like 1, 2, 3, ..., n.

4.3 Recognition of The Viseme for Chinese Mandarin

Every Chinese phonemes consists 20 frames in this experiment. It means that every phoneme contains 20 vectors which corresponds to a number in the numerical sequence of visemes.We used HMM for training the feature model. First,we put the numerical sequence of the viseme into the model ,then we acquired the divert probability of each feature vector. If we want to recognize the viseme, we just calculate the probability of the numerical sequence in every model. The viseme corresponds to the maximum probability of the model is what we are looking for. Recognition result shows that this feature model has good ability to identify the different visual speech.

5 Conclusion

In this paper, we present a novel method for representing the visual speech by combining DCT and LBP. Based on establishing the viseme HMM model, the system realizes the classification of visual speech sequences. The experiment results shows that the video section of single phoneme can be classified into the correct viseme class and error rate in our dataset is less than 10%.

Using the joint visual speech improves the ability to describe the local texture structure of lip contour and inner mouth texture together with global texture distribution low dimension space. Together with the HMM, the dynamic information of visual speech is considered and recognition improves by considering static and dynamic information together. In the following work, we will consider to improve the capability of LBP in describing the local texture structure by adding the weight and the unnecessary texture on the lip for describing the visual structure will be considered to be removed.

Acknowledgment. This work is supported by the Chinese Natural Science Foundation under Grant No. 61070117, No. 60973057 and Ph.D Starter Fund of BJUT under Grant No.X0007013200901.

References

1. Ertan Çetingül, H., Yemez, Y., Erzin, E., Murat Tekalp, A.: Discriminative Analysis of Lip Motion Features for Speaker Identification and Speech-Reading. IEEE Transactions on Image Processing 15 (December 2006)
2. Potamianos, G., Neti, C., Gravier, G., Garg, A., Senior, A.W.: Recent advances in the automatic recognition of audio-visual speech. Proceedings of the IEEE 91 (September 2003)

3. El Aroussi, M., Amine, A., Ghouzali, S., Rziza, M., Aboutajdine, D.: Combining DCT and LBP Feature Sets For Effcient Face Recognition. Information and Communication Technologies (May 2008)
4. Lienharrt, R., Maydt, J.: An Extended Set of Haar-like Features for Rapid Object Detection. IEEE Transactions on Image Processing 1, 900–903 (2002)
5. Ahmed, N., Natarajan, T., Rao, K.R.: Discrete Cosine Transform. IEEE Transactions Computers, 90–93 (1974)
6. Rao, K., Yip, P.: Discrete Cosine Transform - Algorithms, Advantages, Applications. Academic, NewYork (1990)
7. He, J., Zhang, H., Liu, J.: The extraction method in the DCT domain of lip reading for LDA feature. Computer Engineering and Applications (2009)
8. Ojala, T., Pietikainen, M., Harwood, D.: A comparative sludy of texture measures with classification based on feature distributions. Pattern Recognition 29, 51–59 (1996)
9. Tang, H., Sun, Y., Yin, B., Ge, Y.: Expression-robust 3D face recognition using LBP representation. In: IEEE International Conference on Multimedia and Expo (ICME), pp. 334–339 (2010)
10. Ojala, T., Pietikanen, M., Maenpaa, T.: Multire solution gray-scale and rotation invariant texture classification with local binary patterns. IEEE Trans. on PAMI 24, 971–987 (2002)
11. Lu, Q., Ping, Q.: Applying Stochastic Process Tutorial. Tsinghua University Press, Beijing (2004)
12. Wu, Z., Zhang, S., Cai, L., Meng, H.M.: Real-time synthesis of chinese visual speech and facial expressions using MPEG-4 FAP features in a three-dimensional avatar. In: Conference on Spoken Language Processing, pp. 1802–1805 (2006)

Paraphrase of Chinese Sentences Based on Associated Word

Zhongjian Wang and Ling Wang

School of Computer and Information Engineering, Harbin University of Commerce, Harbin 150028, China
wangzhj@hrbcu.edu.cn, wl201011@yahoo.cn

Abstract. Because the processing of complex sentences is more easily generating ambiguity, it is more difficult that deal with a complex sentence than a simple sentence in natural language processing. A method is proposed that make use of structure characteristic of a sentence. The method paraphrases those sentences with associated words by extracting paraphrase template and paraphrase rules, matching paraphrase sentence with paraphrase template by calculating similarity between original sentences with paraphrase template by keyword and restriction words. To evaluate paraphrasing performance of the method, paraphrase experiments have been done and the experiment results are discussed.

Keywords: Complex sentence, associated word, Parapgrase template, Sentence paraphrase.

1 Introduction

With the development of technology of natural language processing and maturity of some foundation processing technology, the research are also from words processing to syntax analysis, semantic analysis, more and more deeply. Although sentence paraphrase is not a new research field, Chinese sentence paraphrase research is placed in the beginning stage comparing with the paraphrase research of English and Japanese.

Paraphrase is an alternative expression of source sentence without changing its meaning. The following examples respectively transfer the same information with different expressions: "A painting hangs on the wall. " , " There is a painting hanging on the wall. " . Application of paraphrase technology is very extensive, it can effectively optimize the performance of various fields, such as, information retrieval, question answering, automatic abstracting, and machine translation. It is very significant to research sentence paraphrase.

The present study on Chinese sentence paraphrase is not abundant. Li et al. proposed a template representation and generalization method, it represent paraphrasing template with semantic code of words and extract slot word by using a dependency parser, using search engine to generalize paraphrasing examples[1]. Zhao et al. proposed a method that acquires context-specific lexical paraphrases

J. Luo (Ed.): Affective Computing and Intelligent Interaction, AISC 137, pp. 109–117.
springerlink.com

automatically, the method obtains paraphrases of a word depend on the specific sentence the word occurs in, extract candidate paraphrase by web mining [2]. Zong et al. proposed a method of spoken Chinese paraphrasing, extracting feather of input utterance by analysis phrase structure and chunk dependency of an utterance, finally got paraphrasing of input based on the analysis result [3]. Wang et al. proposed paraphrase method by analyzing the structure of large amount sentences in Chinese, extracting paraphrase templates based on key items, which can used to paraphrase some sentences with special structure. The method matches paraphrase template with sentences through calculating similarity of paraphrase sentence with paraphrase template. The method uses fixing keywords and structure auxiliary words which reflect the structure of sentences in the templates and combining the qualifier and replacing them by variables, and achieve exactly structural matching in sentence level and enhanced the coverage of the templates[4].

Torisawa proposed a nearly auto-learning method to paraphrase Japanese noun phrase into verb phrase, by using the semantic case frames and word classifications obtained by a statistical learning method [5]. Comparatively, the study on the English and Japanese paraphrase is much rich. Yamamoto proposed a method to get paraphrase acknowledge of content words from a raw text corpus [6]. Richard et al. studied about generating paraphrases of large chunks of texts, not only at the lexical and syntactic levels, but also in document structure [7].

The paraphrasing study in Japanese and English has achieved great progress, and a lot of achievements have been applied to the actual processing systems [8][9].

Chinese language is neither like English which characterizes by tense, person, declension, etc., nor like Japanese which characterizes by case-auxiliary word and declension. It is sentence structure and empty word that plays a key role in Chinese expression. Due to the lack of a strict conformation variety, the study on Chinese paraphrase is relatively difficult. The .means of paraphrasing English and Japanese sentences cannot be directly adopted in Chinese paraphrase.

Based on sentence analysis and paraphrase template extraction, the template-based method combines the advantage of rule approach and example one. The adequate number of concrete words in templates can avoid the shortage that the rule approach is too abstract and easy to make mistakes. The process of template extraction is also the accumulation of paraphrase corpus, which is beneficial to the resources of further study.

2 Our Paraphrase Method

In this paper, we use a method based on template to paraphrase Chinese sentences. At first, we extract paraphrase templates by analyzing structure of large amount of sentences that include associated word. In the templates, some words is as key word which can decide the structure of a sentence, and other variation words which can be abstracted as variables and can be substituted by other words. Then we use similarity calculate to match the paraphrase sentence with paraphrase template, when similarity value between paraphrase sentences with paraphrase templates is equal to the value of setting by experiments the system generates paraphrase sentences automatically. For those generation paraphrase sentences, we evaluate the correct or error by using paraphrase correct rate and paraphrase error rate. At the last, paraphrasing correct rate and template cover rate are calculated and the availability of the method are discussed.

2.1 Outline of the Method

The outline of the method of sentence paraphrase for the sentence that include associated words is as follows.

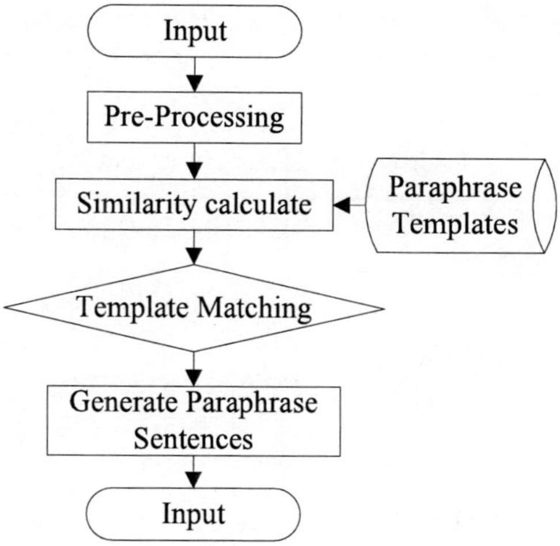

Fig. 1. Outline of the Method

2.2 Analysis of Sentence Structure

The analysis of sentence structure is very important to the establishment of paraphrase template. Expression pattern of a sentence is various with same meaning, i.e., a Chinese sentence can express the same meaning in different syntax. Therefore, there are plenty of synonymous sentences usually. In the process of sentence structure analysis, the substantives words are substituted by variables, while some specific words are maintained, such as, verb, auxiliary word, preposition and empty word. In this way, a paraphrase template is established.

Example 1: "只要坚持锻炼身体，就能增强体质。"
Paraphrase 2: "坚持锻炼身体能增强体质。"
"坚持锻炼身体增强体质。"
"如果坚持锻炼身体，就能增强体质。"
"只要坚持锻炼身体，就能增强体质。"
"......................................."

Example 2: "我们不仅要专心听讲，还要多动脑筋。"
Paraphrase 2: "我们不仅要专心听讲，还要多动脑筋。"
"我们既要专心听讲，也要多动脑筋。"
"我们要专心听讲，多动脑筋。"
"我们专心听讲，多动脑筋。"
"......................................."

We consider those sentences describe same meaning, there maybe a differ in mood or manner, but the facts of narrating are sameness approximately.

In this paper, the object of processing is sentences that include associated word. A associated word is used to associate two or more clauses that there is relation on meaning between clauses. A associated word has the follow characteristics: As a word that associate clause, indicate the relation of clauses. A associated word indicates abstract relation of clauses generally. It is used as a mark of specific complex sentence. Table 1 shows a list of associated words.

Table 1. A list of associated words

	Classification of Associated word	Associated words
1	转折关系 (Transition relation)	尽管~可是~, 虽然~但是~, ~却~, ~然而~
2	假设关系 (Suppose relation)	如果~就~, 即使~便~, 要是~那么~, 倘若~就~, 既然~就~
3	条件关系 (Condition relation)	只要~就~, 只有~才~, 无论~都~, 不管~也~, 即使~也~
4	因果关系 (cause and effect Relation)	因为~所以~, 由于~因此~, 既然~那么~, 既然~就~, 之所以~是因为~
5	并列关系 (Parataxis relation)	不是~而是~, 一边~一边~, 一方面~一方面~, 有时~有时,既~又~
6	承接关系 Continue relation	一~就~, 起先~后面~
7	递进关系 (Progression relation)	不但~而且~, 不光~也~, 不仅~还~
8	选择关系 (Selection relation)	不是~就是~, 是~还是~, 或者~或者, 要么~要么~ 与其~不如~,宁可~也不~

2.3 Paraphrase Rules

Paraphrase rules and paraphrase templates is established by analyzing the grammar structure of sentences that include associated words. Here is an example:

1. 如果我们齐心协力，就能把这件事情办好。
2. 只要我们齐心协力，就能把这件事情办好。
3. 因为我们齐心协力，所以能把这件事情办好。
4. 由于我们齐心协力，因此能把这件事情办好。

Here 1 and 2 describe, 3 and 4 describe almost same meaning. So that, we get rules:

"如果+X……就能→只要+X……就能."
"因为+X……所以→由于+X……因此."

Here use following kind of paraphrase rule:

1. p1+X ……q1 → p2+X……q2
2. p+X……q → X+p… q

In 1,2 rules, p and q are associated words, X and Y are subject of main clause and subject of vice clause respectively.

2.4 Paraphrase Template Extraction

For the template extraction, here is an example : the sentences are processed by part of speech tagging:

```
如果/c我们/rr齐心协力/i，/w就/d能/v把/p这/rz件/q事情/n办/v好/a。/w
如果/c我们/rr齐心协力/i，/w一定/d能/v把/p这/rz件/q事情/n办/v好/a。/w
只要/c我们/rr齐心协力/i，/w就/d能/v把/p这/rz件/q事情/n办/v好/a。/w
只有/c我们/rr齐心协力/i，/w才/d能/v把/p这/rz件/q事情/n办/v好/a。/w
我们/rr齐心协力/i办/v好/a这/rz件/q事情/n。/w
我们/rr齐心协力/i把/p这/rz件/q事情/n办/v好/a。/w
```

Then extracted paraphrase template is as following.

```
如果+[rr]+{*1}，就+[v]+{*2}+[n]+{*3}。
如果+[rr]+{*1}，一定+[v]+{*2}+[n]+{*3}。
只要+[rr]+{*1}，就+[v]+{*2}+[n]+{*3}。
只要+[rr]+{*1}，一定+[v]+{*2}+[n]+{*3}。
只有+[rr]+{*1}，才+[v]+{*2}+[n]+{*3}。
[rr]+{*1}+就+[v]+{*2}+[n]+{*3}。
[rr]+{*1}+[v]+{*2}+[n]+{*3}。
```

In sentences paraphrasing, follow is some sentence paraphrase strategies.

1. XY → YX
2. X →XX
3. X →XY
4. XY →X

In 1, 2, 3 and 4, X and Y are a part of a sentence, or a character string that is insert a sentence, or a part of a sentence that is deleted respectively. We paraphrase a sentence by exchanging the location of part character string, repeat a character string, insert a character string and delete part character string and so on.

For the some associated word, such as "因为…所以", sometime "因为" is used, and sometime only "所以" is used. According to [11], the usage of "因为~所以" is shown in table 2. That is, sometime the associated word can be used partly, and the usage of an associated word is also not fixing, it can be used at begin of a sentence or in the middle of a sentence. This property is used by deletion rule and shift rule. The processing of other associated word also imitate this situation.

Table 2. Usage Example of An Associated Word

Usage of associated word "因为~所以"	Number of Sentences
The number of including "因为"	550
The number of including "所以"	242
The number of including "因为~所以"	78
Total number of sentences	870

2.5 Similiarity Calculate

Paraphrase template matching is an important step in paraphrase processing. In the purpose of achieving a template matching algorithm, which includes few keyword constants and plenty of variables, any symbol string in the source sentence can be automatically replaced by the one in the goal sentence.

We defined a similarity computation method to matching templates with source sentences. The similarity computation is based on key items, restriction items and variety items. Here key items means common words of all extracted templates for same sentence structure, the restriction items means the words in all extracted templates besides the key items and the variety items means the words in all extracted templates variable words. The procedure of paraphrase matching is to extract the structure frame of source sentence, to compute the similarity between the frame and the paraphrasing template, to decide paraphrasing template of using paraphrasing. Similarity calculation of the key item, the restriction item and the variety item are show as (1).

$$\begin{aligned} KIS\ [\%] &= \frac{NKI\ (T,S)}{NTK\ (T)} \times 100 \\ RIS\ [\%] &= \frac{NRI\ (T,S)}{NTR\ (T)} \times 100 \\ VIS\ [\%] &= \frac{NVI\ (T,S)}{NTV\ (T)} \times 100 \end{aligned} \tag{1}$$

Similarity calculation of being paraphrased sentence and template uses formula (2).

$$TemSim[\%] = \begin{cases} 0, & KIS = RIS = VIS = 0 \\ \dfrac{\alpha KIS + \beta RIS + \gamma VIS}{KIS + RIS + VIS} \times 100 \\ & KIS, RIS \ or \ VIS \neq 0 \end{cases} \tag{2}$$

Where: ***NKI*** is the number of key item of the being paraphrased sentence to compare with the template; ***NRI*** is the number of corresponding the restriction item of the being paraphrased sentence to compare with the template; ***NVI*** is the number of corresponding the variety item of the being paraphrased sentence to compare with the template; ***NTK*** is the number of key item in the template; ***NTR*** is the number of restriction item in the template; ***NTV*** is the number of variety item in the template; ***KIS*** is the key item similarity of the being paraphrased sentence; ***RIS*** is the restriction item similarity of the being paraphrased sentence; ***VIS*** is the variety item similarity of the being paraphrased sentence; ***TemSim*** is the similarity of the being paraphrased sentence to match with the template. Where, and are weighted parameter, and.are coefficients, the optimum coefficients are decided by the preliminary experiments using Greedy method, α=10, β=1 and γ=5.

In the template matching procedure, matching of key items and restriction items must consider the order of words, and variety items are calculated only number.

By extracting sentence structure of source sentences and comparing with templates, we can get the number of matching key items, restriction items and variety items in the source sentence, and the number of key items, restriction items and variety items in the corresponding templates. Then, the similarity between the source sentence and the templates is obtained.

3 Experiment and Analysis

3.1 Experiment Data and Results

We extract about 1400 sentences that include associated word from People's Daily Corpus to evaluate the effectiveness of the proposed method to Chinese sentence paraphrase. Those sentences include all kind of associated word. Table 3 shows the number of sentences that include associated words.

Table 3. Associated word Sentence distributing

Associated word	Number of sentence	Associated word	Number of sentence
因为	200	虽然	101
所以	103	然而	121
尽管	95	如果	185
即使	34	由于	309
尚若	2	既然	7

3.2 Experiment Results Analysis

We use the data that extracted from corpus, after pre-processing that segment, label POS, delete some non-Chinese character and non-characters, use those rules and paraphrase templates to carry on paraphrase experiments. For the experiment results, correct or error are evaluated by manual labor. The experiment used 120 rules and templates, original sentences are 1400 and paraphrase result is 2800 sentences. The results is showed in Table 4.

Table 4. Experiment results

Original Sentences	Paraphrase Result	Non-paraphrase	Correct	Error
1400	2685	548	1458	1227

The experiment results indicate that paraphrase correct rate is only 54.3%. The reason of low correct rate is using efficiency too low, a original sentence is too long and include more clauses generally.

4 Conclusion

For the sentence of including associated word, it is consist of several clauses, paraphrase can get a simple sentence. The template based method need to establish templates by manual. The cover rate of template is very important to improve paraphrase rate, but the improvement of paraphrase correct rate need to research much relational problem.

Acknowledgment. The Project Sponsored by the Scientific Research Foundation of the State Human Resource Ministry for Returned Chinese Scholars and the Scientific Research Foundation for the Returned Overseas Chinese Scholars, State Education Ministry.

References

1. Li, W., Liu, T., Zhang, Y., Li, S., He, W.: Automated Generalization of Phrasal Paraphrases from the Web. In: The 3rd International Workshop on Paraphrasing, Jeju Island, South Korea, pp. 49–57 (October 2005)
2. Zhao, S., Liu, T., Yuan, X., Li, S., Zhang, Y.: Automatic Acquisition of Context-Specific Lexical Paraphrases. In: Proceedings of IJCAI, Hyderabad, India, pp. 1789–1794 (January 2007)
3. Zong, C., Zhang, Y., Yamamoto, K., Sakamoto, M., Shirai, S.: Approach to Spoken Chinese Paraphrasing Based on Feature Extraction. In: The 6th Natural Language Processing Pacific Rim Symposium, Tokyo, Japan, pp. 551–556 (November 2001)
4. Wang, Z., Liu, Y.: Paraphrase of Chinese Sentences with Special Structure. In: International Conference on Intelligent Systems and Applications (ISA 2009), Wuhan, China, May 23-24 (2009)

5. Torisawa, K.: A Nearly Unsupervised Learning Method for Automatic Paraphrasing of Japanese Noun Phrases. In: Proceeding of Workshop on Automatic Paraphrasing, Theories and Applications, pp. 63–72 (2001)
6. Yamamoto, K.: Acquisition of lexical paraphrases from texts. In: The 2nd International Workshop on Computational Terminology, pp. 22–28 (2002)
7. Power, R., Scott, D.: Automatic generation of large-scale paraphrases. In: The 3rd International Workshop on Paraphrasing, Jeju Island, South Korea, pp. 66–72 (October 2005)
8. Lepage, Y., Denoual, E.: Automatic generation of paraphrases to be used as translation references in objective evaluation measures of machine translation. In: Third International Workshop on Paraphrasing, IWP 2005 (2005)
9. Rinaldi, F., Dowdall, J., Mollá, D., et al.: Exploiting Paraphrases in a Question Answering System. In: Proceeding of Workshop in Paraphrasing at ACL 2003, Sapporo, Japan, pp. 25–32 (July 2003)
10. Wang, T.: "Yinwei......Suoyi......" of Review. China Academic Journal Electronic Publishing House (April 2004) (in Chinese), all rights reserved, `http://www.cnki.net`

Automatic Registration of Unknown Words for Multi-language Information Retrieval

Zhongjian Wang and Ling Wang

School of Computer and Information Engineering, Harbin University of Commerce, Harbin 150028, China
wangzhj@hrbcu.edu.cn, wl201011@yahoo.cn

Abstract. The development of the internet results in the increasing of electronic text in multi-language. The researches about information retrieval not only aim at single language, study of multi-language information retrieval is paid attention by researchers. To deal with unknown words and ambiguity of query word translation, the existent dictionaries are used. The method can improve search efficiency and search precision.

Keywords: Information retrieval, Multi-language, Query word translation.

1 Introduction

With the development of the internet, electronic text of variety language are increasing ceaselessly, new words are generated more and more. Especially, the development of electronic technology generates large amount of new words. It is not possible that register all of those new words into dictionary. Furthermore, with cultural exchange different country new word are also generated. For example, in Japanese text, large amount katakana words are generated. In news report, one sentence includes one katakana word averagely. Even more in computer related report and magazine, more than half words are katakana words[1]. So that those unregistered words appear high frequency.

When Chinese study Japanese language, they often get the meaning of words by consulting a dictionary, but this situation almost there are not the katakana words in the dictionary. Thus first we use a Japanese-English dictionary and then use an English-Chinese dictionary to get the meaning of a katakana word.

Because there are multi-language text information on the internet, an information retrieval system must find the information of user's longing witch are described by different language. For single language information retrieval method, to extend word gather of query, the thesaurus is used to information retrieval system[2]. The method was proposed for extracting the information of WWW search system users, by analyzing the search keywords list logged by the system. A called after the sequential searching model is proposed, the algorithm calculate the relationship between each search keywords logged, and groups them into several gathers, which is supposed to represents the information needs of users[3]. Paper[4] proposed a method to realize an interactive guidance mechanism for document retrieval system. For multilanguage

J. Luo (Ed.): Affective Computing and Intelligent Interaction, AISC 137, pp. 119–125.
springerlink.com

information retrieval system method, paper [5] proposed a theoretically grounded alternative, which uses sense disambiguation based upon context terms within the source text, introduced the concept of translation probabilities incorporating a context term and extends.

In this paper, we use several different dictionary to deal with problem of unknown words, meanwhile reduce word ambiguity from extending query gather.

2 Automatic Registration of Unknown Words

As an example, we suppose we have not Chinese-Japanese dictionary, because Japanese katakana words are almost unregistered in Chinese-Japanese dictionary. Now we use Japanese-English dictionary and English-Chinese dictionary to get Chinese words, and furthermore we use Chinese-English dictionary to deal with words ambiguity. As an example, Figure 2 gives a demonstration.

1. Input Japanese katakana word
2. Get English words by Japanese-English dictionary
3. Get Chinese words by English-Chinese dictionary
4. Get English words by Chinese-English dictionary

If the English word at step 4 is same with the word at step 2, then translated word of input word can be confirmed.

2.1 Japanese Katakana Words

A Chinese word of the Japanese katakana word can be get by Japanese-English, English-Chinese dictionary. And register it in Chinese-Japanese dictionary.

2.2 Chinese Words

Like figure 2, we use English-Chinese dictionary and Chinese-English dictionary to decide Chinese translated word from English word, deal with ambiguity.

3 Outline of the Method

In this method, we use Japanese-English dictionary, English-Chinese dictionary and Chinese-English dictionary; otherwise we use also Japanese thesaurus and Chinese thesaurus. The usage of those lexicons is showed in Figure 1.

We use those lexicons to decide exact translation word and deal with ambiguity. When a word has multi-translation words, we use two dictionaries to exchange validating, as Figure 2 shows.

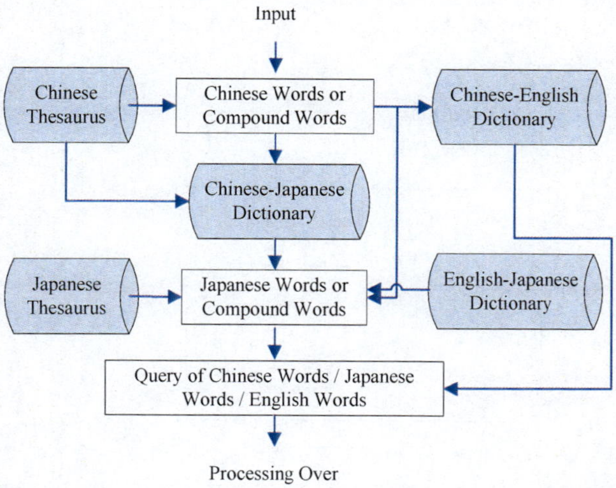

Fig. 1. Outline of the method

3.1 Segmentation Processing

Two exchange dictionaries can be used to decide exact translation words when there are more choice words. Here for the input query sentences, first of the processing is to segment it to words, not only Chinese text but also Japanese text need segment process. For Chinese text, ICTCLAS is used and for Japanese text ChaSen is used[6][7].

3.2 Use of Dictionaries

There is more than one translation word to a source word in a dictionary commonly. Thus we use a method that two dictionaries exchange, reverse to decide correct translation word. Sometimes if we have not a kind of dictionary, we also can get translation words relatively by use this method. Table 1 is results about 20 words selection experiment. In the experiment, correct word is translated by reverse lookup method.

3.3 Use of Thesauruses

To extend query word gather, the thesauruses are used. We use Japanese thesaurus and Chinese thesaurus to extend query.

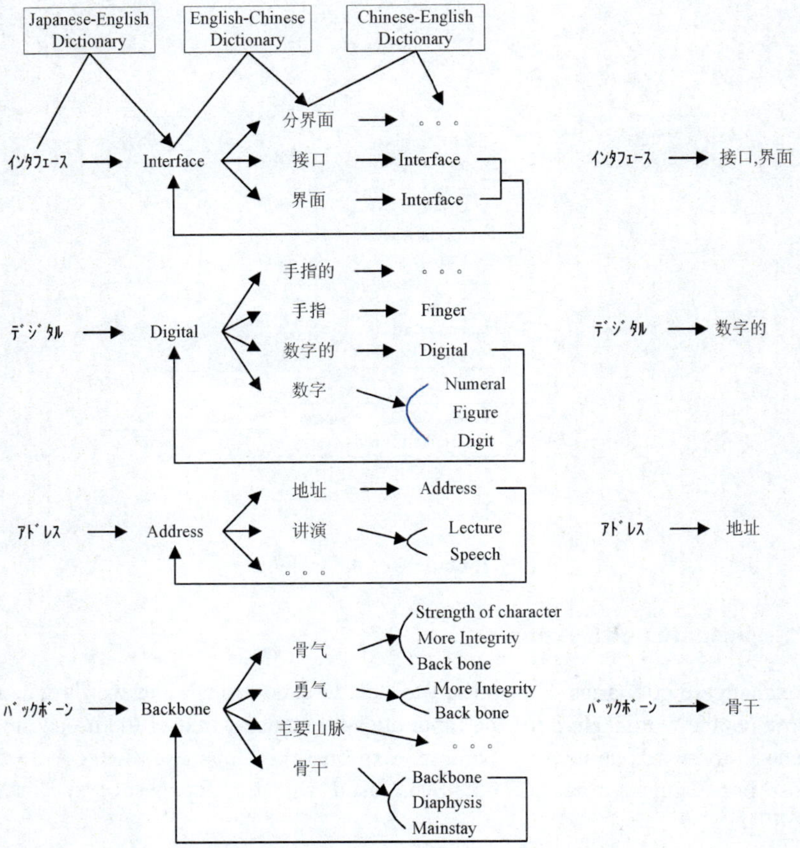

Fig. 2. Translation Processing of Query Words

Table 1. The test results of 20 single words

Japanese	English	Chinese	Correct/ Error
ｱｸﾁｭｱﾘﾃｨｰ	Actuality	现实,现实性	Ok
ｱｽﾀﾘｽｸ	Asterisk	星形符号	Ok
ｱｯﾊﾟｰ	Upper	上面的	Ok
ｱｯﾌﾟﾃﾞｯﾄ	Update	v.使现代化 n. 现代化	Ok
ｱｯﾌﾟﾗｲﾄ	Upright	直立的	Ok
ｲｰｽﾀｰ	Easter	复活的	Ok
ｲｰｾﾞﾙ	Easel	书架	Ok
ｲｴﾛｰ	Yellow	黄色,黄色的	Ok

Table 1. (*continued*)

ｳｲﾝﾄﾞｳ	Window	窗子	Ok
ｳｪｰﾄ	Wait	等待	Ok
ｳｪｰﾊﾞｰ	Weber	韦伯	Ok
ｴﾝﾄﾘ	Entry	进口	Ok
ｵｰｽﾄﾗﾘｱ	Australia	澳洲	Ok
ｵｰﾃﾞｨｵ	Audio	听觉的,声音的	Ok
ｶﾒﾗ	Camera	照相机	Ok
ｼﾝﾀｸｽ	Syntax	语法	Ok
ｺﾝﾊﾟｽ	Compass	指南针	Ok
ﾘﾏｰｸ	Remark	评注	Ok
ﾘﾗｲﾄ	Rewrite	重书写	Ok
ﾘﾎﾟｰﾄ	Report	报告	Ok

4 Experiments

Here the problems of words to words translation are almost one to many, thus how to select exact translation word is very important. We use two mutual dictionary to decide correct selection from multi-translation words. Table 1 and 2 show the test results of 20 single words and 10 compound words respectively.

Table 2. The test results of 10 compound words

Japanese	English	Chinese	Correct/ Error
ｱﾅｺﾝ	analog computer	模拟计算机	OK
ｱﾅﾃﾞｼﾞ	Analog digital	模拟数字	OK
ｲﾝﾌｫﾒｰｼｮﾝｻｲｴﾝｽ	Information science	资料学/信息学	OK
ｲﾝﾍﾞﾝﾄﾘｰｻｲｸﾙ	Inventory cycles	存货周期	OK
ｳｨｰｸﾌﾎﾟｲﾝﾄ	Weak point	弱点	OK
ｳｨﾝﾄﾞｳｼｮｯﾋﾟﾝｸﾞ	Window shopping	溜达着看橱窗	OK
ｴｺﾛｼﾞｰ運動	Ecological movement	---	NO
ｴﾝﾌﾟﾗ	Engineering plastics	---	NO
ｵｰﾄﾘﾋﾟｰﾄ	Automatic repeat	---	NO
ｵﾝﾊﾟﾚｰﾄﾞ	On parade	---	NO

4.1 Extension of Query Word

For input query, first the systems extends the query word gather by thesaurus, then translates the input to other language words and extend the words by thesaurus. At translation, the system uses tow mutual dictionary to deal with word ambiguity.

4.2 Experiment Results

The experiments are carried out under the manual constructed retrieval system to evaluate the availability of the proposed method. The experiment date is download text from the Internet about 20MKB.

At first, for the input query word, the system translates them to other words.

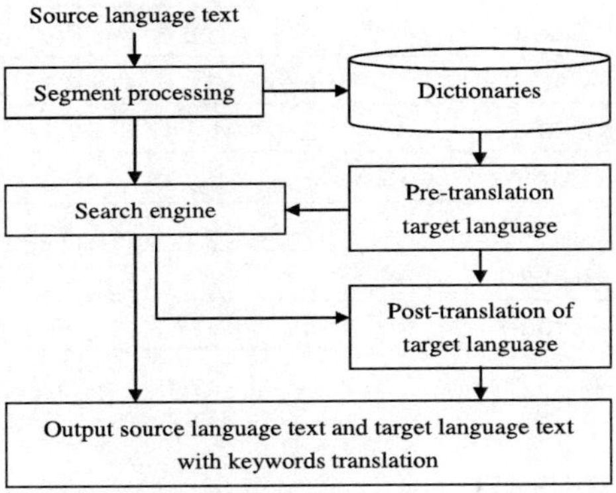

Fig. 3. The information retrieval system

5 Conclusion

To deal with the problem of necessary dictionary lack and word to word translation ambiguity, multi-mutual dictionary are used. The use of multi-mutual dictionary can reduce ambiguity; help the system to select correct translation words that the experiments have demonstrated as table 1 and 2. There are still some problems, such as because query is extended so that too many search results are got. Some search results is not relational with content of user's desirousness. Although the method has a lot of shortage that must be improved, but it still a good method to deal with ambiguity and dictionary lack.

Acknowledgment. The Project Sponsored by the Scientific Research Foundation of the State Human Resource Ministry for Returned Chinese Scholars and the Scientific Research Foundation for the Returned Overseas Chinese Scholars, State Education Ministry.

References

1. Yoshihiro, M., Masahiro, K., Satoshi, S.: Japanese-to-English Translation or 'katakana' Predicates. Information Processing Society of Japan (IPSJ). In: "Share and Reuse of Language Resource" Symposium (1999) (in Japanese)
2. Srinivasanm, P.: Thesaurus Construction. In: Frakes, W.B., Baeza-Yates, R. (eds.) Information Retrieval: Data Structures & Algorithms, pp. 161–218. Prentice Hall, NJ (1992)

3. Yamada, A., Esashi, H.: Proposal of a New Algorithm for Extracting Information Needs on Small Search System. Bull. Tokyo Gakugei Univ. Sect. 6(55), 37–41 (2003) (in Japanese)
4. Yoshiki, N.: Dynamic Co-occurrence Analysis for Interactive Document Retrieval. Information Processing Society of Japan (IPSJ) 96-NL-115, 99–106 (1996) (in Japanese)
5. Kishida, K., Ishita, E.: Translation disambiguation for cross-language information retrieval using context-based translation probability. Journal of Information Science 35, 481–495 (2009)
6. ICTCLAS (Institute of Computing Technology, Chinese Lexical Analysis System), `http://www.ictclas.org/index.html`
7. ChaSen, Japanese Dependency Structure Analyzer, `http://chasen-legacy.sourceforge.jp/`

Study on the Representation and Acquisition Mode of Network Science and Technology Intelligence Based on Multilingual Ontologies

Wu Suyan[1], Li Wenbo[2], and Zhao Xin[1]

[1] Beijing Municioal Institute of Science & Technology Information Beijing, China
[2] Institute of Software, Chinese Academy of Sciences Beijing, China
wusuyan77@163.com, Liwenbo02@iscas.cn, Zhaoxin@163.com

Abstract. This paper studied the representation and acquisition mode of network science & technology intelligence, proposed the representation and acquisition methods based on multi-lingual ontologies and constructed the integral Architecture of the system according to our method.

Keywords: Muti-lingual ontologies, Science & technology intelligence, Representation and acquisition.

1 Introduction

Science and Technology Intelligence(S&T Intelli-gence) currently has two salient features:

1) Based on network. The Internet has been a major source of S&T Intelligence. The features of internet are larger-scale, fast udate and diverse forms, which makes manual information processing methods invalid, Thereby, the information processing must be depend on computer.

2) Multilingual. Most network information exists in their own national language, so automatic cross-language information acquisition must be study multilingual information proceeding technology.

Because of the semi-structured feature of network information, cross-language information acquisition is difficult to get semantic support and only depend on bilingual dictionary and keyword, which lead to the unknown word problem and the semantic ambiguity problem.

a clear concept of formal specification

Ontology technology can well describe the content of the concept and the relationship between the concepts. Multilingual domain ontology is multiple specific representation with multilingual of the same domain ontology representation. Therefore, the multilingual ontology-based information network technology that can provide cross-language semantic information support for solving the problems acquiring Science and Technology Intelligence [1].

J. Luo (Ed.): Affective Computing and Intelligent Interaction, AISC 137, pp. 127–131.
springerlink.com

Multilingual network technology intelligence represented Based on ontology can provide semantic support for cross-language intelligence acquire , In which Ontology technology play a framework role, which makes the line open, loose structure of science and technology information into clear, clear semantic relation ontology instances Information. In additional, it also play a role in navigation for cross-language information retrieval, which Semantic information to increase retrieval of information and reduce the difficulty of cross-language retrieval.

2 The Current Research Both at Home and Abroad

WordNet creaed by Princeton University is a good reference resource for established ontology. Currently, many countries established own corresponding national language ontology referenced Wordnet[2]. In Europe, EuroWortNet was established as multi-ontology[3]. Beginning this ontology contains Dutch, English, Italian and Spanish , they linked by the index between languages. in present, EuroWortNet plans to join other european languages. In china, HowNet is a Chinese ontology referenced WordNet[4-5]. HowNet is an on-line common-sense knowledge base unveiling inter-conceptual relations and inter-attribute relations of concepts as connoting in lexicons of the Chinese and their English equivalents. RussianWordNet created by Russia is a bilingual English and Russian ontology[6]. These multi-language ontology is the good language resources and tools for cross-language information retrieval.

Research in information extraction based ontology is very active, and has also made quite a number of important achievements. For example, some researchers applied it to bio-pharmaceuticals domain and acquire some good benefits. UK's Sheffield University Horacio Saggion and Adam Funk applied this technology to business intelligence to track the business activities of transnational corporations and national information[7-8]. But the study of establishing Science and Technology Intelligence ontology is very little and cross-language network science and technology Intelligence acquisition ontology-based is still no systematic research on it so far. This paper studied how to establish science and technology intelligence ontology and used it to study the effective the multi-language representation model of network intelligence and information and the cross-language acquisition model of network technology information.

3 System Structure

The System structure of the multi-language representation and acquisition model of network Science and Technology Intelligence based on ontology is mainly consisted with three levels.

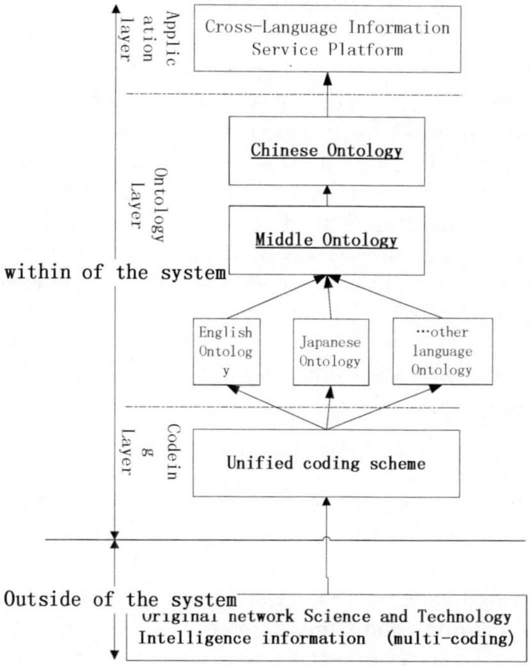

Fig. 1. Structure of the system

The lowest Layer of this system is the coding layer that mainly solved the coding problem about the multi-language representation of network Science and Technology Intelligence, which include automatic coding and Language identification. This layer is to provide services to layers above.

The middle layer is Ontology layer. Which include service users' languages Science and Technology Intelligence middle ontology and multi-language Science and Technology Intelligence ontology? The first ontology belongs to service areas. Its function is providing a united service architecture and evaluated service quality. It is usually a body with specific language. For example, this paper its language is Chinese. The latter ontology belong to Science and Technology Intelligence areas and is composed of many different languages.the key technology problems of this layer is how to establish The two ontology and automatically cross-language information acquisition.

The top layer is the application layer that maily studies the cross-Language Science and Technology Intelligence Retrieval based ontology model. The retrieval models proposed by this paper take advantage of the science and Technology Intelligence middle ontology. The work model is showed in figure 1, which can provide a united service for user and isolate application service with multi-language Science and Technology Intelligence ontology. When service Changes, Rely only on the middle of the escalating body, without changing the multi-language Science and Technology Intelligence ontology.

4 Details of the System Layers

4.1 Coding Layer

The same language has different coding, which is a historical problem. In order to efficiently processing network information must be encoded. Unicode provides a unique number for every character, no matter what the platform, no matter what the program, no matter what the language. The Unicode Consortium is a non-profit organization founded to develop, extend and promote use of the Unicode Standard, which specifies the representation of text in modern software products and standards. The role of this layer is to different network technology intelligence data sources on the local encoding to Unicode. This technique is based on the premise that the computer system knows the type of local codes. But in the case that the type of some system don't decelerate or decelerate error, our system need to automatically determine the encoding type.

4.2 Ontology Layer

Ontology layer is the core structure of the system. The reasonable degree of multiple specific languages Science and Technology Intelligence ontology and the quality and efficiency of acquiring network case automatically direct impact the efficiency of the system. The reasonable degree of service users' languages Science and Technology Intelligence middle ontology and Chinese internet services Ontology determines the upper application service level of cross-language information retrieval.

Constructing multiple specific language science and Technology Intelligence ontology plan is as follows:

Construction of the network more specific language domain science and Technology Intelligence ontology is the definition of the concepts and their relationships of Science and Technology Intelligence in different language, which is a huge workload task and get low efficiency by artificial. It is more reasonable to automatically get ontology from relevant text from each language.

The method of automatically constructing specific language science and technology intelligence ontology is using natural language processing theory and techniques with experts' knowledge in the field of information technology to reconstruction science and technology intelligence. Establishing self-learning mechanism of limited text realize the automatic construction of science and technology intelligence ontology.

Science and technology intelligence is a more specific area. Its information source is clear. The product of scientific and technical information mainly reflected in the patents, monographs, documents, papers, technical reports, products of scientific and technological projects and hi-tech products, awards information of science and technology and other forms. the properties of each scientific and technological achievements are generally more specific. For example belongs to the country, author, level, and so their respective disciplines. Among them, the classification of respective disciplines is a more detailed and has relevant methods, such as Chinese Library Classification. These features make the intelligence of knowledge are basically semi-structured information, so the collection of intelligence and reconstruction of knowledge in other fields of knowledge with respect to said comparison implementation.

4.3 Application Layer

Ontology-based multi-language Science and Technology Intelligence retrieval methods has more advantages, because the introduction of a standardized cross-language synonyms, which makes the concept of different languages to Correspond each other and makes retrieval from lexical match upgrade to the semantic level, namely, between the concept of conversion, eliminating the conversion to bring the different language ambiguity.

5 Conclusion

In this paper, the multi-language representation model of network intelligence and the cross-language acquisition model of network technology information were studied. This paper proposed the multi-language representation and acquisition model of network Science and Technology Intelligence based ontology and given the system structure. This method can provide the semantic level of support for cross-language retrieval.

Type your main text in 10-point Times, single-spaced. Do not use double-spacing. All paragraphs should be indented 1/4 inch (approximately 0.5 cm). Be sure your text is fully justified—that is, flush left and flush right. Please do not place any additional blank lines between paragraphs.

Acknowledgment. This work was supported by the National SoftScience Research Programe of China (Grants No.2009GXQ6D154); the National Science Foundation of China (Grants No. 61003117), and the CAS Innovation Program (Grant No. ISCAS2009-DR12).

References

1. Studer, R., Benjamins, V.R., Fensel, D.: Knowledge engineering; principles and methods. Data Knowledge Engineering 25(1-2), 161–197 (1998)
2. WordNet [EB/OL] (January 18, 2006), http://www.cogsci.princeton.edu/~wn/ (accessed January 2006)
3. EuroWordNet:building a muhilingual database with wordnets for several European languages [EB/OL] (January 18, 2006), http://www.Illc.uva.nl/EuroWordNet/ (accessed January 2006)
4. HowNet. (April 02, 2006), http://www.keenage.com
5. Chen, H.-H., Lin, C.-C., Lin, W.-C.: Building a Chinese English WordNet for translingual applications. ACM Transactions on Asian Language Information Processing 1(2), 103–122 (2002)
6. Loukachevitch, N.V.: Russian language in cross-language information retrieval: Resources and tools in Russia (April 02, 2006), http://www.clef-camPaign.org/workshop2003/Presentations/loukachevitch.ppt
7. Gonzalo, J., Verdejo, F., Peters, C., et al.: Applying EuroWordNet to cross language text retrieval. Computers and the Humanities 32, 185–207 (1998)
8. Honavar, V., Silvescu, A., Reinoso-Castillo, J.: Ontology-Driven Information Extraction and Knowledge Acquisition from Heterogeneous, Distributed Biological Data Sources. In: Proceedings of the LJCAI 2001 Workshop on Knowledge Discovery from Hererogeneous, Distributed Antonomous, Dynamic Data and Knowledge Sources (2001)

Developmental Model of Health Policy: The Third Way for China's Health System Reform*

Xiao Liu

School of Labor Relations and Human Resources, Renmin University of China,
School of Social Welfare, University of California, Berkeley,
59 Zhongguancun St., Beijing, 100872, China, P.R.
Liuxiao8311@yahoo.com.cn

Abstract. This paper involved rethinking of the health system of China. It analyzed key factors that joined health and wealth; pointed out the policy implication that we can promote population health and human well-being without necessarily hampering economic growth; built a developmental model of health policy with Chinese characteristics which identified the juncture of social and economic development. Thus, the author recommended a suitable direction for the health system reform in China through the developmental perspective of view.

Keywords: Developmental model, Social development, Health policy, Health system reform.

1 Introduction

The normative perspective of social policy is often categorized into the residual model (with the typical example of the US) and the institutional model (with the typical example of Western Europe). The former emphasizes limited government intervention and self-responsibility. It leaves social demands to the free market; while government only offers help to the vulnerable who are unable to participate in the market. The latter contends that government intervention should be performed in a more comprehensive way; social policy is a long-term scheme which should not be just residual; people should be universally covered by social welfare. However, this kind of categorization is based on the experience of western industrial countries and is not applicable to developing countries. Stewart MacPherson and James Midgley pointed out in 1987 that in developing countries, which are pursuing economic growth and seeking the coordination between social policy and economic development, neither the residual nor the institutional model could describe their social policy exactly and appropriately [1]. In addition, Midgley put forward the "developmental model of social policy" or "developmental perspective", which emphasized economic policy in accordance with social policy [2]. He believed that social welfare was not always incompatible with economic growth and there was an optimal choice that could promote both social and

* This work is supported by China Scholarship Council, State-Sponsored Study Abroad Grant to Xiao Liu.

J. Luo (Ed.): Affective Computing and Intelligent Interaction, AISC 137, pp. 133–141.
springerlink.com

economic development. He emphasized both social intervention and investment. This model is more applicable to developing countries like China.

The reform of health system in China is now facing multiple choices; however, complete nationalization of health service is not sustainable here, nor is complete marketization. And neither residual nor institutional welfare models can solve the problems in the health system in China. As the third way, the developmental model of social policy provides a better perspective for the transformation. This paper involved rethinking of the health system of China. It developed a developmental model of health policy with Chinese characteristic and recommended a suitable direction and path for the transformation.

2 Variables and Models

2.1 General Model of Health and Well-Being

H. Blum developed a popular model of the force field and well-being paradigms of health (see Fig.1) [3]. As is shown below, health is influenced by four main factors, including population, environment, life styles and medical care services. According

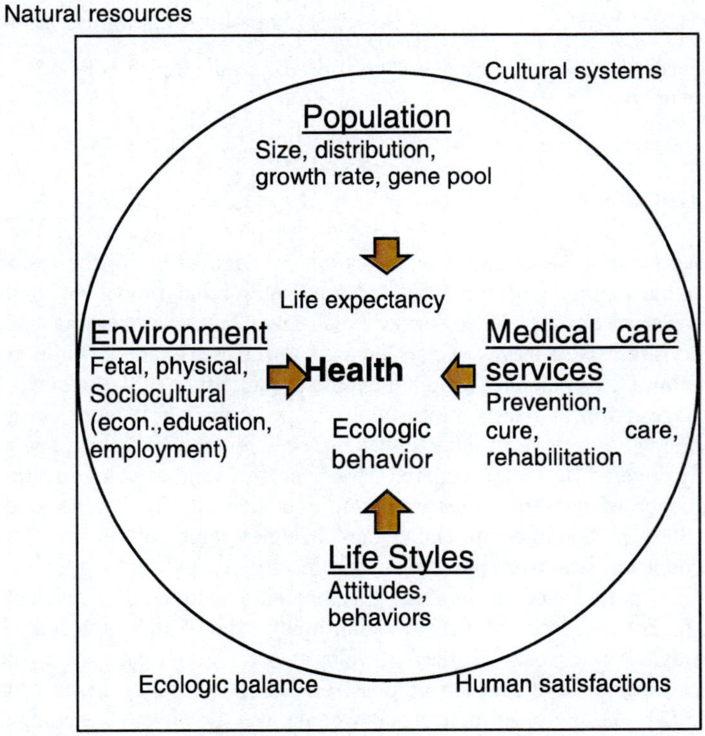

Fig. 1. Blum Model of Health and Well-Being

Source: H. Blum, Planning for Health, New York: Human Sciences Press, 1981.

to Blum, health is a widely defined field with all the forces integrated in a big system. This general model can help us better understand developmental health policy and make comprehensive policy recommendations. Apparently, typical policy makers would not put equal emphasis on these various factors. Population factor is more naturally determined and difficult to be changed by human force. Medical care services, especially curative activities, always attract the most attention because the need of them is often more emergent and the effect of them is always easy to be seen; while environment improvements and life style changes are often overlooked because the effects are long-term (sometimes even longer than the duration of a cabinet) and results are hard to measure. However, investment on those long-tem factors is much more cost-effective than on curative medical care directly. As a developing country, China faces the conflict of limited resource and unmet medical needs. So it is necessary and desirable to build a developmental model that identifies the methods to promote human well-beings through efficient health policies without hampering economic development.

2.2 Developmental Model of Health Policy

The key point for the "developmental model" is to find out the juncture of economic aspect and social aspect of policies. As for health policy, particularly, the objective is to improve health conditions without causing excessive cost expansion; while the measures are to distinguish factors that have most effective and sustainable influence on health results and to choose the most economical ways to improve health conditions. The following factors are picked up from general model and re-organized to build the specific developmental model of health policy: education, public health, primary care and social health insurance (see Fig.2). Through some intermediate variables and processes, these inputs achieve the outputs of increased access, better health condition and lower health expenses. Table I indicates briefly how these factors are applied in different methods to accomplish the expected goals. The following parts of this paper would explain the measures in details and analyze the policy implications within the framework of this model under the given facts in contemporary China.

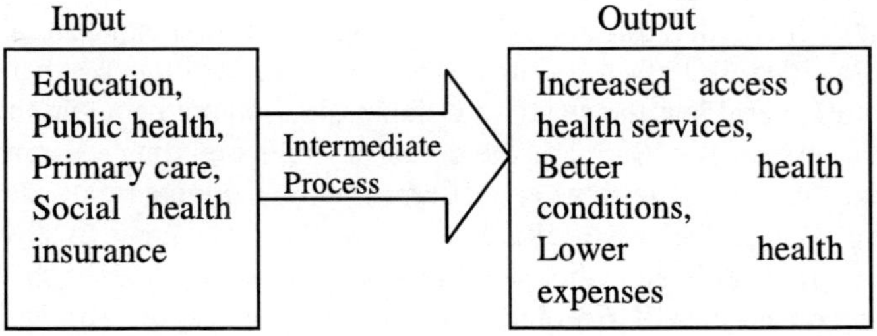

Fig. 2. Developmental Model of Health Policy

Table 1. Influential Factors, Methods and Health Results

Factors	Methods	Intermediate Results	Final Results
Education	Human capital investment in health field	Increase in health human resources	Decrease in health service cost & increase in health service access
Public health	Good life style Encouragement, Prevention	Healthy life style; Lower incidence rate of illness	Better health condition; lower medical expenses
Primary care	Development of primary care in community health service stations	Increased access to health services; Split patient flow from professional hospitals	Increase in access; Decrease in cost; Improvement in services
Social health insurance	Compulsive participation; coverage on basic medical services	Increased access; Lower incidence rate of serious illness	Better health; Lower expenses

3 Policy Implications

3.1 Human Capital Investment in Health Field

It has been proved by Cuban health reform experiences that investment on health education is a cost-effective way to increase health resources and reduce health service expenses since human capital investment often has high returns [4]. As a developing socialist country with the similar social economic situations to Cuba, China can try the same measure of human capital investment in health field to increase its number of health professionals. In addition, with the centralized planning ability in education system, the government can also adjust the course structures and specialist structures to guide the employment of health professionals and to meet the demand of health system. Furthermore, it can adjust human resource distributions in various areas through special incentives in education and/or employment policies, such as free education for those who would like to work in clinics located at remote areas for several years.

It should be noted that different kinds of health education expansion have different policy implications. On the one hand, simple college enrollment expansion in health field tends to lower down the standard of admission requirement and then the quality of future professionals. Suppose the score of college entrance examination is the price of education, with the same education demand, the increase in education supply will definitely cause the price falls down. On the other hand, if government puts more money in health education and sponsors tuitions for medical students, the demand for education will rise and more students will apply for medical schools. Then the score for admission will rise accordingly. At this time, if college expands its enrollment, it has the ability to keep the admission standard unchanged.

3.2 Public Health and Primary Care

As is shown in Fig. 3, "Lower stages tackle problems of higher collective nature, demand lower per capita investments, and have higher yield in terms of LE (life expectancy) [5]."

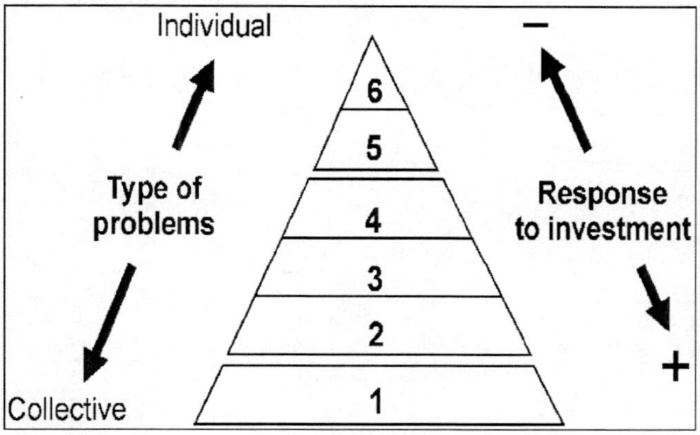

Fig. 3. Schematic illustration of the progressive complexity model of health care

Note: Numbers 1 to 6 enumerate increasing complexity of healthcare.
Source: A. Leirner, "The Health and Wealth of Nations - Coping With Limited Resources," Artificial Organs 30(7): 493–497, 2006.

Level 1 stands for the basic home or institutional care for those who cannot care for themselves and basic treatments for pain or suffering. Level 2 and 3 means preventive and simple curative interventions. Level 6 employs the most advanced technologies for medical therapy such as organ transplant [5]. The lower levels of health services are mostly provided via public health and primary care actions to a large population while the upper levels are provided by highly specialized and professional hospitals to small portion of people in need. The figure above indicates that collective and basic health services have the highest yield, so that public health and primary care are the most cost-effective health interventions which enjoy high investment returns and should be emphasized especially in developing countries like China.

Both involvement and investment in public health are insufficient in China [6]. Since public health has wide extensions including not only disease prevention, control and treatment, but also environmental, behavioural, occupational and social field, it is not merely a product of health sector. "It is to a large extent determined by societal and economic factors, and hence by policies and actions that are not within the remit of the health sector [7]." So the government needs to make comprehensive effort to improve coordination and integration of the field.

Although primary care has not got sufficient emphasis that it deserves, it has been attracting political and public attentions increasingly in China. Responding to the world trend of people-centered primary care (see Table II), China is gradually developing its primary care system based on community health service stations, which are located in communities, serving communities, being responsible for community health conditions, and sharing a close relationship with residents [8]. The development of primary care will not only optimize the structure of health service system, meet more people's health service need, but also save health expenses through better prevention from the starting points, better personal control of diseases and splitting patients flow reasonably from big professional hospitals.

Table 2. Aspects of care that distinguish conventional health care from people-centered primary care

Conventional ambulatory medical care in clinics or outpatient departments	Disease control programs	People-centered primary care
Focus on illness and cure	Focus on priority diseases	Focus on health needs
Relationship limited to the moment of consultation	Relationship limited to Program implementation	Enduring personal relationship
Episodic curative care	Program-defined disease control interventions	Comprehensive, continuous and person-centered care
Responsibility limited to effective and safe advice to the patient at the moment of consultation	Responsibility for disease-control targets among the target population	Responsibility for the health of all in the community along the life cycle; responsibility for tackling determinants of ill-health
Users are consumers of the care they purchase	Population groups are targets of disease-control interventions	People are partners in managing their own health and that of their community

Source: World Health Organization, the World Health Report 2008, pp43.

3.3 Social Health Insurance for Basic Medical Care

Out-of-pocket expenditures on health can exert huge burden on individuals and families. Sometimes patients may choose not to stay in hospital or even not to take any treatment because they are not able to afford it. As is shown in Table III, in 2003 when more than half of the population was not covered by any health insurance, nearly 40% patients out of those who chose not to take any treatment did so due to financial difficulty. While in 2008, almost 85% people were covered by health insurance, the percentage of treatment avoidance due to financial difficulty dropped to 15%. It can be inferred that universal coverage can relieve the burden of health expenditure to some extent for most people. It is also stated in the table that the condition for inpatients kept unchanged from 2003 to 2008. The reason seems to be that the items covered by health insurance, especially for inpatient treatments, is still very limited.

Table 3. Insurance Coverage and Cost Burden of Chinese People in 2003 and 2008

	2003 (%)	2008(%)
Percentage of patients without any health insurance	65.20	15.85
Percentage of patients who choose not to take treatments	48.9	38.2
Percentage of untreated patients who do so because of difficulty in affording medical fees	38.2	14.9
Percentage of patients who choose not to stay in hospital although they should have	29.6	21
Percentage of patients without staying in hospital who do so because of difficulty in affording medical fees	70.0	70.3

Source: Main Results of the 4th National Survey on Health Services, December 2004. Available at http://www.moh.gov.cn/uploadfile/200412/200412314857549.doc
Main Results of the 4th National Survey on Health Services, Feb. 27, 2009. Available at http://www.chinacdc.net.cn/n272442/n272530/n272742/29573.html

If there's no insurance or other social protection, people will be easily exposed to the risk of catastrophic health care expenditures that they are not able to afford. In addition, if individuals give up treatment due to lack of access to health care, when small illness changes into serious diseases, it will cost both the family and the society much more. As is advocated by WHO, universal coverage, the universal access to the full range of personal and non-personal health services with social protection, is not sufficient to guarantee health equality, but it is the fundamental step toward this goal [9]. According to the WHO model, there are three ways of moving toward universal coverage (see Fig. 4), including extending to the uninsured, reducing cost sharing and including other services. From a developmental perspective, although it is not practical for China to afford such a three-dimensional universal coverage system no matter it is tax-based or social insurance based, it is a rational choice to move forward from one or two dimensions.

The way of including other services should be achieved gradually with the economic growing. Moving forward through this dimension in current is apparently unwise. As is already indicated in Fig. 3, preventive health actions and basic medical care have the highest yield in health results, so it is rational to guarantee a basic level of health service universally, that is to extend to the uninsured and cover all the population with the fundamental social protection. With the economic development, the insurance can then include more services, still starting from lower levels and then developing to upper levels. As for cost sharing, its provisions typically include copayment, coinsurance and deductibles. The deductible, or out-of-pocket payment before one can enjoy compensation payment from insurance agency, has only two effects. One is to prevent people from using health treatment when they consider their illness no big deal; however, this action might prevent early diagnosis of serious diseases and induce the risk of larger future expenditures. The other is to encourage people who are pessimistic about their health to take excessive care and to use up deductible amount quickly in order to meet the standard of enjoying insurance compensation. As is proved by research evidences, deductibles are not effective in cost control [10]. So it should be reduced or repealed first.

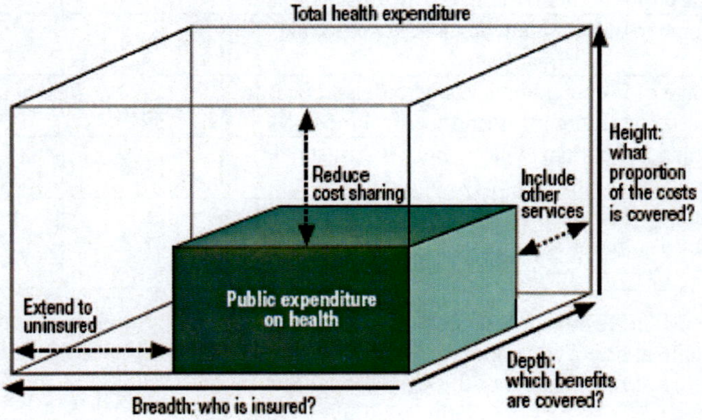

Fig. 4. Three ways of moving towards universal coverage

Source: World Health Organization, the World Health Report 2008, pp26.

Coinsurance, that limits the amount of coverage by the plan to a certain percentage [11], seems to have positive effects in controlling expenditure rising. Furthermore, using different coinsurance rate in different levels of clinics or hospitals can help improve macro-control of patient flows and optimize health resource distribution. As a specified amount that the insured must pay for a specified service or procedure (such as fee for office visit) [11], copayment is not able to have great influence on large expenses and its effect is similar to but weaker than deductible. This provision can be reduced too.

4 Conclusions

Using the theoretical framework of developmental perspective of social policy established by Midgley [1], the author built a developmental model for health policy reform in China. Four factors, including education, public health, primary care and social health insurance, are picked as key points joining the economic and social side of health policy. The findings are: health education investment is an effective way to enlarge health resource and improve health access; public health and primary care investment have higher returns than highly technological health care services; it is practical and desirable to go forward toward universal coverage of social health insurance via the dimension of extending to the uninsured, repealing deductibles and reducing copayment. The policy implications and recommendations were also given in this paper.

Acknowledgment. Xiao Liu thanks Prof. James Midgley for his direct guide on her study in the field of social development as her advisor, and for his enlightening thoughts on the developmental perspective of social policy which inspired the ideas of this paper. She also thanks Prof. Donna Infeld for her course of Health Policy which provided a great amount of knowledge and information that was used in this paper.

References

1. MacPherson, S., Midgley, J.: Comparative Social Policy and the Third World. Wheatsheaf, Brighton (1987)
2. Midgley, J.: Social development: The developmental perspective in social welfare. Sage, Thousand Oaks (1995)
3. Blum, H.: Planning for Health. Human Sciences Press, New York (1981)
4. Waitzkin, H.: Health Policy and Social Change: A Comparative History of Chile and Cuba. Social Problems 31(2), 235–248 (1983)
5. Leirner: The Health and Wealth of Nations - Coping With Limited Resources. Artificial Organs 30(7), 493–497 (2006)
6. Wang, J.: Opening the paths of government buying the public health service goods. Soft Science of Health 16(4) (2002) (in Chinese)
7. World Health Organization, the World Health Report, p. 69 (2008)
8. The State Council of P.R.C., Planning on Recent Steps of Health Reform (2009-2011) (March 18, 2009), http://www.moh.gov.cn/publicfiles/business/htmlfiles/mohzcfgs/s7846/200904/39876.htm
9. World Health Organization, the World Health Report (2008)
10. Liu, X.: Analysis of Social Health Insurance Fraud under Different Payment Models. Population and Economics (2), 64–70 (2008) (in Chinese)
11. Jonas, S.: An Introduction to the U.S. Health Care System, 5th edn. Springer Publishing Company, New York (2003)

Exploratory Comparison between Consumers of China and Developed Countries in Mobile-Commerce

Jinling Lu[1], Xiaolu Cheng[2], and Luzhuang Wang[1,*]

[1] Zhejiang University City College Hangzhou, China
[2] School of Management Zhejiang University Hangzhou, China
julu_1343@163.com, chengxiaolu1224@163.com, zuccjohn@gmail.com

Abstract. This paper is to specify and compare the determinants of individual adoption of mobile commerce between in China and developed countries. An random survey with more than 300 samples was conducted to collect data based on Unified Theory of Acceptance and Use of Technology (UTAUT). This model is tested with mobile device users whoever have used or not used m-commerce before. T test was used for the comparison. It finds perceived innovation is influencing developed countries consumers to adopt m-commerce, but not working in China, even it is an affective factor; and social influence has an opposite effect on these two groups. This comparison with UTAUT is in a cross-cultrue view while introducing new variabilities.

Keywords: Mobile commerce, Technology acceptance model, T test.

1 Background

M-commerce is any transaction, involving the transfer of ownership or rights to use goods and services, which is initiated and/or completed by using mobile access to computer-mediated networks with the help of mobile devices [1]. It provides two-way interactive communications [2] on an anywhere, anytime basis [3]. The activities in the area of m-commerce were guided by a number of trends; that is, localization, instant connectivity and personalization [4]. M-commerce has already been one of the most convenience business model. Driven by a widespread understanding of the Internet's capabilities, the power of electronic commerce, and advances in wireless technologies and devices, mobile commerce is rapidly approaching the business forefront [5].

The Chinese mobile markets, which face approximately a fifth of the world's population, have in recent years been increasing at an amazing pace and already occupied 755 million people. And mobile applications are more acceptable by Chinese consumers to access to Internet. Even 3G in China has just developed for 1 year, accessing to Internet via cell phone occupy 38 percent of Chinese cell phone users, while 27 percent of them in US; accessing to Internet via PCs is still lower than UC, according to Nielsen's report [6]. On the other hand, the population of

* Corresponding author.

J. Luo (Ed.): Affective Computing and Intelligent Interaction, AISC 137, pp. 143–149.
springerlink.com

e-commerce users has researched 15.97 million with 187% adding rate, and 530 million yuan transfer with 248.7% adding rate. However, the growing popularity of cell phone users and growing speed of e-commerce do not keep the same pace of expanding m-commerce [7]. While Amazon, the Wet Seal, Pizza Hut, 1-800-FLOWERS, and Target are retail mobile sites that stand out in the marketplace [8] in US, it is hard to point out some successful retail mobile sites in China. Obviously, the m-commerce model cannot be used in China as same as in developed countries. Different consumer adopting behaviors are affecting the effectiveness of m-commerce.

2 Literature Review and Hypothetical Model

2.1 Related Acceptance Model and Limitations

Studies have given empirical researches in the acceptance of information technology and information system (IT/IS), such as Theory of reasoned action (TRA) [9] [10], Theory of Planned Behavior (TPB) [11], Technology acceptance model (TAM) [12] [13], diffusion of innovations [14 and Task-technology fit (TTF) [15].

TRA was proposed by Ajzen and Fishbein (1980) [10] based on social psychology. TRA is used to predict attitude and behavior, which form behavioral intention. The theory was born largely out of frustration with traditional attitude-behavior research, much of which found weak correlations between attitude measures and performance of volitional behaviors [16].

According to TRB, which was proposed depended on TRA, a high correlation of attitudes and subjective norms to behavioral intention, and subsequently to behavior has been confirmed in many studies [17].

TRA and TAM, both of which have strong behavioral elements, assume that when someone forms an intention to act, that they will be free to act without limitation. In the real world there will be many constraints, such as limit the freedom to act [18] TAM is one of the most influential extensions of TRA. It was developed by Fred DAvis and Richard Bagozzi [18] [19]. TAM replaces many of TRA's attitude measures with two technology acceptance measures-ease of use, and usefulness. Perceived ease of use is the degree to which a person believes that using a particular system would enhance his or her job performance [13]. It was also suggested by diffusion of innovations [14] which defines consumer adoption is a process [20], and considered by TTF [15] to measure the task and technology characteristics. Perceived usefulness is the degree to which a person believes that using a particular system would be free from effort [13]. Lately, a survey conducted by Lu et al. [21] in china in 2005 also strongly supports that usefulness and ease of use are high relevant to behavior intention, and fits on the use of wireless mobile data services.

Criticisms of TAM as a "theory" include its lack of falsifiability, questionable heuristic value, limited explanatory and predictive power, triviality, and lack of any practical value [22]. On the other hand, improvements of TAM are made by scholars, such as UTAUT [23]. The UTAUT aims to explain user's intentions to adopt an IS and following behavior. The theory holds that four key constructs, which are performance expectancy, effort expectancy, social influence, and facilitating condition respectively, are direct determinants of usage intention and behavior, and impacted by four other key (gender, age, experience, and voluntariness) during constructs impact on consumer's intention [23]. And 70% proved to be explained successfully through testing UTAUT empirically [23].

2.2 Research Model

In China, as stated before, consumption of mobile phone and assessing internet via mobile phone is rapid increasing. Consumers having mobile applications have already gotten used to the operation. Thus, facilitating condition needn't treated as a most important factor influencing acceptance of m-commerce.

Perceived innovation is chosen by authors as a variable because consumption group has changed. Youth are growing as a most important group in consumers, and they are more interested in new form of technology. Perceived innovation cannot be ignored now.

Compared with developed countries, m-commerce applications and relative costs may still high for Chinese consumers to afford. So perceived cost suggested by authors to be tested in model.

This study tried to test and modify the constructs of UTAUT model in a sampling groups from China. Based on UTAUT model [23], authors proposed a model constructed by variables of perceived usefulness (PU), perceived ease of use (PEOU), social influence (SI), perceived innovation(PI), and perceived cost(PC), as well as the impacts upon intention.

The detailed discussion about this model is to be published in a separate paper.

3 Methodology and Results

3.1 The Firstsample Group Study

3.1.1 Sampling and Data Collections

The survey was conducted randomly online in China from late in 2009 to August, 2010. 21 survey items for 6 constructs are set in the questionnaire based on prior research empirically. A notation was attached to explain the meanings of the m-commerce, and the survey's purpose to ensure confidentiality. So far, 295 available questionnaires were submitted for analysis. 7 Options are ranging from "strongly disagree" to "strongly agree" for all questions (Seven-point Likert scale).

According to the samples, 56.3% respondents are males and 43.7% are females, and the ages of 21-30 is the largest group (85.4%), and 79.7% has already used m-commerce.

3.1.2 Related Data Analysis

Cronbach α measurement are used for ensure the reliability for questionnaires. The reliability coefficients (α) of each independent variables were PU(0.838), PEOU(0.890), SI(0.859), PI(0.784), PC(0.773), IT(0.875). All the reliability coefficients of independent variables are above 0.70 showing the validity of the model [24]

Structural equation modeling is used for analyzing the data. Structural modeling is used to evaluate the degree of data fitting a theoretical model.

According to t-statistic from measurement model, SI has negative effect on IT, while others have positive effect. From p-statistic, PU-IT, PEOU-IT and PEOU-PU are all showing significant relationship individually. PI has effect on IT. However, SI-IT, PC-IT, and SI-PU do not show significant relevant.

3.2 The Second Sample Group Study

3.2.1 Sampling and Data Collections

Second, the survey was collected in California targets at people from developed countries, including US, German, Japan and UK, during August, 2010. The questionnaire was translated into English and modified by some natives. A notation was also attached to explain the meanings of the m-commerce, and the survey's purpose to ensure confidentiality. So far, 80 questionnaires were submitted. Of the submitted questionnaires, 21 were excluded because of the inconsistency answers, leaving 59 usable responses.

Compared with the first sample group, it can be seen that the utilization rate in second sample group is relatively lower but still be optimistic.

3.2.2 Related Data Analysis

According to the sample demographics shows, 57.6% respondents are males and 42.4% are females, and the group of younger than 20 is the largest group (50.8%), and 72.9% has already used m-commerce.

Obviously, PU was still a important determinants, and PEOU-PU shows significant relation as same as response of the first sample. However, PEOU-IT and PI-IT are showed not so significant as domestic country, while SI-PU presents relationship with each other and is different from that of the statistic of China.

Table 1. T-Statistic

	China		**Deved. Countries**		**Comparison**	
Var.	**Averg**	**STDev**	**Averg**	**STDev**	**Sig**	**Result**
PU	5.466	0.719	5.234	1.315	0.193	-
PEOU	5.117	1.005	5.343	1.290	0.208	-
SI	4.674	1.120	3.898	1.563	0.001	Sig.
PC	4.833	0.928	4.311	1.403	0.104	-
PI	4.992	0.843	5.288	1.339	0.008	Sig.
IT	4.677	1.120	3.898	1.563	0.582	-

SI and IT presents significant differences. SI has much more impact in China, while PI is more influential in developed countries.

4 Discussions and Implications

4.1 Comparisions and Discussion

From analysis above, PU-IT and PEOU-PU are significant relative, and strongly support empirical studies. PU is so important to IT that it could illustrate consumers will adopt a new form of technology; here is m-commerce, when they believe this m-commerce will be beneficial to their daily life. And the degree of PEOU will affect consumers' views toward the usefulness of m-commerce, since the application is easy to use, it will simplify the complex work; but if it is not, it will be more order less for work and resulted as useless. Therefore, simplifying the application and software is still needed to give consumers an image of usefulness, and this image will cause consumers' adoption of m-commerce.

PC-IT and SI-IT are not significant relevant.

Even though some researchers [25] proved that PC is important within the relationship, this model does not support the hypothesis. Applications in China are comparatively high, but Chinese consumers have already gotten used to it. And that is why the population of mobile phone users and assessing internet vie mobile phone users are rapid increasing. And for developed countries, applications with lots of packages, and the cost is more affordable. Trying to offer lower price, thus, will not get more m-commerce consumers in return.

Social influence is not as expect as an important aspect in china, as well as in the developed. Consumers are more confident in them, and are not easily influenced by people around, or kinds of communication channels. It is probably that consumers are tired with too much products have existed in market place, and numberless advertisements in every Conner. Therefore, consumers would better have a try on what they think good, than accept other's suggestion.

PEOU-IT, PI-IT and SI-PU are showing differences between these two samples.

PEOU will not lead to IT in developed countries. Simple and easy application will just let consumers have an idea that it might be useful, but not directly make them adopt it. However, in China, consumers will accept m-commerce if they think it's easy to use. This difference occurred probably because consumers, in developed countries, are get used to new creations, including smart tiny things, while in China, creating new inventions are developed slowly and limited, thus consumers are sensitive to new inventions, and if it is easy to use, consumers will adopt it. It is also the same reason to explain that PI will not lead IT in developed countries, but works in China.

The effect of social influence on perceived usefulness is significant in developed countries. Even they are wide-minded and freely thinking, other's opinion is affecting their feelings about m-commerce. On the contrary, in China, consumers hold their own ideas and not influenced by SI, towards PU, as same as IT. It also can be explained that consumers in China is not believe other's opinion about a product, maybe even other's experience on it. So, in China, selecting a good channel to communicate with consumers will be very harder than in developed countries, or a new channel need to be provided.

SI still an important aspect to influence consumers in China, even it will not cause consumers to adopt m-commerce. On the contrary, SI is not an important factor, but it will lead consumers to accept m-commerce indirectly, in developed countries.

In the opposite, PI is not so influential for consumers in China, but it does lead consumers to adopt m-commerce. However, PI is affective, in developed countries, in consumers' daily life, but has not affect on their behavior towards m-commerce.

4.2 Limitations and Future Study

First, sampling group is limited to present the whole target consumers' mind. 354 respondents are not strong enough to show what all consumers will definitely do, especially, only 59 questionnaires (the second sample) are too limited to represent developed countries people's mind. In another view, the response of the first sample collected online are most from a group who get used to surfing internet or chatting with others, and most of the responders are youth, showing from table 1. So sample needs to be improved in extending responds and variability.

Second, variability is needed to be considered more to explore the case in future. In the further study, authors should put insight in more aspects to improve ability to predict usage intention in m-commerce and find out that their differences between China and developed countries. Then more details can be given to serve study and business.

Acknowledgment. This paper is supported by the construct program of the key laboratory in Hangzhou.

References

1. Wei, T., Marthandan, G., Chong, A.Y., Ooi, K.B., Arumugam, S.: What drives Malaysian m-commerce adoption? An empirical analysis. Industrial Management & Data Systems 109(3), 370–388 (2009)
2. Lee, T.M., Jun, J.K.: Contextual perceived value? Investigating the role of contextual marketing for customer relationship management in a mobile commerce context. Business Process Management Journal 13(6), 798–814 (2007)
3. Balasubramanian, S., Peterson, R.A., Jarvenpaa, S.L.: Exploring the Implications of M-Commerce for Markets and Marketing. Journal of the Academy of Marketing Science 30(4), 348–361 (2002)
4. Lembke, J.: Mobile commerce and the creation of a marketplace, pp. 50–56 (2002)
5. Senn, J.A.: The emergence of m-commerce. IEEE Computer 33(12), 148–150 (2000)
6. Nielson: Chinese mobile phone market insight report, September 29 (2010), `http://cn.nielsen.com/site/0805cn.shtml`
7. iResearch, China Mobile Commerce Research Report, September 29 (2010), `http://www.iresearch.com.cn/Report/Report.aspx?Newsid=1370`
8. Murphy, S.: Focus on: Mobile Commerce Mobile Movers and Shakers (2010), `http://chainstoreage.com`
9. Fishbein, M., Ajzen, I.: Belief, attitude, intention, and behavior: An introduction to theory and research. Addison-Wesley, Reading (1975)

10. Ajzen, I., Fishbein, M.: Understanding Attitudes and Predicting Social Behaviour, pp. 20–35. Prentice - Hall, Englewood Cliffs (1980)
11. Ajzen, I.: From Intentions to Actions: A Theory of Planned Behavior, pp. 11–39. Springer, New York (1985)
12. Davis, F.D.: A technology acceptance model for empirically testing new end- user information system: theory and results, Ph.D. dissertation, MIT Sloan School of Management, Canbridege, MA (1986)
13. Davis, F.D.: Perceived usefulness, perceived ease of use, and user acceptance of information technology. MIS Quarterly, 319–340 (1989)
14. Rogers, E.M.: Diffusion of Innovations. The Free Press, New York (1962)
15. Goodhue, D.L., Thompson, R.L.: Task-technology fit and individual performance. MIS Quarterly 19(2), 213–236 (1995)
16. Hale, J.L., Householder, B.J., Greene, K.L.: Theory of reasoned action. In: Dillard, J.P., Pfau, M. (eds.) The Persuasion Hand Book: Developments in Theory and Practice, pp. 259–286. Sage, Thousand Oaks (2003)
17. Sheppard, B.H., Hartwick, J., Warshaw, P.R.: The theory of resoned action: A meta-analysis of past research with recommendations for modifications and future research. Journal of Consumer Research 15, 325–343 (1988)
18. Bagozzi, R.P., Davis, F.D., Warshaw, P.R.: Development and test of a theory of technological learning and usage. Human Relations 45(7), 669–686 (1992)
19. Davis, F.D., Bagozzi, R.P., Warshaw, P.R.: User acceptance of computer technology: A comparison of two theoretical models. Management Science 35, 982–1003 (1989)
20. Roger, E.M.: New product adoption and diffusion. Journal of Consumer Research 2, 290–310 (1976)
21. Lu, J., Liu, C., Yu, C.S., Wang, K.: Determinants of accepting wireless mobile data services in China. Information & Management 45, 52–64 (2008)
22. Chuttur, M.Y.: Overview of the Technology Acceptance Model: Origins, Developments and Future Directions, Indiana University, USA. Sprouts: Working Paper on Information System, vol. 9(37) (2009)
23. Venkatesh, V., Morris, M.G., Davis, G.B., Davis, F.D.: User Acceptance of Information Technology: Toward a Unified View. MIS Quarterly 27(3), 425–478 (2003)
24. Nunnally, J.C.: Psychometric Theory. McGraw-Hill, New York (1967)
25. Sathye, M.: Adoption of Internet banking by Australian consumers: An empirical investigation. International Journal of Banking Marketing 10, 324–334 (1999)

The Application of Petri Nets to Construction Project Management*

Zhaozhu Chen[1] and Baosen Shan[2]

[1] School Management Engineering, Shandong Jianzhu University Jinan, Shandong Province, China
[2] School Adult Education, Shandong Jiaotong University Jinan, Shandong Province, China
Chenzhaozhu678@163.com, Shanbs@163.com

Abstract. Current construction project management methods are limited in description of the project management process dynamically. Petri nets have been in use as a graphical modeling language in many fields. This paper indicates the advantage of Petri nets in construction project management. The petri nets applications makes it possible to monitor and control the entire management process. A construction project management system based on petri nets is developed to model, simulate and analyze the project. The system makes project running fluently and efficiently by controlling its progress, managing its resources and data, organizing, human resources and equipments. And the modeling can reduce risk and improve efficiency.

Keywords: Construction project management, Petri nets, Application.

1 Introduction

A new project management model with ability to define features accurately and encourage efficient communications between all involved in the planning and execution of the project will yield positive results. Petri net is a graphical and mathematical modeling tool that has been successfully applied extensively in the areas of performance evaluation, communication protocol, legal system and decision making models [1].

Construction Project management contents include the cost management, quality management, project management, human resource management, communication management, risk management, During actual operation, the most difficult problem is cost and time limit for a project. Now the most commonly used methods of project management are Program Evaluation and Review Technique and the Critical Path Method. These methods can be a good way to solve the static planning and scheduling problem. But, these activities often isolated from each other, can't realize the dynamic monitoring management. To realize the construction project management of dynamic monitoring, you must use such as Petri nets as an analysis tool.

---

* This work is partially supported by Jinan University Institutes Independent Innovation Program.

J. Luo (Ed.): Affective Computing and Intelligent Interaction, AISC 137, pp. 151–158.
springerlink.com

2 Modeling Construction Project with Petri Nets

2.1 About Modeling

Petri nets have been applied successfully in the areas of Performance evaluation, communication protocols, legal systems and Decision making models[2]. A variety of Petri nets are reported in literature: untimed[2], timed[3], coloured[4], stochastic[5], predicate[6], priority[7], etc. Analysis methods of High level Petri nets i.e. Petri nets extended with colour, time and hierarchy can be used for the modelling and analysis of many complex systems encountered in industry[8]. These Petri nets are also used for Prototyping of software, design of logistic systems and design of administrative organizations. Stochastic coloured Petri nets for modelling flexible manufacturing systems, material handling systems and machines can be used[8]. Many extensions to Petri nets are suggested to consider issues specific to the problems on hand. Hierarchical timed extended Petri nets, a form of extended Petri net, allow the development of structured MIMO subnets to model complex system functionalities[9]. A class of modelling tools called augmented timed Petri nets has been introduced for the modelling and analysis of robotic assembly systems with breakdown[10]. Project management has also been identified by some researchers[2] as a prospective area, where the modelling power of Petri net could be explored. Petri nets offer many advantages to project managers[12]. Project management approach is drawing increased attention in manufacturing management in recent years and it forms an essential decision making aid as it suits the current trend and characteristics of manufacturing. The mathematical theory of Petri nets is very well developed and the theory of invariants, in particular, is very useful in the analysis and verification of a system modelled by nets[11]. Project with Petri nets: A new project management model with ability to define features accurately and encourage efficient communications between all involved in the planning and execution of the project will yield positive results. Petri net is a graphical and mathematical modelling tool that has been successfully applied extensively in the areas of performance evaluation, communication protocol, legal system and decision making models.

2.2 Software

Petri nets-aided software is developed to deal with project planning with stochastic activity times. It has been completely designed using Visual Basic, so as to make use of its graphical and customized application development tools. The package is developed as menu driven with user-friendly message displays. The software has main menu with Project planning module as sub menu. The planning phase is usually identified with construction of the model graphically, during which specific decisions are made on the method of performing jobs as well as their technological ordering and simultaneously, stochastic times can be assigned. The project planning module has been developed to deal with project modeling and analysis. Extensions to Petri nets are required considering the intricacies of project management [10].

In a Petri net:

A place represents the start or completion of an activity. Circles represent them. A transition represents activity, which is a time /resource consuming part of a project. Bars represent them.

The project planning module has been developed to deal with project modeling and analysis. Extensions to Petri nets is required considering the intricacies of project management [11].

In a Petri nets: A place represents the start or completion of an activity. They are represented by circles (Fig. 1 and 2).

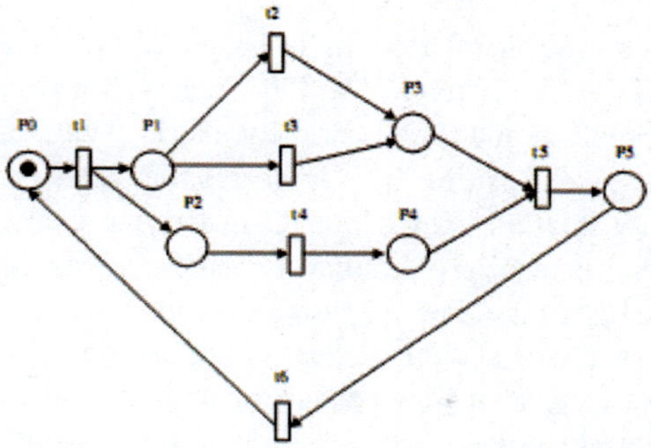

Fig. 1. Petri nets model

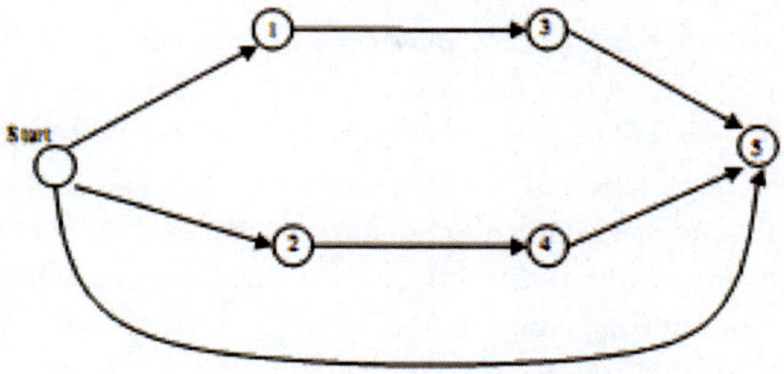

Fig. 2. Construction project management work flow

A transition represents activity, which is a time/resource consuming part of a project. They are represented by bars.

An arc connects places and transitions. It is not allowed to connect two places or two transitions. The arcs are either from a transition to a place or from a place to transition.

Token represent resources of the operation. These are represented by black dot.

A marking is an assignment of tokens to the places of a Petri net. If a place is marked with k, we say that place is marked with k tokens. Pictorially, we place k black dots (tokens) in place. A marking is denoted by M, an m-vector where M is total number places. The p component of the M, denoted by M (p), is the number of tokens in place P.

Token removal from a place pi will be effected, when all or the last activity (transition)emanating from p_i is completed. A token is deposited into an output place p_j after the completion of an activity. When an enabled transition fires, resources are drawn from the buffer and the same are utilized so that the number of resources needed is a starting condition for the transition. When the transition finishes resources are deposited back to the buffer if the resources are of a reusable type. If enough resource is not there for the firing of the enabled transition, then a message insufficient resource will be displayed. Then we have to reallocate the resources so as to fire the transition. We propose token removal or deposition with the help of a PPC matrix.

The removal or deposition of a token is proposed with the help of Precedence Priority Choice matrix (PPC matrix). The PPC matrix is a resultant matrix obtained on summation of Precedence matrix P1, Priority matrix P2 and Choice matrix P3. In a P1-matrix, an element 1 at position (i, j) indicates that activity in the jth column is a predecessor to activity in the ith row. In a P2-matrix, an element 1 at position (i, j) indicates that activity in the jth column is to be considered for the activity in the ith row. In a P3-matrix, an element 1 at the diagonal position is meant to block the transition over its alternatives and an element -1 at position (i, j) indicates that activity in the jth column ceases to be predecessor to the activity in the ith row.

In a PPC matrix, a transition is considered enabled if its row sum is zero. After an enabled transition in the ith row completes its firing, 1 is placed in the position (i, i) and the precedence constraints, namely 1s in the ith column, are removed. A new set of enabled transitions can then be identified, looking for rows with zero sums. The Precedence matrix, Priority and Choice matrices of the project are shown in Table 1. The resultant PPC matrix is shown in Table 2 and it indicates that t1 is enabled. After firing t1, the modified PPC matrix will be as shown in Table 1 and the current row indicate that t2 are enabled after the completion of t1. This process is repeated until all the elements of the leading diagonal contain 1s, indicating that the project is completed.

A token is deposited in the output place of a transition ti, when 1 is placed in its corresponding position in the leading diagonal of the PPC-matrix. A token is removed from place pi when all the emerging activities from it are completed. The completion of emerging activities is established if the corresponding elements of the leading diagonal in the current PPC-matrix contain 1.

Table 1. Precedence Matrix of the Network

Transitions	1	2	3	4	5	6
1	0	0	0	0	0	0
2	1	0	0	0	0	0
3	1	0	0	0	0	0
4	1	0	0	0	0	0
5	0	1	1	1	0	0
6	0	0	0	0	1	0

Table 2. PPC Matrix of the Network before firing t1

Transitions	1	2	3	4	5	6	Rowsum
1	0	0	0	0	0	0	0
2	1	0	0	0	0	0	1
3	0	0	1	0	0	0	1
4	1	1	0	0	0	0	2
5	0	1	1	1	0	0	3
6	0	0	0	0	1	0	1

The PPC-matrix will aid in the traditional computations such as early start, early finish, late start and late finish times. It also assists in determining floats and the critical path. In Petri net analysis, the PPC-matrix can be used to study properties such as deadlocks.

2.3 Construction Project Analysis

Analysis of net models can be based on their behavior (i.e. the set of reachable states) or on the structure of the net, the former is called reach-ability analysis while the latter is structural analysis. Invariant analysis seems to be the most popular example of the structural approach. The power of Petri nets lies in their ability to prove certain qualitative properties of systems. Reach-ability and Invariant analysis is often used to prove the properties of Petri net models.

Reach-ability is a fundamental basis for studying the dynamic properties of any system. The firing of an enabled transition will change the token distribution (marking) in a net. The sequence of firings will result in a sequence of markings. A marking is reachable from another marking if the firing of one or more transitions changes one marking into the other. We can describe the set of all reachable markings from an initial by a reach-ability graph. A related property is aliveness. A transition is live under an initial marking if, for all markings in the reach-ability set there is a sequence of transition firings that enables it. A Petri net is live if all of its transitions are live, otherwise, it is said to be dead lockable. The user constructs a set of equations, which has to be satisfied for all reachable system states. The equations are used to prove properties of the modeled system. The characteristic vector et is a row vector with a 1 in the position of transition t and 0 elsewhere. It simply denotes, by the position of the 1, the transition that is to fire.

The state equation of a net N for transition t is

$$M_{i+1} = M_i + Ce_t$$

where, M_i is the present state, M_{i+1} is the next state and Ce_t is the change in marking when transition t fires.

P-invariants: P-invariants identify sets of places over which the weighted count of tokens always remains constant. They represent system parts which are steady with regard to the amount of involved resources, system parts which will neither lose nor gain resources.

The P-invariants of the net are the integer solutions of the homogeneous linear system.

$$C^T * I = 0$$

Where, I is the P-invariant and C is the incidence matrix of net N.

T-invariants: T-invariants identify sets of transitions with multiplicities which are capable of reproducing a starting marking, provided that- from that starting marking- an occurrence sequence with the indicated multiplicities can actually be realized.

The T-invariants of the net are the integer solutions of the homogeneous linear system

$$C * J = 0$$

where, J is the T-invariant and C is the incidence matrix of net N.

3 CASE Study

The software is validated with a case study of Building built project shown in Table 3.

Table 3. Interpretations of places and transitions for Building built project

Transitions	Interpretation	Places
t1	Signing the contract	P0
t2	Engineering drawing design	P1
t3	Construction preparation	P2
t4	Building and installation project	P3
t5	Municipal engineering	P4
t6	acceptance and handed over	P5

Table 4. Input Data

Transition	Optimistic time	Most time	Pessimistic time	Immediate predecessors	R1	R2
1	1	1	2	-	6	5
2	1	2	3	1	3	4
3	2	3	6	1	4	5
4	4	7	9	1	8	7
5	1	3	5	2,3,4	6	9
6	0	0	0	5	0	0

Table 5. Output Data

Transition	Input Place	Output place	Expected time	Early start	Early finish	Late start	Late finish	total
1	0	1	1	0	1	3	4	3
1	0	2	1	0	1	0	1	0
2	1	3	2	1	3	5	7	4
3	1	3	3	1	4	4	7	3
4	2	4	6	1	7	1	7	0
5	3	5	3	4	7	7	10	3
5	4	5	3	7	10	7	10	0

The time estimates, predecessor relationships and resource requirements are indicted in Table 4 and 5 shows the output based on traditional computations.

The critical indices computed over simulations are shown in Table 6. Table 7 shows incidence matrix of Petri net for Building built project. Reach-ability table of Petri net for Building built project are indicated in Table8.

Table 6. Simulation result: Critical indices after 100 simulations

Transition	Critical indices
1	1.00
2	0.20
3	0.11
4	0.85
5	1.00

Table 7. Incidence matrix of Petri net for Building built project

places	t1	t2	t3	t4	t5	t6
P0	-40	0	0	0	0	40
P1	30	-3	-2	0	0	0
P2	10	0	0	-1	0	0
P3	0	3	2	0	-30	0
P4	0	0	0	1	-10	0
P5	0	0	0	0	40	40

Table 8. Reach-ability table of Petri nets for building built project

Marking	P1	P2	P3	P4	P5	P6
M0	40	0	0	0	0	0
M1	0	30	10	0	0	0
M2	0	25	9	5	1	0
M3	0	20	8	10	2	0
M4	0	15	7	15	3	0
M5	0	10	6	20	4	0
M6	0	5	5	25	5	0
M7	0	0	4	30	6	0
M8	0	0	3	30	7	0
M9	0	0	2	30	8	0
M10	0	0	1	30	9	0
M11	0	0	0	30	10	0
M12	0	0	0	0	0	40
M0	40	0	0	0	0	0

4 Conclusions

Construction Project Management is a complex task. Project managers are on the look out for efficient project management tools to suit specific needs and tackle realistic problems. Though many modeling and controls tools are available, they are limited in application when considering real-life situations. Hence, improved tools to tackle all kinds of situations are called for. Considering the limitations of the traditional project management tools and the benefits offered by the Petri nets, a Petri nets software is developed to model, simulate and analyze a project. Extensions required to Petri nets to help project management is proposed herein. The use of PPC matrix is proposed for the token movement in Petri nets. Analysis of the project network is carried out using reach-ability and invariant analysis. Thus the project details the benefits of Petri nets and attempts to exploits the use of Petri nets in construction project management.

References

1. Murata, T.: Petri nets: Properties, analysis and applications. Proceedings of the IEEE 77(4), 541–580 (1989)
2. Bandelloni, M., Tucci, M., Rinaldi, R.: Optimal resource leveling using non serial dynamic programming. Eur. J. Oper. Res. 78, 162–177 (1994)
3. Kartam, N., Tongthong, T.: Potential of Artificial Intelligence for engineering. Design Analysis and Manufacturing 11, 171–185 (1997)
4. Peterson, J.L.: Petrinet Theory and Modeling of Systems. Prentice Hall, Inc., N.J (1981)
5. Kim, J., Desrochers, A.A., Sanderson, A.C.: Task planning and project management using Petrinet. In: Proc. First Intl. Symp. Assembly and Taskplanning, pp. 265–271. IEEE, Pittsburgh (1995)
6. Kumar, V.K.A., Ganesh, L.S.: Use of Petri net for resource allocation in projects. IEEE Trans. on Engg. Manag. 45, 49–56 (1998)
7. Genrich, H.J.: Predicate/Transition Nets. In: Brauer, W., Reisig, W., Rozenberg, G. (eds.) APN 1986. LNCS, vol. 254, pp. 207–247. Springer, Heidelberg (1987)
8. Jeetendra, V., Kumanan, S., Krishnaiah Chetty, O.V.: Application of petri nets to project management. In: Proceedings of International Conference on CAR and FOF 1998, Coimbatore, India, pp. 853–860 (1998)
9. Wiest, J.D., Levy, F.K.: A management guide to PERT/CPM: with GERT/PDM/DCPM and other Networks. Prentice Hall of India, New Delhi (1977)
10. Martinez, J., Alla, H., Silva, M.: Petri nets for specification of FMS's. In: Kusiak, A. (ed.) Modeling and Design of Flexible Manufacturing Systems, pp. 271–281 (1986)
11. Moore, K.E., Gupta, S.M.: Stochastic colored Petri net models of flexible manufacturing systems: Material handling systems and machining. Comp. Ind. Eng. 29, 333–337 (1995)

Multi-pattern Mobile Collaboration Plotting Framework Model*

Yuhui Cao[1], Fuliang Li[2], Weihong Wang[2], and Lisheng Zhang[2]

[1] Institute of Information and Technology, Hebei University of Economics and Business, Shijiazhuang, Hebei Province, 050061, China
[2] Institute of Information and Technology, Hebei University of Economics and Business, Shijiazhuang, Hebei Province, China
caoyhs@gmail.com, wangwhs@gmail.com

Abstract. Aiming at the problem of mobile collaboration Plotting, this paper proposes a multi-pattern mobile collaboration plotting framework (M2CPF) model. First, in the study for existing mobile collaboration plotting based on common characteristics of the system is given a mobile collaborative system based on plotting the key elements. Subsequently, elements of the systems in place for the model can not fully meet the highly dynamic and complex needs of mobile collaborative environment, and gives a multi-mode flexible mobile collaboration framework model plotting. Furthermore, the introduction of collaborative awareness, concurrency control techniques for the full application of the framework for mobile collaboration in highly dynamic environments. Finally, the framework for Mobile-based plotting system (MCPS) Construction instance. Practice shows that the semantic framework has good accuracy, based on the Framework of the MCPS to fully apply a highly dynamic mobile collaborative environment for the efficient construction of mobile collaborative applications of theory and engineering practice to provide guidance.

Keywords: List key index terms here, No more than 5.

1 Introduction

Nowadays, the collaboration plotting system is widely applied to fields like military, scientific research, education, health care and so on.[1-5] However, the traditional collaboration plotting system based on cable network or single-user mode always fails to meet the demand of plotting for multi-users at anywhere and anytime. With the development of mobile communication technology, the mobile collaboration plotting system[6,7] is created. Whereas. the highly dynamic and complex mobile collaboration plotting computing environment, such as wireless link fluctuations, frequent disconnection, resource-constrained mobile devices and the like[8,9], has made the collaboration plotting faced a more serious challenge.

* This work is partially supported by the Nature Science Fundation of Hebei Province Grant # F2010001497 to Yuhui Cao, and he Nature Science Fundation of Hebei Province Grant #F2008000460 to Weihong Wang.

J. Luo (Ed.): Affective Computing and Intelligent Interaction, AISC 137, pp. 159–167.
springerlink.com

There has been a lot of research done to mobile collaboration plotting system from the perspective of Perceptual System Construction, including Document Six and Document Seven. But less research has been made to Mobile Collaboration Plotting Framework Model. Meanwhile, the existing mobile collaboration plotting model remains single and rigid, causing the system seriously starved of self-adaptation and difficult to satisfy the needs of plotting for multi-users at anywhere and anytime.

So as for the problem mentioned above, this paper puts forward Multi-Patten Mobile Collaboration Ploting Framework(M2CPF) ,full of good semantic accuracy. And Mobile Collaboration Ploting System (MCPS) is built based on it, which can be used to highly dynamic mobile collaboration environment to the fullest, providing good guidance of theory and engineering practice for the construction of effective mobile collaborative applications.

The second part of this paper gives key technology elements in the construction of collaboration plotting system on the basis of the common features of the existing mobile collaboration plotting system; The third part presents us with Multi-Pattern Mobile Collaboration Plotting Framework Model; The fourth part brings us examples based on MCPS, giving an introduction of technologies like collaborative awareness and concurrency control as well as the specific application of the framework model. It also makes engineering verification to semantic accuracy and validity of the framework model; The last part gives not only a summary to this paper but also a brief discription to further research.

2 Common Features and Key Technology Elements of Collaboration Plotting System

2.1 Common Features of Collaboration Plotting System

Features of mobile collaboration plotting system meet for mobile users are as follows:

1) Distributed: Users of mobile collaboration plotting system always uses Cell phone, PDA and other mobile devices without time and space constraints. People in different areas can have access to this system at anytime to collaborate with each other. So this system is distributed on a whole.

2) Real-time: The server of this system responsible for waiting for client connections and receiving data sent by clients and dealing with message forwarding is kept on all the time. When there are three or more terminals participating, real-time communication between them can be achieved through the forwarding server. So real time is another characteristic.

3) Simple interaction: The system is mainly designed for users of mobile phones, which relies on keyboard to interact, thus making mobile users only perform basic graphics rendering to complete collaboration. So it is of simple interaction.

4) Perceptual: Collaboration system users can have access tto other users of collaborative behavior in the sense, that is to say, they can easily know what did other users do at the time, facilitating collaboration efficiency.

5) Multi-user: Collaborative tasks require the involvement of multiple mobile users, making a multi-user nature of the interface in need. in addition , the sharing and operations of data is typical of multi-user.

2.2 Key Technology Elements of Collaboration Plotting System

1) Real-time collaboration communication and information sharing
Using "multi-threading",the mobile collaboration plotting system always runs a number of management threads to handle the user's various requests. Each collaborative group is being considered as a whole, for each of which a collaboration server starts a "group" management thread. The "group" management thread is in charge of maintaining shared data and transmitting the results in the form of a message to all collaborative groups of users within the collaboration,who will do pecific response operations according to these received messages. Besides, there is a thread interacting with the server ,which transmits control information input by users or generated by the logic to the server. At the same time, it can receive data and control information forwarded by server and complete the appropriate action according to this.

2) Collaborative awareness
The client to complete the behavior of group members in the perception of collaboration though list of on-line users and display of messages.When a user participate in the collaboration,the list of on-line users shows the information of the added user and give him the priority of adding messages;When a user exits,information about the user disappeared,displaying the exiting information in the collaboration area. In addition, users can also view all users' information though the list of on-line users,including image information and other data information.

3) Concurrency Control
Supporting multi-user real-time collaborative work, the mobile collaboration plotting system fails to avoid concurrent multi-user operation occurs,bring collaborative conflicts between users. However, because of mobile network's own instability, and unreliability of delay and transmission as well as multi-user collaboration between human factors ,mobile terminal users using wireless networks to share information causes very serious problems.

Thus, locking of data, as a common means to ensure data consistency, is used . It can be divided into "optimistic lock" and "pessimistic lock." The former taking very short time to response is based on less conflicts between shared object manipulation from users and greater probability of success to apply locking.It enables users to obtain experimental lock and operate on shared objects as soon as a request is sent out. The latter not conducive to the interaction could only implement other operations on the condition of being confirmed of further successful lock.

According to user behavior during the approval of waiting for a request, "optimistic lock" can be divided into "full-optimistic lock" and "semi-optimistic lock". "full-optimistic lock" , the common sense of optimistic locking, continues to perform other operations during the waiting period , so a series of operations are undid at the time of unsuccessful requests, allowing users' screen content to return to the previous contents of multiple operations without bring a sense of discontinuity visually. The "semi-optimistic lock" wouldn't implement operations before approved of applications, leaving it withdraw the current operations alone when the request fails.

3 Multi-mode Framework Model

Basic idea of building multi-mode framework model : In order to adapt to highly dynamic and complex mobile collaborative computing environment and overcome the lack of rigidity of a single model, this model uses multi-level, multi-mode flexible adaptive framework for mobile collaborative plotting. As for three levels of coordination mode of the session, the structure for the model and collaborative organizational models, this paper has designed a variety of different mobile collaborative environment for collaborative programs. During running time, it allows users to choose a flexible model to develop mobile collaborative application availability, robustness and efficiency.

Suppose Ar is short for Multi-mode architecture, CO is short for Multi-mode Collaboration and Fl is short for Multi-voice conversation mode. Multi-mode framework model is as follows:

$$\Gamma=(Ar,CO,Fl) \tag{1}$$

In the Equation(1), Ar, CO, Fl are all limited set of patterns. Further elaboration for the pattens are in the following.

3.1 Ar-- Multi-mode Architecture

For mobile collaborative applications, Multi-mode architecture usually has two modes, namely, B/S, C/S mode, in which B/S is also seen as a Web model.

For Web mode, the user access is very simple and convenient if mobile users own Web browser on the device ,allowing you to use Web collaboration model; While the C/S mode could only participate in the adoption of the mobile collaboration client software plotting activities through a mobile collaboration plotting client software installed on the mobile user's mobile device.

However, compared with B/S , C/S mode because of its specific client has more feature-rich, and powerful features. Of course, This also needs strong ability to support mobile devices.

So, Ar can be described as

$$Ar := \{C/S, Web\}.$$

For selection of mobile collaboration architecture model, the user can choose freely according to their ability to portable devices, personal interests and hobbies, etc.

3.2 CO-- Multi-mode Collaboration

Collaborative activities of mobile users are usually organized in groups mode. For the purposes of the group model, some groups correspond to the real world entity organizations, such as a department in the real world working group, a project development team and so on. The user group members know each other or there exits a certain sector organizational relations. But compared to the real world entities organizations, some groups are randomly formed, such as the Forum's discussion groups, a multiplayer game of the group and so on.Usually these groups are often based on personal interest of mobile users and randomly composed of the members of the group familiar with each other. Therefore, from this point of view, mobile collaborative group can be divided into interest-type group and physical-type group.

For some collaborative groups, in the whole process of collaborative activities, the group plays a leader's role to maintain and coordinate group collaborative activities; But some groups were not similar to the role of team leader with all members of full equality. Thus, in this sense, the collaborative model can be divided into leader-group collaborating organizations and non-leader-group collaborating organizations.

So Multi-mode Collaboration is:

Table 1. CO Mode Example

	leader-group type IsGL	No-leader-group type NoGL
Physical-type W	(W, IsGL)	(W, NoGL)
interest-type X	(X, IsGL)	(X, NoGL)

Multi-mode Collaboration is described as:

$$CO := \{(\mathrm{W}, \mathit{Is}\mathrm{GL}), (\mathrm{X}, \mathit{Is}\mathrm{GL}), (\mathrm{W}, \mathit{No}\mathrm{GL}), (\mathrm{X}, \mathit{No}\mathrm{GL})\}$$

3.3 Fl -- Multi-dilogue Conversation Mode

Fl is described as a central model in dialogue mode and no center model, it is as follow.

$$Fl := \{Center, Free\}$$

4 Instance

Construction of the example system set wireless and wired networks in one. Mobile clients (including mobile phone, PDA, laptop, etc.) sends data and requests to communicate to the wired network and the server through a wireless network (CDMA, GPRS, etc.).

The system includes two systems model, namely, the C/S mode and B/S mode. The overall structure of the system shown in Fig 1.

In Fig1, each wireless unit (Wireless Cell) of the wireless communication network has several MU (Mobile Unit, Mobile Unit), which could connect with the wired network through wireless communication interface and wired network server for data communications with the support of the MSS (Mobile Support Station, mobile base station) . Among them, the same MU in the wireless unit can communicate with each other and different wireless units in the MU can also communicate with each other, providing a distributed application system support.

For further instructions on the instance of the system, the following takes the C/S architecture as an example to elaborate in detail.

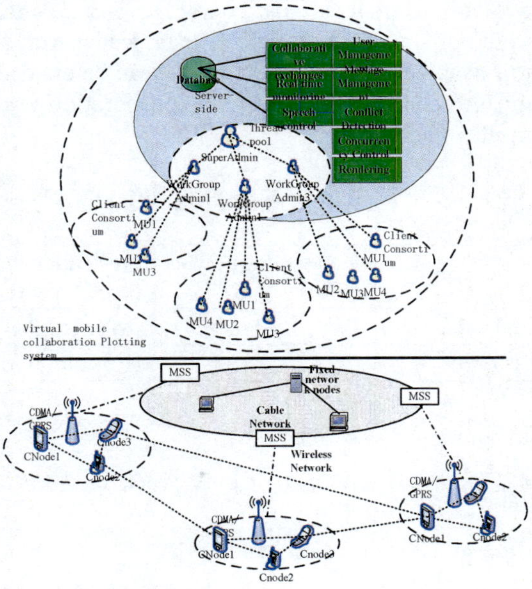

Fig. 1. Architecture of the Instance System

In C/S architecture, the system is divided into the mobile client and server parts. The mobile client are divided into cooperative plotting and collaborative communication. The collaborative plotting supports simplesketching and plotting on the map loads, making it possible for users of collaborative group to share the contents of the plotting screen with each other; collaborative communication provides users within the group with coordination of information sharing and real-time communication, improving efficiency of real-time collaboration.

The server mainly complete support collaboration and coordination from clients, including the management of collaborative exchanges, user information management, real-time monitoring of collaboration information, Collaboration when the voice control, appropriate strategy support from concurrency control and control of plotting shape and other modules, which accordingly does some operations by manipulating the database to complete requests from mobile users.

4.1 Client Side of Instance

The design of instance clints follows the thin client requirements. Most logic deals are transmitted in the form of messages to the server-side, such as the calling program in the corresponding element of the map to add, change, delete operations. In order to reduce the logic burden from the server-side's handling, all will be divided into ordinary users and group-creators who only have the priviliage of management and processing functions. The function of the client is shown in Fig 2 below.

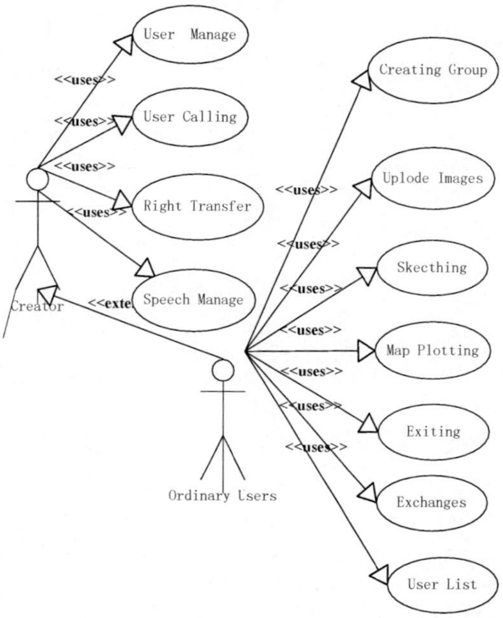

Fig. 2. Client use case diagram of Instance

Fig 2 shows that, when users create collaborative group successfully, the server returns a list of current online users. The creator can choose users to call and the users chosen decide whether to accept the invitation according to the message received. As soon as a user accept the invitation, he will take part in in the current collaboration, otherwise he does not participate in the collaborative.

4.2 Server Side of Instance

The server is mainly responsible for the coordination of group collaboration and message forwarding processing, the former of which includes dealing with concurrency control, conflict detection, speech control in collaborative activities; And message forwarding is in charge of forwarding of all messages and broadcast messages. In fact, all of the processing system is based on the message, the sequence diagram of dealing with messages is shown in Fig3 below.

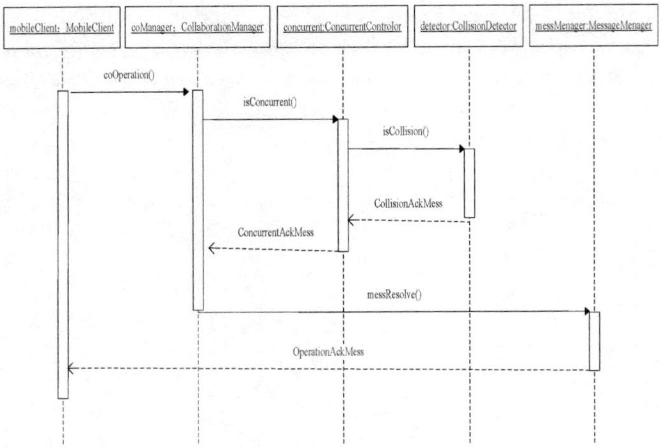

Fig. 3. Server-side message processing sequence diagram

5 Conclusion

In this paper, we put forward Multi-Patten Mobile Collaboration Ploting Framework(M2CPF) First, in the study for existing mobile collaboration plotting based on common characteristics of the system is given a mobile collaborative system based on plotting the key elements. Subsequently, elements of the systems in place for the model can not fully meet the highly dynamic and complex needs of mobile collaborative environment, and gives a multi-mode flexible mobile collaboration framework model plotting. Furthermore, the introduction of collaborative awareness, concurrency control techniques for the full application of the framework for mobile collaboration in highly dynamic environments. Finally, the framework for Mobile-based plotting system (MCPS) Construction instance. Practice shows that the semantic framework has good accuracy, based on the Framework of the MCPS to fully apply a highly dynamic mobile collaborative environment for the efficient construction of mobile collaborative applications of theory and engineering practice to provide guidance.

Acknowledgment. This work is supported in part by the Nature Science Fundation of Hebei Province(Grant # F2008000460), and he Nature Science Fundation of Hebei Province(Grant # F2010001497).

References

1. Zhan, Z., Jin, H., Yuan, P.: Research on instant share locking mechanism in real-time collaborative plotting system. Journal of Huazhong University of Science and Technology (Nature Science) 33(z1), 379–382 (2005)

2. Li, Q., Song, S.: Study of Military Cartography System Architecture. Ship Electronic Engineering 26(6) (2006)
3. Li, L.-X., Xia, X.-Z., Tu, K.: Research and Design of Collaborative Situation Plotting System in Shipborne Command and Control. Computer Engineering 36(16), 240–242 (2010)
4. Yu, W., Jiang, B.: Research on the Problem of Conflict in Realtime Collaborative Pattern Design. Journal of Computer-Aided Design & Computer Graphics 17(7), 1588–1592 (2005)
5. Banerjee, N., Agarwal, S., Bahl, P., Chandra, R., Alec Wolman, M.C.: Virtual compass: relative positioning to sense mobile social interactions. Technical report (2009)
6. Wang, W., Cao, Y.: A Mobile Collaboration Plotting Model Based on Context Awareness. Microelectronics & Computer 23(8), 76–79 (2006)
7. Cao, Y., Wang, W.: A Mobile Collaboration Plotting System Based on Multi-Agent. In: ICICA 2007, pp. 882–885 (2007)
8. Bellotti, V., Bly, S.: Walking away from Desktop Computer: Distributed collaboration and mobility in a Product Design Team. In: Proceeding of the ACM Conference on Computer Supported Collaborative Work, vol. 11, pp. 16–20. ACM Press, Boston (1996)
9. Cui, Y.Q.: How people use the web on mobile devices. In: Proceeding of the 17th International Conference on World Wide Web, vol. 4, pp. 905–914 (2008); Zhui, W., He, W., et al.: Research on Key Technologies in Cooperating Virtual Lab. Computer Simulation 21(12), 252–255 (2004)

Effects of Emotions on Interpersonal Strategies in the Dictator Game

Fang Shuang-hu[1,*], Yao Ben-xian[1], and Ding Hao[2]

[1] Department of Psychology, Anhui Normal University Anhui, Wuhu, 241000, China
[2] Department of Literature and Education, Bengbu College Anhui, Bengbu, 233030, China
fsh9075@163.com

Abstract. By experimental method, this research explored the effects of instant motion, relationship type and previous information of allocation on the selfishness of the interpersonal distribution in the dictator game. The results showed that (1) Emotion had a significant main effect on selfishness of the interpersonal distribution, happy feeling evoked more selfishness than sadness. Under the stranger condition, positive emotions evoked more selfishness than negative ones, but did not reach a significant level. (2)Under unfair condition, the happy mood induced significantly more selfishness than sadness, while under fair condition, happy mood and sad mood induced no significant difference. These results confirmed the mechanisms of emotional impact on selfish, and that negative emotion often produces adaptive and more socially sensitive outcomes.

Keywords: Emotions, Interpersonal strategies, Dictator game, Selfishness, Experimental study.

1 Introduction

If someone gives you 100 yuan, you can take the money that be allocated between self and others in an enjoyable manner. How much money will you leave to yourself? Will good or bad mood affects your final decision? Whether in the private sector or area of work, people often face a similar range of choice between selfish and fairness. Since ancient times, these choices in moral dilemmas are always considered as the eternal concerns by philosophers and writers. Recent studies have shown that evolutionary psychology, humans and other primates evolved a sense of fairness and justice, as an adaptation strategy to restrain selfishness, in order to maintain social cohesion and harmony(Forgas, Haselton, & von Hippel, 2007).

Studies on the relationship between emotion and interpersonal research have achieved the following accomplishments: emotion is one of the important inner drives for interpersonal behavior(Forgas, 2002), particularly low-intensity emotions have lasting effects on people's thinking, judgments and actions(Clore & Storbeck, 2006; Forgas, 2007). A lot of studies about emotional impact on altruism and helping behavior have shown: positive and negative emotions can stimulate altruism and

* Corresponding author.

J. Luo (Ed.): Affective Computing and Intelligent Interaction, AISC 137, pp. 169–178.
springerlink.com

prosocial behavior, but is highly dependent on the environmental variables(Dovidio, Piliavin, Schroeder, & Penner, 2006). But in a simple, minimalist allocation task(dictator game scenario), researchers are available to explore the impact of emotion on selfishness and get out of the environmental dependence. There is only a possible choice for subjects: either to their own advantage or benefit to others.

1.1 The Dictator Game

Making the use of economic strategy game in the context of interpersonal fairness, selfishness, trust and cooperation and other issues(Huangfu Gang & Zhu liqi, 2007,2008; Chen Xi, Ma Jianhong & Shi Zhan, 2007; Yu Jing & Zhu Liqi, 2010) is a reliable method. In the dictator game, the distributor has the absolute power of allocation of rare resources between himself and others(such as lottery winnings or a certain amount of money, securities, etc.). Traditional economic theory predicts rational allocation in the dictator game will leave most of their resources to maximize their own gains. Nevertheless, the actual results shows just the opposite: 30% or even 50% of all the resources are often distributed to others (Bolton, Katok, & Zwick, 1998), suggesting that distributors must take a constructive approach to complete the assigned tasks and they make the final delicate decision basing on comparison between internal conflicts for self-interest demand and external equity norms(Haselhuhn & Mellers, 2005; Pillutla & Murnighan, 1995). The distributors organize and interprete all available information in the subtle way which makes the emotional impact on the selfishness changeable. For example, if the allocation and distribution of those objects is belong to the same organization or group, distributors may have a higher degree of solidarity, emotional impact may be reduced. In the case of all the conditions of equilibrium, members within the group are more likely to be treated fairly than strangers(Forgas & Fiedler, 1996).

1.2 Emotion and Interpersonal Strategies

Moods have impacts on interpersonal strategy from two aspects: (a)affects all the deciding information content (The impact of information); (b)affects the choice of information processing strategies (the impact of treatment strategy).

(1) The Impact of Information Processing

It is easier to recall information consistent with the current mood, which makes the emotional factors affect the choice of interpersonal strategies(Forgas & Bower, 1987). When people are happy to extract more positive message, his behavior is also more optimistic and determined (Forgas, 1999). More negative emotions preferrance leads to the increase of cautious, pessimistic, and social constraints of the reaction(Forgas, 1998,2002). A large number of studies have shown that positive emotions stimulate more active, decisive, confident and optimistic interpersonal style. Delighted people in the social context are more direct and lack of performance etiquette(Forgas, 1999).They are more efficient negotiators (Forgas, 1998) and are more likely to show their mental state in the external world(Bless & Fiedler, 2006). Happy emotions are like incentive resources, allowing happy individuals to accept more risk (Trope, Ferguson, & Raghunanthan, 2001). It is inferred that a happy mood can stimulate more self-confident, self-serving allocation strategies, and negative emotions can lead to more careful and fair distribution.

(2) The Impact of Treatment Strategies

There is much evidence that emotions can affect the way that people respond to social information access(Bless & Fiedler, 2006). Negative emotions seem to promote more adaptive, externally oriented processing style, and positive emotions can lead to more internal-oriented thinking style. Allocation decisions in the dictator game shows the following conflicts: the conflict between the selfish internal impulses and to follow the external social norms(justice). More focus on the external social norms can undoubtedly ease the selfish impulse.

Therefore, the positive emotions stimulates people to care more about the internal impulses, decision-making is more likely prone to selfishness. In contrast, negative emotions can trigger adaptive thinking, focus more on external social norms, which can increase the fairness (Forgas, 2002); apart from this, in bad emotions, people can better identify the fraud(Forgas & East, 2008), and has better witness memories(Forgas, Goldenberg, & Unkelbach, 2009), determining his decision-making dependent on less stereotypes(Unkelbach, Forgas, & Denson, 2008). These studies manifest that bad mood also has a positive adaptive function.

To sum up, experiment 1 was intended to verify the hypothesis: In the dictator game of decision-making context, positive emotions and negative emotions has significant differences on the impact of selfishness, emotions and relationship have the combined effect on the selfishness; experiment 2 was supposes to verify the research hypothesis: the previous information of allocation and emotion have combined effect on selfishness, there are significant differences positive emotions and negative emotions on the impact of selfishness under the condition of previously fixed information.

2 Experiment 1: Emotional and Relational Impact on Selfishness

2.1 Methods

(1) Subjects and Experimental Design

48 university students voluntarily participated in the experiment, including 24 male & 24 female students(M=20.6, SD=1.71).

Use 2(emotions: happy or sad)×2(relationship: members within the group or strangers) between subjects experimental design. Subjects were divided into 4 groups of 12 people (including 6 boys and 6 girls) and were treated by different experimental treatments. The dependent variable is their lottery number assigned by distributors themselves(as the behavior indicators of selfishness of distributors).

(2) Experimental materials and tools

Emotion induction materials. Contains 10 questions graphical reasoning test, firstly subjects were required to select and fill the appropriate blank from the given four graphs based on the regularity of features. Subsequently subjects would be tested on their performance feedback which was secretly controlled by the researcher and unrelated to the actual performance good or bad. Feedback evaluation has two: "wonderful" or "terrible" which causes "good" or "bad" mood (Forgas, 2007).

The emotional self-rating scale. Contains the happy-sad, good-bad, relax-tension 3 terms, 7 point likert scale. 1 represents "sad, bad or tense", 7 represents "happy, well, a feeling of relaxation", was tested from 1 to 7 to select a representative of their current feelings.

The situation of allocation. Dictator game scenario: "You" and one other person are distributed each 10 lottery tickets please, 200 Yuan Securities as the final prize of gold. A lottery more would increase the last chance to win prizes. Other people know the rules, but can only accept any of your decisions, not against it. How many lottery tickets will you give to the other person?

(3)Experimental Procedure

The researcher selected campus open plastic stadium as experimental scenario. Subjects first underwent figure reasoning test. The reseacher pretended to have carried out cautious evaluation and then transmited the false, pre-set evaluation feedback to the subjects. Next, subjects were given each 10 lottery tickets and were informed that they were randomly selected and considered as distributors of this task. Subjects gathered in the classroom waiting for allocation of objects after completion of the task and opened the results of lottery winners. Members Within the group would be specifically told: they and their own distributive objects were belonged to the same group; subjects under strangers experimental condition would be told: their distributive objects are strangers. After the task allocation process, subjects were asked to finish three items 7 point likerts scale to assess their current emotions. After all above procedures, the experimenter uncovered the experimental secret and told the truth by the careful explanation to eliminate any subjects' residual mood effects.

2.2 Results and Analysis

(1) The effect of Mood Induction

The sores on the "happy-sad", "good-bad", "relax- tension" three items were highly correlated. Took "good/bad" information feedback as the independent variable, good or bad scores on 7 points Likert scale as the dependent variable, ANOVA results indicated that the sad group had worse feelings than happy group[(M=4.42; SD=1.37 vs. M= 3.04; SD=1.38), $F(1, 46)=12.02$, $p=0.001$], and proved mood has effective induction effect.

(2)Influence of Emotion on selfishness

2 (Emotion: happy/sad) ×2 (relationship:members within the group/strangers) emotions between subjects analysis of variance showed a significant main effect[$F(1, 44)=19.38$; $p<0.05$], happy students retained more lottery tickets than the grief students(M=5.92 vs. 4.25). Statistical analysis did not prove the existence of other significant effects, but showed a less significant trend: under the strangers condition, positive emotions evoked more selfishness than sad emotions(M=6.42 vs.5.42); while the members' emotions within the group evoked almost no difference on selfishness (M=4.42 vs. 4.08) (Figure 1).

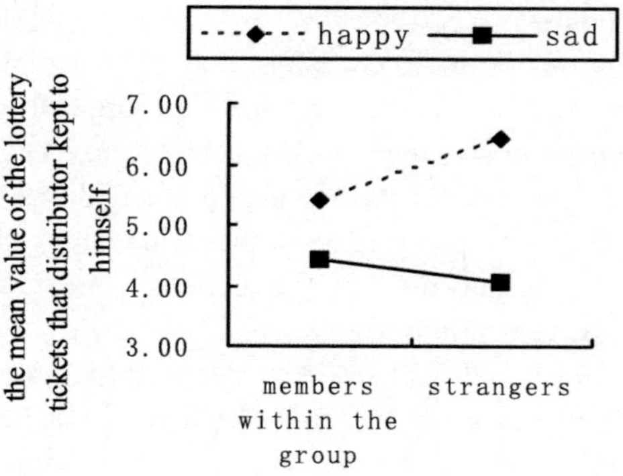

Fig. 1. The average score on the selfishness under the interaction between emotion and relationship

The results confirmed that the immediate mood had an significant impact on selfishness. More distribution plans were around about 50%. This indicated a fair norm doped with selfishness leading decision-making.Emotion has played an subtle role in the allocation process. Designed Experiment 2 further validated and extended findings of Experiment 1.

3 Experiment Two: The Impact of the Emotion and the Previous Information on Selfishness

3.1 Methods

(1) Subjects and Experimental Design
56 university students voluntarily participated in the experiment, including 28 male & 28 female students (M=20.86, SD=1.43).

Use 2(Emotion: happy/sad)×2(previous information of distribution: fair/unfair) between subjects experimental design. Subjects were divided into 4 groups of 14 people (including 7 boys and 7 girls) and were treated by different experimental treatments.The independent variables were the emotion and previous information of distribution, and the dependent variable is their lottery number assigned by distributors.

(2) Experimental Materials and Tools
Emotion induction materials: 2009 Spring Festival Party 2009 short program "Owe No Money" induced positive emotions, the film "warm spring" clips evoked negative emotions (Liu Yongfang, Bi Yufang and Wang Huaiyong, 2010). The above two fragments are about 30 minutes long.

The emotional self-rating scale and distribution scenario arranged the same as Experiment 1.

(3) Experimental Procedure

Each was tested individually in a computer lab to complete all tests. Subjects followed the guidance of the researcher into the experimental procedure, clicked the "enter "button, the screen showed movie clips instructions, and then started playing video clips that evoked positive emotions(or negative emotions).

After the emotion induction procedures, linkage indicator instructed subjects that they would finish the task of allocation in the dictator game. Guidance on the screen: "You then need to complete a task of allocation. Please distribute yourself and another person each 10 lottery tickets, 200 Yuan Securities as the final prize of gold, a lottery more will increase the last chance to win prizes. Other people know the rules, but can only accept any of your decisions, not against it. How many lottery tickets will you give to the other person?" Press "enter" key to view results on the computer,because two students have completed the task of decision-making.

Under fair condition, the computer screen presents the following information: Student A(6:4), student B(5:5) (earlier figure belongs to their own, the latter one is given to the other, the same below). Under unfair conditions, the computer screen presents the following information: Student A(9:1), B students(8:2). In the above context, subjects made allocation decisions.

After the task of allocation was completed, the linkage indicators directed into emotional self-assessment process. That contained three items seven points likert scale to evaluate their current bipolar mood state. After all of the above procedures, the experimenter uncovered experimental secret and told truth. By careful explanation, subjects could eliminate their residual mood effects.

3.2 Results

(1) The Effect of Emotional Induction

The scores on the three items "happy-sad, good-bad, relax-tension" was highly correlated. Took two video clips (happy/sad) and the previous allocation information as independent variables, scores on "happy-sad" item in 7 points scale as the dependent variable, 2×2 between subjects analysis of variance showed a significant main effect of emotion[$F(1, 52)=24.49$; $p<0.005$]. Subjects' emotions in happy group were significantly better than the sad group in the self-rating test($M=4.54$,$SD=1.17$ vs. $M=2.93$, $SD=1.21$), which confirms the validity of mood induction. Previous information of allocation did not affect self-rating emotions[$F(1, 52)=0.012$; $p=0.913$].There was no interaction between previous information of allocation and emotion.

(2) Allocation Strategy

Experiment 3 had no invalid data. Make use of statistical software SPSS13.0 to deal with data processing of the experimental results obtained under different experimental conditions. Selfish allocation mean and standard deviation scores, as shown in Table 1.

Table 1. Numbers mean and standard deviation of lottery tickets that were left to distributors' own under the different experimental conditions (M ± SD)

	Previous information of allocation	
	fair	unfair
happy	5.21 (1.63)	7.86 (1.10)
sad	5.57 (1.34)	4.29 (1.73)

First test of homogeneity of variance [$F(3,52)=0.881$, $P= 0.457>0.05$] showed that it was the appropriate for analysis of variance. Took previous information of allocation and emotions as independent variables, the lottery number distributed to subjects themselves as the dependent variable, 2×2 variance analysis showed that previous information of allocation had no significant main effect on allocation[$F(1, 52)=2.983$; $p=0.09>0.05$], but had significant main effect on emotion[$F(1,52)=6.736$; $p<0.005$], emotions and previous information of allocation had significant interaction[$F(1, 52) =25.000$; $p<0.005$]. To test the specific impact of Interaction of two factors, the researcher further did a simple effect test. According to research needs, researcher studied emotional effect under the condition of fixed previous information of allocation. When the previous information of allocation was fair, the main effect of emotion was not significant, $F<1$; when the previous information of allocation is not fair, emotion had a significant main effect [$F(1,53)=39.83$, $p= 0.000$]. Compared to sad subjects, happy subjects reserved more lottery number with significant selfishness(see Figure 2).

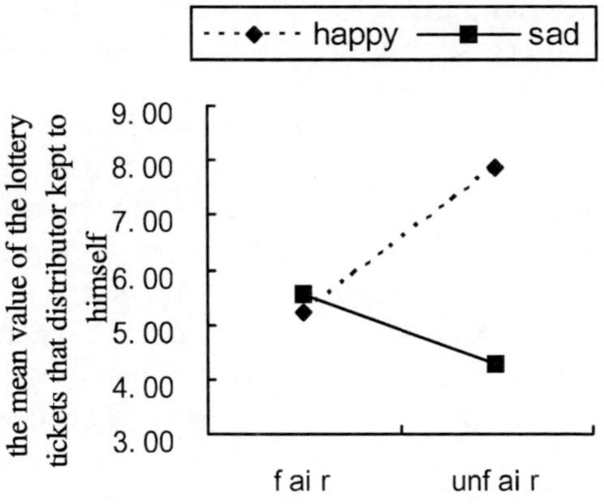

Fig. 2. Selfishness score mean in the interaction between emotion and the previous information of allocation

Experiment 2 results showed that if the researcher emphasized external standard conditions, the emotional impact would be reduced on the allocation, while weakening external norms, the emotion led to larger differences in the allocation. Hypothesis had been confirmed: when the fair norms was destroyed, emotional impact on selfishness was enormous.

4 Discussions

Dictator game is a relatively simple decision-making context, in this context of the study: positive emotions increased significantly selfishness, negative feelings reduced the tendency to act selfishly. This conclusion and functionalist cognitive theory of emotion both believe that emotions have an important signal function for contextual needs(Clore & Storbeck, 2006; Forgas et al., 2007). Negative emotional arousal has an important external focus to adaptive values and functions, positive emotions pay more attention to self-need oriented strategy(Bless & Fiedler, 2006; Fiedler, 2001). Study findings are consistent with theoretical predictions--- emotions can lead to interpersonal differences in treatment strategy(Bless & Fiedler, 2006). The allocation of the dictator game was the hard trade-off betweenthe selfish impulses nd the social norms,distributors in a good mood are more likely to follow the internal selfish desire to make choices; distributors in a bad mood are not too much on the hesitation and pay more consideration of external norms. These above findings and recent research results that negative emotions can increase eyewitness memory, reduce stereotypes and increase courtesy and reduce the error of the trial results are consistent (Forgas, 1998,1999; Forgas et al., 2009; Unkelbach et al., 2008). Experiment 2 controlled the external equity norms, found different emotional impact were weakened or even disappear under fair conditions. This is consistent with the conclusion that emotion uses selective absorption of consistent information to affect the selfishness. Previous information of allocation under a fair decision-making condition can strengthen the presence of social norms, and thus limit the occurrence of emotional impact. While the selfish decision is rejected by social norms and behavior, but the previous information of allocation under selfish decision-making condition can not directly lead to the concerns of fairness norms, on the contrary it would produce a fair standard destructive forces, thus expand the range of emotional impact.

Forgas(2002,2007) studies have shown that the emotional impact on interpersonal behavior, this experimental study of dictator game confirmed that the mechanisms of emotional impact on selfishness, but in real life situations still to be explored further, such as when individual decision-making power(living in a romantic relations, organizational decision-making etc.) dominate over every others' daily situation, emotional impact on selfishness has a wide range of fields. The most common assumption in the application, organizations, clinical and health psychology is: positive emotions have positive and constructive social functions. Results of this study challenge this general assumption, negative emotions also have social value of the same adaptive function, or even a slight negative emotions can increase more attention to the needs of others, resulting in more social equity.

As you see, the emotional impact of the selfish choice is highly dependent on the distributors' information processing strategies determined by a series of situational factors. Dictator game only means measurement of selfishness/ fairness in a simple, minimal the context. future research should consider more types of interpersonal influence and background. For example, the objects of the distribution should be assigned to people who are very familiar with, they should have a right to express their views on the distribution plan(the ultimatum game); they should have the opportunity to return to fair distribution. In other daily life situations, the use of different decision-making tasks and different types of partnerships, continue to validate the study findings are worth waiting. In the future research,we should further concern about the impact of minor emotional state on the decision-making and more intense and specific emotions such as anger, disgust, pride, embarrassment and so on.

5 Conclusions

(1) Emotion had a significant main effect on selfishness of the interpersonal distribution, happy feeling evoked more selfishness than sadness. Under the stranger condition, positive emotions evoked more selfishness than negative ones, but did not reach a significant level.

(2) Under unfair condition, the happy mood induced significantly more selfishness than sadness, while under fair condition, happy mood and sad mood induced no significant difference.

Acknowledgment. This research was support by National Top-quality Curriculums of psychology(2006099) and AHSK05-06D31.

References

1. Bless, H., Fiedler, K.: Mood and the regulation of information processing and behavior. In: Forgas, J.P. (ed.) Affect in Social Thinking and Behavior, pp. 65–84. Psychology Press, New York (2006)
2. Bolton, G.E., Katok, E., Zwick, R.: Dictator game giving: Rules of fairness versus acts of kindness. International Journal of Game Theory 27(2), 269 (1998)
3. Chen, X., Ma, J.H., Shi, K.: The effects of performance, competence and position on organizational distributive justice. Acta Psychologica Sinica 39(5), 901–908 (2007)
4. Clore, G.L., Storbeck, J.: Affect as information about liking, efficacy and importance. In: Forgas, J.P. (ed.) Affect in Social Thinking and Behavior, pp. 123–143. Psychology Press, New York (2006)
5. Dovidio, J.F., Piliavin, J.A., Schroeder, D.A., Penner, L.A.: The social psychology of prosocial behavior. Erlbaum, Mahwah (2006)
6. Fiedler, K., Bless, H.: The formation of beliefs in the interface of affective and cognitive processes. In: Frijda, N., Manstead, A., Bem, S. (eds.) The Influence of Emotions on Beliefs. Cambridge University Press, New York (2001)
7. Forgas, J.P.: Happy and mistaken? Mood effects on the fundamental attribution error. Journal of Personality and Social Psychology 75, 318–331 (1998)
8. Forgas, J.P.: Feeling and speaking: Mood effects on verbal communication strategies. Personality and Social Psychology Bulletin 25(7), 850–863 (1999)
9. Forgas, J.P.: Feeling and doing: Affective influences on interpersonal behavior. Psychological Inquiry 13(1), 1–28 (2002)
10. Forgas, J.P.: When sad is better than happy: Mood effects on the effectiveness of persuasive messages. Journal of Experimental Social Psychology 43, 513–528 (2007)
11. Forgas, J.P., Bower, G.H.: Mood effects on person perception judgements. Journal of Personality and Social Psychology 53, 53–60 (1987)

12. Forgas, J.P., East, R.: On being happy and gullible: Mood effects on scepticism and the detection of deception. Journal of Experimental Social Psychology 44, 1362–1367 (2008)
13. Forgas, J.P., Fiedler, K.: Us and them: Mood effects on intergroup discrimination. Journal of Personality and Social Psychology 70, 28–40 (1996)
14. Forgas, J.P., Goldenberg, L., Unkelbach, C.: Can bad weather improve your memory? A field study of mood effects on memory in a real-life setting. Journal of Experimental Social Psychology 54, 254–257 (2009)
15. Forgas, J.P., Haselton, M., von Hippel, W.: Evolution and the social mind. Psychology Press, New York (2007)
16. Haselhuhn, M.P., Mellers, B.A.: Emotions and cooperation in economic games. Cognitive Brain Research 23, 24–33 (2005)
17. Huang, F.G., Zhu, L.Q.: Cohort difference in the ultimatum and dictator game. Journal of Beijing University of Aeronautics and Astronautics (Social Sciences Edition) 21(2), 17–19 (2008)
18. Huang, F.G., Zhu, L.Q., Dai, H., Zhang, G.Y.: Method effect of ultimatum game and dictator game. Studies of Psychology and Behavior 5(4), 305–308 (2007)
19. Liu, Y.F., Bi, Y.F., Wang, H.Y.: The effects of emotions and task frames on risk preferences in self decision making and anticipating others' decisions. Acta Psychologica Sinica 42(3), 317–324 (2010)
20. Pillutla, M.M., Murnighan, J.K.: Being fair or appearing fair: Strategic behavior in ultamatum bargaining. Academy of Management Journal 38, 1408–1426 (1995)
21. Trope, Y., Ferguson, M., Raghunanthan, R.: Mood as a resource in processing self-relevant information. In: Forgas, J.P. (ed.) The Handbook of Affect and Social Cognition, pp. 256–274. Erlbaum, Mahwah (2001)
22. Unkelbach, C., Forgas, J.P., Denson, T.F.: The turban effect: The influence of muslim headgear and induced affect on aggressive responses in the shooter bias paradigm. Journal of Experimental Social Psychology 44, 1409–1413 (2008)
23. Yu, J., Zhu, L.Q.: The Development of Children's Fair Behavior: Evidence from Experimental Games. Advances in Psychological Science 07, 1182–1188 (2010)

An Empirical Study of China's Stock Market Feedback Trading

Zhang Yanliang and Wang Dongdong

School of Finance and Banking, Shandong University of Finance jinan, shandong Province, China
zhyanliang@sina.com, wdd_ctby@163.com

Abstract. In this paper, based on the theoretical model proposed by Shiller - Sentana – Wadhwani, we use non-symmetric TGARCH model to fit changes in volatility, and carried out an empirical test with feedback trading on China's stock market. The results showed that, Shanghai stock market register as characteristics of negative feedback trading, that is, buy low and sell high. Slightly different from the stock market in Shenzhen, with a lower level of risk, the characteristics of positive feedback trading is more obvious. With the risk increasing, negative feedback traders are dominant. With the volatility increasing, negative feedback trading increases in the Shenzhen stock market more than in the Shanghai stock market. With the high risk, the two stock markets show negative feedback trading.

Keywords: Feedback trading, Empirical study, China's stock market.

1 Introduction

The effective market theory indicated the present stock price already fully responded all related information, so the feedback strategy based on the past income information is non-rational, this will cause the stock price and the foundation value has the great degree deviation, thus aggravating stock price undulatory property. In the US, had many scholars the investment strategy which adopted to the American mutual fund to conduct the research.

Sentana and Wadhwani(1992)[1] analyzed extend the logic of Delong, and saw about the feedback trading, the yield autocorrelation and the relations of fluctuation. They based on Shiller's the results expressed the relationship of the three forms by Shiller-Sentana-Wadhwani model. Bohl and siklos (2004) based on the shiller-Sentana-Wadhwani model, with different GARCH models, tested feedback trading in mature markets and emerging market. The conclusion shows that there are both positive feedback and negative feedback trading in the two markets, but feedback trading is more apparent in emerging markets. In the two markets, with positive feedback are more apparent with the fluctuation augmentation.

---

[1] Sentana,E. and Wadhwani,s. 1992.Feedbaek Traders and Stock Return Autocorrelation: Evidence from a century of Daily Data, Economic Journal, Vol.104, p415-425.

J. Luo (Ed.): Affective Computing and Intelligent Interaction, AISC 137, pp. 179–184.
springerlink.com

Grinblatt, Titman and Wermers(1995)[2] acted according to data of US 274 mutual fund combination the end of 1974 ~the end of 1984, The results suggest that the funds all used the momentum investment strategy (positive investment strategy) to carry on the investment, they favored the purchase display the good stock in the past, but actually has not appeared regarding the past performance unsatisfactory stock systematically sells the behavior. The research indicated that about 77% mutual fund favor in the purchase earlier period display the good stock.

The Wermers(1999)[3] research indicated that what the US securities investment fund usually adopts is the regeneration transaction strategy, favored displays the outstanding stock in the past in purchase same several.

2 The Oretical Model of Feedback Trading

When Sentana and Wadhwani explain stock returns serial correlation with feedback trading, they put forward oundation that contains the value of stocks based on expectations of the traders to invest in (Smart Money) and feedback of the two groups of traders in the market model. The former assumes that the demand function of the assets has the following form: $Q_t^1 = \frac{E_{t-1}(r_t) - \alpha}{\mu_t}$ A

Q_t^1 represents assets ratio which is holden by investors from the first category (Smart Money). $E_{t-1}(r_t)$ represents the return on asset r_t in the moment of t which is expected in the moment of t-1, it is the conditional expectation based on all information in the moment of t-1. α is return on assets if risk-free rate (Merton, 1980), when the expected rate on return is α, these investors do not hold the asset. μ_t represents the risk of assets held by investors in the risk premium, it is the non-descending function of the conditional varianceσ_t^2. Assume : $\mu_t = \mu(\sigma_t^2)$.

μ_t >0,the first category of investors register as the risk aversion preferences, with the increase in risk, investors will reduce the holding the assets. When there is only the first category of investors(Q_t^1 =1)on the stock market, in the equilibrium capital asset pricing formula is $E_{t-1}(r_t) = \alpha + \mu(\sigma_t^2)$.

Feedback trading investors decide the future of the holdings of the assets based on past asset prices rather than expected decide the future of the holdings of the assets. We assume that the current period (t) level of return determine by the holding period (t-1):

$$Q_t^2 = \gamma r_{t-1} \qquad B$$

Q_t^2 represents the proportion of assets held by feedback traders, γ> 0 indicate that the feedback investor is positive feedback trading transaction, that is, "momentum strategy ";

2 Grinblatt, Mark, Sheridan Titman, and Russ Wermers. "Momentum investment strategies, portfolio performance, and herding: A study of mutual fund behavior. American Economic Review".1995;Vol.85, p1088-1105.

3 Wenners. R. "Mutual Fund Herding and the Impact on stock Prices". The Journal of Finance. 1999. No.2. p581-622.

On the contrary, the feedback is negative feedback trading transaction, that is, "buy low and sell high."

The interaction between two types of investors to reach equilibrium, $Q_t^1 + Q_t^2 = 1$. We substitute the formula 1 and 2:

$$E_{t-1}(r_t) = \alpha + \mu(\sigma_t^2) - \gamma\mu(\sigma_t^2)\gamma_{t-1} \qquad C$$

Compared to the standard capital asset pricing model, the model add $\gamma\mu(\sigma_t^2)\gamma_{t-1}$, Because of the existence of feedback traders, The assets in the risk premium held by the first category of investors has changed. When there are feedbacks investors exist, the yield will register as characteristics of first-order correlation. This correlation depends on the type of feedback traders, when the feedback investors is positive feedback type, the yield will register as characteristics of negative serial first-order correlation; when the feedback investors is negative feedback type, the yield will register as characteristics of positive serial first-order correlation. Sentana and others think that there are positive and negative feedback traders at the same time in the market, two types of feedback trading will change with the fluctuations in the rate of intensity change: When the risk is low, the feedback mainly use the negative feedback "buy low and sell high" strategy, the first investors affect market more; when the risk is high, The risk preference of first category of risk-averse investors decide that they demand higher expected revenue, so they will quit from the market partly. The impact of feedback traders will increase. When the risk is high to a certain extent, feedback trading investors show characteristics of risk-aversion, they will adopt the positive feedback strategy “momentum strategy”. Simplified the consideration, we regard the degree of feedback trading volatility as a simple linear function ofσ_1^2, formula 3 simplified as:

$$E_{t-1}(r_t) = \alpha + \mu(\sigma_t^2) - (\gamma_0 + \gamma_1\sigma_t^2)\gamma_{t-1} \qquad D$$

When the risk is low in the market, γ_0 is on behalf of directly affect of feedback towards market returns. When the market risk is high, γ_1 increases, the product of γ_1 and the level of volatility represents directly affect of feedback towards market returns. Sentana and Wadhwani used index from mature stock market which included the United States and England to fit model, the result show that, when the market volatility is low, negative feedback trading dominates, when the fluctuations in securities markets, the positive feedback trading dominates.

The theoretical model is firstly proposed to explain feedback trading gains serial correlation. But the model explains the interaction model of the first category of investors and the feedback trading investor.

3 The Econometrical Model

The accumulation of volatile, non-symmetry and income distribution patterns of peak fat-tail is an important feature of the securities market. Early studies have shown that the prediction error of securities prices will gather arise, larger error for some time, and the other smaller error over a period of time. In this paper, In order to ensure the validity of model parameters estimates and to improve the accuracy of confidence interval, We select TGARCH (1,1) to yield modeling with the conditional variance of the return in this paper. In examination of the existence of China's capital market feedback trading characteristics, joint estimate the following model:

$$r_t = \alpha_0 + \alpha_1 h_1 + (\alpha_2 + \alpha_3 h_t) r_{t-1} + \varepsilon_t \tag{1}$$

Conditional variance equation:

$$h_t = \beta_0 + \beta_1 h_{t-1} + \beta_2 I_{t-1} \varepsilon_{t-1}{}^2 + \beta_3 \varepsilon_{t-1}{}^2 \tag{2}$$

Indicator variables are defined as:

$$I_{t-1} = \begin{cases} 1 \ldots\ldots & \varepsilon_{t-1} < 0 \\ 0 \ldots\ldots & \varepsilon_{t-1} \geq 0 \end{cases} \tag{3}$$

h_t is conditional variance, ε_t ~N(0, h_t). In formula 5 and 6, the conditional variance is the function of sum of residual squares and conditional variance in the past.The smooth of variance equation is required to satisfy: β_1 , β_2 , $\beta_3 \geq 0$, $\beta_1 + \beta_2 + \beta_3$ <1and $\beta_1 + \beta_2 \geq 0$.But combined the regression model, conditional variance is not only function of the sum residual square and the past conditional variance, but also the function of indirectly parameters α_1 . α_2 and α_3 Considering the regression equation, The stable conditions of conditional variance is more complex. β_2 measures the non-symmetrical degree of conditional heteroskedasticity. When $\beta_2 \neq 0$, the residuals and negative residuals last period have different impact to the current period of conditional heteroskedasticity. When $\beta_2 = 0$,conditional heteroskedasticity non-symmetrical phenomenon does not exist, we can use general GARCH model to estimated conditional heteroskedasticity.

4 Data and Empirical Analysis

In this paper, the sample data is from the Shanghai and Shenzhen index. We select the price index of 2554 trading days between January 5, 1998 and August 3, 2008. The data is from the receiving the online data system of "analysts" software. Calculation formula of two market rate of return: $r_t = 100 \times (\ln p_t - \ln p_{t-1})$ p_t represents price index of data period t. Parameter estimation use maximum likelihood methods. Parameter estimation results are presented in Table 1. In Table 1, * denotes parameters in the 1% confidence level is significant. ** denotes the parameters in the 5% confidence level is significant. *** denotes that parameter in the 10% confidence level is significant.

Table 1. Analysis result of Chinese stock market feedback trading existence

Parameter	Shanghai stock market		Shenzhen stock market	
	Estimated value	p value	Estimated value	p value
α_0	-0.0523291	0.061128*	-0.06641895	0.088495*
α_1	0.01526147	0.000009***	0.2294599	<0.000001***
α_2	0.14616414	0.000031**	-0.06813688	0.000171***
α_3	0.04278989	0.00014***	0.09275534	<0.00001***
β_0	0.08357499	<0.00001***	0.09605685	<0.00001***
β_1	0.87073620	<0.00001***	0.91885658	<0.00001***
β_2	0.04995458	<0.00001***	0.01554135	<0.00001***
β_3	0.06330818	<0.00001***	0.02342175	<0.00001***

The results of model fitting are presented in Table 1. ***, ** and * in the table represent significant at 1%, 5% and 10% level of significance. Viewed from results of the model fitting, the coefficient α_2 was not significant in Shanghai stock market. It shows that when stock prices are in decline stage, feedback trading in the Shanghai stock market is not significant. When stock prices are in rise stage, negative feedback trading behavior is more significantly. Slightly different from the stock market in Shenzhen, coefficient α_2 is significant. This shows that when the risk level is low, positive feedback trading is more significantly. With the risk increases, negative feedback trading dominates. With the volatility increases, negative feedback trading increases in Shenzhen stock market more than Shanghai stock market. In high-risk, the two markets both show negative feedback trading in high-risk.

5 Conclusion

Modern financial theory holds that, noise traders not significantly affect the stock price in the assumption of market efficient theory. In the framework of behavioral finance, western financial academic researched and found that the market has a more prominent feature of positive feedback trading. This pattern of behavior promotes the stock price stray from the basis of its market value as an counter-examples. In this paper, based on the Shiller-Sentana-Wadhwani model, we conduct empirical research of feedback trading in Shanghai and Shenzhen stock market. When the risk is low in market, Non-rational investors were majority of negative feedback trading. When the risk is high in market, Non-rational investors were majority of positive feedback trading. At the moment, the positive feedback and negative feedback trading increased the volatility of the market, reducing the effectiveness and predictability of the market.

Investors will not take negative feedback trading; even take positive feedback trading in lower level of volatility in Chinese stock market. So that results the stock ring price volatility. In lower level of volatility, the investors will all take the negative feedback trading to make stock prices return the value of the underlying assets. From this analysis, if feedback trading takes place rightly in right time, it has a positive effect for the stability of the stock market.

References

1. Sentana, E., Wadhwani, S.: Feedback Traders and Stock Return Autocorrelation: Evidence from a century of Daily Data. Economic Journal 104, 415–425 (1992)
2. Shi, D.: Securities Investment Fund and the trading market. The 10th World Economy (2001)
3. Xiang, R., Li, Q.: Empirical analysis of China's institutional investors herding. The First Industrial Economics Research (2006)
4. Zhang, Y., Li, L.: Securities investment fund transactions and the impact of price. The 8th Management Science (2005)
5. Tang, D.: China's securities investment fund and market impact study, pp. 64–75. Economic Science Press (2008)
6. Guo, X.: An empirical study of the characteristics of market volatility of China Securities Investment Fund. The First Management Science of China (2006)
7. Chiang, M.: The efficiency of China's stock market leverage applied empirical research based on the EGARCH model. The Forefront-Seminar of Chinese Financial Sustainable Development (2003)
8. Yao, J., Liu, Z.: Fund investment behavior of a market test – an empirical studies based on the largest of institutional investors in China's stock market. Report on the work of International Business School Nankai University (2005)

Application of the Participatory Method to the Computer Fundamentals Course

Xianyong Fang

School of Computer Science and Technology, Anhui University,
Feixi Road 3#, Hefei, China
xianyong.fang@gmail.com

Abstract. Computer Fundamentals course is widely given to students of different majors in almost all universities. But almost all classes still take the traditional teacher-centered style. This style of course is boring to students who have almost no idea of the computer principles before. In order to solve this problem, this paper introduces the learner-centered participatory method into the course. We present the interaction techniques applied to the course, introduce the course schedule and discuss problem shooting techniques when applying the method. This education method can stimulate the interests of the students and help the student easily understand the fundamentals step by step.

Keywords: Computer Fundamentals course, participatory method, teacher-learned, learner-centered.

1 Introduction

Computer Fundamentals is now a common course for all freshmen of the computer science department and is also widely suggested to study for students from other majors in a university. The widely accepted method of giving such a course is in the traditional teacher-centered style, where the teacher speaks and shows the course material and the students usually just listen to him/her from the beginning to the end. Such kind of teaching is normally boring for the students and, especially for the students have no idea of computer before the course (since there are freshmen or from other majors), it is very difficult to understand some strange concepts of the computer science, such as hardware, pipeline, DMA, etc. Therefore, an improved course style is required to stimulate the interests and help understanding the principles easily. In this paper, we will study this problem and propose a participatory method to cope with it.

There are some studies on the improvement of the learning performance of students. Natvig and Line [1] and Cliburn [2] present gaming methods for introductory courses. Tan et al. [3] concludes that examples and drill-practice method are more interested for students than direct lecturing. This paper also suggests gaming style of teaching. The game-based learning belongs to the broad participatory education method, which has been widely applied in adult education [4]–[9]. However, to our best knowledge, there is no work on how to apply the participatory method to the course of Computer Fundamentals for university students. Therefore, in this paper, we introduce this

J. Luo (Ed.): Affective Computing and Intelligent Interaction, AISC 137, pp. 185–189.
springerlink.com

method into the Computer Fundamentals course and propose practical techniques of the method to the course. We also propose a course schedule using this method. In addition, we introduce techniques to resolve possible problems encountered in the class.

In the following, the basic idea of the participatory method and the possible techniques can be applied to the class of the Computer Fundamentals are introduced in Section 2. Section 3 presents a course schedule for applying the method. Section 4 introduces some problem shooting techniques when give the course with the participatory method. The whole paper is concluded finally in Section 5.

2 The Participatory Method

2.1 What Is the Participatory Method?

Participatory method started in the last century for the community development. It creates a environment that promotes the participants rethink their own experience and ideas about things during different kind of actions and experiences. During participation, they share their experiences and ideas with others, learn from others and thus improve themselves.

The interesting thing about participatory method is that this method does not simply rely on the knowledge of the teacher or veteran but rely on the knowledge discovery of the participants. The participants are not the traditional learners who are passive in learning; they are active in the learning process as participants.

For the students from different background or the freshmen majoring in the computer science, active participation in discovering the unknowns and problems and finding their answers is more attractive and interesting than simply sitting the classroom to listen to the teacher's solo speech. Computer Fundamentals also fits well with the participatory method because it includes some basic knowledge of the computer science which is not difficult for the students to learn it by themselves. Therefore, we propose such a method for the Computer Fundamentals course. Next we show some techniques that can be applied to the class of this course.

2.2 The Techniques Applied to the Computer Fundamentals Course

We now present the techniques in the participatory method that can used in the course. In the class, the students are randomly organized into several groups with each group having 4-6 students. All the methods presented below needs some coordinated actions of each group before the representative of each group shows the results of his/her team to the teacher.

Course calendar: Calendar is to show the progress of the course. The teacher can give the student the book first and then the students will have a general view of the whole content to learn. They make their own decisions on the progressive lessons to study in the future. The teacher only needs to give them some advice after their shows.

Computer game: Game has been advocated by some researches and it can also be adopted in the Computer Fundamentals course. Game can be the one popular in the entertainment market or the one for a teenage. The teacher prepares some questions after the game. The questions will embed the computer technologies inside. Some example questions can be: 1) which character do you like best? 2) how to improve the agent's power? 3) how is the interface and which one do you prefer to? 4) which computer plays the fastest? 5) How to play sound? The teacher collects all answers from students, classifies them with cards and names each card (e. g., sound driver, software, database, user interface and graphics) with explanations.

Experience of buy a computer in the market: After class, each group is required to enter the computer market to buy a computer or see what other customs in the market do when buying a computer. In the class, the teacher asks each group to write down the bargain process from the very beginning to the end buy using a diagram. Then the teacher explains the students with cards on the possible hardware components and their relationship. He/she also shows the students the operation system and its relationship between the software and hardware. The students also need figure out the parameters that may affect the price of a computer in cards. They will understand the architecture better than before after the teacher's explanation.

Diagrams: Each group is asked to draw some diagrams when discussing the architecture, either hardware or software. The students show the relationship of each hard or soft component in the computer and then the teacher helps student clarifying their relationship with a new diagram.

Brainstorm: Brainstorm is widely known for thinking out a number of ideas. For the course, the teacher uses this method for discovering the problems and interests of the students. For example, the teacher may ask students to write down their questions about the computer and its application. He/she can also ask students to give the principles and future developments when he/she is explaining the fundamentals.

Other techniques: There are also some other techniques can be adopted except above five main techniques, such as drawing and thinking. Drawing is to ask the students figure out the computer or software functions and thinking is to ask the students think about the possible problems, principles and relationships when use the computer or software, or what the future computer will be. The teacher can also propose some new techniques to make the participants active in the course.

3 The Course Schedule

After introducing the participatory method and its application techniques to the course, we now show a possible course schedule for the class. Such class can be divided into five parts: introduction, beginning section, course section, end section and summary. Introduction and summary parts are on the introduction and summary of the whole course. Beginning section, course section and end section are the three construction parts of each lesson. We now discuss the possible techniques taken for the five parts.

Introduction: The introduction part only exists in the very beginning of the course, the beginning of the first lesson. In this part, teacher and students can play some games together to get to know each other. He/she may also use the game to separate the students into different groups for future lessons. Each group will draw a logo for themselves. The teacher also need ask the students to set up the rules for the class and fill the calendar for the whole course.

Beginning section: The beginning section will be in the begin of each lesson. In this part, the teacher may assign one group to summarize the contents learned in the last lesson. This summary can be taken by cards. After the summary of the student, the teacher gives thanks to the group and then introduces the content will be taught in this lesson. He/she can ask questions for the introduction of the lesson .

Course section: The course section is on the core content to teach in each lesson. All suggested techniques in last section can be applied. The teacher must carefully select the proper techniques to stimulate the interests of the students. Considering the time limit of each lesson, he/she must control the rhythm of the lesson so that the whole content can be taught in time.

End section: The end section is to conclude the whole lesson and prepare for the next lesson. The teacher can ask questions to the students and each group can show their answers use cards for his summary. He/she also needs to assign a group for next lesson's summary in its beginning part. He/she may also give some experience assignment to students to discussion in the next lesson.

Summary: Summary of the whole course is necessary after the course. In this part, the teacher can use drawings and questionnaires to help students summarize and score the course.

4 The Problem Shooting Techniques

It's normal that there are some problems existed during the participatory course considering many different students participate together. Therefore, the problem solving techniques are very important. We will give some techniques below to cope with possible problems encountered.

I am not really interested in computer.: Nowadays students must get enough course credits for graduation. Some students may choose this course just for the credit but not be really interested in it. In this case, the teacher should try to stimulate their interest considering the importance of the computer in the modern world. Some daily experience may help the students understand its importance and take active altitude to the course.

I don't like to talk.: There are always some students don't like to talk in public, either because they are shy or because they really don't want to think much about the topic under discussion. The teacher need first think whether there are some problems in the course style and try to improve it. He/she may also try to discover the advantages of those students, praise them and encourage them. The teacher can even ask the student directly whether he/she feel boring in the lesson and whether he/she has better suggestions.

Don't stop my speech and I still have something to say.: Generally it is good to have students active in participation. But if there are students like speaking a lot in the class, it definitively costs too much time. In this case, the teacher should not interrupt the students directly but instead ask some other students opinion without interrupting their speech. The teacher can even ask these students to the teacher some time and let them feel the importance of adequate time control of each speech.

5 Conclusions

This paper presents a participatory method to the Computer Fundamentals course for university students, which there is no such study before. We show the possible techniques of the method when applying to the course. A course schedule is presented and some troubleshooting techniques are also introduced. Participatory method has been proved to be very useful in arousing the interests of participants and improve their abilities actively. Therefore, we believe this study will be interesting and constructive to related teachers, department administrators as well as students.

Acknowledgement. The research is partially supported by National Natural Science Foundation of China (No. 61003131) and Education & Research Fund for the Higher Education of Anhui Province, China (No. KJ2010A010).

References

1. Natvig, L., Line, S.: Age of computers: game-based teaching of computer fundamentals. In: ITiCSE 2004: Proceedings of the 9th Annual SIGCSE Conference on Innovation and Technology in Computer Science Education, pp. 107–111. ACM, New York (2004)
2. Cliburn, D.C.: The effectiveness of games as assignments in an introductory programming course. In: 36th Annual Frontiers in Education Conference, pp. 6–10 (October 2006)
3. Tan, P.-H., Ting, C.-Y., Ling, S.-W.: Learning difficulties in programming courses. In: International Conference on Undergraduates Perspective and Perception, Computer Technology and Development, vol. 1, pp. 42–46 (2009)
4. Courtney, S.: Why Adults Learn: Towards a Theory of Participation in Adult Education. Chapman & Hall, New York (1992)
5. Falkenberg, M.G.K., Elwing, B.E., Göransson, A.M., Hellstrand, B.E.S., Riis, U.M.: Problem oriented participatory education in the guidance of adults with non-insulin-treated typeii diabetes mellitus. Scandinavian Journal of Primary Health Care 4(3), 157–164 (1986)
6. Campbell, P., Burnaby, B. (eds.): Participatory practices in adult education. Lawrence Erlbaum Associates, Inc., Mahwah (2001)
7. Francis, C.A., Carter, H.C.: Participatory education for sustainable agriculture: Everyone a teacher, everyone a learner. Journal of Sustainable Agriculture 18(1), 71–83 (2001)
8. Abdullah, M.M.B., Koren, S.F., Muniapan, B., Parasuraman, B., Rathakrishnan, B.: Adult participation in self-directed learning programs. International Education Studies 1(3), 66–72 (2008)
9. Koumlnings, K.D., Brand-Gruwel, S., van Merriënboer, J.J.G.: An approach to participatory instructional design in secondary education: an exploratory study. Educationl Research 52(1), 45–59 (2010)

Cross Layer Algorithm for Lifetime Maximization of Wireless Sensor Network

Ma Jun

Department of Economics and Business Administration, Luo Yang Institute of Science and Technology, Luo Yang, China
mj@lit.edu.cn

Abstract. Wireless sensor networks (WSNs) have been widely studied in many applications such as monitoring environment, embedded system and so on. The nodes of WSN are extremely power constrained, so energy-efficient data management and prolonging the networks lifetime are the major concerns to many scholars in this researching area. The algorithm is proposed using energy-efficient message collection and power management methods to prolong the lifetime of the sensor network. Simulation results show that it is to effectively balance the network energy consumption, improve the network life-cycle and ensure the communication quality.

Keywords: Component, Wireless Sensor Networks, Lifetime maximization diffusion routing protocol.

1 Introduction

The wireless sensor networks can be defined as an autonomous, ad hoc system consisting of a collective of networked sensor nodes designed to intercommunicate via wireless radio. Typically, these nodes coordinate to perform a common task. Wireless sensor network is being used in a wide of variety of commercial and military applications and becomes a new research area of modern technology[1]. In wireless sensor networks, the energy of node is limited and generally could not replenish, so the routing protocol have to use energy efficiently, meanwhile the number of nodes in network are often large, so the node can only acquire the local topology information depend which routing protocol have to select suitable path. In addition, the routing mechanism of WSN is often associated with data fusion technology, by reducing traffic load to save energy. Therefore, the traditional routing protocol for wireless network are not suited to WSN, WSN has its own characteristics. Hence, power efficiency is an important design goal that needs to be enhanced in order to prolong the effective time of a network with a limited power supply. And a lot of researches have been developed today [2].

The key technology based on resource limited is still the bottleneck for WSN to be mature. Designing WSN application algorithm with lower energy cost under microwave wireless communication is an effective way to solve the bottleneck above. Such energy-efficient strategies based on resource limited involve many aspects such

J. Luo (Ed): Affective Computing and Intelligent Interaction, AISC 137, pp. 191–198.
springerlink.com

as density control, energy consumption distribution estimation and node deployment, wireless communication security, cross layer design of wireless communication protocol and so on. Although the main power consumption term in a traditional wireless system is due to the energy required for actual transmissions, this may not be the case in energy-limited wireless sensor networks. In fact, in some cases it is the circuit energy needed for receiver and transmitter processing that is dominant. Network lifetime maximization involves all levels of sensor network hierarchy, from hardware/software design to communication protocols. Recent efforts dealing with communication-related energy costs mainly focus on two separate but equally important fronts: energy-efficient routing and sleep scheduling. In particular, energy-efficient routing aims to balance traffic loads, and hence energy consumption, among sensor nodes across the network. In order to improve the energy-efficiency at the local sensors in an energy-limited wireless sensor network of the above type, in this paper we propose a new.

The organization of this paper is as follows. In Section II, we discuss related work in this area. Section III introduces a cross layer algorithm to extern the lifetime maximization (CALM). In Section IV, includes simulation results and a discussion of various scenarios In Sections VI, presents our conclusions.

2 Related Work

The energy consumption of communication subsystem in the wireless sensor network device can be divided into two parts, the first part is that related with transmitting power, and the second part is the energy consumption in the immunization subsystem circuit.. Let us review some of this work. In [3], the energy consumption of several well-known routing protocols are studied and compared via simulations. Characterization of the topology of a mobile network is a basic component in network-level analysis on energy consumption. Energy allocation or power control can further improve the performance of cooperative communication. Optimal energy distribution among cooperative nodes is studied in [4] to minimize the link outage probability. [5] propose to schedule sensors' activities (Active/Sleep) so that every point in a grid-monitored area is covered at any time. Neighbor sensors exchange a random reference time Tree within cyclic rounds of constant duration T and decide to be active for certain time duration within T. The round period T is equally shared among all the neighbor sensors that cover a common grid. Even though the proposed schedule balances consumed energy over neighbor nodes, it does not take into account the residual energy of sensors. In [6] propose a localized algorithm that finds an optimal subset of sensors ensuring full area coverage or, if not possible, the least uncovered points. If the whole range covered by a sensor is covered by a subset of its neighbors, then the sensor decides to turn off. A random back-off time ensures that two nodes do not make the decision to turn off at the same time. The authors of [7] developed a clustering algorithm that can choose single-hop or multi-hop method adaptively based on the analysis of the energy consumed. This system employed two types of nodes where one type is ordinary sensor node and the other type is complex/high-energy node specially used as the cluster-head. This scheme is not too robust because the system can quit functioning if one of the special cluster-heads dies. There is an extensive literature discussing proactive, reactive, and hybrid routing protocols [8]. However, even though the researches, e.g. [9] begin to formula the energy consumption while re-thinking the states of WNS nodes and the energy consumption both in the steady states and during the transitions, few papers provide a clear mapping between upper layer assumption of device and the feature of communication subsystem in real devices.

The energy consumption of the network, however, is usually not the primary metric for comparison. In fact, on a real wireless channel, because of signal fading and interference, events occurring far away from a sensor are less likely to be detected than events occurring close to it. In other words, an event occurring outside a sensor's range is still detected with a certain probability which decreases when the distance to the sensor increases.

3 CALM Proposed

This paper considers sensor networks where static sensor nodes are randomly located in a given region. Similar to most previous work, this paper assumes that the traffic in the network is light, and hence transmissions are collision free. Data fusion is not considered here, i.e., data will not be aggregated along the routes to the sink nodes. There are multiple sink nodes in the network, and each sensor node is assigned to only one sink node.

Theodore S analysis the underlying link for wireless sensor network communication model of lognormal shadow:

$$PL(d) = PL(d_0) + 10\beta \log(\frac{d}{d_0}) + X_\sigma$$

Where : $PL(d)$ specific location for the path loss, d_0 is the reference distance, β is the path loss index, σ is the standard deviation. X_σ is zero mean Gaussian random variable. From (1) can be received by the receiver of the power:

$$P_{recv} = P_{trans} - PL(d)$$

By equation (2) can be calculated signal to noise ratio in the receiving node (SNR) as:

$$SNR_{(dbm)} = P_{recv(dbm)} - N_{(dbm)}$$

N is the platform noise. Based non-coherent modulation FSK, so the bit error rate formula is:

$$P_e = \frac{1}{2}\exp^{-\frac{1}{2}SNR}$$

Suppose we are given N valid power assignments for the nodes. Denoting by q_{jh} the power assigned to node j in the power assignment, the lifetime maximization task restricted to these power assignments can be represented as the following LP problem over the variables n_t :

Maximize $\sum_t n_t$

Subject to $\sum_t q_{jh} \times n_t \leq w_i, \forall i \in \{1,...,n\}$,

$n_t \geq 0, \forall t \in \{1,...,N\}$,

The associated dual problem over the variables x_j is

Minimize $\sum_i w_i \times x_j$

Subject to $\sum_i q_{jh} \times x_j \geq 1, \forall h \in \{1,...N\}$,

$x_i \geq 0, \forall i \in \{1,...n\}$,

The hardness of the lifetime maximization problem arises from the fact that the power assignments are not given explicitly in the problem instance, but implicitly via the power thresholds d_{ij} , and by specifying the source and sink nodes. In this case, the total number N of different valid power settings can be exponential in , which makes the problem hard to solve.

A node i initializes its RF circuits immediately after it wakes up. Assume that it takes a node T_{rf} time and E_{rf} amount of energy to initialize its RF circuits. If the node has packets to transmit, it first listens to the channel for a period of T_{lis}^i time to see if any of its neighbors are transmitting. We define $T_{\det}^i = T_{rf} + T_{lis}$ the total time from when a node wakes up to when it goes to sleep again. if the receiver happens to be in a sleep mode, the transmitter will resend the RTS preamble after going to a power saving status for a short time T_{sav} .if $T_{\det}^i$ is too short, a node may not be able to capture an RTS preamble transmitted in its RF range. Hence, it requires

$$T_{\det}^i \geq \max_{j \in N_i}(2T_{rf} + 3T_{pre} + T_{sav}^j)$$

Note that $T_{\det}^i$ is dominated by the longest T_{sav} in node i 's neighbor set. This means if different nodes adopt different T_{sav} 's, the ones with smaller T_{sav} 's can always increase their T_{sav} without increasing the idle listening time $T_{\det}^i$ of node i . Hence, it is the most energy efficient way to let all sensor nodes have the same T_{sav} , for otherwise we can always let the ones with smaller T_{sav} 's increase their T_{sav} 's to save energy.

The energy consumption for a node to receive a packet is calculated as

$$\bar{E}_{rx} = (E_{rf} + p_{rx} \times \frac{T_{det}}{2}) + p_{rx}T_{data} + 2 \times p_{tx} \times T_{pre}$$

where the first term on the right hand side is the energy cost to initialize RF circuit and the average energy cost to detect the RF channel before desired RTS is received, the second term is the energy cost to receive the data packet, while the third term is the energy cost to transmit the CTS and ACK packets.

$$E_{det} = E_{rf} + p_{rx}(T_{det} - T_{rf})$$

This is the optimization problem that we will solve in a distributed manner. We can interpret the above problem as minimizing the maximum ratio of power consumption to energy supply at a node.

4 Simulation Test

A. CALM-DD

Directed diffusion [10] is a novel data-centric, data dissemination paradigm for sensor networks. Directed diffusion has some novel features: data-centric dissemination, reinforcement-based adaptation to the empirically best path, and in-network data aggregation and caching. These features can enable highly energy-efficient and robust dissemination in dynamic sensor networks, while at the same time minimizing the per-node configuration that is characteristic of today's networks.

B. CALM for DD

In this paper, using network simulation tool NS-2 to evaluate CALM, In simulation, the value of the sensing range could vary depending on the physical properties of the sensors and on the type of signal they are sensing. To measure the impact of the sensing range, we considered three network sizes of, respectively, 50, 100, 150, 200, 250, 300, 350, 400 nodes randomly placed in 300 × 300 (m) square area. End-to-end packet loss rate, remaining energy En as the evaluation index, to evaluate HOP- DD, ETX- DD and CALM-DD, Then, we draw the network lifetime when, respectively, the sensing range and the cluster size vary. These measures will show how worth it is to invest in sensors with higher sensing range or higher processing capacity.

Multiple simulation experiment is done using the parameters in table 2, in the network topology, choosing two nodes, one as the source node, and the other as the destination node. This paper uses the node id No. 20 as the source node, No. 40 as the destination node. From the simulation experiment, we can make the network throughput curves (Figure 1, Figure2).

Figure 2 shows that from the source node S to destination node D, the average end to end throughput graph; as pre-set conditions, in theory, end to end throughput can up to 25kbps. However, due to Hello, and ACK control packet, the actual throughput was lower than the theoretical value; CALM-DD and ETX-DD take into account the actual traffic situation in the bottom of the communication link, then selecting a relatively reliable link. CALM-DD considers the lower-link communication characteristics of the non-symmetry, so its throughput is even higher than the ETX-DD.

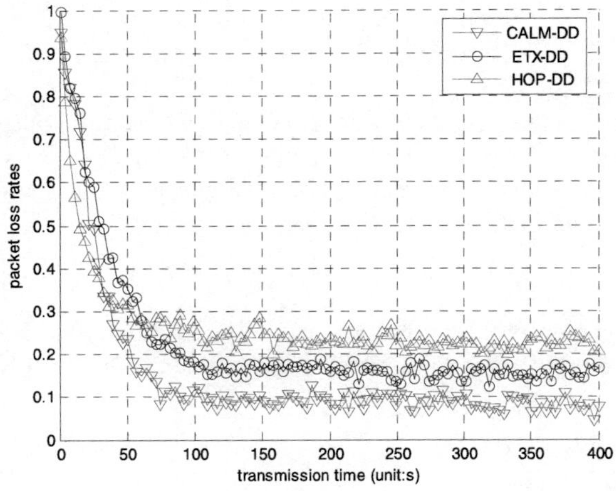

Fig. 1. End to end packet loss rates T vs. simulation time

Figure 2 shows that the simulation results of the experiment: When the end to end route is established, because HOP-DD use the minimum hop count metric method, without considering the actual low-level communication link, Thus, when the transmission of data, the underlying need to re-transmit data, so that the network consumes more energy. CALM-DD, ETX-DD consider the reliability of the underlying link in the establishment of a relatively reliable routing path, comparing the ETX- DD, requires less retransmission, thus reducing the energy overhead. CALM-DD compare with HOP- DD need fewer probe packets to judge the status of the link, so CALM-DD consumes less energy.

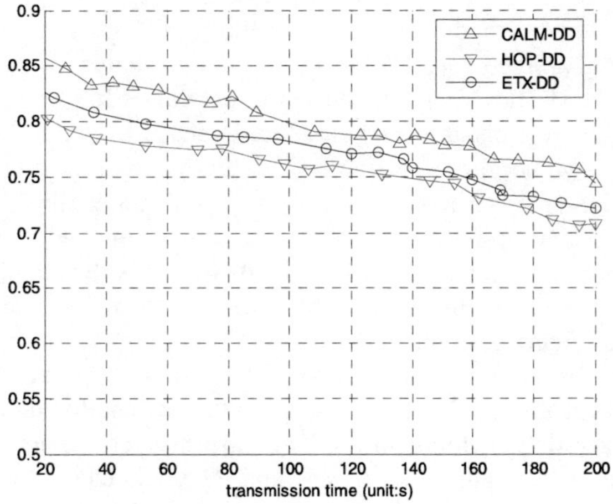

Fig. 2. Remain energy En vs. simulation time t

5 Conclusion

In this paper, an adaptive cross layer algorithm that improves energy consumption and end to end delay simultaneously. Our protocol uses a mechanism of exploration for discovery the network. The selection of routing paths is based on delay and energy metrics. Our approach offers advantages compared with other classical approaches, such as ETX metric. In particular, it reduces much better the energy consumed required for information sent, updates routing and network exploration more reactively. The simulation results have shown the efficiency of the adaptive approaches compared to traditional approaches. In fact, the proposed algorithm takes into account the network state in a better way than the classical approaches do. Also, our protocol is well-suited for a dynamic mobile environment. In particular, adaptive algorithms that simultaneously process an observation and learn to perform better is a good way to consider time evolving networks. In particular, adaptive algorithms that simultaneously process an observation and learn to perform better is a good way to consider time evolving networks. In this paper, we have studied the energy efficient problem in wireless visual sensor networks. we further examined the maximum network lifetime considering transmission errors. We investigated the error remedy techniques in both large-delay WSN applications and small-delay WSN applications, and studied their impacts on maximum network lifetime. the CALM is able to work effectively even in the presence of correlated communication errors which are typical in industrial environments Through extensive numerical simulations, we demonstrated that the proposed algorithm can support a much longer network lifetime for the conventional wireless sensor networks. As future research directions, we intend to develop a more sophisticated heuristic to improve the network lifetime. Furthermore, we intend to consider distance-dependent probabilistic event detection, where the probability that a sensor detects an event is function of the distance of that sensor from the event. Furthermore, we intend to work on distributed algorithms that address energy-efficient clustering under the joint coverage and routing constraint.

References

1. Liu, Y., Liu, K., Li, M.: Passive Diagnosis for Wireless Sensor Networks. IEEE/ACM Transactions on Networking 18(4), 1132–1144 (2010)
2. Masazade, E., Niu, R., Varshney, P.K., Keskinoz, M.: Energy Aware Iterative Source Localization for Wireless Sensor Networks. IEEE Transactions on Signal Processing 58(9), 4824–4835 (2010)
3. Vivekanandan, V., Wong, V.W.S.: Concentric Anchor Beacon Localization Algorithm for Wireless Sensor Networks. IEEE Transactions on Vehicular Technology 56(5), 2733–2744 (2007)
4. Guan, Z.-H., Ding, L., Kong, Z.-M.: Multi-radius geographical spatial networks: Statistical characteristics and application to wireless sensor networks. Physica A: Statistical Mechanics and its Applications 1(15), 198–204 (2010)
5. Wu, Y., Zhang, L., Wu, Y., Niu, Z.: Motion-Indicated Interest Dissemination With Directional Antennas for Wireless Sensor Networks With Mobile Sinks. IEEE Transactions on Vehicular Technology 58(2), 977–989 (2009)

6. Fu, T.-Y., Peng, W.-C., Lee, W.-C.: Parallelizing Itinerary-Based KNN Query Processing in Wireless Sensor Networks. IEEE Transactions on Knowledge and Data Engineering 22(5), 711–729 (2010)
7. Boukerche, A., Samarah, S.: A Novel Algorithm for Mining Association Rules in Wireless Ad Hoc Sensor Networks. IEEE Transactions on Parallel and Distributed Systems 19(7), 865–877 (2008)
8. Liu, Y., Liu, K., Li, M.: Passive Diagnosis for Wireless Sensor Networks. IEEE/ACM Transactions on Networking 18(4), 1132–1144 (2010)
9. Kaur, T., Baek, J.: A strategic deployment and cluster-header selection for wireless sensor networks. IEEE Transactions on Consumer Electronics 55(4), 1890–1897 (2009)
10. Le, T., Sinha, P., Xuan, D.: Turning heterogeneity into an advantage in wireless ad-hoc network routing. Ad Hoc Networks 1(8), 108–118 (2010)

Design for the System of Music Learning and Accompanying

Chen Xuehuang[1], Lin Qian[2], and Yan Guilin[1]

[1] Physics Department of Qinghai Normal University, Xining, Qinghai, China
[2] College of Physics and Electronic Information Engineer of Qinghai University for Nationality, Xining, Qinghai, China

Abstract. Designing a system which based on the new high-speed MCU with the accompaniment of music learning system, to make up the capabilities of the device has a single voice accompanied by defects. Dual-core structure, the image microprocessor and the microprocessor of the respective audio and video information processing and analysis of voice detection processing tasks. Major hardware elements of the system, including MCU, VGA controller, standard MIDI sound source characteristics and the use of such methods to study the system in the main application software, including human-computer interaction, sound detection, VGA driver and other design methods to make the system to normal operation.

Keywords: Music learning machine, Audio\Video technology, Microprocessor System.

1 Sound Accompaniment System Status

Existing voice and video machine automatic accompaniment music devices, voice-activated technology solutions, using computer technology to produce beautiful music. electronic music, singing melody with automatic recognition function, as the singer s voice can be issued for a harmonious musical accompaniment . Although these devices have good performance, but because of high user demand music theory as well as accompaniment in the form of Kara OK there, so that it can not be effectively promoted. Therefore, changes in thinking, using an interactive way to increase the appropriate hardware and software, music, learning and accompaniment system design, its main function will be located in the music school supplies, so that it has a new space for development, for improving the students and lovers of music quality has a positive meaning.

2 System Design

Music learning and accompaniment system structure shown in Figure 1. It can use hardware circuitry to detect human voice in real-time of the fundamental frequency, or the voice to do A / D converter with a software baseband after testing, and then based

J. Luo (Ed.): Affective Computing and Intelligent Interaction, AISC 137, pp. 199–206.
springerlink.com

on music theory, the analysis by determining the fundamental frequency to which it belongs scale, to drive the audio tone sounds in harmony with the issue of electronic tones, and tones of the musical alphabet and other information will be displayed. In order to integrate audio and video technology, a dual-core processor system structure, shall external memory (SD card) in the stored images and music material processed by the microprocessor to give the video image output unit, respectively, in an external monitor and speakers generated images and sound, or sight-singing mode for users to sing. Baseband by the software to determine the valuation and analysis of fundamental frequency belongs to the scale, gives the material provided with the standard deviation of pitch and displayed on the monitor or through the voice prompts to complete a practice session. At the same time the microprocessor according to the calculation of the relationship between music and sound, the output associated with the musical scale information (instructions) to the standard sources (uses MIDI sound source), but also the information given to the image microprocessor, the scale information converted to the tone, timbre, intensity and other relevant images, the output video signal to the monitor [1].

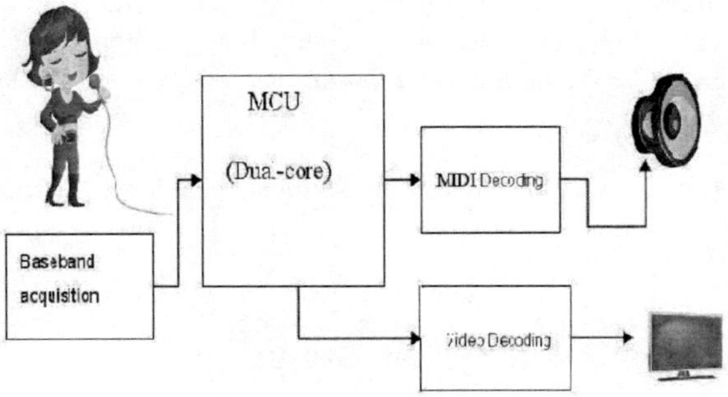

Fig. 1. With the accompaniment of music learning system architecture diagram

3 Display Control and MIDI Sound Source

Systems use two new high-speed MCU - STC12C5AxxS2, for the system's audio and video signal processing foundation. Two MCU, the main microprocessor to complete the image from the external memory (SD card) for ear training with reading material and pictures of the MIDI sound material, and after an external display card output video signal. Audio microprocessor complete the user's valuation of the fundamental frequency sounds and produce relevant musical information sent to standard MIDI sound source.

Video output interface using FX-VXC256 series VGA controller (also known as VGA card), the controller has two channels: analog video signal (CVBS) channel and MCU digital display channels. MCU-channel is 8-bit data bus mode and master MCU interface, can display 256 colors. The controller is able to directly accept the ranks of the coordinate values, no calculation of memory addresses, thus reducing the burden on the MCU, the small size of the image display can achieve satisfactory speed. Figure 2 shows the VGA card and MCU interface connection diagram.

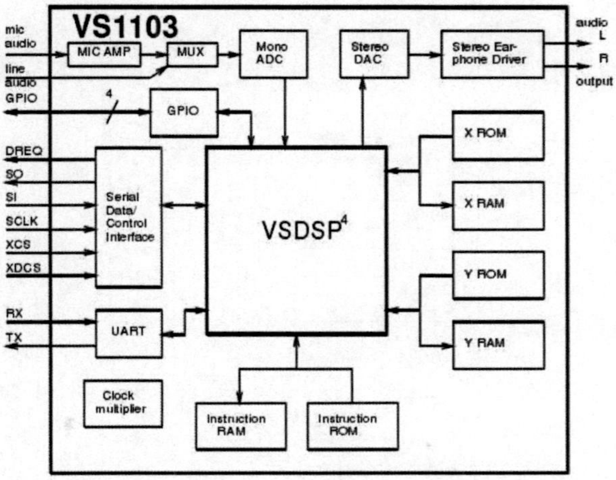

Fig. 2. MIDI sound source within the structure

Standard audio use a single VLSI chipWMA / MIDI audio decoder and ADPCM encoder chip VS1103B, produced by a Finland company. it has A high-performance low-power DSP processor core VS_DSP, 5K instruction RAM, 0.5K data RAM, serial control and data input interface, its internal structure.

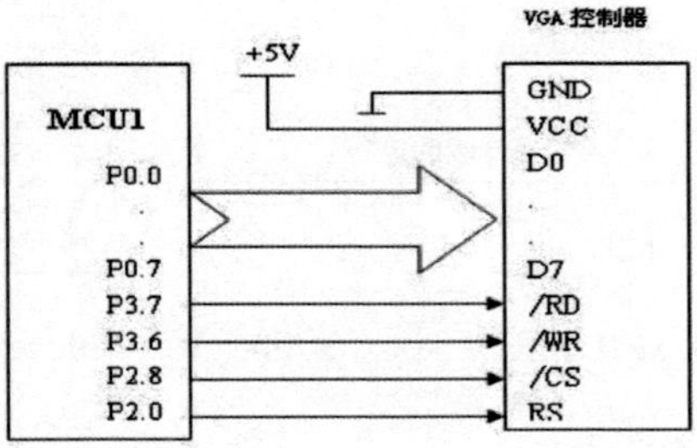

Fig. 3. VGA controller and MCU interface diagram

In Figure 3, it has many connections with the MCU, in doing MIDI decoding is the most convenient way serial communication, VS1103B the first 26, 27 pins RX, TX pins for serial communication, use the serial connection to the processor [2].

4 Software Design

System software design includes the program for MCU2 MCU1 running the program as two relatively independent parts, the serial communication between the two parts through the exchange of information. Program subroutine by the main program and a number of functional form, the need for real-time response functions such as human-computer interaction, real-time notes display, pitch evaluation is used to complete the interrupt service routine, Figure 4 shows the main program flow chart MCU1.

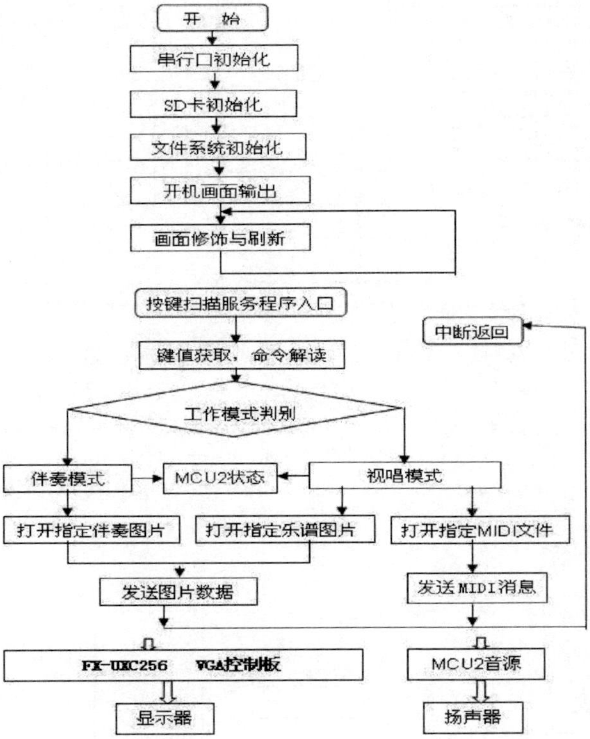

Fig. 4. MCU1 main program flow chart with key scan

4.1 Pitch Detection Software Algorithms

Pitch detection using software methods include time-domain method, frequency-domain method and time-frequency mixing method. This method should be used in the computational domain small average magnitude difference method and the autocorrelation function method, the processor speed requirements can be reduced. As the voice signal has good stability, and its detection accuracy to meet the basic requirements.

For the system through the A / D sampled discrete signal, the signal delay with its own degree of similarity that is the autocorrelation function.

$$R(k) = \frac{1}{N}\sum_{n=0}^{N-1} x(n) \bullet x(n-k)$$

Where N is a signal of the sample points. Periodic signal autocorrelation function is equal to the delay period where it functions great value, that is, the autocorrelation function of the cycle and the cycle of the same signal. Convolution by doing so find that the correlation function can be obtained from the signal of the pitch period (frequency). As the unused digital signal processor DSP, in the high speed in MCU2 case, by proper control of computation can be applied (based on experiments in music learning within 300ms delay is acceptable). Autocorrelation function method has the advantage of simple algorithm and have more documentation (including software), some of which fast algorithms can significantly reduce the number of multiplications easy programming, and practical it is more valuable [3].

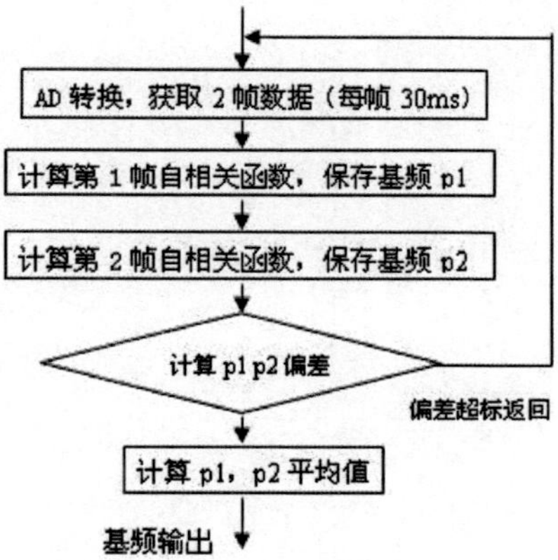

Fig. 5. Bad value filter algorithm flowchart

The method also has shortcomings: high accuracy by the noise and the impact of a larger resonance peak. If the software filter to remove noise and high frequency resonance peak will affect the real-time. Therefore, the hardware signal conditioning part of the signal must first be leveled, filter, to ensure that voice signal contains the fundamental frequency characteristics and noise reduction. Desirable band-pass filter band 80Hz-900Hz (including F2-a2 sound range), so do not have to do post-filtering software. In addition, appropriate to reduce the number of samples (lower sampling rate, but the same frame length 30ms) can be reduced by increases in the number of process operations. To reduce the error shown in Figure 5 can be added to filter out bad value algorithm can improve the detection accuracy.

4.2 SD Card SPI Communication Software

MCU and the external memory SD card for data exchange, to complete the initialization, including SD card reset, system initialization file, where file using the FAT32 file system management, the specific use to help some of the basic file operations and management functions to complete, more relevant references [4] [5]. As the microprocessor STC12C5AxxS2 do not have the SPI interface, the system uses the common I / O port for data communication.

There are two communication modes SD card SD mode and SPI mode, in which SPI mode is generally used for less demanding applications read and write speed, compatibility with the hardware interface is relatively high, the hardware connection is relatively simple, this uses the SPI mode. SPI uses HOST / SLAVE structure, in bytes transmission unit; interface definition has four pins CS, SI (DI), SO (DO), SCK, is in Mode 0, the pin function as follows:

1) CS is chip select pin, low effective;
2) SI output for the Slave Host input, idle is high, SCK rising effective;
3) SO output for the Slave Host input pin, SCK down effectively;

SCK is the clock synchronization;
According to the microprocessor STC12C5AxxS2 and SD card interface (as Figure 2), the I / O ports are defined as follows:
sbit SD_SCL = P1 ^ 7; / / SD card synchronization clock input
sbit SD_SI = P1 ^ 5; / / SD Card synchronous data input
sbit SD_CS = P1 ^ 4; / / SD card input selection
sbit SD_SO = P1 ^ 6; / / SD Card synchronous data output

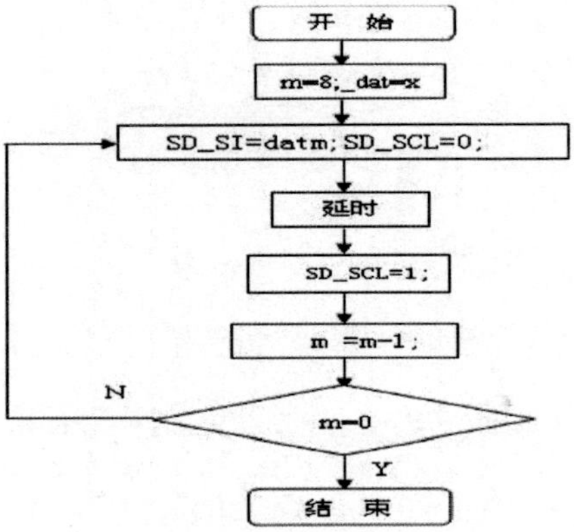

Fig. 6. I / O ports to simulate SPI communication flow

According to SPI communication timing diagram can be simulated communication instructions, such as the sending process subroutine shown in Figure 6.

4.3 VGA Card Driver

The use of the VGA card with simple script format, just by sending the appropriate commands to control the display memory , script format see the instructions with the card. The system MCU1 by 8-bit data bus and control signal lines, the use of I / O device in the form of access control VGA card. For example, specify the data to the monitor output (the background color) of the method shown in Figure 7. Reference code assignment cycle is as follows:

```
        void clear (void)
{uint data i;       / / row loop variable definitions
uchar data j;    / / column loop variable definitions
WCMD = 0x00; WDAT = 0x00; / / VGA variable assignment
WCMD = 0x01; WDAT = 0x00;
WCMD = 0x02;
for (i = 0; i <640; i) {   / / row loop variable
for (j = 0; j <480; j) {   / / column loop variable
WDAT = charbkcolor;}
}
}
```

charbkcolor variable is stored in memory in the background color image data or data, in order to read it to WDAT variables.

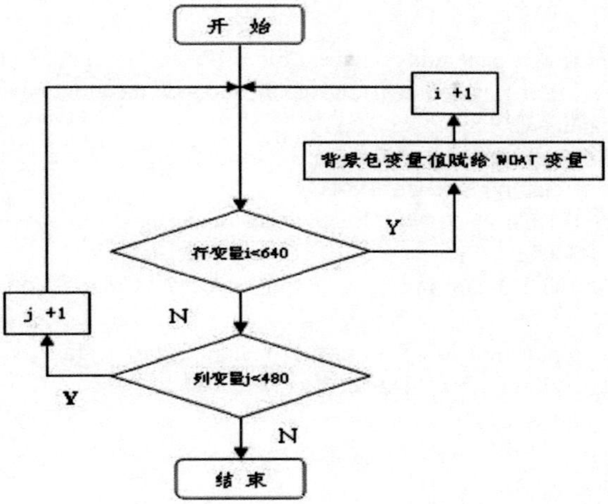

Fig. 7. Flow chart of VGA card driver

5 Examples of the Work System

Various parts of the system can work smoothly , Figure 8 is the state of the signal flow accompaniment. The problem currently remained for System is the screen display is slow, the color and clarity of lower value is similar to the picture given in Figure 7 displays the flow loop to read the image data to WDAT variables. At the same time make the resolution of VGA 640 * 480 standard, external expansion RAM capacity must be greater than 64KB, fortunately, after reducing the resolution of this application for a simple screen display resolution can still be met. Therefore, the display further improve related technologies to reduce the high cost of the system (mainly the high cost of an external VGA card) and lack of application of the system to the key.

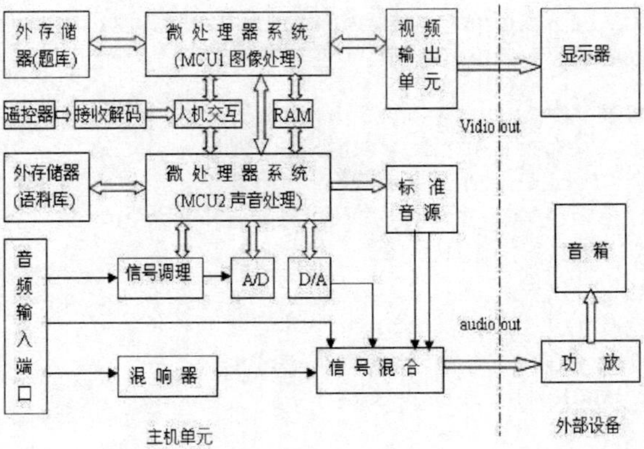

Fig. 8. Accompaniment status signal flow diagram

References

1. Music video output voice generating device. Chinese Patent Gazette, Invention 44(11) (2009)
2. Chen, X., et al.: MIDI development and application of the principle. National Defense Industry Press (January 2008)
3. Ding, Z., Yi, M.: Autocorrelation function of the valuation of the fast algorithm. Anhui Normal University (Natural Science) 19(4)
4. Yu, Z.: SPI protocol based on MMC card reader mechanism implementation. Computer Simulation 1(22) (2005)
5. Yao, F.: 51 SCM and SD card interface design. Industrial Control Computer 21(9), 60–62 (2008)
6. Chen, X.: The application of multimedia material, processing and material library construction. Jiangxi Radio & TV University (3) (2008)

Study on Service Modeling Approach for Regional Logistics Information Platform

Wang Qifeng[1], Lin Qing[1], and Huang Yongbin[2]

[1] Modern Logistics School, Zhejiang Wanli University, Ningbo, P.R. China
lhywqf@163.com, linq@126.com
[2] Business School, Zhejiang Wanli University, Ningbo, P.R. China
Huangyb@zwu.edu.cn

Abstract. Service modeling is one of the key issues for constructing the regional logistics information platform (RLIP). This paper put forward a service modeling approach for regional logistics information platform based on the presently service-oriented system modeling approach. The service modeling approach for RLIP is composed by four steps. Firstly, the function model is constructed based on the RLIP's function requirements. Secondly, the business process model of RLIP is established based on the function model, and the business object model is constructed based on the business process model. On the basis, the service model of the RLIP is established based on the above business object model and process model. Finally, the service interactive model of RLIP is realized by service choreography.

Keywords: Service-Oriented architecture, Regional Logistics Information Platform, Service Modeling.

1 Introduction

Regional logistics information platform is a regional oriented logistics management information system, which is the important component for presently logistics informatization construction and important measures to improve the development of modern logistics industry for local government. System modeling is the key technology for construction the regional logistics information platform. As a typical information system, the system modeling approach for RLIP is similar to other information system modeling, and which can be referenced in the process of RLIP system modeling. System modeling approach and corresponds technologies are research focus in information system research fields. At present, service-oriented system modeling approach is studied in academy and industry fields. Ali Arsanjani[1] provided a integrated system modeling approach with the integration of top-down, business-driven and down-top approaches based on the analysis of service conception model and service-oriented architecture. Wu Ying[2] provided a UML based system modeling approach for distribution center information system based on the analysis of UML modeling language. Ge Lei[3] discussed the service modeling process from the view of service extracting, service encapsulation and service realization, and put forward the steps of service extracting, service encapsulation principle and

J. Luo (Ed.): Affective Computing and Intelligent Interaction, AISC 137, pp. 207–213.
springerlink.com

mathematical model for service encapsulation. Rahmani Adel Torkaman[4] and Oliver[5] provided a MDA-based SOA system modeling and design approach for information system based on the analysis of the insufficient in traditional object-oriented and component-based software design approaches in complex distribution information system modeling and design. The information system modeling approaches are widely studied and have gained plenty of achievements, while the system approaches for the information system lifecycle are still to be further studied. Presently, with the needs of logistics information construction, the modeling approach for logistics information system is need to be further studied and practices.

2 The Analysis of the Service Modeling Process for RLIP

The service modeling and construction process mainly includes three steps, which are service recognition, service classification, service design and realization[1].

(1) Service recognition
The process of service recognition is combined with the sub process of domain decomposition, analysis of legacy systems, and target modeling approaches. The domain decomposes from the service domain composes to its function region and subsystem's decomposition, contains its flow or the process decomposes the process, uses the example from the process and the high rank service. The legacy system analysis is important for service recognition form legacy systems by API encapsulation, transaction and function model encapsulation, etc. View from the middle of the field composed of the target service modeling to validate and found that top-down or bottom-up service does not capture the means of identification to the other services. It will serve to link the goals and sub-goals, key performance indicators and criteria.

(2) Service classification
Service classification begins from the point when the service is assigned. Service classification is very important for service design. The service can be designed with simple function or combined function, which means with the service classification, the services can be designed in coarse granularity or in thin granularity. The coarse granularity services are well defined to design, deploy the system function and help to realize the business collaboration.

(3) Service design realization
The main steps of service design and realization is the assignment of components to services and the decision of service realization. The assignment of services includes the process of assigning the services to the logistics sub-system, which including the open function components. While the decision of service realization determines the realization mode of services, such as encapsulate from legacy system or develop as required.

3 The Service Modeling Realization for RLIP

Based on above service-oriented RLIP modeling framework, four steps for RLIP service modeling is put forward. Firstly, the function model is constructed based on the RLIP's function requirements. Secondly, establishing the business process model of RLIP based on the function model, and constructing the business object model based on the business process modeling. On the basis, the service model of the RLIP is established based on the above business object model and process model. Finally, the service interactive model of RLIP is realized by service choreography.

(1) Function model modeling for RLIP
The main task of function model construction is the decomposition of business domain for regional logistics system. With the business domain decomposition to analyze the logistics business process and recognize the business processes and business user cases, which will be the candidates for the business services. The domain decomposition includes two steps. Firstly, the function of logistics business is decomposed with different users. On the basis, user case analysis is fulfilled to each business function. The user case for RLIP is shown in figure 1.

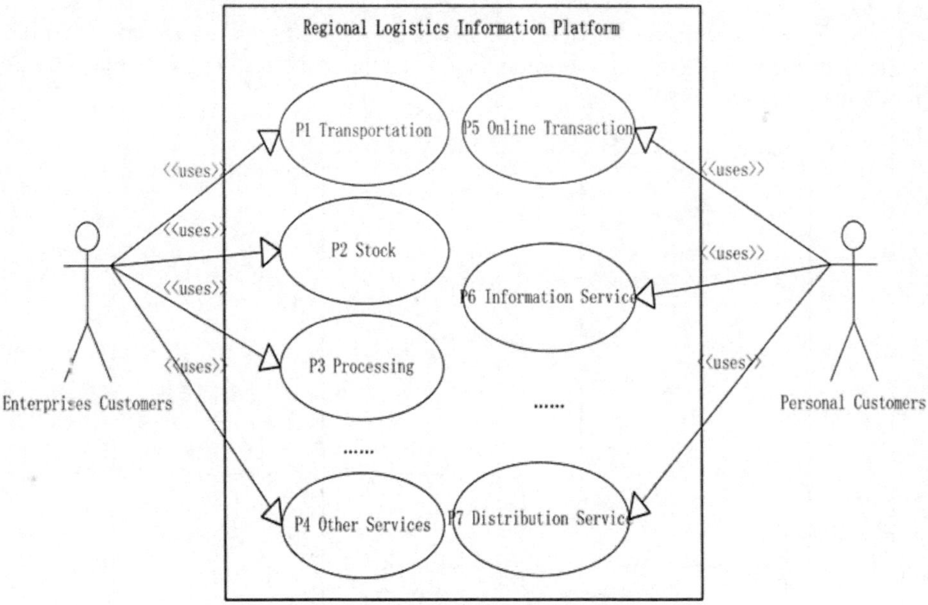

Fig. 1. Function User Case for RLIP

(2) Process model modeling for RLIP
Based on the function modeling for RLIP, using UML activity diagram, and BPM process analysis methods to analyze the operation business process of regional logistics system, including logistics flow, information flow and control flow, etc, and analysis the business rules, to design and optimize the business process to establish the process model of regional logistics information platform. In the process of process modeling for RLIP, the process model for each function is established, considering the process optimization and the integration between logistics business processes.

1) Analyzing the each user case deeply based on the result of function modeling, and refining the business process list of the regional logistics system. The result will be the basis for RLIP process model modeling.

2) The activity diagram of UML is used to analyze and design the business process based on the combing of business processes of RLIP. Figure 2 shows the process model of logistics order processing with UML activity diagram.

(3) Servicemodel modeling for RLIP

Service is the basic elements for RLIP. Based on the above function model and process model, and combined with the business objects of logistics operation, the function components can be classified. The services are encapsulated with the function components with coarse granularity, which is described with WSDL and registered in UDDI. Other logistics service can search the service in UDDI and called via SOAP protocol. According to practices requirements, the logistics order services, the information publish services, the member registration services, the stock services, the logistics cost services, etc, can be established. On the same time, the services can be combined into synthesis Services with coarser granularity via service combination, described and published with the same way as simple services. The concrete service modeling process is as followings.

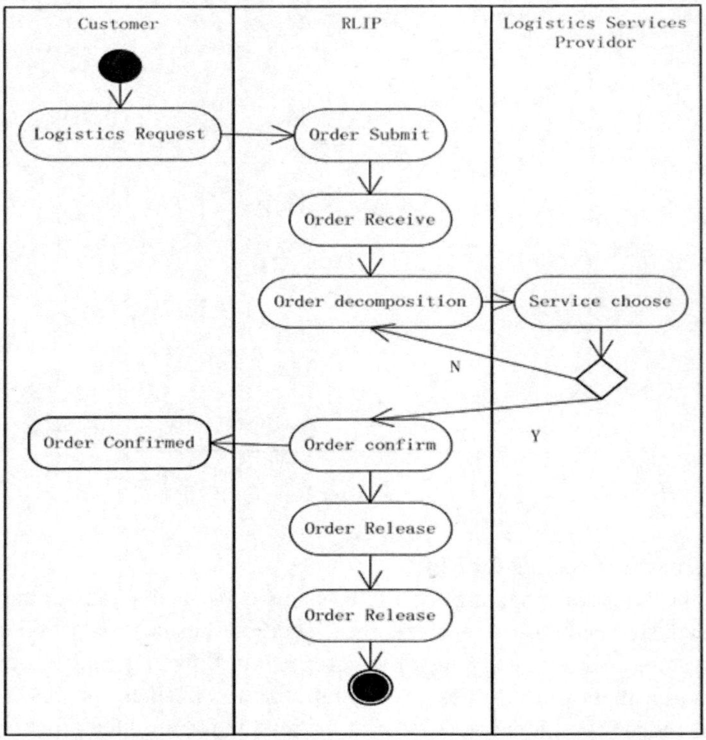

Fig. 2. Process Model for Logistics Order Processing

1) Based on the above function model and process model of RLIP, the services are recognized with the integration of top-down, down-up and align center approaches. And the candidates services lists for RLIP will be generated.

2) With the difference of logistics services providers and requestors, the logistics services can be classified and to know which function model provides the services and who are the targets users of the logistics services. This paper users the approach of integration the target users and function to classify the candidates logistics services. For example, the logistics services can be classified into enterprises required logistics services, personal required logistics services, government required logistics services and common technology services for the RLIP, etc.

3) Designing the service with the service classification of step 2. Service designing is the key issue in RLIP system design. The services of RLIP include three types of services, which are basic services, synthesis services and composite services. According to the basic services, the service function property and non-function properties, such as security, performance, etc, can be described with WSDL. The component model of basic logistics service is shown in figure 3. While according to the synthesis services and composite service, which can be described with BPEL based on the establishing of process model via sequence diagram of UML, BPMN, etc.

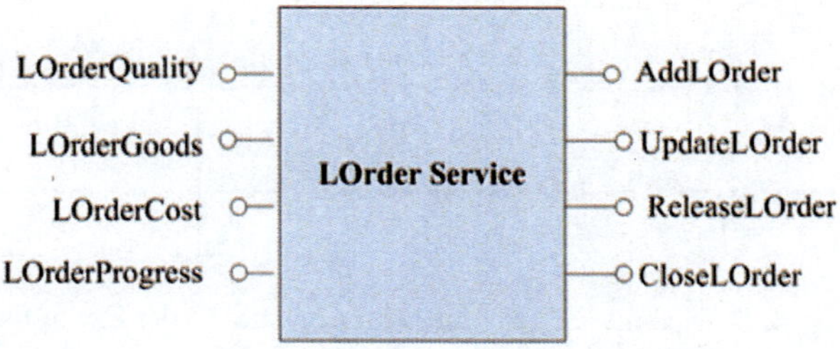

Fig. 3. Component Model for basic Logistics Service

(4) Service interactive model modeling for RLIP

The service-oriented RLIP is a process-oriented and business-driven logistics information management system. Based above service modeling, the next step is to establish the service interactive model via service combination. The service interactive model can be modeled with sequence diagram of UML and described with BPEL, which is a XML-based business process execution language. Figure 4 shows an example of logistics order execution service interactive model expressed by sequence diagram.

With the modeling of UML sequence diagram, each logistics business process can be described as the interaction of logistics services. Take the logistics process shown in figure 4 for example, the execution process is as follows.

1) When the customer submits the requirement to RLIP, the customer requirements receiving service will activated and save the customer's requirements to RLIP.

2) The Logistics order create service create the order after the RLIP receives the customer's requirements.

3) The logistics order service searches and maps the suitable logistics services and generates the logistics service chain with the requirements of the customer.

4) The BPEL execution engine of RLIP called the stock service via ESB and fulfills the stock service.

5) The BPEL execution engine of RLIP called the transportation service via ESB and fulfills the transportation service.

6) The service execution information feedbacks to the customer via RLIP in the end of logistics services. By the way, the customer can query the execution information via the query services of RLIP on demand.

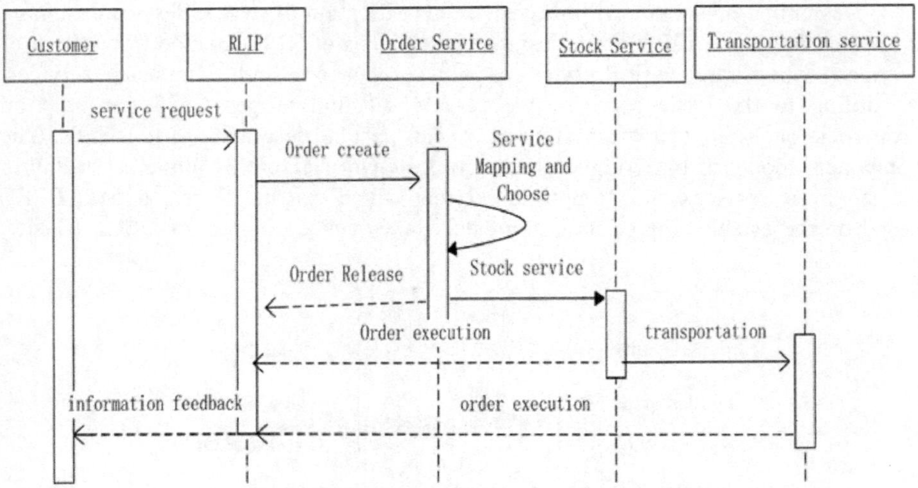

Fig. 4. Service Interactive Model for Logistics Order Execution

4 Conclusions

Regional logistics information platform is an important platform for improving the regional logistics management level and operation efficiency. This paper is focus on the approach of service modeling for RLIP construction. A service modeling approach for service-oriented RLIP is established based on the service-oriented RLIP system architecture. The service modeling approach is composed by four steps, which are function modeling, process modeling, service modeling and service interactive modeling. The service-oriented computing technologies and UML system modeling approach are integrated in the service modeling approach, which provides a feasible service modeling solution for RLIP construction.

Acknowledgment. This Research is supported by the Ningbo Natural Science Foundation Program (No.2008A610023).

References

1. Arsanjani: Service-Oriented Modeling and Architecture (SOMA), IBM developer Works (2004), http://www.ibm.com/developerworks/webservices/library/ws-soa-design1
2. Wu, Y., Zhang, Q.H., Cheng, G.Q.: Modeling Method of Distribution Center Information System Based on UML. Logistics Technology 1, 80–83 (2006)
3. Ge, L., Huang, S.X., Fan, Y.S.: Servie Modeling in SOAD. Manufacturing Automation 30, 27–29 (2008)
4. Rahmani, A.T., Vahid, R., Saeed, S.: An MDA-based modeling and design of service oriented architecture. In: Proc. of the 6th International Conference on Computer Science, December 10, pp. 578–585. IEEE Press (2006)
5. Oliver, I.: Applying UML and MDA to real systems design. In: Proceedings of the Conference on Design, Automation and Test in Europe, pp. 70–71. IEEE Press (May 2005)

A Scheme of Test Data Compression and Power Reduction Based on Common-Run-Length Coding (CRLC)

Wenfa Zhan, Bing Shi*, and Huaizhi Zha

Dept. of Educational Technology, Anqing Normal College
zhanwenfa@gmail.com

Abstract. A new scheme of test data compression and power reduction, namely common-run-length coding (CRLC) scheme is presented, which is based on run-length. This scheme uses two shorter codewords to represent the whole second run of two consecutive runs, the lengths of which are the same. Compared with other already known schemes this scheme has some characteristics, such as high compression ratio, low power dissipation, easy control and implementation. The performance of the algorithm is mathematically analyzed and its merits are experimentally confirmed on the larger examples of the ISCAS89 benchmark circuits.

Keywords: Test data compression, Low power testing, Run-length coding, Frequency-directed run-length code.

1 Introduction

The complex functionality and size caused by increasing integration levels of VLSI chips makes testing for these chips more and more difficult. Testing remains a dominant cost factor in VLSI design and reducing the testing cost is one of the objectives which VLSI producers are devoted to achieve vigorously. The most two important sources of the test cost are test data volume and test power.

Test data compression can reduce test data effectively and increases the transferring speed from ATE to chip, which makes full use of limited bandwidth and memory of ATE. Moreover, the hardware overhead is very low and acceptable. Thus this technology is widely used.

A popular class of compression schemes relies on the use of a linear decompressor. These compression schemes exploit the fact that scan test vectors typically contain a large fraction of unspecified bits even after compaction. However, the compression effect of these schemes is limited by the Smax, where Smax is the maximum number of specified bits in the test vector. That is to say, the compression effect of these schemes is related with the worst case in the test set.

* Corresponding author.

J. Luo (Ed.): Affective Computing and Intelligent Interaction, AISC 137, pp. 215–223.
springerlink.com

Another category of compression methods is based on coding. These methods exploit the regularity inherent in test data to achieve high compression. They involve partitioning the original data into symbols, and then replacing each symbol with a codeword to form the compressed data. According to the volume of the symbol in original test data and codeword in compressed data, the coding schemes can be divided into five types[1][2].

Conventional test data compression schemes generally increase test power. A large number of test pattern bits (don't care bits) being assigned randomly result in a large number of transitions in the scan chains thereby increasing power dissipation during test.

A new scheme, namely common-run-length coding (CRLC), is proposed. This scheme first considers both types of runs of 0's and 1's, so the total number of runs and transitions during scan-in will both decrease, which result in better compression effect and lower test power. Then it further explores the relationship between two consecutive runs on the basis of traditional coding characteristic which uses shorter codeword to represent longer symbol. This presented scheme uses two shorter codewords to represent the whole second run of two consecutive runs, the lengths of which are the same. Theoretical analysis and experimental results show that the compression effect of this scheme is higher than that of previously proposed schemes.

2 Algorithm of CRLC

2.1 CRLC

From FDR coding, it can be found that the prefix of every codeword is in the form of 1…10 from the group A2. So another form of 0…01 can also be used in the prefix of codeword. It has been observed that test sets contain a large number of runs of 1's in addition to runs of 0's. By considering both types of runs, the total number of runs will decrease, which could result in higher test data compression. Hence, two types of prefix can be used to represent the types of runs. That is to say, the prefix in the form of 1…10 can be used to represent the runs of 1's and the prefix in the form of 0…01 can be used to represent the runs of 0's. So a new coding table is constructed and named as CRLC, which is shown in Table 1.

The coding table of CRLC is a variable-to-variable-length coding way. The size of the groups is increasing as the number of groups increasing, which is in accordance with the distribution of the run-length in actual test data and this characteristic is similar to FDR[3].

It is found that the tails of those two codewords are very similar. The only difference between them is that of the run-length of FDR is longer by 2 than that of CRLC for the same tail of the codeword. From Table 1, it is also found the tails are the same if the run-length of 0's and 1's are the same. The only difference between the same run-length of 0's and 1's is that the prefix for the runs of 0's is in the form of 0…01, but the prefix for the runs of 1's is in the form of 1…10. So the codeword of CRLC can simply be obtained from the codeword of FDR.

Table 1. CRLC

Run-length	Group	Prefix for runs of 0's	Tail for runs of 0's	Codeword for runs of 0's	Prefix for runs of 1's	Tail for runs of 1's	Codeword for runs of 1's
1			01	0101		01	1001
2	A_1	01	10	0110	10	10	1010
3			11	0111		11	1011
4			000	001000		000	110000
5			001	001001		001	110001
……	A_2	001	……	……	110	……	……
10			110	001110		110	110110
11			111	001111		111	110111
12			0000	00010000		0000	11100000
14			0001	00010001		0001	11100001
……	A_3	0001	……	……	1110	……	……
26			1110	00011110		1110	11101110
27			1111	00011111		1111	11101111
……	……	……	……	……	……	……	……

It needs to be pointed out that for the runs of 0's this coding approach will not increase the length of codeword compared with the codeword of FDR if the original run-length and later run-length obtained from original run-length plus 2 belongs to the same group of FDR coding table. Therefore, for the runs of 0's this coding approach will get almost compression effect compared with FDR. In addition, this coding approach can also encode the runs of 1's. Compared with EFDR coding this coding approach can also encode both runs of 0's and runs of 1's. The characteristic of this approach is that the type of runs can be told from the prefix, but the coding approach of EFDR needs an additional bit to represent the type of runs. Furthermore, CRLC scheme explores the relationship between the two consecutive runs. If the length of a consecutive run is the same as that of the former, then the whole second run can be represented by a shorter codeword. Two codewords (0100 and 1000) are used to represent a repeated run. The run-length of encoded run is the same as the former run. Codeword 0100 represent the encoded run is runs of 0's and 1000 represent the encoded run is runs of 1's. This further improves the compression ratio.

From the above analysis, it can be concluded that the scheme of CRLC uses two different coding ways: (1) Type I. This coding way uses coding table of CRLC scheme to code the run (symbol) in the original test data. This coding way is similar to that of the traditional run coding. (2) Type II. This coding way uses two shorter codewords (0100 and 1000) to represent the second run of two consecutive runs, the lengths of which are the same. That is to say, if the lengths of two consecutive runs are the same, then the later run is represented by the codeword 0100 or codeword 1000, or else the later run should be coded by using coding table.

Original test data typically have only 1-10% of the bits specified, while the rest are don't-cares [1]. Filling those don't-care bits appropriately can increase the same length probability of two consecutive runs. If the two consecutive runs lengths are equal, then the later can be represented by one of the two shorter codewords. Thus, the whole later run can be represent by 4 bits, which further improves the compression ratio.

In order to improve the effectiveness of proposed scheme, the coding way of Type II can be further explored. Namely, the don't-care bits should be appropriately filled to make the length of consecutive two runs the same. Then the first run is encoded by the way of Type I and the second run is encoded by the way of Type II. In this way compression effect can be improved further. For example, let take test vector slice of 000000X000000X000000X XX X1X111XXX111XXX1110 (X represents don't care bit) into account. The don't-care bits can be filled by 0 and 1 appropriately and two consecutive runs 000000000 000000000001 and 11111111111111111110 can be gotten. Then the second run can be encoded by 1000. The length of original test vector slice is 42 and that of corresponding compressed data is 12. If the don't-care bit is filled into 000000000000000000 000000111111 111111111110 and Type I coding way is used, then the compressed data is 00011000 11100100, whose length is 16. From this, conclusion can be made that appropriately filling the don't-care bits can decrease the volume of test data.

In a word, type I coding approach can encode the runs of 0's and runs of 1's by using different prefix, which gets a good compression effect. And type II coding way will further compress the test data. Hence, CRLC scheme can get a better compression effect than any already known schemes.

An example of coding using CRLC scheme is shown in Figure 1. This example codes the length of runs 0s. Let's take the test vector slice of 000000XX XXXXX110 into account. The lengths of those two runs equal to 7. So the later run can be encoded by 1000. The results encoded of the slice is 001011 1000. The length coded is shorter by 6 than that of the original slice.

Original test vector slice: 000000XX XXXXX110 000000000001 000000000001 1111X1XXXXXX 000000000001

Encoded test vector: T_E=001011 1000 001111 0100

Length of original test vector: 64 bits; T_E: 20 bits; reduce: 44 bits.

Fig. 1. An Example for CRLC Encoding

2.2 Power Estimation

We use the weighted transitions metric(WTM) introduced in [4] to estimate the power consumption due to scan patterns. The WTM metric models the fact that scan power for a given vector depends not only on the number of transitions in it but also on their relative positions. WTM is also strongly correlated to the switching activity in the internal nodes of the circuit under test during scan operation. It was shown experimentally in [4] that scan vectors with higher weighted transition metric dissipate more power in the circuit under test. In addition, in [5] experimental results for an industrial ASIC confirm that the transition power in the scan chains is the dominant contributor to test power.

Consider a scan chain of length l and a scan vector $t_j = t^*_{j,1} t^*_{j,2} \cdots t^*_{j,l}$, with $t^*_{j,1}$ scanned in before $t^*_{j,2}$, and so on. The weighted transitions metric for tj, denoted WTMj, is given by $WTM_j = \sum_{i=1}^{l-1} (l-i)\cdot(t^*_{j,i} \oplus t^*_{j,i+1})$. If the test set TD contains n vectors t1, t2, …, tn then the average scan in power Pavg and peak scan in power Ppeak are estimated as $P_{avg} = \sum_{j=1}^{n} \frac{WTM_j}{n}$ and $P_{peak} = \max_{j\in\{1,2,\ldots,n\}}\{WTM_j\}$.

In our proposed scheme, both types of runs are considered. Hence, the filling of don't care bits is selective to decrease the number of transitions during scan-in.

3 Design of Decoder

In this section, the design of the decoder of the CRLC coding scheme is illustrated.

As mentioned above, the real run-length (i) of FDR is the sum of prefix prefix (a) and tail (b), namely i=a+b. For example, (13)10 = (110)2+(111)2. Further analyzing the prefix, we find that it will appear in the form of 111……1110. By transforming the equation i=a+b, we can get

$$i=(a+2+b)-2=((\underbrace{11\cdots\cdots110}_{k})_2+(\underbrace{10}_{2})_2)+(\underbrace{b'}_{k})_2)-2$$
$$=((\underbrace{100\cdots\cdots00}_{k+1})_2+(\underbrace{b'}_{k})_2)-2=(\underbrace{1b'}_{k+1})_2-2 \quad (10)$$

That is to say, if we increase 1 bit to the tail of FDR codeword and put the value of 1 in this bit, then we get a new code which represents the run-length that is longer by 2 than that of actual run. As mentioned above, for the same tail the run-length of FDR is longer by 2 than that of CRLC. So we can increase 1 bit to the tail of CRLC codeword and put the value of 1 in this bit, then we get a new code which represents the run-length that is longer by 4 than that of actual run. That is to say, the length information of CRLC is hided in the tail of codeword. So we can use this discipline to decompress. This discipline can be implemented by a special counter. The special counter has two characteristics: (1) setting the lowest bit of the counter to number 1. When the data needing to be decoded is shifted into the counter, the number 1 and other data are together shifted to high bits of the counter. (2) using non-0 low bound. Traditional counter counts down to 0, but this counter counts down to 4 (the content of counter is "100"). This function can be implemented easily by some combinational circuits, which will not introduce high hardware overhead.

The design of this decoder is similar to that of the FDR using the FSM-based design. This decoder is embedded in chip, which has some characteristic of small size and simple structure. This decoder does not introduce hardware overhead obviously, and is independent of not only DUT, but precomputed test data as well. The block diagram of the decoder is shown in Figure 4. The decoder decodes the precomputed test data TE to output TD. The structure of this decoder is simple, which only needs a FSM, a special k+1 bit counter, a log2k bit counter and a k+1 bit register.

The block diagram of the proposed CRLC decoder is given in Figure 2.

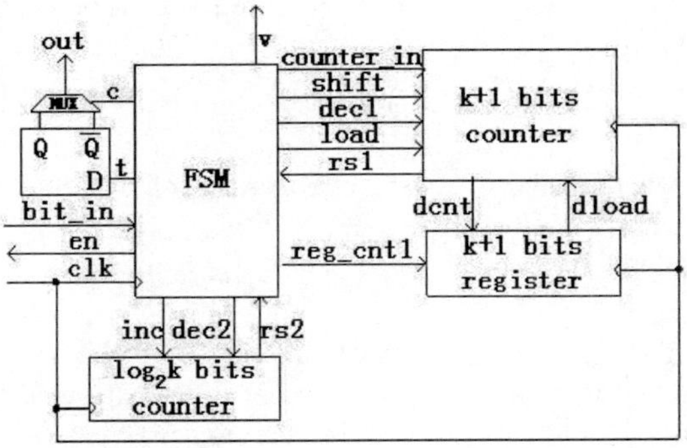

Fig. 2. Block Diagram of CRLC Decoder

4 Experimental Results

In this section we will verify the effectiveness of the proposed CRLC scheme by using experiment results. For comparison with other schemes, we adopted first the MinTest test sets provided by Duke University of America, which are the same as [3]. Experiments were performed on the several largest ISCAS 89 benchmark circuits [6].

Table 2. Compression Obtained by Using Different Codes Methods (No Difference)

Circuit	Size of T_D bit	Scheme presented		Golomb		FDR	EFDR	VIHC
		Size /bit	α %	m	α %	α %	α %	α %
s5378	23754	14176	50.01	4	40.70	48.02	**51.93**	51.78
s9234	39273	23659	**47.37**	4	43.34	43.59	45.89	47.25
s13207	165200	31324	81.26	16	74.78	81.30	81.85	**83.51**
s15850	76986	27111	66.83	4	47.11	66.22	**67.99**	66.23
s35932	28208	14716	**82.91**	4	4.69	10.19	45.63	56.08
s38417	164736	88686	**61.07**	4	44.12	43.26	60.57	43.26
s38584	199104	80812	61.82	4	47.71	60.91	**62.91**	60.92
Avg.			**64.47**		43.21	50.50	59.53	58.43

The compression ratio of this scheme compared with other schemes is shown in Table 2. The first column of Table 2 is the circuit name. The second column is test size of original test data. The compression effect is shown in third column of Table 2. The other columns are data size of compressed data and compression ratio of Golomb,

FDR, EFDR[6], and VIHC[7] coding. The compression ratio can be determined by $\alpha = (T_D - T_E)/T_D \times 100\%$, of which, T_D is original length of test data and T_E is the length of compressed data. In Table 2 the original test data did not do the operation of difference. The filling of don't-care bits is to make the length of consecutive runs the same as possible. The remaining of don't-care bits are filled randomly.

From Table 2, as can be seen, the proposed scheme achieves higher compression than the Golomb, FDR, EFDR, VIHC technique for some of the circuits. The scheme achieves higher compression ratio than Golomb for all of the compared test sets. It also achieves higher compression ratio than FDR for six out of the seven compared test sets. It achieves higher compression ratio than EFDR for three out of seven compared test sets, almost the same compression ratio for three out of seven compared test sets (about less than 1%). It also achieves higher compression ratio than VIHC for five out of seven compared test sets. On average, the percentage compressions of the proposed scheme are 21.26%, 13.97%, 4.94% and 6.04% higher than those of Golomb, FDR, EFDR and VIHC techniques.

Table 3. Experimental Results on Peak and Average Scan-in Power consumption

Circuit	Mintest		Golomb		Proposed	
	Peak power	Average Power	Peak power	Average Power	Peak power	Average Power
s9234	17494	14630	12994	5692	13720	4534
s13207	135607	122031	101127	12416	94886	8190
s15850	100228	90899	81832	20742	70908	13815
s35932	707280	583639	172834	73080	107226	40395
s38417	683765	601840	505295	172665	438005	118628
s38584	572618	535875	531321	136634	481185	86415
Avg.	369499	324819	234233	70205	200988	45329

Next we present the experimental results on the peak and average power consumption during the scan-in operation. These results show that test data compression can also lead to significant savings in power consumption. We estimate power using the weighted transitions metric. Let P_{peak}^{C} (P_{avg}^{C}) be the peak (average) power with compacted test sets obtained using Mintest. Similarly, let P_{peak}^{G} (P_{avg}^{G}) be the peak (average) power when Golomb coding is used by mapping the don't cares in TD to zeros and let P_{peak}^{SRLC} (P_{avg}^{SRLC}) be the peak (average) power when CRLC coding is used by mapping the don't cares in TD to zeros or ones according to our program. Table 3 compares the average and peak power consumption for Mintest test sets when Golomb coding and our proposed scheme. The first column of Table 3 is the circuit name. The second and third columns are the peak power and average power if original Mintest test sets are used. The fourth and fifth columns are the peak power and average power if Golomb coding is used. The last two columns are peak power and average power if our proposed scheme is used. Table 4 shows that the peak power and average power are significantly less if proposed scheme is used for test data compression and the decompressed patterns are applied during testing. On average, the peak (power) is 33.41% (12.43%) less in our proposed scheme than for the Golomb coding.

In a word, what can be concluded from the experimental results is that our proposed scheme can achieve better compression result, lower power, and shorter application time at lower hardware overhead, which is acceptable.

5 Conclusion

Test data compression is an effective solution to reduce the increasing volume of test data. A new test data compression scheme (CRLC) is proposed. This scheme first considers both types of runs of 0's and 1's, then further explores the relationship between two consecutive runs on the basis of traditional coding characteristic which uses shorter codeword to represent longer symbol in the original test data. This proposed scheme uses two short codewords to represent the whole second run of two consecutive runs, the lengths of which are the same. It is shown that the presented scheme decreases the ATE memory and channel capacity requirements by obtaining good compression ratios. Moreover, this scheme can reduce power dissipation. Thus, it is an effective solution for test data compression/decompression of IC design.

Acknowledgment. This work is supported in part by the Natural Science Foundation of Educational Department of Anhui Province of China under Grant KJ2010A230 and the Natural Science Foundation of Anhui Province of China for young teachers in high colleges under the Grant No. 210SQRL110.

References

1. Touba, N.A.: Survey of Test Vector Compression Techniques. IEEE Design & Test of Computers 23(4), 294–303 (2006)
2. Zhan, W., Liang, H., Shi, F., Huang, Z.: Test data compression scheme based on variable-to-fixed-plus-variable-length coding. Journal of Systems Architecture 53(11), 877–887 (2007)
3. Chandra, A., Chakrabarty, K.: Test data compression and test resource partitioning for system-on-a-chip using frequency-directed run-length (FDR) codes. IEEE Transactions on Computers 52(8), 1076–1088 (2003)
4. Sankaralingam, R., Oruganti, R., Touba, N.A.: Static compaction techniques to control scan vector power dissipation. In: Proc. of VLSI Test Symp., pp. 35–40 (2000)
5. Saxena, J., Butler, K., Whetsel, L.: An analysis of power reduction techniques in scan testing. In: Proc. Int. Test Conf., pp. 670–677 (2001)

6. El-Maleh, A., Al-Abaji, R.: Extended frequency-directed run-length code with improved application to system-on-a-chip test data compression. In: Proc. 9th International Conference Electronics Circuits and Systems, pp. 449–452 (2002)
7. Paul, T., Al-Hashimi, B., Nicolici, N.: Variable-Length Input Huffman Coding for System-on-a-Chip Test. IEEE Transactions on CAD of Integrated Circuits and Systems 22(6), 783–796 (2003)
8. Chandra, A., Chakrabarty, K.: System-on-a-Chip test data compression and decompression architectures based on Golomb codes. IEEE Transitions on CAD of Integrated Circuits and System 20(3), 355–368 (2001)

J2ME-Based Mobile Applications' Development Mode

Han Jianmin

Henan Vocational College of Economics and Trade
Zhengzhou, China 450053
1979hjm@163.com

Abstract. With the popularity of wireless communication networks, and common use of mobile phone, PDA and other mobile devices, the demand for mobile devices accessing Web is increasing, and thus the demand for mobile applications also increases. Meanwhile J2ME as a flexible, practical one of the main support platform has become increasingly popular. This paper proposes a J2ME-based mobile application development model.

Keywords: Mobile applications, software development framework, J2ME, security.

1 Introduction

With the development of mobile communication technology and the increasing popularity of mobile devices, people increasingly prefer using wireless devices to access Internet. More and more people prefer using mobile phones to access internet anytime and anywhere through HTTP protocol. Such as, entertainment, web browsing, e-commerce. So, the development of mobile applications is urgent, and J2ME is the best development platform.

2 J2ME Architecture and Its API

Java-based J2ME is an open, standards-based development and deployment platform, and has the portability of java technology, deployment flexibility, security and so on. Currently, the field of mobile application development has the J2ME architecture as the standard in fact. J2ME's API available to developers are shown in Figure 1[1]. MIDP(Mobile Information Device Profile) divides user interface into high-level user interface and the underlying user interface. High-level user interface's abstraction level is higher, it encapsulates user interface components' high-level objects, and object's perception are not under its control [2]. Device use user's own local interface perception, so portability is good, for the end users, this achieves a seamless user interaction.

3 System Model's Analysis and Design

3.1 System Model's Analysis

The client of mobile application system uses phone, PDA and other wireless devices in general. Due to wireless devices' fewer resources, so its processing power is weak,

J. Luo (Ed.): Affective Computing and Intelligent Interaction, AISC 137, pp. 225–233.
springerlink.com

so the form of "thin client" is well-received, and business processing is mainly on server-side. J2ME provides excellent support for multi-level structure, and each layer is transparent to other layers, so is the choice for using J2ME [3].

Mobile application system, in addition to selecting framework, it should also adapt its own uniqueness. Currently, there are many kinds of wireless devices, they have different functions, and wireless communication technology develops rapidly, a variety of wireless networks coexist in markets, such as RS, GSM, etc. Therefore, the development of mobile application system needs to consider an important issue is how to make use of mobile application system without the constraints of wireless devices, wireless network.

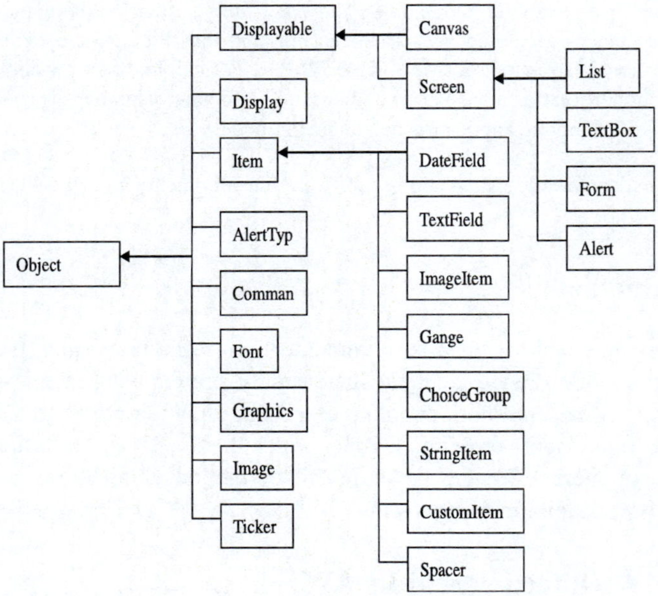

Fig. 1. User interface class' inheritance tree

3.2 System Model's Design

J2ME-based mobile application development model can be divided into four layers: client layer, presentation layer, business logic and data layer [4], as shown in Figure 2.

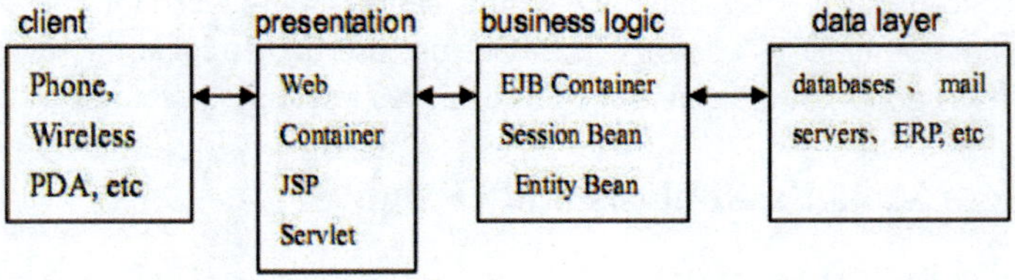

Fig. 2. The framework of J2ME application system

Client layer runs on mobile phones, PDA and other mobile devices and terminals, provides user interface in user's terminal devices, implements user interaction with other layers, and can also be used to implement some business logic, but only necessary and simple business logic. Interaction manner provided by client layer is generally traditional Web browser, but can also be a dedicated non-Web-based application.

Presentation layer implements interaction with client layer and business logic layer. Presentation layer accepts user's request sending from client layer, calls the appropriate module through Servlet and JavaBean technology to communicate with the business logic layer. Business logic processes, and generates dynamic responses content and returns the content to the client layer in order to achieve interactivity [5].

Business logic layer is composed by a number of EJB components running in the EJB container / server to implement the system's business logic. EJB components can be used repeatedly and many times, also can be easily ported to other servers. EJB receives client layer's request data, and does logic processing, then sends data to data layer for saving. EJB accessing the data layer's database is generally through traditional JDBC or ODBC[6].

Data layer contains varieties of servers for permanent storage, such as database servers, mail servers, etc. database systems, in addition to storing data record, may also store coarse-grained logic blocks such as the stored procedures and triggers. Business logic layer reads the data and calls services through a variety of standard protocols.

This structural design has its original advantages. Presentation layer separates from business logic layer, specializes as a separate layer to response to varieties of mobile devices, wireless networking and communication protocols, and adapts and meets the diversity of mobile devices and wireless network in system model design. This hierarchical structure makes the application logic clarity, simple, easy development, maintenance and management; this architecture uses the middleware technology, by means of middleware technology's stability, maturity and development's simplicity, we can significantly reduce development risk and lower development costs. Through middleware platform's clustering technology, we can be very easy to obtain high performance.

4 System Model's Implementation

Mobile devices have different hardware configuration, and these hardware features such as power supply capacity, memory size, have seriously affected the java program running in this device. In order to facilitate application development in the mobile device, we urgently need to develop a specification to unify these mobile devices. Because there are various kinds of mobile devices and we can not use a unified way to specify these devices, so we can only use the J2ME model to achieve our goal.

The definition of mobile java J2ME platform is divided into four layers: the host operating system, configuration, summary tables, and optional package vendors. As the java application is independent of the operating system, therefore, J2ME specification contains only three layers: configuration, summary tables, and optional package vendor. Through configuration, summary tables and optional package vendor, J2ME defines a complete runtime environment for java application in mobile devices. Because the vendor optional package is optional, J2ME mainly depends on configuration and summary table.

4.1 Configuration

Device configuration is the minimum java platform for a certain kind of device; it includes the minimal subset to meet such device's virtual machine running and minimal subset for such device's core class libraries.

J2ME specifies a large number of vastly different hardware capabilities of mobile devices through configuration. Configuration specification defines the ability that hardware must possess. For example, how much ROM and RAM does hardware possess, the minimum CPU clock cycles, the minimum bandwidth when connect network, etc. J2ME configuration defines the specific equipment's available virtual machine and low-level API, and meanwhile provides the appropriate basis for the development. Currently J2ME's available configuration has two kinds: CDC (Connected Device Configuration) and CLDC (Connected Limited Device Configuration).

4.2 Summary Tables

Summary table structures above the configuration, so each summary table will simply declare the configuration type which it is rooted. Summary table defines advanced API for different hardware features; these high-level API is expansion of libraries which are not related to other hardware devices. These advanced API determines the compilation method of java program in such device. And the API which can be called are defined in a mobile information device summary table (MIDP).

Usually, how the user interface of java program present in a variety of mobile devices is defined in the summary table. Such as MIDP's LCDUI (limited configure the device user interface) defines the mobile information device's user interface. The summary table also defines how the program interacts with the user or device. Expansion libraries defined in summary table is built based on the underlying core class libraries defined in configuration, In order to better meet the needs of special device, summary table will add new method for the class in configuration, or make the method throw a new exception.

4.3 Handling Process

The processing flowchart is as follows, shown in figure 3:

Due to adopting "Thin client" principle, the client only provides simple human-computer interaction and the necessary, simple processing; the main processing logic is executed on the server side. The communications mechanism of Client and server is HTTP protocol. Because the traditional HTTP communication protocol has been prevalent for a long time, HTTP communication's needed infrastructure has also been established completely, and mobile application can just take advantage of these existing facilities. Also it is important that HTTP protocol is supported by almost all MIDP devices, so it is not difficult to use HTTP protocol.

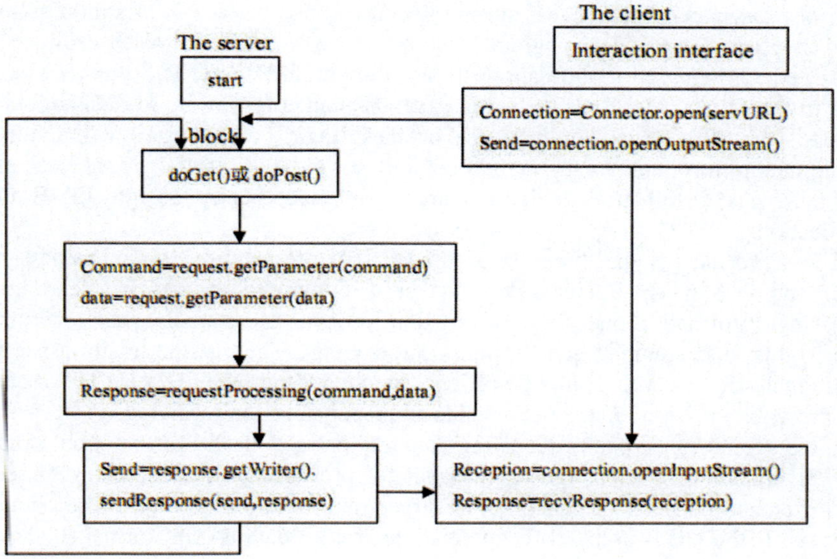

Fig. 3. The processing flowchart

After the application program starts, the server will start automatically and be at the block state; listen to the client's request at any time. When the client needs the server's processing capacity, the client will establish connection, and send request. During the connection process, the way of HttpConnection interface is often used, which defines a comprehensive approach to easily achieve the network connection. For example, in the following table, Connection = Connector.open (servURL), in the program source code it is similar to the form of "httpConnection = (HttpConnect) Connector.open (" http://www.baidu.com ")", and if the application program establishes the connection by connecting framework (GCF), then it can only open the connection by the CLDC configuration. The server calls the doGet () or the doPost () method to accept the client's connection requests. When accepted, the server extracts processing instruction and processing data respectively, calls the server-side business processing logic to process the request. Processing also supports database access. Because of the existing of database accessing class in JavaBean, so you can easily operate the database, such as connect, query, add, delete, modify and other operations. After processing, the server will generate response content, and connect the client once again, send the response content to the client and display the content in user's browses.

4.4 Example Scenarios

An application example of mobile application system is the mobile office system; we are familiar with automatic office system in front of computer, Mobile office system will extend the traditional automatic office system to the mobile terminal devices, so As long as users of handling mobile devices are in the mobile communication network's coverage, he/she can handle official business anytime and anywhere. Users can send and receive messages, query information and release information as usual. The client runs a MIDlet program, customers use a traditional Web browser or a dedicated J2ME-based mobile application, through HTTP / HTTPS protocol to send data to the server side. Because xml has self explanatory power, so data adopts xml format, when the server complete the business logic processing, it will return the result data to the client. As the mobile devices' resource is constrained themselves, so we must make full use of its resources and specifically design J2ME client application.

The server adopts the three-tier architecture of presentation layer, business logic layer and data layer. Business logic layer is composed of message sending and receiving, information querying, information retrieval and statistics, data encryption, user rights management and business processing. This is the main aspects of traditional office as well as mobile office. The client sends a HTTP / HTTPS requests, the presentation layer of the server side is responsible for receiving, in accordance with the agreed rules, parses receiving data and extracts useful information, sends to the EJB module in the business logic layer for processing. After handling, based on the difference between the clients, the server returns different data to the client for analysis. If the client uses J2ME applications, return data as xml format. If the user uses a traditional Web browser, return the data as dynamically generated JSP page form. EJB is mainly responsible for business logic processing, but also responsible for transmitting data to the data layer for permanent storage; transmitting data can use JDBC, ODBC to communicate with database server, you can also use SMTP, POP3 protocol to communicate with the mail server.

4.5 Implementation Instance of Mobile Phone Operating System

Android operating system is a popular operating system on mobile phone, in the following, through a music player example on Android operating system to further elaborate the implementation of mobile applications development. Open Core is the core of Android multimedia framework, which provides media player function, and media streaming records function. MediaPlayer class can be used to play audio, video and streaming media, MediaPlayer include Audio and Video playback capabilities. In Android interface, Music and Video two applications are achieved by called MediaPlayer. This mini version music player is implemented by Open Core and MediaPlayer. This player need pause, stop, previous track, next track, and other basic button. And able to play more than a music format, it can automatically recognize which suffix files in the specified file path can play, which are not supported format. The following lists brief code.

```
Public class musicActivity extends ListActivity {
    private MediaPlayer myMediaPlayer;
private List<String> myMusicList=new ArrayList<String>();
private int currentListItem=0;
private static final String MUSIC_PATH=new String("/sdcard/");
Public void onCreate(Bundle savedInstanceState) {
   super.onCreate(savedInstanceState);
         setContentView(R.layout.main);
        myMediaPlayer=new MediaPlayer();
```

```
        findView();
        musicList();
        listener();
}
void musicList(){
    File home=new File(MUSIC_PATH);
        if(home.listFiles(new MusicFilter()).length>0){
            for(File file:home.listFiles(new MusicFilter())){
             myMusicList.add(file.getName());
            }
            ArrayAdapter<String> musicList=new ArrayAdapter<String>
            (musicActivity.this,R.layout.musicitme, myMusicList);
            setListAdapter(musicList);
        }
    }
void findView(){}
void listener(){
    viewHolder.stop.setOnClickListener(new OnClickListener() {
    public void onClick(View v) {
if(myMediaPlayer.isPlaying()){
      myMediaPlayer.reset();
            }
        }
    });

        viewHolder.start.setOnClickListener(new OnClickListener() {
    public void onClick(View v) {
      playMusic(MUSIC_PATH+myMusicList.get(currentListItem));
        }
    });
    viewHolder.next.setOnClickListener(new OnClickListener() {
    void onClick(View v) { nextMusic();   }
    });
        viewHolder.pause.setOnClickListener(new OnClickListener() {
    public void onClick(View v) {
if(myMediaPlayer.isPlaying()){
                myMediaPlayer.pause(); }
else{ myMediaPlayer.start();
            }
        }
    });
  void playMusic(String path){
        try {
    myMediaPlayer.reset();
    myMediaPlayer.setDataSource(path);
    myMediaPlayer.prepare();
    myMediaPlayer.start();
    myMediaPlayer.setOnCompletionListener(new OnCompletionListener() {
      public void onCompletion(MediaPlayer mp) {
```

```
                nextMusic();
            }
        });
    } catch (Exception e) {
        e.printStackTrace();
    }
}
```

5 The Security ofJ2ME Mobile Application System

First of all, java language does its best in the security of the network. Java strictly complies with object-oriented specification, encapsulates the details of the data, only provides the interface to users, and increases data-level security. Java has no pointer operations. Java can not directly manipulate the memory address, thereby increasing memory-level security. Java checks on the array bounds, and forces type conversion. Java provides thread-safe support.

In traditional network transmission, when the user inputs important personal information, the net uses SSL protocol to protect user data safety. Also in mobile communications, MIDP applications adopt SSL protocol to protect the communications security, and end to end security is ensured by KSSL protocols. KSSL agreement is a "thin client" model, the main calculated is on the server, the client only do simple and necessary processing. Using of KSSL to build wireless mobile applications is based on HTTPS.

In the J2ME application interaction, through the integrity check, authentication, encryption, security audit to ensure security. Due to increasing security on the need for additional resource consumption, so according to the importance degree of resource, the security can be ensure separately and graded.

6 Conclusion

Practice has proved that, mobile applications based on J2ME development model, make the system can plug and play, and has good scalability, security and stability. As a result of "thin client" model, the clients can choose many kinds of devices, and the traditional J2EE- based applications, just do a little modification and can easily be extended to the mobile edge. With the popularity of mobile applications, The J2ME model in the development will become a mainstream.

References

1. Wells, M.J.: J2ME game programming, translate by Peng Li et al. Tsinghua University Press, Beijing (2005)
2. Riggs, R., Peursem, J.V., Patel, M., Huopaniemi, J., Uotila, A., Taivalsaari, A.

3. Translate by Wei Xiao et al. J2ME wireless devices programming. Electronics Industry Press, Beijing (2004)
4. Hu, X., Yang, Z., Li, H.: J2ME Mobile Applications development. Tsinghua University Press (2005)
5. Hao, Y., Li, X.: J2ME Mobile-Application development. Tsinghua University Press (2006)
6. Feng, M.: The analysis of EJB JPA database persistence layer's development practices. Electronic Industry Press (2008)

A Lightweight Encryption Algorithm for Images

Zhen Chen

Department of Computer Science and Technology
Hunan International Economics University, Changsha 410205, Hunan, China
chenzhenp@163.com

Abstract. In order to prevent images from unauthorized access and distribution, the images should be encrypted before transmitting to different customers. The traditional encryption algorithms such as RSA, DES and AES are not suitable for encrypting images due to their high computational complexity. To this end, this paper proposes a new image encryption algorithm that has a lower computational complexity. Compared to those traditional encryption algorithms, the proposed scheme can be considered as a lightweight encryption algorithm.

Keywords: image encryption, selective encryption, DCT, secure content distribution.

1 Introduction

The rapid development of computer networks and digital technology allowed large files, such as digital images, to be easily distributed and shared over the internet. Images are widely used in different scenes and the protection of image data from unauthorized access and distribution is very important. To prevent the unauthorized user from accessing the image data, the image data should be encrypted before sending to the corresponding customer. Image encryption is a hot research topic and a lot of image encryption schemes have been proposed in recent years [2-4].

Images are different from text and hence, the traditional encryption algorithms such as RSA, DES and AES are not suitable for encrypting images due to their high computational complexity and some other constrains. For example, the data size of the images is almost always greater than that of text data. In addition, in text, data, the decrypted text must be equal to the original text. However, this requirement is not necessary for image data. According to the human visual system (HVS) model [5, 6], the decrypted image has a small distortion is usually acceptable.

The public-key encryption algorithms such as have a high computational complexity and hence, they are usually used to key exchange and data signature, not used to encrypt the multimedia data directly. Compared to the public-key encryption algorithms, the symmetric encryption algorithms have a relatively lower computational complexity and used to encrypt the multimedia data. In these existing image encryption schemes with the symmetric encryption algorithms, the image data is presented as a vector of real numbers, and each element of this vector is encrypted

J. Luo (Ed.): Affective Computing and Intelligent Interaction, AISC 137, pp. 235–241.
springerlink.com

by a different symmetric key. Since the number of the sampling coefficients in each image is very large, the key has a very long length [8].

Designing an efficient image encryption scheme is very important for protecting and distributing the images. The aim of this paper is to introduce a new image encryption algorithm that has a low computational complexity and a desired security. The remainder of this paper is organized as follows. In section 2, we introduce some related work. In section 3, we introduce the proposed encryption algorithm in detail. Some experimental results are presented in Section 4, and in Section 4, we analyze the security of the proposed scheme. A short conclusion is given in Section 5.

2 Related Work

The discrete cosine transform (DCT) helps separate the image into parts (or spectral sub-bands) of differing importance (with respect to the image's visual quality). Similar to the discrete Fourier transform, the DCT is also a mathematical transformation, which transforms a signal or image from the spatial domain to the frequency domain. Because the DCT can minimize the amount of data needed to recreate a digitized image, the DCT is a common transformation of image [1]. Let $f(x, y)$ denote a $n \times n$ image, then the following equation shows the discrete cosine transform

$$F(u,v)=\frac{2}{n}c(u)c(v)\sum_{x=0}^{n-1}\sum_{y=0}^{n-1}f(x,y)I(u,v), \tag{1}$$

where

$$I(u,v)=\cos\left(\frac{\pi u(2x+1)}{2n}\right)\cos\left(\frac{\pi v(2y+1)}{2n}\right) \tag{2}$$

and

$$c(x)=\begin{cases} 1/\sqrt{2} & x=0 \\ 1 & x=1,2,\ldots n-1 \end{cases} \qquad x\in\{u,v\} \tag{3}$$

The corresponding inverse discrete cosine transform (IDCT) is shown as follow

$$f(x,y)=\frac{2}{n}\sum_{u=0}^{n-1}\sum_{v=0}^{n-1}c(u)c(v)F(u,v)I(u,v). \tag{4}$$

In [1], the authors proposes a well known watermarking algorithm (secure spread spectrum watermarking), where a watermark consists of a sequence of real numbers $X=(x_1, x_2,\ldots, x_n)$. Each element x_i is chosen independently according to a probability distribution, such as standard normal distribution whose mean is 0 and variance is 1. To embed the watermark signal into the original image P, the distributor first performs the DCT on P, and selects n low frequency different DCT coefficients $P=(p_1, p_2, \ldots, p_n)$ as the candidate coefficients to embed the watermark. The secure spread spectrum watermark embedding process can be denoted by

$$p_i' = p_i + \theta x_i \quad i = (1, 2, ..., n), \tag{5}$$

where θ is a scaling factor that determines the extent to which x_i alters v_i.

Secure spread spectrum watermark embedding algorithm as a well known watermarking algorithm has been used widely. In [7], the author used the spread spectrum water-mark algorithm to propose an asymmetric digital fingerprinting scheme, where the encryption key generated from the customer's fingerprint and Gaussian vector is considered as a high strength watermark signal to be embedded into the original image. In [9], the authors proposed an asymmetric digital fingerprinting scheme by utilizing the selective (partial) encryption algorithm. The main goal of selective encryption is to reduce the amount of data to be encrypted by taking the different characteristics of the multimedia data into account while achieving a required level of security. In [8], the original image is presented by a vector of real numbers (selected DCT coefficients) and the length of the encryption key is equal to the length of the vector.

3 Proposed Encryption Scheme

Taking the underlying characteristics of the image into account, it is not necessary to encrypt the image bit by bit. We assume that an image P can be denoted as a vector $P = (p_1, p_2, \ldots, p_n)$, where the elements of this vector denote the perceptually significant parts of P. Here, we assume that each element p_i represents a DCT coefficient in low frequency.

We assume that the encryption key $ek = (e_1, e_2, \ldots, e_n)$ can be considered as a high strength watermark signal, where p_i is encrypted by e_i. In this paper, for simplify, we assume that each element of ek is draw from a standard normal distribution with the generator g. The encryption operation can be considered as the watermark embedding process with the secure spectrum watermarking algorithm. The following equation shows the encryption operation

$$c_i = p_i + \alpha e_i \quad i = (1, 2, ..., n), \tag{6}$$

where α is a real number used to control the encryption strength. A larger α means a higher encryption strength.

The decryption operation is the reversion of the encryption operation, which can be denoted by

$$p_i = c_i + \alpha dk_i \quad i = (1, 2, ..., n). \tag{7}$$

Here $dk = ek$ is the corresponding decryption key (symmetric encryption). In order to overcome the problem that the encryption (decryption) key has a long length, the encryption key is transferred into a function f. Since each ek is determined by g and α, f is a function of g and α, where g and α are combined together so that try to separate g and α is very difficult.

4 Discussion

As an image encryption scheme, the following three aspects should be considered.

(1) Computational complexity. Compared to the existing scheme, the proposed scheme has a lower computational complexity as the encryption is only simple addition.

(2) The size of the ciphertext. The encrypted image should be not extended too much. In the proposed scheme, the encryption operation equals to the watermarking embedded operation and hence, the encrypted image is equal to the original image.

(3) The security. As an image encryption algorithm, the security of it includes two aspects: cryptographic security and perceptual security [10]. The cryptographic security means the security against the cryptographic attacks, while the perceptual security means the perceptibility of the encrypted image. For the cryptographic security, performing the cryptanalysis on the different images is the common technique. We can achieve the cryptographic security by selecting different g and α (different encryption function) for different images. Without the original image, the attacker cannot deduce the encryption key and encryption algorithm from the encrypted image. The perceptual security depends on the number and the encryption strength of the encrypted components, the more the encrypted components, the higher the perceptual security, the high the encryption strength, the higher the perceptual security.

5 Experimental Results

In this section, we show some experimental results to examine the proposed encryption algorithm. We take a 512 × 512 gray scale image as the test image. The original image and the encrypted image are shown in Fig. 1 and Fig. 2 respectively.

Fig. 1. The original image

Fig. 2. The encrypted image

In order to show the impact of that the length of the encryption key and the encryption strength to the encrypted image, we did two comparative experiments.

In the first simulation experiment, I keep the encryption strength (α) unchanged and use the encryption key with different lengths. The corresponding encrypted images are shown in Fig.3, Fig.4 and Fig.5 respectively, where $L_3 = 3L1$, $L_2 = 2L1$. From the experimental results, we can see that the longer the length of the encryption key, the higher the perceptual security.

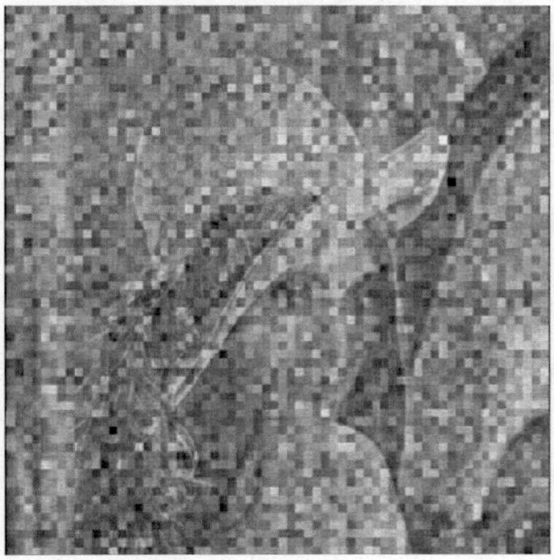

Fig. 3. The encrypted image ($l = L_1$)

Fig. 4. The encrypted image ($l = L_2$)

Fig. 5. The encrypted image

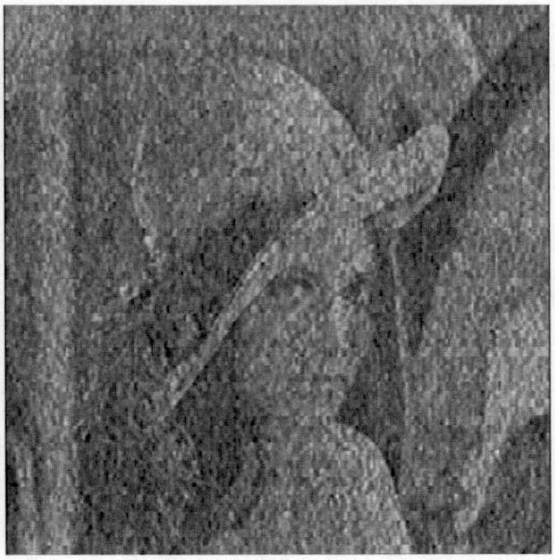

Fig. 6. The encrypted image ($\alpha = \alpha_1$)

In the second simulation experiment, I keep the length of the encryption key unchanged and use different encryption strengths (α). The corresponding encrypted images are shown in Fig.6, Fig.7 and Fig.8 respectively, where $\alpha_3 = 3\alpha_1$, $\alpha_2 = 2\alpha_1$. From the experimental results, we can see that the larger the value of α the higher the perceptual security.

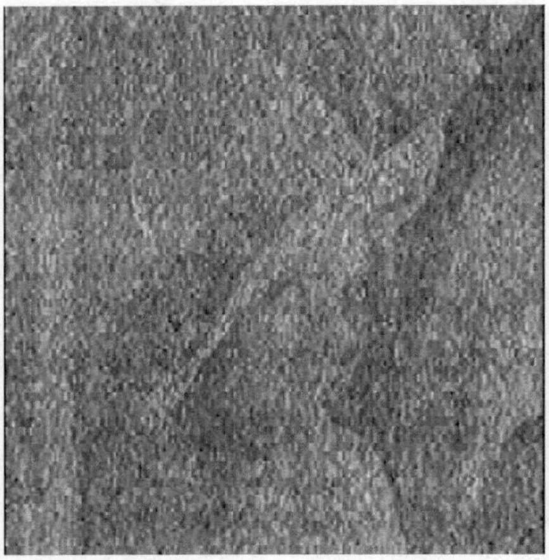

Fig. 7. The encrypted image ($\alpha = \alpha_2$)

Fig. 8. The encrypted image ($\alpha = \alpha_3$)

6 Conclusion

An image encryption scheme is proposed in this paper, where the underlying characteristics of the image are taken into account during encryption. Compared the traditional encryption, it has a low computational complexity. In addition, its encryption key has a short length, which is more practical.

References

1. Cox, I., et al.: Secure spread spectrum watermarking for multimedia. IEEE Trans. Image Process. 6(12), 1673–1687 (1997)
2. Kamali, S.H., Shakerian, R., Hedayati, M., Rahmani, M.: A new modified version of Advanced Encryption Standard based algorithm for image encryption. In: International Conference on Electronics and Information Engineering, pp. 141–145 (2010)
3. Kumar, G.M.B.S.S., Chandrasekaran, V.: A novel image encryption scheme using Lorenz attractor. In: IEEE Conference on Industrial Electronics and Applications, pp. 3662–3666 (2009)
4. Zhou, Y., Panetta, K., Agaian, S.: Image encryption using binary key-images. In: IEEE International Conference on Systems, Man and Cybernetics, pp. 4569–4574 (2009)

5. Koz, A., Alatan, A.A.: Oblivious Spatio-Temporal Water-marking of Digital Video by Exploiting the Human Visual System. IEEE Transactions on Circuits and Systems for Video Technology 18(3), 326–337 (2008)
6. Podilchuk, C., Zeng, W.: Image adaptive watermarking using visual models. IEEE J. Select. Areas Commun. 16(4), 525–540 (1998)
7. Hu, D., Li, Q.: Asymmetric Fingerprinting Based on 1-out-of-n Oblivious Transfer. IEEE Communications Letters 14(5), 453–455 (2010)
8. Massoudi, A., Lefebvre, F., De Vleeschouwer, C., et al.: Overview on selective encryption of image and video: challenges and perspectives. EURASIP Journal on Information Security, 1–18 (2008)
9. Hu, D., Li, Q.: Bandwidth efficient asymmetric fingerprinting based on one-out-of-two oblivious transfer. Int. J. Information and Computer Security 4(2), 152–163 (2010)
10. Lian, S., Liu, Z., Ren, Z., Wang, H.: Joint Fingerprint Embedding and Decryption for Video Distribution. In: IEEE International Conference on Multimedia and Expo., pp. 1523–1526 (2007)

Face Recognition Using Block-Based DCT and Weighted Generalized KFD

Jin Zou and Feng Sun

College of Mathematics and Information Science,
Leshan Normal University, Leshan, China
zoujin@lstc.edu.cn, sunfeng1005@163.com

Abstract. An improved feature extraction for face recognition is presented in this paper. In the proposed technique, the input face image is divided into blocks and two-dimensional Discrete Cosine Transform (DCT) approach is applied to each block. Then the low frequencies of all two-dimensional DCT coefficients from each block are extracted and combined to form a feature vector. Thereafter, weighted generalized kernel Fisher discriminant is performed on these vectors. Experimental results on the ORL face database demonstrate the effectiveness of the proposed method.

Keywords: Face recognition, Block-based discrete cosine transform, Weighted generalized Kernel Fisher discriminant.

1 Introduction

Face recognition has been a very active research area in pattern recognition because of its numerous technical challenges and commercial applications. Extracting proper features to reduce the amount of data and provide discrimination power is important in face recognition. The two most well-known methods for linear feature extraction and dimensionality reduction are principal component analysis (PCA) and linear discriminant analysis (LDA). Over the past decade, many variants of the original LDA algorithm have been proposed for face recognition, with most of them trying to overcome the problem due to under sampling, i.e. the Small Sample Size (SSS) problem. Some of these methods perform PCA first before applying LDA in the PCA-based subspace, as is done in Fisherface [1] (also known as PCA+LDA). However, Fisherface loses effective discriminant features while discarding the null space of within-scatter matrix to guarantee the successful use of LDA. Thus many improved algorithms based on null spaces have been developed [2, 3, 4, 5, 6]. Although algorithms based on LDA or PCA of face recognition have been proved to be both efficient and effective for linear problems, they are inadequate to represent the complex and nonlinear variations of real face images, such as illumination, facial expression and pose variations. To overcome the limitation of PCA and LDA, Yang et al. [7] proposed kernel Fisher discriminant (KFD) in which kernel principal component analysis (KPCA) is first performed and then LDA is used for a second feature extraction in the

J. Luo (Ed.): Affective Computing and Intelligent Interaction, AISC 137, pp. 243–251.
springerlink.com

KPCA-transformed space. Despite the SSS problem, traditional KFD-based algorithms have another drawback since the Fisher separability criterion is not directly related to classification rate. That is, the classes with larger distance to each other in feature space are more emphasized while the Fisher criteria is optimized, which leads the resulting projection preserves the distance of already well-separated classes, causing a large overlap of neighboring classes. To improve Fisher criterion, weighted schemes were introduced [8, 9].

Recall that PCA-based transforms are optimal in the sense of energy compact. However, the PCA is data dependent and computationally expensive transform. As this is the case, the Discrete Cosine Transform (DCT), a very useful tool for signal representation both in terms of information packing and in terms of computational complexity due to its data independent nature, has been shown to approach the optimality of PCA [10]. Combing null space method, improved Fisher criterion and DCT, we propose a new feature extraction method using block-based DCT and Weighted generalized KFD to improve the recognition results. In Section 2, we present the background of Weighted KFD and DCT. Section 3 describes the procedures of our approach. Experimental results are presented and discussed in Section 4. Finally, conclusions are drawn in Section 5.

2 Brief Review of Weighted KFD and DCT

2.1 Weighted Kernel Fisher Discriminant

For a given nonlinear mapping ϕ, the input data space $\mathcal{D}$ can be mapped into the feature space $\mathcal{F}$: $X \in \mathcal{D} \mapsto \phi(X) \in \mathcal{F}$. The idea of KFD is to solve the problem of LDA in the feature space $\mathcal{F}$. The optimal projection matrix V^T can be achieved by maximizing the following Fisher criterion in the feature space $\mathcal{F}$:

$$J(V) = \arg\max_V \frac{V^T S_b^\phi V}{V^T S_w^\phi V},$$

where S_b^ϕ and S_w^ϕ are the between-class scatter matrix and within-class scatter matrix defined in feature space $\mathcal{F}$:

$$S_b^\phi = \frac{1}{n}\sum_{i=1}^{c} n_i (m_i^\phi - m_0^\phi)(m_i^\phi - m_0^\phi)^T, \tag{1}$$

$$S_w^\phi = \frac{1}{n}\sum_{i=1}^{c}\sum_{j=1}^{n_i} (\phi(X_i^j) - m_i^\phi)(\phi(X_i^j) - m_i^\phi)^T, \tag{2}$$

where X_i^j denote the jth sample in ith class X_i containing n_i samples, c is the class number and n is the number of all samples, $m_i^\phi = \frac{1}{n_i}\sum_{j=1}^{n_i}\phi(X_i^j)$ and $m_0^\phi = \frac{1}{n}\sum_{i=1}^{c}\sum_{j=1}^{n_i}\phi(X_i^j)$.

As has been stated, the Fisher separability criterion is not directly related to classification rate. To improve existing Fisher criterion, Zhou et al. [9] introduced the following weighted schemes:

$$\widehat{J}(V) = \arg\max_{V} \frac{V^T \widehat{S}_b^{\phi} V}{V^T S_w^{\phi} V},$$

where $\widehat{S}_b^{\phi} = \sum_{i=1}^{c-1} \sum_{j=1+1}^{c} \frac{n_i}{n} \frac{n_j}{n} w(d_{ij}^{\phi}) \, (m_i^{\phi} - m_j^{\phi}) \, (m_i^{\phi} - m_j^{\phi})^T$,
$d_{ij}^{\phi} = \sqrt{(m_i^{\phi} - m_j^{\phi})^T (m_i^{\phi} - m_j^{\phi})}$
and $w(d_{ij}^{\phi}) = (d_{ij}^{\phi})^{-q}$, $q \geq 2$.

2.2 Discrete Cosine Transform

The two-dimensional DCT (2D-DCT) coefficients matrix C of an $M \times N$ image X is defined as follows [11]: $C_{pq} = \alpha_p \beta_q \sum_{m=0}^{M-1} \sum_{n=0}^{N-1} X_{mn} \cos \frac{(2m+1)\pi p}{2M} \cos \frac{(2n+1)\pi q}{2N}$,

where $\alpha_p = \begin{cases} \sqrt{\frac{1}{M}}, & p = 0, \\ \sqrt{\frac{2}{M}}, & 1 \leq p \leq M-1, \end{cases}$ and $\beta_q = \begin{cases} \sqrt{\frac{1}{N}}, & q = 0, \\ \sqrt{\frac{2}{N}}, & 1 \leq q \leq N-1. \end{cases}$

The values C_{pq} are called the DCT coefficients of X. Though the DCT coefficients matrix C has the same size as X, the energy distribution changes with most energy being compacted to the low frequencies coefficients which mainly located in the upper left corner of the DCT coefficients matrix. The Direct Current (DC) coefficient C_{00} holds most of the image energy and represents the average energy or intensity of the image. The remaining $MN - 1$ coefficients, called Alternate Current (AC) coefficients, represent some different pattern of image variation and directional information. For most facial images, much of the signal energy lies at low frequencies, corresponding to large DCT coefficient magnitudes, which are located to the upper left corner of the DCT coefficients matrix. Conversely, the lower right values of the DCT coefficients matrix represent higher frequencies, and turn out to be small enough to be truncated or removed with little visible distortion, especially as p and q approach the image width and height, respectively.

3 Procedures of Our Approach

3.1 Theoretical Analysis

Block-Based 2D-DCT. It is known that variations on the facial appearance caused by occlusion, illumination and expression can lead to modifications on the entire representation coefficients in a holistic representation scheme. In this respect the approach of analyzing faces locally is believed to perform superior to the holistic appearance-based approaches, where a local change affects only the corresponding part of the representation and does not modify the representation vector as a whole. Rather than taking the transformation of the image as a whole, the block-based DCT is applied separately to blocks of the image. It first divides a face image into $N \times N$ blocks, performs the $N \times N$ 2D-DCT on each block to get DCT coefficients matrices of all blocks. Then, uses a selection of the 2D-DCT coefficients from each block to construct the feature vector for an image. In dividing a face image into $N \times N$ blocks, if the image dimensions do not divide by the block size, the image is zero-padded to the next multiple that will allow for division. Zero-padding is considered a pre-processing step to the proposed recognition algorithm.

For an $N \times N$ block M, the 2D-DCT coefficients matrix Q of M can be computed as $Q = DMD^T$, where D is the $N \times N$ DCT matrix with $D_{ij} = \begin{cases} \sqrt{\frac{1}{N}}, & i = 0,\ 0 \le j \le N-1 \\ \sqrt{\frac{2}{N}} \cos \frac{(2j+1)i\pi}{2N}, & 1 \le i \le N-1,\ 0 \le j \le N-1 \end{cases}$ and D^T is the transpose of D.

Since D is a real orthonormal matrix, its inverse is the same as its transpose. Therefore, the inverse 2D-DCT of Q is given by $M = D^T QD$. With the help of the inverse 2D-DCT, we can reconstruct an image from its 2D-DCT coefficients of each block. Figure 1 shows the original image, its 2D-DCT coefficients and the reconstructed image. Fig.1(a) is a 92×112 image taken from ORL database, and it is divided into 4×4 blocks, after performing 2D-DCT on each block, we get their 2D-DCT coefficients shown in Fig.1(b). Retaining the DC coefficient of each block while zero setting all AC coefficients, we reconstruct image listed in Fig.1(c). On the contrary, retaining all AC coefficients of each block while zero setting the DC coefficient, we get the reconstructed image presented in Fig. 1(d).

For the best identification rates, the coefficients selected from each block should be the most significant in representing important and distinguishing image details. Using some effective coefficients to represent each block and combining selected coefficients of all blocks from left to right then top to down, all images of the database are represented as feature vectors for further extraction.

Weighted Generalized KFD. Similar to KFD, Weighted KFD also suffer from the SSS problem. To over come this problem, Zhou et al. [9] proposed to discard the null space, spanned by the eigenvectors corresponding to the zero eigenvalues, of between-scatter matrix. However, this may lose some significant discriminant information. To avoid loss of the most discriminant information, we suggest to ignore the null space of the weighted total-scatter matrix $\widehat{S}_t^\phi = \widehat{S}_b^\phi + S_w^\phi$. Then we get the generalized and weighted Fisher criterion:

$$\widetilde{J}(V) = \arg\max_V \frac{V^T \widehat{S}_b^\phi V}{V^T \widehat{S}_t^\phi V}.$$

The following two lemmas ensure the rationality of the above suggestion.

(a) Original image

(b) 2D-DCT coefficients

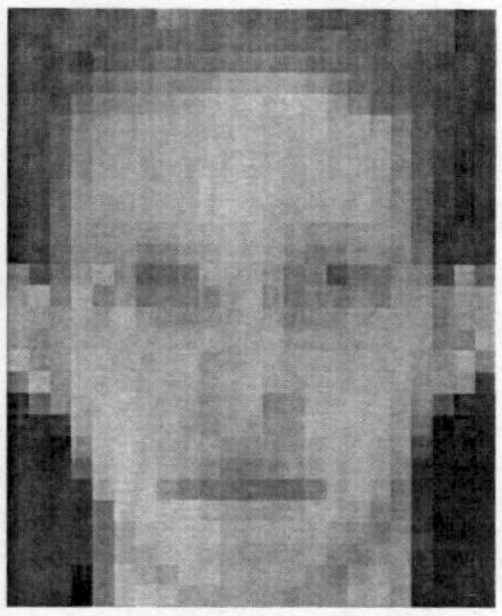

(c) Reconstructed image from DC coefficient

(d) Reconstructed image from AC coefficients

Fig. 1. Original image, DCT and Reconstructed image from DCT coefficients

Lemma 1. *The null space $\mathcal{N}(\widehat{S}_t^{\phi})$ of $\widehat{S}_t^{\phi}$ is equivalent to the space $\mathcal{N}(\widehat{S}_b^{\phi}) \cap \mathcal{N}(S_u^{\phi})$.* *None*

Proof. For any $v \in \mathcal{N}(\widehat{S}_t^{\phi})$, $\widehat{S}_t^{\phi} v = 0$, then $(\widehat{S}_b^{\phi} + S_w^{\phi})v = 0$. Since $\widehat{S}_b^{\phi}$ and S_w^{ϕ} are nonnegative definite, we have $\widehat{S}_b^{\phi} v = 0$ and $S_w^{\phi} v = 0$. That is $v \in \mathcal{N}(\widehat{S}_b^{\phi})$ and $v \in \mathcal{N}(S_w^{\phi})$, therefore $v \in \mathcal{N}(\widehat{S}_b^{\phi}) \cap \mathcal{N}(S_w^{\phi})$. Conversely, for any $v \in \mathcal{N}(\widehat{S}_b^{\phi}) \cap \mathcal{N}(S_w^{\phi})$, we have $\widehat{S}_b^{\phi} v = 0$ and $S_w^{\phi} v = 0$. Hence $(\widehat{S}_b^{\phi} + S_w^{\phi})v = 0$, i.e. $v \in \mathcal{N}(\widehat{S}_t^{\phi})$. □

Lemma 2. *The optimal projection matrix with respect to $\widehat{J}(V)$ are also optimal with respect to $\widetilde{J}(V)$, and vice versa.* *None*

Proof. Suppose that V is optimal with respect to $\widehat{J}(V)$, i.e. V maximizes $\frac{V^T \widehat{S}_b^{\phi} V}{V^T S_w^{\phi} V}$, then minimizes $\frac{V^T S_w^{\phi} V}{V^T \widehat{S}_b^{\phi} V}$, which is equivalent to minimize $\frac{V^T S_w^{\phi} V}{V^T \widehat{S}_b^{\phi} V} + 1 = \frac{V^T S_w^{\phi} V + V^T \widehat{S}_b^{\phi} V}{V^T \widehat{S}_b^{\phi} V} = \frac{V^T \widehat{S}_t^{\phi} V}{V^T \widehat{S}_b^{\phi} V}$. Hence V is optimal with respect to $\widetilde{J}(V)$. Vice versa, if V is optimal with respect to $\widetilde{J}(V)$, then V is optimal with respect to $\widehat{J}(V)$ since $\widehat{S}_t^{\phi} = \widehat{S}_b^{\phi} + S_w^{\phi}$. □

The optimal projection matrix with respect to $\widetilde{J}(V)$ is equivalent to solve the eigenvalue problem:

$$\widehat{S}_b^{\phi} V = \Lambda \widehat{S}_t^{\phi} V, \tag{3}$$

where Λ is a diagonal eigenvalue matrix.

Denote $\Phi = (\phi(x_1^1), \cdots, \phi(x_1^{n_1}), \cdots, \phi(x_c^1), \cdots, \phi(x_c^{n_c}))$, $U_i = \frac{1}{n_i}(0_{1\times(n_1+\cdots+n_{i-1})}, 1_{1\times n_i}, 0_{1\times(n_{i+1}+\cdots+n_c)})^T$, and $U_i^j = (0_{1\times(n_1+\cdots+n_{i-1}+j-1)}, 1, 0_{1\times(n_i-j+n_{i+1}+\cdots+n_c)})^T$, with $0_{p\times q}$ and $1_{p\times q}$ are $p \times q$ matrices consist of 0 and 1, respectively. Then the kernel matrix $K = \Phi^T\Phi$ with its element $K_{ij} = \phi(x_i)^T\phi(x_j) = k(x_i, x_j)$, $m_i^{\phi} = \Phi U_i$, $\phi(x_i^j) = \Phi U_i^j$ and $w(d_{ij}^{\phi}) = (d_{ij}^{\phi})^{-q} = (\sqrt{(m_i^{\phi} - m_j^{\phi})^T(m_i^{\phi} - m_j^{\phi})})^{-q} = [(U_i - U_j)^T K (U_i - U_j)]^{-q/2}$. Thereby,

$$S_w^{\phi} = \frac{1}{n}\Phi W \Phi^T \text{ and } \widehat{S}_b^{\phi} = \frac{1}{n}\Phi U \Phi^T,$$

where $W = \text{diag}(I_{n_1} - \frac{1}{n_1}1_{n_1\times n_1}, \cdots, I_{n_c} - \frac{1}{n_c}1_{n_c\times n_c})$ with I_m be an $m \times m$ identity matrix, and $U = \sum_{i=1}^{c-1}\sum_{j=i+1}^{c} \frac{n_i n_j}{n} w(d_{ij}^{\phi})\ (U_i - U_j)(U_i - U_j)^T$.

According to the reproducing kernel theory, any eigenvector of (3) can be expressed by a linear combination of the observations in feature space, we have:

$$V = \Phi A. \tag{4}$$

Substituting (4), $\widehat{S}_t^{\phi} = \widehat{S}_b^{\phi} + S_w^{\phi}$, $S_w^{\phi} = \frac{1}{n}\Phi W \Phi^T$, $\widehat{S}_b^{\phi} = \frac{1}{n}\Phi U \Phi^T$ and $K = \Phi^T\Phi$ to (3), we have:

$$KUKA = \Lambda K(U + W)KA. \tag{5}$$

The optimal projection matrix A is actually the eigenvector matrix of $(K(U + W)K)^{-1}(KUK)$. Without losing discriminant information, we propose removing the null space of $K(U + W)K$ by Lemma 1. Performing eigenvalue analysis on $K(U + W)K$, we have

$$P^T(K(U + W)K)P = \Lambda_t, \tag{6}$$

$$\Lambda_t^{-1/2}P^T(K(U + W)K)P\Lambda_t^{-1/2} = I_r, \tag{7}$$

where r is the rank of $(K(U+W)K)$, P is the eigenvector matrix corresponding to the nonzero eigenvalues of $K(U + W)K$ and $\Lambda_t = \text{diag}(\lambda_1, \cdots, \lambda_r)$ is the corresponding eigenvalue matrix with diagonal elements in decreasing order. Project KUK to the complementary space of $\mathcal{N}(K(U + W)K)$:

$$\widetilde{S}_b^{\phi} = \Lambda_t^{-1/2}P^T(KUK)P\Lambda_t^{-1/2}. \tag{8}$$

Calculating the eigenvector matrix Q of $\widetilde{S}_b^{\phi}$, we have the optimal projection matrix with respect to (5):

$$A = P\Lambda_t^{-1/2}Q. \tag{9}$$

Thereby, the optimal projection matrix with respect to $\widetilde{J}(V)$ is defined by:

$$V = \Phi A = \Phi P \Lambda_t^{-1/2} Q. \tag{10}$$

3.2 Proposed Algorithm

In summary of the discussion above, our proposed algorithm is given as follow:

Training: given a set of n training samples having c classes with ith class has n_i samples.

Step1. Choose the block size $N \times N$ before performing block-based 2D-DCT.

Step2. Pre-processing: in dividing a face image into $N \times N$ blocks, if the image dimensions do not divide by the block size, the image is zero-padded to the next multiple that will allow for division.

Step3. Perform block-based 2D-DCT to get training vectors: divide each face image into $N \times N$ blocks, perform the $N \times N$ 2D-DCT on each block to get DCT coefficients matrices of all blocks. Then, use a selection of the 2D-DCT coefficients from each block to construct the feature vector for each image. The obtained DCT features are used for further training.

Step4. Calculate the kernel matrix K: $K_{ij} = \phi(x_i)^T \phi(x_j) = k(x_i, x_j)$.

Step5. Calculate $\widehat{S}_b^\phi = KUK$ and $\widehat{S}_t^\phi = K(U+W)K$.

Step6. Remove the null space of $K(U+W)K$: $P^T(K(U+W)K)P = \Lambda_t$ according to (6).

Step7. Project KUK to the complementary space of $\mathcal{N}(K(U+W)K)$: $\widetilde{S}_b^\phi = \Lambda_t^{-1/2} P^T (KUK) P \Lambda_t^{-1/2}$ according to (8).

Step8. Find the eigenvector matrix Q of $\widetilde{S}_b^\phi$.

Step9. Get the optimal projection matrix $A = P\Lambda_t^{-1/2} Q$.

Feature extraction: for any input vector x, its low dimensional feature y representation extracted by our proposed method can be defined as: $y = V^T \phi(x) = (\Phi P\, \Lambda_t^{-1/2} Q)^T\, \phi(x) = Q^T \Lambda_t^{-1/2} P^T (k(x, x_1^1), \cdots, k(x, x_c^{n_c}))^T$.

4 Experimental Results

The proposed method is tested on the ORL face image database which contains a total of 40 people, 10 per person a total of 400. The images are taken at different times and contain various facial expressions and facial details. The size of image is 92×112 pixels with 256 gray levels. In our experiment, we choose Gaussian kernel function, i.e. $k(x,y) = e^{\frac{-||x-y||^2}{2t^2}}$. We then tested the proposed algorithm with different block sizes and weighting functions over different training numbers. For classification, we choose nearest neighbor classifier with Euclidean distance. When $w(d_{ij}^\phi) = \frac{1}{10^6}(d_{ij}^\phi)^{-4}$, the recognition rates over different training numbers are as follows:

Note TNPP means training numbers per person and BS means block size. For each person the first TNPP images are used for training and the rest are used for testing.

Table 1. Recognition rate(%) when $q = 4$

TNPP	1	2	3	4	5	6	7	8	9
BS=6	68.89	84.38	87.14	92.08	95.00	96.25	96.67	96.25	95.00
BS=8	67.50	80.94	84.64	89.58	95.50	95.63	95.83	96.25	95.00

When $w(d_{ij}^{\phi}) = \frac{1}{130}(d_{ij}^{\phi})^{-5}$, the recognition rates over different training numbers are as follows:

Table 2. Recognition rate(%) when $q = 5$

TNPP	1	2	3	4	5	6	7	8	9
BS=6	68.61	82.81	85.71	92.50	96.50	95.63	96.67	95.00	95.00
BS=8	66.94	80.00	83.93	90.42	95.50	95.63	98.33	95.00	95.00

We also compared our proposed method with others, as is shown in Table 3.

Table 3. Recognition rate(%) of different methods when TNPP=7

Methods	Recognition rate	Remarks
Our method	98.33	BS=8, $k(x,y) = e^{\frac{-\|\|x-y\|\|^2}{2000^2}}$, $q = 5$
No DCT	94.17	$k(x,y) = e^{\frac{-\|\|x-y\|\|^2}{2000^2}}$, $q = 5$
No kernel trick	95.83	BS=8, $q = 5$
No weight	96.67	BS=8, $k(x,y) = e^{\frac{-\|\|x-y\|\|^2}{2000^2}}$

5 Conclusions

In this article new face recognition method was examined. It exploits two approaches namely block-based DCT and weighted generalized KFD trying to avoid using PCA because of its data dependent nature, improve Fisher criterion and overcome SSS problem without losing effective discriminant features. Experimental results showed that our proposed method achieves better performance than some other linear and nonlinear dimensionality reduction methods without DCT or weighted scheme.

Acknowledgment. This research was supported by the 2009 third application technology research and development project in Sichuan Province (No.2009JY0077).

References

1. Belhumeur, P.N., Hespanha, J.P., Kriegman, D.J.: Eigenfaces vs. Fisherfaces: recognition using class specific linear projection. IEEE Trans Pattern Analysis and Machine Intelligence 19, 711–720 (1997)
2. Chen, L.F., Liao, H.Y.M., Ko, M.T., Lin, J.C., Yu, G.J.: A new LDA-based face recognition system which can solve the small sample size problem. Pattern Recognition 33, 1713–1726 (2000)
3. Chen, W.S., Huang, J., Zou, J., Fang, B.: Wavelet-face based subspace LDA method to solve small sample size problem in face recognition. International Journal of Wavelets, Multiresolution and Information Processing 7, 199–214 (2009)
4. Huang, R., Liu, Q.S., Lu, H.Q., Ma, S.D.: Solving the small sample size problem of LDA. In: Proceedings of the 16 th International Conference on Pattern Recognition (ICPR 2002), vol. 3, pp. 29–32. IEEE Press (2002)
5. Lu, J., Plataniotis, K., Venetsanopoulos, A.: Regularized discriminant analysis for the small sample size problem in face recognition. Pattern Recognition Lett. 24, 3079–3087 (2003)
6. Yu, H., Yang, J.: A direct LDA algorithm for high-dimensional datawith application to face recognition. Pattern Recogntion 34, 2067–2070 (2001)
7. Yang, J., Jin, Z., Zhang, D., et al.: Essence of kernel Fisher discriminant: KPCA plus LDA. Pattern Recognition Letters 37, 2097–2100 (2004)
8. Lotlikar, R., Kothari, R.: Fractional-step dimension reduction. IEEE Trans. Pattern Anal. Mach. Intell. 22, 623–627 (2000)
9. Zhou, D., Tang, Z.M.: Kernel-based improved discriminant analysis and its application to face recognition. Soft Comput. 14, 103–111 (2010)
10. Hafed, Z.M., Levine, M.D.: Face recognition using the discrete cosine transform. International Journal of Computer Vision 43, 167–188 (2001)
11. Schwerin, B., Paliwal, K.: Local-DCT features for facial recognition. In: Proceedings of the 2nd International Conference on Signal Processing and Communication Systems (ICSPCS 2008). IEEE Press (December 2008)

A Hybrid Active Filter Based on Current Detection for Harmonic Suppression

Liu Yunjing[1] and Wang Fengwen[2]

[1] Department of Automation Engineering, Northeastern University at Qinhuangdao, Qinhuangdao, China
yunjingliu@126.com
[2] Department of Automation Engineering, Northeastern University at Qinhuangdao, Qinhuangdao, China
yjliu66@163.com

Abstract. Power-electronics circuits are widely used in industrial equipment, such as frequency changers, motor-drive systems, etc. Such equipment presents nonlinear impedance to the utility, generating large harmonic currents with well known adverse effects, such as low power factor, low efficiency and destruction of other equipment. Also, some precision instruments and communication equipment will be interfered. Therefore, utility power quality has become an important issue recently. This paper presents a hybrid active filter based on current detection for harmonic suppression. The compensating current flows into the power feeder in order to suppress the harmonic currents generated by nonlinear loads. A hybrid active power filter is developed to demonstrate the performance of the proposed method. The system is modeled in MATLAB SIMULINK to consist of the non-linear loads, with a hybrid active filter to compensate for the harmonic current injected by the nonlinear loads. The simulation results show that the proposed hybrid active power filter has the expected performance.

Keywords: Harmonic, current detection, power systems, hybrid active filters.

1 Introduction

Modern power industries are using more and more semiconductor-based devices, such as adjustable speed drive, furnaces, computer power supplies, uninterruptible power supplies, etc. Non-linear loads such as diode or thyristor rectifiers draw non-sinuscidal currents from utility grids, thus contributing to the degradation of power quality in utility or industrial power systems. Harmonic current in power system is time-varying. Harmonics can not only influence the normal work state of electric equipments, but also interfer communication equipments. To improve the power quality, some methods for harmonic suppression have been used in power systems. Compensation techniques based on active and passive filters to eliminate current harmonics and to compensate reactive power have already been presented and published in the technical literature.

J. Luo (Ed.): Affective Computing and Intelligent Interaction, AISC 137, pp. 253–259.
springerlink.com

Conventionally, passive LC filters have been used to eliminate line current harmonics and to increase the load power factor. However, in practical applications these passive filters present the following disadvantages:

1) The source impedance strongly affects filtering characteristics.

2) As both the harmonic and the fundamental current components flow into the filter, the capacity of the filter must be rated by taking into account both currents.

3) When the harmonic current components increase, the filter can be overloaded.

4) Shunt resonance between the power system and the passive filter causes amplification of harmonic currents on the source side at a specific frequency.

5) The passive filter may fall into series resonance with the power system, so that voltage distortion produces excessive harmonic currents flowing into the passive filter.

Though traditional passive filter (PF) is cheap and simple, the effect of harmonic suppression is greatly influenced by impedance and it's own parameters, and resonance may occur between power impedance and passive filter. The PF can't satisfy the needs of modern industry and power system.

An active filter (APF) basically consists of a power electronic inverter circuit and a control circuit. The common operation of the active filter is as follows. First the harmonic detecting unit in the control circuit samples the instantaneous waveform of the power-line voltage and current and then computes the required compensating current. Second, based on the computed amount of current, the control unit generates gating signals to drive the inverter power circuit. Third, according to the gating signals, the inverter generates the compensating current and injects it into the power line to cancel its harmonic distortion.

Active power filter has good harmonic suppression effect, but is very expensive. So hybrid active power filter which combines low-capacity active power filter with high-capacity passive filter becomes a "must-be"choice in application.

Active power filter (APF), if designed properly, can eliminate harmonic disturbances caused by non-linear loads on power lines in the industrial and commercial power distribution systems. This paper presents a shunt active filter based on current detection for harmonic suppression.

2 Analysis of Active Power Filter

Compared with the passive power filter technology, active power filter technology has the advantages of small impact of power grid impedance, dynamic tracking, and integrated compensation for power quality. It plays a leading role in controlling the pollution of power supply system.

According to the power circuit configurations and connections, the active power filter can be divided into series active filters, shunt active filters and hybrid active filters.

The shunt active power filter is connected in shunt to the load and the generated compensation current opposes to the load harmonic current to being injected into the power feeder. The shunt active power filter has many configurations. Among these configurations, the standard inverter type is widely used and discussed. Fig. 1(a) shows the system configuration of the standard inverter type shunt active power filter. The conventional shunt active power filter can perform the harmonic current suppression, reactive power compensation and balancing three-phase currents.

Fig. 1(b) shows the system configuration of the series active filter. The major advantages of the series active filter over the shunt active power filter are that it can maintain the output voltage waveform to be sinusoidal and balance the three-phase voltages. However, the series filter is less popular in the industrial applications due to the inherent drawbacks of series circuits, namely it must handle high load currents, which increases their current rating compared with the shunt active power filters.

Hybrid active power filter is a new type of filter, it can compensate harmonic which has the varied frequency and size and realize the dynamic harmonic suppression. Compared to the traditional LC filter composed with capacitors, inductors and resistances and shunt active power filter, it has more flexible application and smaller physical size.

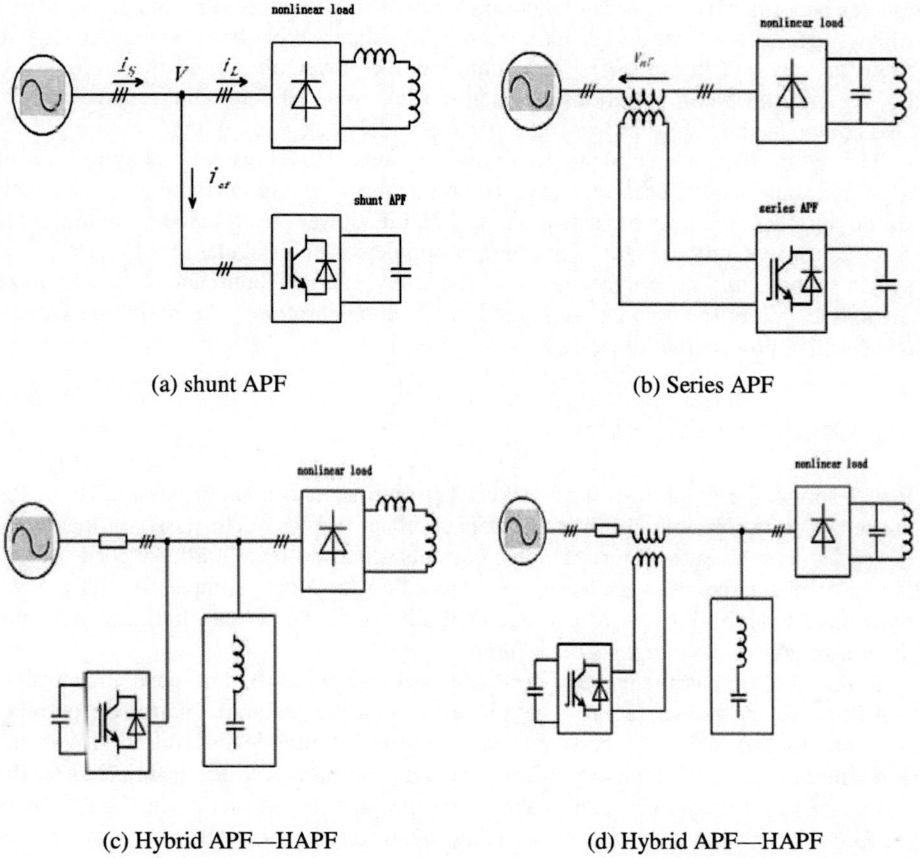

(a) shunt APF

(b) Series APF

(c) Hybrid APF—HAPF

(d) Hybrid APF—HAPF

Fig. 1. System configuration of active power filter, (a) standard inverter type shunt active ower filter, (b) series active power filter, (c)-(f) hybrid power filter

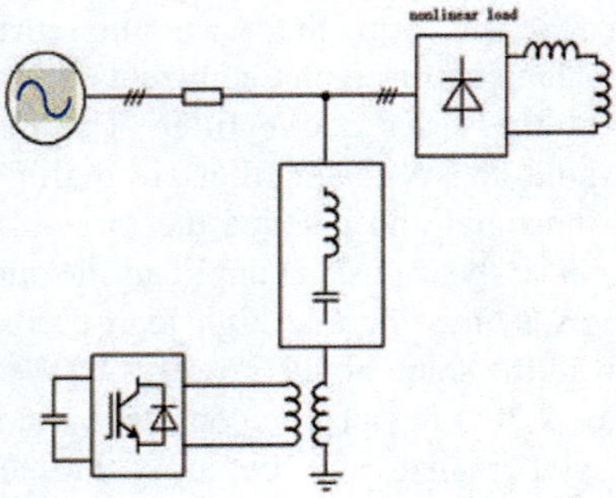

(e)Hybrid APF—HAPF

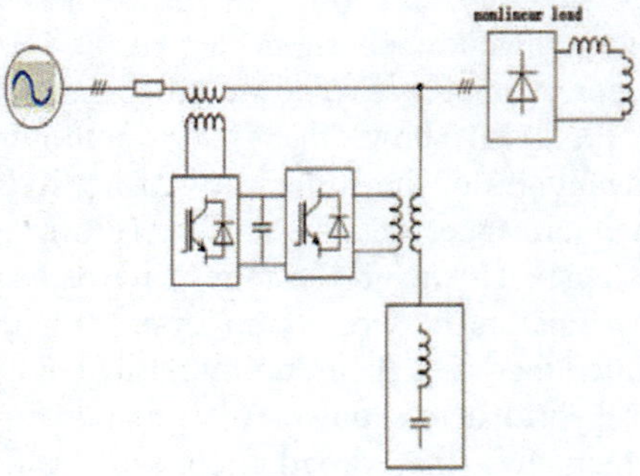

(f) Hybrid APF—HAPF

Fig. 1. *(continued)*

In some applications, the combinations of several types of filters can achieve greater benefits. The major combinations include shunt active filter and series active filter, series active filter and shunt passive filter, shunt active filter and shunt passive filter, and active filter in series with shunt passive power filter as shown in Fig1.(c)-(f). Among these configurations, the active filter in series with shunt passive filter, also known as the hybrid power filter, is more widely discussed in the literature.

The hybrid filter is suitable to high-power applications. However, the hybrid power filter requires a bulk passive power filter set and a voltage-matching transformer. Based on the type of compensation, the active power filter can be divided into reactive power compensation, harmonic compensation, balancing of three-phase systems and multiple compensations. The conventional shunt active power filter belongs to multiple compensations, and it can compensate for the harmonic current and reactive power simultaneously.

3 Operation Principles

Fig. 2 shows the combination of a shunt active filter and a shunt passive filter, the entire circuit is composed of inductor, capacitor and resistor. The passive filters were devised to compensate major harmonic currents produced by a nonlinear load. Active power filters have been an effective way since they can compensate harmonics dynamically. Hybrid active filters retain the advantages of active filters and have not the drawbacks of passive and active filters.

In this system, passive power filters are tuned to suppress part of harmonic currents and improve the power factor. The active power filter generates harmonic currents opposite to the harmonic currents caused by non-linear loads and improves the performance of passive power filters. The compensating current is injected to the power grid through power feeder. The active power filter shown in the fig.2 focuses on suppressing harmonic currents and improving the performance of passive power filters, it is not necessary to compensate reactive power.

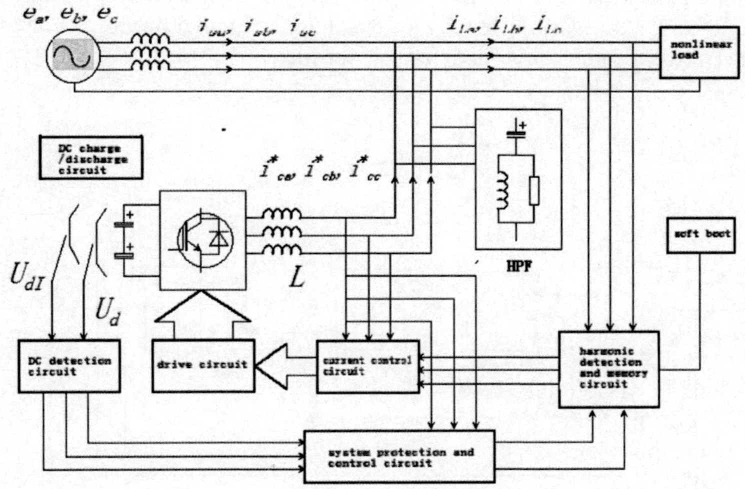

Fig. 2. Configuration of the active power filter

The harmonic current can be determined by using the synchronous reference frame (SRF) transformation as shown in Fig.3, where phase-locked loop (PLL) is required for grid synchronization.

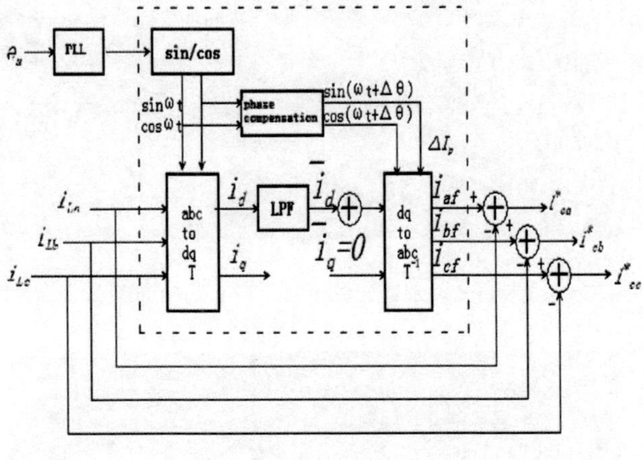

Fig. 3. Harmonic detection based on phase compensation

4 Simulation

To verify the performance of the proposed high-capacity hybrid active power filter, simulation as shown in Fig.4 has been carried out in the simulation environment MATLAB.

The comparison between before-compensation and after-compensation of the current waveforms are shown in Fig.5 and Fig.6. Simulation results show the effectiveness of the proposed hybrid active power filter.

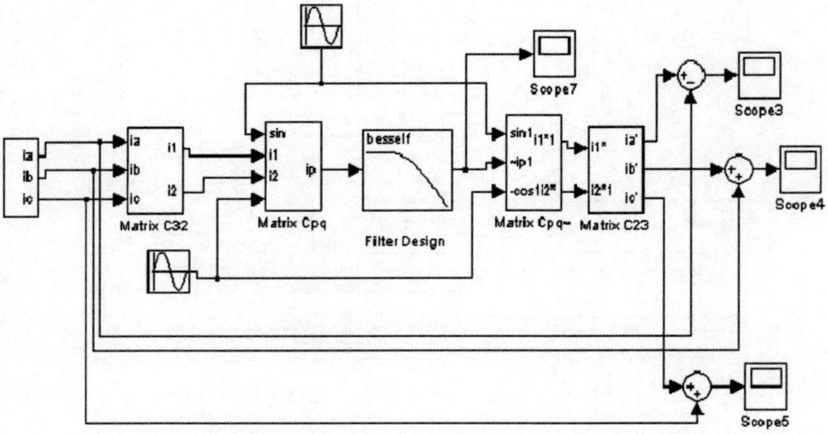

Fig. 4. Simulation model of the hybrid active power filter

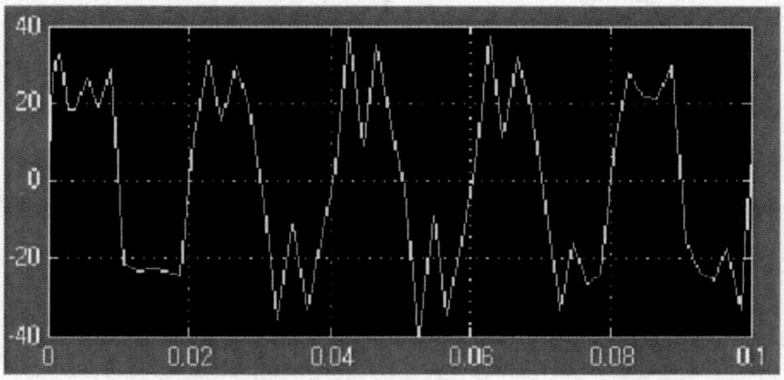

Fig. 5. Current waveform produced by the nonlinear load

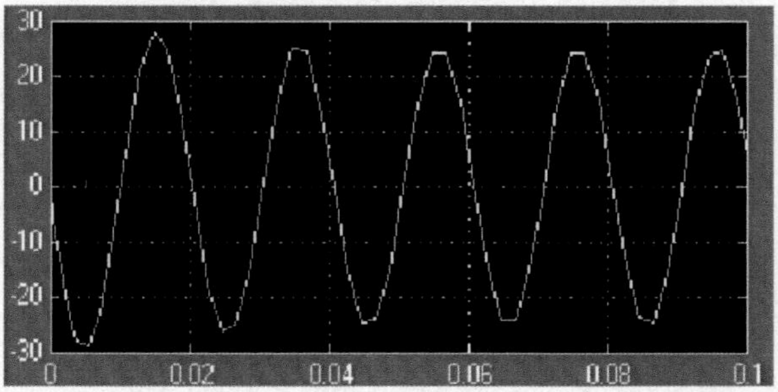

Fig. 6. output current waveform produced by the active portion of the hybrid filter

5 Conclusions

This paper presents harmonic current detection and regulation for the hybrid active filter to enhance the harmonic suppression capability in industrial power system. Simulation results indicate that we can achieve satisfactory consequence by adopting hybrid active power filter to achieve harmonic suppression and improve the power quality. This paper has led to the following conclusions: the detecting method is much better in stability and detection accuracy than the other methods; the hybrid filter can reduce the harmonic current as low as the passive filter wed alone.

References

1. Cheng, P.T., Bhattacharya, S., Divan, D.: Control of square-wave inverters in high power hybrid active filter systems. IEEE Trans. Ind. Applicant. 34, 458–472 (1998)
2. Jintakosonwit, P., Fujita, H., Akagi, H.: Control and Performance of a Fully-Digital-Controlled Shunt Active Filter for Installation on a Power Distribution System. IEEE Trans. Power Electron. 17, 132–140 (2002)
3. Hirofumi, A.: Active Harmonic Filters. Proceedings of the IEEE 93(12), 2128–2141 (2005)
4. Nunez-Zuniga, T.E., Pomilio, J.A.: Shunt Active Power Filter Synthesizing Resistive Loads. IEEE Trans. Power Electron. 17, 273–278 (2002)
5. Zhou, L., Li, Z.: A Novel Active Filter Based on the Least compensation Current control Method. IEEE Trans. Power Electronics 15(4), 655–659 (2000)
6. Li, D., Chen, Q., Jia, Z.: A High-Power Active Filtering System With Fundamental Magnetic Flux Compensation. IEEE Trans. Power Delivery 21, 823–830 (2006)
7. Jintakosonwit, P., Akagi, H., Fujita, H., Ogasawara, S.: Implementation and performance of automatic gain adjustment in a shunt active filter for harmonic damping throughout a power distribution system. IEEE Trans. Power Electron. 17(3), 438–447 (2002)
8. Bhende, C.N., Mishra, S., Jain, S.K.: TS-Fuzzy Controlled Active Power Filter for Load Compensation. IEEE Trans. Power Delivery 21, 1459–1465 (2006)
9. Wu, T.-F., Nien, H.-S., Shen, C.-L.: A single-Phase Inverter System for PV Power Injection and Active Power Filtering With Nonlinear Inductor Consideration. IEEE Trans. Ind. Applicant. 41, 1075–1083 (2005)
10. Tan, P.C., Morrison, R.E., Holmes, D.G.: Voltage form factor control and reactive power compensation in a 25kV electrified railway system using a shunt active filter based on voltage detection. IEEE Trans. Ind. Appl. 39, 575–581 (2003)
11. Wu, J.-C., Jou, H.-L., Feng, Y.-T.: Novel Circuit Topology for Three-Phase Active Power Filter. IEEE Trans. Power Delivery 22, 444–449 (2007)
12. Chang, G.W., Yeh, C.-M., Chen, W.-C.: Meeting IEEE-519 Current Harmonics and Power Factor Constraints With a Three-Phase Three-Wire Active Power Filter Under Distorted Source Voltages. IEEE Trans. Power Delivery 21, 1648–1653 (2006)

Paraphrasing Approach of Chinese Based on Seed Sentence

Lingfen Meng and Zhongjian Wang

School of Computer and Information Engineering
Harbin University of Commerce
Harbin 150028, China
menglingfen1984@163.com, wangzhj@hrbcu.edu.cn

Abstract. How to construct paraphrasing corpus and to provide language resource for the further language information processing research are one of the new research topics. The seed sentence was used in sentence paraphrasing, paraphrasing phrases was found by similarity calculation, and paraphrases were created by combining different seed phrase. In order to improve the paraphrasing correct rate, rich language knowledge in words or phrases between sentences and part of speech or the properties of phrases were used. Experimental results show 70.91% for precision, 63.67% for recall and 67.1% for F. The effectiveness of method was approved by paraphrasing experiment results.

Keywords: paraphrasing, seed sentence, complex sentence, similarity calculation.

1 Introduction

The development and mature of the underlying technology (such as participle, speech marking, syntactic analysis) of natural language processing make a foundation for deeper technology. Currently, the advanced technology (such as paraphrasing, etc.) in natural language processing has got a widespread concern by many researchers, and paraphrasing has been successfully applied to information extraction, search engines, machine translation and other fields.

Paraphrasing is a phenomenon, in which we can express the meaning of language in a different way [1] without changing the intent; paraphrasing can fully reflect the diversity of language. The purpose of the research on approaches of seed sentence is mainly to achieve the following objectives:

The first, to build a more comprehensive large-scale paraphrasing corpus: currently, most of the corpus used in paraphrasing is small-scaled; furthermore, its content is singular. So, through extracting formal sentences from some kinds of relevant corpus, we can build a large-scale corpus which could reflect the fact of the language and contain abundant knowledge of the language, this corpus also provides substantial corpus foundation;

The second, to construct reasonable seed sentence expressions: nowadays, we are short of a tool that can analyze and comprehend sentence in-depth, and there is no

J. Luo (Ed.): Affective Computing and Intelligent Interaction, AISC 137, pp. 261–268.
springerlink.com

study about extracting seed sentence, therefore, building reasonable seed sentence expressions and obtaining seed sentence automatically make the study of paraphrase a major challenge;

The third, to generate a variety of paraphrasing sentences: generation is the final stage of the paraphrasing process; this process will make use of various language knowledge and some kinds of underlying technology of natural language processing, so, how to put multi-lingual language in a comprehensive model effectively is a long-term goal, from this, we can handle the generation of paraphrasing efficiently and produce a large number of expressions that contain a wider range;

The fourth, to give a reasonable assessment of paraphrasing: the evaluation of paraphrasing will promote the development of paraphrasing technology effectively. To deal with paraphrasing evaluation, the key is to raise a reasonable evaluation model which can simulate manual evaluation well.

For the paraphrasing, domestic and foreign researchers conducted some studies; relatively speaking, the study abroad earlier and more mature.

For example, Yves Lepage and Etienne Denoual [2] proposed an approach of generating paraphrasing set automatically from seed sentence in the objective evaluation of machine translation, the set gained can be used as a reference set in MT evaluation;

Ali Ibrahim, Boris Katz and Jimmy Lin [3] proposed an approach of acquiring automatic paraphrase from singular corpus, and this conquers some shortcomings of previous approaches;

Based on feature extraction, Kazuhide Yamamoto [4] and other scholars proposed an approach of Chinese spoken language paraphrasing and language generation technology, this can help to reduce system processing workload.

2 Outline of Proposed Approach

2.1 Approach Description

In order to improve the accuracy of paraphrasing, the paraphrasing approach that based on seed sentence is not only use abundant of knowledge of the language in the sentence between words and phrases, but also the knowledge between part of speech and phrases. In the seed sentence paraphrasing approach, seed sentence is the criteria when we are paraphrasing the training corpus. In paraphrasing process, we can generate paraphrased sentence according to the matching of training corpus and seed sentence. The detailed implementation steps of seed sentence approach are as follows: acquisition of seed sentence; the abstraction of complex seed sentence; seed sentence matching and similarity computation of seed sentence.

2.2 Acquisition of Seed Sentence

According to the structure of sentences, we divided sentences into two categories: single sentence and complex sentence, then, we selected the following two representative grammatical means that can applied to single sentence and complex sentence, and from this, the diversity of sentence can be seen.

1) The mean of exchanging word order:

In the original sentence and paraphrased sentence, the same components do not appear. Acquiring a pair of seed sentences which are different in word order but there are no changes in components.

Example 1: Getting a single seed sentence

[Original] 在黑板上写字. (Write on the blackboard.)

[Paraphrase] 字写在黑板上. (Characters are written on the blackboard.)

2) The mean of changing components:

In the original sentence and paraphrased sentence, the same components can appear. Acquiring a pair of seed sentences in which the components in the original sentence and paraphrased sentence are not all the same.

Example 2: Getting a complex seed sentence

[Original] 如果他没有病，也就不会来医院.

(If he does not get sick, he could not come to the hospital.)

[Paraphrase] 他没有病也不会来医院.

(He is not sick and will not come to the hospital.)

2.3 The Abstraction of Complex Seed Sentence

As the complex sentence involving too much detailed knowledge, containing several words or phrases that are irrelevant to paraphrasing, and the transformation of complex sentence structure is mainly reflected in its conjunction and clause word order. therefore, through carefully research, we proposed a method of abstracting complex seed sentence to elevate coverage rate of complex sentences.

Example 3: Complex seed sentence

[Original] 即使下雨，我们也要干. (Even if it rains, we have to do it.)

[Paraphrase] 雨我们也要干. (Rain, we have to do it.)

Example 4: Abstracted complex seed sentence

[Original] [即使/ c] + <P> + {，/w} +1 [r / 也 / c] + <Q>

[Paraphrase] <P> +1 [r / 也 / c] + <Q>

The description of symbols used in complex sentence abstraction is as follows:

- n[ ]n: This symbol is used to mark the conjunction in the complex sentence and the pronouns that are previous or subsequent the conjunction. In the abstracting process, the morphology of the conjunction is kept down; the morphology of the pronoun is replaced by its part of speech. The figure appearing before represents the pronoun before conjunction, otherwise, the pronoun after conjunction. If the content which is marked by figure is the same in the pair of seed sentences, it can be omitted.

- { }: This symbol is used to mark the comma appears in the complex sentence. In the abstracting process, its morphology will be replaced by part of speech.

- n< >: This symbol is used to mark the other components in complex sentence. In abstracting process, the components that are before the final comma marked P, after the comma, marked Q, the figure represents the number that other components are divided by conjunction or comma.

2.4 Seed Sentence Matching

When we are paraphrasing, the pair of seed sentences we choose should generate an expression that meets our requirement. In this situation, we select the sentence which has a high similarity with original sentence as far as possible. The simplified implementation process is as follows:

- As to the original sentence, we conduct phrase dividing and phrase marking, meanwhile, we record the information contained in the phrase;
- Scanning the sentence from left to right. If there is a conjunction, at least a comma and two predicate components, then, we classify the sentence into complex sentence and implement the step three; otherwise, we classify the sentence into single sentence and implement the step four;
- Acquiring the complex seed sentence which has the same conjunction with the original sentence; then, calculating the similarity of the original sentence and the abstracted complex seed sentence. If they are similar, make use of this seed sentence to paraphrase, else enter into the next seed sentence matching;
- Acquiring the seed sentence with the use of the phrase and properties of phrase; if the phrase and properties of phrase appear in one seed sentence, use this seed sentence to paraphrase the original sentence, however, if the phrase and properties of phrase appear in more than one seed sentence, conduct calculating similarity of single sentence; if the similarity is higher than threshold, use this seed sentence to paraphrase, else enter into the next seed sentence matching until there is no seed sentence meets requirement.

2.5 Similarity Computation of Seed Sentence

1) Similarity Computation of Single Sentence
Given the original sentence $I=w_1w_2...w_n$, retrieving the phrase it can matching in the candidate unit. The similarity of original sentence $I=w_1w_2...w_n$ and seed sentence $A=s_1s_2...s_n$ depends on:

a) The similarity of phrase
If the original sentence $I=w_1w_2...w_n$ matching more than one seed sentences, use this formula:

$$SimPhrase_i = \frac{2\times match\ (w_i, s_i)\times map\ (w_i, s_i)}{len\ (w_i) + len\ (s_i)} \tag{1}$$

In this formula, $match(w_i, s_i)$ represents the number of same words in the phrase; $map(w_i, s_i)$ represents whether the position of the phrase in original sentence and seed sentence is the same or not, the same, its figure is one, else is zero; $len(w_i)$ represents the length of w_i, $len(s_i)$ represents the length of s_i.

b) The number of matching phrases
The number of matching phrases represents the number of matching phrases in the original sentence and seed sentence; if a seed sentence has more matching phrases, this means it has a high probability to be used in paraphrasing.

To summarize, the similarity of phrase and the number of matching phrases are core factors of the similarity of sentence. The similarity of the original sentence and seed sentence can be calculated from the following formula:

$$Sim\,(I, A) = \sum_{i=1}^{n} SimPhrase_{\,i} \tag{2}$$

2) The similarity of complex seed sentence
Reflecting the correlation of the meaning of two sentences, the computation method of the similarity of complex sentence is mainly through the conjunction of original sentence and seed sentence.

The similarity of sentence I and A is calculated from the following formula, here, $Sim(I, A)$ represents this similarity.

$$Sim(I, A) = \frac{samePA(I, A) + \beta \cdot sameSA(I, A)}{LenPA(A) + \beta \cdot sameSA(A)} \tag{3}$$

In this formula, $samePA(I, A)$ represents the number of same primary attributes between original sentence and seed sentence; $sameSA(I, A)$ represents the number of same secondary attributes between original sentence and seed sentence; $LenPA(A)$ represents the number of primary attributes in original sentence; $LenSA(A)$ represents the number of secondary attributes in original sentence; β is a figure between zero and one, representing the weight of secondary attribute.

3 Experiments

3.1 Method of Experiment

The purpose of researching paraphrase is mainly to generate expressions as many as possible. To verify the feasibility and stability of our approach, we used experimental data which is selected randomly from training corpus as input, and conducted the test of seed sentence approach.

3.2 Experiment Data

With the help of Chinese grammar book and internet, we collected 6000 sentences, of which 5000 for training data and about 1000 as seed sentence.

As the input of seed sentence paraphrasing, training corpus is to train the stability and feasibility of seed sentence paraphrasing; furthermore, for evaluating precision and recall rate, we randomly selected 300 sentences as experimental data in training corpus. In the experimental data, the longest sentence is 38 characters; the shortest one is 4 characters, the average length is 13 characters. Because there are no tools to phrase dividing, for part of the phrases, we marked them by hand.

3.3 Results of Experiment

Through seed sentence paraphrasing, the results show that of 300 sentences, there are 191 sentences were paraphrased; from 191 sentences, we obtained 262 results, of which 186 results are correct. The results of the experiment are shown in table 1.

Table 1. Results of experiment

number of experiment sentences	300
paraphrased number/per sentence	1.37
the most paraphrased number/per sentence	4
number of un-paraphrased sentences	109

4 Discusion

4.1 The Analysize of Experiment

For the 262 results, we conducted manual evaluation. Evaluation criteria is that the correct results must meet two conditions [5], the first one is paraphrased result must be a correct Chinese sentence; the other is paraphrased result must has the same meaning with the original sentence. In this experiment, we used precision rate, recall rate and F metric as evaluation parameters of the paraphrasing results. Through making the use of evaluation parameters, we can optimize seed sentence. When retrieving more than one seed sentence matching with the original sentence, sort out the F metrics of evaluation parameters of seed sentences from high to low, we prefer the higher F metric; from this step, we could find out the best seed sentence to paraphrase.

Assuming S as the total number of sentences, S_u as the total number of sentences that can be paraphrased, S_t as the total number of paraphrased sentences, S_p as the total number of correct paraphrased sentences.

In the seed sentence paraphrasing approach, precision rate refers to the percentage that correct paraphrased sentences account for the total number of the paraphrased sentences. Its computation formula is as follows:

$$precision = \frac{S_P}{S_t} \tag{4}$$

In the seed sentence paraphrasing approach, recall rate refers to the percentage that sentences can be paraphrased account for the total number of sentences. Its computation formula is as follows:

$$recall = \frac{S_u}{S} \tag{5}$$

The formula of how to calculate F metric is as follows:

$$F = \frac{2\times precision \times recall}{precision + recall} \tag{6}$$

The result shows that there are 186 correct sentences in 262 paraphrased sentences; these 186 correct sentences are paraphrased sentences of the original sentences. The evaluation results are shown in table two.

Table 2. Evaluation of test results of paraphrasing

number of experiment sentences	300
Precision rate[%]	70.91
Recall rate[%]	63.67
F metric[%]	67.10

The following is an example of a successful paraphrase.

Example 7

[Original] 乌云遮住了**太阳**. (Dark clouds cover the sun.)

[Paraphrase 1] **太阳被**乌云遮住了.

(The sun is covered by clouds.)

[Paraphrase 2] **太阳**让乌云遮住了.

(The dark clouds hid the sun.)

[Paraphrase 3] **太阳叫**乌云遮住了.

(The dark clouds let the sun to hide.)

[Paraphrase 4] 乌云把太阳遮住了.

(Dark clouds hid the sun.)

Example 7 explains the way seed sentence paraphrasing acts. Paraphrase 1 comes from inserting "被 (by)" into the original sentence by the means of changing components; Paraphrase 2 and 3 come from the paraphrase 1 by changing word order; paraphrase 4 comes from the paraphrase 3 by changing components: first delete "被 (by)", then add "把(let)". These four paraphrased sentences are correct ones, and they have the same meaning with original sentence, so, they are paraphrased sentence of the original sentence.

4.2 The Analysis of Errors of Experiment

The result that evaluation gave shows that about 29.09 percent of paraphrase is incorrect. Through careful analysis of the wrong paraphrased sentences, we got the main reasons as follows:

- Participle and part of speech marking errors account for 19.37 percent of total errors;

- Because of so many modifiers in the complex sentence, the information unit after the phrase is too much; it is beyond the seed sentence can describe, this situation accounts for 41.25 percent of total errors;
- Errors caused by wrong recognition of the complex sentence symbol account for 14.46 percent of total errors;
- Errors caused by wrong seed sentence matching are divided into two conditions: errors caused by seed sentence extraction account for 10.14 percent of total errors; errors caused by imprecise of the similarity formula account for 14.78 percent of total errors.

5 Conclusion and Prospect

This paper proposed a paraphrasing approach based on seed sentence. Compared with the traditional paraphrasing approach based on templates, our approach makes full use of part of speech and the properties of phrase, this can better reflect the features of the sentence and make seed sentence matching more accurate. Using small-scale data, we conducted experiment and evaluation; for some sentences with special grammar means, we acquired the paraphrased results. Furthermore, the analysis of errors of experiment pointed out the problems in our approach. Future research includes the improvement of the abstraction of complex seed sentence and the computation method of similarity; expanding abstraction range of seed sentence with different sentence structures; doing experiment with large-scale data and so on.

Acknowledgment. The Project Sponsored by the Scientific Research Foundation of the State Human Resource Ministry for Returned Chinese Scholars and the Research Foundation of Ministry of Education of PRC for the returned overseas Chinese scholar.

References

1. Xu, S.: Research and Implementation of Paraphrase Recognition for Question Answering. Master Harbin Institute of Technology Engineering Papers (2009)
2. Lepage, Y., Denoual, E.: Automatic generation of paraphrases to be used as translation references in objective evaluation measures of machine translation. In: Third International Workshop on Paraphrasing (IWP 2005) (2005K); Elissa, K.: Title of paper if known (unpublished)
3. Ibrahim, A., Katz, B., Lin, J.: Extracting Structural Paraphrases from Aligned Monolingual Corpora. In: Inui, K., Hermjakob, U. (eds.) Second International Workshop on Paraphrasing, pp. 57–64 (2003)
4. Zong, C., Zhang, Y., Yamamoto, K., Sakamoto, M., Shirai, S.: Approach to Spoken Chinese Paraphrasing Based on Feature Extraction. Journal of Chinese Language and Computing 12(1), 63–77
5. Zhang, Y., Yamamoto, K.: Paraphrasing of Chinese Utterances. In: Proc. of the 19th International Conference on Computational Linguistics, pp. 1163–1169. ICCL, Taipei (2002)

A Spectral Graph Approach to Object Recognition

Xiang Liu[1] and Haishan Wu[2]

[1] College of Electrical and Electronic Engineering
Shanghai University of Engineering Science, Shanghai, China
[2] School of Computer Science, Fudan University, Shanghai, China

Abstract. A spectral graph approach to 3D object recognition from its 2D images is presented. Each image of an object is represented by a weighted graph and the feature is extracted by the eigenvectors of the affinity matrix of this graph. The proposed approach is validated on COIL data set. Experimental results show that this method is efficient for object recognition.

1 Introduction

Object recognition is a subject that has been studied in computer vision for decades. Given a query image of an object, a recognition algorithm is to determine which, if any, of a given set of objects appear in a given image database. Thus the object recognition is a problem of matching models from a database with representations of those objects. In this paper, we are concerned with model-based recognition methods in which an object is represented using its 2D images [1].

A variety of object recognition algorithms have been proposed and an extensive survey of object recognition in object recognition can be founded in [2] and [3]. These methods will be discussed briefly in Section 2.

Since the cognitive process of object recognition by human has not yet completely understood, it is still difficult for us to prove theoretically that one particular object recognition algorithm is the best. In practice, several benchmark data sets and their associated performance evaluation are available to empirically measure the quality of existing object recognition algorithms.

In this paper, we propose an object recognition method based on spectral graph, which can be viewed as a combination of feature-based and brightness-based method. Each image of a object can be represented by a weighted graph and each vertex of the graph corresponds to a pixel of the image and the weight value of each edge is determined by the affinity of two incident vertices. Then the eigenvalues of Laplacian matrix of the graph can be used as feature vectors. It is worth noting that with the aid of sub-sampling and interpolation via Nyström approximation [5], the spectral embeddings and eigenvalue-based feature are quite efficient to compute.

The rest of the paper is organized as follows. After briefly discussing previous work on object recognition in Section 2, we describe the details for constructing the weighted graph for an image. In section 4, we test our method in COIL data sets. The experiment results and discussions are also presented in this section. Finally, we conclude in Section 5 and suggest possible future work.

J. Luo (Ed.): Affective Computing and Intelligent Interaction, AISC 137, pp. 269–276.
springerlink.com

2 Previous Work

Many methods have been proposed for object recognition, and broadly speaking, these methods can be classified into two categories: 1)feature based, which involve the use of spacial arrangement of extracted features such as edge elements or junctions, and 2) brightness-based, which make more direct use of pixel brightness.

For feature-based methods, a great deal of research on shape similarity has been done using the boundaries of *silhouette* images. Since silhouettes do not have holes or internal markings, the associated boundaries are conveniently represented by a single-closed curve which can be parameterized by arclength. Many feature-based methods using silhouette have been proposed, a comprehensive comparison of different shape descriptor for comparing silhouettes was done as part of the MPEG-7 standard activity. However, silhouettes are fundamentally limited as shape descriptors for general objects; they ignore internal contours and are difficult to extract from real images. More promising are approaches that treat the shape as a set of points in the 2D images [6][7].

Brightness-based (or appearance-based) methods offer a complementary view to feature-based methods. Instead of focusing on the shape of the occluding contour or other extracted features, these approaches make direct use the gray values within the visible portion of the object. [8] classified the framework using brightness into two categories. In the first category, we have the methods that explicitly find correspondences/alignment using grayscale values[9]. The second category includes those methods that build classifiers without explicitly finding correspondences. In this approach, one relies on a learning algorithms having enough examples to acquire the appropriate invariance[10].

Recently, spectral graph method has been widely used in computer vision area. The idea of spectral graph, used for data clustering [11], dimension reduction, is not new. On successful example of spectral graph in computer vision is normalized cut [4] used for image segmentation. As for supervised learning, it also performs better than other methods. For example, Locality Preserving Projection (LPP)[12], which finds an embedding that preserving local information, achieves better results than PCA and LDA for face recognition. With development of range data acquisition device, spectral graph method is also used for 3D model retrieval [13] and 3D face recognition[14]. The excellent experimental results motivate us to construct spectral graph-base algorithm for 3D object recognition from its 2D images.

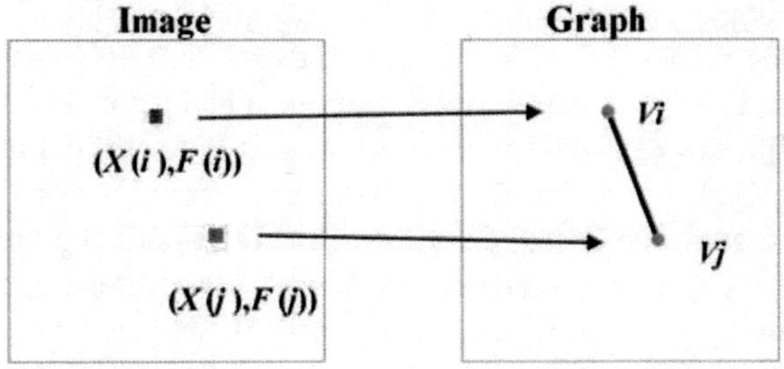

Fig. 1. Construction of weighted grpah from an image.*F* and *X* denote the intensity and distance of two pixels on image. *V* denotes the vertex of the graph.

3 Construction of Spectral Graph

3.1 Affinity Matrix and Weighted-Graph

Given an 2D image with n pixels, construct a weighted graph **G=(V,E)** by taking each pixel as a node connecting each pair of pixels by an edge. Then weight on that edge should reflect the similarity between the two pixels of an object. We then form a $n \times n$ affinity matrix W such that the ij^{th} entry of W is the affinity between the ith and the jth vertices. Figure 1 illustrates this procedure. Like normalized cut, using the brightness value and the spacial location of the pixels, we define the graph edge weight connecting the two nodes i and j as:

$$W_{ij} = e^{\frac{\|F_{(i)} - F_{(j)}\|^2}{\sigma_I^2}} \times e^{\frac{\|X_{(i)} - X_{(j)}\|^2}{\sigma_X^2}} \tag{1}$$

Thus the affinity matrix represents the edge structure of the graph. Clearly, if the graph is undirected, the matrix W is symmetric.

Note that we do not use a sparse, approximate version of W in which each element is connected only to a few of its nearby neighbors in the image plane and other connections are assumed to be zero[4]. While this makes it possible to employ efficient, sparse eigensolvers (e.g., Lanczos), the effects of this process are not well understood. In this paper, we adopt the modified version in [5]. It is based on sampling and allows all affinities to be retained at the expense of some numerical accuracy in their values.

3.2 Spectral Decomposition of Laplacian Matrix

First, we compute the diagonal degree matrix whose entries are column (or row, since W is symmetric) sum of W, that is:

$$D_{ij} = \sum_j W_{ij} \tag{2}$$

From the affinity matrix W and diagonal degree matrix constructed above, we obtain the Laplacian matrix L,i.e. the degree matrix minus the affinity matrix:

$$L = D - W \tag{3}$$

The normalized Laplacian matrix is given by:

$$\hat{L} = D^{-\frac{1}{2}} L D^{-\frac{1}{2}} \tag{4}$$

Then the spectral decomposition of the normalized Laplacian matrix is:

$$\hat{L}\Phi = \Lambda\Phi \tag{5}$$

Since $\hat{L}$ is symmetric and positive semi-definite, it has n nonnegative real eigenvalues. And it is easy to verify that 0 is an eigenvalue of $\hat{L}$. We denote the eigenvalues of $\hat{L}$ by $0 = \lambda_0 \le \lambda_1 \le \cdots \le \lambda_{n-1}$. [15] offers more properties of Laplacian matrix. According to Eq. (3) and (4), (5) can be transformed into:

$$D^{-\frac{1}{2}} W D^{-\frac{1}{2}} \Phi = (I - \Lambda)\Phi \tag{6}$$

Therefore, we can perform the spectral decomposition of normalized affinity matrix *W* instead of $\hat{L}$.

3.3 Nyströ m Approximation

The time complexity for constructing the full affinity matrix *W* for an image with n pixels is $O(n^2 logn)$. Moreover, the eigen-decomposition of an $n \times n$ matrix takes $O(n^3)$ time; $O(kn^2)$ if only the first *k* eigenvectors are computed. To speed things up, we use Nyströ m approximation [5] to efficiently approximate the eigen-values and eigenvectors of *W*.

This method can be described briefly as follows:assume *m* randomly chosen samples come first and the remaining *n-m* samples come next. Now, partion the affinity matrix *W* as :

$$W = \begin{bmatrix} A & B \\ B^T & C \end{bmatrix} \tag{7}$$

with $A \in R^{m \times m}$, $B \in R^{(n-m) \times m}$, and $C \in R^{(n-m) \times (n-m)}$. Here, A represents the subblock of weights among the random samples with diagonalization $A = U \Lambda U^T$, B contains the weights from the random samples to the rest of the pixels, and C contains the weights between all of the remaining pixels. In the case of interest, m<<n, so C is huge. Letting the $\bar{U}$ denote the approximate eigenvectors of *W*, the Nyströ m extension[5] gives:

$$\bar{U} = \begin{bmatrix} U \\ B^T U \Lambda^{-1} \end{bmatrix} \tag{8}$$

and the approximation of W, which we denote $\bar{W}$, then takes the form:

$$\bar{W} = \bar{U} \Lambda \bar{U}^T \tag{9}$$

$$= \begin{bmatrix} U \\ B^T U \Lambda^{-1} \end{bmatrix} \Lambda \begin{bmatrix} U^T & \Lambda^{-1} U^T B \end{bmatrix} \tag{10}$$

$$= \begin{bmatrix} U \Lambda U^T & B \\ B^T & B^T A^{-1} B \end{bmatrix} \tag{11}$$

$$= \begin{bmatrix} A & B \\ B^T & B^T A^{-1} B \end{bmatrix} \tag{12}$$

From equations above, one can see that Nyström extension implicitly approximate C using $B^T A^{-1} B$, thus can approximate the eigenvectors of *W* by only knowing the sampled sub-blok *A* and *B*. The overall complexity is thus reduced from O(n3), without sub- sampling, down to $O(ln \log n+l^3)$, where $l<<n$, in practice.

3.4 Spectral Feature

While the eigenvectors of the affinity matrix form a spectral embedding which is a normalized representation of the object, the eigenvalues specify the variation of the 2D image along the axes given by the corresponding eigenvectors. This leads us to consider eigenvalues as spectral shape descriptors. Recall that for efficiency, we only compute a sampled affinity matrix, required by Nyström approximation. Thus with the same number of samples taken on each image, the eigenvalues of the sampled affinity matrices are comparable. Similar as in [8], we have chosen χ^2- distance to measure dissimilarity so as to remove bias towards any particular eigenvalue. Specifically, given two images *P* and *Q*, let λ_i^P and λ_i^Q be *i-th* the eigenvalues of their respective affinity matrices. We define:

$$Dist(P,Q) = \frac{1}{2}\sum_{i=1}^{N}\frac{[\lambda_i^P - \lambda_i^Q]^2}{\lambda_i^P + \lambda_i^Q} \tag{13}$$

where *N* is the number of eigenvalues. Based on feature distance, we then use the KNN classifier to realize object recognition. For simplicity in computation, 1NN is adopted in our experiment.

4 Experiment Results

We now test our method on COIL (Columbia Object Image Library) database. COIL-100 is a database of color images of 100 objects. The objects were placed on a motorized turntable against a black background. Images of the objects were taken at pose interval 5 degrees. This corresponds to 72 poses per object. In our experiment, we choose 20 objects with an interval of 20 degrees, that is 20×18 images. Figure 2 illustrates some prototypes in our experiments. The recognition results in this data set are illustrated by their Precision-recall (PR) curve, which is widely used to measure the recognition effectiveness.

In this experiment, one image was selected as a query image each time, and the remaining 359 images were used to find which one is the closet to the query one in feature space. First, the sensitivity parameters, σ_I^2 and σ_X^2, were selected with scrutiny, then finally fixed at 40 and 7 , respectively. Based on the proper parameters, we then construct the weighted graph and obtain the affinity matrix. Then by using Nyström approximation, we can get the eigenvalues of affinity matrix. We finally use the eigenvalues of a 50×50 sampled affinity matrix for each image. The reason for choosing 50 samples and thus 50 eigenvalues is experimental. We have found that the recognition result improve as the number of samples/eigenvalues increase, but only to a point, after which no conceivable improvement is obtained by using more eigen-values. Obviously, at this point, more samples/eigenvalues would only induce higher computational cost. At last, we compute the χ^2-distance between the query image and images in database and sort the retrieved images from database in descending order by their distance.

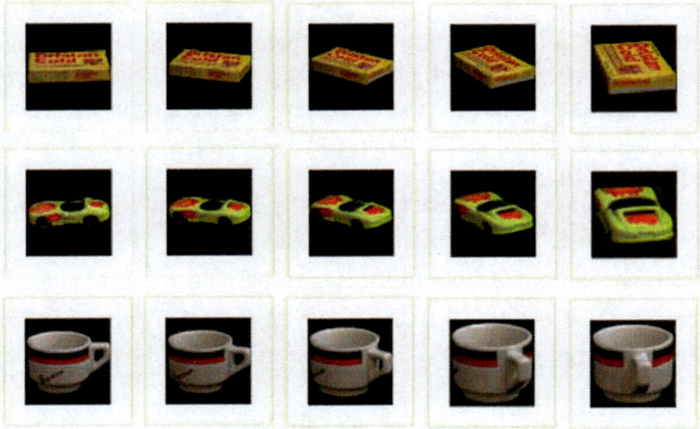

Fig. 2. Prototype views selected for two different 3D objects from the COIL data set

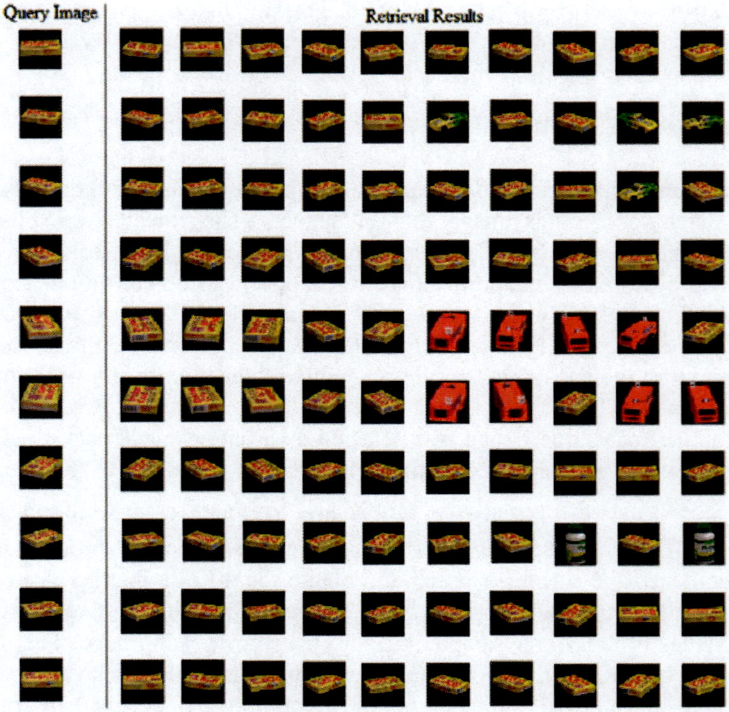

Fig. 3. Some retrieval results from COIL database. First Column is the query image.

The recognition result is shown in Figure 3, from which one can observe that: top 5 retrieval images are in the same category as each query image. Actually, with the parameter selected above, the average correct recognition ratio is 96.6%. The PR curve is illustrated in Figure 4, which also verifies the effectiveness of our method.

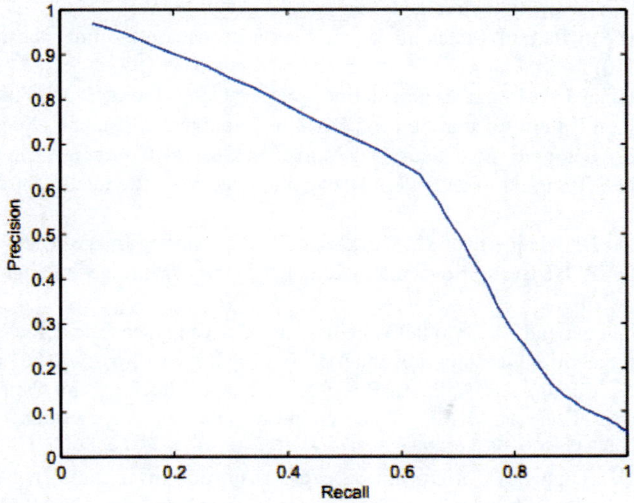

Fig. 4. Precision-recall(PR) plots for our method in COIL database

5 Conclusion and Future Work

In this paper, we consider the problem of 3D object recognition by using its 2D images. By using the brightness and location information, an image can be viewed as a weighted graph by constructing the affinity matrix. Because the spectral of Laplacian matrix processes the intrinsic properties of a graph, we can use the spectral as feature for object recognition.

By using Nyström approximation, the eigenvalues of the affinity can be obtained efficiently. In our experiment, spectral method achieved high recognition accuracy.

In our future work, we would like to explore more ways to define affinities that are insensitive to the parameters selection and/or robust to object recognition in complex scene. Another interesting study is the design of novel spectral features that can be efficiently for 3D object recognition. We have suggested to use the eigenvalues of the affinity matrix as the feature vector, which performs apparently better. However, issues such as, the number of eigenvalues to choose and the distance norm to use, all require further investigation. Finally, we would like to improve the applicability of our method to larger or complex data sets.

References

1. Pope, A.R.: Model-based Object Recognition, A Survey of Recent Research. Technical Report, University of Columbia (1994)
2. Veltkamp, R.C., Hagedoorn, M.: State of Art in Shape Matching. Technical Report UU-CS-1999-27, Universiteit Utrecht (1999)
3. Hagedoorn, M.: Pattern Matching Using Similarity Measures. PhD thesis, Universiteit Utrecht (2000)
4. Shi, J., Malik, J.: Motion Segmentation and Tracking Using Normalized Cut. IEEE Transaction on Pattern Recognition and Machine Intelligent 22(8), 888–905 (2000)
5. Fowlkes, C., Belongie, S., Chung, F., Malik, J.: Specral Grouping Using the Nyström Method. IEEE Transaction on Pattern Recognition and Machine Intelligent 23(3), 861–869 (2004)
6. Huttenlocher, D., Klanderman, G., Rucklidge, W.: Comparing Images Using the Hausdorff Distance. IEEE Transaction on Pattern Recognition and Machine Intelligent 15(9), 850–863 (1993)
7. Sclaroff, S., Pentland, A.: Modal Matching for Correspondence and Recognition. IEEE Transaction on Pattern Recognition and Machine Intelligent 17(6), 545–561 (1995)
8. Belongie, S., Chung, F., Malik, J., Puzicha, J.: Shape Matching and Object Recogniton Using Shape Contexs. IEEE Transaction on Pattern Recognition and Machine Intelligent 24(24), 509–522 (2002)
9. Lades, M., Vorbuggen, C., Buhmann, J., et al.: Distortion Invariant Object Recognition in teh Dynamic Link Architechture. IEEE Transaction on Computers 42(3), 300–311 (1993)
10. Oren, M., Papageorgiou, C.,Sinha, P., Osuna, E., et al.: Pedestrian Detection Using Wavelet Templates. In: Proc. IEEE Conf. Computer Vision and Pattern Recognition, pp. 191–199 (1997)
11. Ng, A.Y., Jordan, M.I., Weiss, Y.: On Specral Clustering: Analysis and an algorithms. In: Proc. Neural Information Processing System (NIPS), pp. 857–864 (2002)
12. He, X.F., Yan, S.C., Niyogi, P.: Face Recognition Using Laplacianface. IEEE Transaction on Pattern Recognition and Machine Intelligent 27(3), 328–340 (2005)
13. Jain, V., Zhang, H.: Shape-based Retrieval of Articulated 3D Models Using Spectral Embbedings. In: Kim, M.-S., Shimada, K. (eds.) GMP 2006. LNCS, vol. 4077, pp. 299–312. Springer, Heidelberg (2006)
14. Elad, A., Kimmel, R.: On Embedding Invariant Signature of Surfaces. IEEE Transaction on Pattern Recognition and Machine Intelligent 25(10), 1285–1295 (2003)
15. Chung, F.R.K.: Spectral Graph Theory. Am. Math. Soc. (1997)

The Study on Image Compression Methods

Zhi Chen[1,2] and Yuehong Qiu[1]

[1] Xi'an Institute of Optics and Precision Mechanics of CAS Xi'an, China
[2] Graduate University of Chinese Academy of Sciences Beijing, China

Abstract. Because of the higher resolution, the data volume of remote sensing image is very large. To communicate and use them on network, efficient compression is very important. In this study, we present the main subjects in the theory of data compression and some image compression methods.

Keywords: image compression, JPEG2000, remote sensing image, CCSDS.

1 Introduction

In the transmission of radio and television field, such as through the satellite remote sensing technology, aircraft, radar, sonar, meetings, computer communication, fax, compression techniques are constrained by real time and on-line considerations that limit the size and complexity of the hardware. There are three main reasons, current multimedia system needed data must be compressed: 1) plenty of multimedia data storage requirements, 2) comparatively slow storage devices, are not allowed to play multimedia data (especially video), (3) the current network bandwidth, don't allow video data transmission of real-time.

2 Image Compression Techniques

It's classified into two groups as lossy and lossless compression.

2.1 Lossy Techniques

Data compression is formed in very high compression ratios. It's separated into two groups as block compression and full-frame compression.

2.1.1 Block Compression

Each part of the image is compressed with the opinion of that the image is separated in parts. The compression and accelerate the process of parallel provides the time [1].

2.1.2 Full-Frame Compression

Here, the compression is thought as an entire process instead of partly compression in the previous one. Although the block compression seems available, full-frame compression is accepted as much more available since high ratio is needed for the radiological images [2].

J. Luo (Ed.): Affective Computing and Intelligent Interaction, AISC 137, pp. 277–283.
springerlink.com

Each two techniques include three steps: transformation, quantization and encoding lastly.

Transformation: first, the original image is converted into the area intended. For example, the Fourier transform and discrete cosine transform [1].

Quantization: It provides the relationship of data matrix formed on the important coefficient and the transformation of the ratio than ordinary rate.

Encoding: The coefficients transferred are coded by using zigzag ordering of each 8x8 images and showed with the number flows including lots of zeros. It's necessary to pay attention that the frequency varies from the smallest ratios to the biggest ratios. Zigzag ordering starts from the top left 8x8 block and finishes in the bottom right 8x8 block [2].

2.2 Lossless Techniques

Lossless image compression techniques typically consider images to be a sequence of pixels in row major order. Each pixel treatment consists of two independent operations. The first step formed a forecast for the next pixel values. A typical linear combination forecast factors include adjacent pixels, in conjunction with edge detection, and tried to make the strength of heuristic.

The second step, the differences between actual pixel values and predict future pixels intensity of entropy coding using probability distribution and selected an encoder.

In high-performance mechanisms such as JPEG-LS, the error differential is coded in a conditioning context established by a possibly-different set of neighboring pixels, and multiple different probability distributions are maintained in parallel as the image is coded, one for each possible value arising in the context evaluation heuristic.

2.2.1 GIF (Graphics Interchange Format)

GIF, developed by CompuServe in 1987, is one of the algorithms that work with LZW. It allows the images, and had eight bytes of depth (28 = 256 colors), most. Although it is incapable of color image compression for this reason, it is used for compressing images, including some color, some simple Numbers. Another reason is that it USES is low, it is in the patent protection [5].

2.2.2 TIFF (Tagged-Image File Format)

It is by Aldus Company. Aldus since joining adobe system, its copyright belongs to the adobe system. It has successfully using compressed RGB bitmap data released LZW algorithm to 5.0 editions in 1988. The JPEG lossy compression to version 6.0 published in 1992. Because of the copyright problem, develop TIFF - LZW ZIP compression TIFF under some entrepreneurs.

2.2.3 PNG (Portable Network Graphics)

Since the GIF and LZW algorithm in its infrastructure by copyright protection, a free compression standard is necessary. In 1996, Thomas Boutell PNG, Scott elliott, mark, Tom and many entrepreneurs in 1995. After the version 1.2 published in 1999, utility edition increases so fast. Results show that it is standard and practical and TIFF GIF 48-byte gray color and tone.

2.2.4 JPEG

The human eye cannot detect from a proper level. Therefore, the color values in the original image can be moved to the frequency base. A coefficient will appear in the transformation of the frequency of use. It's original image. However, it is unnecessary benefited from infinite frequency components. Coefficients can be abandoned adventure partial loss. The digital display frequency abandoned image quality. In application, although very little mass loss, can be in proportion of 1:0 smaller image.

3 JPEG2000 Coding Flow

The basic coding flow scheme of JPEG2000 consists of the following parts (as Fig.1):

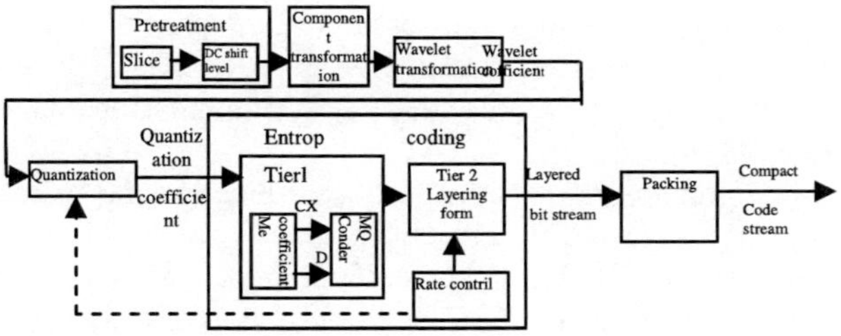

Fig. 1. JPE2000 Coding Flow Diagram

1) Pretreatment: mainly to divide image into rectangle area with same size and not overlapping, that is image patch, do independent coding and decoding process on each image patch. The purpose is to decrease the internal memory resources required in compression process.

2) Component transformation: refer to that the image with many components firstly go through some kind of transformation to decrease the correlation of each component.

3) Discrete wavelet transformation (DWT): do discrete wavelet transformations to each component of every image patch, that is to transform from two dimensional space to resolution space, and the purpose of transform is to increase the redundancy of data so as to increase compression efficiency significantly. The discrete wavelet transformation can be nonreversible; it also can be reversible wavelet transformation. The nonreversible wavelet transformations default to use Daubechies9/7 filter to realize, that is lossy coding. The reversible transform defaults to use LeGal5/3 filter to realize, that is lossless coding. The JPEG2000 uses the hoisting algorithm of wavelet.

4) Quantization: The quantization process can further improve the image compression efficiency. The quantization is that wavelet coefficient is divided by quantization step length and the result is represented by integer. This can reduce the data accuracy, increase compression efficiency and enforce lossy compression.

5) Entropy coding: Embedded Block Coding with Optimized Truncation (EBCOT) algorithm is entropy coding method put forward by David Taubman based on context. EBCOT coder mainly consists of two tiers (Tierl and Tier2): Tier 1 consists of inter-embedded bit plane coding and self adapting arithmetic coding MQ, which mainly

does arithmetic coding based on context, while Tier2 part is composed of layering form, it intercepts applicable bit stream in every independent coding block and code stream to form the final image compact code stream.

When choosing image compression algorithm, JPEG2000 standard is considered at first. This standard is a new still image compression criterion established by ISO and IEC. Compared with JPEG, JPEG2000 Standard shows more advantages as [1]: 1) better - bit rate - the squeezing properties. It can provide better visual effect and the peak value signal-to-noise ratio in 0.25 BPP bitrates. At the same rate, it can provide better quality image, is suitable for low rate of image transmission. (2) Fitting precision pixel image compression. Now it is the only compression arithmetic which supports pixel precision over 16 bits. 3) Supporting larger image size. The maximum size of the image can reach $(2^{32}-1)\times(2^{32}-1)$ and the pixel number per bit can be 2^{14}. This feature especially suits for the compression of satellite and multi-spectrum image. 4) Supporting loss and lossless compression at the same time. 5) Gradual transmission. 6) User-interested region code (ROI). 7) Resistance of error code. 8) Image encrypting. Therefore, JPEG2000 Compression Standard will have wider application field [3].

Because of the complexity of JPEG2000 arithmetic, the present JPEG2000 compression arithmetic hardware system constructed by DSP+FPGA and double DSP only carries out reduced JPEG2000 image compression, that is to say, it uses no-list SPHIT coding as a substitute for EBCOT coding. Based on image processing and constructing an embedded system is the core of compress storage power components, ADV212 FPGA, although the FPGA responsible for preparing satellite remote sensing image data compression technology and production control signals, and the power PC is used to load allocation and management ADV212 image files[4].

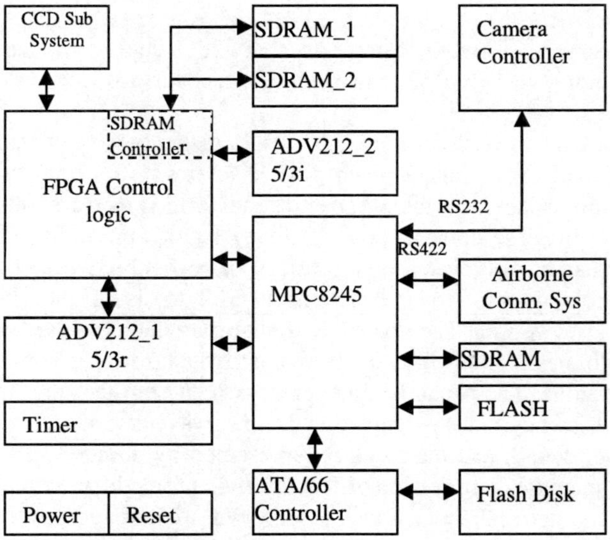

Fig. 2. Compression system block diagram

4 CCSDS Algorithm

The compressor of CCSDS algorithm consists of two functional parts, depicted in Fig.3, a Discrete Wavelet Transform module that performs decorrelation on input images, and a Bit-Plane Encoder which encodes the decorrelated data.

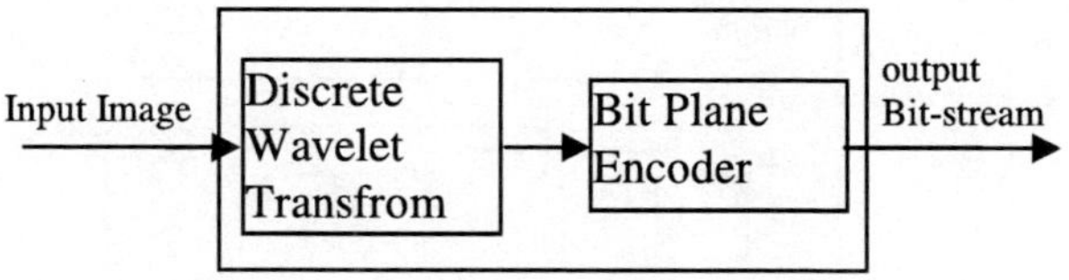

Fig. 3. General schematic of CCSDS

4.1 Discrete Wavelets Transform

The CCSDS standard for the decorrelation module makes use of a three-level, two dimensional (2-d), separable DWT with 9/7 taps for low- and high-pass filters, respectively. Two specific wavelets are specified with Float DWT and Integer DWT. The Float DWT generally exhibits superior compression efficiency in the lossy domain, whereas only the Integer DWT supports strictly lossless compression. Float DWT needs floating operation which is complex for hardware implementation. So 9/7 Integer DWT is adopted in this paper. The Integer DWT is defined by equation (1) and (2). The implementation of Integer DWT is very simple, just needs addition and shift operations.

$$D_j = x_{2j+1} - \left[\frac{9}{16}(x_{2j} + x_{2j+2}) - \frac{1}{16}(x_{2j-2} + x_{2j+4}) + \frac{1}{2} \right] \tag{1}$$

$$C_j = x_{2j} - \left[-\frac{D_{j-4} + D_j}{4} + \frac{1}{2} \right] \tag{2}$$

4.2 Bit Plane Encoding

After DWT, the Bit Plane Encoder (BPE) processes wavelet coefficients in groups of 64 coefficients referred to as blocks. An example of a block is illustrated in Fig.4 as comprised of shaded pixels. Information pertaining to a block of coefficients is jointly encoded by the BPE. The adjacent pixels have high correlations, so do the adjacent blocks in transformed domain. So in CCSDS standard a segment is defined as a group of consecutive blocks. The DWT coefficients coding proceeds segment by segment and each segment is coded independently with the others. The less the value of S is, the less resources is consumed, and then, the deblocking is more obvious [5].

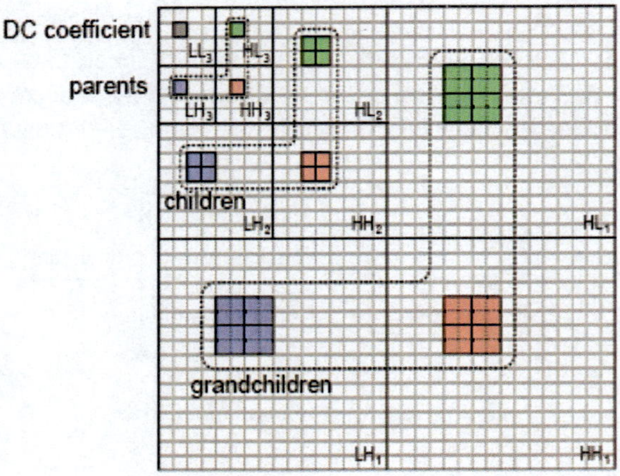

Fig. 4. The relationship between coefficients

4.3 The Hardware Implementation

The proposed top-level architecture ideal for the compressor core is illustrated in Fig.5. It consists of two main modules, i.e. DWT and BPE. The DWT Module is used to perform the three-level 2D-DWT on the input image, and the BPE scans and encodes the resulted wavelet coefficients. The ByteBulilder module then rearranges the bit-stream as a stream composed of a proper number of bytes. The pipeline technology can be used to achieve high throughput capacity. However more levels pipeline need more hardware costs especially memory size. So the scheme of two level pipeline between module DWT and module BPE is adopted in this paper. Thus, the throughput capacity of this compression core is determined by the module BPE which consumes more time in the experiment simulation. To improve the throughput capacity of the core, the optimization of the BPE architecture is needed[6].

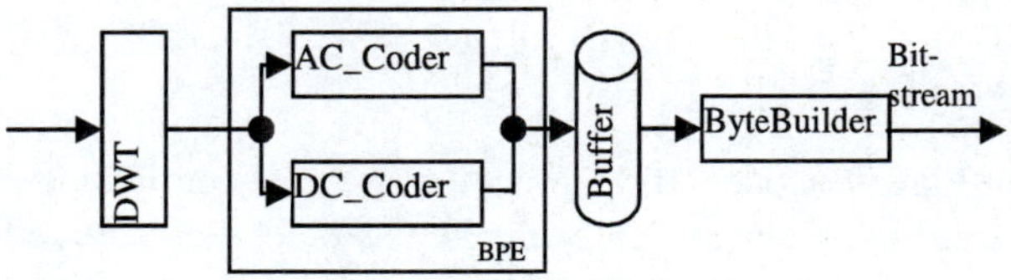

Fig. 5. The proposed processor

5 Conclusion

This paper presents an overview of image coding. In this paper we have discussed a CCSDS algorithm-based image compression and JPEG2000. We motivate the use of transform coding in practical settings, and describe the properties of various wavelet transforms. We motivate the use of the wavelet transform in coding using rate-distortion considerations as well as approximation-theoretic considerations.

References

1. Chaudhuri, D., Samal, A.: An automatic bridge detection technique for multispectral images. IEEE Trans. Geosci. Remote Sens. 46(9), 2720–2727 (2008)
2. Sirmacek, B., Unsalan, C.: Urban-area and building detection using SIFT key points and graph theory. IEEE Trans. Geosci. Remote Sens. 47(4), 1156–1167 (2009)
3. Hou, P., Hojjatoleslami, A., Petrou, M., Underwood, C., Kittler, J.: Improved JPEG coding for remote sensing. In: Proc. 14th Int. Conf. Pattern Recog., vol. 2, pp. 1637–1639 (August 1998)
4. Pelon, H., Spiwack, R.: Advanced image compressor for SPOT-5 satellite. In: Proc. Round Table Data Compression Techn. Space Appl., pp. 23–27 (October 1996)
5. Jiang, H., Zhou, X., Li, B.: Texture-resembling-based image real-time compression. In: Proc. Int. Conf. Commun. Technol., pp. 1–4 (November 2006)
6. Abrardo, A., Barni, M., Magli, E., Nencini, F.: Error-resilient and low-complexity onboard lossless compression of hyperspectral images by means of distributed source coding. IEEE Trans. Geosci. Remote Sens. 48(4), 1892–1904 (2010)

The Building of University Science Education Platform Based on Web

Yu-You Dong[1] and Shuang Chen[2]

[1] School of Information Technology
Hebei University of Economics & Business, Shijiazhuang, China
[2] School of Public Administration
Hebei University of Economics & Business, Shijiazhuang, China

Abstract. From the investigation, the demand in the information age and the power of the new curriculum reform, analyzes the deficiency of scientific inquiry, then introduces the concept and characteristics of cognitive tool. Finally, puts forward a design of web quest learning platform for university education, scientific research, and can provide reference for similar.

Keywords: university science education platform, scientific inquiry, web.

1 Introduction

It points out that developing students' scientific literacy, without scientific learning process, in the University of Science Curriculum Standard. Explore the science and education for the core, is one of the important targets in order to promote the development of students. Therefore science research should be highlighted in science classes. This new concept of university teachers is supporting the research-type learning and obtaining scientific methods and techniques through the process of scientific inquiry. However, many studies have shown that the traditional teaching method has been unable to effectively support the development of scientific inquiry [1]. Therefore this paper presents web quest learning platform for scientific education institute of design based on the cognitive tool to promote the development of universities and universities of science education of science inquiry.

2 Literature

Scientific research (SI) is the core of scientific study. Application of scientific research and teaching practice must rely on a good learning environment. National science curriculum standard "(3-6 with clear: to achieve the goals of science curriculum" scientific literacy for all children. "However, the researchers carried out between 2000 and 2001 team shows the overall level of students' scientific literacy is not satisfactory qualification for less than 50% of the traditional scientific teaching. Reason is too seriously, but ignore the knowledge teaching, how to cultivate students' scientific methods and spirit [1]. Standard solutions are put forward: it with scientific research as the core of the scientific study. In education, scientific orientation

J. Luo (Ed.): Affective Computing and Intelligent Interaction, AISC 137, pp. 285–291.
springerlink.com

scientific accomplishment of science teachers must adopt a new method of teaching, scientific research, rather than simply guide the textbook knowledge education target, including the diversity of the scientific method and skill training, training and development ability of science and scientific attitude and value of another textbook knowledge.

However, the current scientific education, primary always lack of good learning environment for implementing inquiry-based learning. This textbook is the guidance of teachers and students to participate. Although some experiments in the classroom, and most of them are in accordance with the experiment, the description of the process of textbook [2].

The foregoing scientific inquiry environment usually has the following characteristics:

(1) Supply with tools for acquiring scientific data.

(2) Presuppose the themes of scientific inquiry: part of inquiry themes are designed beforehand so that teachers can use them directly.

(3) Support various steps and management of scientific inquiry projects via net technologies, such as put forwarding questions, presupposing, enacting research plans, collecting evidences, communicating and cooperating.

(4) Coordinating many learning tools, such as cooperating tools, inquiry maps, sharing works, etc.

3 Scientific Inquiry

3.1 What Is Scientific Inquiry?

It can be described as follows: "science" refers to various ways and methods, the use of natural scientists and interpretation of the facts. It also refers to the activities of students, the construction of scientific knowledge, and understanding of the concept of scientific research method. In the national science education standards[2], Institute of scientific inquiry learning, refers to the students build scientific knowledge, scientific concept, and formed about scientific research methods in various activities, including the process of learning, and puts forward a problem, forecast and assumptions, planning and designing experiment, experiment of facts and evidence, testing and evaluation and expression and sharing.

3.2 Deficiencies of University Scienctific Inquiry Learning

Inquiry learning is becoming more and more important in the university science class, but there exist some defects from classroom observation, interview, including the following four aspects:

•Unable to track the students' thinking process

Teachers usually take some method to arouse the students' thinking in classroom teaching, but it is difficult to detect the learning process problems should be paid attention to, and make the appropriate support and guidance.

• Lack of proper scientific methods and tools

Due to the lack of appropriate scientific methods and tools, students don't know where to start in some inquiry activity. They can make a definite plan and observation, or to make myself clear. All these show that students are lack of planning, observation, reasoning, communication and assess ability; this is necessary for cultivating ability, and the accomplishment of science research.

• Results-oriented rather than process-oriented in evaluation

At present, most of the evaluation of science and education is oriented. Although some teachers try to use the combination, they cannot increase students' participation. This is a single form of evaluation, individuation, so it is difficult to evaluate students' comprehensive utilization. It is very difficult, stimulates the student to study the interest of research-oriented learning.

• No sufficient time for students in classroom inquiry learning

Most of the students in the team can test in the limited class time. They need extra time. Extracurricular research-oriented learning requires teachers to guide and support, this is far from enough.

In a word, the student investigative study without inspiration, also there are some deficiencies of teaching method and evaluation, the inquisition ability and developing scientific literacy.

How to provide strong support and the study-based learning? How to evaluate the students more comprehensively? Cognitive tool is a perfect answer with its popularization and broad application prospect.

4 Cognitive Tools

According to the Jonassen (1996), the famous experts in the teaching design, cognitive tools are generalizable computer tool, is to participate in and improve cognitive processing. Help learner's cognitive tool and complex cognitive learning activities and critical thinking.

There are many kinds of cognitive tool. Jonassen (1996) put forward eight kinds of cognitive tools, tools, dynamic semantics organization modeling tools, social common cognitive tools [3]. Cognitive tools adopted in this system are: semantic network tools, spreadsheets, computer communications tools, and hypermedia tools, as shown in table 1:

Table 1. Cognitive tools used and their features

Type	Features
Semantic network tools	Visualization of the internal schema cognitive structure.(mind map, K-W-L-H chart)
Spreadsheets	Three primary functions: storing, calculating, and presenting information. (Excel)
Computer communication tools	Asynchronous communication and synchronous communication to Support students to construct their knowledge.(chat room)
Hypermedia tools	Manage information in a nonlinear way(Powerpoint, PPT as abbreviation)

Cognitive tool can support research-oriented learning, and their own advantages, not only in helping students to collect information, the experimental observation, record the data, analyzing and express the result, but also provide individuation in the evaluation. As a result, it is the important significance of inquiry-based learning platform based on cognitive tool design, to apply it to practice.

5 Design of a Web-Based Inquiry Learning Platform for University Science Education

Its purpose is to support the whole process of scientific inquiry, tracking and record the process of scientific inquiry of comprehensive evaluation, and provide sufficient guidance.

5.1 To Support the Whole Process of Scientific Inquiry

5.1.1 Design of the Inquiry Learning Mode Based on Cognitive Tools

There are various ways to scientific research. Generally speaking, the most basic process has six factors: put forward scientific problems, forecast and assumptions, planning and designing experiment, experiment of facts and evidence, testing and evaluation and expression and sharing.

Some contain only a few of these elements investigation, and don't take orders above. The main purpose of scientific research is to help students understand and Master of Science inquiry skills need to understand scientific research method and the process of scientific inquiry. According to the complete process of scientific inquiry, the inquiry learning mode based on Cognitive tools is as shown in Figure 1.

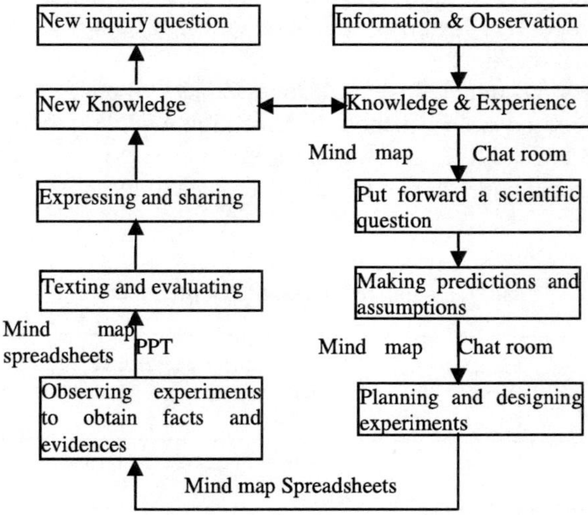

Fig. 1. Inquiry learning mode based on Cognitive tools

5.1.2 According to the Mode above, the Platform will Provide Various Cognitive Tools to Support the Whole Inquiry Process. Making Predictions and Assumptions

a) Putting forward scientific questions

Generally speaking, the scientific problems for students are through the observation and research, reading, etc. However, science can by teachers or teachers and students from the discussion. In this platform, the teachers and students, through the discussion questions think map and chat rooms. Teachers guide students through think map, discussed the concept and knowledge, and that of the students, the decision problem for inquiry.

b) Making predictions and assumptions

Students can use the mind map and chat rooms do individual think map can describe them on this question, or do knowledge and understanding of the group think map assume this scientific problems through discussion. To know the problem, students can use K - W - L - H, including four aspects: we know (" know "), we will find (" hope "), we have got the information (" "), how can we learn more (" how to ").Students can draw this chart on the paper by themselves or using the mind map software, such as Mind manager, as shown in Figure 2.

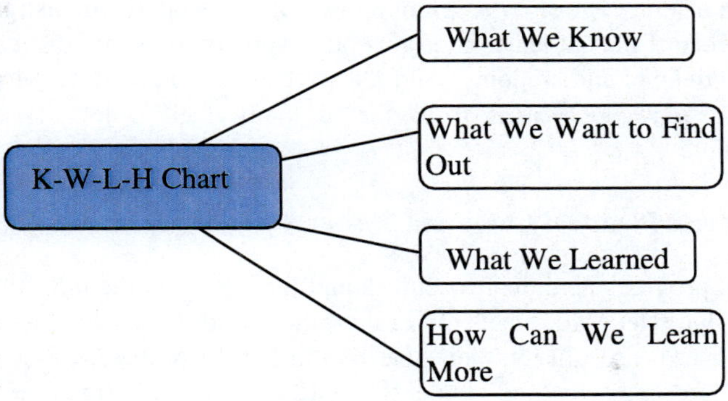

Fig. 2. K-W-L-H Chart

Students can be summed up their knowledge, they are K W - - - L diagrams and investigation method, explore clear goal. Let students themselves or group of prediction, the answers to these questions by discussions between the teachers and students to discuss and students or write them down in different ways.

c) Planning and designing experiment.

Students of the planning and design experiments are through think map, spreadsheet and chat rooms. This process may need to teachers' participation and guidance. The observation and experiment, before the students should know the three problems that clear. What is the experiment? What is the sequence of the experiment? What is the detail, every program? Clear record experiment program flowchart and mental map and determine the plan inquiry experiment.

d) Observing experiments to obtain facts and evidences
Students use the related experimental equipment and materials research, retrieval, observation and measurement. In addition, ensure the safety of operation, the student's main task is to record the observation and measurement results. The mind map, spreadsheets forms can be used to provide teachers or inquiry here to record the process and result.

e) Testing and evaluating
This course aims to give explanation, and through the analysis results of the observation, measuring and test results, and assumptions and assumptions. At the same time, collect more evidence to support the explanation and check if there is some problems, process and method. If necessary, give some improvement measures. Mind mapping and spreadsheets are the main tools.

f) Expressing and sharing
Its main purpose is to write the inquiry report and share the results through Mind Map, PPT, chat room, etc.

5.2 To Track and Record the Process of Scientific Inquiry for Comprehensive Evaluation

This platform offers a variety of cognitive tool support students investigation process, tracking and record the student's mind mapping, spreadsheets and discussion between teachers and students and students. And the students can upload scanning think map, experiment, it may be written form for they provided a lot of comprehensive evaluation.

5.3 To Provide Plenty of Cases and Resources

This platform provides abundant resources and the case. On the other hand, they can acquire more knowledge of science. On the other hand, they can learn the scientific method from these cases, use in their investigation of the process. Cases and resources are provided for teachers and students. Teachers play an important role in the selection, organization and delivery of the learning resources. Therefore, it is necessary for all kinds of resources pre-alpha - they teachers.

5.4 To Provide Sufficient Guidance

These platform extended students learning science courses. Teachers can give all the students enough support and guidance of the platform outside.

6 Conclusion

The application of information technology education in university is an important research field, education technology development. This paper introduces web quest learning platform design based on cognitive science education tools, in order to promote the development of science and education of university science inquiry. And this may be a good attempt in reality of teaching practice. It will be a development for a very long time, so we have to make further efforts.

References

1. Fu, Z.-W., Wang, X.-W., Ke, Q.-C.: Design of Webbased Science Inquiry Learning System for K12 Science Education. Modern Educational Technology, 56–59 (September 2008)
2. Center for Science, Mathematics, and Engineering Education (CSMEE), The National Science Education Standards, http://www.nap.edu/openbook.php?record_id=4962&page=11 (access time: May 2009)
3. Zhu, Z.-T.: Modern education technology, p. 106. Higher Education Press, Beijing (2001)
4. Chinese Teachers Journal. American Scientists Appeal to strengthen the Spatial Thinking Education, http://www.Chinateacher.com.cn/news/2006/0419/880.asp
5. Slotta, J.: Designing the 'Web-Based Inquiry Science Environment (WISE). Educational Technology 42(5), 15–20 (2002)
6. Jin, Y.: Inquiry Learning, vol. (11). Sichuan Education Press, Sichuan (2005)
7. Fan, W., Feng, K.: The Comparative Study on Different Courses in Inquiry Learning. IT Education in Primary and Secondary Schools (12), 33–35 (2004)
8. Martin-Hansen, L.: Defining Inquiry: Exploring the Many Types of Inquiry in the Science Classroom. The Science Teacher 69(2), 34–37 (2002)
9. Zhao, J.: Research on the Evaluation Tools of Pupils' Science Literacy. Suzhou University (2003)
10. http://www.worldwatcher.northwestern.edu/
11. http://www.wise.berkeley.edu/

A Novel Image Semantic Annotation Method Based on Image-Concept Distribution Model

Ma Ying, Zhang Laomo, and Gao Jixun

School of Software, Henan Institute of Engineering, Henan, China
myfirst@163.com, cyechina@163.com, cyechina@126.com

Abstract. The optimal solution to support a great annotator is very difficult to derive. Up to the present, little work can precisely tag the images since the so-called semantic gap is not easy to reduce. This paper constitutes the approach of approximating the solution to discover the visual-concept associations from the image-concept distribution. The experimental results show that our proposed annotation approach is very effective in facing the diverse relations between visual features and human concepts.

Keywords: image semantic annotation, entropy, image-concept distribution model.

1 Introduction

Advances on digital capturing technologies lead to the explosive growth of image data. For a user retrieving the desired images from a huge amount of image data, textual query can represent her/his interest more precisely than visual query.

Fig. 1. Example of diverse visual-concept associations

To aim at this problem, in this paper, we propose an innovative annotation approach, namely Annotation by Image-Concept Distribution Model (AICDM), to discover the visual-concept associations from image-concept distribution. Based on Figure 1, Figure 2 depicts an example for the basic idea of our proposed AICDM. In Figure 2, we further group the visual-relevant images as the descriptive genes of

J. Luo (Ed.): Affective Computing and Intelligent Interaction, AISC 137, pp. 293–301.
springerlink.com

images using visual features. Our intent behind this idea is to identify the conceptual distinctness of genes. According to the discriminative genes, the image-concept distribution model is constructed.

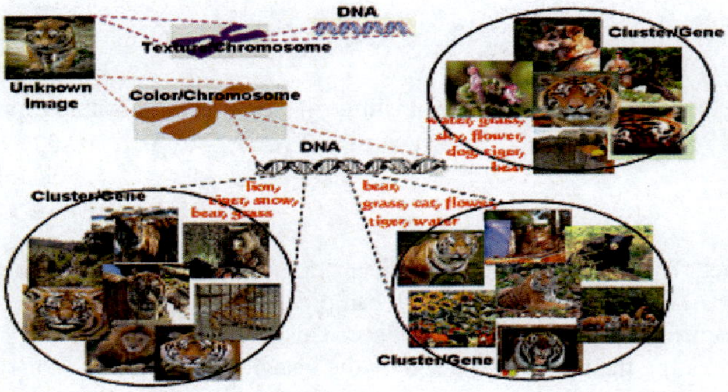

Fig. 2. The basic idea of the proposed AICDM

2 The Proposed Approach: AICDM

As shown in Figure 3, the whole procedure can be decomposed into two phases, namely offline learning and online prediction. For offline learning, the annotated images are grouped into a set of visual clusters feature by feature.

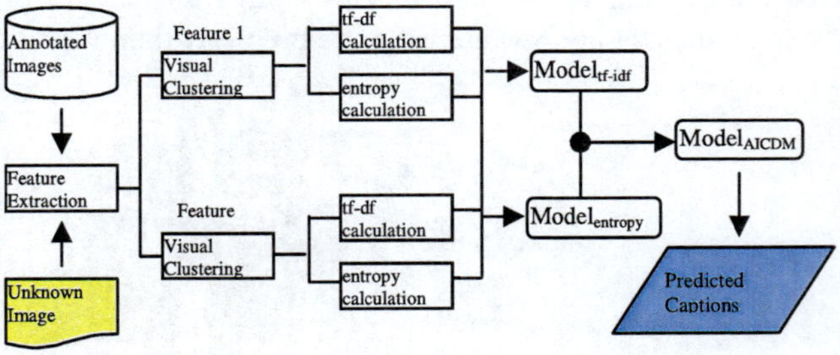

Fig. 3. The framework of the proposed AICDM method

2.1 Offline Learning

Before constructing the proposed models, the annotated images are grouped by calculating visual distances. Then we can obtain a set of clusters for each visual feature. A cluster contains a set of images and an image is annotated by a set of keywords. For example, assume that the images are grouped into 5 clusters $\{C_1, C_2, C_3, C_4, C_5\}$ by Scalable Color, and each image is projected as a set of keywords, as shown in Figure 4. Note that, the best cluster cardinality might be different for different visual features.

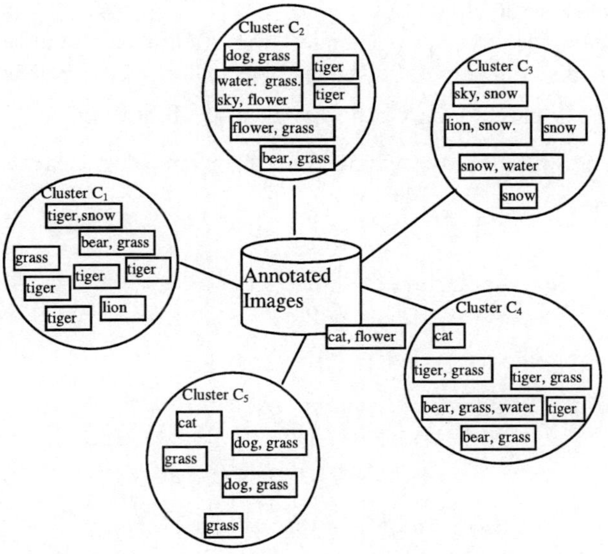

Fig. 4. Example of visual clusters containing the related concepts

2.1.1 Construction of Model$_{entropy}$

From the clusters, we can obtain an aspect that, the images in a cluster are very similar on the visual features but containing a number of concepts. This is the problem called diverse associations we mention in above. If the cluster is regarded as a kind of visual genes, our intent is to calculate the caption frequency and the gene entropy. The major idea behind this model is that, the higher the frequency of a caption, the more representative the caption. In contrast, if a large number of captions occur in a cluster, the related entropy would be too high to disambiguate visual concepts. That is, entropy can be viewed as a local weight for a gene. According to this notion, two related measures, called Term_Frequency (tf) and Entropy, are defined as follows.

Definition 1. Consider that a training data set IDB is divided into t clusters, $IDB = \{C_1.C_2,\dots,C_t\}$, and there are y unique captions $\{cp_1, cp_2, \dots, cp_{y_t}\}$. Assume that a cluster contains several images and each image is assigned several captions. Hence a cluster can be viewed as a collection of captions, $C_J = \cup cp_i$. The entropy of the jth cluster/gene can be defined as:

$$Entropy^j = \sum_{cp_i \in C_j} \left[\frac{tf^j_{cp_i}}{\sum_{cp_v \in C_j} tf^j_{cp_v}} \log\left(\frac{\sum_{cp_v \in C_j} tf^j_{cp_v}}{tf^j_{cp_i}} \right) \right], \tag{1}$$

Where $tf^j_{cp_i}$ stands for the frequency of caption cp_i in the jth cluster. For example, for C_5 in Figure 4, the frequency set of {grass, dog, cat} is {4, 2, 1} and Entropy^5 is 4/7*log(7/4)+ 2/7*log(7/2)+ 1/7*log(7/1)= 0.415.

2.1.2 Construction of Model$_{\text{tf-idf}}$

Similar to above model, tf is also generated first. Additionally, another issue of how to determine the discrimination of a gene is inverse document frequency, namely idf. The idf of the jth cluster/gene can be defined as the following.

Definition 2. Following the above definitions, the inverse document frequency for the jth cluster/gene is

$$Idf^{\,j} = \log\left(\frac{y}{\sum_{1\le g\le y} df^{\,j}_{cp_g}}\right) \tag{2}$$

where

$$df^{\,j}_{cp_g} = \begin{cases} 1, & \text{if } C_j \text{ contains caption } cp_g \\ 0, & \text{otherwise} \end{cases}$$

Table 1. The relations of captions and clusters

Caption	Related Cluster			
tiger	C_1	C_2	C_4	
grass	C_1	C_2	C_4	C_5
bear	C_1	C_2	C_4	
lion	C_1	C_3		
snow	C_1	C_3		
sky	C_2	C_3		
flower	C_2	C_4		
eater	C_2	C_3	C_4	
cat	C_4	C_5		
dog	C_2	C_5		

In this model, if the gene contains most of unique captions, it would be a general gene. Therefore its discrimination with respect to idf is low. Let us take Table 1 as an example based on Figure 4. The set of unique captions is {tiger, grass, bear, lion, snow, sky, flower, water, cat, dog}. For cluster C_1, it contains the caption set {tiger, grass, bear, lion, snow}. The idf of C_1 is log(10/5) = 0.301. For cluster C_2, the idf of C_2 is log(10/7) = 0.155. In this case, C_1 is more distinguishable than C_2 for our proposed approach. Overall idf can be viewed as the global weight for a gene. The final idf set for {C_1, C_2, C_3, C_4, C_5} is {0.301, 0.155, 0.398, 0.222, 0.523}.

2.2 Online Prediction for Annotation

For example, Figure 5 shows an unknown image and its three genes on color features. Once the genes of the unknown image are determined, three prediction models are triggered to predict the potential captions.

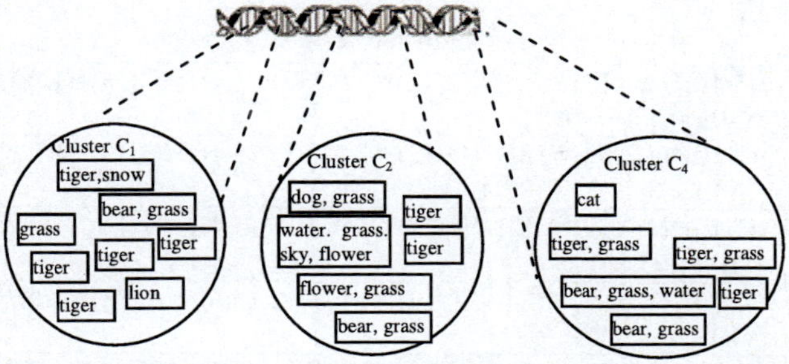

Caption cp_i	Frequency $tf^1_{cp_i}$
tiger	5
grass	2
snow	1
bear	1
lion	1
total	10

Caption cp_i	Frequency $tf^2_{cp_i}$
grass	4
tiger	2
flower	2
bear	1
dog	1
water	1
sky	1
Total	12

Caption cp_i	Frequency $tf^4_{cp_i}$
tiger	3
grass	3
bear	2
water	2
cat	2
flower	1
Total	13

Fig. 5. Example of three relevant clusters to the unknown image

2.2.1 Prediction by Model$_{entropy}$

In this model, our intent is to weight captions by using gene entropy and caption frequency. The major notion is that, if the caption occurs frequently in a gene with low entropy, its referred degree would be high. Otherwise, its degree would be low. Finally, the captions are ranked by the related degrees. In this paper, the entropy-based degree is defined as:

$$ED_{egree_{cp_x}} = \sum_{C_j \in CS} \left[\left(\frac{tf^{\,j}_{cp_x}}{\sum_{cp_v \in C_j} tf^{\,j}_{cp_v}} \right) \times \frac{1}{Entropy^{\,j}} \times \left(\frac{\sum_{1 \le k \le |IDB|} dis^k}{dis^{\,j}} \right) \right] \quad (3)$$

where $1 \le x \le y$. Finally, the EDegrees referred to color and texture are aggregated and normalized as the final degrees for a caption. Let us take an example based on Figures 2 and 4. In this case, there are 5 clusters and top 3 most-relevant clusters are identified as the genes, with respect to {C_1, C_2, C_4}. Figure 5 shows that three most-relevant clusters contain 21 images and 10 unique captions. Example of the resulting entropy-based degrees.

Caption	EDegree
tiger	((5/10)*(1/0.59)*(18))+((2/12)*(1/0.778)*(5.4))+((3/13)*(1/0.755)*(10.8)) = 19.714
grass	((2/10)*(1/0.59)*(18))+((4/12)*(1/0.778)*(5.4))+((3/13)*(1/0.755)*(10.8)) = 11.717
bear	((1/10)*(1/0.59)*(18))+((1/12)*(1/0.778)*(5.4))+((2/13)*(1/0.755)*(10.8))=5.831
lion	((1/10)*(1/0.59)*(18))+((0/12)*(1/0.778)*(5.4))+((0/13)*(1/0.755)*(10.8)) = 3.051
snow	((1/10)*(1/0.59)*(18))+((0/12)*(1/0.778)*(5.4))+((0/13)*(1/0.755)*(10.8))= 3.051
sky	((0/10)*(1/0.59)*(18))+((1/12)*(1/0.778)*(5.4))+((0/13)*(1/0.755)*(10.8)) = 0.578
flower	((0/10)*(1/0.59)*(18))+((2/12)*(1/0.778)*(5.4))+((1/13)*(1/0.755)*(10.8)) = 2.257
water	((0/10)*(1/0.59)*(18))+((1/12)*(1/0.778)*(5.4))+((2/13)*(1/0.755)*(10.8))= 2.780
cat	((0/10)*(1/0.59)*(18))+((0/12)*(1/0.778)*(5.4))+((2/13)*(1/0.755)*(10.8))= 2.202
dog	((0/10)*(1/0.59)*(18))+((1/12)*(1/0.778)*(5.4))+((0/13)*(1/0.755)*(10.8)) = 0.578

2.2.2 Prediction by $\text{Model}_{\text{tf-idf}}$

To weight captions by considering the global weight, we adopt idf to reveal the degrees of the captions in the most-relevant genes. The major notion behind this model is that, if the caption occurs frequently in a gene with high idf, its related degree would be high. Otherwise, its degree would be low. At last, the captions are ranked by the related degrees. In this paper, the idf-based degree is defined as

$$FD_{egree_{cp_z}} = \sum_{C_j \in CS} \left[\left(\frac{tf_{cp_z}^{\,j}}{|C_j|} \right) \times idf^{\,j} \times \left(\frac{\sum_{1 \le k \le IDB} dis^{\,k}}{dis^{\,j}} \right) \right] \tag{4}$$

where $1 \le x \le y$. At last, the FDegrees referred to color and texture are aggregated and normalized as the final degrees for a caption.

2.2.3 Prediction by Fusion Model $\text{Model}_{\text{AICDM}}$

To this end, 70% of annotated images are used as the learning set to construct $\text{Model}_{\text{entropy}}$ and $\text{Model}_{\text{tf-idf.}}$ Then the EDegrees and FDegrees of the remaining annotated images are generated by the above two models, respectively. Eventually the derived caption degrees are used as feature vectors to build the fusion model ($\text{Model}_{\text{AICDM}}$) by SVM, as shown in Figure 6.

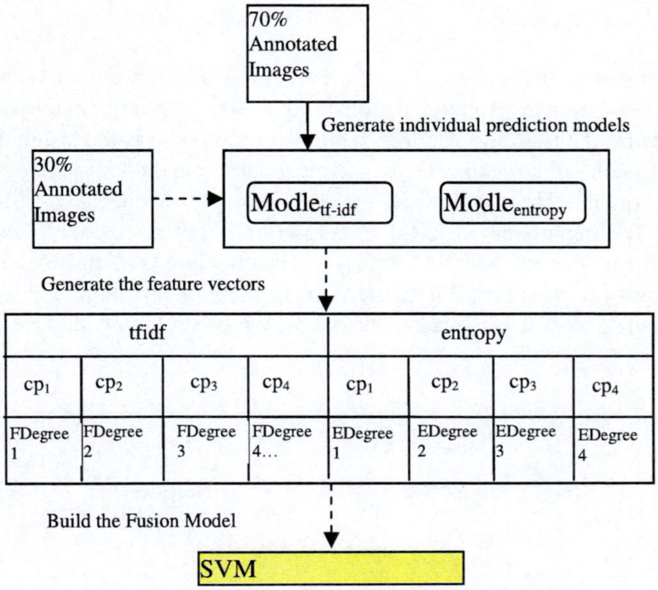

Fig. 6. Procedure of fusing $\text{Model}_{\text{entropy}}$ and $\text{Model}_{\text{tf-idf}}$

3 Empirical Evaluations

3.1 Experimental Data and Parameter Settings

To make the experiments complete, the experimental data came from the collections of TREC 2002 videos and web images. For TREC videos, the total duration of the experimental data is around 525 minutes and the data size is about 6 Giga Bytes. Moreover, there are 3090 shots generated in this experimental dataset. As for keyword setting, we selected 133 basic vocabularies regulated by National Institute of Standards and Technology (NIST) as the relevant keywords. For web images, the images are gathered from Google by using 10 categories. Each category contains 100 unique images. Note that, clustering is performed by K-means and the number of clusters is determined manually in this paper.

Table 2. The parameter settings for the experiments

Parameter	Data Set	Model	Default Value
Number of clusters, t	TREC Videos	entropy	color:87,texture:52
		tf-idf	color:90,texture:50
	web images	entropy	color:64,texture:42
		tf-idf	color:64,texture:65
Number of the most-relevant clusters, $\lvert CS \rvert$	TREC Videos	entropy	2
		tf-idf	2
	web images	entropy	2
		tf-idf	2

3.2 Experimental Results

In our experiments, 70% images are selected as the training data and the others are testing data. Furthermore, for the fusion model, 70% training images are chose as learning set from the training data and 30% of training data is the validation data. On basis of precision and recall, the experiments are conducted in two aspects: 1) comparisons of the three proposed models, namely $Model_{entropy}$, $Mode_{ltf\text{-}idf}$ and $Model_{AICDM}$, 2) comparisons of CRM (Continuous Relevance Model) [2], MMG [5], SVM [1] and our fusion model $Model_{AICDM}$. From Figures 7 and 8, we can obtain some important observations. First, the precision for web image set is worse than TREC video data set. It is because the web images are more diverse than TREC videos.

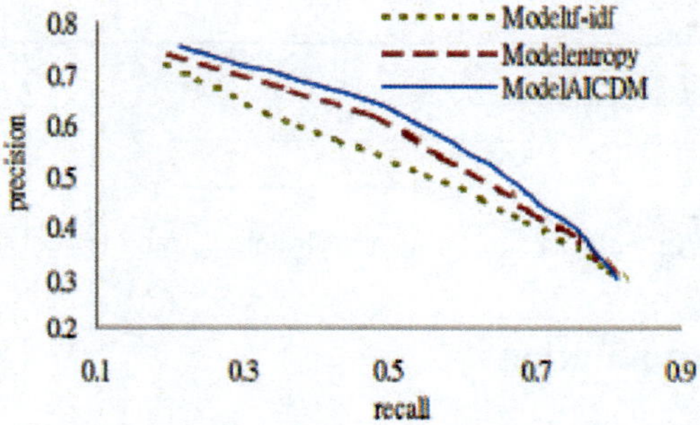

Fig. 7. Precision-recall of Modelentropy, Modeltf-idf and ModelAICDM for TREC vid

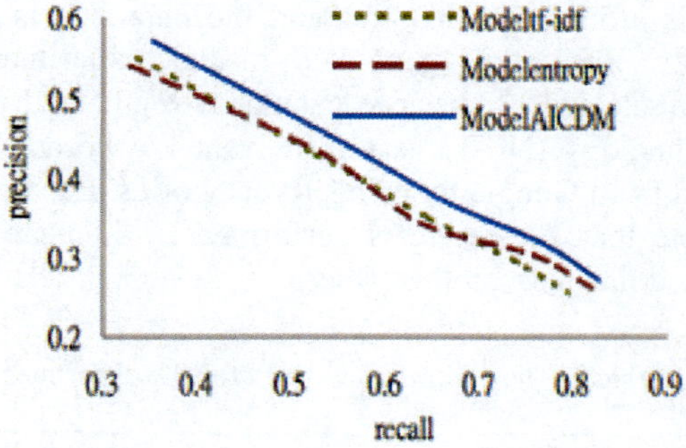

Fig. 8. Precision-recall of $Model_{entropy}$, $Model_{tf\text{-}idf}$ and $Model_{AICDM}$ for web images

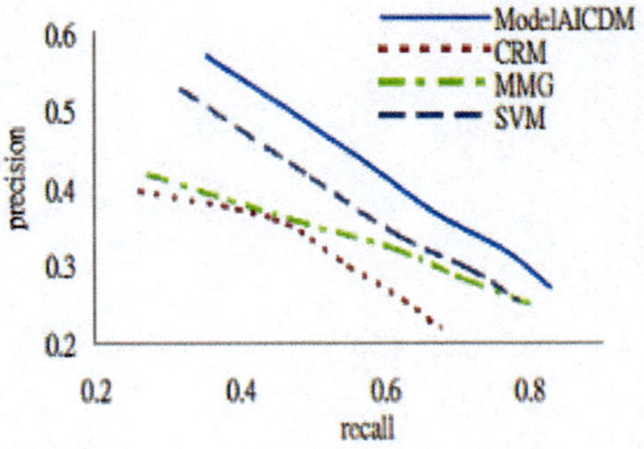

Fig. 9. Precision-recall of CRM, MMG, SVM and Model$_{AICDM}$ for web images

4 Conclusions

In the future, there remain some issues to further investigate. First, we will use more useable features to enhance the annotation quality. Second, we will try to mine the rules from the image-concept distribution by other data mining methods. Third, the involved ideas will be applied to other multimedia applications.

References

1. Feng, H., Shi, R., Chua, T.S.: A bootstrapping framework for annotating and retrieving WWW images. In: Proc. of the 12th Annual ACM International Conference on Multimedia (2009)
2. Lavrenko, V., Feng, S.L., Manmatha, R.: Statistical models for automatic video annotation and retrieval. In: Proc. of IEEE International Conference on Acoustics, Speech, and Signal Processing (2004)
3. Liu, J., Li, M., Ma, W.Y., Liu, Q., Lu, H.: An Adaptive Graph Model for Automatic Image Annotation. In: Proc. of ACM International Conference on Multimedia Information Retrieval (2006)
4. Li, W., Sun, M.: Automatic Image Annotation Using Maximum Entropy Model. In: Dale, R., Wong, K.-F., Su, J., Kwong, O.Y. (eds.) IJCNLP 2005. LNCS (LNAI), vol. 3651, pp. 34–45. Springer, Heidelberg (2005)
5. Li, J., Wang, J.Z.: Automatic Linguistic Indexing of Pictures by a Statistical Modeling Approach. IEEE Transactions on Pattern Analysis and Machine Intelligence 25(9), 1075–1088 (2003)
6. Li, J., Wang, J.Z.: Real-Time Computerized Annotation of Pictures. IEEE Transactions on Pattern Analysis and Machine Intelligence 30(6), 985–1002 (2008)

Chinese Text Speech Recognition Derived from VQ-LBG Algorithm

Xiaojun Wang and Weidong Lai

Science & Technology College
North China Electric Power University, Baoding, Hebei, P.R. China
laiwd@hbu.cn

Abstract. Speech recognition has been drawn extensive attentions for identity verification, man-machine interactive control, security appliance and so on, while the recognition rate of the speech text is essentially affected by the pronounce mode of different language. In order to recognize the Chinese text, which is syllabic in nature with severe word boundary uncertainty problem, a combined algorithm has been established based on *Linde–Buzo–Gray* (LBG) method to achieve vector quantization (VQ) for the MFCC processed Chinese speech text. Simulation indicates that the proposed algorithm is effective to construct codebooks for each solicited Chinese text word constricted by the minimum average distortion measure criterion, and the speech recognition is achieved through comparing between the codebooks and the unknown speech signal. Results also show that the codebook template dimension plays important role to affect the recognition rate, with 32 dimensions as the optimal size which can result in higher recognition rate and few computational resources occupancy. The proposed algorithm can be integrated in Chinese speech analyzing software to achieve voice signal recognition.

Keywords: Speech Recognition, VQ-LBG method, Vector quantization, Chinese word, Signal processing.

1 Introduction

During biological feature recognition and identity verification for human beings, speech is always regarded as the most powerful information resource due to its rich dimensional character such as the speech text, gender, attitude, emotion, health situation and so on. Speech recognition has drawn interests in security appliance, man-machine interactive control, and so on [1]. Valin et al have set up a mobile robot which can achieve robust recognition of simultaneous speech, and demonstrate an average reduction in relative recognition error rate of 42% when combined the missing features approach with the post-filter [2]. Since the speech signal can not take as its original form for processing, the feature abstracting methods include the Mel-frequency cepstral coefficients (MFCC), the Line spectra pairs (LSP), the Wavelet packet parameter (WPP), or PCA and ICA [3]. When it comes for feature recognition, algorithms such as hidden Markov models (HMM), vector quantization (VQ), dynamic time wrapping (DTW), artificial neural networks (ANN) etc are established in recent reports [4]. Sha and Saul studies the parameter estimation in

J. Luo (Ed.): Affective Computing and Intelligent Interaction, AISC 137, pp. 303–311.
springerlink.com

continuous density hidden Markov models (CD-HMMs) for automatic speech recognition (ASR), and a learning algorithm based on the goal of margin maximization is proposed [5]. H. Jiang reviews the confidence measures (CM) of speech recognition results, and divides the custom CM as a combination of predictor features, a posterior probability or utterance verification kind [6]. J. He et al reports novel method of group vector quantization (GVQ) to train VQ codebooks for closed-set speaker identification, in which speaker codebooks are optimized for vector groups rather than for individual vectors [7].

Vector quantization (VQ) oriented speech recognition is derived from the idea that is to compress large number of short-term spectral vectors into small set of code vectors. For many VQ applications, the information directly provided by speech signal is trained as codebooks for further unknown speech recognition. VQ method has been verified as the most commercially viable and useful technology to provide effective means for compressing the short-term spectral representation of speech signals [8]. In this paper, the *Linde–Buzo–Gray* (LBG) method is selected to achieve vector quantization for the MFCC processed speech signal. Focused on the Chinese text, which is syllabic in nature, the algorithm is verified and factors influencing speech recognition rate is discussed.

For clarity, Sections II focuses on MFCC transformation and VQ-LBG method for voice signals. Section III is devoted to the introduction of the combined algorithm, which is decomposed into three parts: firstly, the MFCC is achieved to voice signals; secondly, VQLBG algorithm is applied to obtain codebooks and achieve recognition; thirdly, the compressing efficiency and recognition rate of algorithm is discussed.

2 LBG Principle and Speech Recognition Method

LBG algorithm, which can minimize the quantization error, is selected to train codebooks in this article.

2.1 LBG Traning Principle for Chinese Speech

The codebook based on *LBG* algorithm leads to the density and clustering of the training data constricted by the minimum distortion measure value. When it comes to the training data abstracted from speech signal, the Mel-frequency cepstrum method is applied. The Mel-frequency cepstrum coefficient (MFCC) represents the short-term power spectrum of a sound, based on a linear cosine transform of a log power spectrum on a nonlinear Mel scale of frequency, with its nth coefficient $c(n)$ just as

$$c(n) = \sum_{m=1}^{M-1} S(m) \cos\left(\frac{\pi n(m+1/2)}{M}\right) \qquad (0 \le m < M) \tag{1}$$

In which,

$$S(m) = \ln\left(\sum_{k=0}^{N-1} \left|X(k)\right|^2 H_m(k)\right)$$

The $X(k)$ presents the kth Fourier transformation coefficient of the original speech signal, with the $H_m(K)$ as the Mel band-pass filters.

The *Linde–Buzo–Gray* algorithm (*LBG*) is a kind of vector quantization algorithm to derive codebook, which is similar to the *k*-means method in data clustering. With $d(x_j, y_l)$ defined as the distortion measure between the training data x_j, which is derived from the MFCC from Equation (1), and the code y_l, the average distortion measure of the training data set is

$$D^{(n)} = \frac{1}{J}\sum_{j=0}^{J-1} d(x_j, \hat{x}_r) \tag{2}$$

In which,

$$\hat{x}_r \in Y_R$$

$$\hat{x}_r = \arg\min_{y_l \in Y_R} d(x_j, y_l)$$

The iteration computation to the y_l can result in optimal codebook $D^{(n)}$ with the minimum criterion as

$$\frac{D^{(n-1)} - D^{(n)}}{D^{(n)}} \le \varepsilon \tag{3}$$

In this paper, the *LBG* Algorithm is practiced to train MFCC processed Chinese speech signal with the ε=0.01.

2.2 Recognition Method Based on VQLBG Codebook

The unknown speech signal is also MFCC processed to obtain a set of coefficients and then compared with the codebooks, as shown in Fig.1.

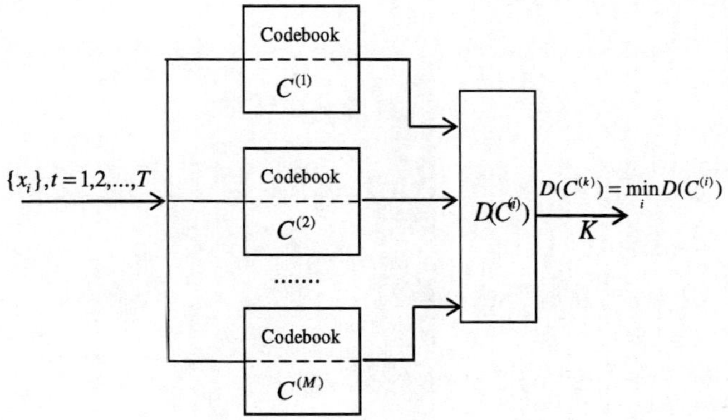

Fig. 1. Recognition Method for the unknown speech vector set $\{x_i\}$based on VQLBG codebook

In Fig. 1, the average distortion measure corresponding to the *codebook i* is defined as

$$D(C^{(i)}) = \frac{1}{T}\sum_{t=1}^{T} d(x_t, \hat{x}_t^{(i)}) \quad (4)$$

In which,

$\hat{x}_t^{(i)} \in C^{(i)}$ and

$$\hat{x}_{t^{(i)}} = \arg\min_{y^{(i}{}_j \in c^{(i}} d(x_t, y_j^{(i)})$$

The speech sample is recognized as the k class when

$$D(C^{(k)}) = \min_i D(C^{(i)}) \quad (5)$$

has the minimum average distortion measure.

3 Results and Discussions

The speech signals are selected from our database of the Chinese number and text pronounced by a male in laboratory quiet condition, with the sampling frequency of 22050 per second. The algorithm is integrated and simulated in GUI platform.

3.1 Waveform of Chinese Speech

The Chinese language is syllabic in nature with frequent homonym phenomena and severe word boundary uncertainty problem. But in given situation, such as secret code inputting, telephone number recording, text input for special human beings who can not input number and text information by hands, the isolating Chinese number and text signal are important for recognition. In Fig. 2, the number of 7 'qi' is pronounced in common spoken Chinese tone.

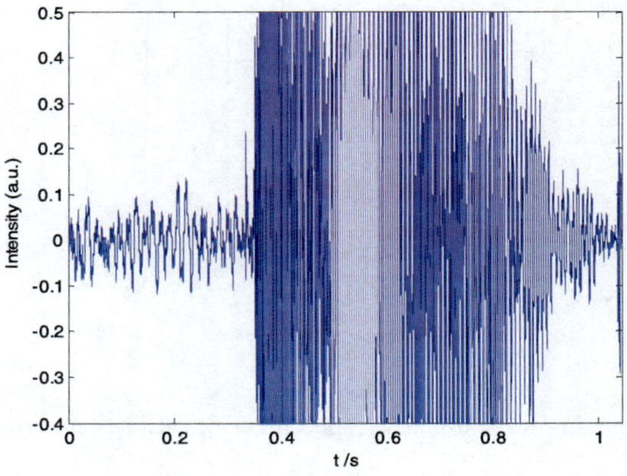

Fig. 2. Waveform of pronounced Chinese word of 7 'qi' versus time

The waveform has not clear variance versus time. Such phonetics phenomenon is ascribed that the Chinese tone has the distinct characteristic by its pronounced method from front mouth cavity that is different from the *Latin* language which presents less syllabic. To abstract feature from Chinese tone, there should develop efficient method.

3.2 Mel Filter Construction

The *Mel* filters is composed of triangular bands. The spacing of filter follows the Mel frequency scale, which is inspired by critical band measurements of the human auditory system, and the bandwidth of each filter is chosen by aligning the triangle base with the center frequencies of the neighboring filters. In Fig.3, the MFCC filter number is setting as 20.

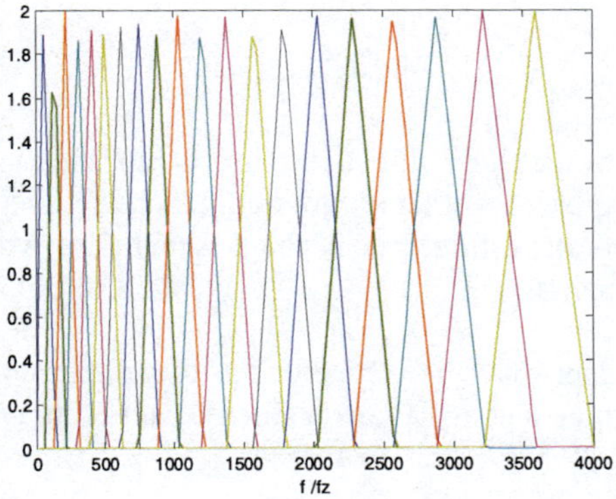

Fig. 3. The Mel filter structure construction

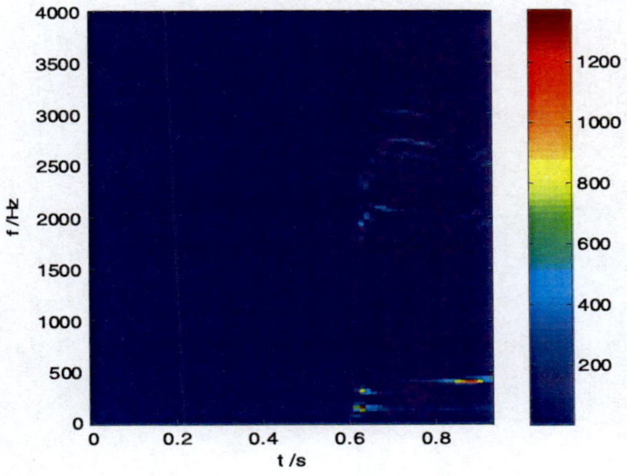

(a)

Fig. 4. Mel filer effect on the voice signal. (a)unmodified; (b) Mel filter modified.

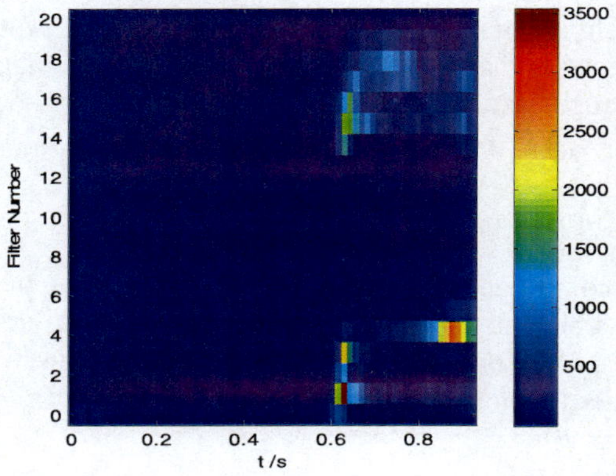

(b)

Fig. 4. *(continued)*

After filtering, the voice signal of Chinese text is modified in its power spectrum as shown in Fig.4. Compared with the unmodified power spectrum in Fig 4(a), the constructed Mel filters have filtered the voice signal. The MFCC coefficients have been established as vectors.

The MFCC coefficients have been regarded as two-dimensional vectors. By MFCC transformation, the voice signal has been filtered and compressed into quantization form.

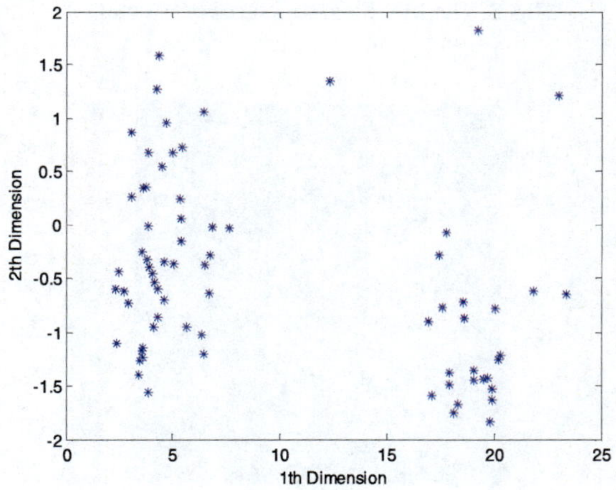

Fig. 5. MFCC vectors estblished in this article

3.3 VQLBG Vector Quantization to Construct Codebook

Furthermore, the VQ-LBG algorithm is put forward with iteration to achieve the minimum average distortion measure between code and the training data. The optimal 1th and 2th part of codebook for Chinese text 7 'qi' is shown in Fig. 5. As seen, there has established two-dimensional vector set as the template.

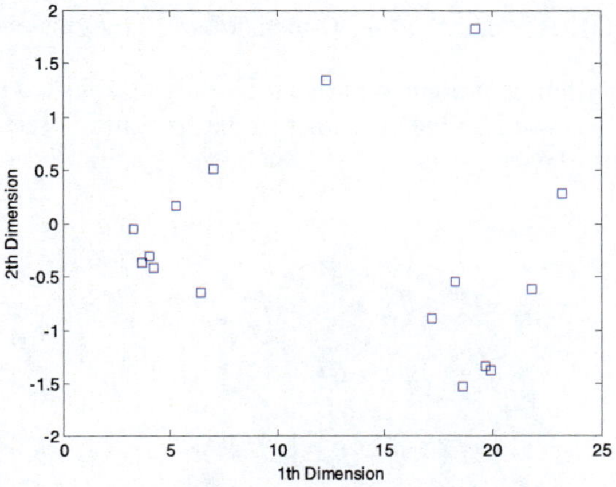

Fig. 6. Two- dimensional templet codebook with 16 cells of Chinese number 7

The VQ codebook is consisting of a small number of highly representative vectors, with the dimension of 16 cells that can represent the MFCC in Fig. 5.

3.4 Solicited Chinese Speech Recognition

The unknown signal is compared in the set of the codebook templates. With the extreme value criterion of average distortion measure indicated in Equation 5, the most suitable template with the minimum Euclidean distance to the unknown signal is the recognition result, according to the principle of Fig. 1.

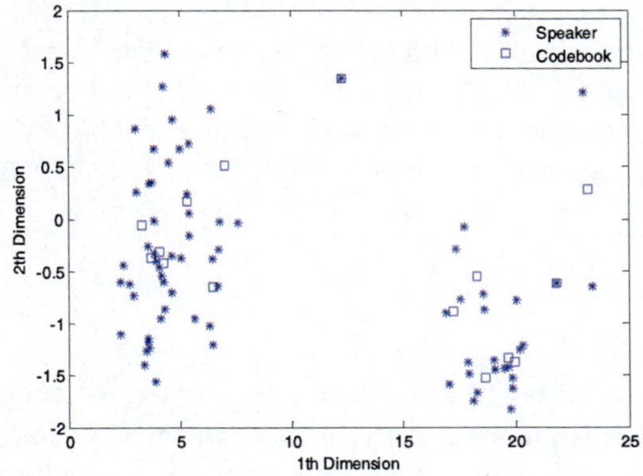

Fig. 7. The matching method between unknown signal and the codebook template

Such result verifies that the algorithm established in this paper is suitable to recognize the Chinese text speech. The entire algorithm is integrated in GUI platform for convenience in order to integrate with other software. For certain person, the combined algorithm of 16 dimensional codebook have achieved good recognition rate, though the pronunciation of Chinese number 'yi' and 'qi' are in some extent syllabic.

3.5 Effect of VQLBG Quantization Dimension on Recognition Rate

The combined algorithm is presented to the artificially established database including Chinese numbers and words. When it comes to the recognition rate, the dimension of the trained template plays important roles, which is present in Fig. 8.

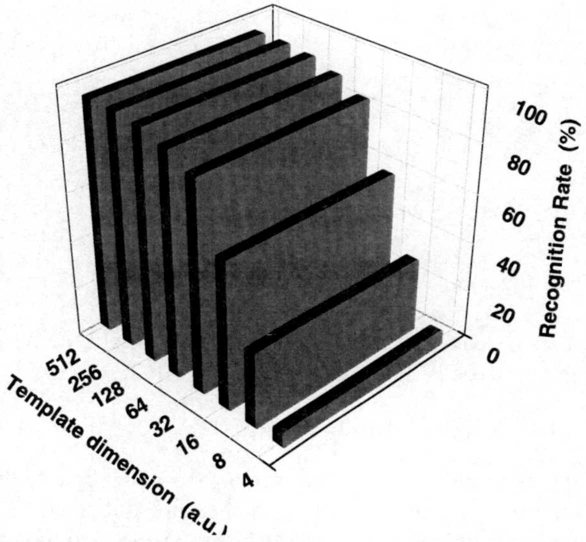

Fig. 8. Recognition rate under different template dimension

It is obvious that the template with dimension more than 32 has not great influence on the recognition rate for the Chinese texts. To heighten the operation velocity and lessen the algorithm complexity, the dimension of 32 is chosen for the codebook size, which can achieve optimal recognition result with few computational resource requirements.

4 Conclusion

The VQ algorithm based on LBG method is performed to recognize Chinese text speech. The Mel filter is proposed firstly in this article for decompressing the voice signal and abstracting features. The LBG algorithm is effective to construct codebooks for each solicited Chinese text word constricted by the minimum average distortion measure criterion, and the speech recognition is verified through comparing between the codebooks and the unknown speech signal. Results also indicate that the codebook template dimension plays important role to affect the recognition rate, with 32 dimensions as the optimal size which can result in higher recognition rate based on few computational resources occupancy.

The algorithm proposed in this article can be integrated into Chinese text detection system to obtain higher recognition accuracy of the Chinese speech.

Acknowledgment. This work is financially supported by Foundation of Hebei Educational Department of China Z2009305.

References

1. Zue, V.: The use of speech knowledge in automatic speech recognition. Proceedings of the IEEE 73, 1602–1615 (2005)
2. Valin, J., Yamamoto, S., Rouat, J., Michaud, F., Nakadai, K., Okuno, H., Zue, V.: Robust recognition of simultaneous speech by a mobile robot. IEEE Transactions on Robotics 23, 742–752 (2007)
3. Reddy, D.R.: Speech recognition by machine: A review. Proceedings of the IEEE 64, 501–531 (2005)
4. Benzeghiba, M., De Mori, R., Deroo, O., Dupont, S., Erbes, T., Jouvet, D., Fissore, L., Laface, P., Mertins, A., Ris, C., Rose, R., Tyagi, V., Wellekens, C.: Automatic speech recognition and speech variability: A review. Speech Communication 49, 763–786 (2007)
5. Sha, F., Saul, L.: Large margin hidden Markov models for automatic speech recognition. In: Advances in Neural Information Processing Systems, vol. 1, pp. 1249–1256 (January 2007)
6. Jiang, H.: Confidence measures for speech recognition: A survey. Speech Communication 45, 455–470 (2005)
7. He, J., Liu, L., Palm, G.: A discriminative training algorithm for VQ-Based speaker identification. IEEE Transactions on Speech and Audio Processing 7, 353–356 (1999)
8. Kinnunen, T., Kärkkäinen, I.: Class-Discriminative Weighted Distortion Measure for VQ-based Speaker Identification. In: Caelli, T.M., Amin, A., Duin, R.P.W., Kamel, M.S., de Ridder, D. (eds.) SPR 2002 and SSPR 2002. LNCS, vol. 2396, pp. 681–688. Springer, Heidelberg (2002)

Pronunciation Quality Scoring for Single Syllable Word in PSC

Long Zhang[1,2], Haifeng Li[1,*], and Jianhua Wang[2]

[1] School of Computer Science and Technology, Harbin Institute of Technology, Harbin, Heilongjiang, China
zlwalkman@sina.com, haifengli@hit.edu.cn

[2] College of Computer Science and Information Engineering, Harbin Normal University, Harbin, Heilongjiang, China
wjh@vip.163.com

Abstract. This paper discusses pronunciation quality scoring for single syllable word in Putonghua Shuiping Ceshi (PSC) that is a nationwide spoken test to evaluate the standard level and the practical ability that an individual uses the mandarin in china. This study mainly includes some algorithms about the syllable separation, acoustic units selection, posterior probability scoring, threshold values setting and neural network combination. Experiment shows that the proposed approach achieves a high correlation of 0.731, a value very close to 0.786 between human experts. It is also observed that the combination method of neural network gets the best evaluation result. This method is eligible for the automatic scoring in PSC.

Keywords: pronunciation quality Scoring, single syllable word, PSC, neural network, HMM, automatic machine scoring.

1 Introduction

China is a multinational country with many different languages and dialects, so it is important to popularize the common speech (Putonghua) for promulgating cultural education, developing the science and technology, raising working efficiency and meeting the need of the development of market economy and information society. Putonghua Shuiping Ceshi (PSC) is a spoken test throughout the country to evaluate the standard level and the practical ability that an individual uses the mandarin. It is an important approach to popularize Putonghua. Now, PSC is performed through the manual test, and there are many disadvantages, such as great expense, long time, lower efficiency and subjective feeling. So, it is an active demand and inevitable result to develop an automatic PSC testing system using the computer.

PSC consists of four parts: reading single syllable word, reading multi syllable parse, reading short text and speaking with a special subject. The detailed standards of grading can prefer to the literature [1]. This paper mainly researches the methods to evaluate the single syllable word. This part needs a candidate to read 100 single syllable words in three minutes and aim to evaluate the standard level of the initials,

* Corresponding author.

J. Luo (Ed.): Affective Computing and Intelligent Interaction, AISC 137, pp. 313–319.
springerlink.com

finals and tones of these syllable words. Scoring criterion lists as follow: (1) speech error, 0.1 point penalty per syllable; (2) speech defect, 0.05 point penalty per syllable; (3) overtime in 1 minute, 0.5 point penalty; overtime over 1 minute, 0.8 point penalty.

This paper is organized as follows. Section 2 introduces the system structure. Section 3 discusses the traditional phonetic scoring algorithm and improves it on its shortcomings. Some experiments and results are given in Section 4. Conclusions are drawn in Section 5.

2 System Structure

Phone is the most fundamental assessment unit of our scoring system. The phonetic evaluation is carried out by using speech recognition techniques of HMM and Viterbi decoding (Force Align). And then calculate the posterior probability, compare with a threshold value and detect the error (or defect) of pronunciation. Finally computing the machine score and comparing the human score in order to evaluate the preferment of System. Block diagram of the system is shown in Fig. 1.

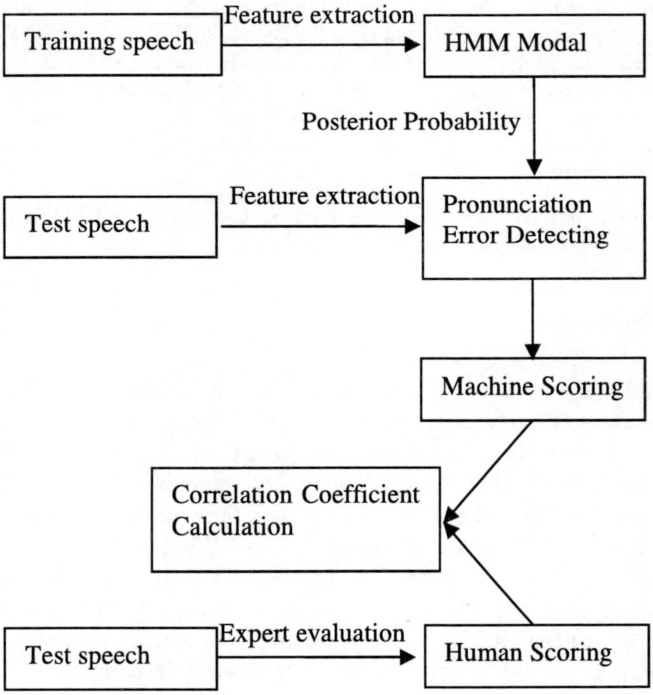

Fig. 1. Scoring System Structure

3 Scoring Algorithm

According the scoring criterion, the key of study includes two parts. One is the syllable separation; the other is the initials evaluation, finals evaluation and tones evaluation.

3.1 Syllable Separation

There are many methods to separate the syllables from a speech. The common method is that training firstly the initials and finals based HMM model or syllables based HMM model and then according the given transcription feeding into HMM model net to do one-pass Viterbi decoding. The Viterbi decoding is a force alignment between the speech frames and the HMM models in the net. But the problem is when the speech and the given transcription are disagreement, the force alignment boundaries will be not right. For example, missing or reading again is enviable in a real speech in a test. Some traditional syllable separation method adapts the features of short energy, zero-crossing rate and their variations, such as energy-zero-product, energy-zero-ratio and so on. When the SNR is low, the efficient of these methods will become bad. So, a frame based smooth accumulated energy features will been proposed.

X_i is a speech signal, and its smooth accumulated energy is defined as follow:

$$E_i = 0.99 * E_{i-1} + X_i * X_i$$

3.2 Initials, Finals and Tones Evaluation

1) Acoustic Units Selection

According to Mandarin phonology, each syllable is divided into there parts: initials, finals and tones. Initials are consonants beginning a syllable and finals which consist of head (glide), body (main) and tail (coda) of finals follow close behind. Body of finals is necessary and the other parts are optional in an available syllable. Tones mainly have an effect on finals. The phonetic symbols of putonghua comprises 22 initials (including a zero initial), 31 finals, and 4 tones (excluding the soft voice tone). Totally, there are about 420 base syllables and more than 1200 tonal syllables in the whole pronunciations of Mandarin.

In order to integrating initials, finals and tones into an acoustic model, we modify the segmental tonal modeling which is proposed in [2] by Chao Huang. Basic idea consists of three parts. Modeling tone types by combinations of segmental 2-scale of pitch with high/low (H/L). For Example, tone 1, 2, 3, and 4 are realized as high-high (HH), low-high (LH), low-low (LL) and high-low (HL) respectively. Assigning the glide to initial parts and forming an extended initial set. Dividing the rest of finals into two parts and each of them can carry the tone characteristics respectively through the index of pitch. So, there are three parts, that is initial, final1 and final2, in a syllable. E.g. syllable /Huang1/ -> {hu aaH ngH } and /tiao3/ -> {ti aaL oL}.

extended initial set (57): b、bi、bu、c、ch、chu、cu、d、di、du、f、fu、g、ga、ge、go、gu、h、hu、ji、jv、k、ku、l、li、lu、lv、m、mi、mu、n、ni、nu、nv、p、pi、pu、qi、qv、r、ru、s、sh、shu、sil、su、t、ti、tu、wu、xi、xv、yi、yv、z、zh、zhu、zu segmental tonal final set(26) : aH、aL、aaH、aaL、 ehH、ehL、elH、elL、erH、erL、iH、iL、ibH、ibL、ifH、ifL、ngH、ngL、nnH、nnL、oH、oL、uH、uL、vH、vL.

2) Scoring Algorithm
To assess the pronunciation quality of a speaker, many measures have been proposed before [3-6]. In this paper, posterior probability (PP) is chosen since it is less affected by the changes in spectrum due to speaker specific characteristics or acoustic channel variations. It measures the phonetic quality more directly than other measures. We firstly extract the features from the input speech waveforms, which are 13-dementional MFCCs, and their first and second order derivatives. In an HMM-based speech recognizer, given an isolated syllable or phone acoustic observation sequence O and its corresponding transcription T, we can calculate the output probability $P(O|T)$. $P(O|T)$ is likelihood of given HMM model T to given observation vector O. The most useful evaluation algorithm is posterior probability, Which is $P(T|O)$. When each phone in a syllable has the same duration, using Bayes' rules, $P(T|O)$ can be written as:

$$
\begin{aligned}
P(T|O) &= \sum_{i=1}^{N} |\log(P(T_i \mid O^{(T_i)}))| / N \\
&\approx \sum_{i=1}^{N} |\log(\frac{P(O^{(T_i)} \mid T_i)P(T_i)}{\sum_{q\in Q} P(O^{(T_i)} \mid q)p(q)})| / N \\
&\approx \sum_{i=1}^{N} |\log(\frac{P(O^{(T_i)} \mid T_i)}{\sum_{q\in Q} P(O^{(T_i)} \mid q)p(q)})| / N
\end{aligned}
$$

Q is the model set, and q is the phone that T_i could be misread as. Ti is a phone and T is a syllable that consists of T_i. $P(O(T_i)|T_i)$ is the likelihood of given HMM model T_i to the observation vector $O(T_i)$. N is the number of phones in a syllable, so N is 3 in our modified segmental tonal model.

When the weighted duration of phones in a syllable, $P(T|O)$ can also be calculated as:

$$
\begin{aligned}
&P(T|O) \\
&= \sum_{i=1}^{N} \frac{NF(T_i)}{NF(T_1)+...+NF(T_i)+...+NF(T_N)} |\log(P(T_i \mid O^{(T_i)}))| \\
&= \sum_{i=1}^{N} \frac{NF(T_i)}{NF(T_1)+...+NF(T_i)+...+NF(T_N)} |\log(\frac{P(O^{(T_i)} \mid T_i)P(T_i)}{\sum_{q\in Q} P(O^{(T_i)} \mid q)p(q)})| \\
&\approx \sum_{i=1}^{N} \frac{NF(T_i)}{NF(T_1)+...+NF(T_i)+...+NF(T_N)} |\log(\frac{P(O^{(T_i)} \mid T_i)}{\sum_{q\in Q} P(O^{(T_i)} \mid q)p(q)})|
\end{aligned}
$$

$NF(T_i)$ is the total frame of phoneme T_i.

According to the Scoring criterion in PSC, speech defect (or error) derives the defect (or error) of the initials or finals or tones, so we can set two threshold values for each phone respectively. The threshold value can be calculated as:

$$T_{q_i} = \mu_q + \alpha_i \delta_q + \beta_i$$

μ_q and δ_q is the mean and variance of the posterior probability of phone q occurred in the training corpus and pronounced right. The value of i is either 0 (respecting a error) or 1 (respecting a defect) . α_i and β_i are experience values.

So, we can use calculated posterior probability to evaluate phone. After posterior probability of the pronounced phone T_i is derived, if its value is less the threshold Tq_i, T_i is an error (or defect). If it matches or exceeds Tq_i, T_i is judged as right. When the values of T_i of each phone composed a syllable word are all right, a neural network training will be performed in order to getting a fine result. Otherwise, the syllable word is error (or defect) if any of T_i is a error (or defect).

3) Neural Network Training

Neural network is one of the main tools of machine learning. A neural network can be capable of implementing arbitrary maps between input and output spaces. To get a fine result, we put the machine scores of all phones that compose a syllable and are right pronunciation through comparing to some certain threshold values as the inputs of the neural network; the predicted syllable grade are the output value of the network; the actual human grades provide the targets for the training of the network. The number of input units corresponds to the number of machine scores combined. Neural network parameters, the weights, are adjusted by the training algorithm to minimize the mean square error criterion.

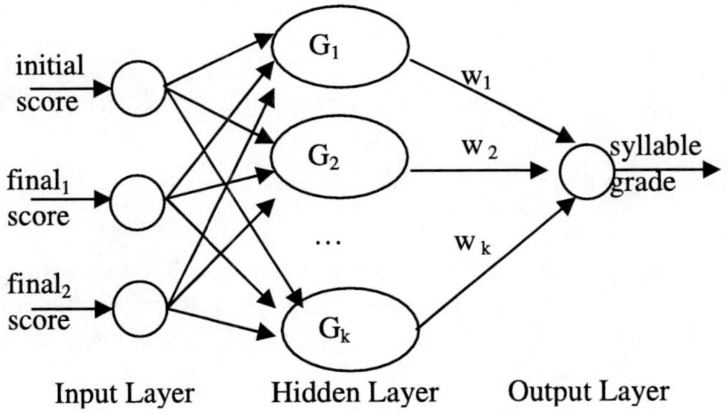

Fig. 2. Neural Network Structure

After some preliminary experiments with different network architectures, we choose the two-layer back propagation (BP) network with a single linear output unit and a hidden layer of log-sigmoid units to estimate the syllable grade. We try to vary the number of hidden layer units, and the best performance is obtained with 7 hidden units.

4 Experiments and Results

4.1 Corpus

All of the speech waveforms used by the following experiments come from a PSC test. A PSC test set has 100 mono-syllable words. We totally collect 100 sets of test samples from 100 test attendees, among whom half are male and half are female. The test attendees are all students in the third year of our normal university. They come from the 18 provinces in china and study in 12 different majors. Each syllable word in the test is graded on a 0-2 scale by two expert raters with national certificate. A rating of 0 indicates pronunciation right, a rating of 1 indicates pronunciation defect and a rating of 2 indicates pronunciation error. 80% of the collected samples are used as training set and the other 20% are used as testing set. We select speech waveforms of the right pronunciation syllable that all tagged 0 by two experts, and use them to train the HMM model.

4.2 Correlation Coefficient

The popular way to evaluate the performance of one pronunciation assessment system is to calculate the correlation between machine grades and human expert's subjective grades. The correlation coefficient (CC) is defined as:

$$CC_{J1,\,J2} = \frac{X_{J_1}^T X_{J_2}}{\left\|X_{J_1}\right\|_E \left\|X_{J_2}\right\|_E}$$

Where, X_{J_1} and X_{J_2} is the machine grade vector and the corresponding human rating vector respectively. $\left\|X_{Ji}\right\|_E = \sqrt{\sum_{j=0}^{N-1} x^2(j)}$ is norm of vector X_{Ji}.

In scoring system, when machine grade correlates positively with corresponding human score, $CC_{J1,\,J2}$ gets the maximum value 1. Otherwise, the more different between the two scorings, the smaller the value of $CC_{J1,\,J2}$ will become.

4.3 Results

At last we evaluate the method used the neural network to combine phone based posterior probability scores and map to based syllable posterior probability scores, and calculate the correlation between human and machine scores. The parameters of the regression and estimation models are trained on the training set and evaluated on the testing set. The evaluation results are shown in Table 1.

Table 1. Correlation between human and machine scores

	Rater1	Rater2
Machine	0.713	0.748
Rater1		0.786

5 Conclusion

The main focus of this study is the methods to evaluate the single syllable word. Experiments show that the proposed approach achieves a high correlation of 0.731, a value very close to 0.786 between human experts. It is also observed that the combination method of neural network gets the best evaluation result. This method is eligible for the automatic scorings in PSC.

Acknowledgment. This study is supported by the Youth Science and Technology Special Foundation of Heilongjiang Province of China (QC2008C27), the science and technology project of the Education Department of Heilongjiang Province (11541093, 11551125) and the Heilongjiang Provincial Key Laboratory of Intelligence Education and Information Engineering.

References

1. Meng, G.Z., An, Y.H., Zheng, W.F.: The guidebook of Putonghua Shuiping Ceshi. Heilongjian Education Press, Harbin (2001)
2. Huang, C., Shi, Y., Zhou, J.L., Chu, M., Wang, T., Chang, E.: Segmental tonal modeling for phone set design in Mandarin LVCSR. In: Proc. ICASSP 2004, pp. 901–904 (2004)
3. Zheng, J., Huang, C., Chu, M., Soong, F.K., Ye, W.: Generalized Segment Posterior Probability for Automatic Mandarin Pronunciation Evaluation. In: Proc. ICASSP 2007, pp. IV201–IV204 (2007)
4. Witt, S.M.: Use of Speech recognition in Computer assisted Language Learning. PhD Thesis, University of Cambridge (1999)
5. Franco, H., Neumeyer, L., et al.: Automatic pronunciation Scoring for Language Instruction. In: Proc. ICASSP 1997, pp. 1471–1474 (1997)
6. Neumeyer, L., Franco, H.: Automatic Scoring of Pronunciation Quality. Speech Communication 30, 83–93 (2000)

Applications of the Fingerprint Identification Technology in the Fire Protection Monitoring System

Yongqiang Wang, Hao Li, and Guiying Ma

Department of Computer Science and Technology
Tangshan College, Tangshan, China
wangyongqiang12332@163.com

Abstract. Aiming at drawbacks existed in means of the traditional ID authentication and the authority control in the fire protection monitoring system, this paper proposes solutions for the ID authentication in the monitoring system based on the Fingerprint Identification Technology, describes the working principle of the fingerprint identification in the fire protection monitoring system, introduces the methods of the fingerprint identification, the realization modes of the ID authentication, and the design ideas of the program, and analyzes the applications of the fingerprint identification technology in the fire protection monitoring system, providing a new way of thinking for the safe operation of the fire protection monitoring system.

Keywords: fingerprint identification, safety monitoring, fire protection monitoring system.

1 Introduction

The urban fire safety monitoring system as an important means to improve the rapid response capacity of fire forces and put out the starting fire hazard, is playing an increasingly important role in real work, and is of great significance to enhance the fire safety management and to timely detect and handle fire hazards and fire situations [1]; more and more large buildings are all equipped with complete fire protection monitoring system, which includes a variety of fire detectors, fire fighting equipments, alarm devices, and camera devices, etc.; management staffs in the central control room can conduct the unified management and allocation on these devices, when conduct the system configuration, records checking, video checking and other system operations, the ID authentication and the authority control on operators are completed usually in the form of user name and password, while due to frequent operations and relatively simple passwords, which operators know one another, passwords lose the confidential role, which is prone to occur operators assume the administrator' s name to operate, making the centralized control center monitoring system exist potential safety hazards [2]. In addition, for some devices that can be controlled on the spot, such as shutter doors, fire water monitors and so on, of which the on-site control panel generally adopts traditional key as the method of the authority authentication, operators must bring keys with them, which also bring

J. Luo (Ed.): Affective Computing and Intelligent Interaction, AISC 137, pp. 321–329.
springerlink.com

inconveniences and potential safety hazards for handling dangerous cases of fire hazards. To solve drawbacks brought from the "user name + password" or the traditional key-authentication way, thus it proposes to apply the fingerprint identification technology into the fire centralized control center monitoring system, in order to achieve the ID authentication before users controlling the fire fighting equipments.

2 System Compositions and Functions

The fire safety monitoring system approximately can be divided into three sections of detection, linkage, and monitoring [3], while the detection section includes ball machines video cameras and other devices used for video detections, and conventional devices of fire detectors used for fire detections; these devices are deployed in each space of a building that needs the protection monitoring to conduct the real-time monitoring for uncommon conditions within the region. The linkage section primarily used in the alarming and fire fighting measures after the conditions of fire takes place, is an important implementing agency in the system, mainly including fire broadcasting, shutter doors, and alarm calls, etc. associated equipments. The monitoring section possesses functions like viewing live videos, recording fire events, picture recording of fire hazards and other functions, of which the mainly composing equipments include the television wall, the hard disk recorders, and fire host, etc. The system conducts the ID authentication to operators by means of the fingerprint identification; the fingerprint identification module mainly completes collections and verifications of fingerprints, registers users' fingerprints, cancels fingerprints and resets to use, as well as sets administrators' authority. The structured chart of the system is shown in Figure 1.

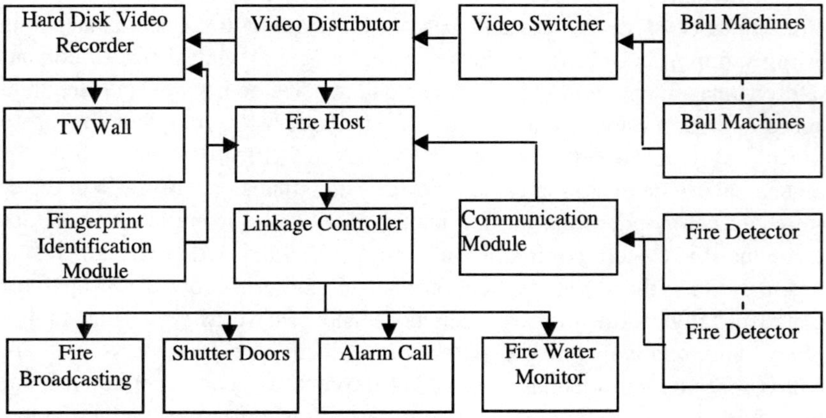

Fig. 1. The Structured Chart of the Fingerprint Identification Monitoring System

3 Fingerprint Identification Technology

Fingerprints are patterns showed on the finger skins of the body; these patterns are composed of ridge line and valley line and diverse from one another in the patterns, breakpoints and cross points, which are called "characteristics" in the information processing. These characteristics remain unchanged for a lifetime after the formation in the early growth stage of human, and are unique. According to the characteristic of uniqueness, it can correspond a person with his fingerprints; through comparing the characteristics of the collected fingerprints and the pre-stored fingerprints, it can verify the identity of a person; the whole process known as the fingerprint identification, is a high-new technology concentrating the sensor technology, the biotechnology, the electronic technology, the digital image processing, and the pattern identification in one [4].

At the beginning, through the fingerprint reading device reading the fingerprint image of the human body, after extracting the fingerprint image, it needs to conduct the preliminary processing on the original image to make it more clear[6].

Next, the fingerprint identification software establishes the figure representations of fingerprints – characteristic data, a one-way conversion, which can convert fingerprints to characteristic data, but cannot convert characteristic data to fingerprints, and two different fingerprints will not produce same characteristic data. The software finds the data point of the so-called "nodes" (minutiae) from fingerprints, that is, those coordinate positions of bifurcations, terminations or circular areas in fingerprints patterns; these points also have more than seven unique characteristics. Usually there is an average of 70 nodes in fingers therefore this method will produce approximately 490 data[7].

Some algorithms combine nodes and direction information to produce more data; these direction information indicate the relationship between each node, and some algorithms also handle the whole fingerprint images. In a word, these data, often referred to as templates, save records with the size of 1k. No matter how they are composed, there is still no one standard for templates until now, nor one published abstract algorithm, instead each manufacturer has its own way.

Finally, by means of the computer fuzzy comparison, compare templates of the two fingerprints, calculate the degree of their similarity, and ultimately obtain matching results of the two fingerprints.

In the process of the fingerprint identification of this system in this paper, we design four steps for the identification algorithm: the assessment of the image quality, the pre-processing of images, the extraction of fingerprint characteristics and the matching of fingerprints. The pre-processing link mainly includes the segmentation, the enhancement, the binarization, the thinning, and the repairing, etc. operations of fingerprint images; the segmentation, the binarization and the thinning, etc. operations, have been studied a lot by many scholars and proposed various different methods; here we integrate the previous study results. The enhancement of fingerprint images is a very critical part of the automated fingerprint identification system, in which it will inevitably encounter poor-quality fingerprint images; the fingerprint enhancement adopts a certain algorithm on low-quality fingerprint images to make their pattern structures clarify, highlighting and preserving the original characteristic information to avoid the generation of false information [5]; this paper adopts the Gabor function filter to enhance images.

The Gabor function filter is the band-pass filter selective in the frequency, playing strong roles of growth and enhancement in the pattern directions of fingerprints, and having good attenuation role in the gradient direction of fingerprints; it can well filter out noises between fingerprint patterns, enhance the contrast degree of the spine, and improve the quality of images. The Gabor function filter in the space of frequency domains taking the (x, y) point as the center is defined as:

$$G(x,y,\omega,\sigma_x,r,\theta)=\frac{1}{(\pi r\sigma_x)^{1/2}}e^{-1/2}[(R_1/\sigma_x)^2+(R_2/(r\sigma_x)^2)]e^{i\omega R_1}$$

In the formula: $R1=x\cos\theta+y\sin\theta$, $R2=-x\sin\theta+y\cos\theta$;$\theta$ is the two-dimensional rotation angle of the Gaussian envelope, that is the included angle between the pattern gradient direction of fingerprint images and the horizontal direction, also the angle of extracting fingerprint characteristics. The range of the ratio between the length of the Gaussian envelope and r is among $1.5\leq r\leq 2.0$.

The algorithm of extracting the fingerprint direction information is summarized as follows:

Step One: divide the collected fingerprint images (the pixel size in this paper as 300 × 300) into w × w quadrels (w in this paper taking the pixel size of 15).

Step Two: Calculate a step partial derivative of each pixel point in fingerprint images; the paper adopts the Sobel operator as the template of discrete convolutions, and the dimension of the template is 3 × 3.

Horizontal and vertical templates separately are $\begin{bmatrix}-1 & 0 & 1\\ -2 & 0 & 2\\ -1 & 0 & 1\end{bmatrix}$ and $\begin{bmatrix}-1 & -2 & -1\\ 0 & 0 & 0\\ 1 & 2 & 1\end{bmatrix}$. The original fingerprint images respectively conduct the discrete convolutions with the two templates, then it can obtain a step partial derivatives.

Step Three: Calculate directions of sub-blocks in each image:

$$V_x(x,y)=\sum_{u=x-\frac{w}{2}}^{x+\frac{w}{2}}\sum_{v=y-\frac{w}{2}}^{y+\frac{w}{2}} 2\partial_x(u,v)\,\partial_y(u,v)$$

$$V_y(x,y)=\sum_{u=x-\frac{w}{2}}^{x+\frac{w}{2}}\sum_{v=y-\frac{w}{2}}^{y+\frac{w}{2}} (\partial_x^2(u,v)-\partial_y^2(u,v))$$

$$\Theta(x,y)=\frac{1}{2}\tan^{-1}(V_y(x,y)/V_x(x,y))$$

Step Four: As the existence of noises during inputting fingerprint images, the direction information obtained above has a certain inaccuracy; according to the characteristics of fingerprints, except individual areas, within the local small-block areas, the changes in trends of the fingerprint patterns generally are relatively slow; using a low-pass filter to horizontally slide on the direction information of each block of fingerprint images, to amend inaccurate results of calculations. The specific method is to firstly calculate the projection component of θ (x, y) in the x-axis and y-axis:

$$\xi_x(x, y) = \cos[\theta(x, y)]$$
$$\xi_y(x, y) = \sin[\theta(x, y)]$$

Low-pass filtering is expressed as:

$$\xi_x'(x, y) = \sum_{u=-wl/2}^{wl/2} \sum_{v=-wl/2}^{wl/2} H(u, v) \cdot \xi x(x-uw, y-vw)$$

$$\xi_y'(x, y) = \sum_{u=-wl/2}^{wl/2} \sum_{v=-wl/2}^{wl/2} H(u, v) \cdot \xi y(x-uw, y-vw)$$

In which H (u, v) is the two-dimensional low-pass filter, $w_l \times w_l$ as the filter dimension of 5 × 5, w as the dimension of sub-blocks in images (this paper taking the dimension as w = 15 pixels).

The direction of sub-blocks after sliding is expressed as follows:

$$\overline{\theta}(x,y) = \frac{1}{2}\tan^{-1}(\xi_y'(x, y) / \xi_x'(x, y))$$

Step Five: In order to make the obtained directional diagram have a certain degree of robustness to the appropriate rotation, use directions of the surrounding eight blocks to adjust the direction of the current block, of which the specific method is to use the average value of the direction sum of eight blocks in the neighborhood of 3 × 3 of the current block as the direction of the current block.

Filtering steps are as follows:

(1) Decompose the low-frequency signal L and the high-frequency signal H of images;

(2) Decompose the low-frequency signal L further into the low-frequency signal LL1 and the high-frequency signal LH1; similarly decompose the high-frequency signal H further into the high-frequency signal HH1 and the low-frequency signal HL1;

(3) Repeat the process according to the actual needs so that the resolution factor of the frequency becomes higher and higher, and select the appropriate energy threshold for the decomposed frequency in the frequency range to conduct the threshold quantification, to achieve the predetermined target of finally reducing noises.

Subsequently, use the dynamic threshold method to conduct the binary processing, and then conduct thinning on binarized images to obtain the skeleton image; in the following stage of the characteristic extraction, use the template matching method to obtain the information about locations, directions and types of the characteristic points of minutiae (endpoints and bifurcation points); finally, the characteristic matching adopts the matching algorithm based on the characteristic points of minutiae.

4 The Design and Realization of the ID Authentication in the System

4.1 The Design of the Program

The system realizes the information security of the database system through two aspects: one is the management measures provided by the database management system itself like the user name / password identification, the control of authorization to use, and audit, etc.; the other is the management control set by the application program. As a database user, the security of the data information is the most concerned item for him, especially the issue of the user's authorization. Monitoring the security management of the system database is mainly divided into the following several categories:

1) *The user's authorization.* As a database user, the security of the data information is the most concerned item for him, especially the issue of the user's authorization. Different types of users are granted with different data management authorizations. Here the authorizations are divided into three categories: the system login authorization, the record viewing authorization and the system setting authorization.

2) *Data classification.* The data in the system include two categories: the first category: the configuration of information, such as the location and the attribute of detection equipments, the location and motion parameters, etc. of fire fighting equipments; the second category: records of the system, such as the time, site, location, and on-duty staff of the alarm and other written records and the on-site video record. The system administrator has all authorizations to be able to conveniently configure the system and to view and amend the records. While on-duty staffs only have partial authorizations such as the system login authorization, the record viewing authorization and the linkage alarm controlling authorization and so on.

3) *Designs of the means for ID authentications.* Designs of the means for ID authentications are the regular method of "the user name + password" and the method of the fingerprint authentication, and the system uses the method of combining the two types, which not only facilitates operations, but also ensures the security of the system.

4) *The audit function.* There are two ways that is users audit and systems audit. When users conduct the audit, the audit system records all attempts to access its tables or views, including successful and unsuccessful accesses, and the user's name, time, the operation code and other information in each operation. Such information usually is recorded in the data dictionary (or namely system tables); users can use such information to conduct the audit analysis. The system audit is conducted by the system administrator, in which the audit contents are primarily the first-level commands of the system and the using conditions of the database object.

4.2 Designs of the Means for ID Authentications

The means for ID authentications can be designed as the following two categories, and the means for authentications can be any of the following categories, while the defaulted first choice is the fingerprint identification.

1) The knowledge-based means to determine the identity, such as the password or PIN; the security of such ID authentication means is based only on the confidentiality of the user's password or PIN.

2) On the basis of his or her unique physiological behavioral characteristics to determine the identity that is the biometric authentication, such as the fingerprint authentication.

4.3 The Realization of the ID Authentication Function by Fingerprint Identifications

1) Functions to be Realized

Use the database of the monitoring system to unify the management of the fingerprint information, realizing the following several functions in the aspect of ID Authentications based on fingerprint authentications:

a) It can identify the user through his fingerprint, and allocate the authorization to operate the database according to the user's authorization he belongs to;

b) Directly input the fingerprint registration records into the database of the monitoring system; when maintain the database of the monitoring system, it can maintain the fingerprint database at the same time, no need to separately maintain the fingerprint database any more;

c) ID authentications in all operations of the monitoring system can pass fingerprint authentications, and record operation logs in detail;

d) After reading the operator's fingerprint information, compare with the documented information in the database, and after the comparison is correct, read again the guardian's fingerprint information; only when the operator's and his guardian's fingerprints both pass authentications, and they are not the same person, the remote-controlled operation can be continued, thus ensuring the security of the controlling operations.

2) The Realization of the Fingerprint Identification Function

The system on the basis of the interface program of the PIS2004 fingerprint input device, develops in the Visual C + + integrated development environment. The system includes the equipment management, the linkage control, the alarm display, the record checking, the user management, etc. functions, and the fingerprint identification function and the monitoring system function are fully integrated.

a) Fingerprint Image Acquisitions

Through the fingerprint sensor in the USB interface acquire the fingerprint images of living bodies. According to the above analysis on the fingerprint image acquisition, adopt the plane stamping of fingerprints, to improve the speed of identifications. The interface of fingerprint acquisitions is shown in Figure 2:

Fig. 2. The interface of fingerprint acquisitions

b) Fingerprint Preservations
The acquired fingerprint information is preserved in the application database of the monitoring system, achieving the unified management and maintenance of the database in the monitoring system. At the same time, compress the fingerprint images, and preserver the large-capacity fingerprint database after compressing through JPEG to reduce the memory space.

c) Users' ID Authentications
According to users' information obtained in the fingerprint identification system, complete the ID authentications on operators to ensure operators in the monitoring system of the centralized control center are legal on-duty staffs; through the permission authentication determine whether on-duty staffs have the required permissions, and ensure on-duty staffs can only use the allocated operated authorization.

d) Record Operation Logs
The database in the monitoring system of the centralized control center operates the automatic transference of the fingerprint ID authentication program, meanwhile preserving the password input window; when the fingerprint ID authentication fails, it can enter the password input window; passing the authorizations obtained in the fingerprint ID authentication equals to pass the password ID authentication, and record detailed operation logs in the application database of the monitoring system.

5 Conclusion

This paper applies the fingerprint identification technology to the fire monitoring system to realize that users conduct the ID authentications before the controlling on the fire fighting equipments in buildings. The implementation of the system greatly improves on-duty staffs' work efficiency in the centralized control center and shortens the operating time, especially plays an active role in emergent situations; meanwhile, it effectively prevents the non-authorized personnel from maliciously disrupting operations, provides the technical support for the safe operation of the fire monitoring system, and obtains users' unanimous praise.

References

1. Chen, S.: Remote Monitoring System. Electronic Products World (10), 18–20 (2000)
2. Bo, L.: Jinbin Light Rail Control Center Automatic Fire Alarm and the Design of the Linkage System. Standard Design of Railways 49(5), 89–92 (2005)
3. Bao, M., Huang, Y.: Computer Linkage Control System of Fire Alarms Based on Windows. Fire-fighting Technology and Products Information 15(12), 67–71 (2002)
4. Jie, J., Lijie, Hu, X.: Study on Images in the Automatic Fingerprint Identification and the Processing Algorithm. Micro Computer Information 21(36), 169–171 (2005)
5. Fu, J.: Study on Several Key Algorithms in the Fingerprint Identification. Full-text Database of China's Outstanding Doctorate's and Master's Theses, 18–25 (2003)
6. Lu, C., Zhang, G., Lin, L.: Evaluation Methods of Fingerprint Identification System Performances. Journal of Xi'an Electronic Science and Technology University (Natural Science Edition) 29(6) (2002)
7. Dai, P.: Research Progresses of the Fingerprint Identification Technology. Journal of Xiamen University (Natural Science Edition) 41(6) (2002)

Complex Scenes Modeling and View-Dependent Simulation Roaming Optimization Technology Based on VRML

Zhang Ruixia[1] and Yang Guozeng[2]

[1] School of Information Engineering
North China University of Water Conservancy & Electric Power,
Zhengzhou, China
[2] Department of Mathematics, ZhengZhou Normal University, Zhengzhou, China
zhangruixia@ncwu.edu.cn, ygz_0907@163.com

Abstract. Terrain scene modeling has become a research hotspot in virtual reality applications. However, due to the huge amount of 3D spatial data, there are still many problems unresolved such as low scene rendering frame rate, scene image popping. In this paper, a new simulation roaming optimization algorithm is proposed based on dynamic simplification which is implemented on VRML technology platform. Firstly, the appropriate terrain nodes are selected from the original DEM data according to the position and orientation of the viewpoint. Then on the basis of the selected nodes, data interpolation is carried out to form regular grid model data structure. Thirdly, the grid model data is loaded into elevation grid node of VRML to create terrain surface model. Finally, visualization rendering technologies such as illumination, texture, material are used to make the scene model realistic. The results of simulation experiment show that this algorithm improves scene rendering frame rate greatly and eliminates popping effect effectively and has a wide application prospect.

Keywords: 3D modeling, scene roaming, virtual reality, VRML, simplication algorithm.

1 Introduction

The rapid development of computer hardware and software technology provides a powerful technical platform for the construction of the virtual reality system. Currently, virtual reality technology has been used widely in urban planning, visual simulation, geographic information systems, simulation of virtual battlefield and so on. In recent years the virtual reality technology has become the research hotspot in related fields. The complex three-dimensional scenes include not only buildings, natural environment, but also large-scale terrain and so on. To achieve interactivity, immersion and imagination of virtual reality system, we need consider many factors such as model quality, scenes integration, scenes roaming, VR system web publishing

J. Luo (Ed.): Affective Computing and Intelligent Interaction, AISC 137, pp. 331–339.
springerlink.com

and so on. For the model quality, we hope the models possess not only high precision but also small volume (that is the models require little storage space). The modeling tools commonly used currently is 3D MAX. However, these type 3D models require more storage space which makes it is difficult to publish the model system on the web. For the scenes roaming, we hope the model render system possesses high frame rates and the scenes with different resolution switch smoothly. But in the actual scene rendering, huge 3D spatial data and attribute data leads to low frame rates and the popping effect of switching among levels of detail. For the VR system web publishing, we hope the rendered scenes can be published on the web to achieve the purpose of resource sharing. Therefore, based on the previous studies, we proposed a new scene modeling method and scene roaming optimization technology based on VRML technology platform which realize terrain scene simplification modeling. Results of simulation experiments demonstrate that the method proposed in this paper eliminates popping effect effectively, and has a higher frame rate compared with other methods. At the same time the modeling process is simple and easy to implement.

2 View-Dependent and View-Independent

Terrain scene modeling is a very important part in virtual reality system. For its huge spatial data, it is difficult to describe large-scale terrain accurately. Therefore, terrain simplification algorithm has become the key in the current research works. The current simplification algorithms includes static simplification (view-independent) and dynamic simplification (view-dependent).

2.1 View-Independent Simplification Algorithm

Early simplification algorithms mostly belong to static simplification which construct simple model from complex model based on some compact principles. These algorithms consider only the information themselves and have nothing to do with the viewpoint and cannot recover the original information. Static simplification may be used to construct multiple levels of detail models. However, it is necessary to pre-store many approximation models with different resolutions which require more storage space. At the same time, the models switching between the adjacent resolution may leads to image pop effect. Static simplification mainly includes vertex clustering, edge folding method, regional consolidation and so on. The commonly used method is vertex clustering. In vertex clustering algorithm, firstly the bounding box of the original model is divided into many small regional by uniform spatial subdivision. Then in each regional, the representative vertexes are selected to replace other vertexes. Finally, we can obtain the geometric model of low complexity by re-triangulating the selected representative vertexes. Figure 1 shows the basic principle of vertex clustering simplification algorithm.

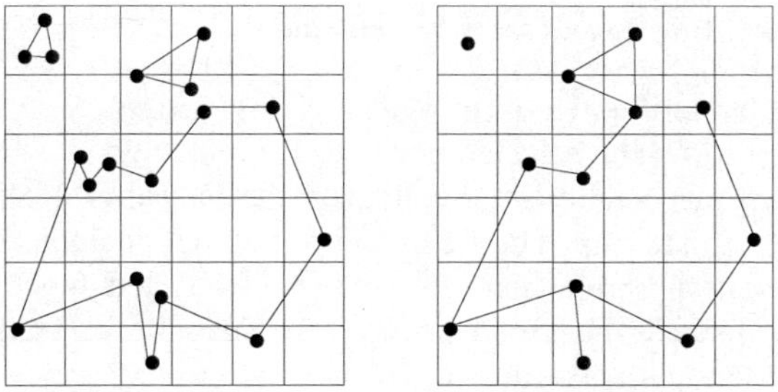

(a) before simplification (b) after simplification

Fig. 1. Vertex clustering simplification

2.2 View-Dependent Simplification Algorithm

The basic idea of dynamic simplification algorithm is as follows: in the simplification procession, we can obtain the model with appropriate resolution. The simplification degree of each model is decided by the factors out of the model itself——viewpoint. Generally the dynamic simplification is achieved by a number of local geometric transformations to generate approximate model with continuous different resolution. The main problem is how to select appropriate terrain nodes based on the position of viewpoint. In this paper, the idea of nodes selection is to define each node's height difference firstly based on the variations of other nodes surrounding the node. If height difference is larger than allowable error, the node will be selected. The calculation method of height difference in the height field dataset is defined as follows:

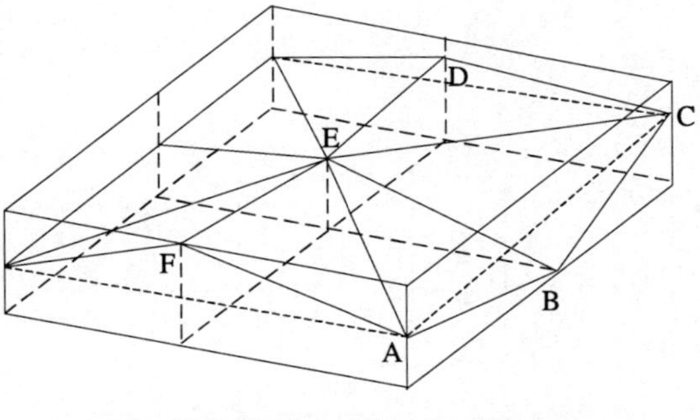

Fig. 2. Definition of height difference

We use the amount of change in slope between two triangles (co-triangle pair) to describe height difference [4]. For example, in Figure 2, node B is the common vertex of co-triangle ΔABE and ΔBCE in the height field dataset. The conditions under which node B can be selected are primarily decided by the amount of change in slope when ΔABE and ΔBCE is combined into a single triangle ΔACE. Therefore, the height difference of terrain node is the minimum distance from the node to the plane formed by the single triangle. The height difference reflects influence of nodes simplification over terrain mesh model. As the height difference increases, the chance of terrain node being selected increases [6].

2.3 Data Interpolation Based on Selected Nodes

To construct the simplified terrain model, it is necessary to implement data interpolation for based on the selected nodes to form new grid model data. The main idea of spatial data interpolation is to calculate the elevation of other region nodes based on the given conditions. The Calculation methods include distance weighted, polynomial interpolation, spline interpolation, multi-faceted function and so on. In this paper we use distance weighted method. By spatial data interpolation, we can obtain regular grid model data structure which can be used to construct terrain model with the help of VRML technology.

3 Work Principle and Framework of VRML

The basic objective of VRML is to create three-dimensional interactive Internet multimedia. It takes the internet as its internet-application platform and basic structure. Since its appearance, it has changed the network's two-dimensional scene to three-dimensional space. Like HTML, the VRML is a description language by browser too. Compared with other languages the VRML the following advantages such as simple syntax, platform-independent, fast network transmission speeds, powerful 3D modeling capabilities, openness and so on. The access method of VRML is based on B/S architecture. The server provides the scene model files and all kinds of resources. Customer terminals send request command through networks, then servers send the objectives scene model fields based on the request command to the customer terminal. Finally the customer terminal is responsible to render the 3D terrain scene by its browser. As the browser is provided by the local terminal, the rendering architecture is platform independence. Moreover, the transmission objects through networks are text file describing the scene model. Therefore, the data volume is very small and transmission efficiency is very high. Figure 3 shows the work principle of VRML[5].

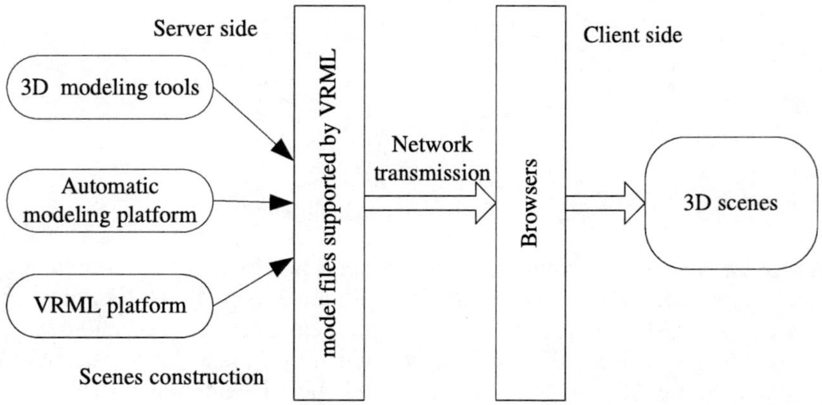

Fig. 3. Frame and work priciple of VRML

The latest version of VRML (VRML2.0, SO / IEC 14772-1:1997) added a new node to build terrain surface model with grid data——ElevationGrid. With this node, designers can specify the elevation of each cell within the given area. The syntax is as follows:

```
Elevation Grid {
   eventIn  MFFloat    set-height
   exposedField  SFNode  Color        NULL
   exposedField  SFNode normal      NULL
   exposedField  SFNode texcoord   NULL
   field MFFloat  height [ ]
   field SFBool    CCW TRUE
   field SFBool    colorRerVetex   TRUE
   field SFFloat   creaseAngle  0
   field SFBool    normalPerVetex   TRUE
   field SFBool    solid TRUE
   field SFInt32   xDimension   0
   field SFFloat   xSpacing   1.0
   field SFInt32   zDimension   0
   field SFFloat   zSpacing   1.0
   }
```

The principle and structure of VRML ElevationGrid is shown in figure 4.

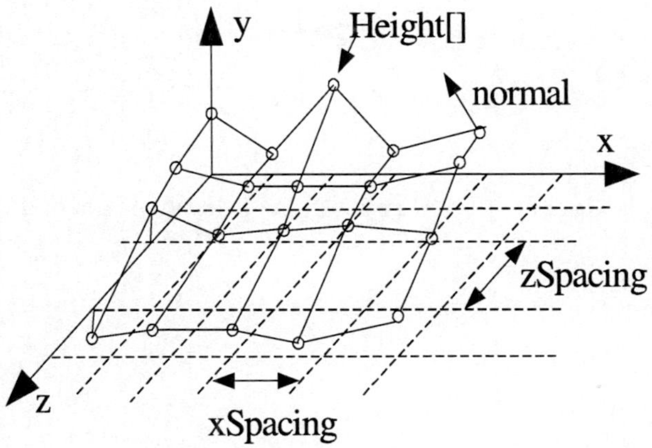

Fig. 4. Modeling priciple of elevation grid

Parameter description:

xDimension: the number of vertices in x direction;
xSpacing: the distance between the vertices in x direction;
zDimension: the number of vertices in z direction;
zSpacing: the distance between the vertices in z direction;

height [ ] : elevation value array, each value corresponds to a grid vertex.
normalPervertex: vector coordinates
texCoord : texture mapping
creaseAngle: smooth angle threshold
normal: normal vector of node

Grid model data structure provides effective support for building terrain surface model by VRML ElevationGrid node. We can load the regular grid model data into the domain "height []" of ElevationGrid node. For example, At some moment, suppose that the file name of the standard grid data processed through height difference calculation and spatial data interpolation is "meshdata.dem" in which there are 64x64 terrain nodes, zSpacing and xSpacing are both 1 unit length and other parameters take default values. The pseudo-code for terrain model can be like this:

```
#VEML V2.0 utf8
NavigationInfo {
type EXAMINE
}
Shape {
     appearance Appearance {
          material Material {
               diffuseColor 0.5 0.5 5.0 5.0
          }
     }
    geometry ElevationGrid {
        xDimension    64
zDimension    64
xSpacing        1.0
zSpacing        1.0
height ["meshdata.dem" ]  # (Here insert the #elevation data of grid node by Java
script)
solid     false
creaseAngle   5
……
}
}
```

This pseudo-code template possesses some universality. For the different input mesh data, different terrain model can be generated easily. Based on above analysis, we can obtain the optimization algorithm framework (as shown in figure 5).

To control the accuracy of the model, we can adjust the threshold of node height difference dynamically. That is the higher the threshold is, the more accurate the terrain model is.

To construct realistic three-dimensional scene, it is necessary to add lighting in the scene, set the material properties, map surface textures and set interaction. Under the VRML technology platform, these functions can be realized easily.

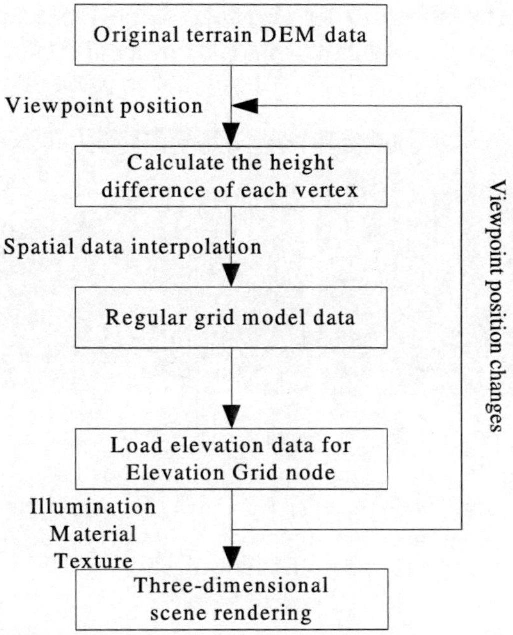

Fig. 5. Optimization algorithm framework for scene rendering

4 Simulation Experiment Analysis

To test the effect of the algorithm proposed in this paper, we developed a prototype experimental system with based on VRML technology platform. The area to be modeled lies in a mine in Hebei province. We select an area in which there are 256x256 measuring elevation data nodes based the original terrain data. The static grid model of this area is shown in figure 6.

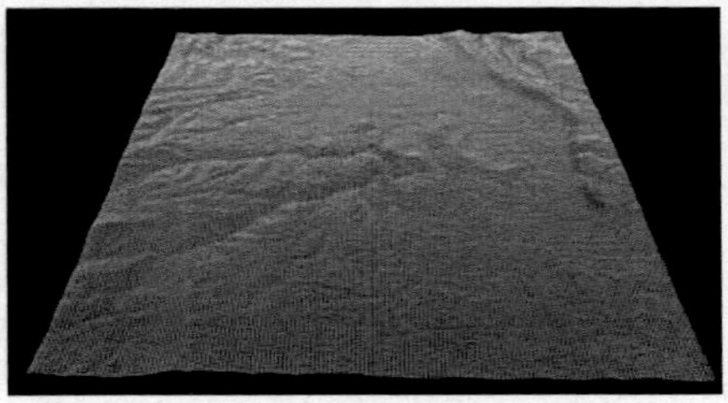

Fig. 6. Terrain grid model

With the VRML texture mapping technology, we can implement texture mapping for the given area. Then the simulation roaming experiment can be made on graphics workstation. The configuration of graphics workstation is as follows: CPU: Intel

E5502, 1.86GHz, 4GB (2x2GB) DDR3 RDIMM, nVidia Quadro FX 1800(768M). The movement speed of viewpoint is controlled under 70-100m / s. Figure 7 is a roaming scene.

Fig. 7. Terrain roaming scene

Terrain roaming simulation results show that the algorithm overcomes some shortcomings of static optimization algorithms. It improves the frame rate of scene rendering and eliminates the popping effect effectively.

5 Conclusions

The virtual reality technology has been used more and more wildly in many fields. And scene modeling optimization has become one of the focuses in VR study. This paper proposed an optimization algorithm based on node height difference selection of terrain nodes. With the VRML technology platform, a prototype experimental system is developed. The results of simulation experiment show that this algorithm has a wide application prospect in terrain scene modeling.

References

1. Hoppe, H.: New quadric metric for simplification meshes with appearance attributes. In: Proceedings of the IEEE Visualization 1999, pp. 59–66 (1999)
2. Wu, Y.-G., Du, Y., Wang, X.-M.: Lod-based Algorithm of Tin Model for Large Terrain Simulation. Journal of System Simulation 17(3), 665–669 (2005)
3. Wei, D., Huang, Y.-Q.: Research of Visual Builder of Virtual Scene Based on VRML. Journal of System Simulation 17(1), 72–74 (2005)
4. Wu, H.-X., Su, J.-Q., Xue, H.-F.: View-dependent real-time rendering algorithm for large-scale terrain. Application Research of Computers 25(10), 235–237 (2008)
5. Dong, H., Tang, J.: Three-Dimensional Multi-Resolution Terrain Model Rendering Under Network Platform Based on LOD unpublished
6. Li, X., Wu, H., Xue, H.: A New Large-Scale Terrain Real-Time Simplification Algorithm Based on Delaunay Triangulation. In: The Second International Conference on Space Information Technology, Wu Han, China (November 2007)
7. Liu Ch, Y.: The Construction Method of Geometric Objects in 3D Terrain. Computer Applications (21), 29–31 (2001)

8. Garland, M., Heckbert, P.S.: Simplifying surfaces with color and texture using quadric error metrics. In: Proceedings of IEEE Visualization 1998, pp. 263–269 (1998)
9. Morackot, P., Tempfli, K., Molenaar, M.: A Tetrahedron-Based 3D Vector Data Model for Geoinformation. In: Molenaar, M. (ed.) Advanced Geographic Data Modelling, pp. 129–140. Geodesy Press, Sylvia De Hoop (1994)
10. Losa, A.D.E., Cervelle, B.: 3D Topological Modeling and Visualisation for 3D GIS. Computers & Graphics (23), 469–478 (1999)
11. Cui, H.-G., Yan, C.-H.: Research on 3D topography construction algorithm based on VRML & Java. Journal of Naval University of Engineering 14(5), 34–36 (2002) (Chinese)

Consensus Problems of DM System Network with Different Time-Delays

Wei-Long Shan, Wu-Neng Zhou, and Ming-Hao Li

College of Information Science and Technology,
Donghua University, Shanghai 201620 P.R. China
Shanweilong0725@yahoo.com.cn, wnzhou@dhu.edu.cn,
lmh@mail.dhu.edu.cn

Abstract. As we all know, consensus problems of multi-agent networks have attracted more and more attention recent years. In these related papers, many cases of networks have been studied, including directed and undirected networks, fixed and switching topology, with communication time-delay and no time-delay. However, those papers have not discussed the real systems. And in this paper, we will discuss the DM（Delta Modulation） system network, a real complex system with encode and decode modules in each agent. DM system is widely used in the A/D and D/A modules, and the signal encoded is transmitted during the communication channels. From a practical system, I try to extract the abstract signal relationship, and find the consensus problems in real system. Following, through a new method, main point we discussed is actually the undirected discrete-time DM networks with different time-delays created by the communication constraints.

Keywords: Communication, discrete-time (DT) model, DM (Delta Modulation) system, Distributed consensus.

1 Problem Formulation

Delta modulation is referred to as ΔM or DM (Delta Modulation). It is another method for analog signal to transmit with digital transmission method, which can be seen as an important special case of DPCM (Differential Pulse Code Modulation). Its purpose is to simplify the speech coding method. During the digital communication, we do not directly transmit the original voice signal sequence, but the difference sequence, which can replace the original sequence of effective information. And the difference signal energy is much smaller than the original value which can reduce the number of quantization levels, so this method can greatly compress digital rate. At the receiving end, as long as we add the difference sequence to the forecast sequence stacking sequence, we can reconstruct the original sequence.

Compared with the PCM (Pulse Code Modulation) encoding method, DM has the advantage of simple equipment for coding and decoding, high SNR (Signal to Noise Ratio) with low bit rate quantization, good characteristics in anti-error and so on. It has been widely used in the military and the industrial sector of private communication

J. Luo (Ed.): Affective Computing and Intelligent Interaction, AISC 137, pp. 341–349.
springerlink.com

networks and satellite communications, and in recent years used as A/【坡12123D converter in high-speed ultra-large-scale integrated circuits.

Concerning the DM system, we consider the average-consensus control for a network of agents with the dynamics, which can be used in the DM system. An iterative form of the consensus algorithm can be stated as follows in discrete-time format:

$$x_i(k+1) = x_i(k) + \varepsilon u_i(k) \tag{1}$$

$$\begin{cases} u_i(k) = \sum_{j \in N_i} a_{ij}\left(x_j(k) - x_i(k)\right), \\ k = 0,1,2\cdots, i = 1,2,\cdots N, \end{cases} \tag{2}$$

where $x_i(k)$ represents the state of the i th agent, ε is the control gain, and $u_i(k)$ is the control gain. Usually we represent the communications by undirected graph $G = \{\nu, E, \mathrm{A}\}$, where $\nu = \{1,2,\cdots,N\}$ is the set of nodes, η_G is the set of edges, $\mathrm{A} = \left[a_{ij}\right]$ represents the weighted adjacency matrix. (j,i) shows the communication channel connection from the i th agent to the j th agent. And the neighborhood of the i th is showed by $N_i^{[1]} = \{j \in \nu / (i,j) \in E\}$.

When $j \in N_i$, $i, j \in \nu$; $a_{ij} = a_{ji} \geq 0$, and we suppose $a_{ii} = 0$. $\deg_i = \sum_{j=1}^{N} a_{ij}$ is called the degree of node i, $d^* = \max_i \deg_i$ is degree of G. The Laplacian matrix of G is defined as $L = D - \mathrm{A}$, $D = diag(\deg_1, \deg_2, \cdots, \deg_N)$. If there is a communication channel for any $(i,j) \in \eta$, we call G is connected.

We usually call the dynamics (1)-(2) together with graph G a dynamic network. $\mu = \{u_i : i = 1,2,\cdots,N\}$ is the distributed protocol, and the so-called average-consensus control means to design a distributed protocol for the dynamic network, such that for any initial values $x_1(0),\ldots,x_N(0)$, the states of all the agents converge to $\frac{1}{N}\sum_{j=1}^{N} x_j(0)$ when $t \to \infty$.

2 Protocol Design

Network with continuous-time (CT) integrator agents is discussed in Paper [2][3]. About the DT model [4] has discussed the distributed average consensus protocol:

$$\begin{cases} u_i(k) = \sum_{j \in N_i} a_{ij}\left(x_j(k) - x_i(k)\right), \\ k = 0,1,2\cdots, i = 1,2,\cdots, N. \end{cases} \tag{3}$$

Node i depends on the states of its neighbors. Concerning the DM system we discussed, whose structure is stated as bellow including the encoding and decoding modules in each agent. After the encoding module of one agent, the discrete time state will be transmitted through the communication channel. Then the neighbor agents will receive the discrete state and reconstruct the discrete state into the continuous state through the decoding module:

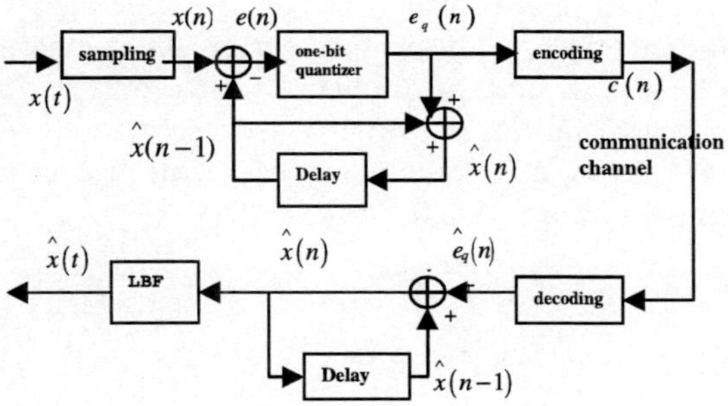

Fig. 1. The undirected topology of DM network

Remark 1: From the graph below, $m(t)$ represents analog signal with continuous time-change. we can use a step waveform with time interval Δt and adjacent magnitude difference is $+\sigma$ or $-\sigma$ to approach $m(t)$. as long as Δt is small enough, which is $f_s = {}^{1}\!/_{\Delta t}$ is high enough, and is small enough, step waveform $m'(t)$ can approximately represent $m(t)$. where σ is the quantitative level, $\Delta t = T_s$ sampling interval.

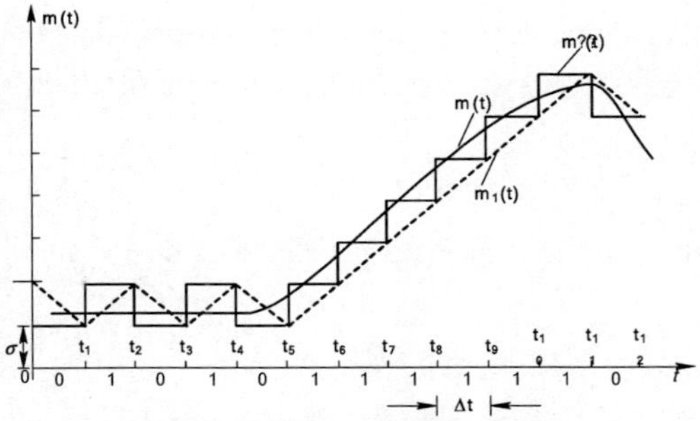

Fig. 2. DM waveform diagram

Remark 2: As there exists certain relevance between adjacent sampling rate of the voice signals, we can predict the moment value based on previous value of time. As long as there exists the difference between the actual value and the prediction value, without the need for transmission of each value. This method is DPCM.

Remark 3: The value of voice signals can be divided into predictable and unpredictable parts. The predictable part(relevant part) is the weighted value past ; some unpredictable (non-relevant parts) parts can be seen as a prediction error. In this way, during the digital communication, we do not directly transmit the original voice signal sequence, but only send the difference sequence. Because the difference sequence information can replace the original sequence of effective information, and the difference signal energy is much smaller than the original value which can reduce the number of quantization levels, so this method can greatly compress digital rate. At the receiving end, as long as we add the difference sequence to the forecast sequence stacking sequence, you can reconstruct the original sequence.

From the structure of this DM system we can write the coder and encoder representation according the information flow[5]:

(a): encoder

$$\begin{cases} e(k) = x_i(k) - \hat{x}_i(k-1), \\ \hat{x}_i(k) = \hat{x}_i(k-1) + e_q(k) \\ e_q(k) = q(e(k)), \end{cases} \tag{4}$$

(b): decoder

$$\hat{x}_{ji}(k) = \hat{x}_{ji}(k-1) + \hat{e}_q(k) . \tag{5}$$

For each communication channel $(j,i) \in \eta$, the i th agent receives the output of jth agent whose encoder output is $c(n)m'(t)\Delta t = T_s - \sigma f_s = {}^1\!/_{\Delta t}{}'$, actually the structure of this DM system is convenient for the analysis in theory, but in the real structure $c(n)$ is equal to $e_q(n)$, so we use the decoder to estimate $x_j(k)$, here $\hat{x}_{ji}(k)$ is the output of decoder.

From the above equations, we can conclude that there exists a functional connection between $x_i(k)$ and $\hat{x}_i(k)$, and we can represent it from (4) and (5) as

$$x_i(k) = f\left[\hat{x}_i(k), \hat{x}_i(k-1)\right]. \tag{6}$$

In this paper we discussed the consensus problems of DM system, which can be stated as :

$$x_1(k) = x_2(k) = \cdots = x_N(k) = s(k) \ \ when \ \ k \to \infty .$$

As long as the state $x_i(k)$ achieve consensus, $\hat{x}_i(k)$ will in-phase achieve consensus from the equation (6). So the protocol can be represented by :

$$\begin{cases} u_i(k) = \sum_{j \in N_i} a_{ij}\left(\hat{x}_{ji}(k) - \hat{x}_i(k)\right), \\ \quad k = 0,1,\ldots,i = 1,2,\ldots,N. \end{cases} \tag{7}$$

Otherwise, we can get $\hat{x}_{ji}(k) = \hat{x}_i(k)$ from the coding and decoding equations (4) and (5). Consensus protocol (7) can be changed into the following formation:

$$u_i(k) = \sum_{j \in N_i} a_{ij}\left(\hat{x}_j(k) - \hat{x}_i(k)\right) \tag{8}$$

and

$$\hat{x}_i(k+1) = \hat{x}_i(k) + \varepsilon u_i(k). \tag{9}$$

In the actual system, decoder module will receive delayed- signal because of kinds of communication channel transmission reasons from i to j, just like different channels, different lengths, different terrains and different noises. In order to research conveniently, we suppose the self-delay τ_{ij} is equal to the communication delay. About the CT model, [4][6][7]discussed the consensus with time delays. Then the communication protocol design with time delay τ_{ij} is

$$\hat{x}_i(k+1) = \hat{x}_i(k) + \varepsilon \sum_{j \in N_i} a_{ij}\left(\hat{x}_j(k-\tau_{ij}) - \hat{x}_i(k-\tau_{ij})\right),$$

$$k = 0,1,\ldots,i = 1,2,\ldots,N. \tag{10}$$

3 Consensus Analysis

When the above system (10) achieves consensus, which is also called the balanced system, we can get

$$\begin{cases} \hat{x}^* = \left(\hat{x}_1^*, \hat{x}_2^*, \cdots, \hat{x}_n^*\right)^T, \\ \hat{x}_1^* = \hat{x}_2^* = \cdots = \hat{x}_n^*. \end{cases} \tag{11}$$

As $u_i(k) = \varepsilon \sum_{j \in N_i} a_{ij}\left(\hat{x}_j(k-\tau_{ij}) - \hat{x}_i(k-\tau_{ij})\right)$, we get $\sum_{i=1}^{n} u_i(k) = 0$. Suppose

$$\alpha(k) = Ave(x) = \sum_{i=1}^{n} \hat{x}_i(k) \Big/ n,$$

we get

$$\alpha(k+1) - \alpha(k) = \sum_{i=1}^{n} u_i(k) \Big/ n = 0.$$

From the above expressions, $\alpha(k)$ is invariable, and the system converges to $\alpha(0)$.

Let $\xi_i(k) = \hat{x}_i(k) - \hat{x}^*$, together with (10) we can get

$$\xi_i(k+1) = \xi_i(k) + \varepsilon \sum_{j \in N_i} a_{ij} \left(\xi_j(k - \tau_{ij}) - \xi_i(k - \tau_{ij}) \right). \tag{12}$$

Lemma 1 [8] :

Suppose the network G is fixed undirected graph, $\varepsilon \in (0, 1/\Delta)$,if $\rho(\varepsilon L) < 2\sin \pi/2(2\tau+1)$ the system will converges to zero.

Remark 4: In this equation, we want to insist that τ_{ij} in this discrete system are random integers, which is to say that it differs in different channels. And in order to make research convenient in the following deduction, τ_{ij} is just a symbol indicating the random integers, not a variable depending on i and j.

Concerning (12), using Z-transform, and letting

$\zeta_i(z) = Z(\xi_i(k))$, $\zeta_i(0) = 0$,

we have

$$z\zeta_i(z) = \zeta_i(z) + \varepsilon z^{-\tau_{ij}} \sum_{j \in N_i} a_{ij} \left[\zeta_j(z) - \zeta_i(z) \right],$$

$$\zeta_i(z) = \varepsilon \frac{z^{-\tau_{ij}}}{z-1} \sum_{j \in N_i} a_{ij} \left[\zeta_j(z) - \zeta_i(z) \right], \tag{13}$$

$$\det \left[I + diag\left(\frac{z^{-\tau_{ij}}}{z-1} \right) \varepsilon L \right] = 0. \tag{14}$$

Let $z = e^s$, and we have $\det[I + G(s)] = 0$,

$$G(s) = diag\left(\frac{e^{-s\tau_{ij}}}{e^s - 1} \right) \varepsilon L. \tag{15}$$

From the Lemma 1, substituting τ with τ_{ij}, we can see that the system is approximately stable.

4 Simulation Result

The undirected topology of DM system is stated as below and we suppose the weighted value is 1, the communication time-delay is random. Let $\varepsilon = 0.05$, and $\varepsilon \in (0, 1/\Delta)$. We can get $\rho(L) = 5.9$, and from the Lemma 1, have $\tau_{\max} = 4.045$.

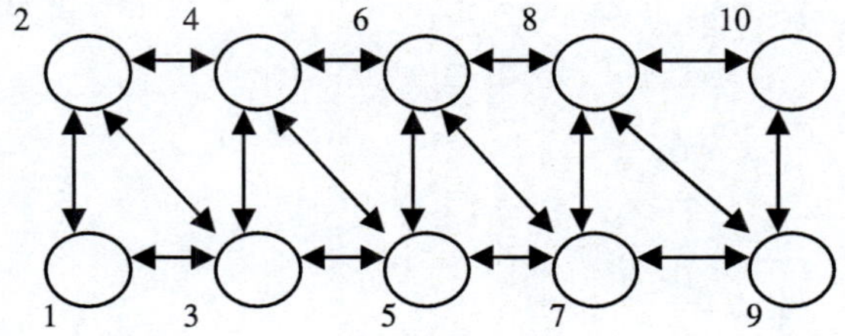

Fig. 3. The undirected topology of DM network

Suppose the initial state of each node is -5,-4,-3,-2,-1,1,2, 3,4,5. And the simulation graph with no time-delay is:

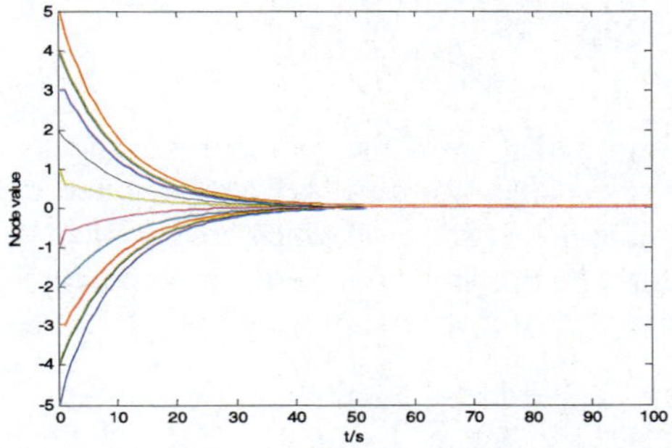

Fig. 4. The simulation with no time-delay

Suppose the time-delay is random at [1, 4], $\tau_{ij} \leq 4$, simulation graph is as following Fig. 4. We can see that with time-delay system converges slower at t=500s compared with t=50s with no time delay.

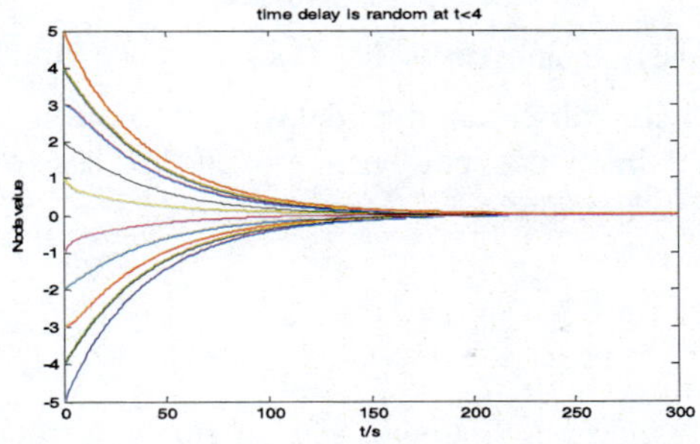

Fig. 5. The simulation with delay $\tau_{ij} \leq 4$

Suppose the time-delay is random at [5, 10], $\tau_{ij} > 4$, the simulation graph is as follows:

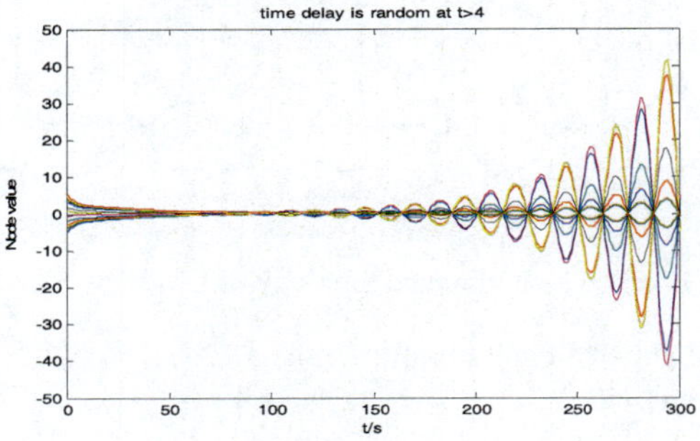

Fig. 6. The simulation with delay $\tau_{ij} > 4$

We can see from Fig. 5 that, with the protocol designed, the DM system could achieve consensus satisfying the lemma 1. Of course, there exists another complex situation with random time-delay, some of which are less than 4, others greater. It can produce different results and this situation needs further research.

5 Conclusion

In this paper, we have discussed the distributed consensus problems of DM system, which combines the knowledge about both controlling and communication. From the structure of each node, we can see the connection between $x(k)$ and $\hat{x}(k)$, so the consensus problem is proved in a new way. Basing on the relation between $x(k)$ and $\hat{x}(k)$,I transmit the real complex DM system into the certification of consensus of $\hat{x}(k)$ in order to prove the consensus of $x(k)$.And we discussed the distributed consensus protocol with different time-delays to confirm that the system is asymptotically stable, which has not been proved. At last we use the computer simulation to prove the efficiency.

References

1. Jadbabaie, A., Lin, J., Morse, A.S.: Coordination of groups of mobile agents using nearest neighbor rules. IEEE Trans. on Automatic Control 48(6), 988–1001 (2003)
2. Moreau, L.: Stability of multi-agent systems with time-dependent communication links. IEEE Trans. on Automatic Control 50(2), 169–182 (2005)
3. Lin, Z., Broucke, M., Francis, B.: Local control strategies for groups of mobile autonomous agents. IEEE Trans. on Automatic Control 49(4), 622–629 (2004)
4. Olfati Saber, R., Murray, R.M.: Consensus problems in networks of agents with switching topology and time-delays. IEEE Trans. on Automatic Control 49(9), 1520–1533 (2004)

5. Fax, J.A., Murray, R.: Information flow and cooperative control of vehicle formation. IEEE Trans. on Automatic Control 49(9), 1465–1476 (2004)
6. Yu, H., Jian, J.G., Wang, Y.J.: Weighted average consensus for networks of multi-agent with time delays. Control and Decision 22(5), 558–561 (2007)
7. Liu, C., Tian, Y.: Consensus of multi-agent system with diverse communication delays. Journal of Southeast University (Natural Science Edition) 38(1), 170–174 (2008) (in Chinese)
8. Yang, H., Zhang, S.: Consensus of multi-agent system with discrete time. Control and Decision 24(3), 413–416 (2009) (in Chinese)

Design and Implementation of Management and Query Center of Log Auditing System Based on SSH2*

Zhengjun Cheng, Weiming Wang, and Ming Gao

Institute of Network and Communication Engineering, Zhejiang Gongshang University, No.18, Xuezheng Str., Xiasha University Town, Hangzhou, P.R. China, 310018
{chengzhengjun,gmyyqno1}@pop.zjgsu.edu.cn, wmwang@zjgsu.edu.cn

Abstract. As an important component of the Log Audit System, Management and Query Center is mainly responsible for log query and analysis and displays results in a web browser. For the system should have a good Maintainability and scalability to support different log types, through a comparative study of two design options, we decided to implement Management and Query Center based on an integration of Struts2, Spring and Hibernate framework (SSH2) and the system implemented have a very good scalability and maintainability.

Keywords: Log Audit, Struts2, Spring, Hibernate.

1 Introduction

Almost all of the equipments in computer network system can generate logs to record their own behavior and relevant network events. Log management and analysis is beneficial for indentifying security incidents, policy violation and operational problems, and also Logs are useful when performing auditing and forensic analysis, supporting internal investigations, establishing baselines, and identifying operational trends and long-term problems1. Yet there have some logs analysis tools (Log Parser, Logwatch). But many of them can only simply analyze several types of logs on specific equipment or computer2. For a lot of systems are distributed, It is impractical and immensely time consuming to examine the contents of each security log on every server. Furthermore, indentifying some kinds of attacks or security incidents may need association analysis of the logs from different servers3. So we need to collect the logs from different hosts in one place and Log Auditing System works in a process as generating, transmitting, storing, analyzing, and disposing of computer security log data 4.

As an important component of the Log Audit System, Management and Query Center is mainly responsible for log query and analysis, Agent and Auditing Center's configuration and display the reports in a web browser5. To design and implement Management and Query Center, we need solve these problems: First, the system should

* Supported by the National Natural Science Foundation of China under Grant No. 60903214, 60970126, the National High Technology Development 863 Program of China under Grant No. 2008AA01A323, Zhejiang Provincial NSF China No. Y1080078, Y1090452, Zhejiang Sci & Tech Project No.2009C31066, 2008C21093,2009c11050 and Zhejiang Gongshang University NSF.

J. Luo (Ed.): Affective Computing and Intelligent Interaction, AISC 137, pp. 351–357.
springerlink.com

have a good scalability to support new types of logs8. Second, the system is implemented in Java and Auditing Center is implemented in c language, so we need to realize the communication between this two systems.Third, for there are so many logs that we need to process, we should think about that we may change a new database management system in the future.

To make our system having a good Maintainability and scalability, we decided to implement Management and Query System based on SSH2 integrated framework6.

2 Comparative Study of Implementations

We decided to realize the system in B/S model based on J2EE so that the user could access the system on any host. There are many WEB application development methods and we proposed two different implementations such as Java Applet + Servlet7 and SSH2.

2.1 Java Applet + Servlet

An applet is a program written in the Java programming language that can be included in an HTML page. When you use a Java technology-enabled browser to view a page that contains an applet, the applet's code is transferred to your system and executed by the browser's Java Virtual Machine (JVM).A servlet is a small Java program that runs within a Web server. Servlets receive and respond to requests from Web clients, usually across HTTP.

So we use applet technology to realize our client program and servlet technology to realize server program. Actually our system based on Java Applet and Servlet shown in Figure 1 can be seen as a four-layer structure: the User Interface Layer (GUI), the Message Management Layer (MML), the Daemon Layer (DL) and the Plug-in Layer(PL).

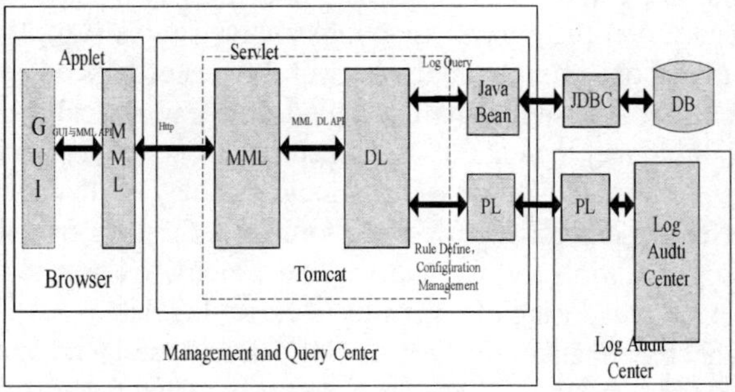

Fig. 1. Applet and Servlet system

A Ftp logs query operation in this system are shown in Figure 2: GUI map the query operation which contains many parameters such as the log type ,the logs generated date and the page number to Applet client's MML message. And the client's MML transfer its message to the server's MML across HTTP. Servlets explain the message and identify the operation as an Ftp log query operation based on the specific content of the message. Then servlets call the corresponding Daemon handler procedures to process the query and the Daemon will search the logs from the logs database and return logs list to the GUI in an inverse way.

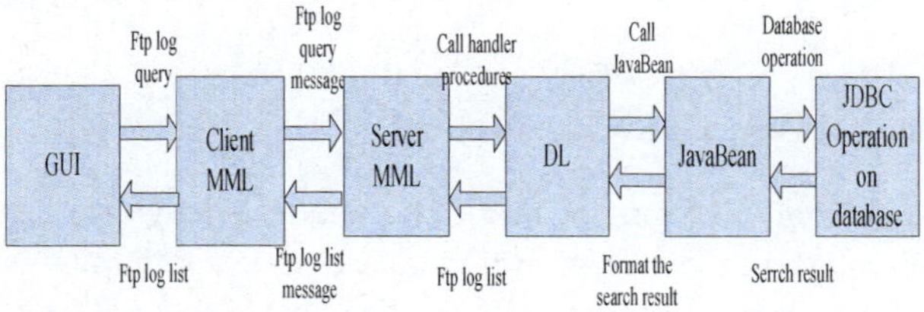

Fig. 2. A Ftp logs query operation

2.2 SSH2

Model-View-Controller (MVC) model is a software architecture which recommends programmers to design their software system in three layers7. The separation of View layer and Logic layer can make the program have low-coupling, maintainability, high reusability and to be conducive to software engineering management.

Struts2 was developed from Struts and Webwork. As a popular framework, Struts2 cleanly separate the model (application logic that interacts with a database) from the view (HTML pages presented to the client) and the controller (instance that passes information between view and model)9.The Spring Framework features its own MVC framework and also can integrate with a number of well-known frameworks such as Strust2 and Hibernate . But compared with Spring Framework, Struts2 have a more outstanding performance in View Layer. So we decided to implement our system based on an integration of these two frameworks. Object-relational mapping (ORM) is a programming technique for converting data between incompatible type systems in relational databases and object-oriented programming languages. Hibernate is an ORM library and provide a framework for mapping an object-relational domain model to a traditional relational database.

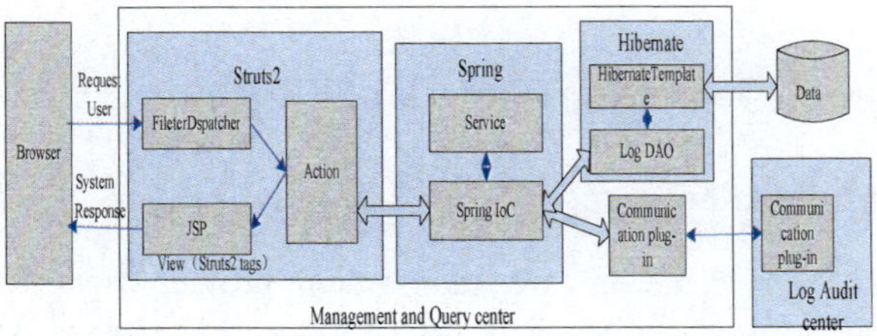

Fig. 3. SSH2 System architecture

We implement our system based on a integration of Struts2, Spring and Hibernate (SSH2), because these three frameworks have a good performance in different layers.

The system architecture was shown in Figure 3. The user's request (logs query or system configuration) will be sent to the FilterDispatcher of Struts2 and FilterDispatcher will pass it to the Action class. Then Action class call service class in service layer through Spring IoC container and service class can access to communication plug-in or log DAO class.

2.3 Difference between These Two Methods

Applet program is only written in Java programming language, so it's easy for programmer to implement the system. But there are some shortcomings for the Applet client program. First, Applet program runs on the client side, so it needs some time to download the program to the client. And if the program is large, the client should wait for a long time. Also the client host needs to install Java Virtual Machine (JVM). Second, We need to design a Message Management Layer (MML) to transfer messages between the client host and the server host. The messages contain a lot of information such as the client operation types and we should know the meaning of these messages. For the applet program can not access to the database to search the logs directly, so we should know how to transfer the search result from server host to client host. Third, There are so many log types and also a lot of new log types will be defined in different network device or application program, so our program should have a good scalability and maintainability. But it's difficult to design an applet program to have such a great performance.

Well we resolve those problems based on the SSH2 method. SSH2's hierarchical design method makes the program have a clear structure. Struts2 achieve a separation of the View layer and Controller layer. And the JSP pages can use the Struts2 tags so that JSP pages do not contain any program logic but rather a simple display. This make the interface art to be easy and the code have a good readability. Also, Struts2 can provide a good data authentication and rights management implementation. Spring can realize a management of Java object, For example, when we modify the code of class A, we can change the Spring configuration file in accord and don't need to change the code of all the other classes which contain class A as before. Hibernate can support different databases which is a very good solution to the problem of the database portability.

3 System Implementation Based on SSH2

3.1 System Introduction

Management and Query center of Log Audit system provide users with an interaction platform based on B/S mode on which can we configure and manage system, search and analyze the logs and generate reports. System's location of the Log Audit System was shown in Figure 4.

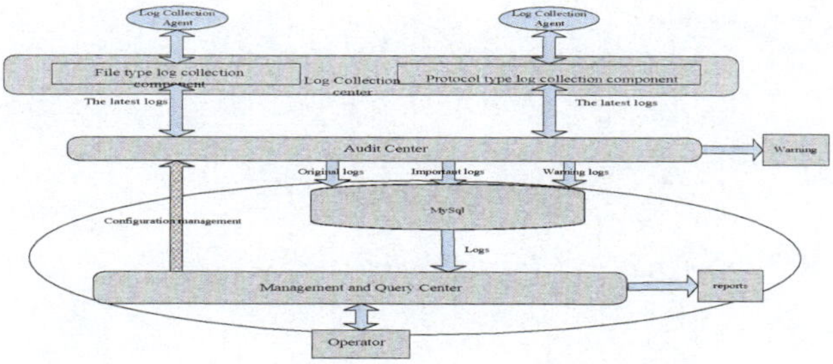

Fig. 4. Log Audit System

3.2 Implementation of System Based on SSH2

The system has seven main components: real-time audit, log query, data management, rule definition, object monitor, system management and behavior audit. We check the log query component as an example to introduce our system implementation briefly.

Log query component allow us to search latest logs or combined search based on fields of logs. But There are so many log types and also a lot of new log types will be defined in different network device or application program, we need find a way to support this new log types easily. We can define our log type in a XML file. For example, a http type log can be defined in XML file which can be named http.xml as shown in Figure 5. This file contains the http type log's format and search condition of each field of logs.

```
- <class name="Http">
  + <property name="id" type="int">
  - <property name="sourceip" type="string" length="20">
      <operation name="large" />
      <operation name="small" />
      <operation name="between" />
      <operation name="equal" />
      <operation name="unequal" />
    </property>
  + <property name="visiip" type="string" length="20">
  + <property name="time" type="java.util.Date">
  + <property name="aproach" type="string" length="10">
  + <property name="visitObject" type="string" length="256">
  + <property name="visitPro" type="string" length="20">
  + <property name="resp_code" type="int">
  - <property name="xfer_size" type="int">
      <operation name="include" />
      <operation name="equal" />
      <operation name="unequal" />
    </property>
  + <property name="refer" type="string" length="256">
  + <property name="info" type="string" length="256">
  </class>
```

Fig. 5. http.xml

Struts2's function in log query component was shown in Figure 6. For example, when client host make a http type log query, the request will be submitted to a series of filters (such as ActionContextCleanUp, FilterDispatcher). FilterDispatcher is the center of the controller of MVC. FilterDispatcher find the class LogSearchAction and then pass all the datas to the LogSearchAction. LogSearchAction can call Service Layer program to get the query results and then pass these results to user interface.

Spring is mainly responsible for dealing with application logic and transaction management. Spring's IoC container can mange the dependencies between objects very well. Figure 7 is the Spring's function in log query.

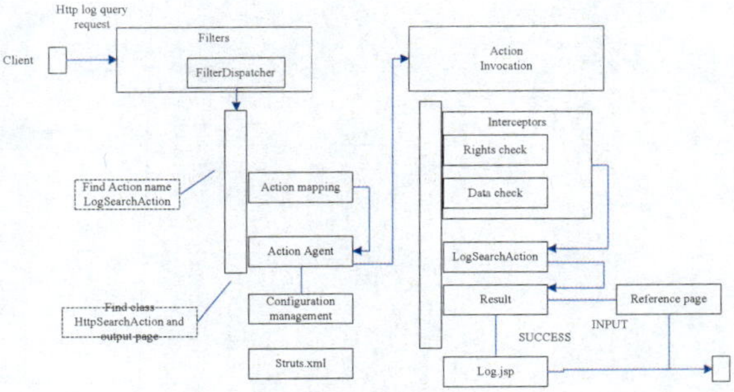

Fig. 6. Struts2's function of the system

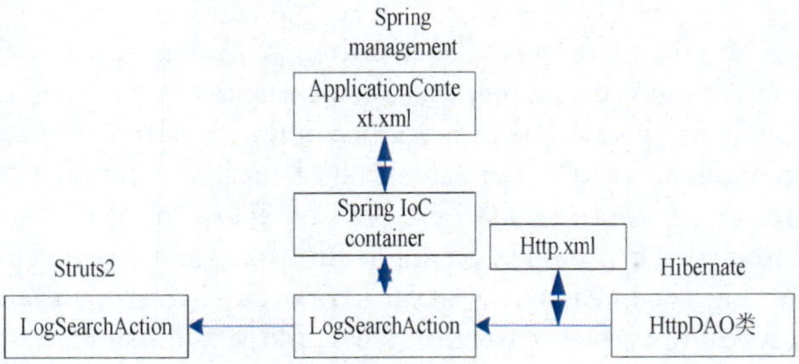

Fig. 7. Spring manages objects

A log query request will be processing in the system as follows: First, the user sends a log query request., for example, the latest http type logs (LogSearchAction?logtype="http"&keyword="latest"). Second, Struts2's Interceptors get the request and check the request's access rights. Then they find class LogSearchAction to process the request. Third, LogSearchAction will call Spring IoC container to find class LogSearchService to do a further processing. Fourth, LogSearchService will find http.xml and parse the file, then it will call HttpDao to access database to get logs. The results was shown in Figure 8.

2	1	2008-11-04 14:02:16	10.240.46.51	10.20.0.59	http	GET	200	17443	/ HTTP/1.1"
2	2	2008-11-04 14:02:16	10.240.46.51	10.20.0.59	http	GET	200	17443	/ HTTP/1.1"
2	3	2008-11-04 14:02:16	10.240.46.51	10.20.0.59	http	GET	200	1051	/images/image_engli...
2	4	2008-11-04 14:02:16	10.240.46.51	10.20.0.59	http	GET	200	1193	/images/image_conta...
2	5	2008-11-04 14:02:16	10.240.46.51	10.20.0.59	http	GET	200	2609	/images/image2.jpg ...
2	6	2008-11-04 14:02:16	10.240.46.51	10.20.0.59	http	GET	200	5602	/images/logo2.jpg H...
2	7	2008-11-04 14:02:16	10.240.46.51	10.20.0.59	http	GET	200	2359	/images/image4.jpg ...

Fig. 8. Http logs list

So if we want our system to support a new kind of logs, for example A, we can write a XML file A.xml first, Then write ADao.java and modify Hibernate configuration file.

4 Conclusion

Management and Query Center is very important to Log Audit system and it should have a good scalability and maintainability. We try to implement our system based on Applet+Servlets at first, but then find to make our system have a good performance. Well the integration of Struts2, Spring and Hibernate can reduce the implementation complexity and also the system can be easy to develop, maintain and upgrade.

References

1. Kent, K., Souppaya, M.: Guide to Computer Security Log Management (Recommendations of the National Institute of Standards and Technology). NIST Special Publication (September 2006)
2. Crocker, N.: Intelligent Security Reporting: Auditing Security Logs. Computer Audit Update 1997(2), 8–17 (1997)
3. Roger, M., Goubault-Larrecq, J.: Log Auditing through Model Checking. Research Report LSV-01-3 (April 2001)
4. Chuvakin, A.: Advanced Log Processing. Security Focus Infocus (August 1, 2002), http://online.securityfocus.com/infocus/1613 (April 30, 2003)
5. Huang, Y., Hu, J.: Design and Implementation of Log Audit System. Computer Engineering 32(22), 67–68 (2006)
6. Liu, B., Xu, R.: The Design of the Integrative Log Pre-processing Model Based on XML. Journal of Putian University 14(5), 65–69 (2007)
7. Du, X., Li, Z.: Architecture of J2EE Web Development Framework. Computer Science 33(8), 236–239 (2006)
8. Wang, W., Li, C., Li, J.: An approach to network based Security Auditing System. Computer Application and Software 19(11), 24–26 (2002)
9. Li, G.: Definitive Guide to Struts 2:MVC development based on WebWork. Publishing House of Electronics Industry, Beijing (2007)
10. Walls, C., Breidenbach, R.: Spring in Action, 2nd edn. Post & Telecom Press, Beijing (2008)

Development on Intelligent Controller of Automobile ABS Based on the Slip Ratio

Wanxin Kang[1], Renyun Sun[2], Xiucheng Dong[1], and Yunyun Xu[1]

[1] Sichuan Provincial Key Lab on Signal and Information Processing,
Xihua University, Chengdu 610039, China
[2] School of Transportation & Automotive Engineering, Xihua University,
Chengdu 610039, China
dxc136@163.com, Kangwx@mail.Xhu.Edu.Cn, xuyy19851223@sina.com

Abstract. For the optimal slip ratio, this paper adopted intelligent control technology to realize automobile Anti-lock control system. First, the two dimensional fuzzy Anti-lock Braking System(ABS) controller was designed, followed by design of the fuzzy neural network ABS controller. Finally the genetic algorithm was adopted to optimize the weights and thresholds of fuzzy neural network. Simulation results show a good anti-lock braking effect can be obtained by adopting the compound intelligent control technology.

Keywords: Automobile ABS, slip ratio, fuzzy control, neural network, genetic algorithm.

1 Introduction

Antilock braking system (ABS) is a nonlinear and uncertain time variant system, the model's establishment and the simulation accuracy of the actual working conditions will determine the control effect of control theory. It's difficult to establish its precise mathematical model with modern control theory, therefore, control theory without modeling could be considered. Artificial intelligence control which mainly contains fuzzy control technology, neural networks and evolutionary technology, uses all kinds of automatic machines or intelligent machines to imitate, extend and expand human capabilities, achieving the judgment, understanding, reasoning, recognition and learning of knowledge. Fuzzy control technology is good at expressing and dealing with uncertain knowledge, neural network is good at learning and expanding knowledge, both don't require the mathematical model of complex controlled object. The combination of those two control technologies can compensate each other and exert their merits, can effectively solve the impact of changing working conditions in the process of automobile braking and non-linear characteristics, can increase automobile braking system's security and stability, bringing the ABS into the age of intelligent control.

J. Luo (Ed.): Affective Computing and Intelligent Interaction, AISC 137, pp. 359–367.
springerlink.com

2 Establishment of the Mathematical Model of Automobile ABS

The wheel's force status of ABS vehicle development model[1] based on single-wheel when braking is shown in Figure 1.According to Figure 1,we can establish the motion equation of the vehicle as the expression (1) :

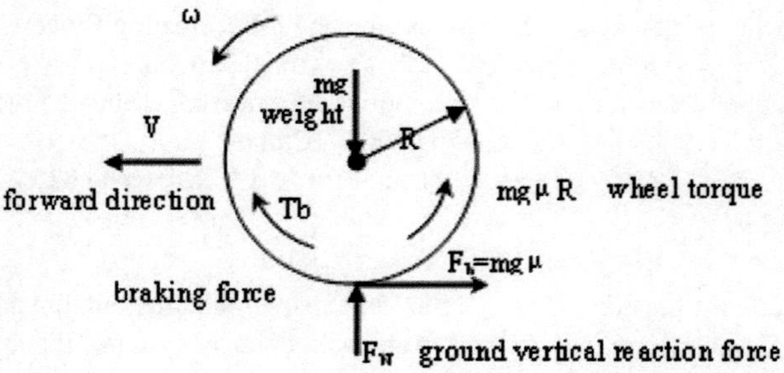

Fig. 1. Force model of single-wheel braking

$$m\frac{dv}{dt} = -F_b = -mg\mu \tag{1}$$

wheel motion equation is as in (2)

$$I\frac{d\omega}{dt} = T_b - F_b R = T_b - mg \cdot \mu R \tag{2}$$

Braking torque acting on the wheel and hydraulic pressure acting on the braking wheel cylinder can be simplified as a first-order delay component as in (3) :

$$T_b = K \cdot P \tag{3}$$

the wheel slip ratio is defined as in (4)

$$\lambda = \frac{v - \omega R}{v} \tag{4}$$

The tire model can be simplified by using a double linear model, and can be expressed by the following two linear equations in (5).

$$\begin{cases} \mu = \dfrac{\mu_k}{\lambda_0}\lambda, \lambda \le \lambda_0 \\ \mu = \dfrac{\mu_k - \mu_g \lambda_0}{1 - \lambda_0} - \dfrac{\mu_k - \mu_g}{1 - \lambda_0}\lambda, \lambda > \lambda_0 \end{cases} \tag{5}$$

The function of ABS is to control the slip ratio between 10% and 30%, to make full use of surface friction coefficient and improve braking performance, achieving the best braking effect.

The optimal slip ratio of the current road surface which the ground recognition system gives can be seen as our control objective. The wheel speed sensor measures current wheel speed. After calculating, we can get the actual slip ratio.

The difference between them is defined as slip ratio error and then it is input into the controller which obtains the control variable according to the control algorithms designed. Control variable acts on the vehicle dynamics system. Again we could obtain the new actual slip ratio. Compared with the optimal slip ratio of this time, the slip ratio error is obtained by re-calculating. Thus, the control forms a cycle. When the speed is zero, the simulation ends.

The parameters of car model[1]: single-wheel car's quality: 450M/Kg; braking force factor: 0.02; initial braking speed: 80km/h; wheel scroll inertia: 20kg·m^2; the wheel, S effective radius: 0.3m; gravity acceleration: 10m/s^2; the peak value of high friction coefficient road surface parameters: 0.9; slip value: 0.6.

3 Development of Intelligent ABS Controller

3.1 Design of Fuzzy ABS Controller (FC)

In the MATLAB / SIMULINK, we could build a ABS closed-loop system model. As it is shown in Figure 2.

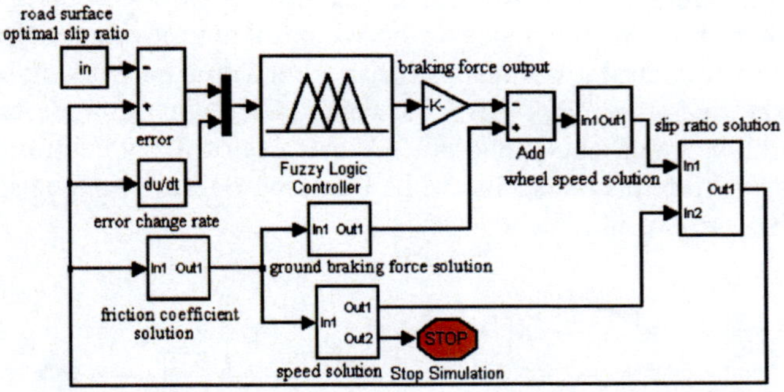

Fig. 2. ABS System Model

Table 1. Seven-grade fuzzy control rule table

Rules		E						
		NL	NM	NS	ZE	PS	PM	PL
EC	NL	PF	PF	PF	PF	PF	PF	ZE
	NM	PF	PF	PF	PF	PF	ZE	NL
	NS	PF	PF	PF	PF	ZE	NL	NL
	ZE	PF	PF	PF	ZE	NL	NL	NL
	PS	PF	PF	ZE	NL	NL	NL	NL
	PM	PF	ZE	NL	NL	NL	NL	NL
	PL	ZE	NL	NL	NL	NL	NL	NL

Parameters of fuzzy controller in this system are designed as follows. The slip ratio error is the input signal E, which has seven grades after fuzzification--negative large NL, negative medium NM, negative small NS, zero ZE, positive small PS, positive medium PM, positive large PL, whose universe is [-0.2,0.2]. The slip ratio error rate of change is seen as the input signal EC, which has 7 grades after fuzzification--negative large NL, negative medium NM, negative small NS, zero ZE, positive small PS, positive medium PM, positive large PL, whose universe is [-0.2,0.2]. The hydraulic wheel cylinder pressure control variable is seen as the output signal U, which has three-grade fuzzification, whose language values are NF, ZE and PF which respectively represents the situation of error decompression, zero and pressurized; the universe is [0,6]. Changing fuzzy interval of the Gauss membership function, we can obtain a greater membership degree when the fuzzy hierarchy gets near to zero[2]. Control rules are as follows, if E=A and EC=B then U=C. Reasoning type selects mamdani; anti-fuzzification uses the center of gravity method. From Table 1,we can know that control rules contain 49 pieces.

3.2 Design of Fuzzy Neural Network ABS Controller (FNC)

After having mature expert rules, using neural network to replace fuzzy inference machine can enhance the generalization ability and improve control effect; can eliminate the influence brought by the shape of membership function; also, can improve the interference suppression capability of the controlled object. Fuzzy neural network has many combination modes, whose control algorithms are also various[3] . In this paper, fuzzy control is our main structure and neural network embeds in it, thus we get a fuzzy neural network ABS controller. Through sample's training to the neural network, we could obtain and amend the rules of fuzzy control, this structure can be seen as the inference machine while inputting online, completes the operation and enters the control results.

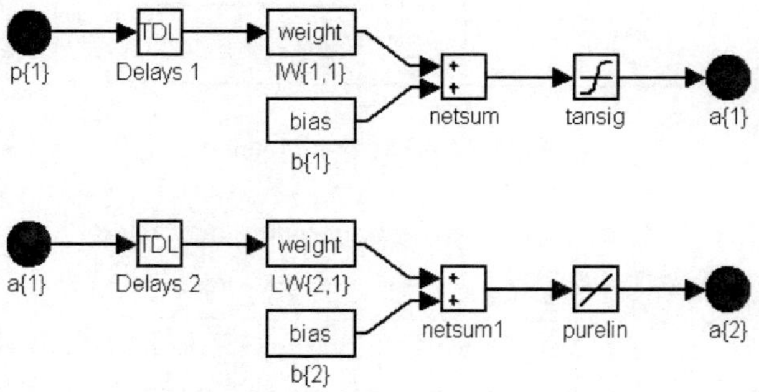

Fig. 3. The internal structure of fuzzy neural network controller

The model uses three-layer BP network to form a fuzzy neural network controller. The input layer inputs vector p{1}—(E EC). Respectively, E is the element obtained after the input error signal *e* is quantified to the universe X; EC is the element obtained after the input error change rate *ec* signal is quantified to the universe Y.

Hidden layer has 7 neurons; activation function(transfer function) uses tansig; its output vector is a {1}. Output layer has 1 neuron; the activation function is purelin; its output vector is y{1}.As it is shown in Figure 3.

The fuzzy rules are transferred into input-output sample couples. Using BP algorithm to train the network, thus the network can remember those rules. Training goal sets the output error for the sample to 0.01. The trained neural network will generate simulation module through gensin (net, st), which will replace this fuzzy inference machine part of Figure 2 system, thus forming a closed-loop fuzzy neural network ABS system.

3.3 Design of Fuzzy Neural Network ABS Controller (FNC)

The fitting degree of weights of fuzzy neural network to the rules would seriously affect the network's generalization ability. Using genetic algorithm to optimize the weights, the network can learn and understand the rules much completely[4].

The global search method which adopts a minimum error as the objective function, generates a new population through selection, cross and mutation. For those characteristics of BP network, this paper selects multi-parameter cascade encoding which is divided into four levels, any one group of complete [IW1 B1 IW2 B2] is equivalent to a chromosome. Each parameter employs float to encode, which can avoid discrete errors caused by binary-encoding. Real-encoding can also avoid the design of decoding process. Population size is 60. The objective function: error sum of squares as in (6),

$$E = \frac{1}{2}\sum_{i=1}^{n}[t_i - a_i]^2 \tag{6}$$

our aim is to find the minimum of objective function. Where: t_i =target value; a_i =the network output value; n =the number of samples. Fitness function is expressed as in (7)

$$Fitness = M / E \tag{7}$$

4 Simulation and Performance Analysis

Figure 4 to Figure 6 show three key data curves got from the simulation of the FC. From these figures, we can achieve these data: braking distance S=34.6m; braking time t=3.0s; speed from v to vb = 0.8v (km/h) driving distance Sb=13.7m; speed from v to ve=0.1v (km/h) driving distance Se=34.2m; braking average deceleration MFDD is calculated by this formula as in (8):

$$MFDD = \frac{v_b^2 - v_e^2}{25.92(S_e - S_b)} = \frac{64^2 - 8^2}{25.92 \times (34.2 - 13.7)} \tag{8}$$

$$= 7.59\,\mathrm{m}/s^2$$

As it is shown that braking distance, braking time and the average deceleration are all within the range of GB12676-1999. So, the control achieves a good effect.

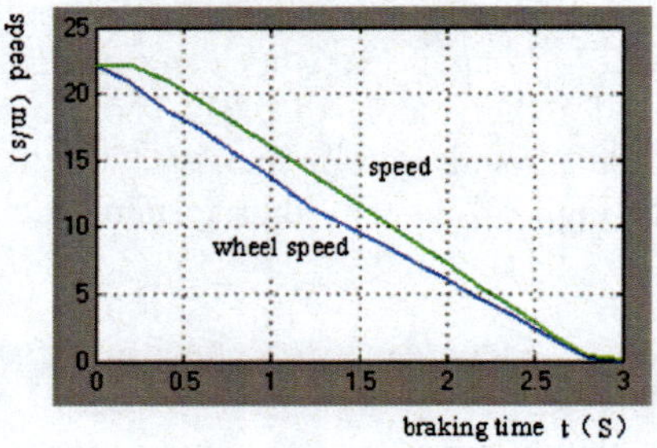

Fig. 4. FC speed and wheel speed curve

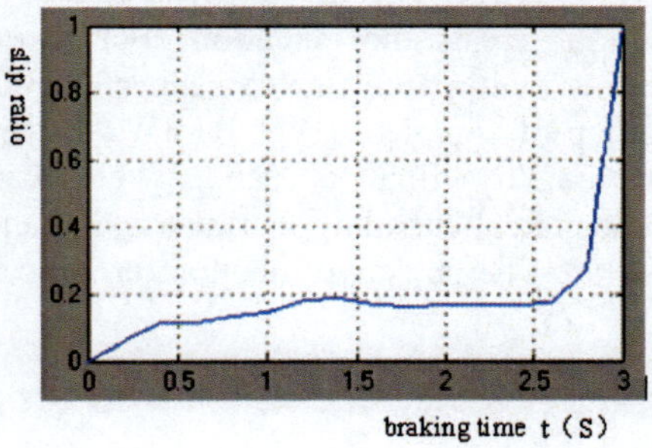

Fig. 5. FC slip ratio curve

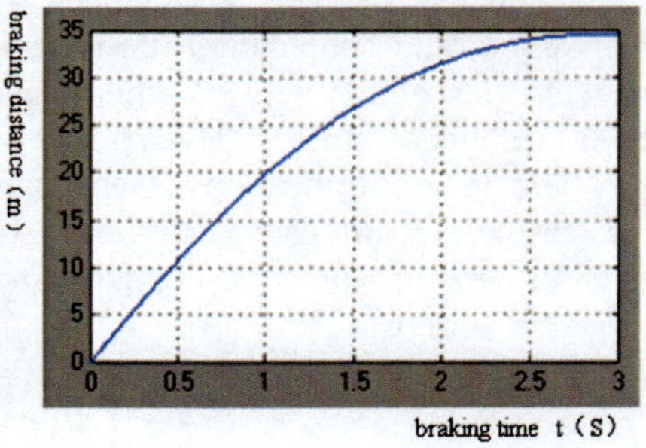

Fig. 6. FC braking distance curve

Figure 7 to Figure 9 show the results of FNC's simulation. From these figures, we can obtain these data: braking distance S=31.27m; braking time t=2.85s; speed from v to vb=0.8v (km/h) driving distance Sb= 2.64m; speed from v to ve=0.1v (km/h) driving distance Se=30.97m; braking average deceleration rate is 8.49 m/s^2.. Contrast to a simple fuzzy intelligent control, the braking effect of multiplex intelligent control has a significant improvement: braking distance shortens by 2.33m; braking time reduces by 0.15s, the braking average deceleration also increased. Slip ratio changes near the optimal value.

The weights and thresholds optimized with GA are written into seven-grade fuzzification network. When input value universe is the same to the FNC design's universe, we can get these simulation results as Figure 10 to Figure 12 show. From these figures, we can get these data: braking distance S=30.0m; braking time t=2.8s; speed from v to vb = 0.8v (km/h) driving distance Sb = 12.6m; speed v to ve = 0.1v (km/h) driving distance Se = 29.2m; braking average deceleration MFDD is 9.37 m/s^2.

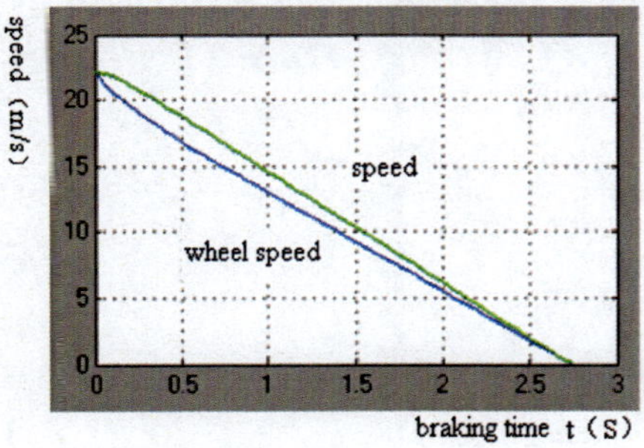

Fig. 7. FNC speed andwheel speed curve

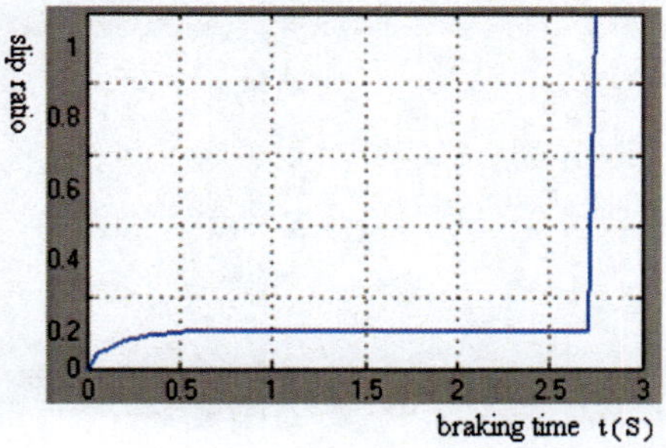

Fig. 8. FNC slip ratio curve

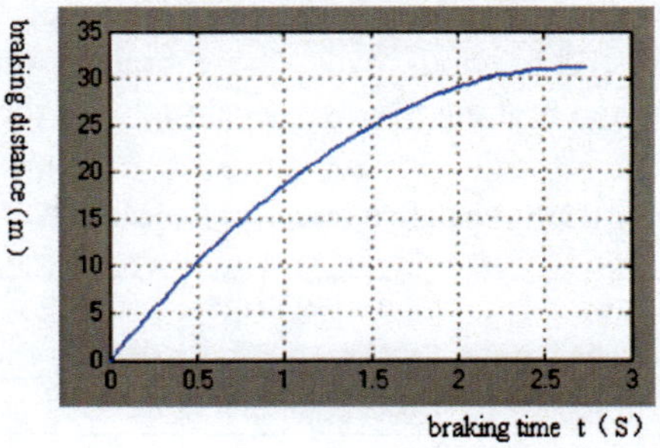

Fig. 9. FNC braking distance curve

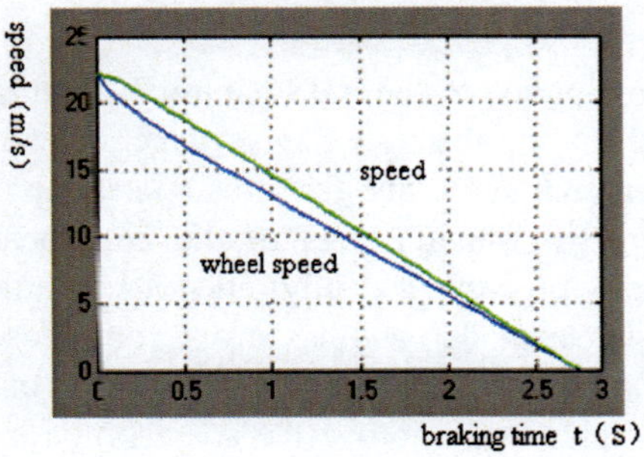

Fig. 10. GA-FNC speed and wheel speed curves

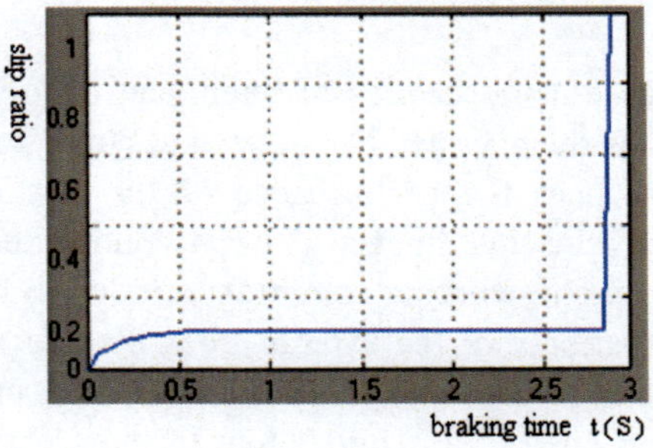

Fig. 11. GA-FNC slip ratio curve

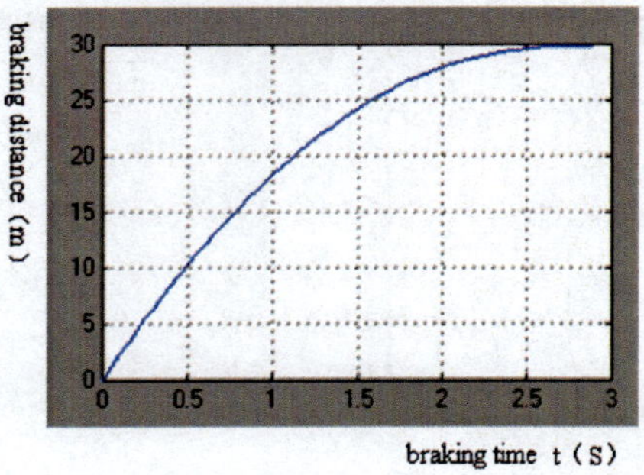

Fig. 12. GA-FNC braking distance curve

Table 2 sums up the three simulation results of typical ABS intelligent control, from which we could draw the conclusion: the effect of ABS braking process optimized is better than the previous two, whose braking distance is the shortest, braking time is the least, and braking average deceleration is the maximum.

Table 2. Comparing three simulation results of typical ABS intelligent control

	braking distance (m)	braking time (t)	braking average deceleration (m/s^2)
FC	34. 6	3. 0	7. 59
FNC	31. 27	2. 85	8. 49
GA-FNC	30. 0	2. 8	9. 37

5 Conclusion

Three main theories of Intelligent control are applied to the ABS control algorithm, play their respective advantages and make up for their shortages, which greatly enhances the performance of the controller and plays out the great potential of artificial intelligence. Fuzzy control technology which adapts to the non-linear, changing working condition, and lag in braking process, can fully reflect the control experience of the experts, does not depend on car model, and has strong robustness. Combined with neural network, fuzzy control system can have learning function, and thus can get a better output in interpolation input and improve the adaptability and coverage of the control system. Optimization of genetic algorithm to neural network weights, thresholds promotes the rules' learning and understanding, thereby improving the accuracy and control effect of ABS braking control.

Combined with road surface recognition system, automobile intelligent ABS controller will be further improved. We could build a sound, concise ABS control rule base, find a more suitable neural network for ABS control[5], and transform the model to make it have the predictive and adaptive capacity.

Acknowledgment. This project is supported by the fund of Sichuan Provincial Key Subject and Key Lab on Vehicle Engineering (SZD0410-1). Special thanks should go to it.

References

1. Zhao, S., Sun, T.-J., Kang, W.-X.: Automotive ABS Hybrid modeling and simulation Based on Simulink / Stateflow. Bus Technology and Research (January 2006)
2. Zhao, Z.G., Yu, Z.P., Sun, Z.C.: Research on fuzzy road surface identification and logic control for anti-lock braking system. In: IEEE International Conference, America (2006)
3. Mo, Y.-S., Zhu, R., Li, S.-E., Shen, L.: Automotive ABS fuzzy neural network control system. Journal of Shanghai Jiaotong University 33(5) (May 1999)
4. Lu, L.-Y., Zhu, W.-X.: The application of Fuzzy Intelligent Control in ABS. Computer Measurement and Control (November 2003)
5. Mohammadi, K., et al.: Fault Diagnosis of Analog circuits with Tolerances By Using RBF and BP Neural Networks. In: Student Conference on Research and Development Proceedings, vol. 12(2) (2002)
6. Euktai, K., Park, M., Sevnghwan, J., et al.: A new approach to fuzzy modeling. IEEE Trans.on Fuzzy System 15(3), 1029–1039 (1997)
7. Wai, R., Chu, C.: Robust Petri fuzzy-neural-network control for linear inductionmotor drive. IEEE Transactions on Industrial Electronics 54(1), 177–189 (2007)

Higher-Order Sampled-Data Iterative Learning Control for Nonlinear Systems

Xu Hongwei, Huang Yanyan, and Gao Yanfeng

College of Mechanical Engineering, China Jiliang University, Hangzhou, China
xhw@cjlu.edu.cn, hyy9791@cjlu.edu.cn, gaoyanfeng@cjlu.edu.cn

Abstract. In this paper, a higher-order sampled-data iterative learning control method is presented for nonlinear system with relative degree and output time delays. The higher-order learning algorithm does not include numerical differentiations of any error order. A sufficient condition is proposed that the iterative learning control converges to the desired trajectory at each sampling instant if some Lipschiz conditions are satisfied and the sampling period is small enough. A simulation example shows that higher-order algorithm has larger convergence rate.

Keywords: Sampled-data iterative learning control, Relative degree, Higher-order learning algorithm, Bellman lemma, convergence rate.

1 Introduction

Iterative Learning Control (ILC for short) theory has made great development and taken many successful applications in recent years since Japanese scholar proposed ILC in [1] in 1980s. The control algorithm take into application by data sampling and computer system, and iterative learning control theory paper on sampled systems, especially in nonlinear systems, far less than on continuous systems.

ILC algorithm convergence and robustness in sampled systems and discrete-time systems has been studied in some paper as in [2][3][4][5].Chiang JuChien in [6][7] presented a sampling ILC algorithm in the systems with initial state migration, state disturbances and measurement noise, proved the system output converges to the desired output bounded neighborhood when the iteration parameters meet some conditions. The convergence condition is closely related to the definition of the relative order which different from the definition of relative degree in continuous and discrete systems. Mingxuan Sun in [8] presents a sample definition of relative degree on nonlinear sampled system. With the definition of relative degree, first-order iterative learning control algorithm and convergence conditions are given. In this paper, based on the definition of relative degree in [8], study the convergence of higher order ILC algorithm of nonlinear sampled system with output delay.

J. Luo (Ed): Affective Computing and Intelligent Interaction, AISC 137, pp. 369–377.
springerlink.com

2 High-Order Sampled-Data Iterative Learning Control Algorithm

Consider the nonlinear systems with output delay of the form

$$\begin{cases} \dot{x}_k(t) = f(x_k(t)) + g(x_k(t))u_k(t) \\ y_k(t) = h(x_k(t-\tau)) \end{cases} \quad (1)$$

Where $x_k(t) \in R^n, u_k(t) \in R^r, y_k(t) \in R^m$ are, respectively, the state, input and output of the kth iteration, k=0,1,2,…, $t \in [0,T]$; $g(x_k(t)) = [g_1(x_k(t)), \cdots, g_r(x_k(t))]^T$, $h(x_k(t)) = [h_1(x_k(t)), \cdots, h_m(x_k(t))]^T$, $f(x(t))$, $h(x(t))$ exist higher-order derivatives, τ is the output delay time.

Assuming the followed (A1)-(A4) are met :

(A1) The system have relative degree $\eta = [\eta_1 ; \cdots , \eta_m]$, as to say when $n \le \eta_i - 1, i \in [1,m]$,we have

$$L_g L_f^n h_i(x) = 0 \quad (2)$$

it can be rewritten as

$$y_i^{(n)}(t+\tau) = L_f^n h_i(x(t)) + L_g L_f^{n-1} h_i(x(t)) \cdot u(t) \quad (3)$$

(A2) $f(x(t))$, $h_i(x(t))$ and Higher order Li derivatives $L_f^c h_i(x(t))$, $L_g L_f^{c-1} h_i(x(t))$, $c \in [1, \cdots, \eta_i]$, $i \in [1, \cdots, m]$ satisfy Lipschitz condition, Lipschitz content is $l_f, l_h, l_{fh}, l_{gfh}$ respectively

$$\|w(x_1(t)) - w(x_2(t))\| \le l \cdot \|x_1(t) - x_2(t)\| \quad (4)$$

Where $w(\bullet) \in \{f(x(t)), h(x(t)), L_f^c h(x(t)), L_g L_f^{c-1} h(x(t))\}$, $l \in \{l_f , l_h , l_{fh} , l_{gfh}\}$.

(A3) Sampling period $h = T/N$, where [0,T] is time range, N is the number of sampling. The control input sampling by zero order holder $u_k(t) = u_k(nh)$, $nh \le t < nh + h$ .The output delay time τ is integer multiple of the sampling period, $\tau = Zh$.

The sampled systems can be rewritten as

$$\begin{cases} x_k(nh+t) = x_k(nh) + \int_{nh}^{nh+t} (f(x_k(\tau)) + g(x_k(\tau))u_k(nh))d\tau \\ y_k(nh+t) = h(x_k((n-Z)h+t)) \end{cases} \quad (5)$$

(A4) Let $y_d(nh)$ be a desired reference trajectory. Assuming there be an unique input $u_d(nh)$, this implies that

$$\begin{cases} x_d(nh+t) = x_d(nh) + \int_{nh}^{nh+t} (f(x_d(\tau)) + g(x_d(\tau))u_k(nh))d\tau \\ y_d(nh+t) = h(x_d((n-Z)h+t)) \end{cases} \quad (6)$$

The purpose is to find a input updating serial $u_k(jh), j=[0, N-1]$, such that the output $y_k(jh) \to y_d(jh), k \to \infty$.

First of all, supplement the definition. As the system with output delay, to add the definition interval of the control input, system output range. Adding the desired trajectory

$$y_d(t) = y_d((N-1)h), t \in [Nh, (N+Z)h]$$

Firstly let's cite two lemmas.

Lemma 1. If the function $h(x(t))$ have n order derivative in an open interval (t_0, t_T), then $h(x(t))$ can be expressed as sum of an n-1 polynomial of $t - t_0$ and a remainder $R_n(t)$

$$h(x(t)) = h(x(t_0)) + h'(x(t_0))(t-t_0) + \frac{h''(x(t_0))}{2!}(t-t_0)^2 + \cdots + \frac{h^{(n-1)}(x(t_0))}{(n-1)!}(t-t_0)^{n-1} + R_n(t) \qquad R_n(t) = \int_{t_0}^{t}\int_{t_0}^{t_1}\cdots\int_{t_0}^{t_{n-1}} h^{(n)}(x(t_{n-1}))dt_{n-1}\cdots dt_2 dt_1 \tag{7}$$

Where $t_{n-1} \in (t_0, t_{n-2}), \cdots, t_2 \in (t_0, t_1), t_1 \in (t_0, t_T)$

It can be seen that each element of output vector can be expressed with high order integral points of the series expansion form.

Lemma 2. Non-negative real number sequence $\{\alpha_i, i=1,\cdots n\}$ satisfy the following inequality

$$\alpha_i \le \vartheta\alpha_{i-1} + \sum_{j=i-a}^{i-2} l_j \alpha_j$$

Where $\vartheta < 1, l_i \ge 0$. If $\vartheta + \sum_{j=i-a}^{i-2} l_j < 1$, then $\alpha_i \to 0, i \to \infty$.

The output of the iterative decomposition of Lemma 1, using Lemma 2, Bellman lemma and Lipschiz conditions, we have Theorem 1.

Theorem 1. A nonlinear continuous system (1) with relative degree $\eta = [\eta_1; \cdots, \eta_m]$, by sampled with the h period first order sample and hold device ,the use of high order iterative learning algorithm

$$u_{k+1}(nh) = u_k(nh) + \sum_{p=k-a}^{k} \Gamma_p(nh) \cdot e_p((n+Z+1)h) \tag{8}$$

Where a is iteration order, $a \in [0,1,\cdots,k], n \in [-Zh, (N-1)h]$.

① $$\rho = \|I - \Gamma_k(n)D_k(n)\| < 1 \tag{9}$$

$$D_k(n) = \begin{bmatrix} \int_n^{n+1}\int_n^{t_1}\cdots\int_n^{t_{\eta_1-1}} [L_{g_1}L_f h_1^{(\eta_1-1)}(x_k(t_{\eta_1-1})), L_{g_2}L_f h_1^{(\eta_1-1)}(x_k(t_{\eta_1-1})), \cdots, L_{g_r}L_f h_1^{(\eta_1-1)}(x_k(t_{\eta_1-1}))] dt_{\eta_1-1}\cdots dt_2 dt_1 \\ \vdots \\ \int_n^{n+1}\int_n^{t_1}\cdots\int_n^{t_{\eta_m-1}} [L_{g_1}L_f h_m^{(\eta_m-1)}(x_k(t_{\eta_m-1})), L_{g_1}L_f h_m^{(\eta_m-1)}(x_k(t_{\eta_m-1})), \cdots, L_{g_r}L_f h_m^{(\eta_m-1)}(x_k(t_{\eta_m-1}))] dt_{\eta_m-1}\cdots dt_2 dt_1 \end{bmatrix}$$

② Iterative learning initial state is strict repetition.

When assumptions ① and ② hold and the sampling period h is small enough, the system output at sampling points converges to the desired output.

Proof: Study the kth iteration output at sampling time (n+Z+1)h, we have

$$y_i((n+Z)h+h)=h_i(x_k(nh+h))$$

For convenience, the following $h_i(x_k(nh+h))$ expressed as $h_i(x_k(n+1))$, the others expressed as the same.

According to Lemma 1, it's easy to know

$$h_i(x_k(n+1))=h_i(x_k(n))+h_i'(x_k(n))h+\cdots+\frac{h_i^{(\eta_i-1)}(x_k(n))}{(\eta_i-1)!}h^{\eta_i-1}+\int_n^{n+1}\int_n^{t_1}\cdots\int_n^{t_{\eta_i-1}}h_i^{(\eta_i)}(x(t_{\eta_i-1}))dt_{\eta_i-1}\cdots dt_2dt_1 \tag{10}$$

Under the definition of relative order, we have $L_gL_f^jh_i(x)=0, j<\eta_i-1$.

$$\begin{aligned}h_i(x_k(n+1))=&h_i(x_k(n))+L_fh_i(x_k(n))h+\frac{h^2}{2!}L_fh_i^2(x_k(n))+\cdots+\frac{h^{(\eta_i-1)}}{(\eta_i-1)!}L_fh_i^{(\eta_i-1)}(x_k(n))+\\&\int_n^{n+1}\int_n^{t_1}\cdots\int_n^{t_{\eta_i-1}}L_fh_i^{(\eta_i)}(x_k(t_{\eta_i-1}))dt_{\eta_i-1}\cdots dt_2dt_1+\int_n^{n+1}\int_n^{t_1}\cdots\int_n^{t_{\eta_i-1}}[L_{g_1}L_fh_i^{(\eta_i-1)}(x_k(t_{\eta_i-1}));\cdots,L_{g_m}L_fh_i^{(\eta_i-1)}(x_k(t_{\eta_i-1}))]dt_{\eta_i-1}\cdots dt_2dt_1\cdot u_k(n)\end{aligned}$$

The kth iterations error of the ith outputs

$$\begin{aligned}e_{ik}(n+Z+1)=&h_i(x_d(n))-h_i(x_k(n))+L_fh_i(x_d(n))h-L_fh_i(x_k(n))h+\cdots+\\&\frac{h^{(\eta_i-1)}}{(\eta_i-1)!}L_fh_i^{(\eta_i-1)}(x_d(n))-\frac{h^{(\eta_i-1)}}{(\eta_i-1)!}L_fh_i^{(\eta_i-1)}(x_k(n))+\\&\int_n^{n+1}\int_n^{t_1}\cdots\int_n^{t_{\eta_i-1}}L_fh_i^{(\eta_i)}(x_d(t_{\eta_i-1}))dt_{\eta_i-1}\cdots dt_2dt_1-\int_n^{n+1}\int_n^{t_1}\cdots\int_n^{t_{\eta_i-1}}L_fh_i^{(\eta_i)}(x_k(t_{\eta_i-1}))dt_{\eta_i-1}\cdots dt_2dt_1+\\&\int_n^{n+1}\int_n^{t_1}\cdots\int_n^{t_{\eta_i-1}}[L_{g_1}L_fh_i^{(\eta_i-1)}(x_d(t_{\eta_i-1})),L_{g_2}L_fh_i^{(\eta_i-1)}(x_d(t_{\eta_i-1})),\cdots,L_{g_m}L_fh_i^{(\eta_i-1)}(x_d(t_{\eta_i-1}))]dt_{\eta_i-1}\cdots dt_2dt_1\cdot u_d(n)-\\&\int_n^{n+1}\int_n^{t_1}\cdots\int_n^{t_{\eta_i-1}}[L_{g_1}L_fh_i^{(\eta_i-1)}(x_k(t_{\eta_i-1})),L_{g_2}L_fh_i^{(\eta_i-1)}(x_k(t_{\eta_i-1})),\cdots,L_{g_m}L_fh_i^{(\eta_i-1)}(x_k(t_{\eta_i-1}))]dt_{\eta_i-1}\cdots dt_2dt_1\cdot u_k(n)\end{aligned}$$

Estimating the state's bias

$$\delta x_k(n)=x_d(n)-x_k(n),\delta u_k(n)=u_d(n)-u_k(n)$$

the state function can be written

$$\delta x_k(n+t)=\delta x_k(n)+\int_n^{n+t}(f(x_d(\tau))+g(x_d(\tau))u_d(n)-f(x_k(\tau))-g(x_k(\tau))u_k(n))d\tau$$

Norm on both sides

$$\|\delta x_k(n+t)\|\le\|\delta x_k(n)\|+\int_n^{n+t}(\alpha\|\delta x_k(\tau))\|+b_g\cdot\|\delta u_k(n)\|)d\tau$$

Where $b_d=\sup_{0\le t\le N}(u_d(t))$, $\alpha=l_f+l_g\cdot b_d$, $b_g=\|g(x_k(\tau))\|$

According to Bellman lemma

$$\|\delta x_k(n+t)\|\le e^{\alpha t}\cdot\|\delta x_k(n)\|+\int_n^{n+t}e^{\alpha(t-\tau)}\cdot\alpha\cdot\|\delta u_k(n)\|d\tau+(b_g-\alpha_n)\cdot\|\delta u_k(n)\|\cdot h \tag{11}$$

$$\|\delta x_k(n+t)\|\le\beta(t)\cdot\|\delta x_k(n)\|+\gamma(t)\cdot\|\delta u_k(n)\| \tag{12}$$

$\beta(t)=e^{\alpha(n+t)h}$ $\gamma(t)=e^{\alpha nh}(e^{\alpha(n+t)h}-1)\cdot\alpha+(b_g-\alpha)\cdot h$

If t=1, then

$$\|\delta x_k(n+1)\| \leq \beta \cdot \|\delta x_k(n)\| + \gamma \cdot \|\delta u_k(n)\| \tag{13}$$

Where $\beta_n = e^{\alpha(n+1)h}$, $\gamma_n = e^{\alpha nh}(e^{\alpha(n+1)h} - 1) \cdot \alpha + (b_g - \alpha) \cdot h$

So (12) can be rewritten

$$\|\delta x_k(n+1)\| \leq \beta^{n+1} \cdot \|\delta x_k(0)\| + \sum_{i=0}^{n} \beta^{n-i} \cdot \gamma \cdot \|\delta u_k(i)\|$$

Supposed iterative initial state strict repetition

$$\|\delta x_k(n+1)\| \leq \sum_{i=0}^{n} \beta^{n-i} \cdot \gamma \cdot \|\delta u_k(i)\| \tag{14}$$

Control input error norm iterative relationship can be expressed as follow

$$\begin{aligned} \delta u_{k+1}(n) &= \delta u_k(n) + \sum_{p=k-a}^{k} \Gamma_p(n) \cdot e_p(n+Z+1) \\ &= (I - \Gamma_k(n) D_k(n)) \delta u_k(n) - \sum_{p=k-a}^{k} \Gamma_p(n) \omega_p(n) - \Gamma_k(n) v_k(n)) - \sum_{p=k-a}^{k-1} \Gamma_p(n) \mu_p(n) \end{aligned} \tag{15}$$

Where

$$D_k(n) = \begin{bmatrix} \int_n^{n+1}\int_n^{t_1}\cdots\int_n^{t_{\eta_1-1}} [L_{g_1}L_f h_1^{(\eta_1-1)}(x_k(t_{\eta_1-1})), L_{g_2}L_f h_1^{(\eta_1-1)}(x_k(t_{\eta_1-1})), \cdots, L_{g_r}L_f h_1^{(\eta_1-1)}(x_k(t_{\eta_1-1}))] dt_{\eta_1-1}\cdots dt_2 dt_1 \\ \vdots \\ \int_n^{n+1}\int_n^{t_1}\cdots\int_n^{t_{\eta_m 1}} [L_{g_1}L_f h_m^{(\eta_m-1)}(x_k(t_{\eta_m-1})), L_{g_2}L_f h_m^{(\eta_m-1)}(x_k(t_{\eta_m-1})), \cdots, L_{g_r}L_f h_m^{(\eta_m-1)}(x_k(t_{\eta_m-1}))] dt_{\eta_m-1}\cdots dt_2 dt_1 \end{bmatrix}, \omega_p(n) = [\omega_{1p}(n); \cdots, \omega_{mp}(n)]^T,$$

$$v_k(n) = [v_{1k}(n); \cdots, v_{mk}(n)]^T, u_p(n) = [u_{1p}(n); \cdots, u_{mp}(n)]^T$$

Where

$$\begin{aligned} \omega_{jp} &= h_j(x_d(n)) - h_j(x_p(n)) + L_f h_j(x_d(n))h - L_f h_j(x_p(n))h + \cdots \\ &+ \frac{h^{(\eta_j-1)}}{(\eta_j-1)!} L_f h_j^{(\eta_j-1)}(x_d(n)) - \frac{h^{(\eta_j-1)}}{(\eta_j-1)!} L_f h_j^{(\eta_j-1)}(x_p(n)) \end{aligned} \tag{16α}$$

(16β)

$$\begin{aligned} v_{jk}(n) = &\int_n^{n+1}\int_n^{t_1}\cdots\int_n^{t_{\eta_j-1}} (L_f h_j^{(\eta_i)}(x_d(t_{\eta_j-1})) - L_f h_j^{(\eta_i)}(x_k(t_{\eta_j-1}))) dt_{\eta_j-1}\cdots dt_2 dt_1 + \int_n^{n+1}\int_n^{t_1}\cdots\int_n^{t_{\eta_j-1}} [L_{g_1}L_f h_j^{(\eta_j-1)}(x_d(t_{\eta_j-1})) - L_{g_1}L_f h_j^{(\eta_j-1)}(x_k(t_{\eta_j-1})) \\ &\cdots, L_{g_m}L_f h_j^{(\eta_j-1)}(x_d(t_{\eta_j-1})) - L_{g_m}L_f h_j^{(\eta_j-1)}(x_k(t_{\eta_j-1}))] dt_{\eta_j-1}\cdots dt_2 dt_1 \cdot u_d(n) \end{aligned}$$

$$\begin{aligned} \mu_{jp}(n) = &\int_n^{n+1}\int_n^{t_1}\cdots\int_n^{t_{\eta_j-1}} (L_f h_j^{(\eta_i)}(x_d(t_{\eta_j-1})) - L_f h_j^{(\eta_i)}(x_p(t_{\eta_j-1}))) dt_{\eta_j-1}\cdots dt_2 dt_1 + \\ &\int_n^{n+1}\int_n^{t_1}\cdots\int_n^{t_{\eta_j-1}} [L_{g_1}L_f h_j^{(\eta_j-1)}(x_d(n)) - L_{g_1}L_f h_j^{(\eta_j-1)}(x_p(n)); \cdots, \\ &L_{g_m}L_f h_j^{(\eta_j-1)}(x_d(t_{\eta_j-1})) - L_{g_m}L_f h_j^{(\eta_j-1)}(x_p(t_{\eta_j-1}))] dt_{\eta_j-1}\cdots dt_2 dt_1 \cdot u_d(n) + \\ &\int_n^{n+1}\int_n^{t_1}\cdots\int_n^{t_{\eta_j-1}} [L_{g_1}L_f h_j^{(\eta_j-1)}(x_p(t_{\eta_j-1})); \cdots, L_{g_m}L_f h_j^{(\eta_j-1)}(x_p(t_{\eta_j-1}))] dt_{\eta_j-1}\cdots dt_2 dt_1 \cdot \delta u_p(n) \end{aligned}$$

(16χ)

Where $j \in [1, m]$ and $p \in [1, a]$ is the high iterative order.

Norm on both sides in (14)

$$\|\delta u_{k+1}(n)\| \leq \rho \cdot \|\delta u_k(n)\| + \sum_{p=k-a}^{k} b_p \|\omega_p(n)\| + b_{\eta k}\|v_k(n)\| + \sum_{p=k-a}^{k-1} b_{\Gamma p}\|\mu_p(n)\| \tag{17}$$

Where $\rho = \|I - \Gamma_k(n) D_k(n)\|$, $b_p, b_{\eta k}, b_{\Gamma p}$ are the corresponding coefficients of the maximum norm.

$$\|\omega_{jp}\| \le (l_{fh}(1+h+\frac{h^2}{2!}+\cdots+\frac{h^{(\eta_j-1)}}{(\eta_j-1)!}))\|\delta x_p(n)\|$$

\We have

$$\|\omega_p\| \le b_1\|\delta x_p(n)\| \tag{18}$$

Where $b_1 = \sup_{j=[1,m]} (L_{fh}(1+h+\frac{h^2}{2!}+\cdots+\frac{h^{(\eta_j-1)}}{(\eta_j-1)!}))$

$$\|v_k(n)\| \le b_2 \left\| \begin{bmatrix} \int_n^{n+1}\int_n^{t_1}\cdots\int_n^{t_{\eta_1-1}}\|\delta x_k(t_{\eta_1-1})\|dt_{\eta_1-1}\cdots dt_2 dt_1 \\ \vdots \\ \int_n^{n+1}\int_n^{t_1}\cdots\int_n^{t_{\eta_j-1}}\|\delta x_k(t_{\eta_j-1})\|dt_{\eta_j-1}\cdots dt_2 dt_1 \end{bmatrix} \right\| \tag{19}$$

Where $b_2 = l_{fh} + rb_d l_{ghf}$

$$\|\mu_p(n)\| \le b_2 \left\| \begin{bmatrix} \int_n^{n+1}\int_n^{t_1}\cdots\int_n^{t_{\eta_1-1}}\|\delta x_p(t_{\eta_1-1})\|dt_{\eta_1-1}\cdots dt_2 dt_1 \\ \vdots \\ \int_n^{n+1}\int_n^{t_1}\cdots\int_n^{t_{\eta_m-1}}\|\delta x_p(t_{\eta_m-1})\|dt_{\eta_m-1}\cdots dt_2 dt_1 \end{bmatrix} \right\| + b_3\|\delta u_p(n)\| \tag{20}$$

Where $b_3 = rb_4$,

$b_4 = \max\{\frac{h^{\eta_i}}{\eta_i!}, i\in[1,r]\}\cdot\max(\|[L_{g_1}L_f h_i^{(\eta_i-1)}(x_p(t_{\eta_i-1}));\cdots, L_{g_m}L_f h_i^{(\eta_i-1)}(x_p(t_{\eta_i-1}))]\|, i\in[1,r])$ Bundles (13), (19), (20), (21) into (18)

$$\|\delta u_{k+1}(n)\| \le \rho'\cdot\|\delta u_k(n)\| + \sum_{p=k-a}^{k} b_p b_1\|\delta x_p(n)\| + b_6\|\delta x_k(n)\| + \sum_{p=k-a}^{k-1} b_7\|\delta u_p(n)\| + \sum_{p=k-a}^{k-1} b_8\|\delta x_p(n)\| \tag{21}$$

Where $\rho' = \rho + b_{\eta k}b_5 b_2\gamma$, $b_5 = \max(\frac{h^{\eta_i}}{\eta_i}, i\in[1,r])$, $b_6 = b_{\eta k}\beta$,

$b_7 = b_{\Gamma p}b_3 + b_{\Gamma p}b_5 b_2\gamma$, $b_8 = b_{\Gamma p}b_5\beta$

Bundles $\|\delta x_k(n)\|$ into (21)

$$\|\delta u_{k+1}(n)\| \le \rho'\cdot\|\delta u_k(n)\| + \sum_{p=k-a}^{k-1}(b_9\sum_{j=0}^{n}\|\delta u_p(j)\|) + b_9\sum_{j=0}^{n-1}\|\delta u_k(j)\| \tag{22}$$

Where $b_9 = \max(b_6\beta^{n-i}\gamma, i\in[1,n])$

From equation (12) we can see, when $\rho<1$ and sampling period h small enough, makes γ small enough, we have $\rho' = \rho+\varepsilon<1$, $\varepsilon = b_{\eta k}b_5 b_2\gamma$

The convergence of input λ norm. Take λ norm on both sides on equation (22)

$$\|\delta u_{k+1}(n)\|_\lambda \le \rho''\|\delta u_k(n)\|_\lambda + b_{10}\sum_{p=k-a}^{k-1}\|\delta u_p(n)\|_\lambda \tag{23}$$

Where $\rho'' = \rho' + \frac{1-e^{-(n-1)\lambda}}{1-e^{-\lambda}}$, $b_{10} = b_9\frac{1-e^{-n\lambda}}{1-e^{-\lambda}}$. If λ is sufficient small,

$\rho'' < 1$ is set up, and $\rho'' + \sum_{p=k-a}^{k-1} b_{10} < 1$. According to lemma 2, $\|\delta u_k(n)\|_\lambda \to 0, k \to \infty$. Proof is completed.

Iterative learning control converges to desired control, at the same time, in that the output expression satisfied Lipschitz condition, we can get the convergence characteristics of the system output. In fact:

$$\|e_k(n+Z)\|_\lambda \le L_h \|\delta x_k(n)\|_\lambda \le L_h \gamma \frac{1-(\beta e^{-\lambda})^n}{1-\beta e^{-\lambda}} \|\delta u_k(n)\|_\lambda$$

When $\|e_k(n+Z)\|_\lambda \to 0, \|\delta u_k(n)\|_\lambda \to 0$, iterative output converges to desired output.

From (8), we known iterative convergence condition are related to the relative order and the delay time in the system.

3 The Simulation

Example: using model in paper [8]

$$\dot{x}_1(t) = x_2(t)$$
$$\dot{x}_2(t) = 0.1 \cdot \sin(x_2(t)) + (1 + 0.1 \cdot \cos(x_1(t))) \cdot u(t)$$
$$y(t) = x_1(t - 0.1)$$

Where $t \in [0,1]$, the relate order is 2,sampling period is 0.1,N=10.The desired output:

$$y_d(t) = 4(nh)^3 - 3(nh)^4, 0 \le n \le 10$$

Additional control input $u(t), t \in [-h,(N-1)h]$, additional desired output $y_d(11h) = 1$.

Iterative initial state strict repetition and $x_k(0) = 0$. Stop conditions for the output error is less than 0.001.

When 1 order iterative learning algorithm, select the increment parameter $\Gamma_k = 70$. From figure 1 it can be obtained at 49 iterations times the convergence of the error norm is less than 0.001.

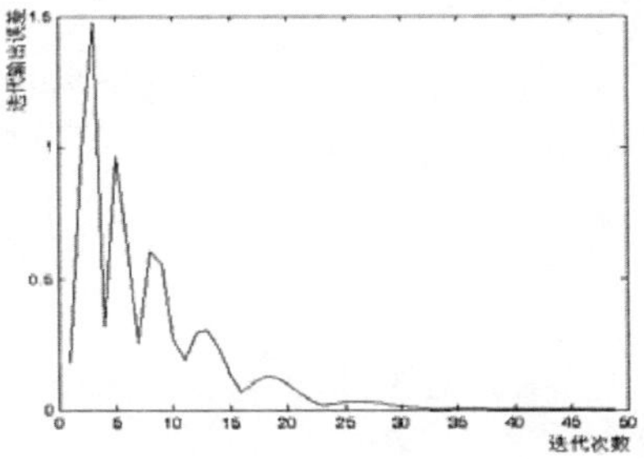

Fig. 1. Convergence curve of the 1-order iterative error by $\|e_k\|_\infty$

When 2-order iterative learning algorithm, select the increment parameter $\Gamma_k = 70$, $\Gamma_{k-1} = 10$. From figure 2 it can be obtained at 38 iterations times the convergence of the error norm is less than 0.001.

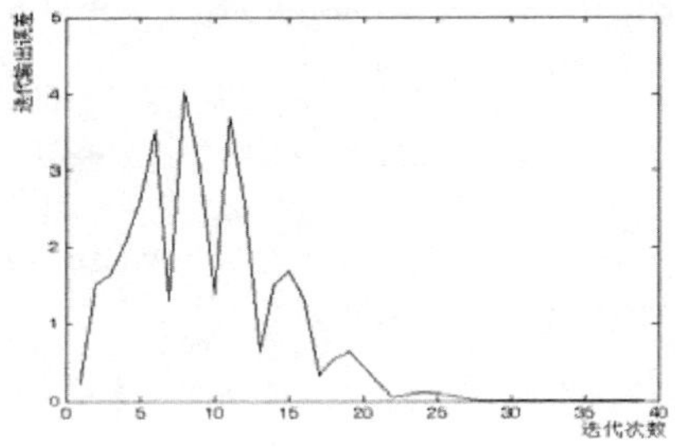

Fig. 2. Convergence curve of the 2-order iterative error by $\|e_k\|_\infty$

Comparison can be seen, second-order learning algorithm converges faster than first order convergence speed significantly improved. Side effect of the convergence process is slight increase volatility, but does not affect the output convergence.

4 Conclusion

In this paper, a higher-order learning algorithm convergence is studied of nonlinear discrete by sampling a continuous system with a relative degree. Simulation results show that high order convergence speed is much better than the first order algorithm. At the same time the high-order algorithms cause large oscillation. It remains to be further research that how to improve the convergence rate and reducing volatility of the sampling system using higher order derivative information, that the relations between sampling period h and the convergence of iterative learning algorithm.

Acknowledgment. I would like to show my deepest gratitude to my supervisor. I would also like to thank all my teachers who have helped me to develop the fundamental and essential academic competence.

This paper is supported by the Science Foundation of Zhejiang Province Y1080950 and Zhejiang Provincial Education Department projects Y200909258.

References

1. Arimoto, S., Kawamura, S., Miyazaki, F.: Bettering operation of robots by learning. Journal of Robotic Systems 1(1), 123–140 (1984)
2. Amann, N., Owens, D.H., Rogers, E.: Iterative learning control for discrete-time systems with exponential rate of convergence. IEE Proc.-Part, Control Theory Appl. 143(2), 217–224 (1996)
3. Fang, X., Chen, P., Shao, J.: Optimal higher-order iterative learning control of discrete-time linear systems. IEE Proc.-Control Theory Appl. 152(1), 43–48 (2005)

4. Norrlöf, M.: An Adaptive Iterative Learning Control Algorithm With Experiments on an Industrial Robot. IEEE Transactions on Robotics and Automation 18(2), 245–251 (2002)
5. Tayebi, A.: Adaptive iterative learning control for robot manipulators. Automatic (40), 1195–1203 (2004)
6. Chien, C.-J.: A Discrete Iterative Learning Control of Nonlinear Time-Varying Systems. In: Proceedings of the 35th Conference on Decision and Control, pp. 3056–3061 (1996)
7. Chien, C.-J.: The Sampled-Data Iterative Learning Control for Nonlinear Systems. In: Proceedings of the 36th Conference on Decision and Control, pp. 4306–4311 (1997)
8. Sun, M., Wang, D.: Sampled-data iterative learning control for nonlinear systems with arbitrary relative degree. Automatica 37, 283–289 (2001)

DNA Sequences Analysis Based on Classifications of Nucleotide Bases

Long Shi[1,*] and Hailan Huang[2]

[1] School of Science, Central South University of Forest and Technology, Changsha, Hunan 410004, China
chu_9@163.com
[2] School of Mathematics and Computational Science, Xiangtan University, Xiangtan, Hunan 411105, China
huanghailan_1980@sina.com

Abstract. Four bases A, C, G and T of DNA sequences are divided into three kinds of classifications according to their chemical properties. We convert a DNA primary sequence into three symbolic sequences. The frequencies of group mutations of three symbolic sequences have been grouped into a twelve-component vector to represent the DNA sequence. The Euclidean distances among introduced vectors are applied to characterize and compare the coding sequences of the first exon of beta globin gene of 11 different species.

Keywords: bases classification, symbolic sequence, group mutation, similarity analysis.

1 Introduction

With the development of the sequencing technique, a large number of DNA primary sequence data are collected and become available in public databases. It is one of the challenges for bio-scientists to mathematically analyze the large volumes of biological data to find biological interest. In recent years, several authors outlined different graphical representations of DNA sequences based on 2D or 3D space[1-19]. The advantage of graphical representation of DNA sequences is that they allow visual inspection of data, helping in recognizing major differences among similar DNA sequences. But some graphical representations of DNA sequence are accompanied with some loss of information due to overlapping and crossing of the curve representing DNA with itself[2-5]. To avoid the limitations associated with crossing and overlapping some authors provided many novel graphical representation methods recently[6-19]. These techniques provide useful insights into local and global characteristics along a sequence which are not easily obtained by other methods. Based on these graphical representations, several authors developed some numerical methods, for example E matrix, D/D matrix, M/M matrix, L/L matrix and their leading eigenvalues and so on, to compare DNA sequences[6-12]. But when the length of the

* Corresponding author.

J. Luo (Ed.): Affective Computing and Intelligent Interaction, AISC 137, pp. 379–384.
springerlink.com

sequences under studying is long the computation will be complicated and time-consuming.

In the past years the dinucleotide analysis has also been tried by several authors[18-20]. This approach can offer fast, qualitative comparisons and reveal the biology information of DNA sequences. In this paper, motivated by works on dinucleotide analysis, according to bases' chemical properties, we divide four bases into classes and use three mappings to convert a DNA primary sequence into three symbolic sequences. We construct a 12-component vector consisting of twelve frequencies of group mutations to represent a DNA sequence. The Euclidean distances between the end points of the vectors are applied to characterize and compare the coding sequences of the first exon of beta globin genes of 11 different species in Table 1.

Table 1. The coding sequences of the first exon of beta globin gene of 11 different species

Species	Coding sequences
Human	ATGGTGCACCTGACTCCTGAGGAGAAGTCTGCCGTTACTGCCCTGTGG GGCAAGGTGAACGTGGATTAAGTTGGTGGTGAGGCCCTGGGCAG
Goat	ATGCTGACTGCTGAGGAGAAGGCTGCCGTCACCGGCTTCTGGGGCAAG GTGAAAGTGGATGAAGTTGGTGCTGAGGCCCTGGGCAG
Opossum	ATGGTGCACTTGACTTCTGAGGAGAAGAACTGCATCACTACCATCTGG TCTAAGGTGCAGGTTGACCAGACTGGTGGTGAGGCCCTTGGCAG
Gallus	ATGGTGCACTGGACTGCTGAGGAGAAGCAGCTCATCACCGGCCTCTGG GGCAAGGTCAATGTGGCCGAATGTGGGGCCGAAGCCCTGGCCAG
Lemur	ATGACTTTGCTGAGTGCTGAGGAGAATGCTCATGTCACCTCTCTGTGGG GCAAGGTGGATGTAGAGAAAGTTGGTGGCGAGGCCTTGGGCAG
Mouse	ATGGTTGCACCTGACTGATGCTGAGAAGTCTGCTGTCTCTTGCCTGTGG GCAAAGGTGAACCCCGATGAAGTTGGTGGTGAGGCCCTGGGCAGG
Rabbit	ATGGTGCATCTGTCCAGTGACGAGAAGTCTGCGGTCACTGCCCTGTGG GGCAAGGTGAATGTGGAAGAAGTTGGTGGTGAGGCCCTGGGC
Rat	ATGGTGCACCTAACTGATGCTGAGAAGGCTACTGTTAGTGGCCTGTGG GGAAAGGTGAACCCTGATAATGTTGGCGCTGAGGCCCTGGGCAG
Gorilla	ATGGTGCACCTGACTCCTGAGGAGAAGTCTGCCGTTACTGCCCTGTGG GGCAAGGTGAACGTGGATGAAGTTGGTGGTGAGGCCCTGGGCAGG
Bovine	ATGCTGACTGCTGAGGAGAAGGCTGCCGTCACCGCCTTTTGGGGCAAG GTGAAAGTGGATGAAGTTGGTGGTGAGGCCCTGGGCAG
Chimpanzee	ATGGTGCACCTGACTCCTGAGGAGAAGTCTGCCGTTACTGCCCTGTGG GGCAAGGTGAACGTGGATGAAGTTGGTGGTGAGGCCCTGGGCAGGTT GGTATCAAGG

2 Proposed Method

For a DNA primary sequence, there are three kinds of classification methods to divide four bases A, C, G and T according to their chemical properties, i.e., purine group R={A,G} and pyrimidine group Y={C,T}; amino group M={A,C} and keto group K={G,T}; weak H-bond group W={A,T} and strong H-bond group S={C,G}. We call them RY classification, MK classification and WS classification correspondingly. Let $X = s_1 s_2 \cdots s_n$ be a DNA primary sequence with length n. we have three maps ϕ_{RY}, ϕ_{MK}, ϕ_{WS}, which map X into three symbolic sequences respectively. Explicitly,

$$\phi_{RY}(X) = \phi_{RY}(s_1)\phi_{RY}(s_2)\cdots\phi_{RY}(s_n), \quad (1)$$

where $\phi_{RY}(s_i) = \begin{cases} R & \text{if } s_i \in R \\ Y & \text{if } s_i \in Y \end{cases}$, i=1,2,…,n

$$\phi_{MK}(X) = \phi_{MK}(s_1)\phi_{MK}(s_2)\cdots\phi_{MK}(s_n), \quad (2)$$

where $\phi_{MK}(s_i) = \begin{cases} M & \text{if } s_i \in M \\ K & \text{if } s_i \in K \end{cases}$, i=1,2,…,n

$$\phi_{WS}(X) = \phi_{WS}(s_1)\phi_{WS}(s_2)\cdots\phi_{WS}(s_n), \quad (3)$$

where $\phi_{WS}(s_i) = \begin{cases} W & \text{if } s_i \in W \\ S & \text{if } s_i \in S \end{cases}$., i=1,2,…,n.

One example is three symbolic sequences corresponding sequence $X = ATGGTGCACCTGACT$ as follows:

$\phi_{RY}(X) = RYRRYRYRYY\ YRRYY$
$\phi_{MK}(X) = MKKKKKMMMM\ KKMMK$
$\phi_{WS}(X) = WWSSWSSWSS\ WSWSW$

We call obtained symbolic sequences RY sequence, MK sequence and WS sequence respectively. Then we analyze mutation information of above three symbolic sequences. In each classification we focus on group mutation information. From above symbolic sequences we find twice $R \to R$, five $R \to Y$, four $Y \to R$, and three times $Y \to Y$ group mutations; four $M \to M$, three times $M \to K$, twice $K \to M$ and five $K \to K$ group mutations; once $W \to W$, five $W \to S$, five $S \to W$ and three times $S \to S$ group mutations. It is difficult to obtain the information from DNA primary sequence directly. The frequency of group mutation will be applied to make comparisons among coding sequences belonging to eleven Species in Table 1 for analysis of similarity in next section.

3 Application

3.1 Mutation Analysis

For a symbolic sequence identified with a word over an alphabet {R,Y}, we define the frequency of group mutation as follows:

$$f_{UV} = \frac{\text{the number of word } UV}{n-1}, \quad (4)$$

where $U,V \in \{R,Y\}$, n is the length of the symbolic sequence. So we have four frequencies of group mutations in RY classification, denoted by $f_{RR}, f_{RY}, f_{YR}, f_{YY}$ respectively. Similarly, we can compute the frequencies of group mutations in MK classification and WS classification. In Table 2 we list frequencies of group mutations based on three classifications of the coding sequences of the first exon of beta globin gene of eleven species in Table 1. From Table 2, we conclude:

1) The frequencies of $R \to Y$ group mutations and $Y \to R$ group mutations of all listed species are identical, although the lengths of coding sequences are different.

2) In RY sequences, (human, gallus) and (goat, bovine) have the same frequencies of group mutation respectively.

3) The frequencies of $R \to R$ group mutation and $K \to K$ group mutation are larger than 0.3.

3.2 Similarities and Dissimilarities

We construct a twelve-component vector consisting of twelve frequencies of group mutations to represent a DNA sequence. The analysis of similarity and dissimilarity between two DNA sequences represented by the twelve-component vectors is based on the assumption that two sequences are similar if the corresponding twelve-component vectors point to a similar direction in the 12D space and have similar magnitudes. The similarities between such vectors can be computed by calculating the Eucilidean distance between their end points. The smaller Euclidean distance is, the more similar are the DNA sequences.

In Table 2, we list the similarity/dissimilarity matrix for the coding sequences of Table 1 based on the Euclidean distances between the end points of the twelve-component vectors.

Table 2. Twelve frequencies of group mutation based on three classifications of the coding sequences of Table 1

Species	f_{RR}	f_{RY}	f_{YR}	f_{YY}	f_{MM}	f_{MK}	f_{KM}	f_{KK}	f_{WW}	f_{WS}	f_{SW}	f_{SS}
Human	0.3407	0.2308	0.2308	0.1978	0.1978	0.1978	0.1868	0.4176	0.0989	0.3077	0.2967	0.2967
Goat	0.3882	0.2118	0.2118	0.1882	0.1647	0.2353	0.2235	0.3765	0.1059	0.2941	0.2824	0.3176
Opossum	0.3077	0.2308	0.2308	0.2308	0.2308	0.2198	0.2088	0.3407	0.1319	0.3407	0.3297	0.1978
Gallus	0.3407	0.2308	0.2308	0.1978	0.2418	0.2308	0.2198	0.3077	0.0989	0.2747	0.2637	0.3626
Lemur	0.3626	0.2198	0.2198	0.1978	0.1319	0.2418	0.2308	0.3956	0.1429	0.3187	0.3077	0.2308
Mouse	0.3152	0.2283	0.2283	0.2283	0.1957	0.2065	0.1957	0.4022	0.1087	0.3152	0.3043	0.2717
Rabbit	0.3736	0.2308	0.2308	0.1648	0.1758	0.1978	0.1868	0.4396	0.0989	0.3187	0.3077	0.2747
Rat	0.3297	0.2418	0.2418	0.1868	0.1978	0.2198	0.2088	0.3736	0.1758	0.2747	0.2637	0.2857
Gorilla	0.3478	0.2283	0.2283	0.1957	0.1957	0.1957	0.1848	0.4239	0.0978	0.3043	0.2935	0.3043
Bovine	0.3882	0.2118	0.2118	0.1882	0.1765	0.2118	0.2000	0.4118	0.1294	0.2824	0.2706	0.3176
Chimpanzee	0.3462	0.2308	0.2308	0.1923	0.1923	0.1923	0.1827	0.4327	0.1250	0.2981	0.2885	0.2885

Observing from Table 3, we find the smallest entries are associated with the pairs (Human, Gorilla), (Gorilla, Chimpanzee) and (Human, Chimpanzee). On the other hand the largest entries in the similarity/dissimilarity matrix appear in the row belonging to Opossum (the most remote species from the remaining mammals) and Gallus (the only non-mammalian representative). We believe it is not an accident, but shows their relationship in evolutionary sense.

Table 3. Similarity/dissimilarity matrix for the coding sequences of Table 1 based on the Euclidean distances between the end points of the twelve-component vectors

Species	Human	Goat	Opossum	Gallus	Lemur	Mouse	Rabbit	Rat	Gorilla	Bovine	Chimpanzee
Human	0	0.0974	0.1523	0.1507	0.1263	0.0531	0.0622	0.1077	**0.0143**	0.0814	**0.0357**
Goat		0	0.1859	0.1287	0.1121	0.1172	0.1071	0.1159	0.0948	0.0576	0.1059
Opossum			0	0.2014	0.1424	0.1128	0.1744	0.1523	0.1653	0.1960	0.1616
Gallus				0	0.2097	0.1596	0.1941	0.1373	0.1527	0.1483	0.1683
Lemur					0	0.1130	0.1142	0.1263	0.1316	0.1241	0.1215
Mouse						0	0.0976	0.1070	0.0652	0.1129	0.0683
Rabbit							0	0.1355	0.0597	0.0900	0.0595
Rat								0	0.1125	0.1028	0.0962
Gorilla									0	0.0736	**0.0344**
Bovine										0	0.0729

4 Conclusion

Mutation analysis and similarity analysis of DNA sequences are still important subjects in bioinformatics. According to chemical properties of nucleotide bases we divide four bases into two classes in each classification. For a DNA primary sequence we obtain three symbolic sequences and we compute the frequencies of group mutations of each symbolic sequence. From Table 2 we find some useful conclusions about group mutation. The frequencies of group mutations are applied to construct a 12-component vector to represent the sequence. The Euclidean distances among the introduced vectors are applied to compare the similarity and dissimilarity of the coding sequence of the first exon of beta globin gene of 11 different species. The results are coincident with what we expect in the evolutionary sense. Method we introduced is simple and clear so that we can quickly make comparisons of the similarity/dissimilarity among DNA sequences and its computation is easy, even for a long sequence.

Acknowledgment. This research is supported by Scientific Research Fund of Hunan Provincial Education Department (No.09C1015, No.08C882).

References

1. Hamori, E., Ruskin, J., Curves, H.: A Novel Method of Representation of Nucleotide Series Especially Suited for Long DNA Sequences. J. Biol. Chem. 258, 1318–1327 (1983)
2. Gates, M.A.: A Simple way to look at DNA. J. Theor. Biol. 119, 319–328 (1986)
3. Nandy, A.: A new graphical representation and analysis of DNA sequence structure: I. Methodology and application to globin genes. Curr. Sci. 66, 309–314 (1994)
4. Leong, P.M., Morgenthaler, S.: Random walk and gap plots of DNA sequences. Comput. Appl. Biosci. 11, 503–507 (1995)
5. Guo, X.F., Randic, M., Basak, S.C.: A novel 2-D graphical representation of DNA sequences of low degeneracy. Chem. Phys. Lett. 350, 106–112 (2001)

6. Randic, M., Vrakoc, M., Lers, N., Plsvsic, D.: Novel 2-D graphical representation of DNA sequences and their numerical characterization. Chem. Phys. Lett. 368, 1–6 (2003)
7. Randic, M., Vrakoc, M., Lers, N., Plsvsic, D.: Analysis of similarity/dissimilarity of DNA sequences based on novel 2-D graphical representation. Chem. Phys. Lett. 371, 202–207 (2003)
8. Wu, Y.H., Liew, A.W., Yan, H., Yang, M.S.: DB-Curve: a novel 2D method of DNA sequence visualization and representation. Chem. Phys. Lett. 367, 170–176 (2003)
9. Liao, B., Wang, T.M.: New 2D graphical representation of DNA Sequences. J. Comput. Chem. 25, 1364–1368 (2004)
10. Liao, B., Wang, T.M.: Analysis of similarity/dissimilarity of DNA sequences based on 3-D graphical representation. Chem. Phys. Lett. 388, 195–200 (2004)
11. Liao, B., Wang, T.M.: 3-D graphical representation of DNA sequences and their numerical characterization. J. Mol. Struct. Theochem. 681, 209–212 (2004)
12. Yao, Y.H., Wang, T.M.: A class of new 2-D graphical representation of DNA sequences and their application. Chem. Phys. Lett. 398, 318–323 (2004)
13. Liao, B., Tang, M.S., Ding, K.Q., Wang, T.M.: Analysis of similarity /dissimilarity of DNA sequences based on a condensed curve representation. J. Mol. Struct. Theochem. 717, 199–203 (2005)
14. Song, J., Tang, H.W.: A new 2-D graphical representation of DNA sequences and their numerical characterization. J. Biochem. Biophys. Methods 63, 228–239 (2005)
15. Li, C., Tang, N.N., Wang, J.: Directed graphs of DNA sequences and their numerical characterization. J. Theor. Biol. 241, 173–177 (2006)
16. Yao, Y.H., Nan, X.Y., Wang, T.M.: A new 2D graphical representation—Classification curve and the analysis of similarity/dissimilarity of DNA sequences. J. Mol. Struct. Theochem. 764, 101–108 (2006)
17. Liao, B., Ding, K.: A 3D graphical representation of DNA sequences and its application. Theor. Comput. Sci. 358, 56–64 (2006)
18. Liu, X.Q., Dai, Q., Xiu, Z.L., Wang, T.M.: PNN-curve: A new 2D graphical representation of DNA sequences and its application. J. Theor. Biol. 243, 555–561 (2006)
19. Qi, Z., Qi, X.: Novel 2D graphical representation of DNA sequence based on dual nucleotides. Chem. Phys. Lett. 440, 139–144 (2007)
20. Qi, Z., Fan, T.: PN-curve: A 3D graphical representation of DNA sequences and their numerical characterization. Chem. Phys. Lett. 442, 434–440 (2007)

An Approach to Support Evolution of Software Architecture and Its Formalization

Li Xiaojian[1] and Zheng Ying[2]

[1] International Software School, Wuhan University, Hubei, P.R. China
[2] Economics and Management School, Wuhan University, Hubei, P.R. China
Lxj0713@tom.com, Butterfly0601cn@yahoo.com.cn

Abstract. This paper presents a novel reflective approach to support evolution of software architecture. Based on the approach, Pi calculus model was established. And what' more, through case study, the evolution process of software architecture was formalized by Pi calculus description.

Keywords: Evolution, reflective, formalization, pi calculus.

1 Introduction

Recent years, the evolution of software architecture is getting a concentrated topic. For a running software system, how can we add a component, delete a component, or combine two sub-architecture into a whole architecture dynamically? Here, we propose a reflective approach to achieve the goal of supporting architectural dynamical evolution.

Reflection is a technology which is relevant with meta information, through meta information, reflection manages and controls information. Meta information is a kind of information about information, it can be used to describe the structure、semantic and usage of information. Meta information technology provides an effective approach for managing, controlling and using complicated information. Through reflection mechanism from meta information to information, users can use information in a more simple、flexible and automatic method. Reflection technology is getting more and more attractive for researchers, some program languages(CLOS, Smalltalk and Java) , operating systems(Apertos), CORBA and OLE technologies adopt reflection technology. In the code layer of object-oriented, reflection technology provides effective method for code reusing. For example, JavaBean is a code component reusing technology based on reflection mechanism, BeanBox makes use of reflection mechanism to provide a simple and efficient visual method for programmer reusing Bean components. All these are achievements for reflection technology. Here, we will use reflection in the evolution of software architecture, it should be a novel trying. And the more important thing is that, we will give the formal description in Pi calculus for the process of dynamical evolution.

J. Luo (Ed): Affective Computing and Intelligent Interaction, AISC 137, pp. 385–392.
springerlink.com

The next section is about reflective mechanism. Section 3 is about the basis of Pi calculus. And section 4 is case study, section 5 is about related work, and the last section is the conclusion.

2 Reflective Mechanism

Implementation of reflection is based on meta information, we separate software architecture into two parts: meta-level and base-level. Meta-level includes meta objects which embody meta information, and base-level includes base objects – every component. The scheme is as the following figure:

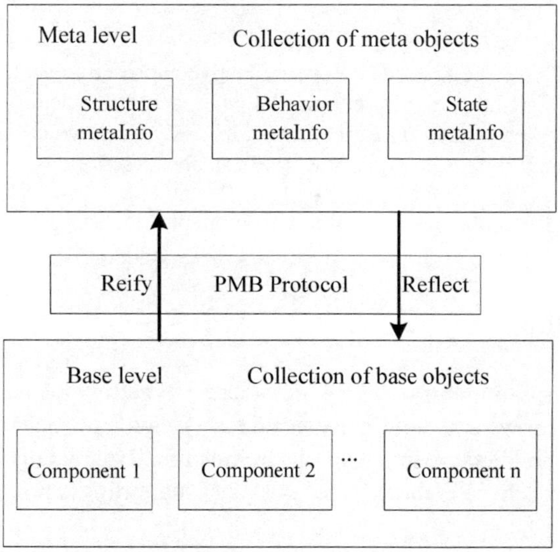

Fig. 1. Architectural reflection

In this figure, “PMB protocol” is a Protocol for connecting Meta-level architecture and Base-level architecture, through the PMB, any change of meta object can be reflected to base-level, that is, components can be isolated or replaced, meanwhile , any change of base-level can also be reified to meta-level. From this figure, we can find there are three main parts: meta-object, object, PMB. In fact, if we want to change the architecture dynamically, there should be a user end. So we can establish the π calculus model for the reflective mechanism as follows.

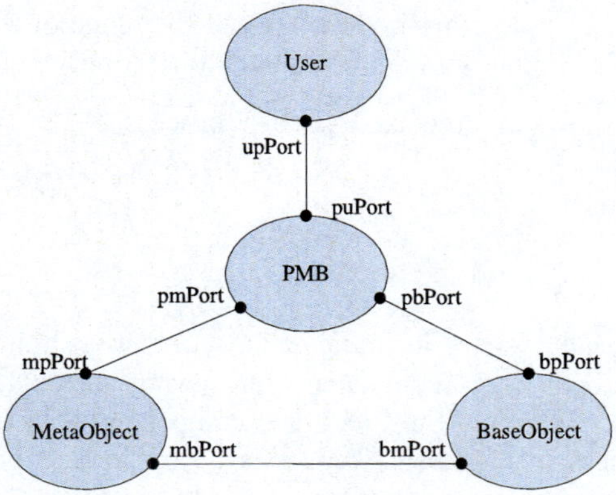

Fig. 2. Pi calculus model for reflective architecture

In the above figure, there are four basic processes, they are user process, PMB process, Meta Object process, Base Object process. upPort and puPort are the ports between user and PMB process. mpPort and pmPort are the ports between PMB and Meta Object processes.

3 Foundation of Π Calculus

Pi calculus is an important concurrent computing model, which is proposed through expanding CCS by Milner and the others[1,2,3,4]. Process is a kind of entity which run concurrently. Pi calculus uses name to define channel and the object which is transferred in the channel, each process has a number of channels associated with other processes, the processes interact through their shared channels. High-order multi-type Pi calculus is developed by Sangiorgi[5,6] on the basis of one-order Pi calculus, its aim is to describe concurrent system whose structure or behavior change dynamically. Pi calculus is very suitable for describing the dynamical evolution of software architecture.

In pi calculus, process is the basic unit for a running system. a process t can be defined as one of these form:

$$t ::= 0 \mid \overline{a} < x > .P \mid a(x) \mid \tau.P \mid P + Q \mid P \mid Q \mid (va)P \mid !P \mid [x = y]P$$

- 0: blank process. This process have not any operations.
- $\overline{a} < x > .P$: output prefix expression. It shows that the process is P after sending x through channel a.
- a(x).P: input prefix expression. It shows that the process is P after receiving x through channel a.
- τ.P: mute prefix expression. It shows that the process can be P without perform any operations.
- P+Q: summation expression. It shows that the process should be P or Q.
- P|Q: concurrently composition expression. It shows that P and Q can be executed concurrently.

- [x=y]P: match expression. It shows that if the name of x and y is same, then the process is P, otherwise there is no any operation.

In Pi calculus, fn(P) shows the free name set of P process.

4 Case Study

4.1 Case

There is a existed bank system, its basic architecture is as follows. There are five components. SavAccount is a component for saving account, BusAccount is a component for business account, SimClock is a component of system timing, BM is a component for business management, People is a component for client information, ADTConn is a component for transferring messages from one component to other component.

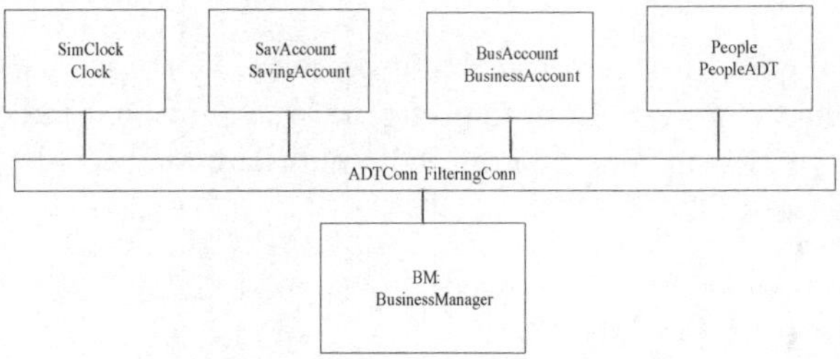

Fig. 3. The architecture of a bank system

Now, we hope that the bank system can handle check account, so we need to add a component for handling check account. We name the component as CheckingAccount. First, we establish the Pi calculus model for adding CheckingAccount component.

4.2 Formalization Expression

Before adding CheckingAccount component, the process in the meta level of the bank system can be described by:

$$\begin{aligned}&MetaBankAccountArch(Ports, msgs)\\&= MSavingAccount(Ports, msgs) \mid MBu\sin essAccount(Ports, msgs)\\&\mid MPeopleADT(Ports, msgs) \mid MBu\sin essManager(Ports, msgs)\\&\mid MADTConn(Ports, msgs) \mid MWrite(Ports, msgs)\end{aligned}$$

the process of base level can be described by:

$$\begin{aligned}&BaseBankAccountArch(Ports, msgs)\\&= BSavingAccount(Ports, msgs) \mid BBu\sin essAccount(Ports, msgs)\\&\mid BPeopleADT(Ports, msgs) \mid BBu\sin essManager(Ports, msgs)\\&\mid BADTConn(Ports, msgs) \mid BWrite(Ports, msgs)\end{aligned}$$

the whole process can be described by:

$$BankAccountArch(Ports, msgs)$$
$$= User(Ports, msgs)$$
$$PMB(Ports, msgs)$$
$$MetaBankAccountArch(Ports, msgs)$$
$$BaseBankAccountArch(Ports, msgs)$$

The above formula show that the whole architecture is composed by user process, PMB process, meta level process and base level process.

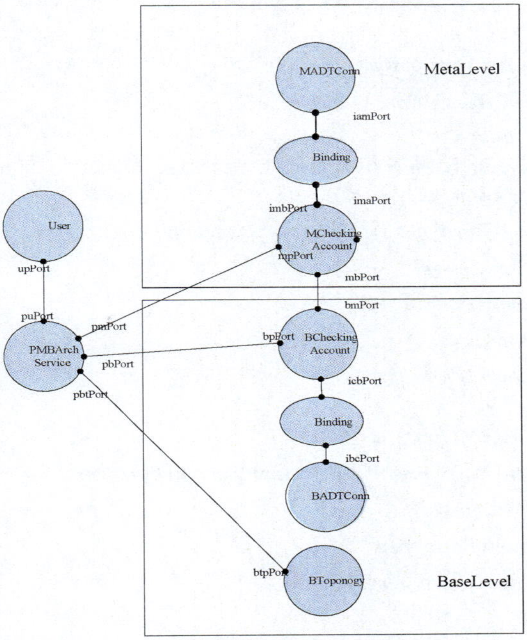

Fig. 4. Pi calculus model for adding checkAccount component

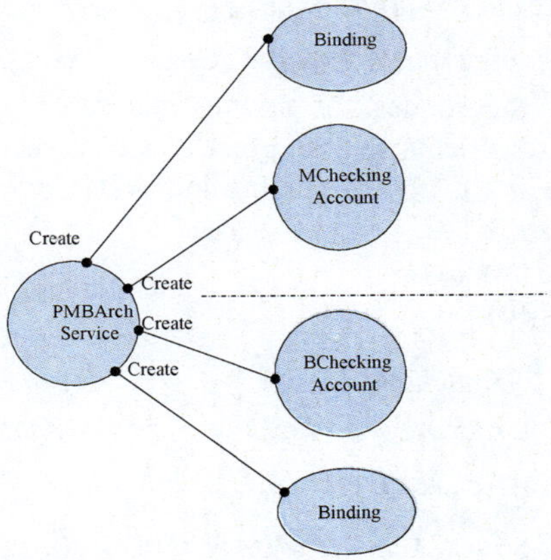

Fig. 5. The construction of CheckingAccount process

Now we need to add checkAccount component to current architecture, we can show the whole process in two parts when we make use of Pi calculus. The first part is the construction of process when user render the request of adding component, and the other part is the process of interchanging between different processes.

(1) Construction of new process

After user submit the request of adding component, PMBArchService process will construct MCcheckingAccount process in the meta level and BCcheckingAccount process in the base level, and what's more, it will construct a binding process for connecting MCheckingAccount and MADTConn.

$$PMBArchService(fn)$$
$$=[msg=add\,\mathrm{Re}\,quest].\overline{Create}<MCheckingAccount>.$$
$$\overline{Create}<BCheckingAccount>.\overline{Create}<Binding>$$
$$Binding<icmPort,imcPort>$$

After processes have been constructed, the meta level can be expressed by:

$$MetaBankAccountArch(Ports,msgs)$$
$$=MSavingAccount(Ports,msgs)\mid MBu\sin essAccount(Ports,msgs)$$
$$\mid MPeopleADT(Ports,msgs)$$
$$\mid MCheckingAccount(Ports,msgs)$$
$$\mid Bingding(Ports,msgs)$$
$$\mid MBu\sin essManager(Ports,msgs)$$

the base level can be expressed by:

$$BaseBankAccountArch(Ports,msgs)$$
$$=BSavingAccount(Ports,msgs)\mid BBu\sin essAccount(Ports,msgs)$$
$$\mid BPeopleADT(Ports,msgs)$$
$$\mid BCheckingAccount(Ports,msgs)$$
$$\mid BBu\sin essManager(Ports,msgs)$$

(2) User process

User send its request to PMB interface, it can be expressed by:

$$User(fn)$$
$$=puPort<addCA\,\mathrm{Re}\,quest>.upPort(msgs).$$
$$[msg=addCA\,\mathrm{Re}\,questAck]puPort<addCALink>.User(fn)$$

user process send the request of adding checkAccount to PMBArchService process, after the request is dealt, if user process accept success information, then it send a connection request for checkAccount and ADTConn , if succeed, then user process enter the wait state.

(3) PMBArchService process

$$PMBArchService(fn)$$
$$=puPort(addCA\,\mathrm{Re}\,quest).\overline{mpPort}<addCA>.pmPort(msg).$$
$$[msg=addCAAck].mpPort<addMLink>.pbPort(msg).$$
$$([msg=addBaseCAAck].bpPort<addBLink>.$$
$$\overline{upPort}<addCA\,\mathrm{Re}\,questAck>.PMBArchService(fn)$$

After PMBArchService process accept the request of adding CheckingAccount, it will construct MCheckingAccount process in the meta level, and then, through mpPort port, addCA message was sent. After dealt, if success message return, it will send addMLink message to meta level process through mpPort port. And then, PMBArchService process accepts the acknowledge message from BCheckingAccount process, if the acknowledge message is a success message, then send addBLink message to base level process. If succeed, PMBArchService will return a addCARequestAck message to user process. Finally, PMBArchService process enter wait state.

(4) MCheckingAccount process

$$\begin{aligned} &MCheckingAccount(fn)\\ &= mport(addCA).\overline{Create} < Verify > .\tau.(MCheckingAccount(fn) +\\ &\overline{bmPort} < addCANotify > .\overline{bmPort} < addCALinkNotify >).\\ &MCheckingAccount(fn) \end{aligned}$$

Once the MCheckingAccount process in the meta level accept the message of adding checkingAccount component, it will set values for the meta object, and then, construct a verify process to check whether these value is legal, if illegal, then the process state is wait. Otherwise, the process in the meta level will send messages of adding component and link to the process in the base level. Finally, the whole process enter wait state.

(5) BCheckingAccount process

$$\begin{aligned} &BCheckingAccount(fn)\\ &= bmPort(addCANotify).\tau.\\ &(\overline{pbPort} < addBaseCAAck > .\overline{pbPort} < addBLinkAck >\\ &.BCheckingAccount(fn) \end{aligned}$$

the base level process BCheckingAccount accepts add- CANofity message from meta level process, after internal dealing, BCheckingAccount process will send success message of adding component and adding link to PMBArchService process. Finally, BCheckingAccount process enter the wait state.

5 Relate Work and Discussion

The most impressive work on architectural reflection maybe is from J-C- Fabre's team LAAS. They implemented a MOP-based system *Friends*[7,8,9] for fault-tolerance CORBA system, and made a detailed test on reflective mechanism. But they did not declare their work is for the evolution of software architecture. Their main object is to achieve a fault-tolerance system. Our work is aimed to support the evolution of software architecture, and obviously presents the basic evolution unit-component.

6 Conclusion

This paper presents a reflective approach for evolution of software architecture, and gives formal description in Pi calculus, in the future, we will do some formal verification for our approach.

References

1. Milner, R., Parrow, J., Walker, D.: A calculus of mobile processes. Information and Computation 100(1), 1–40 (1992)
2. Milner, R., Parrow, J., Walker, D.: A calculus of mobile processes. Information and Computation 100(1), 41–77 (1992)
3. Milner, R.: Communicating and Mobile Systems: The Pi Calculus. Cambridge University Press (1999)
4. Milner, R.: The polyadic-calculus: a tutorial. Logic and Algebra of Specification 94, 91–180 (1993)
5. Sangiorgi, D.: From π-Calculus to Higher-Order π-Calculus and Back. In: Gaudel, M.-C., Jouannaud, J.-P. (eds.) CAAP 1993, FASE 1993, and TAPSOFT 1993. LNCS, vol. 668, pp. 151–166. Springer, Heidelberg (1993)
6. Sangiorgi, D., Walker, D.: The π-calculus: A Theory of Mobile Processes. Cambridge University Press (2001)
7. Ruiz-garcia, J.C., Thevenod-fosse, P., Fare, J.C.: A Strategy for Testing MetaObject Protocols in Reflective Architectures. In: International Conference on Dependable Systems and Networks (2001)
8. Ruiz, J.C., Killijian, M.-O.: Reflective Fault Tolerant Systems from Experience to Challenges. IEEE Transactions on Computers 52(2) (February 2003)
9. Killijian, M.-O., Fabre, J.C.: Implementing a reflective fault-tolerance CORBA System. In: Proceedings of the 19th IEEE Symposium on Reliable Distributed Systems, Nuremberg, Germany, pp. 154–163 (October 2000)

On an Intelligent Fault-Tolerant Algorithm in CAN Network Control System

Cao Xiaohua and Zhou Yong

Logistics Department, Wuhan University of Technology, Wuhan, China
Tomm_cao@163.com, Zhoyo@163.com

Abstract. Network control system is applied widely in the industrial field. However, the reliability and real-time cannot meet the system's requirements in the overloading system. A new idea called as Active-Request-And-Dynamic-Load-Distribution was introduced into redundant fault-tolerant services through analysis and comparison to several fault-tolerant algorithms. An effective algorithm was presented for CAN fault-tolerant services. According to the system's load, CAN nodes decided whether to request actively the task as a redundant server. Considering the load and the current status of node hosts, the management system distributed dynamically its load to each redundant server and changed the number of redundant servers. Finally, the model of new algorithm was created. The new algorithm was compared with other fault-tolerant algorithms and tested throught some examples. The results showed that the new algorithm was superior and feasible.

Keywords: Dynamic Load Distribution, Active Request, Fault-Tolerant Algorithm, CAN.

1 Summary

The redundant fault-tolerant technologies have been proposed to enhance network reliability. Their core is the fault-tolerant algorithm[1-2].

A number of scholars dedicated to fault-tolerance technologies and got a series of achievements. A primary-backup method was presented in the reference paper [3]. The main server realized its fault-tolerant functions through the twin servers. When the primary server was failed, the backup server would replace it. This algorithm was simple, but when the system load was heavy, the server processing ability would become the bottleneck in the entire system and might cause system to collapse. It is difficult to meet the requirements of real-time and reliability. An active replication algorithm was presented in the reference paper [4]. All the servers were applied as a primary servers in this algorithm. Simultaneously, they worked and responded to customer requests. It belonged to a kind of spatial redundancy fault-tolerant algorithm. This algorithm could respond quickly to the requests, but it consumed vast system resource. Therefore it debased the system performance. A RAWA (Read-Any-Write-Any) algorithm was proposed in the reference paper [5]. It adopted a broadcast-styled load-balanced mothod. In this mothod the load distribution was determined by the

J. Luo (Ed.): Affective Computing and Intelligent Interaction, AISC 137, pp. 393–398.
springerlink.com

service side. The task manager sent the same requests to all the redundant servers. When a redundant server completed the request, the task manager would remove the same request in other redundant servers. Although Its execution speed might been improved through the multi-thread program, this algorithm still have some disadvantages. 1) It is focused on the balance distribution of tasks requests. If the current node collapsed, the node processing tasks would be lost. 2)It completed the tasks through a multi-thread program, but the requests were executed ordinally and could not been skipped. So the system had to wait and wast vast free time. 3)The task manager broadcasted the requests to all servers. When the execution results returned, the task manager would cancel the same request in other servers. So the system would bring large communication data which easily led to system congestion.

2 Structural Design of ARADLD Fault-Tolerant System

ARADLD fault-tolerant system consists of Administrator, OM(Object–Manager), HM (Host-Manager) and SA (Server-Agent). Its framework is shown in Figure 1.

In figure 1, The administrator is a global manager. The OM owns an information aggregate of service object, including a Request_Queue according to FIFO principle, the Instance_List and other management information. The Request_Queue consists of the workers' request information. The Instance_List is the address information of instance object. The instance_status points out the working state of instance object (normal or failure). Each OM manages a type of service. The HM owns an information aggregate of CAN node servers managed by Administrator, which contains various server receiving identifier, the ability to process the requests and other management information, such as the server's current load, the value of max Load, an instance queue waiting for execution and the current host node state (good or bad) and so on. The HM and SA reside on each CAN host. The SA check the load status of the host from HM to decide whether to apply for a task from the OM.

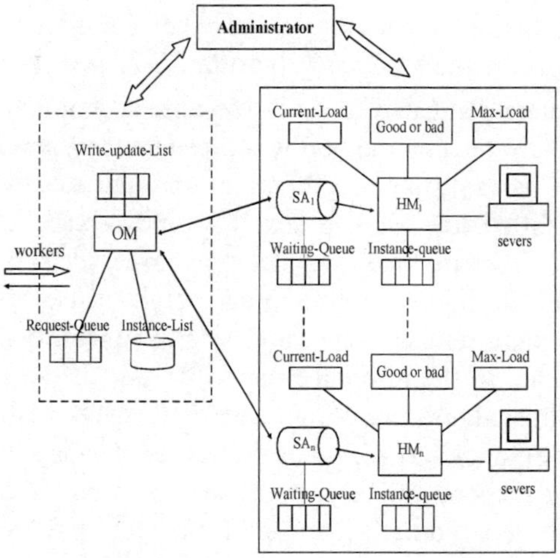

Fig. 1. System Framework of ARADLD Algorithm

3 ARADLD Fault-Tolerant Algorithm

3.1 Core Idea of Algorithm

The core idea of ARADLD fault-tolerant algorithm is shown as follows. Firstly each SA inquires the cuurent load of corresponding CAN node, then the CAN node with less service actively applies for the task from OM as a redundant server. Through the active load request way, the system realizes the dynamic load distribution to achieve load balance. Simultaneously the ARADLD algorithm greatly reduces the OM management task so that the OM management functions become simple and the processing bottlenecks is avoided in the system.

3.2 Algorithm Process

According to FIFO principle, the OM puts the worker's requests into the request_Queue, adjusts the maximum number of workers' requests in the request_Queue and updates the serial number number of next active request in the Instance_List through replacing null next_requests by the newly arriving requests, and then waits for the arrival of some SA's active application for the task.

The SA in CAN node checks whether the node host is normal with T cycle. if the host is normal, the SA will test the load status in the host. If the host overloading, the SA won't send the request message to the OM, otherwise the SA will actively send a request message to the OM. After the OM receives the request message from the SA in CAN node, firstly looks for the current SA's request serial number in the Instance_List, the value of next_request, then tests this request and make the following treatment according to the testing results.

If the number of the current redundant servers is less than the maximum number of redundant servers, the OM will add 1 to the number of current redundant servers and assigns the first task in the Instance_List to the SA. Because this task has been served in the Instance_List, the next_request address of Request_Queue will be updated.

If the number of the current redundant servers has reached the maximum number, the OM will test the next request in the Request_Queue. If there is the request which number of needful redundant servers is less than the maximum, the following operation is the same with the first step, otherwise the OM sends non-task message to the SA, then send non-task message to the SA, and updates the SA's applying value with the NULL, indicating that there is no new tasks assigned to the SA.

When a CAN node fails, the waiting queue in the SA will be sent to the OM, and the OM will re-placed this request to the Request_Queue.

After the CAN node as redundant server has completed the customer requests, the execution results will return to the OM, then the OM sends the results to the worker and removes this worker's request from the Instance_List, and updates the next_request with the request address of next worker.

4 Performance Analysis and Test

4.1 Analysis on Average Response Time

About the average response time, Some researchers have compared among the RAWA algorithm, the active-replication algorithm and primary-backup method, and got the results. The average response time increases by degrees with RAWA, Active replication, primary backup[5]. Therefore, the average response time of ARADLD algorithms will be only compared with RAWA.The average response time of two algorithms is

$$T_{DY} = T_{work} + T_{wait} \tag{1}$$

Twait is the average waiting time for the request, Twork is the service time for the requet, TDY is the average response time for the request. Twait = L / k, L is the average waiting queue length, k is the average arrival rate of the request.

Assume that two algorithms' number of redundant servers is m, for the RAWA algorithm, the average service time is tRA, then Twork = tRA, Twait = L / k, the average response time of the RAWA algorithm is

$$T_{DY} = t_{RA} + L/k \tag{2}$$

In the ARADLD algorithm, the average period that SA applies for a task on its own is T, the average number of redundancy server is N, the average parallel degree is m/N, so the average request service time is

$$T_{work} = t_{RA} + TN/(2m) \tag{3}$$

Assume that two algorithms' average arrival rate is k, the actual length of waiting queue in ARADLD algorithm should be L / (m / N), the average response time is

$$T_{DY} = T_{work} + LN/(mk) \tag{4}$$

Synthetizeing the formula 3 and the formula 4,

$$T_{DY} = t_{RA} + TN/(2m) + NL/(mk) \tag{5}$$

By analyzing the formula 2 and formula 5, when $T \leq \frac{2L}{k}(\frac{m-N}{N})$, the average response time of the ARADLD algorithm is less than one of the RAWA algorithm.

4.2 System Response Time Test

In the test, there is only one type of service object in the system, so the system just owns an OM which resides on one CAN host node. The service time of each instance is 5 seconds. There are eight test nodes in the test. The instance can be executed by any node and its redundancy may be random one from 1 to 8. The Administrator sends the requests to the OM. The Performance Comparison Test is done between RAWA and ARADLD algorithm based on CORBA platform.

Firstly, Suppose that the application period T is 0.1 second, the request frequency is 256 in one second, the request number is 512. The response time of the two algorithms is got by changing the number of redundant severs from 1 to 8. The result is shown in Figure 2.

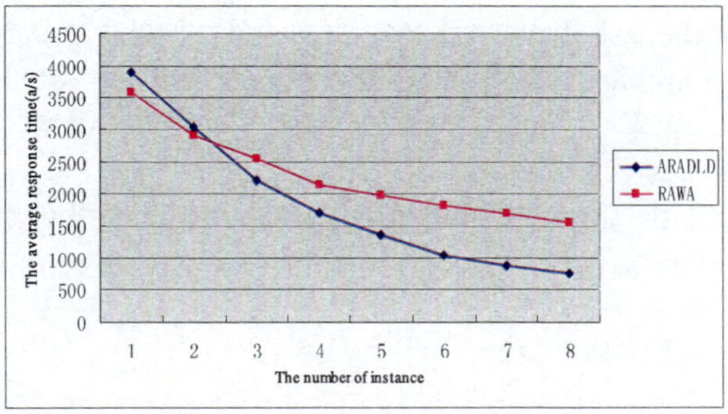

Fig. 2. Change Curves of Response Time with the Number of Redundant Severs

From Figure 2, with the increasing of the number of redundant severs, the response time of the two algorithms gradually reduced, so the system realtime increases with the increasing of the redundant severs. When the redundancy is 1 or 2, the realtime of ARADLD algorithm is less than the realtime of RAWA because ARADLD algorithm has cycles T to apply for the task. But with the increase of redundancy, the response time of ARADLD algorithm is improved more than RAWA algorithm.

Secondly, Suppose that the redundancy is 5 and the application period T is 0.1 second. The response time of the two algorithms is got by changing the number of request arrical from 1 to 200 in one second. The result is shown in Figure 3.

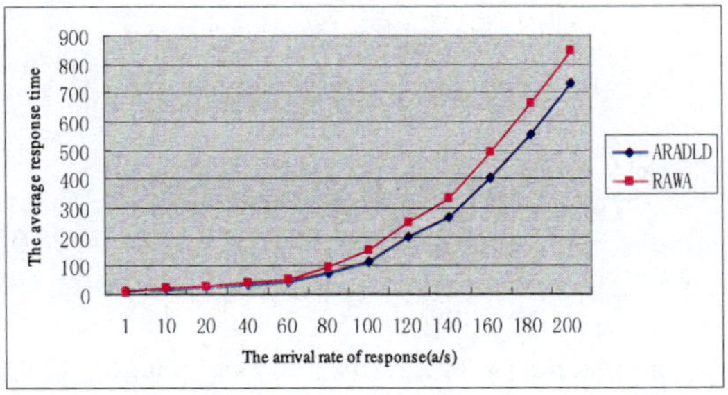

Fig. 3. Response time change curves with the arrival number of request in one second

As can be seen from Figure 3, with the increasing of request number in one second, the response time is gradually increasing in two algorithms. When the request number in one second is more than 10, the response time of ARADLD is less than that of RAWA. When the request number in one second is more than 80, its response time is much less than that of RAWA.

In a general way, the request number is vast in CAN network control system, therefore the ARADLD is a more suitable fault-tolerant algorithm for CAN network system. Its advantages are mainly included in the following.

First, the system Load is dynamically allocated to the less-load nodes by the less-load nodes applying for task initiatively, which largely alleviates the burden of task manager and avoids the task manager becoming a bottleneck of the system.

Second, when the number of redundant servers and requests are much in the system, the realtime of ARADLD is much better and the efficiency of entire system may be ensured.

Third, Because of the introduction of dynamic load distribution mechanism, the process of application and execution is parallel. Reading requests, reading and writing requests, writing requests can be executed in parallel. So the execution efficiency of requests is greatly improved.

5 Conclusion

All the nodes in CAN network system can be used as a redundant host. According to this characteristics, the ARADLD fault-tolerant algorithm based on Active-Request-And- Dynamic-Load-Distribution was presented. The test results show that the performance of ARADLD algorithm is better than several other fault-tolerant algorithms. It has less average response time of the request and better realtime, reduced the system jam-up and improved the system reliability. It can be used in a control system which requires higher reliability and real-time.

Acknowledgment. This work is supported by National Nature Science Foundation under Grant 50975216 and by Guangdong Production and Education Foundation (2009B090200059).

References

1. Hu, X.-M., Zhai, Z.-J., Cai, X.-B.: Active Dynamic Load Balancing Algorithm for Collaborative Virtual Environments. Computer Engineering 33(20), 104–106 (2007)
2. Yu, C., Xu, H., Liu, G.: Research of the Redundant Methods of CAN. Measurement & Control Technology 22(1), 28–30 (2003)
3. Zhou, W.: Detecting and tolerating failures in a loosely integrated heterogeneous database system. Computer Communication 22(11), 1056–1067 (1999)
4. Raehid, G., Andre, S.: Software-Based replication for fault tolerance. IEEE Computer 30(4), 68–74 (1997)
5. Qian, F.: A Dynamic Fault Tolerant Algorithm for Improving Performance of Redundant Services. Journal of Software 12(6), 928–935 (2001)
6. Chen, J., Hu, J.: Distributed fault-tolerant system based on CAN bus. Journal of Tsinghua University (Science and Technology) 49(7), 1007–1011 (2009)
7. Avizienis, A.: Toward Systematic Design of Fanlt-Tolerant Systems. IEEE Computer 30(4), 51–58 (1997)

Rebuilding the Competitive Equilibrium Theory with Differentiable Dynamic Systems

Li Yingjuan[1] and Meng Ying[2]

[1] School of Business Management, Shenyang University, Shenyang, China
[2] School of Economics, Xiamen University, Xiamen, China
lyjmeng@tom.com

Abstract. We rebuilt and proved the competitive equilibrium theory with theories and methods of the differentiable dynamic systems in this paper. We construct the state equations of the economic behavior, and get rid of the market economy assumes that each consumer is endowed with a nonnegative amount of every commodity. We established equivalent relationship between variant economical conduct. The results may have more universal significance and practicability than that of Arrow Debreu. It is possible for us to explain debt consumption with the new competitive equilibrium theory, and therefore to extend the applied area of the dynamical economic equilibrium.

Keywords: Competitive equilibrium, Economical behavior, Distributed parameter control, Behavior equation.

1 Introduction

A complete different means is offered to the reconstruction of the competitive equilibrium theory [1] and the proof of the existence of the competitive equilibrium. First we define some conceptions and explain special properties of the competitive equilibrium. On the basis of these conceptions and properties, some behavior state equations are constructed for the consumers and firms with the differentiable dynamic systems. Then we prove that the variant behaviors are equivalent by the operator semigroup theory and the infinitesimal analytical craftsmanship. In the whole process, we eliminate the assumption that every consumer initially possesses a nonnegative amount of some commodity, and therefore extend the applied area of the competitive equilibrium theory.

2 Definitions, Assumptions and Notation

We consider economy E that consists of m consumers, n firms, and l commodities. We denote by N_k the set that is equal to the finite nature number set $\{1,2,\cdots k\}$, and denote the space of endowments by Ω. Each consumer i, for $i \in N_m$, has an initial endowment ω_i, a consumption set $X_i \in R^l$, and ownership shares $\theta_{ij} \geq 0$,

J. Luo (Ed.): Affective Computing and Intelligent Interaction, AISC 137, pp. 399–407.
springerlink.com

$j \in N_n$, in the n firms. There is a continuous utility function $u_i : X_i \to R, x_i \in X_i$ is a feasible commodity bundle. Each firm $j, j \in N_n$, has a production possibility set $Y_j \subset R^l$, an element $y_j \in Y_j$ is a netput vector, and $\sum_{i=1}^{m} Q_{ij} = 1$, for $j \in N_n$. We define $\omega = \sum_{i=1}^{m} \omega_i$ as the aggregate endowment, $Y = \sum_{j=1}^{n} Y_j$ as the aggregate production possibility set. Let $(a_i)_k$ be the vector $(a_1, a_2, \ldots, a_k)$, then a production plan is a collection $(y_i)_n$ of netput vector with $y_j \in Y_j$ for $j \in N_n$. An allocation is a collection $(x_i)_m$ of feasible commodity bundles, with $x_i \in X_i$ for $i \in N_m$. To simplify notation, sometimes $\sum_{i}$ will be subsequently used instead of $\sum_{i=1}^{m}$ to indicate summation over a complete index set, and $\sum_{i \in N_m}$ will be used to indicate summation over a subset N_m of a complete index set. Commodities are infinitely divisible.

We assume that the economy E operates under perfect competition, so that both consumers and firms are price takers. Prices are denoted by a price vector $P \in R^l$. Give a price vector P, the profit of firm j operating at netput vector y_j is denoted by $\pi_j(P, y_j), \pi_j(P, y_j) = P' y_j$. The profit share of consumer i at price P is $\sum_{j=1}^{m} \theta_{ij} \pi_j$. An attainable state of E is a triple $((x_i)_m, (y_j)_n, P)$ which consist of an allocation $(x_i)_m \in \prod_{i=1}^{m} X_i$, a production plan $(y_j)_n \in \prod_{j=1}^{n} Y_j$ and a commodities price vector P such that market clears i.e., $\sum_i x_i = \omega + \sum_j y_j$. We denote yet by E the aggregate attainable state set of economy E when understand, that satisfies:

$$E = \{((x_i)_m, (y_j)_n, P) \mid ((x_i)_m, (y_j)_n, P) \in (\prod_{i=1}^{m} X_i) \times (\prod_{j=1}^{n} Y_j) \times R^l, \ \sum_{i=1}^{m} x_i = \omega + \sum_{j=1}^{n} y_j\}$$

The aggregate attainable state of economy is also said to be economy.

We assume that firms act to maximize utility, subject to budget constraints. Let E^* be a subject of E, if it satisfies that $E = \{((x_i^*)_m, (y_j^*)_n, P^*) \mid ((x_i^*)_m, (y_j^*)_n, P^*) \in E\}$

and $P^* \neq 0$ and

C1: $\sum_{i=1}^{m} x_i^* = \omega + \sum_{j=1}^{n} y_j^*$, market clearing.

C2: $y_j^* \in Y_j$, profit maximizing for each $j \in N_j$.

C3: $x_i^* \in X_i$, unitility maximizing under the budget constraints for each $i \in N_m$, then E^* is called an economic equilibrium set. The triple $((x_i^*)_m,(y_j^*)_n,P^*) \in E$ is called an equilibrium of E. Therefore the existence of equilibrium is that E^* is non-empty i.e. $E^* \neq 0$ and the uniqueness of equilibrium is that E^* is set consist of one point $((x_i^*)_m,(y_j^*)_n,P^*)$. I.e.,

$$E^* = \{((x_i^*)_m,(y_j^*)_n,P^*)\}$$

Furthermore, we define an excess demand z that satisfies $z = \sum_{i=1}^{m} x_i - \sum_{j=1}^{n} y_j - \omega$, the total excess demand set Z with $Z = X - Y - \Omega$. The condition $Z = \{0\}$ is a necessary condition for economical equilibrium.

3 Dynamical Characteristic of Competitive Equilibrium

First, Let [0,T] $\subseteq R$ be the time area in which economy operating, $z(t,P)$ be an excess demand at time t and price P. We define an excess demand matrix $sign$ function $M(z)$ that satisfies $M(z) = diag[sign(z_1), \cdots, sign(z_l)] = diag[sign(z)]$.

The $sign(z)$ is the $sign$ vector of z such that $\forall k \in N_l, sign(z_k) = -1,0,+1$, if and only if $z_k < 0, = 0, > 0$ respectively. $diag[sign(z)]$ denote the diagonal matrix that diagonal elements are the components of $sign(z)$. In the analysis of Walrasian equilibrium of the dynamic economy E, the commodities prices are the function of the excess demand z. Now we denote by emphatic "·" the derivative of price P with respect to the time t, i.e., $\dot{P} = \frac{dp}{dt}$. According to Walrasian's rule, if the excess demand z_k is great than zero, the demand is great than supply in the commodity market k, the related price P_k tend to increase. Therefore P_k has positive varied rate on time t, that is $\dot{P} > 0$. To the contrary, if the excess demand z_k is negative i.e., $z_k < 0$, the demand is small than supply in the market k, the related price P_k tent to decreasing, so that P_k has a negative changed rate about time t, that is $\dot{P} < 0$. If $z_k = 0$, it is shows that the demand is equal to supply in commodity market k, the price P_k is constant and $\dot{P} = 0$. Let $\eta(t,z)$ be an adjustment rate for the commodities prices when excess demand is z at time t, Δt is a very small variation about time t. In the light of discuss above, the prices should satisfy the formula as follows:

$$P[t + \Delta t, z + \Delta z] = P(t,z) + \Delta t M(z)\eta(t,z)$$

$\eta(t,z)$ is nonnegative, and is also called control vector of price. We suppose that price and excess demand are all continuous and differentiable, when $\Delta t \to 0$, we have

$$\dot{P} = -\frac{\partial}{\partial P} P\dot{z} + M(z)\eta \tag{3.1}$$

This is the commodities prices state equation. Note that $z = z(t,P)$, and substitute it into (3.1), we obtain

$$\dot{P} = M(z)\frac{1}{3}\eta \tag{3.2}$$

Model (3.2) coincides with Walrasian's price equation.

Now we consider firms behavior state equation. When price P_k of the commodity k tend to rising, the individual rationality manner makes firm j turn to increasing its supply of commodity k for all $j \in N_n$. On the other hand, the firm j turns to decreasing its supply of that. The supply of commodity k is not only the function of the price P_k, but also that of all commodity prices P. The direction of the supply development is the same as the price. Let $\zeta_j(t,P)$ be an adjustment rate of the supply of the firm j, and $\zeta_j(t,P)$ is nonnegative. Let Δt be a very little increase of time t. When commodity prices are P at time t, the productive supplies $y_j(t,P)$ of the firm j satisfies the equation as follows:

$$y_j(t + \Delta t, P + \Delta P) = y_j(t,P) + \Delta t M(\Delta P)\zeta_j(t,P)$$

Where $\Delta P = P(t + \Delta t) - P(t)$. Because ΔP and gradient of P has the same direction, $sign(\Delta P)$ is equal to $sign(\dot{P})$, this also implicate that $M(\Delta P)$ is equal to $M(\dot{P})$. Due to the continuity of production, we have

$$\dot{y}_j = -\frac{\partial}{\partial P} y_j \dot{P} + M(\dot{P})\zeta_j \tag{3.3}$$

Where ζ_j is also called productive control variate of firm j.

Next we discuss consumers' behavior state equations. Note that the action of the consumers are contrary to that of the firms with regard to the announced price vectors, using the same analytical process for that to consumers, we can obtain the state equation of the consumers manner as follows:

$$\dot{x}_j = -\frac{\partial}{\partial P} x_j \dot{P} + M(\dot{P})\xi_j \tag{3.4}$$

For all $i \in N_m$, where $\xi_i(t,P)$ is nonnegative, it is the consumption adjustment rate of the consumer i, and it is also called consumption control variate vector of the consumer i.

We establish the commodity equations of the capital accumulation for consumer. Let $\omega_i(t)$ be the capital accumulation of the consumer i at time t, $\mu_i(t)$ be an increasing rate of the capital accumulation for consumer i at time t. The commodities of consumer i should be

$$\omega_i(t+\Delta t) + \sum_{j=1}^{n} \theta_{ij} y_j(t+\Delta t, P+\Delta P) - x_i(t+\Delta t, P+\Delta P)$$

at time $t+\Delta t$, where Δt is a very small increasing quantum of time. This is consumptive remains $\omega_i(t) + \sum_{j=1}^{n} \theta_{ij} y_j(t,P) - x_i(t,P)$ at time t and the growing quanta $\Delta t \mu_i(t)$ of the capital during Δt . So the consumer's capital accumulation is contented with the following formula:

$$\begin{aligned}&\omega_i(t+\Delta t) + \sum_{j=1}^{n} \theta_{ij} y_j(t+\Delta t, P+\Delta P) - x_i(t+\Delta t, P+\Delta P)\\ &= \omega_i(t) + \sum_{j=1}^{n} \theta_{ij} y_j(t,P) - x_i(t,P) + \Delta t \mu_j(t)\end{aligned}$$

Where $P(t+\Delta t) = P(t) + \Delta P(t)$ and $\mu_i(t)$ is also called capital accumulation control of the consumer i. From this we obtain the capital accumulation state equation for consumer i:

$$\dot{\omega}_i = -\sum_{j=1}^{n} \theta_{ij} \left(\frac{\partial}{\partial P} y_j \dot{P} + \dot{y}_j\right) + \left(\frac{\partial}{\partial P} x_i \dot{P} + \dot{x}_i\right) + \mu_i \tag{3.5}$$

4 Differentiable Dynamic Systematic Description of the Competitive Economical Action

Before We assume without loss of generality that $u_i : (0,T) \times x \to R, t \in (0,T), x_i \in X_i$ is a utility function of consumer i, a function that is smooth, smoothly increasing (i.e. $Dv_i \in R_{++}$) and smoothly strictly concave (i.e. the quadratic form defined by the Hessain matrix of second derivatives $D^2 v_i$ is negative definite), $\phi_i : \{0,T\} \times x_i \to R, t=0$, or $T, x_i \in X_i$, is the boundary function of u_i. Let

$\pi_j : (0,T)\times Y_j, t\in(0,T), y_j \in Y_j, \pi_j(t,y_j) = P(t)y_j$, be profit function of the firm j, $\psi_j : \{0,T\}\times Y_j \to R, t=0,T, y_j \in Y_j$ be the boundary function of π_j. Given a price vector P, consumer i act to maximize utility subject to budget constraints. Therefore under the announced price vectors, the consumptive behavior of consumer i is as follows:

$$[CB_i]\quad Maximize\quad \{J = \varphi_i(t,x_i)\Big|_0^T + \int_0^T u_i(t,x_i)dt\}$$
$$subject\quad to\quad \begin{aligned} \dot{x}_i &= -\frac{\partial x_i}{\partial P}\dot{P} - M(\dot{P})\xi_i \\ x_i &\le \omega_i + \sum_{j=1}^{n} \theta_{ij} y_j \end{aligned}$$

In which $x(0)$ is given and is initial consumption of consumer i, $i \in N_m$, y_j is the production of firm j. In the same way for given a price P, the firm j act to maximize profits subject to its economical and technical constraints. Therefore the productive behavior of firm j is as follows under the pronounced vectors:

$$[FB_j]\quad Maximize\quad \{J = \psi_j(t,y_j)\Big|_0^T + \int_0^T \pi_j(t,y_j)dt\}$$
$$subject\quad to\quad \dot{y}_j = -\frac{\partial y_j}{\partial P}\dot{P} - M(\dot{P})\zeta_j$$

Where $y_j(0)$ is given, for all $j\in N_n$, and it is the initial production.

Definition 4.1. For a given commodities price vector P, if the consumption behavior is [CB_i] about consumer i, $i\in N_m$, it is called that the consumption of consumer i is an equilibrium consumption with regard to the announced price vectors.

Definition 4.2. For given commodities prices vector P, if the productive behavior is [FB_j] in respect of firm j, $j\in N_n$, it is called that the production of firm j is an equilibrium production with regard to the announced price vectors.

In order to discuss relationship between different equilibrium, we now consider the economical manner on over economy. The consumptive action of E is defined by the sum of [CB_i] on I, i.e.

$$[CB]\quad Maximize\quad \{J = \sum_{i=1}^{m} \varphi_i(t,x_i)\Big|_0^T + \int_0^T \sum_{i=1}^{m} u_i(t_i,x_i)dt\}$$
$$sunject\quad to\quad \begin{aligned} \sum_{i=1}^{m} \dot{x}_i &= -\frac{\partial}{\partial P}\Big(\sum_{i=1}^{m} x_i\Big)\dot{P} - M(\dot{p})\sum_{i=1}^{m} \xi_i \\ \sum_{i=1}^{m} x_i &\le \sum_{i=1}^{m} \omega_i + \sum_{j=1}^{n} y_j \end{aligned}$$

Where $x_i(0)$ is given, so as $\sum_{i=1}^{m} x_i(0)$. Similarly, the productive action of E is defined by the sum of [FB_j] on j as follows:

$$[FB]\quad Maximize\ \{J=\sum_{j=1}^{n}\psi_j(t,y_j)\Big|_0^T+\int_0^T\sum_{j=1}^{n}\pi_j(t,y_j)dt\}$$
$$subject\ to\quad \sum_{j=1}^{n}\dot{y}_j=-\frac{\partial}{\partial P}(\sum_{j=1}^{n}y_j)\dot{P}-M(\dot{P})\sum_{j=1}^{n}\zeta_j$$

Where $y_j(0)$ is given, for all $j\in N_n$, so as $\sum_{j=1}^{n}y_j(0)$.

Let
$$\varphi:\{0,T\}\times\prod_{i=1}^{m}x_i\to R,\varphi=\sum_{i=1}^{m}\varphi_i,\ u:(0,T)\times\prod_{i=1}^{m}x_i\to R,$$
$$\psi:\{0,T\}\times\prod_{j=1}^{n}y_j\to R,\ \psi=\sum_{j=1}^{n}\psi_j,\ \pi:(0,T)\times\prod_{j=1}^{n}y_j\to R$$

Then
$$[FB]_a\quad Maximize\ \{J=\psi[t,(y_j)_n]\Big|_0^T+\int_0^T\pi[t,(y_j)_n]dt\}$$
$$subject\ to\quad \dot{y}=-\frac{\partial y}{\partial P}\dot{P}-M(\dot{P})\zeta$$

$$[CB]_a\quad Maximize\ \{J=\varphi[t,(x_i)_m]\Big|_0^T+\int_0^T u[t,(x_i)_m]dt\}$$
$$subject\ to\quad \begin{aligned}\dot{x}&=-\frac{\partial x}{\partial P}\dot{P}-M(\dot{P})\xi\\ x&\le\omega+y\end{aligned}$$

Where $y=\sum_{j=1}^{n}y_j$ is the aggregate production of E, $x=\sum_{i=1}^{m}x_i$ is the aggregate consumption of E, and $\omega=\sum_{i=1}^{m}\omega_i$ is the aggregate endowment of the economy E. We define $\xi=\sum_{i=1}^{m}\xi_i$ as the aggregate consumptive control, $\zeta=\sum_{j=1}^{n}\zeta_j$ as the aggregate productive control of the economy E.

Definition 4.3. Let P be the systems of the commodities prices. If the economical consumption action is [CB], the economical consumption action is called equilibrium consumption. If the economical productive action is [FB], it is called equilibrium production.

At the last of this section, we consider the actions of production and consumption on the aggregate production and consumption sets. Let $u:(0,T)\times X\to R$ be the continuous and differentiable consumptive utility function that is smooth, smoothly increasing and smoothly strictly concave. We can then assume without loss of generality that $\varphi:\{0,T\}\times X\to R$ is the boundary function of u. We defined the economical consumption action on the aggregate consumption sets as follows:

$$[CB]_s\quad Maximize\ \{J=\varphi(t,x)\Big|_0^T+\int_0^T u(t,x)dt\}$$
$$subject\ to\quad \begin{aligned}\dot{x}&=-\frac{\partial x}{\partial P}\dot{P}-M(\dot{P})\xi\\ x&\le\omega+y\end{aligned}$$

Where $x(0)$ is given. In the meanwhile, we furthermore assume that $\pi : (0,T) \times Y \to R$ is the continuous and differentiable production profit function, $\psi : \{0,T\} \times Y \to R$ is the remnants function of u. The productive action of E on the aggregate production set can be written as

$$[FB]_s \quad Maximize \quad \{J = \psi(t,y)\Big|_0^T + \int_0^T \pi(t,y)dt\}$$

$$subject \quad to \quad \dot{y} = -\frac{\partial y}{\partial P}\dot{P} - M(\dot{P})\zeta$$

Where $y(0)$ is given.

5 The Main Definitions and Theorem

Definition 5.1. Let economy E be consisting of l commodities, m consumers, and n firms. If there is a price systems P such that the consumption behavior of consumer i, for all $i \in N_m$, is [CB $_i$], the production behavior of firm j, for all $j \in N_n$ is [FB$_j$], and the excess demand $z(t,P)$ is equal to zero, we call economy E is competitive equilibrium. The competitive equilibrium of economy E is also called dynamic equilibrium or Walras equilibrium of economy E.

Theorem 5.1. Let P be the price systems of economy E. If each consumption x_i of i, $i \in N_m$, is equilibrium consumption, and each production y_j of firm j, $j \in N_n$ is equilibrium production, we call the economy E is Walras equilibrium under the price systems P.

Theorem 5.2. Let P be the given commodities price systems of economy E. If the consumption and the production are respectively the equilibrium consumption and production on the economy E, the economy E is Walras equilibrium in respect of the given commodities price.

Theorem 5.3. If there is a commodities price systems P such that the consumption and production are respectively the equilibrium consumption and production, the economy E has equilibrium.

In the light of structures of [FB$_j$], [FB] and [FB]$_a$, and of [CB$_i$], [CB] and [CB]$_a$, we obviously have the follows theorem:

Theorem 5.4. The consumer behavior [CB$_i$] $i \in N_m$, is equivalent to [CB] (or [CB]$_a$), i.e. [CB$_i$] $\Leftrightarrow$ [CB] $\Leftrightarrow$ [CB]$_a$, for all $i \in N_m$. The firm behavior [FB$_j$], $j \in N_n$ is equivalent to [FB] and [FB]$_a$, i.e.[FB$_j$] $\Leftrightarrow$ [FB] $\Leftrightarrow$ [FB]$_a$.

Note that the consumer behaviors [CB]$_a$ and [CB]$_s$ have the same restricted conditions, but the various sets of definition with respect to the objective functions.

The former is the set $(0,T)\times\prod_{i=1}^{m} X_i$, and the latter is $(0,T)\times X$. On the behavior $[CB]_a$ and $[CB]_s$, we have the following result:

Theorem 5.5. The consumptive behavior $[CB]_a$ is equivalent to $[CB]_s$, that is $[CB]_a \Leftrightarrow [CB]_s$.

Theorem 5.6. In the economy E, the firm behavior $[FB]_a$ is equivalent to $[FB]_s$.

Theorem 5.7. Between the economical actions, there are the equivalent relationship

$$\forall j, j\in N_n \qquad [FB_j]\Leftrightarrow[FB]\Leftrightarrow[FB]_a\Leftrightarrow[FB]_s$$
$$\forall i, i\in N_m \qquad [CB_i]\Leftrightarrow[CB]\Leftrightarrow[CB]_a\Leftrightarrow[CB]_s$$

6 Conclusion

We rebuild economical equilibrium analysis and some relative concepts, and study the characteristics of commodity prices and economical actions with the theory and method of differentiable dynamic systems. This is more practical and operable than Arrow and Debreu's explanation[2,3] which is expressed by set theory. It is convenient and advantageous to apply the competitive equilibrium theory to the macro-adjustment in the development of national economies and the optimal decision in the operational management. We have proven that the economical actions are equivalent. This results guarantee that the individual and aggregate economical actions are identical when we explain competitive equilibrium with control theory.

References

1. Maxifield, R.R.: General equilibrium and theory of directed graphs. Journal of Mathematical Economics 27, 23–51 (1997)
2. Arrow, K.J., Hahn, F.H.: General competitive analysis. North-Holland, Amsterdam (1971)
3. Debreu, G.: New concepts and techniques for equilibrium analysis. International Economic Review 3, 257–273 (1962)

An Improved Acquisition Algorithm Based on Power Spectrum Analysis

Zhonghui Chen, Xinxin Feng, Lingxiao Huang, and Xin Chen

College of Physics and Information Engineering, University Of Fuzhou, Fuzhou, China
czh@fzu.edu.cn, huang_lxiao@163.com,
Fxx1116@fzu.edu.cn, Chen-xin@fzu.edu.cn

Abstract. A new synchronization acquisition method based on the power spectrum analysis was studied and the theory analysis proved that the new algorithm has better synchronization performance than the traditional algorithm of sliding correlation acquisition. Also, the best band was got by discussing the center frequency of band-pass filter. At last, the whole simulation was given and the result showed that the new algorithm is fit for the real Spread Spectrum System.

Keywords: Power spectrum analysis, synchronization acquisition, band.

1 Introduction

Spread spectrum communication is a modulation mode[1], which spreads the base-band signal into wideband signal to send. By using the PN code with many times the rate of base-band signal, the transmitter spreads the power spectrum of base-band signal into low power spectral-density and wideband signal, at the same time, the receiver uses the synchronized PN code to demodulate and regain the sending signal, only to reduce the effect of noise.

Spread spectrum communication has been widely used in many communication systems because of its advantages of noise immunity, code division multiple access, low power spectral density and so on. And how to design a spread spectrum system with good performance becomes the research hotspot. Generally, the acquisition performance of PN code decides the performance of spread spectrum receiver to a certain extent, thus, the most important problem in spread spectrum system is the PN code acquisition and it's also the theme of this paper.

The traditional acquisition algorithm mainly includes sliding correlation acquisition and digital matched filter algorithm. The digital matched filter algorithm has shorter acquisition time but higher complexity implementation [2], and when it's used in low signal-to-noise ratio, the length of PN code is limited. On the contrary, the sliding correlation acquisition algorithm has longer acquisition time but it's easier to realize and has lower cost. Considering the advantage of sliding correlation acquisition algorithm, this paper studies an improved sliding correlation acquisition algorithm and deeply discusses its synchronous acquisition performance.

J. Luo (Ed.): Affective Computing and Intelligent Interaction, AISC 137, pp. 409–416.
springerlink.com

2 Improved Sliding Correlation Acquisition Algorithm

Fig.1 shows the configuration of the traditional sliding correlation algorithm [3](except the dashed line box), the received signal is first multiplied by PN code, filtered by BPF1 , then squared, detected and sent into integrator (integration time is T_D), which outputs a related value every T_D and reset. When the related value of integrator is lower than the fixed threshold of comparator, the spread spectrum system outputs a signal to the phase control circuit to control the state of clock circuit, and then changes the phase state of PN code. After that, the system repeats the course above until the related value of integrator is more than the fixed threshold, which means the synchronization acquisition is realized. At this time, the system keeps the clock circuit state unchanged and output a signal to the following circuit.

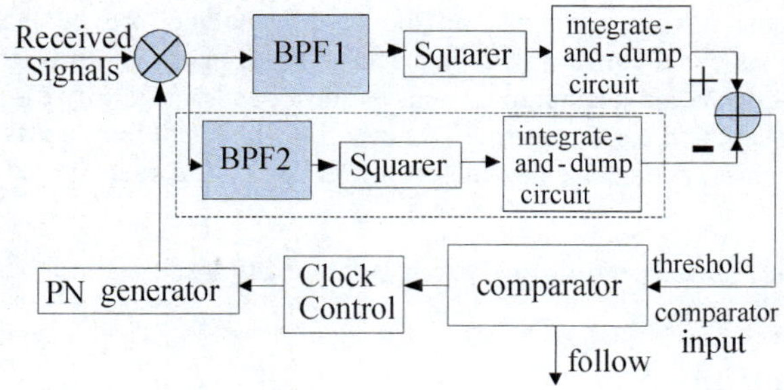

Fig. 1. Diagram of improved sliding correlation acquisition algorithm

When the traditional acquisition algorithm is adopted, the sum of in-band and out-band power of BPF1 is certain. When synchronization acquisition is achieved[4], the power mainly concentrates in the band of BPF1 and the power out of band is close to zero, otherwise, the out-band power of BPF1 may be more than or equal to the power in the band of BPF1.

This paper studies an improved acquisition algorithm based on power spectrum analysis. This algorithm adds another branch circuit in the traditional sliding correlation algorithm as shown in Fig.1 (including the dashed line box). The only difference between the new branch and the original branch is the filter. In the new branch, the filter BPF2 is used to filter a part of power out of filter BPF1 [5] to analyze. This algorithm uses the power difference between BPF1 and BPF2 as the input of comparator instead of that of BPF1 in the traditional algorithm.

As we know, the correlator outputs to the maximum when the received signal is completely aligned with the local PN code, otherwise, the output of correlator decreases. In the improved acquisition algorithm, when the receiving system is close to synchronization, the power of the original branch becomes more and more big and that of the new branch becomes more and more small.

Suppose spread spectrum and BPSK modulation are adopted in the spread spectrum system, then the received signal can be expressed as:

$$s(t) = Ad(t)c(t)\cos(2\pi f_0) + n(t) \tag{1}$$

Here, the received signal $s(t)$ is described by two components: the desired signal and the noise $n(t)$. A is the amplitude of the desired signal. $d(t)$ is the sending base-band signal, $c(t)$ is the sending PN code, f_0 is the carrier frequency.

Suppose T is the actual delay of the received signal, and T_1 is the estimation of T. In the case of not synchronization, $T - T_1 \neq 0$, the function $c(t-T)c(t-T_1)$ contains direct current and some interference noise.

Let $|T - T_1| = \varepsilon T_c$, when $|\varepsilon| > 1$, the output of correlator is nearly white noise, and the difference between BPF1 and BPF2 is close to 0, that is to say, the comparator input is nearly zero, which is definitely lower than the threshold of comparator. Thus, this paper supposes $0 \leq |\varepsilon| \leq 1$ as the relative delay of local PN code and the received PN code.

Through deduction [6], the function of power spectral density of $C(t,\varepsilon) = c(t-T)c(t-T_1)$ can be expressed as:

$$\varphi_c(f) = \left(1 - \frac{N+1}{N}|\varepsilon|\right)^2 \delta(f) + \frac{N+1}{N}\varepsilon^2 \sin c^2(\pi f \varepsilon T_c) \sum_{\substack{k=-\infty \\ k\neq 0}}^{\infty} \delta\left(f + \frac{k}{T_c}\right) + \frac{N+1}{N^2}\varepsilon^2 \sin c^2(\pi f \varepsilon T_c) \sum_{\substack{k=-\infty \\ k\neq 0}}^{\infty} \delta\left(f + \frac{k}{NT_c}\right) \tag{2}$$

In the function, $\sin c(x) = \dfrac{\sin(x)}{x}$, T_c is the PN code chip period, N is the length of PN code, when $|\varepsilon| = \dfrac{1}{2}$, the function of power spectral density of $c(t,\varepsilon)$ is shown as Fig.2.

In this algorithm, the power difference between BPF1 and BPF2 is used as the input of comparator.

Suppose the band of BPF1 is: $f_0 - f_b \leq f \leq f_0 + f_b$, which is shown as Fig.2, f_0 is the carrier frequency, f_b is the rate of base-band signal, f_c is the rate of PN code and $f_c = Nf_b$, then the output of BPF1 can be calculated by formula (2) [7] [8]:

$$S_u = \left(1 - \frac{N+1}{N}|\varepsilon|\right)^2 + 2\frac{N+1}{N^2}\varepsilon^2 \sin c^2\left(\frac{\pi\varepsilon}{N}\right) \tag{3}$$

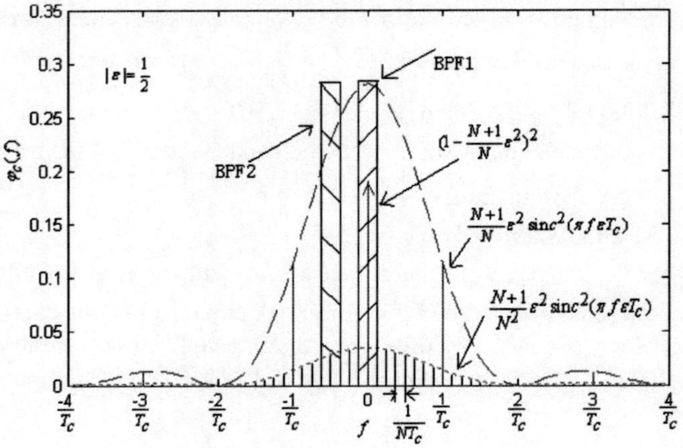

Fig. 2. Power spectrum function of $C(t,\varepsilon)$

Suppose the band of BPF2 is: $f_0 - 3f_b \le f \le f_0 - f_b$, which is also shown as Fig.2. By the same rule, the output of BPF2 can be described as:

$$S_d = \frac{N+1}{N^2}\varepsilon^2[\sin c^2(\frac{\pi\varepsilon}{N}) + \sin c^2(\frac{2\pi\varepsilon}{N}) + \sin c^2(\frac{3\pi\varepsilon}{N})] \tag{4}$$

3 The Performance Comparison of Improved and the Traditional Algorithm

Let $\varepsilon_1 < \varepsilon_2$, ε_1 and ε_2 are two relative delay of local PN code and received PN code. In the following, this paper discusses the magnitude of the comparator input of ε_1 and ε_2, by respectively using traditional and improved acquisition algorithm.

① The traditional sliding correlation algorithm: it only uses the output power of BPF1 as the comparator input.

Suppose $S_u(\varepsilon_1)$ and $S_u(\varepsilon_2)$ are respectively the output power of BPF1 of ε_1 and ε_2, and their difference can be expressed as:

$$\Delta S_u = S_u(\varepsilon_1) - S_u(\varepsilon_2) \tag{5}$$

Owning to $\varepsilon_1 < \varepsilon_2$, ε_1 is closer to synchronization than ε_2, then the output power of BPF1 of ε_1 is bigger than that of ε_2. So $S_u(\varepsilon_1) > S_u(\varepsilon_2)$ can be got, that is, $\Delta S_u > 0$.

② The improved acquisition algorithm: it uses the output power difference between BPF1 and BPF2 as the comparator input.

Suppose $S_u(\varepsilon_1)$ and $S_d(\varepsilon_1)$ are respectively the output power of BPF1 and BPF2 of ε_1, $S_u(\varepsilon_2)$ and $S_d(\varepsilon_2)$ are respectively the output power of BPF1 and BPF2 of ε_2 .Then the output power difference between BPF1 and BPF2 of ε_1 and ε_2, that is, the comparator input of ε_1 and ε_2 are respectively $\Delta S(\varepsilon_1)= S_u(\varepsilon_1)- S_d(\varepsilon_1)$ and $\Delta S(\varepsilon_2)= S_u(\varepsilon_2)- S_d(\varepsilon_2)$, and their difference can be expressed as:

$$\begin{aligned}\Delta S &= \Delta S(\varepsilon_1)- \Delta S(\varepsilon_2)\\ &= S_u(\varepsilon_1)- S_u(\varepsilon_2)+S_d(\varepsilon_2)- S_d(\varepsilon_1)\\ &= \Delta S_u +S_d(\varepsilon_2)- S_d(\varepsilon_1)\end{aligned} \tag{6}$$

Owing to $\varepsilon_1 < \varepsilon_2$, ε_2 is father to synchronization than ε_1, then the output power of BPF2 of ε_2 is bigger than that of ε_1. So $S_d(\varepsilon_2)- S_d(\varepsilon_1)>0$ can be got, that is, $\Delta S>\Delta S_u$.

From the deduction above, we can get: comparing with the traditional acquisition algorithm, when adopting the improved acquisition algorithm, the input to comparator has larger change before and after the synchronization and it's much easier to judge whether it's synchronous or not.

4 The Choice of the Best Band

Suppose the band of BPF1, ε_1 and ε_2 are unchanged, then ΔS_u is also fixed, and formula (6) shows that: in the improved acquisition algorithm, the magnitude change of the comparator input (ΔS for short) is related to the band of BPF2. This paper aims at finding the band of BPF2 whose ΔS is the biggest as the best band. The following two band of BPF2 both have a bandwidth of $2f_b$, and the band are:

Band a: $f_0 - kf_b < f < f_0 - (k-2)f_b$,

Band b: $f_0 - (k+1)f_b < f < f_0 - (k-1)f_b$, $k \ge 3$.

From formula (2) above, the output power of band a and band b can be expressed as:

$$S_{da} = \frac{N+1}{N^2}\varepsilon^2\{\mathrm{sinc}^2[\frac{(k-2)\pi\varepsilon}{N}]+\mathrm{sinc}^2[\frac{(k-1)\pi\varepsilon}{N}]+\mathrm{sinc}^2(\frac{k\pi\varepsilon}{N})\} \tag{7}$$

$$S_{db} = \frac{N+1}{N^2}\varepsilon^2\{\mathrm{sinc}^2[\frac{(k-1)\pi\varepsilon}{N}]+\mathrm{sinc}^2[\frac{k\pi\varepsilon}{N}]+\mathrm{sinc}^2(\frac{(k+1)\pi\varepsilon}{N})\} \tag{8}$$

From the deduction above, when band a is used, ΔS can be expressed as:

$$\begin{aligned}
\Delta S_a &= \Delta S_u + S_{da}(\varepsilon_2) - S_{da}(\varepsilon_1) \\
&= \Delta S_u + \frac{N+1}{N^2}\{\varepsilon_2^2 \sin c^2[\frac{(k-2)\pi\varepsilon_2}{N}] \\
&- \varepsilon_1^2 \sin c^2[\frac{(k-2)\pi\varepsilon_1}{N}] + \varepsilon_2^2 \sin c^2[\frac{(k-1)\pi\varepsilon_2}{N}] \\
&- \varepsilon_1^2 \sin c^2[\frac{(k-1)\pi\varepsilon_1}{N}] + \varepsilon_2^2 \sin c^2(\frac{k\pi\varepsilon_2}{N}) - \varepsilon_1^2 \sin c^2(\frac{k\pi\varepsilon_1}{N})\}
\end{aligned} \tag{9}$$

When band b is used, ΔS can be expressed as:

$$\begin{aligned}
\Delta S_b &= \Delta S_u + S_{db}(\varepsilon_2) - S_{db}(\varepsilon_1) \\
&= \Delta S_u + \frac{N+1}{N^2}\{\varepsilon_2^2 \sin c^2[\frac{(k-1)\pi\varepsilon_2}{N}] \\
&- \varepsilon_1^2 \sin c^2[\frac{(k-1)\pi\varepsilon_1}{N}] + \varepsilon_2^2 \sin c^2(\frac{k\pi\varepsilon_2}{N}) \\
&- \varepsilon_1^2 \sin c^2(\frac{k\pi\varepsilon_1}{N}) + \varepsilon_2^2 \sin c^2[\frac{(k+1)\pi\varepsilon_2}{N}] - \varepsilon_1^2 \sin c^2[\frac{(k+1)\pi\varepsilon_1}{N}]\}
\end{aligned} \tag{10}$$

Comparing ΔS_a and ΔS_b, we can get:

$$\begin{aligned}
&\Delta S_a - \Delta S_b \\
&= \frac{N+1}{N^2}\{\varepsilon_2^2 \sin c^2[\frac{(k-2)\pi\varepsilon_2}{N}] - \varepsilon_1^2 \sin c^2[\frac{(k-2)\pi\varepsilon_1}{N}] \\
&- \varepsilon_2^2 \sin c^2[\frac{(k+1)\pi\varepsilon_2}{N}] + \varepsilon_1^2 \sin c^2[\frac{(k+1)\pi\varepsilon_1}{N}]\} \\
&= \frac{N+1}{N^2}\{\varepsilon_2^2(\sin c^2[\frac{(k-2)\pi\varepsilon_2}{N}] - \sin c^2[\frac{(k+1)\pi\varepsilon_2}{N}]) \\
&- \varepsilon_1^2(\sin c^2[\frac{(k-2)\pi\varepsilon_1}{N}] - \sin c^2[\frac{(k+1)\pi\varepsilon_1}{N}])\} \\
&= \frac{N+1}{N^2}[S(\varepsilon_2,k) - S(\varepsilon_1,k)]
\end{aligned} \tag{11}$$

In the formula above,

$$S(\varepsilon,k) = \varepsilon^2(\sin c^2[\frac{(k-2)\pi\varepsilon}{N}] - \sin c^2[\frac{(k+1)\pi\varepsilon}{N}]),$$

let N=31, and this function can be shown in Fig.3:

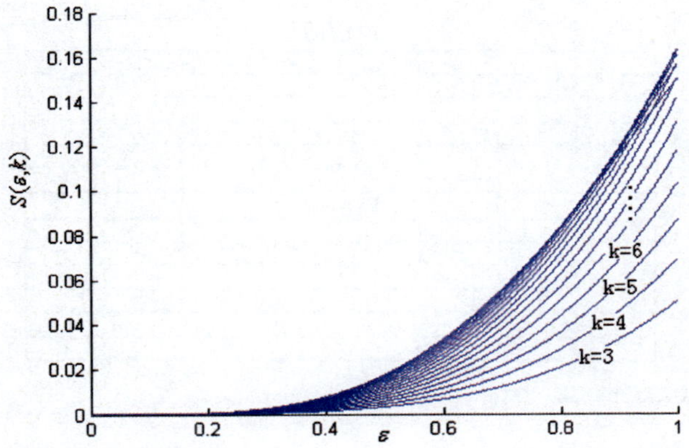

Fig. 3. Function $S(\varepsilon,k)$

As shown in Fig.3 above, when the value of k is determined, $S(\varepsilon,k)$ is a monotone increasing function. Therefore, when $\varepsilon_1 < \varepsilon_2$, $S(\varepsilon_1,k) < S(\varepsilon_2,k)$, then according to the formula (11), $\Delta S_a - \Delta S_b > 0$, that is, $\Delta S_a > \Delta S_b$, which means that it's better to choose band a than band b. Also, it can be analogized that the larger value of k, the smaller the value of ΔS . Therefore, the band that is closest to the band of BPF1 can be judged as the best band of BPF2, that is, when k=3, $f_0 - 3f_b \le f \le f_0 - f_b$ is the best band of BPF2.

5 Simulation and Results Analysis

The spread spectrum communication simulation system was built in MATLAB/SIMULINK. The base-band signal rate is 1.2Kbps, the length of PN code is 15, the PN code rate is 18Kbps. According to the experience value, the comparator threshold was set to 0.005, that is, when the comparator input is above this value, the system will be considered as synchronization. The simulation results are shown in Fig.4, which describes how the comparator input changes with the time (output a value every symbol period) in traditional and improved algorithms.

Fig.4 (a) expresses the curve that the output power of BPF1 is changing with time in traditional acquisition algorithm, Fig.4 (b) expresses the curve that the output power difference between BPF1 and BPF2 is changing with time in the improved acquisition algorithm, the horizontal axis represents time t and the vertical axis represents the comparator input.

From the Fig.4 above, the system synchronization is achieved at 0.012s or so. Comparing with Fig.4(a) and Fig.4(b), before and after synchronization, the change value in Fig.4(b) is larger than that in Fig.4(a), which makes it easier to judge whether it's synchronous or not. To sum up, the improved algorithm has better acquisition performance than the traditional one.

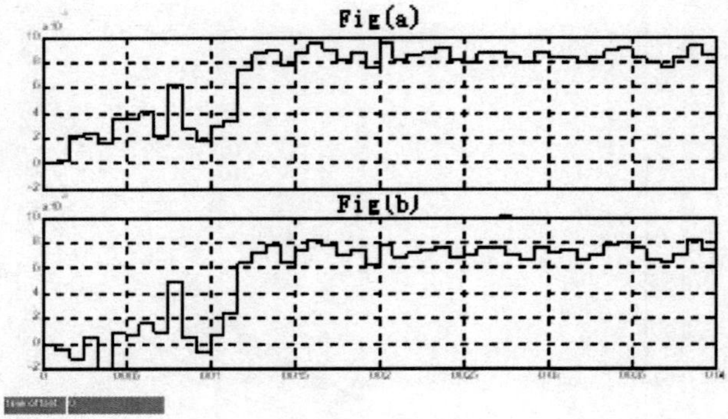

Fig. 4. Comparator input in traditional and improved algorithms

6 Conclusion

This paper studied an improved acquisition algorithm based on power spectrum analysis. With formula deduction and system simulation, it also proved that the improved algorithm has better acquisition performance than the traditional one, and has actual research meaning.

References

1. Wei, H.: Spread Spectrum communication technology and application. Xidian University Press, Xi'an (2007)
2. Shang, X., Hu, X., Yang, L.: Comparison of two methods' performance in PN code acquisition. Communications Technology 9, 90–93 (2001)
3. Zhu, J.: CDMA communication technology. Post& Telecom Press, Beijing (2001)
4. Yin, W., Hu, X., Yang, L., et al.: The acquisition of PN code of DSSS-slippage correlation and its improvement. Journal of Huazhong University of Science and Technology 31(3), 28–30 (2003)
5. Hong, H., Chen, Z., Chen, X.: Spread spectrum signals synchronous algorithm based on power spectrum estimation and adaptive threshold. Applied Science and Technology 35(9), 39–42 (2008)
6. Gill, W.J.: Effect of Synchronization Error in Pseudorandom Carrier Communications. In: First Annunal IEEE Comm. Conv. Conference Record, Denver, June 7-9, pp. 187–191 (1965)
7. Holmes, J.K., Liang, Z.X., Chai, K.J. (trans.).: Coherent spread spectrum systems. Defense Industry Press, Beijing
8. Tian, R.C.: Spread spectrum communication. Tsinghua university press, Beijing (2007)

Study on the Influence of Organizational Slack on Firm Growth

Yuerui Wu, Chen Zhang, and Yi Cui

School of Business Administration, South China University of Technology, Guangzhou, China
wuyuerui@yahoo.cn, bmyicui@scut.edu.cn

Abstract. Base on Penrose's (1959) theory of firm growth, effect of organizational slack on performance is warmly discussed. And the conflicting views from different scholars, different theories call for exploiting current theories and exploring the black box between organizational slack and firm growth. This paper reviews the more recent study, and establishes a clear research framework including organizational slack's concept, nature, and the path of its effect. Further, we provide several possible research trends to service in later study.

Keywords: Organizational slack, firm growth, research trend.

1 The Penrose's Theory of Firm Growth

The issue of firm growth has always been the objectives and outcome variables in the organizational management research field. In 1959, Penrose's 'theory of firm growth' first established the resource-based theory on firm growth. She believes that enterprise is a collection of resources; organization growth mainly depends on more effective use of existing resources and to face new unused resources; excess resources – organizational slack is an engine of firm growth; the growth of enterprise is an unlimited process of continuous mining of unused resources under the accumulation of management capacity.

Later, some scholars(e.g.B. Wernerfet, 1984; Curtis, Pisano, and Schon, 1990; Pan Dian, 1992. etc) rich and perfect Penrose's theory, form the resource-base firm growth academic school, the release of B.Wernerfet's classic paper 'the resource-based view of the firm' represent the formation of this theory. The main point of this view is that a corporation is formed by a series of resource, competitive advantage derived from the enterprises owned resources, especially the heterogeneity resources. Firm growth is a dynamic process, through innovation, changing and strengthening the management, to save, integrate and enhance resources, and thus pursuing sustainable growth of enterprises.

Such studies led to establish 'Enterprise Resource–organizational capability – firm growth' analytical framework and firm growth mechanism based on internal resources, organizations slack (slack resources) as an important variable on theory of firm growth, begin to be recognized and researched by scholars and enterprise managers.

J. Luo (Ed.): Affective Computing and Intelligent Interaction, AISC 137, pp. 417–426.
springerlink.com

2 Concept and Definition of Organizational Slac

As early as 1938, Barnard in his book talked about the effect of organizational slack, after that lots of scholars gave their definition and concept. Cyer & March (1963) took organizational slack as the disparity between the resources available to the organization and the payments required to maintain the coalition. Child (1972) defined the organizational slack as the organization actually purchased resources exceeded the practical needs, which are organized in the rights groups of the organization to meet in their preferences for structural adjustment. It is worth noting, in the year of 1981, Bourgeois defined organizational slack as: a cushion of actual or potential resources which allows an organization to adapt successfully to internal pressures for adjustment or to external pressures for change in policy, as well as to initiate changes in strategy with respect to the external environment. This definition has been generally recognized. However, since Nohria & Gulati (1996) argue some of the specific characteristic of organizational slack, this trend has changed. Nohria & Gulati concluded slack as the pool of resources in an organization that is in excess of the minimum necessary to produce a given level of organizational output. In recent important research on slack, scholars defined slack are affected to varying degrees by Nohria & Gulati's study. Table 1 lists several key definitions; the evolution on slack concept can be seen from it.

According to some needs of study, researchers often organize slack as single dimension construct in definition and measure. Growing number of scholars put it now as a multidimensional construct, and researchers define three categories of slack: available slack, recoverable slack, potential slack (Bourgeois & Sigh, 1983; Sharfman, Wolf, Chase, & Tansik, 1988). Available slack refer to resources that have not been put into the organization or specific program design; recoverable slack refer to resources which have been incorporated into the production system,cause higher operating costs (e.g., beyond the ordinary running costs), but can be recovered to the normal cost by organization re-designed; potential slack refer to the future resources - such as resources gathered from additional debt or equity capital as well as other resources generated from the environment(Bourgeois & Sigh, 1983).As potentially utilizable resources can be diverted or redeployed to achieve enterprise goals (George, 2005), organizational slack can be broadly divided into absorbed slack which is not easy to redeploy and unabsorbed slack which is more easily redeployed (Singh, 1986). Sharfman et al (1988) divided slack into high discretionary slack and low discretionary slack on whether can be used in various aspects. They consider slack is the exceeded investing resources of minimum required input under given level of output in the organization. It's the over-investment concludes redundant staff, unused capital and unnecessary capital costs; also concludes opportunities undeveloped and increased output, such as, profits can be obtained from customers and return from innovation close to technology frontier.

Table 1. Several major definitions of organizational slack

Researcher	Definitions of Organizational Slack
1. Cyer & March (1963)	• The disparity between the resources available to the organization and the payments required to maintain the coalition.
2. March(1979)	•Since organizations do not always optimize, they accumulate spare resources and unexploited opportunities which then become a buffer against bad times. Although the buffer is not necessarily intended, slack produces performance smoothing, reducing performance during good times and improving it during bad times.
3.Boumeois(1981)	• A cushion of actual or potential resources which allows an organization to adapt successfully to internal pressures for adjustment or to external pressures for change in policy, as well as to initiate changes in strategy with respect to the external environment.
4.Nohria & Gulati(1996)	•The pool of resources in an organization that is in excess of the minimum necessary to produce a given level of organizational output.
5.Runsheng(2004,2005)	• Enterprises in the production output, given level of technical possibilities beyond minimum input generated in resources and its function and potential value.
6.Geiger & Makri(2006)	• Resources readily available to an organization that are in excess of the minimum necessary to produce a given level of organizational output as well as the resources that are recoverable from being embedded in the firm.

Source: Relevant literatures reviewed.

3 Conflicting Views and Evolution in Organizational Slack

Based on organizational slack, firm growth theory suggest that organizational slack can contribute to competitive advantage, the organization have the power to search it and thus enhance firm performance, therefore the slack seen as a positive role in enterprise development by early scholars. From the 50s and 60s of the 20th century, a wealth of literatures have been accumulated, and the effect of this special resource is no longer has a single aspect, mainly include the proponents, opponents, as well as the combination of the former two's new ideas.

3.1 Positive Views of Proponents

Organizational slack has long been regarded as one of the key building blocks of the behavioral theory of the firm. In organization theory, slack has been suggested to perform four major functions. First, slack acts as an inducement, which represents `payments to members of the coalition in excess of what is required to maintain the organization (Cyert and March, 1963). Second, slack can become a resource for conflict resolution (Tan & Pang, 2003). The upshot is that, with sufficient slack, there can be a solution for every problem. Third, slack may be employed as a buffer, which insulates the technical core of the organization from environmental turbulence (Tan & Pang, 2003). Finally, slack can be a facilitator of strategic behavior, which allows the firm to experiment with new strategies such as introducing new products and entering new markets (Thompson, 1967). Moreover, from the real options perspective, McGrath

(1999) argue that organizational slack helps the firm to manage the cost of failure by limiting exposure to the downside while preserving access to attractive opportunities and maximizing gains.

3.2 Negative Views of Opponents

Scholars, especially organizational economists such as Leibenstein (1969) and Williamson (1963, 1964), have adopted a more hostile view of slack. They view it as synonymous with waste and as a reflection of managerial self-interest, incompetence, and sloth rather than as a buffer necessary for organizational adaptation. Leibenstein (1969) even coined the felicitous term X-inefficiency to highlight the discrepancy that slack creates between actual output and maximum output for a given set of inputs. Agency theorists often suggest that slack is a source of agency problems, which breeds inefficiency, inhibits risk-taking, and hurts performance (Fama, 1980). In addition, other opponents argue that organizational slack may lead to organizational inertia (Leonard-Barton, 1992). The firm with more slack would be unwilling to develop and learn new knowledge to renew its existing capabilities (Leonard-Barton, 1992). This would make it difficult for a resource-rich firm to find a fit with the environmental changes, and further lower performance. In sum, these theorists have suggested that although excess slack undoubtedly spurs R&D expenditures that lead to the pursuit of many new projects, very few of these projects actually translate into value-added innovations for firms, because the loose controls placed on these projects allow decision makers to make choices that "accord better with their own preferences than with economic considerations" (Child, 1972).

3.3 Curvilinear Theory of Others

Form a dynamic process standpoint, some researchers believes that slack's performance-enhancing effect—is nonlinear. This notion rests on the argument that slack can he both good and bad for performance. And the relationship may well be an inverse parabola. In an effort to reconcile existing arguments, researchers have speculated that 'there is an optimal level of slack for any given firm. If the firm exceeds that level, performance will go down'(Sharfman et al, 1988; Bourgeois, 1981). Concerning the relationship between slack and innovation, Nohria and Gulati (1996) sought to synthesize the literature—The basis for whether slack has positive or negative impacts on innovation was argued to be a function of how much slack existed within the organization. Overall, scholars have suggested that it is U-shaped or inverted U-shaped relationships (Bromiley, 1991; Nohria & Gulati, 1996; Geiger & Cashen, 2002; Tan & Peng, 2003), but no one know what is the right amount. Interestingly, Chunyan & Shuming (2004) showed a model of three stages between slack and performance, and using the data of Chinese listed companies to test the relationship. More important is they suggest a way to find out the ideal slack level.

3.4 Recent Research Trend

Enterprises only can develop after survival. H anbrick& Daveni (1988) using a 10-year comparative study of data, the results showed: compare to enterprises faced bankruptcy, enterprises able to survive owned significantly higher average organizational slack. In the year 2008,seriously affected by the financial crises, many companies went bankrupt, with the famous collapse of Lehman Brothers, one of the main reasons we study professional managers often do fictitious transactions, that is , the enterprise has little or no slack. As the increasing uncertainty of the external environment, the positive effect of slack is also more and more prominent. From the mainstream related articles in recent years published in international journals, more and more attention has been put on the focus of the organizational slack, and tends to study its positive effect in both internal and external environment of the enterprise, with the role of such actives, which have very real significance to the operation of enterprises.

4 Explore the Black Box in Slack-Performance Link

Organizational slack is the core concept in firm growth theory, and provide an effective analytical framework about firm growth, scholars have tried to explain the performance of slack. However, the theoretical explanation and empirical study results are not consistent on the relationship between organizational slack and organization performance; even have some conflicting conclusions, as discussed above. This give rise to another important organizational slack branch to explore the internal mechanisms of organizational slack and performance, namely, organizational slack opened the black box in firm growth.

4.1 The Mediating Role of Organizational Capacity

Penrose is the first person put organization capacity as the dynamic mechanism in the firm growth under effect of organizational slack. Penrose argue, in the traditional economics theory, firm is the “black box”, actually, it is the internal resources of the enterprise and organization capacity lay a solid foundation of the firm performance and development direction. Penrose deepen the outstanding features and functionality of the basis of firm growth-enterprise resources and capabilities, and consider the resources and capabilities as a source of sustainable competitive advantage in the enterprises. Specific resources are owned and produced by the excess capability is the driving force for firm growth. So an organizational slack – management capability – growth performance framework is established.

In 2003, Runsheng F. systematic research the enterprise resource and organizational capabilities, in the vision of enterprise resources and integration, he build a new model consist of slack, knowledge absorptive capacity and innovation output, in order to explain how does the enterprise establish and maintain a competitive advantage under intensifying environmental change, and, use empirical study to test the authenticity of the model.

As March (1991) dichotic the organizational learning into exploitation of knowledge and exploration of knowledge, scholars take more time on these interesting contractive variable. And exploitation capability and exploration capability become the new hot pot.(Jianli, 2009; Yuerui, 2009; Jansen, Bosch & Volberda, 2006, etc.). They think that enterprise should not only explore but also exploit, only to maintain the balance between these two may promote the progressing and sustained growth of the enterprise.

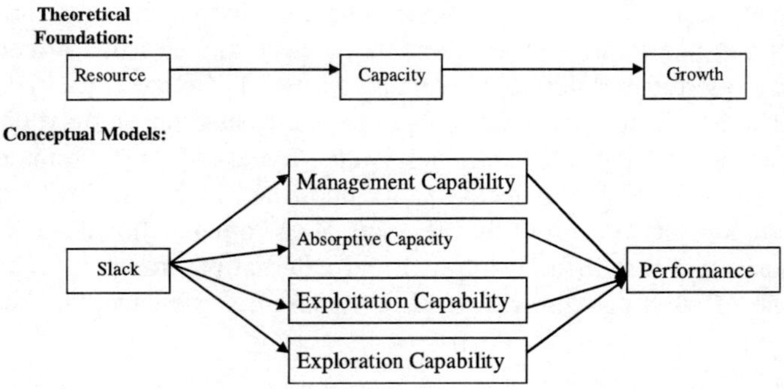

Fig. 1. The model on the mediator of organizational capacity

4.2 The Mediating Role of Organizational Strategy

The resources based firm growth theory argue, enterprises may have more benefit and lower risks if they focus resources on its existing product, however, the pursuit of growth is likely to expand into new areas. Expansion has become a necessary business means to achieve growth. There are many ways for expansion, such as innovation, mergers and form an alliance with other enterprises, through these strategies, companies can gain access to adequate growth.

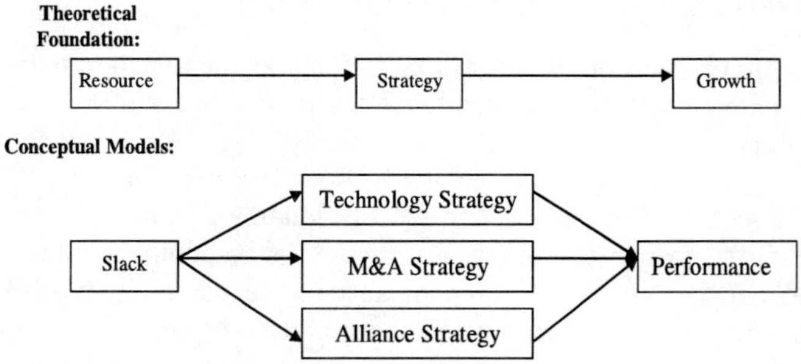

Fig. 2. The model on the mediator of organizational strategy

Enterprise often develops appropriate business strategy according to the resources capacity. The term strategic flexibility has been widely used by strategy researchers to denote the ability of firms to respond to various demands from dynamic environments, characterized by high levels of uncertainty, where slack is the means for achieving flexibility (Evans, 1991; Genus, 1995; Wernerfelt and Karnani, 1984). Ujari (2002), Vaara & Monin(2010), Marino et al (2008), and Frank & David (2004) studied the effect of slack on technology strategy, M&A strategy and alliance strategy, and then influence to the firm growth process. This is give a more and more clearly presentation in the important effect of organizational slack in organizational adaptation and strategic flexibility. In the process of strategic adaptation or strategic flexibility, it is showed more and more clearly to organizational slack's importance.

4.3 The Moderating Role of External Environment

The external environment is a very familiar investigation variable, every enterprise is surviving in a certain environment, and they do business plan and make adjustment to the original plan after observe and forecast the environment. Organizational slack is considered as a buffer to environmental changes (Bourgeois, 1981), researchers study the relationship between slack and environment in different aspect. Thompson (1967) argued that the primary use of slack is to protect the firm from its environment. Sharfman et al. (1988) identify three specific environmental forces that help shape a firm's level of slack. And they believe that firms will have forms of slack, dependent on the amount and types of resources in their environment. Xiaozhen (2009) suggest that the financial crisis affected enterprise generally reduce the slack resources, thus inhibiting innovation activities.Glenn, Deepak & Zannie (2008) use 214 U.S. nonprofit theaters data in empirical tests showed that the perception of external environment crisis, unabsorbed slack will lead exploratory behavior.

According to the present environment of China, the institutional environment of transition economies has two significant characteristics: munificence and dynamism (Peng, Zhang, & Li, 2007). Zhongfeng, En & Yuan (2009) argue that the two characters may play critical roles on the slack–performance link. Based on the data of Chinese listed firms, they find 'that unabsorbed slack is critical for firms to sustain their competitive advantages. And such a positive impact is especially likely to be profound for firms confronting resource scarcity and environmental dynamism caused by institutional transitions.'

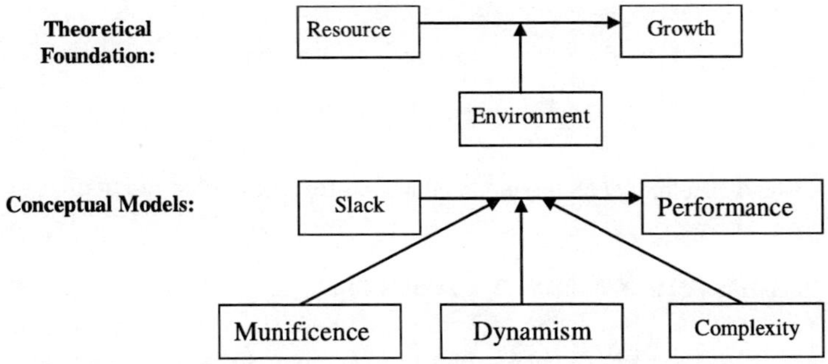

Fig. 3. The model on the moderator of external environment

4.4 The Moderating Role of Entrepreneurial Orientation

The motivation to transform slack resources quickly into growth is rooted in what Penrose called the 'entrepreneurial ambition' of the top management team: management's desire for growth and its propensity for taking risks to ensure that growth occurs (Mishina et al., 2004). Entrepreneurial orientation is a core concept in the study of corporate entrepreneurship, and become a hot spot in the academic research. Some scholars believe that entrepreneurial orientation and entrepreneurial spirit is the same term with different interpretations, and other people feel that entrepreneurial orientation is the process of entrepreneurial performance.

Entrepreneurial orientation includes three dimensions: innovativeness, risk-taking and proactiveness (Miller & Friesen,1982; Lumpkin & Covin, 1997; Barringer & Bluedorn, 1999). Mostly used dimension which linked resources with entrepreneurial orientation is the risk-taking and innovativeness(Singh, 1986; Steensma & Corley,2001; Franquesa & Brandyberry,2009; Ujari,2002). Steensma & Corley (2001) provide two different views about the relationship of slack-risk taking. On one hand, they believe, 'according to an abundance-driven view, the scarcity of slack leads to reduced risk-taking, As slack increases, managers accept more risk'. On the other, a hunger-driven view suggests the opposite relationship, wherein higher levels of slack result in decreased managerial risk taking. As slack decreases, managers become more accepting of risk (Steensma & Corley, 2001). Singh (1986) argue that organizations with more slack engage in greater risk taking such as the innovative activities. Managers with entrepreneurship will seek to extract growth from resources, and thus prefer instead to invest slack resources to expand a firm's market or product position (Mishina et al., 2004).

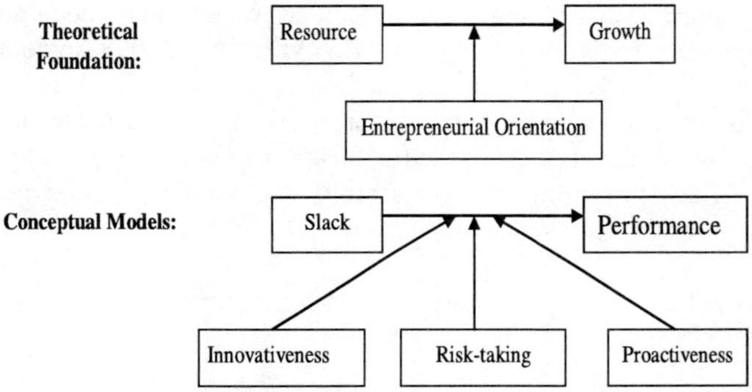

Fig. 4. The model on the moderator of entrepreneurial orientation

5 Conclusions and Research Prospects

As a core concept in the study of firm growth mechanism, organizational slack provide an effective analytical framework, the researchers tried to explain the performance using slack. However, organizational slack and firm performance in the test results are not consistent and there was even conflicting conclusions. As previously mentioned Singh (1986), Hambrick & D' Aaveni(1988), Bromiley (1991), Geiger & Cashen(2002), and Zhongfeng, En & Yuan (2009),the empirical results show that there is a positive relationship between firm performance and slack, the enterprise should reserve and be good at using these resources. However, some other scholars' showed slack was negatively related to performance (Davis & Stout, 1992; Oktemgil & Greenley, 1996), and they argued that as soon as possible to eliminate slack, to ensure corporate efficiency. Further, researchers have demonstrated the curvilinear relationship of slack–performance. Cheng & Kesner (1997), Tan & Peng (2003) found that there is an optimal level of slack, too much or too little will do no good on firm performance; Chunyan & Shuming (2004) showed a three stages of relationship between slack and performance, and suggested that 'the ideal range of slack isn't really optimal, the corporate should shorten this stage as possible, and to enter the third stage for higher performance return'.

The inconsistency conclusions between slack and performance make scholars begin to use contingency perspective to study the path of this relationship, and attempt to open the black box. They try to use mediator or moderator into the direct relationship, and explore the complex process that slack affect the performance, thus establishing the inherent mechanism of slack's effect on growth. Penrose as a representative use organizational capability to explain the bridge of the slack-performance link; Marino et al (2008), Frank & David (2004) and Ujari (2002) introduce various strategies into it, establishing the path based on resources-strategy. Aim at China's situation, Zhongfeng, En & Yuan (2009) argue the transition economies play an important role in the slack-performance link; greater number of researchers of entrepreneurship theory find that characteristics and situation of entrepreneurship act on the slack-performance link, promote the progress of firm growth theory and enrich entrepreneurship research.

Although Barnard (1938) had discussed the role of slack in his early work, and today the heat of researching slack has not reduced, from recent literature, research trends can be summarized as follows:

Research Trends 1. Classify the organizational slack. Meyer (1982) pointed out that slack can exist in different forms, different forms of slack could provide the different buffer on the impact of environmental. G Geiger & Makri (2006) show that available slack, but not recoverable slack, positively affects the process of innovation in the R&D intensity firms. Glenn, Deepak, & Zannie (2008) believe that absorbed slack can increase product exploitation and decrease exploration, but unabsorbed slack is just the opposite when perceived environmental threat is high. Zhongfeng, En & Yuan (2009) find that unabsorbed slack is critical for firms to sustain their competitive advantages.

Research trends 2. Diversification study. More and more study focus on characteristics on particular industry or organization. Sharfman et al.(1988) suggest that different types of industries will affect the level of discretion slack. For example, service industries have positive relationship with the level of high discretion slack, and manufacturing industries have positive relationship with the level of low discretion slack. Glenn, Deepak,& Zannie (2008) research 214 nonprofit professional theaters in the U.S.. Geiger & Makri(2006) investigate U.S. manufacturing. Herold et al.(2006) use the high-tech industry data. The research, using the data of China or other developing countries, is also more and more. And they have many interesting results.

Research trends 3. Use organizational slack theory to explain and find out other research topic, and verify this theory in turn. Organizational slack has accumulated a wealth of literature, recently, scholars no longer only study some features of its own, but use the theory to obtain more meaningful and extensive research phenomenon. Dovev & Lori (2005) use the organizational slack perspective to explain the positive effect of financial success on function-based exploration, and provide valuable conclusions.

References

1. Bourgeois, L.J.: On the Measurement of Organizational Slack. Academy of Management Review 6(1), 29–39 (1981)
2. Bourgeois, L.J., Singh, J.V.: Organizational slack and political behavior among top management teams. In: Academy of Management Proceedings, pp. 43–47 (1983)
3. Wernerfelt, B.: A Resource-Based View of the Firm. Strategic Management Journal 5(2), 171–180 (1984)
4. Cheng, J., Kesner, I.: Organizational slack and response to environmental shifts: The impact of resource allocation patterns. Journal of Management 23, 1–18 (1997)
5. Child, J.: Organization structure, environment and performance: The role of strategic choice. Sociology 6, 1–22 (1972)
6. Chunyan, J., Shuming, Z.: Relationship between organizational slack and performance: Empirical research on Chinese listed firms. Management World 5, 108–115 (2004)
7. Cyert, R.M., March, J.D.: A behavioral theory of the firm. Prentice-Hall, Englewood Cliffs (1963)
8. Glenn, B.V., Deepak, S., Zannie, G.V.: The effects of slack resources and environmental threat on product exploration and exploitation. Academy of Management Journal 51(1), 147–164 (2008)
9. Geiger, S.W., Makri, M.: Exploration and exploitation innovation processes: The role of organizational slack in R&D intensive firms. Journal of High Technology Management Research 17, 97–108 (2006)
10. George, G.: Slack resources and the performance of privately held firms. Academy of Management Journal 48, 661–676 (2005)
11. Leonard-Barton, D.: Core capabilities and core rigidities: A paradox in managing new product development. Strategic Management Journal 13, 111–125 (1992)
12. March, J.G.: Exploration and exploitation in organizational learning. Organization Science 2(1), 71–87 (1991)
13. Mishina, Y., Pollock, T.G., Porac, J.F.: Are more resources always better for growth? Resource stickiness in market and product expansion. Strategic Management Journal 25, 1179–1197 (2004)
14. Nohria, N., Gulati, R.: Is slack good or bad for innovation. Academy of Management Journal 39, 1245–1264 (1996)
15. Penrose, E.: The Theory of the Growth of the Firm. Oxford University Press, Oxford (1959)
16. Runsheng, F., Yi, G.: Organizational slack and its potential value. Economic Survey 6, 85–87 (2003)
17. Sharfman, M.P., Wolf, G., Chase, R.B., Tansik, D.A.: Antecedents of organizational slack. Academy of Management Review 13, 601–614 (1988)
18. Singh, J.V.: Performance, slack, and risk taking in organizational decision making. Academy of Management Journal 29, 562–585 (1986)
19. Steensma, H.K., Corley, K.G.: Organizational context as a moderator of theories on firm boundaries for technology sourcing. Academy of Management Journal 44(2), 271–291 (2001)
20. Tan, J., Peng, M.W.: Organizational Slack and Firm Performance During Economic Transitions: Two Studies from An Emerging Economy. Strategic Management Journal 24, 1249–1263 (2003)
21. Gong, X.: Effect of the financial crisis on organizational slack and technological innovation in enterprise. Modern Finance & Economics 29(9), 19–22 (2009)
22. Su, Z., Xie, E., Li, Y.: Organizational slack and firm performance during institutional transitions. Science 26, 75–91 (2009)

Application of Process Neural Network on Consumer Price Index Prediction

Li Ge[1,2] and Guisheng Yin[2]

[1] School of Computer and Information Engineering,
Harbin University of Commerce, Harbin 150028, P.R. China
[2] College of Computer Science and Technology,
Harbin Engineering University, Harbin 150001, P.R. China
gl2728@163.com

Abstract. In this paper we present a prediction method of consumer price index (CPI) based on process neural network (PNN). In order to reduce errors, after the raw data was directly expressed as a set of orthogonal basis expanded form, we made use of time-varying input function feature of process neural network and trained process neural network with combined type improved BP algorithm. We achieved a multi-variable CPI prediction with non-linear model of process neural networks gotten by above-mentioned result and illustrated the advantage of process neural network compared to traditional neural network in economic time series prediction. We provide a new method for economic time series prediction in this paper.

Keywords: CPI, process neural network, time series, prediction.

1 Introduction

As an important indicator reflecting inflation (or deflation) degree, consumer price index is an important basis to analyze and formulate national economic policies and national economic accounting. Therefore, it has great significance to monitor in real-time and predict accurately CPI. CPI prediction not only depends on a country's overall macroeconomic situation, but also its own inherent characteristics of time series. However, the entire macroeconomic system is a complex and elusive nonlinear system; it is difficult to predict accurately CPI. Therefore, it is the key of economic time series prediction to simulate accurately the characteristics of the above two strands. Neural network is an effective prediction method, which has a good nonlinear mapping ability to simplify modeling process of time series. It is widely used in various fields. In particular, process neural network technology, the input is time-varying function; process neuron not only includes the space aggregate computing functions of traditional neuron, but also includes the extraction function of time cumulative effects. It coincides with time-related features of time series. Therefore, compared to the previous prediction method, it has its own unique advantages to fit the inherent laws between eight categories data of China's household consumption and CPI, and establish non-linear model of CPI time series using process neural network.

J. Luo (Ed.): Affective Computing and Intelligent Interaction, AISC 137, pp. 427–432.
springerlink.com

2 Process Neural Network Model

2.1 Process Neuron

The structure of process neuron is composed of weighted, aggregation and activation operator of three parts. The difference from the traditional neuron is that the input, output, and weights of process neuron can be time-varying, and its aggregation operations is composed of multi-input aggregation in space and cumulative aggregation of time. A process neural network is a network composed of some neurons in the certain topology structure. The structure of a process neuron as shown in Fig. 1, where $X(t) = (X_1(t), X_2(t), ... X_n(t))^T$ is input function vector of process neuron, $w_1(t), w_2(t), ... w_n(t)$ is weight function, $f(\cdot)$ is the activation function, which may take linear function, Sigmoid function, Gauss-type function, and so on.

The relationship between input and output of process neuron is

$$Y = f((w(t) \oplus X(t)) \otimes K(\cdot) - \theta) \tag{1}$$

Where " $\oplus$ " indicates a sort of spatial aggregation operation," $\otimes$ " indicates a sort of time (process) aggregation operation, " θ " indicates threshold value of process neuron, $K(\cdot)$ indicates an integrable function of [0, T] [1].

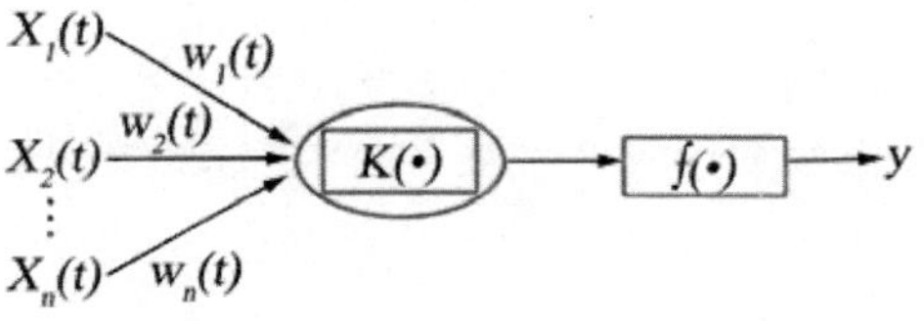

Fig. 1. Process neuron

2.2 Topology of Process Neural Network

Fig. 2 shows a topology structure of feedforward process neural network: the input is a time-varying function, the output is a constant, and the topology structure of network is n $-$ m $-$ L $-$ 1 .Input layer has n nodes used for n time-varying function input to the network. The first hidden layer is a process neuron hidden layer composed of m nodes to complete the n input function weighted aggregation in space and the aggregation operation of the time course, as well as the character extraction of sample process model. The second hidden layer is a non-time-varying hidden layer (Also known as normal neuron hidden layer), which have *L* nodes. The fourth layer is the output layer, whose output is y.

The input-output relationship of the network as follows:

$$y = \sum_{l=1}^{L} f_2\left(\sum_{j=1}^{m} v_{jl} f_1\left(\sum_{i=1}^{n} \int_0^T w_{ij}(t) x_i(t) dt - \theta_j^{(1)}\right) - \theta_l^{(2)}\right) \mu_l \tag{2}$$

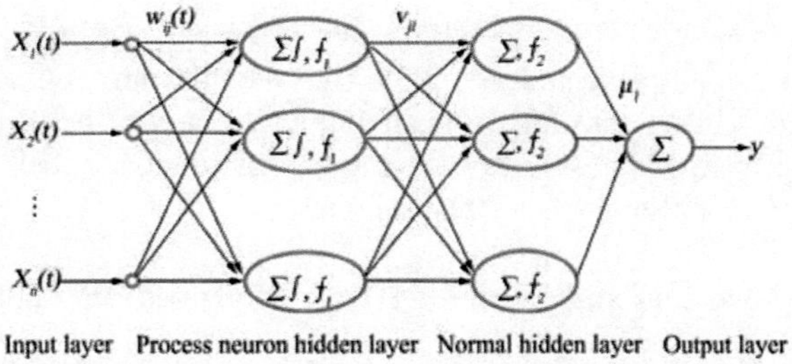

Fig. 2. Process neural network

Where $x_i(t)$ is system input, $w_{ij}(t)$ is the connection weight function between input layer and the first hidden layer, $\theta_j^{(1)}$ is the output threshold of j-neuron of the first hidden layer, $[0\ T]$ is sampling interval, f_1 is the activation function of the first hidden layer. v_{jl} is the connection weight between j-neuron of the first hidden layer and l-neuron of the second hidden layer, $\theta_l^{(2)}$ is output threshold of j-neuron in the second hidden layer, f_2 is the activation function of the second hidden layer, which can be taken as a linear function or S function. y is final output of network, μ_l is the connection weight between neuron of the second hidden layer and l-neuron in output layer [2].

3 CPI Prediction Based on Process Neural Network

3.1 Datasets

China's CPI data are composed of the eight categories residents' basic consumption data, and has very strong nonlinear characteristics. As a general rule, monthly economic time series is affected not just by the seasonal but also non-seasonal factors. In order to achieve a more accurate prediction result of CPI and extract sufficiently its non-linear variation characteristics, process neural network is introduced into CPI prediction. Making use of extraction capabilities of process neurons to time accumulative feature, the next month CPI data is gotten by prediction, after training set and test set are composed of historical data fully reflecting the characteristics of seasonal and non-seasonal factors, and make up for the deficiencies in past prediction methods. A sample structure is as follows:

$$\{x_1(t), x_2(t), \dots, x_i(t), \dots, x_n(t), d\} \tag{3}$$

Where $x_i(t)$ indicates an arbitrary input function, which is fitted by a series of consecutively discrete data, d is network expected output. In order to represent seasonal and non-seasonal characterizations, as well as time accumulate feature more fully, finally, CPI data of 10 consecutive months was selected as the data to fit input function, after repeated comparison and analysis. CPI data in the eleventh month was selected as expected output of PNN. Namely:

$$x_i(t) \approx \varphi(a_0, a_1, \dots, a_m) \tag{4}$$

Where $(a_0, a_1,..., a_m)$ indicates consecutively discrete data, $m = 9$ in CPI prediction.

Data were specifically selected as follows: 22 samples are composed of the eight categories residents' basic consumption data published in National Bureau of Statistics of china web site from May 2007 to September 2009, then selected 17 of which as a training sample and 5 of which as the test samples.

3.2 Concrete Measures and Combined Type Improved BP Algorithm

In addition, in order to improve overall prediction accuracy, the following measures were adopted in CPI prediction:

1) The raw data were normalized.

2) The raw data normalized was directly expressed as a set of orthogonal basis expanded form so as to reduce errors and speed up network convergence. Thereinto, network orthogonal basis function selected Legendre polynomial [3], number of basis functions is 8 (The value is determined based on the status of network convergence at last).

3) Combined type improved BP algorithm was adopted to train PNN.

The concrete implements of combined type improved BP algorithm in PNN training are followed as:

There are three main measures of improving BP Algorithm: the momentum method [4], adaptive learning rate method and steepness factor method [5] [6]. But it often happens that the three methods used simultaneously can not get the best convergence results in the actual application, whereas using the above three ways alone or combined can get better convergence results in connection with different data characteristics. Therefore, using random combinations of the above three measures, PNN is trained by the above random combinations respectively, and finally decides the concrete improving measures of BP algorithm by convergence accuracy and test errors, which is a feasible improvements in the actual application. Specifically, Adaptive learning rate method is realized by using equation (5).

$$\eta(k+1)=\begin{cases}\alpha_1\eta(k), E(k+1)>\lambda_1 E(k)\\ \alpha_2\eta(k), E(k+1)<\lambda_2 E(k)\end{cases} \tag{5}$$

Where k is training number, η is learning rate, E is error function, α_1 is incremental factor used to increase learning rate of network, and α_2 is as reduction factor used to decrease learning rate. λ_1, λ_2 are amplitude control factors new added, which are used to indicate the changing magnitude of two successive error function and accelerate the network convergence speed.

In CPI prediction, weight functions in process neural hidden layer were firstly expressed as the same set of orthogonal basis expanded form (which is identical to orthogonal basis in input function expansion), so weight function training is transformed into a set of weights (orthogonal basis parameters) training problem. Moreover, all parameters of orthogonal basis and the weights in normal neuron hidden layer were initialized in (-1, 1).

Secondly, PNN used in CPI prediction was trained by combination of the above three measures. After repeated combination training and comparative analysis, finally, identify the momentum method to train PNN, momentum factor is 0.8, learning rate is 0.8, and steep factor is 1.

The correlation parameters in CPI prediction were selected as follows: 8 input nodes, 15 hidden layer neurons in process neuron hidden layer, 1 normal neural node, and 1 output node. The largest iterative number is 50, and learning accuracy is 0.01.The activation function in the first hidden layer is tangent sigmoid function. The activation function in the second layer is a linear function.

3.3 Prediction Result and Comparison Analysis

Network has convergence, after it operated 3 generations. The error during the training stage for sample was shown in Fig. 3, prediction values and real data of test samples as shown in Table 1.The average of relative errors is 0.3623%, the average of relative errors absolute value is 0.6842%.

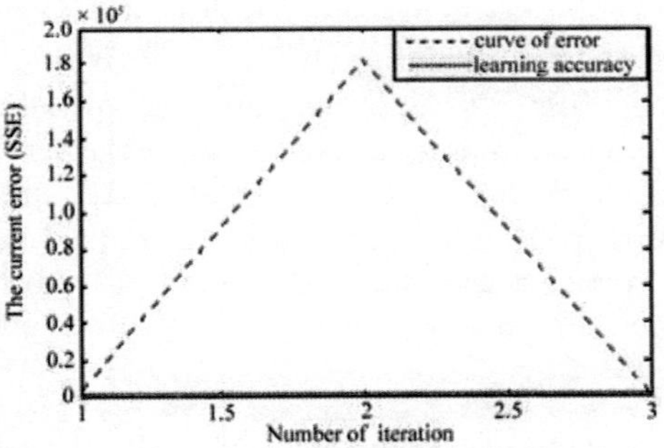

Fig. 3. The error value during the training stage for sample

For the same training samples and test samples (Discrete real data were used as input), traditional neural network method was adopted to make model. After repeated adjust the network structure and repeated training using the same learning algorithm a model with the best test error was choose as the end result. Its test errors are shown in Table 1. Training parameters selected are as follows: 8 input nodes, 100 the first hidden layer nodes, 1 the second hidden layer nodes, learning rate is 0.01, the largest iteration number is 10000, learning accuracy is 0.001, momentum factor is 0.8,activation function of the first hidden layer is the tangent sigmoid function, activation function of the second layer is a linear function, the minimum run gradient is 1e-010.The network was terminated after 305 generations, because the gradient does not meet the minimum run gradient value.

Table 1. Prediction values of test samples on different prediction ways

Test Samples	Real Data	Prediction Values of Traditional Neural Network	Prediction Values of Process Neural Network
1	98.8	103.7647	98.7438
2	99.5	103.7647	98.9253
3	98.3	103.7647	98.6142
4	101	103.7647	99.3142
5	98.5	103.7647	98.666

Can be seen from Table 1, the average relative error of test samples in the traditional neural network model is -4.591%, the average of relative errors absolute value is 4.591%. Thus, under the premise of the same learning algorithm, the test result of PNN is superior to the traditional neural network result. This shows that PNN with time-varying input has advantage in CPI short-term prediction.

4 Conclusions

Aiming at CPI multivariable prediction problem in economic domain, a short-term prediction method of economic time series based on process neural network is presented. Its concrete realization process is given in this paper. A multivariable and nonlinear PNN prediction model of CPI was established, and was compared to traditional neural network perdition model using the same learning algorithm, the result illustrates that process neural network method has advantage in short-term prediction of time series. It provides a new way for other prediction problem in economic domain.

Acknowledgment. This work was supported by Harbin Technology Innovation Research Special Fund (2008RFQXG072) and Heilongjiang Province Science and Technology Project (GC05A118).

References

1. He, X., Liang, J.: Some theoretical issues on process neural networks. Engineering Science 2, 40–44 (2000) (in Chinese)
2. Xu, S., He, X.: Learning algorithms of process neural networks based on orthogonal function basis expansion. Chinese Journal of Computers 27, 645–649 (2004) (in Chinese)
3. Liu, C.: Orthogonal functions and applications, pp. 40–49. National Defense Industry Press, Beijing (1982) (in Chinese)
4. Zhong, S., Piao, S., Gang, D.: Application research of improved BP algorithm in process neural networks. Journal of Harbin Institue of Technology 6, 840–842 (2006) (in Chinese)
5. Han, L.: Artificial neural network theory, design and application, pp. 55–69. Chemical Industry Press, Beijing (2002) (in Chinese)
6. Jiang, Z.: Introduction to artificial neural networks, pp. 86–98. Higher Education Press, Beijing (2001) (in Chinese)

Design and Realization of an Embedded Gateway Base on 3G

Shuang Wu and Gui-gen Zeng

Nanjing University of Posts and Telecommunications, NUPT, Nanjing, China
1987ws@gmail.com, kingmuddy@163.com

Abstract. This paper introduced an embedded access platform, which took OMAP3530 as the center microprocessor and was built upon Linux2.6 operating system. By adopting SIM5218A and transplanting the PPP software package, the wireless data communication link between the embedded gateway and 3G network was set up. Also, by transplanting the Iptables tool and setting up a DHCP server, communication among the local equipments and the data conversion task between 3G network and Ethernet was completed. Results show that 3G remote control and data communication of the local equipments are realized. Compared with other access gateway, this system possesses the advantages of high-speed, stability and reliability.

Keywords: 3G, Gateway, PPP, NAT, DHCP.

1 Introduction

Based on the extensive coverage of 3G (the abbreviation of the 3rd Generation) network and high performance of 3G technology in China, it is a new, advanced and revolutionary technology program when using 3G as the remote access network. Meanwhile, in the industrial area, Ethernet has become a trend because more and more products and solutions have already supported the Ethernet standard. Therefore, how to integrate the advantages of 3G network and the user's requirements of remote control and data communication has become an urgent problem.

This system uses WCDMA (Wideband Code Division Multiple Access) [1] technology to access this embedded gateway to 3G network, and then to the Internet. And local users connecting the gateway through Ethernet interface can communicate with each other and access the Internet by WCDMA mode through the gateway, thus a embedded gateway based on 3G is set up. The combination of 3G and Ethernet has a very important significance in the embedded gateway system.

2 Hardware Design Scheme

The hardware design of this gateway is based on ARM CortexTM-A8. Through expanding the 3G and Ethernet modules, the platform can realize the remote control and data communication function. Structure is shown in Fig. 1.

J. Luo (Ed.): Affective Computing and Intelligent Interaction, AISC 137, pp. 433–439.
springerlink.com

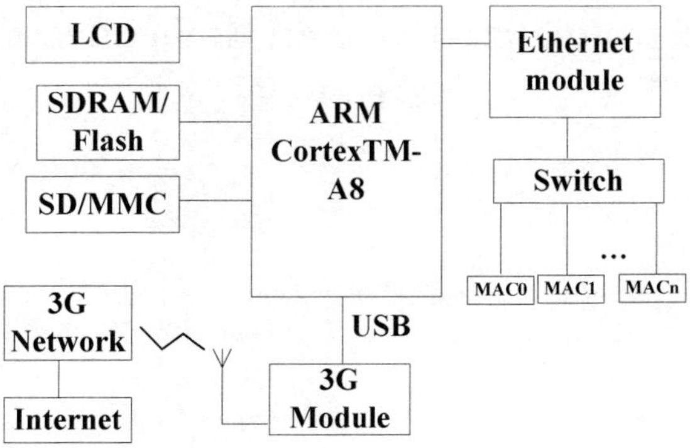

Fig. 1. Structure of the system

2.1 CPU

In the system, embedded gateway is responsible for data communication between Ethernet and 3G network, and also responsible for local network management, all of tese make demands on high reliability and powerful processing ability. Thus, we adopt TI's OMAP3530 [2] microprocessor as the main chip. Based on ARM CortexTM-A8 core, OMAP3530 can operate in a high-speed (up to 1GHz frequency above). Its off-chip memory using 256MByte 32bit mDDR and 512Mbyte 16bit NAND Flash integrated chip to support the powerful processing function. With its good scalability as well as the rich peripheral interfaces, including 10/100M adaptive Ethernet interface、serial and USB interface, OMAP3530 can communicate with 3G module and Ethernet module expediently. And this processor can also be embedded in a tailored Linux operating system, it will be very convenient for the PPP(Point-to-Point Protocol)[3]dial-up software, Iptables tool and DHCP (Dynamic Host Configuration Protocol) server to be transplanted.

2.2 Ethernet Module

In the system, local devices need a high-speed Ethernet interface controller to access the embedded gateway, so this system adopts the DM9000 Ethernet controller chip, which is a fully single integrated fast MAC controller chip launched by China Taiwan DAVICOM company. It follows the promulgation of the IEEE 802.3 Ethernet protocol, provides 10/100Mbps Ethernet access and supports TCP / IP acceleration (IPV4 check sum offload) to reduce the CPU burden and improve machine performance.

2.3 3G module

For the high reliability and performance-price ratio, we use SIM5218A chip in WCDMA mode. This module supports data communication service with downstream rate up to 7.2Mbps and uplink speeds up to 5.76Mbps; and it has rich interfaces including UART, USB2.0, GPIO, etc, and also offers a wealth of AT commands. In this system, we use USB port to link OMAP3530 and SIM5218A for full-duplex communication by AT commands.

3 Software Design Scheme

Embedded Linux is a widely used embedded operating system, for its embedded feature, we modify and recompile it's kernel, retain the system's original powerful network function, high stability, and ample documentation system support function.

A. As the embedded gateway connects the 3G and Ethernet by USB interface, therefore, it is necessary to add DM9000 Ethernet controller driver, retain the support of USB, and reserve the support of the network ("network support") as well as the TCP / IP protocol stack.

B. Configure the kernel

1) Support the PPP function: Select "network device support" in the kernel configuration, and choose as follows:

<*> PPP (point-to-point protocol) support
<*> PPP support for async serial ports
<*> PPP support for sync tty ports
<*> PPP over Ethernet (EXPERIMENTAL)
<*> PPP over L2TP (EXPERIMENTAL)

Here, we choose to incorporate it into the core, because it generates the executable program by default, instead of ". ko" module file.

2) Support firewall and NAT(Network Address Translation) function: Select "networking support"→ "networking options"→ "network packet filtering framework (Netfilter) " and choose according to the kernel hints, of which some are loaded as the module while the others are incorporated into the core.

3) Set up a DHCP server: Select "networking support"→ "networking options", and choose as follows:

[*] IP: DHCP support
[*] Network packet filtering framework (Netfilter)

C. Configure the file system

Apart from some basic commands and directories, some specific configurations also needed to be done to make the embedded gateway work properly.

4 Realization of the Embedded Gateway

4.1 3G Access

Data transmitted by 3G network requires Internet networks to transfer, so it has to be packaged by TCP / IP protocol. This paper uses software to package data. The transmitting data is firstly packaged by UDP Layer protocol header, and then packaged by IP protocol header, at last packaged by PPP protocol.

PPP software package works on embedded Linux2.6 operating system, and controls 3G module to complete network access, the software structure is shown in Fig. 2.

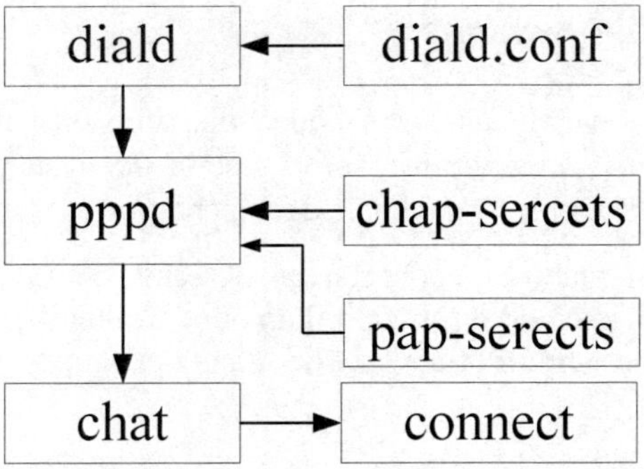

Fig. 2. Software structure diagram of PPP

The chat dial-up files can be placed anywhere in the system, its position can be designated by the pppd options, but usually in "/etc/ppp" directory, as follows[4]:

'' 'AT'
OK 'AT+CGDCONT=1,"IP","3gnet", 0, 0'
OK 'ATDT*99#'
CONNECT

AT is the string that send AT, expected to be OK, and then set China Unicom's 3G access point to be 3gnet, use "ATDT*99#"dialing-up to visit GGSN (Gateway GPRS Supporting Node), afterwards, it would achieve "CONNECT" string.

In the end, type the command line: $ pppd

Run pppd, and dial-up Internet by ppp.

After the success of dial-up, the server's dynamic IP address and DNS addresses should be allocated in dial-up information.

4.2 Firewall and NAT

Firewall is a protective mechanism, which is located in a transit point between the specific and strictly limited network (such as local area network) and the adjacent network (usually the Internet) to protect local devices will not be threatened by the external risks. The security management mechanism of firewall determines the kind of the data that can be passed through.

NAT is one of IETF (Internet Engineering Task Force) standards, allowing the whole devices appears a public IP address in the Internet, that's to say, NAT translates the private source IP address to the available legal IP address[5]. Applying the technology of NAT, the large numbers of private IP addresses that can be reused inside local domain satisfy the increasing demand of IP address. In fact, NAT is a technology of IP address reusing.

There is a firewall tool, Iptables, in the Linux 2.6 kernel, which contains three tables, NAT tables, mangle table and filter table. There are chains in each table. Rules and policies which used to decide the doom of the packets are embodied in these chains. The structure is shown in Fig. 3.

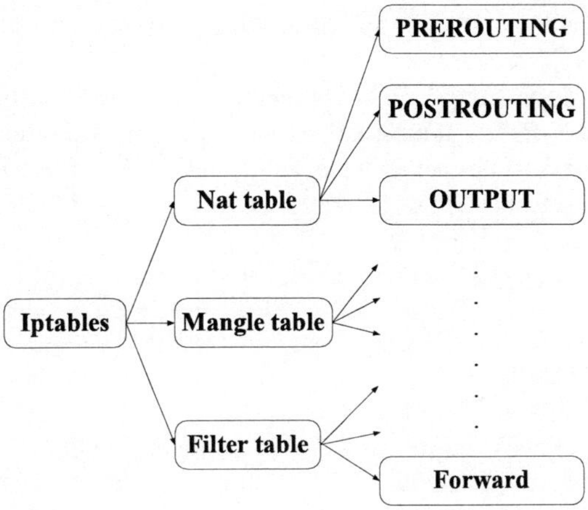

Fig. 3. The structure of Iptables

Here, the NAT table is adopted. In NAT table, there are three chains, PREROUTING, POSTROUTING and OUTPUT. In this paper, a realization of NAT under embedded Linux 2.6.28 system is involved, the following lines come from the NAT configure file [6]. (Characters following '#' are notations, and following '$' is a command)

#Load modules, which are encapsulating the realization of the iptable and its sub-chains

modprobe ip-tables
modprobe ip-conntrack
modprobe ip-conntrack-ftp
modprobe ipt-state
modprobe iptable-nat
modprobe iptable-filter
#Enable IP forwarding for NAT, so that the legal packets can pass through.
$echo 1 >/proc/sys/net/ipv4/ip-forward

To enable the hosts assigned with private IP address in the local network to attend the public network

$iptables –t nat –A POSTROUTING –o eth0 –j SNAT –to 192.168.1.100-192.168.1.200

To allow the internal devices with no known address outside their network, to communicate with the outside through the ppp0 interface.

$iptables –t nat –A POSTROUTING –o ppp0 –j MASQUERADE

4.3 DHCP Server

DHCP is proposed by the IETF, for automating the configuration of computers that use TCP/IP. DHCP is widely used by Internet service providers and network administrators for network management, particularly in dial-up or mobile environments. A commonly used DHCP service is IP address allocation. When joining the network, a client cannot connect to other services (e.g., file server and mail server) until after it obtains an IP address from the DHCP server.

In the system, for supporting DHCP server function, some configurations have been made to the kernel and file system. Besides, we deploy the file udhcpd.conf, including IP address, subnet mask, gateway address, DNS and DHCP leases information.

In the end, type the following command line:

$ udhcpd –fS /etc/udhcpd.conf &.

Run udhcpd in the backyard and get a allocated IP address for each local device.

5 Network Testing

As is shown in Fig. 1, OMAP3530 gateway communicates with the 3G module via USB interface and Ethernet module via eth0. At first, gateway successfully accesses the Internet through WCDMA network, and gets a ppp0 interface with a IP address 172.17.16.23 and the DNS address 58.242.2.2, as shown in Fig. 4. Then, with the NAT and DHCP technology, this system set a legal IP address pool of the local hosts which connect the gateway through Ethernet from 192.168.1.100 to 192.168.1.200 while the gateway's eth0 is 192.168.1.1. Network testing result in any local host is shown in Fig. 5.

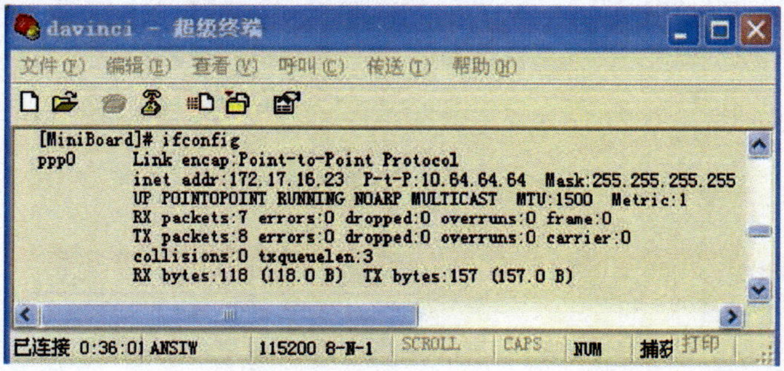

Fig. 4. Example of the network testing 1

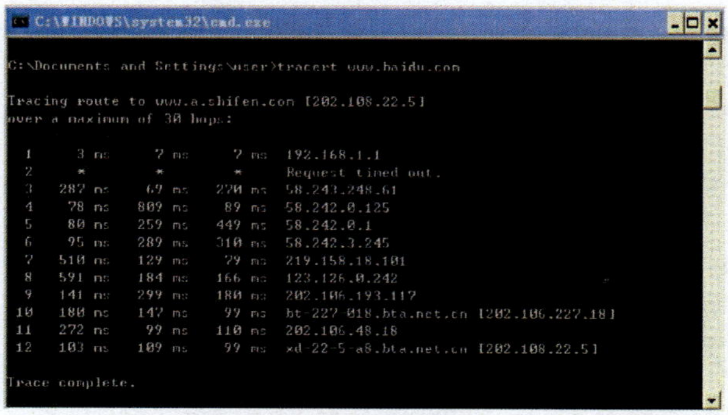

Fig. 5. Example of the network testing 2

6 Conclusion

The embedded gateway described in this article is based on 3G network while using the Linux2.6 operating system with a powerful network function and ARM CortexTM-A8 processor with high-performance, which greatly simplifies application development, and has good scalability, combined with PPP, NAT, DHCP technology, a 3G gateway with multiple functions was build up successfully. Staff can log on the Internet to check the working conditions of the gateway's internal devices, thus the 3G remote control and data communication is realized. With the development of 3G, this gateway will have a promising application foreground.

References

1. Holma, H., Toskala, A. (eds.): WCDMA for UMTS: Radio access for third generation mobile communications. Wiley (2000) ISBN0-471-72051-8
2. ICETEK-OMAP3530-Mini Manual Rev3.0 (2009)
3. Simpson, W.: The Point to Point Protocol. RFC1661 (1994)
4. SIMCOM Limited, SIM5218A_ATC_V1.07.AT Command set (2009)
5. Rekhter, Y., Moskowitz, B., Karrenberg, D., de Groot, G.: Address Allocation for Private Intemets, RFC1597, T.J. Watson Research Center, IBM Corp., Chrysler Corp., RIPE NCC (March 1994)
6. Marie, F.: Netfilter Extensions HOWTO, http://samba.org/ftp/unpacked/netfilterihomepage/output/HOWTO/

Parameter Optimization for Machined Round Parts by Using Grey Relational Analysis

Hong Zhennan, Liu Chunbao, and Li Junhong

School of Electrical Engineering, University of South China, Hengyang, China

Abstract. The roundness and surface roughness are important quality performances for machined round parts. In order to get the best quality performance by less experiment search for optimum parameters, an orthogonal array was used for the experimental design, optimal machining parameters were determined by the grey relational grade obtained from the grey relational analysis, and the main machining parameters that affect the machining performance can be found by the difference between the maximum and minimum value of the grey relational grade of the machining parameters. Experimental results have shown that the roundness and surface roughness of machined round parts can be improved effectively by the proposed approach.

Keywords: Roundness Surface roughness, Grey relational analysis, Optimization.

1 Introduction

In machining operation, it is an important task to select machining parameters for achieving high machining performance. Usually, the desired machining parameters are determined based on experience or by use of a handbook. However, this does not ensure that the selected machining parameters have optimal or near optimal machining performance for a particular machine and environment. To select the machining parameters properly, several mathematical models [1–5] based on statistical regression techniques or neural computing have been constructed to establish the relationship between the machining performance and the machining parameters. Then, an objective function with constraints is formulated to solve the optimal machining parameters using optimization techniques. Therefore, considerable knowledge and experience are required for using this modern approach. Furthermore, a large number of cutting experiments has to be performed and analyzed in order to build the mathematical models. Thus the required model building is very costly in terms of time and materials. In this paper, an effective approach based on the grey relational analysis is used to determine the desired machining parameters.

The grey system theory is a new technique for performing prediction, relational analysis and decision making in many areas. With regard to the quality characteristics of machined round parts, some of the problems encountered include the roundness and surface roughness. Among two characteristics, the roundness plays better important role than the surface roughness in the performance of a round part. The vertical

J. Luo (Ed.): Affective Computing and Intelligent Interaction, AISC 137, pp. 441–448.
springerlink.com

machining center(model:A55) use circular contour cutting of a mechanical parts is research object. There is seven machining factors affecting performance characteristics in machined parts. such as acceleration time, load inertion rate, pre-control coefficient, pre-control time constants, cutting speed, feed rate and cutting depth. In order to improve machining efficiency, reduce the machining cost, and improve the quality of machined parts, it is necessary to select the most appropriate machining conditions. In a system that is complex and multivariate, the relationship between various factors such as those described above is unclear. Such systems often are called as "grey" implying poor, incomplete, and uncertain information. Their analysis by classical statistical procedures may not be acceptable or reliable without large data sets that satisfy certain mathematical criteria.

The purpose of the present work is to introduce the use of grey relational analysis in selecting optimum machining conditions on multi-performance characteristics, namely, the roundness and the surface roughness. The setting of machining parameters was accomplished using the Taguchi experimental design method, the effect of parameters is achieved by using grey relational analysis. In addition, the most effective factor and the order of importance of the controllable factors to the multi-performance characteristics in the machined process were determined. Thus, by properly adjusting the control factors, we can improve work efficiency and produce quality.

2 Experimental Work

The experimented machine(model:A55) is vertical machining center use circular contour cutting of a mechanical parts. The experimental material is aluminium alloy T6061, material size is Φ32mm*80mm, and cutting size isΦ30mm*10mm. cutting tool material is tungsten carbide, and cutting tool size isΦ12mm.

Classical experimental design methods are too complex and are not easy to use. A large number of experiments have to be carried out when the number of process parameters increases. To solve this problem, the Taguchi method uses a special design of orthogonal arrays to study the entire parameter space with only a small number of experiments. According to the Taguchi method, a robust design and an L_{18} orthogonal array are employed for the experimentation. Seven machining parameters are considered as controlling factors, and each parameter has three levels. The experimental results after machined parts were evaluated in terms of the following measured machining performance: (1) surface roughness (*Ra*); (2) roundness (h). The experimental results are summarized in Table 1.

Table 1. Experimental design using L_{18} orthogonal array

Exp.no	A	B	C	D	E	F	G	Ra(μm)	h(μm)
1	1	1	1	1	1	1	1	2.09	8.03
2	1	2	2	2	2	2	2	2.18	12.2
3	1	3	3	3	3	3	3	2.15	33.6
4	2	1	1	2	2	3	3	2.11	30.1
5	2	2	2	3	3	1	1	2.2	6.58
6	2	3	3	1	1	2	2	2.24	17.7
7	3	1	2	1	3	2	3	2.14	24.7
8	3	2	3	2	1	3	1	2.35	23.2

Table 1. (*continued*)

Exp.no	A	B	C	D	E	F	G	Ra(μm)	h(μm)
9	3	3	1	3	2	1	2	2.15	4.01
10	1	1	3	3	2	2	1	2.28	9.21
11	1	2	1	1	3	3	2	2.30	70.9
12	1	3	2	2	1	1	3	2.26	9.06
13	2	1	2	3	1	3	2	2.21	8.71
14	2	2	3	1	2	1	3	2.23	6.77
15	2	3	1	2	3	2	1	2.03	26.85
16	3	1	3	2	3	1	2	2.24	6.55
17	3	2	1	3	1	2	3	2.58	4.75
18	3	3	2	1	2	3	1	2.45	61.05

3 Grey Relational Analysis

In grey relational analysis, black represents having no information and white represents having all information. A grey system has a level of information between black and white. In other words, in a grey system, some information is known and some information is unknown. In a white system, the relationships among factors in the system are certain; in a grey system, the relationships among factors in the system are uncertain [7,8]. Grey relational analysis is an impacting measurement method in grey system theory that analyzes uncertain relations between one main factor and all the other factors in a given system. In the case when experiments are ambiguous or when the experimental method cannot be carried out exactly, grey analysis helps to compensate for the shortcomings in statistical regression. Grey relational analysis is actually a measurement of the absolute value of the data difference between sequences, and it could be used to measure the approximate correlation between sequences.

3.1 Data Pre-processing

Data pre-processing is normally required since the range and unit in one data sequence may differ from the others. Data preprocessing is also necessary when the sequence scatter range is too large, or when

Table 2. Experimental design using L_{18} orthogonal array

Exp.no	A	B	C	D	E	F	G	Ra(μm)	h(μm)
1	1	1	1	1	1	1	1	2.09	8.03
2	1	2	2	2	2	2	2	2.18	12.2
3	1	3	3	3	3	3	3	2.15	33.6
4	2	1	1	2	2	3	3	2.11	30.1
5	2	2	2	3	3	1	1	2.2	6.58
6	2	3	3	1	1	2	2	2.24	17.7
7	3	1	2	1	3	2	3	2.14	24.7
8	3	2	3	2	1	3	1	2.35	23.2
9	3	3	1	3	2	1	2	2.15	4.01
10	1	1	3	3	2	2	1	2.28	9.21
11	1	2	1	1	3	3	2	2.30	70.9
12	1	3	2	2	1	1	3	2.26	9.06

Table 2. *(continued)*

Exp.no	A	B	C	D	E	F	G	Ra(μm)	h(μm)
13	2	1	2	3	1	3	2	2.21	8.71
14	2	2	3	1	2	1	3	2.23	6.77
15	2	3	1	2	3	2	1	2.03	26.85
16	3	1	3	2	3	1	2	2.24	6.55
17	3	2	1	3	1	2	3	2.58	4.75
18	3	3	2	1	2	3	1	2.45	61.05

$$x_i^*(k) = \frac{x_i^0(k) - \min x_i^0(k)}{\max x_i^0(k) - \min x_i^0(k)} \quad (1)$$

When the "lower is better" is a characteristic of the original sequence, then the original sequence should be normalized as

$$x_i^*(k) = \frac{\max x_i^0(k) - x_i^0(k)}{\max x_i^0(k) - \min x_i^0(k)} \quad (2)$$

However, if there is a definite target value (desired value) to be achieved, the original sequence will be normalized in form:

$$x_i^*(k) = 1 - \frac{\left|x_i^0(k) - x^0\right|}{\max x_i^0(k) - x^0} \quad (3)$$

where $i = 1, \ldots, m$; $k = 1, \ldots, n$. m is the number of experimental data items and n is the number of object parameters. $x_i^0(k)$ denotes the original sequence, $x_i^*(k)$ the sequence after the data pre-processing, $\max x_i^0(k)$ the largest value of $x_i^0(k)$ $\min x_i^0(k)$ the smallest value of $x_i^0(k)$ and x^0 is the desired value.

3.2 Grey Relational Coefficient and Grey Relational Grade

In grey relational analysis, the measure of the relevancy between two systems or two sequences is defined as the grey relational grade. When only one sequence, $x_0(k)$, is available as the reference sequence, and all other sequences serve as comparison sequences, it is called a local grey relation measurement. After data pre-processing is carried out, the grey relation coefficient $\xi_i(k)$ for the kth performance characteristics in the ith experiment can be expressed as:

$$\xi_i(k) = \frac{\Delta\min + \rho\Delta\max}{\Delta_{0i}(k) + \rho\Delta\max} \quad (4)$$

where Δ_{0i} is the deviation sequence of the reference sequence and the comparability sequence.

$$\Delta_{0i} = \left\| x_0^*(k) - x_i^*(k) \right\|$$

$$\Delta\min = \min_k \min_j \left\| x_0^*(k) - x_j^*(k) \right\| \quad \Delta\max = \max_k \max_j \left\| x_0^*(k) - x_j^*(k) \right\|$$

$x_0^*(k)$ denotes the reference sequence and $x_i^*(k)$ denotes the comparability sequence. ρ is distinguishing or identification coefficient: $\rho \in [0,1)$ (the value may be adjusted based on the actual system requirements). A value of ρ is the smaller and the distinguished ability is the larger. $\rho = 0.5$ is generally used.

After the grey relational coefficient is derived, it is usual to take the average value of the grey relational coefficients as the grey relational grade [12]. The grey relational grade is defined as follows:

$$\gamma_i = \frac{1}{n}\sum_{k=1}^{n} \xi_i(k) \tag{5}$$

However, in a real engineering system, the importance of various factors to the system varies. In the real condition of unequal weight being carried by the various factors, the grey relational grade in Eq. 5 was extended and defined as:

$$\gamma_i = \sum_{k=1}^{n} \omega_k \xi_i(k) \quad \sum_{k=1}^{n} \omega_k = 1 \tag{6}$$

where ω_k denotes the normalized weight of factor *k*. The grey relational grade γ_i represents the level of correlation between the reference sequence and the comparability sequence. If the two sequences are identical, then the value of grey relational grade is equal to 1. The grey relational grade also indicates the degree of influence that the comparability sequence could exert over the reference sequence. then the grey relational grade for that comparability sequence and reference sequence will be higher than other grey relational grades.

4 Analysis and Discussion of Experimental Results

In the present study, the surface roughness and roundness values in different machined parameters are listed in Table 2. Lower surface roughness and roundness are indications of better performance. For data pre-processing in the grey relational analysis process, both surface roughness and roundness are taken as the "lower is better" (LB). Let the results of 18th experiments be the comparability sequences $x_i^0(k)$ i = 1–18, k = 1,2. All the sequences after data pre-processing using Eq. 2 are listed in Table 3 and denoted as $x_0^*(k)$ and $x_i^*(k)$ for reference sequence and comparability sequence, respectively. For example, $x_1^*(1)$, $x_1^*(2)$, $\Delta_{01}(1)$, $\Delta_{01}(2)$ can be calculated as follow:

$$x_1^*(1) = \frac{\max x_i^0(1) - x_1^0(1)}{\max x_i^0(1) - \min x_i^0(1)} = \frac{2.58 - 2.09}{2.58 - 2.03} = 0.891$$

$$x_1^*(2) = \frac{\max x_i^0(2) - x_1^0(2)}{\max x_i^0(2) - \min x_i^0(2)} = \frac{70.9 - 8.03}{70.9 - 4.01} = 0.940$$

$$\Delta_{01}(1) = \left|x_0^*(1) - x_1^*(1)\right| = \left|1 - 0.891\right| = 0.109$$

$$\Delta_{01}(2) = \left|x_0^*(2) - x_1^*(2)\right| = \left|1 - 0.940\right| = 0.06$$

The other $x_i^*(k)$ and $\Delta_{0i}(k)$ were calculated using the same method. Using Table 3, Δmax and Δmin can be found as follows:Δmax = $\Delta_{17}(1) = \Delta_{11}(2) = 1$, Δmin = $\Delta_{15}(1) = \Delta_{09}(2) = 0$. The distinguishing coefficient $\rho = 0.5$ can be substituted into Eq. 4 to produce the grey relational coefficient. Because the roundness plays better important role than surface roughness in the performance of a round part. So where $\omega_1 = 0.3, \omega_2 = 0.7$. The grey relational grade values for each experiment of the L_{18} orthogonal array were calculated by applying Eqs. 6.

Table 3. The response table for grey relational grade

parameters	Level 1	Level 2	Level 3	Max-min
A	0.6907	0.7452*	0.7009	0.0545
B	0.7569*	0.7079	0.6721	0.0848
C	0.7158	0.7008	0.7203*	0.0195
D	0.6310	0.7253	0.7806*	0.1496
E	0.7594*	0.7103	0.6672	0.0922
F	0.8398*	0.7245	0.5726	0.2672
G	0.7040	0.7236*	0.7092	0.0196

According to the performed experiment design, it is clearly observed from Table 2 and Fig. 1 that the machining parameters'setting of experiment no. 9 has the highest grey relational grade. Therefore, experiment no. 9 is the optimal machining parameters' setting for minimum surface roughness and minimum roundness simultaneously (i.e. the best multi-performance characteristics) among the 18 experiments.

In order to the determination of optimum machined parameters for surface roughness and roundness, the response table for the Taguchi method was used to calculate the average grey relational grade for each level of the machined parameters. The procedure is: i) group the grey relational grades by factor level for each column in the orthogonal array, ii) take the average of them. For example, the grey relational grade for factor A at level1 can be calculated as follow:

$$\gamma_{A1} = \frac{1}{6}(\gamma_1 + \gamma_2 + \gamma_3 + \gamma_{10} + \gamma_{11} + \gamma_{12}) = 0.6907$$

The grey relational grade values for each level of the machining parameters were calculated using the same method. The grey relational grade values are shown in Table 4. Since the grey relational grade represents the level of correlation between the reference sequence and the comparability sequence, the greater value of the grey relational grade means that the comparability sequence has a stronger correlation to the reference sequence. In other words, regardless of category of the performance characteristics, a greater grey relational grade value corresponds to better performance. Therefore, the optimal level of the machining parameters is the level with the greatest grey relational grade value. An asterisk (*) indicates that the level value results in a better performance. Based on the grey relational grade values given in Table 3, the optimal machining performances for both the roundness and the surface roughness were obtained for load inertia rate (level 2), acceleration time (level 1), pre-control time constants (level 3), pre-control coefficient (level 3), cutting speed(level 1), feed rate(level 1) and cutting depth. (level 2) combination.

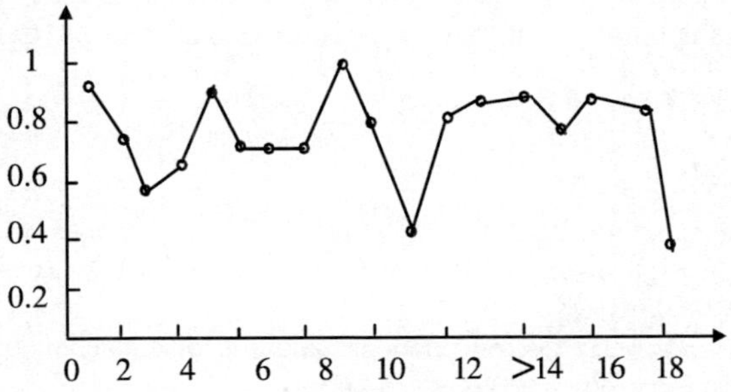

Fig. 1. Grey relation grades for the minimum Ra and h

As listed in Table 3, the difference between the maximum and the minimum value of the grey relational grade of the machining parameters is as follow: 0.0545 for load inertia rate, 0.0848 for acceleration time and so on. The most effective factor affecting performance characteristics is determined by comparing these values. This comparison will give the level of significance of the controllable factors over the multi-performance characteristics. The most effective controllable factor was the maximum of these values. Here, the maximum value is 0.2672, it indicates that the feed rate has the strongest effect on the multi-performance characteristics among the other machining parameters. On the other hand, the significance of role that every controllable factor plays over the multi-performance characteristics can be obtained by examining these values. The order of importance of the controllable factors to the multi-performance characteristics in the machining process, in sequence can be listed as: factor F>D>E>B>A>G>C. Factor F,D,E and B were clearer, but Factor F(feed rate) was the most effective factor to the performance. This indicates that the machining performance is strongly affected by the feed rate.

5 Conclusion

In this paper, the optimal parameters were determined for the roundness and surface roughness in the machined round parts by using the grey relational analysis.

The grey relational analysis, based on the Taguchi method's response table, was proposed as a way of studying the optimization of machined process factors. The roundness and surface roughness were selected to be the quality targets. From the response table of the average grey relational grade, the largest value of grey relational grade for the machining parameters was found. These values are the recommended levels of controllable parameters for the multi-performance characteristics. It was found that the feed rate has the strongest effect among the other parameters used on the multi-performance characteristics. In other words, the most influential factor is feed rate. The order of importance of the controllable factors to the multi-performance characteristics is feed rate, pre-control coefficient, cutting speed, acceleration time, load inertia rate, cutting depth and pre-control time constants. Experimental results have shown clearly that the roundness and surface roughness in the machined round parts can be improved effectively through the proposed approach. This study indicated that grey relational analysis approach can be applied successfully to other operations in which performance is determined by many parameters at multiple quality requests.

References

1. Chua, M.S., Rahman, M., Wong, Y.S., Loh, H.T.: Determination of optimal cutting conditions using design of experiments and optimization techniques. Int. J. Mach. Tools Manuf. 33(2), 297–305 (1993)
2. Li, P., Tan, T.C., Lee, J.Y.: Grey relational analysis of amine inhibition of mild steel corrosion in acids. Corrosion 53(3), 186–194 (1997)
3. Tosun, N., Ozler, L.: Optimization for hot turning operations with multiple performance characteristics. Int. J. Adv. Manuf. Technol. 2, 777–782 (2004)
4. Wang, C.C.L., Chen, S.F., Yuen, M.M.F.: Fuzzy part family formation based on grey relational analysis. Int. J. Adv. Manuf. Technol. 8, 128–132 (2001)
5. Deng, J.L.: Introduction to grey system theory. J. Grey System 1, 1–24 (1989)
6. Lin, Z.C., Ho, C.Y.: Analysis and application of grey relation and anova in chemical–mechanical polishing process parameters. Int. J. Adv. Manuf. Technol. 21, 10–14 (2003)
7. Fung, C.P.: Manufacturing process optimization for wear property of fibre-reinforced polybutylene terephthalate composites with grey relational analysis. Int. J. Adv. Manuf. Technol. 254, 298–306 (2003)
8. Lo, S.P.: The application of an ANFIS and grey system method in turning tool-failure detection. Int. J. Adv. Manuf. Technol. 19, 564–572 (2003)

The Online System for Measuring Sediment Concentration Based on Data Fusion Technology

Liu Mingtang[1], Yu Hui[2], Fu Libin[3], and Li Li[4]

[1] Department of Information Engineering, North China University of Water Conservancy and Electric Power, Beihuan Road 36, Zhengzhou City, Henan Province, 450011, China
[2] Henan Press and Publishing School, Zhengzhou City, Henan Province, China
[3] College of Civil Engineering, Huanghe S&T University, Zhengzhou, Henna 450006, China
[4] Yellow River Institute of Hydraulic Research, Zhengzhou, Henan 450003, China
liumt99@163.com

Abstract. In the Yellow River, the sediment concentration is one key parameter of the water information detection. An online system is designed for measuring sediment concentration based on data fusion technology. This paper particularly introduces principle of the capacitive differential pressure sensor, PLC hardware design and multi-channels data fusion technology. This system will focus on a variety of information gathering, such as sediment concentration, water temperature, water depth, and flow rate. The data fusion technology is used though multi-sensor to adapt to environmental changes and increase the sediment concentration detection accuracy and stability. A large number of calibration test have been done to find the proportion of bottles and traditional value, and this system will improve the measurement accuracy and stability. Through the " Xiaolangdi Reservoir Model " simulation, this system is more suitable to measure sediment concentration in the reservoir dredging, water quality management and slurry treatment etc.

Keywords: Online system, sediment concentration, data fusion capacitive differential pressure, PLC design.

1 Introduction

The regression analysis is one of the simplest and most intuitive data fusion methods. The basic idea of it is to create multi-dimensional regression equation with many parameters of the sensors. The classic sensor input and output relationship is one-dimensional regression equation to describe. Multi-dimensional use the least square error method from experimental data to calculate the coefficients of the regression equation .Thus, we can know the measured value from the multi-dimensional regression equation. Multi-sensor device can measure many parameters and can have a variety of fusion algorithm. The surface fitting algorithm is one kind of multi-dimensional regression analysis [1] [2].

In the Yellow River, the sediment concentration is one key parameter of the water information detection. Sediment concentration not only can provide scientific guidance of the governance in the Yellow River, but also can prevent and control soil erosion

J. Luo (Ed.): Affective Computing and Intelligent Interaction, AISC 137, pp. 449–455.
springerlink.com

caused by natural disasters [3] [4]. In recent years, the modern methods of measuring sediment have been actively explored both at home and abroad. However, there are some limitations of these methods in the practical application, such as expensive equipment, single function and instability [5]. As the flow rate, water temperature, depth and other factors have an impact on the accuracy of silt content, we must consider data fusion processing when measuring sediment concentration [6].

To resolve the problems, we have adopted a capacitive differential pressure method to measure sediment content in the Xiaolangdi reservoir area. The main system is composed of the water samples pretreatment, signal acquisition part, Programmable Logic Controller (PLC), processing of the data, configuration software and so on. The system is based on differential capacitive principle. The system can detect not only sediment, but also the depth of the measuring point, the water temperature and gradient to enhance detection. The system uses electromotor to drag sensors and the sediment content can be measured at different depths. The system comprehensive analyses kinds of factors, such as water temperature, depth, tilt, etc, and made up of the changes caused by circumstance. Through data fusion, the system will increase detection accuracy and stability of sediment concentration.

2 Principle of Capacitive Differential Pressure Sensor

In the absence of differential pressure, the initial capacitance of the moving plates on both sides is C_0.When there exist differential pressure, the moving plate distort to the dotted line, and capacitor between dotted line and the initial location is C_A. The capacitor between dotted line and the low pressure side is C_L, and high-pressure side of the capacitance is C_H [7]. Working principle of capacitive differential pressure sensor can be shown in Fig.1.

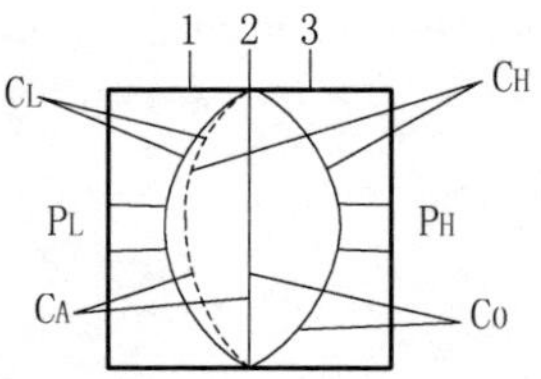

Fig. 1. Capacitance change

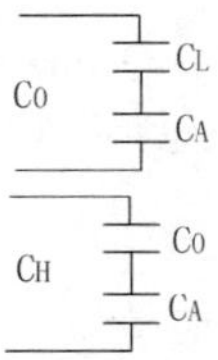

Fig. 2. The relationship of differential pressure

The four capacitors C_0, C_A, C_L and C_H have the equivalent relationship as shown in the Fig. 2. Therefore, the equation (1) can be written based on series capacitor-type.

$$\frac{1}{c_0}=\frac{1}{c_L}+\frac{1}{c_A},\frac{1}{c_H}=\frac{1}{c_0}+\frac{1}{c_A} \tag{1}$$

The following results can be deduced as equation (2):

$$\frac{c_A}{c_0} = K_L(P_H - P_L) = K_L \Delta P \tag{2}$$

K_L is a coefficient with differential coefficient of the capacitance. By equation (2) we can educe: C_0/C_A is directly proportional to the differential pressure, that is, changes in differential sediment pressure sensor will be converted into the changes in capacitance.

Capacitive differential pressure sensor requests the liquid to completely cover the two capacitors, and the minimum water level is higher than 100mm to differential pressure sensor. If the measured liquid is fluid, it must be less than or equal to 0.4m/s.Testing sediment pressure is one of the important ways of measuring. Liquid pressure can be expressed as the equation (3):

$$\Delta P = P_B - P_A = \rho g \Delta h \tag{3}$$

By the pascal liquid static theorem, the density is available between two points A, B as show in the equation (4).

$$\rho = \frac{\Delta P}{g\Delta h} = \frac{1}{g\Delta h}\frac{1}{k_l}\frac{C_A}{C_0} \tag{4}$$

Where h is a high level, ΔP for the pressure difference, ρ for the liquid density, but here for sediment concentration, g is the acceleration due to gravity. By equation (4) can be seen: when two differential pressure sensors are at a certain height, the problem of sediment testing converse differential pressure measurement problems [8].

3 Experiment Process

We carry out a large number of calibration test to find the laws and traditions. The experiment uses the bottles to calibrate in order to improve measurement accuracy and stability. We have joint with Research Institute of Yellow River Conservancy to do the test. The capacitive differential pressure sensor is installed on the test platform in the "Model Yellow River". All completely simulate the working conditions of the actual situation. The mixture is from 0kg/m3 to 500kg/m3 according to the actual requirements.

The information of sand content is converted to the current 4-20mA analog signal by the capacitive differential pressure sensor. This current 4-20mA analog signal is converted to digital signal by DVP06XA, using the K value in the PLC [9] [10].

Figure 3 is shown the model of K value and sediment content measured by the bottles actually. The equation (5) indicates a simple regressive equation and correlation coefficient of the K value of sediment concentration and sediment concentration in the bottles. From Figure 14 and the regression equation (6) can be seen, K value and the actual measured value by bottles has basic linear relationship.

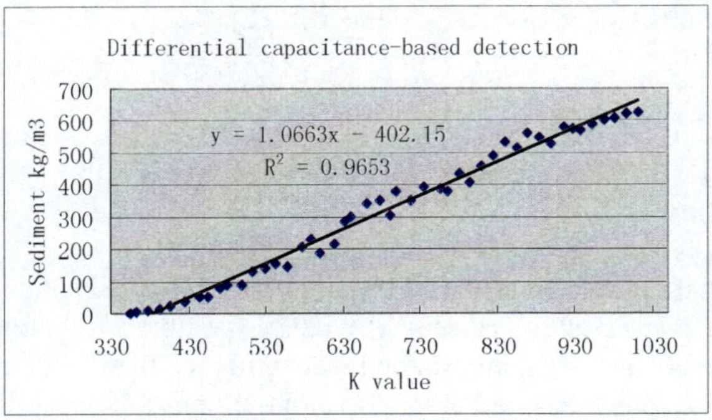

Fig. 3. The relationship between K and sediment more than 65kg/m^3

$$y = 1.0663x - 402.15 \tag{5}$$

$$R^2 = 0.9653 \tag{6}$$

We have carried out a test according to equation (5) for measuring sediment using capacitive differential pressure detection method. As shown in Figure 4, we can see that the average measurement error is in the ± 0.25 kg / m^3 and the maximum one-time measurement error is less than ± 8.5 kg / m^3. The reason why lead to large error at the beginning is that we had non-uniform mixed sand and water when sediment concentration at the low level.

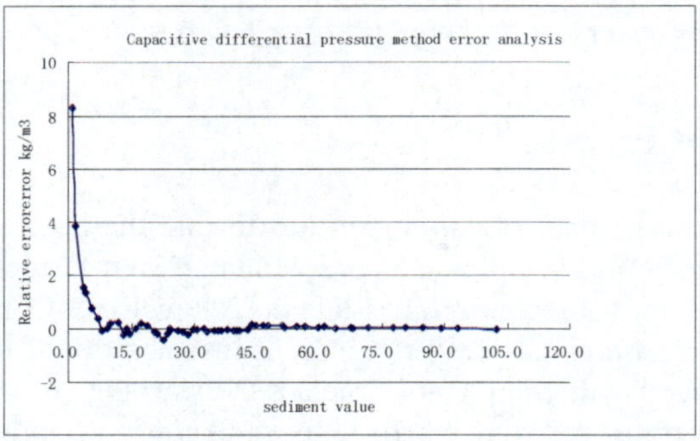

Fig. 4. Capacitive differential pressure error analysis

4 Data Fusion Technology

The capacitance sensor circuit output current as the transformation k, and there existence the temperature sensitivity. Therefore, only one-dimensional capacitive sensor will have greater error for the reason that the output S is not a unary function of the value of K. Now the output S can be solute by the information t of the water temperature, the S information of silt content and the H detected by the depth sensors, as show in equation (7).

$$S = g(K, H, T) \tag{7}$$

For S-plane determined by (H, K, T), we can make use of quadratic surface fitting equations, namely three-dimensional quadratic regression equations. Taking into account the impact of temperature is larger, the information on the temperature T and sediment content of the information a second fitting, fitting equation as shown in equation (8).

$$S = \alpha_0 + \alpha_1 H + \alpha_2 K + \alpha_3 T + \alpha_4 K^2 + \alpha_5 KT + \alpha_6 T^2 + \varepsilon \tag{8}$$

In the equation 14, α0 ~ α6 are coefficients; ε for the higher-order infinitesimal. If the constant coefficients in the equation 14 are known, the S of capacitance sensor output is set. When the T value, the K value and the H value, the sensor output S can be measured. Therefore, we should calibration three-dimensional firstly, and then determine the coefficients α0 ~ α6 by the least square method. Though a large of experiments, we find that the K value, the T value and the H value have a good linear relationship with the output S. To simplify the calculation, three ternary with one fitting is adopted, as shown in equation (9):

$$S = \alpha_0 + \alpha_1 H + \alpha_2 K + \alpha_3 T \tag{9}$$

5 Surface-Fitting Equation to Determine

In generally speaking, we use the least square method to determine constant coefficients in equation (9), and meet the conditions for the minimum mean square error. Coefficient α0 ~ α3 of the equation 14 are solute by the following steps:

Firstly, there is a error ΔK between S (K, T) values and calibration values of S_k. The variance of ΔK^2 will be expressed in equation(10).

$$\Delta_k^2 = [S_k - S(H,K,T)]^2 \quad k=1,2,\cdots,m*n \tag{10}$$

There are $m*n$ fixed-point, which should be the minimum mean square error R_1 in equation (11).

$$R_1 = \frac{1}{m*n}\sum_{k=1}^{m*n}[S_K - (\alpha_0 + \alpha_1 H + \alpha_2 K + \alpha_3 T)]^2 = R_1(\alpha_0, \alpha_1, \alpha_2, \alpha_3) \tag{11}$$

According to the demanding multi-function extreme condition, the following partial derivatives are zero in equation (12).

$$\frac{\partial R_1}{\partial \alpha_0} = 0;\ \frac{\partial R_1}{\partial \alpha_1} = 0;\ \frac{\partial R_1}{\partial \alpha_2} = 0;\ \frac{\partial R_1}{\partial \alpha_3} = 0 \tag{12}$$

Equation (11) is added in formula (12), α0 ~ α3 function will be available. Its coefficient can be test by the calibration points with the input value S_k, and finally determine the fitting equation.

6 Data Fusion

There are 10 points to calibrate Silt content test, and n=10. The number of the temperature reference points to take 10, as 10°C, 15°C, 23°C, 36.6°C, 39°C, and verify the fusion result with the value 12°C, 17°C, 24.5°C, 36°C, 37.7°C,so m=5. Depth of H is 10.5, 11, 11.5, 12, 12.3, 12.5, 12.8 (cm).

S（kg/m3）:12.58, 18.84, 19.45, 36.42, 57.78, 67.92, 93.11, 93.4, 95.68, 98.61
T (°C) : 10, 11, 15, 17, 23, 24.5, 36, 36.6, 37.7, 39
H(cm) : 10.5, 11, 11.5, 12, 12.3, 12.5, 12.8

Table 1. The fitting data

K	T	H	S
350	10	10. 5	12. 58
389	15	11	18. 84
450	23	12	57. 78
493	36. 6	11. 5	93. 11
511	35	12. 5	91. 8

According to the data in Table1, using matrix decomposition method calculates the value of α0 ~ α3, so fitting equation 13 will determined to be:

$$S = -54.1124 + 32.1185 * H + 0.9820 * K + 7.1191 * T \tag{13}$$

$$R = 0.9947 \tag{14}$$

From the equation (14) can be seen, S value and the actual measured value by bottles has more linear relationship than in equation (6).The system input silt content of K, the depth H and temperature T. The output of S is silt content though data fusion. What kind of output format will be accordance with the fitting formula may be chosen. Table 2 shows the results of the data fusion and error.

Table 2. The fitting data

K	H	T	S of calculation	S of measurement	error	Relative error
355	10. 5	11	12. 83195	15. 11	-2. 27805	0. 17752
405	11. 5	17	38. 56505	36. 42	2. 14505	0. 05562
480	12. 8	24. 5	60. 06235	60. 92	-0. 857565	0. 01427
494	11. 5	37. 7	98. 53242	96. 68	1. 85242	0. 01880
514	12. 3	36	92. 48475	93. 40	-0. 91525	0. 00989

We can see that the average relative error of measurement is ± 0.055kg/m3 from the figure 5, and a single absolute error is less than ± 2.28 kg/m3. From Table 2 and Figure 5 can be seen that when considered in conjunction with the silt content and temperature, the error of detection is relatively small, and the accuracy of the system is higher.

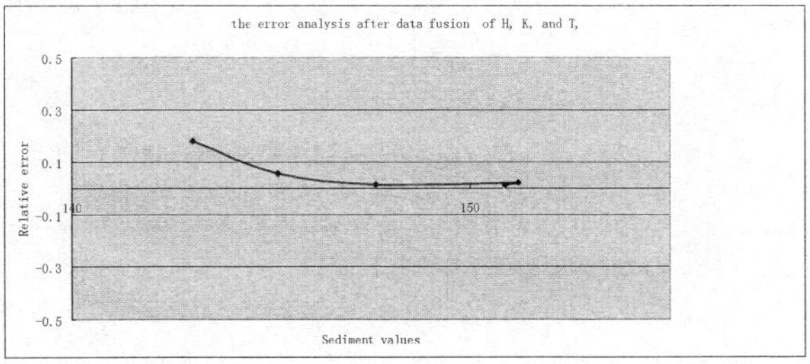

Fig. 5. The error analysis after data fusion of H, K, and T

7 Conclusion and Outlook

By theoretical analysis and experimental validation, this system will improve the measurement accuracy and stability with the data fusion technology. The equation (9) regard as a measure formula between the K value and sediment concentration. The error of measurement is in the range of 3%, to meet the requirements of engineering measurement. The range of the Sediment content measured in this system is $0kg/m^3$ to $500kg/m^3$. As compared to similar products, this system is more suitable for the detect sediment concentration in the Xiaolangdi reservoir area. This system is also suitable to measure sediment concentration in the reservoir dredging, water quality management and slurry treatment etc.

References

1. Zhang, M.: The information fusion research of capacitive sensors measuring the sediment content. Northwest A & F University, a master's degree thesis 6, 13–43 (2003)
2. Pan, Q., Yu, X.: Essential Methods and Progress of Information Fusion Theory. Acta Automatica Sinica 4, 599–615 (2003)
3. Hong, S., Zhang, T., Li, L.: Model Yellow River needs analysis. Yellow River 3, 14–15 (2004)
4. Zhu, Q.: Model Yellow River project Summary. Yellow River 3, 1–3 (2004); Progress in Geophysics 9, 420–425 (2003)
5. Zhang, P., Zheng, C.: Differential Pressure seabed mud density measurement. Shanghai Water Conservancy 2, 17–22 (1997)
6. Chen, S., Li, H., Tian, Y.: List instrument and its testing sand. Water Resources and Hydropower Express (20), 6–18 (2001)
7. Wang, Y., Wang, X., Wang, Y.: The intelligent measurement system in reservoir based on capacitive sensor. Micro-computer Information 7(1), 150–178 (2007)
8. Jiao, R., Zhuang, Q., Wang, Y.: Tank Liquid Level Measurement Intelligent System. Sensors and Microcomputer 3, 53–55 (2007)
9. Delta DVP Series Programmable Logic Controller Manual. Delta Power Company Ltd. (2003)
10. Jing, W., Wang, P., Wu, W.: Computer Monitoring System for test bed Based on PLC and Kingview Software. Micro-computer Information 6(1), 55–56 (2008)
11. Liu, M.-T.: Male, Huangchuan County, HeNan Province, lecturer of the department of information of engineer of North China University of Water Conservancy and Electric Power, Research area focus on intelligent detecting and control technology (1974)

Parallel Implementation of Motion Parameters Estimation Algorithm

Feng Han, Guohua Wei, and Siliang Wu

School of Information and Electronics,
Beijing Institute of Technology, Beijing 100081, China
hanfengtea@bit.edu.cn

Abstract. The motion parameters estimation of the moving target is a key problem in radar signal process. Recently, in order to increase the accuracy of estimation parameters, the technology of modern signal process is used in radar signal process with the cost of increase of computational complexity. So the technology of distributed technology based on the multiprocessor computers is introduced into the field of radar signal process, which improves the compute performance greatly. The design details of the parallel algorithm of motion parameters estimation are addressed from the design, software structure of and adaption. And it was realized with the help of MATLAB distributed computing toolbox.

Keywords: Distributed computing, motion parameter estimation, signal process.

1 Introduction

Parameter estimation of the motion trajectory is a classical problem of radar signal process, which is used to describe the relative position of moving target and radar. The accuracy of parameters is constrained by the algorithm of the signal process. To obtain precise measurements, more sophisticated algorithms and data processing flow are applied, which involve the theory of mathematical statistics, time series analysis, time-frequency analysis, wavelet transform etc. All above lead large amount of calculation. Parallel method was considered in signal process to satisfy the demand of processing time limitation, which is concrete modeled in the math problem of non-linear spectral estimation, multi-dimensional nonlinear optimization. Through the excavation and development of inherent parallelism, the parallel algorithm is designed on multiple servers in high-speed computing combined with the help of distributed parallel computing toolbox (DCT) [1], which is a commercially available solution for parallel compute by Math. Works, Inc. The DCT may let designers distribute and control a set of computational tasks across several computational workers running under the Distributed Computing Engine (DCE). The toolbox also contains additional job-control functions and supports high level constructs such as parallel for loops and distributed arrays.

With the application of parallel processing, it is a difficult task of efficient mappings of algorithm to multiprocessor computers in engineering realization of

J. Luo (Ed.): Affective Computing and Intelligent Interaction, AISC 137, pp. 457–463.
springerlink.com

high-performance computing. It is a detailed problem of determine the particle size, which is defined as an assignment of blocks of data to processing elements. The DCT based on multi-processor structure provides a simple parallel computing interface engine, which increases the ease of programming and productivity and allows the designers only focus on the implementation of parallel algorithms. Inefficient duplication of labor of program is avoided and more energy and more time are centralized on the physical nature of the algorithm, such as decomposition of tasks and integrated, parameter settings, rational division of the particle size etc. to improve efficiency.

2 Theoretical Background

In radar array signal processing, we can estimate a complete set of target motion parameters using a small antenna array with the information of the Doppler frequency and phase differences among the sensors[2]. Fig. 1 shows the general geometrical configuration for a moving target. Consider the target flies at a straight line path with a constant velocity v. Assign an initial time $t=0$ when the distance between the source position and the point of the closest point of approach (CPA) is L_0. Coordinates of CPA is $\{x_0, y_0, z_0\}$ and scalar miss distance is r_0,

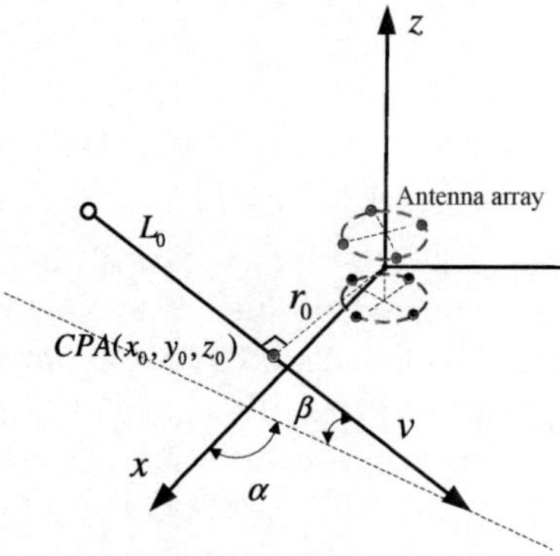

Fig. 1. Sensor and target geometry

$$r_0 = \sqrt{x_0^2 + y_0^2 + z_0^2} .$$

The variables $\{r_0, v, L_0\}$ of scalar score may be modeled as a mathematical problem of nonlinear optimization, formulated as follows:

$$\hat{r}_0, \hat{v}, \hat{L}_0 = \arg \min_{r_0, v, L_0} \sum_{n=1}^{N_0} \left[\hat{f}_d(t_n) - \frac{L_0 - vt_n}{\sqrt{r_0^2 + (L_0 - vt_n)^2}} \right]^2 \tag{1}$$

t_n : Discrete observed time;

The variables $\{(x_0, y_0, z_0), \alpha, \beta\}$ of vector miss distance may be modeled as a mathematical problem of constrained nonlinear optimization, formulated as a five-dimensional minimization as follows:

$$\min J_0(\alpha, \beta, x_0, y_0, z_0)$$
$$s.t.\begin{cases} x_0^2 + y_0^2 + z_0^2 = r_0^2 \\ x_0 \cos\alpha \cos\beta + y_0 \sin\alpha \cos\beta - z_0 \sin\beta = 0 \end{cases} \tag{2}$$

Where,

$$J_0(\alpha, \beta, x_0, y_0, z_0) = \sum_{j=1}^{M-1} \sum_{i=1}^{N} \left| e^{j\hat{\phi}_j(t_i)} - e^{j\varphi_j(t_i)} \right|^2 \tag{3}$$

$\hat{\phi}_j$: The observed noisy wrapped phase differences;

φ_j : The theory phase differences computed from the parameters.

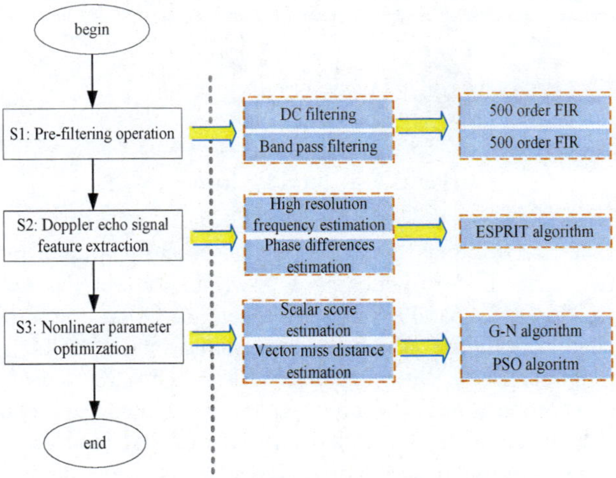

Fig. 2. Flowchart of signal process

The flow of parameters estimation algorithm may be divided into three stages, which is show in figure 2. Firstly, operation of improving the signal to noise ratio (SNR) is done, including narrow-band filter, dc filtering etc. Secondly, using some modem spectral analysis method, such as ESPRIT [3] or MUSIC algorithm to estimate Doppler frequencies and the phase differences among the receiving antennas. Then, the estimated Doppler frequencies and the phase differences among the antennas can be curve fitted to the ideal Doppler frequencies and the ideal phase differences respectively to get the target motion parameters using the equation (1) (2). The random Gauss-Newton method and particle swarm optimization algorithm (PSO) are considered to get the parameter of the nonlinear optimization and give flight position, velocity and other parameters.

3 Realization

Fig. 3 describes the layers of the signal process system, which is an open system platform, including open system software, support software, user interface and so on. The first layer of user interface use the multi-media technology to complete database operations, process prompts and graphic display of the target trajectory.

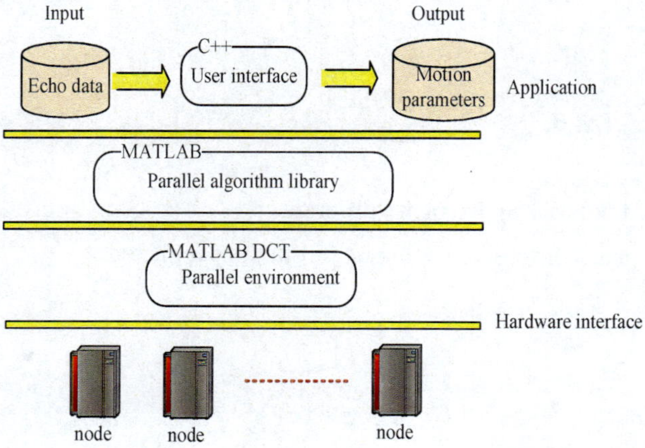

Fig. 3. Data process platform

The second layer is the library of parallel signal processing algorithm based on the computing software MATLAB, which is a powerful tool in the field of scientific research. The methods in parallelizing a sequential program consist of function decompose and data decompose, which are divided into smaller tasks.

Usually the radar echo data is huge such as SAR data and Phased Array Radar data. The echo data decomposition is often used by the distance, frame, or snapshot. Assigning the data in equal length is taken to balance the load on each computing node. And there is no communication among the nodes at work which is a peer-to-peer model. It is not necessary between node number and efficiency. During the task division, it should pay attention to the balance of task time and communication time between job manager and works. As far as possible to increase the ratio of the task implements time to communication time. Thus the efficiency of the software can be improved [4].

The third layer is parallel environment base on the application of the DCT toolbox, which map the processes to computational nodes. Parallel tasks are managed through manager, whose mission is mainly responsible for the decomposition and synthesis; during the process of the division of mission, the size of the particle size need to be consideration carefully. If necessary, experiment should be done to determine the particle size. In order to reduce the complexity of parallel computing, parallel computing structure is used mainly by the data parallel.

The fourth layer is the hardware including the computer nodes. The computers are connected by high speed data bus.

3.1 Pre-filtering Operation

The Doppler echo signals are collected by the score radar system and stored in the computer as data files. It contains 8 antennas echo signal and each antenna echo signal has two orthogonal I and Q channels. Firstly, the signal is divided by frame (512 points × 8 antennas). Through period gram spectrum estimation and 8 channels no coherent integration, the Doppler frequency of each frame is detected and the ineffective data frames are abandoned. Narrow-band digital filtering is designed by frame (512 points × 8 antennas) according to the detected Doppler frequency, because the band pass is difference frame by frame and the order is 500. After filtering, SNR is improved, which will make better parameter estimation performance of precise frequency and phase difference. Designate the number of tasks to be scheduled in the simulation with the method of DC removal. The divisible task may be a session of the block data collected from one antenna or a session of segment data collected from all the antennas. Be sure of the mean execution time for each task is equation and the stand deviation of execution time is least.

3.2 Doppler Echo Signal Feature Extraction

The motion of the target leads to the change of the Doppler frequency. The power spectrum method only got coarse spectrum estimation. To obtain more accurate Instantaneous frequency, the length of frame should become shorter. High-resolution frequency estimation method of TLS-ESPRIT is used. Dividing the signal by the frame (64 points × 8 antennas）, there is no relationship among frames. So the data can be used in parallel processing algorithms for echo signal to estimate the frequency and phase difference.

3.3 Nonlinear Parameter Optimization

The scalar parameters estimation uses the random Gauss-Newton optimization. Initial seeds are generated randomly. Each seed correspond to one thread and each thread corresponds to once compute of G-N algorithm.

Objective function of the vector miss distance is the form of exponential function with a multi-peak characteristic. The traditional G-N algorithm is hard to get the global optimal value with multi-dimensional parameters and constants. PSO algorithm is a swarm intelligence based on an effective global optimization algorithm. The concept is simple and easy to achieve on the project. And it is more conducive to the realization of parallel algorithms. Relative work on PSO may refer to 5.

The optimization infrastructure is organized into a coordinating node and several computational nodes. PSO algorithm functions and task orchestration are performed by the coordinating node, which assigns the seeds to the computational nodes. With this approach, no communication is required among computational nodes and individual fitness evaluations are independent of each other. The only necessary communication is between the coordinating node and the computational nodes. Fitness values reported from slave nodes to coordinating node.

4 Performance

The parallel experiment environment on which we verify the parallel algorithm is given in table 1.

Table 1. The characteristics of the system used for the parallel experiments

Server type	IBM X3950
Compute node	8
Network	Fast Ethernet 1-Gbps
Memory	4GB per Server
Operating System	Windows Server2003
Soft	MATLAB 2007b

Parallel performance is quantized by calculating speed up. Speed up is the ratio of sequential execution time to parallel execution time and ideally should equal the number of processors. Fig. 4 shows the time comparison of each step for serial program and parallel program. A high speed up is obtained by the implementation of the parallel environment.

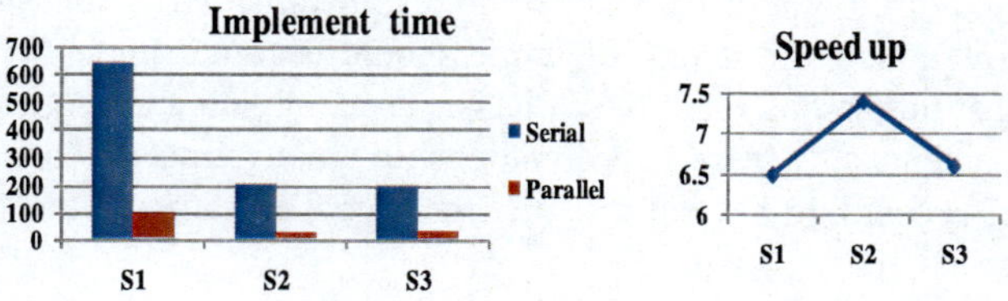

Fig. 4. Implement time and speed up for different steps

5 Conclusions

The echo data collected by radar system is large. So it is an inevitable trend to imply the parallel technology, which will have a strong impact on the radar signal process. Parallel environment discussed here provide a solution to parallel algorithm of radar signal process.

References

1. Krishnamurthy, A., Nehrbass, J., Chaves, J.C., Samsi, S.: Survey of Parallel MATLAB Techniques and Applications to Signal and Image Processing. In: IEEE International Conference on Acoustics, Speech and Signal Processing, vol. 4, pp. 1181–1184 (2007), doi:10.1109/ICASSP.2007.367286
2. Chan, Y.T., Towers, J.J.: Sequential localization of a radiating source by Doppler-shifted frequency measurements. IEEE Transactions on Aerospace and Electronic Systems 28, 1084–1090 (1992), doi:10.1109/7.165370

3. Roy, R., Kailath, T.: ESPRIT – estimation of signal parameters via rotational invariance technique. IEEE Transiton on Acoust. Speech Signal Process. 37, 984 (1989), doi:10.1109/29.32276
4. Zhang, B.L., Gu, T.X., Mo, Z.Y.: Principles and Methods of Numerical Parallel Computation. National Defence Industry Press, Beijing (1999)
5. Schutte, J.F., Reinbolt, J.A., Fregly, B.J., Haftka, R.T., George, A.D.: Parallel Global Optimization with the Particle Swarm Algorithm. Int. J. Num. Methods Eng. 61(13), 2296–2315 (2004)

Formal Modelling of Immune Complex Network Based on B Method

Yujing Peng[1], Hao Sun[2], Chunqiu Liu[3], and Shengrong Zou[3]

[1] Department of Computer Science and Technology Yangzhou Vocational, College of Environment and Resources, YangZhou, China
124807475@qq.com
[2] Comprehensive Management Office of Jiang Su Oil Field, YangZhou, China
[3] Yangzhou UNiversity, Yangzhou, China
124807475@qq.com, pyj_2007@qq.com

Abstract. We attempt to use formal B methods to study complex network. Experts of complex network only can study the static feature of the system as a whole. We use B methods to describe bipartite graph of immune network and simulate the function of the immune system. We present a formal specification and try to find some rules.

Keywords: B method, formal method, complex network, immune system, bipartite graph.

1 Introduction

The last few years have witnessed a tremendous activity devoted to the characterization and understanding of networked systems. Indeed large complex networks arise in a vast number of natural and artificial systems. Ecosystems consist of species whose interdependency can be mapped into intricate food webs. Social systems may be represented by graphs describing various interactions among individuals. The Internet and the World-Wide-Web (WWW) are prototypical examples of self-organized networks emerging in the technological world. Large infrastructures such as power grids and the air transportation network are critical networked systems of our modern society [1]. Finally, the living cell is not an exception either, its organization and function being the outcome of a complex web of interactions among genes, proteins and other molecules.

For a long time all these systems have been considered as haphazard set of points and connections, mathematically framed in the random graph paradigm. Researchers thus have started to have a systematic look at these large data sets, searching for hidden regularities and patterns that can be considered as manifestations of underlying laws governing the dynamics and evolution of these complex systems. The attempt to model and understand the origin of the observed topological properties of real networks has led to a radical change of perspective, shifting the focus from static graphs, aiming to reproduce the structure of the network in a certain moment, to modeling network evolution. we currently view networks as dynamical systems that evolve through the subsequent addition and deletion of vertices and edges. The set of dynamical rules

J. Luo (Ed.): Affective Computing and Intelligent Interaction, AISC 137, pp. 465–473.
springerlink.com

defining these processes thus outlines the dynamical theory required in order to understand the macroscopic properties of networks.

However, experts of complex network are usually study static feature of the overall network, it is difficult to use a technique to describe the function of the network. In addition to the complexity of these systems, it is difficult to use traditional software engineering methodologies to model these complex systems. Therefore, we try to use formal B method to study complex network.

Formal specifications can be used to provide an unambiguous and precise supplement to natural language descriptions and can be rigorously validated and verified. This method can detect early specification errors.

B is a formal software development method that covers software process from specification to implementation. The B abstract machine notation (B/AMN) is based on set theory, the generalized substitutions language of and first order logic.

B benefits from commercial tool support for proof, Atelier-B [2] and the B-Toolkit [3].

2 B Method

The B/AMN [4] is a state model-based, formal specification notation, designed to support formally verified development by refinement from specification through to implementation. Formal verification of proof obligations ensures that a specification is consistent throughout its refinements. B, like its predecessor, Z [5], is based on set theory and first-order predicate logic. For refinement, B requires a 'refinement relation' as part of its invariant predicate, which is analogous to an 'abstraction relation' schema in Z. An invariant in B is a property of the state that operations are expected to maintain. B provides facilities for generation of executable code, it can also be used as an abstract specification language similar to Z. However, proof is difficult for inexperienced practitioners. A more accessible, automated form of verification is model checking[6]. A B model checker, ProB, has been developed at the University of Southampton [7]. It is equally important to ensure that the specification is a useful one. ProB includes an animator, which allows us to observe and validate the simulated behaviour of a specification.

A B module consists of a number of B components from the most abstract specification, through possibly many refinements. B provides a structuring mechanism to decompose a component so that parts of the state can be encapsulated and segregated, making them easier to comprehend, reason about and manipulate.

3 Case Study

Many biological systems can be expressed as networks. The immune cytokine regulating network is a typical example of biological networks. Immune system is the most important defense system to resist human pathogens. This system consists of immune organs, immune cells and immune molecules. Each cell plays an important role in immune system. The function of many cells in immune system mediated by a group of protein called cytokines. Cytokines are rapidly produced by immune cells in response to tissue injury, infection, or inflammation. The overproduction of cytokines mediates tissue damage and physiological and molecular mechanisms have evolved to control their production and to prevent the injury during the host response. In addition to regulate cellular interactions, cytokines are the molecular players that signal the brain to respond to the danger of viruses, bacteria, fungi and parasites through an elaborated coordination[8-9].

In our previous work, we have already described the network relation of the cell and cytokine by bipartite graph. Now, we use a portion of cells and cytokines as an example to build a direction bipartite graph. As is shown in Figure 3 [10].Bipartite graph have two types of vectors. The edges connect the two types vectors.

Figure 3 have two types nodes, the top nodes are the cells, and the bottom node are cytokines, the directed edges from cells to cytokines indicate the cytokines secreted by the cell, the directed edges from cytokines to cells indicate the cells affected by the cytokines.

According to the above bipartite graph, we try to use B method to describe the bipartite graph.

The cells in the bipartite graph was packaged in a machine of B method. In this type of machine, we need to define three sets.

The first set is CELL.It is an extension set. It indicates the present or the future cells.

The second set is CATEGORY. It indicates the types of the cells.

The third set is STATUS. It indicates the state of the cells.

So the machine cell are as follows:

```
    MACHINE
  Cell
  SETS
     CELL;
     STATUS={secret}
  CATEGORY={B-CELL,   T-CELL  ,   Mast-CELL,   Basophil,
Eosinphil,Neutrophil,Macrophag}
  VARIABLES
  Cell,status,category
  INVARIANT
  cell⊆CELL∧
    category∈CELL→CATEGORY∧
  status∈CELL→STATUS
  INITIALIZATION
     Cell,status,category:=Φ,Φ,Φ
```

These are the initialization of variables . Now we try to describe the operation of the Cell machine. If we need to increase a cell in the bipartite graph, then we can create a new cell(create_cell) by B methods .The edges from the cells to cytokines can expressed by secretion operation(secret_cytokine).The operation description by B methods are as follows:

OPERATION

```
   i  ←  Creat_cell (c) ≙
  PRE
   CELL-cell≠Φ∧
   c∈CATEGORY
 THEN
 ANY j WHERE
 j∈CELL-cell
 THEN
   i:=j||
  cell:=cell∪{j}||
 category(j):=c
 END
 END;
 i←get_cell≙
 PRE
 cell≠ΦTEHN i: ∈cell
 END;
 Secret_cytokine(c) ≙
 VAR c IN
   c← Get_cell( )
 WHILE((status(c)=secret) ∧(c≠Φ)) DO
 SELECT T-Cell  THEN
 Produce_IL-2;
 Produce_IL-4;
 Produce_IL-6;
 Produce_IL-8;
 Produce_IL-9;
 Produce_IL-16;
 Produce_IL-17;
 Produce_IL-1;
 Produce_IL-10;
 Produce_IL-13;
 Produce_IL-14;
 Produce_IL-3;
 Produce_IL-5;
 Produce_IFN-α;
 Produce_IFN-β;
 Produce_IFN-γ;
 Produce_TNF-γ;
 Produce_TFN-α;
 Produce_TNF-β
 SELECT B-Cell THEN
 Produce_IL-1;
 Produce_IL-6;
 Produce_IL-10;
 Produce_IL-12;
 Produce_TFN-α;
```

```
Produce_TNF-β;
Produce_IFN-α;
Produce_IFN-β
SELECT Macrophage THEN
Produce_IFN-α;
Produce_IFN-β;
Produce_TFN-α;
Produce_TNF-β;
Produce_IL-1;
Produce_IL-6;
Produce_IL-10;
Produce_IL-12;
Produce_IL-15;
Produce_IL-18;
Produce_IL-8
SELECT Mast-Cell THEN
Produce_IL-4;
Produce_IL-5
SELECT Neutrophil THEN
Produce_TFN-α
SELECT Eosinophil THEN
Produce_IL-5
END
END;
END;
```

For cytokines, we use the same method to describe them. We need to create a cytokine machine. In this machine, we define two sets. One is an extension set CYTOKINE, the other is a type set CATEGORY. Because the cytokine machine need to use the variable of cell, we must include cell machine. The description of cytokine machine are as follows:

```
      MACHINE
Cytokine
USES Cell
SETS
   CYTOKINE;
CATEGORY={IL-1,IL-2,IL-3, IL-4, IL-5, IL-6, IL-8, IL-9,
IL-10, IL-12, IL-13, IL-14, IL-15, IL-16, IL-17, IL-18, TFN-α,
TNF-β, TNF-γ,IFN-α,IFN-β, IFN-γ}
                 VARIABLES
                        cytokine,actobject,category
                                 INVARIANT

                              cytokineCYTOKINE  ⊆
                 category∈CYTOKINE→CATEGOR∧
     actobject∈(cytokine,cell)
INITIALIZATION
   Cell,actobject,category:=Φ,Φ,Φ
```

In this machine, there are three variables. The variable actobject indicates the cells affected by the cytokine. In this way, we can describe the nodes with machine.

The edges from cytokines to cells indicates the cells affected by the cytokines. The same cytokine may affect many cells and the same cell may be affected by many cytokines. Now we use B method to describe the relation. First we need to create a new cytokine (create_cytokine), and then we need to get the name of the cytokine. At last we use operation to describe the relation of the cytokines and the cells.

```
   OPERATION
  k  ←  Creat_cytokine (c) ≙
PRE
  CYTOKINE-cytokine≠Φ∧
  c∈CATEGORY
THEN
ANY j WHERE
j∈CYTOKINE-cytokine
THEN
  k:=j||
 cytokine:=cytokine∪{j}||
category(j):=c
END
END;
i←get_cytokine()≙
PRE
cytokine≠ΦTEHN k: ∈cell
END;
Function_cytokine(k) ≙
VAR K IN
k  ←  Get_cytokine()
WHILE (k∈(CYTOKINE→CATEGORY)) DO
  SELECT IL-1 THEN
   Act_B-CELL;
   Act_T-CELL;
   Act_Macrophage;
   Act_Neutrophil;
   Act_Basophil
SELECT IL-2 THEN
   Act_B-CELL;
   Act_T-CELL
SELECT IL-3 THEN
   Act_Macrophage;
   Act_Neutrophil;
   Act_Basophil;
   Act_Mast-cell;
    Act_Eosinophil
SELECT IL-4 THEN
    Act_B-CELL;
    Act_T-CELL;
```

```
    Act_Macrophage;
    Act_Neutrophil;
    Act_Basophil;
    Act_Eosinophil
SELECT IL-5 THEN
   Act_Eosinophil
SELECT IL-6 THEN
   Act_B-CELL;
    Act_T-CELL
SELECT IL-8 THEN
    Act_T-CELL;
    Act_Neutrophil;
    Act_Basophil
SELECT IL-9 THEN
      Act_T-CELL;
     Act_Mast-cell
SELECT IL-10 THEN
    Act_B-CELL;
    Act_T-CELL;
     Act_Macrophage;
    Act_Mast-cell
SELECT IL-12 THEN
     Act_B-CELL;
    Act_T-CELL
SELECT IL-13 THEN
    Act_B-CELL
SELECT IL-14 THEN
Act_B-CELL
SELECT IL-15 THEN
Act_B-CELL
SELECT IL-16 THEN
Act_B-CELL;
    Act_T-CELL
SELECT IL-17 THEN
    Act_T-CELL
SELECT IL-18 THEN
    Act_T-CELL
SELECT TNF-αTHEN
Act_B-CELL;
    Act_T-CELL;
  Act_Macrophage;
Act_Neutrophil
       SELECT TNF-βTHEN
Act_B-CELL;
     Act_T-CELL;
  Act_Macrophage;
Act_Neutrophil
SELECT TNF-γ THEN
Act_Neutrophil
SELECT IFN-αTHEN
```

```
        Act_T-CELL;
         Act_Macrophage
   SELECT IFN-β THEN
        Act_T-CELL;
         Act_Macrophage
   SELECT IFN-γ THEN
    Act_B-CELL;
        Act_T-CELL;
         Act_Macrophage
   END
   END;
   END;
```

Here we have described all the nodes and the edges using B method.

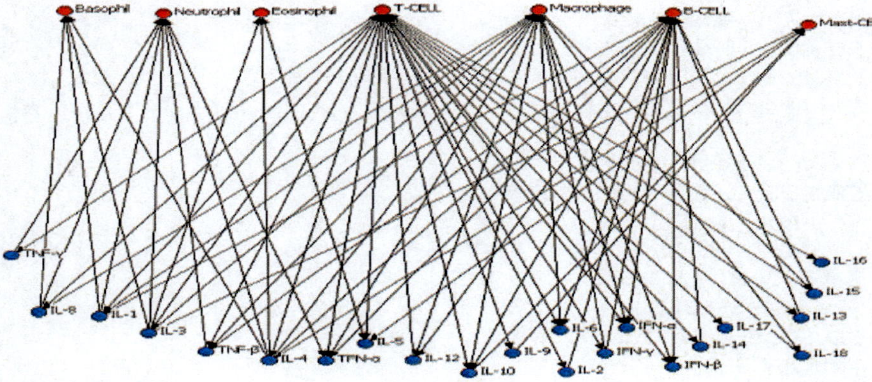

Fig. 3. The immune system network

4 Conclusions and Further Work

In this paper, we present an idea of connecting complex network with B method. The next step is to complete the verification and refinement of the machine, and generates executable code. We try to use executable code to display the dynamic process of the immune system, and explore more ways to solve complex network problems.

Acknowledgement. The research is supported by National Natural Science Foundation of China under the grant No. 70671089 and 10635040.

References

1. Barabási, A.-L.: Virtual round table on ten leading questions for network. The European Physical Journal
2. ClearSy. Atelier-B User Manual V3.6 (2003)
3. B-Toolkit User's Manual, Release 3.2. B-Core(UK) Ltd., Oxford, UK (1996)

4. Abrial, J.R.: The B Book—Assigning Programs to Meanings. Cambridge University Press, NY (1996)
5. Spivey, J.M.: Understanding Z a specification language and its formal semantics. CambridgeUniversity Press, NY ClearSy System Engineering, Aix-en-Provence, F (1988)
6. Clarke, E.M., Grumberg, O., Peled, D.: Model Checking. MIT Press, Cambridge, MA (1999); ClearSy. AtelierB Training Course Level 2. ClearSy System Engineering, Aix-en-Provence, F. (2000)
7. Leuschel, M., Butler, M.: ProB: A Model-Checker for B. In: Araki, K., Gnesi, S., Mandrioli, D. (eds.) FME 2003. LNCS, vol. 2805, pp. 855–874. Springer, Heidelberg (2003)
8. Abbas, A.K., Lichtman, A.H., Pober, J.S.: Cellular and Molecular Immunology, W. W. B. Saunders Company, Philadelphia (2003)
9. Janeway, C.A., Travers, P., Walport, M., Shlomchik, M.: Immunobiology. Garland Publishing, New York (2001)
10. http://www.copewithcytokines.de/

Experimental Study on Web2.0 Blending Learning

Rui Cheng

JiangHan University, Wuhan, China
carycheng78@gmail.com

Abstract. The objective of this study is to bring forward an applicable instructional model of blending learning based on web2.0 technologies and validate it by experiments. General guidelines for applying the model to practice are also explored.

Keywords: Web2.0, Blending learning, Learning model.

1 Introduction

Blending learning generally represents a learning mode integrating diverse technologies and approaches for instructional purposes. In the web2.0 era, many web2.0 tools can be applied to practical activities based on this mode so as to perform online learning.

2 Web2.0 Blending Learning Model

An experimental study on web2.0 blending learning model has been carried out in our university, and several web2.0 tools has been applied to our online learning procedures for different purposes simultaneously: (1) a blog website was used for storing and transferring text and image learning contents, (2) a video website provided us with bigger space to store video, audio and animation learning contents, (3) a web bookmark tool made it easy for instructor or students to register and retrieve useful web resources,(4) a miniblog was used for publishing notice, and (5) an rss tool was recommended to the students for assembling contents dispersed in different websites.

Furthermore, some instant message tools such as MSN and QQ and E-mails were used for feedback or interactivities between the instructor and students as well.

2.1 Web2.0 Blending Learning Procedure

As is shown in Fig. 1, the learning procedure comprises 3 stages, 2 branches and several steps.

The first is instructional design stage, in which the instructor prepares for the work to be carried out in both class room settings and online environments. In this stage, the instructor need to: (1) analyze learning needs, (2) analyze the learners, (3) analyze learning contents, (4) set down learning objectives, (5) develop learning strategies and approaches and (6) select evaluation instruments.

J. Luo (Ed.): Affective Computing and Intelligent Interaction, AISC 137, pp. 475–478.
springerlink.com

The second is performance stage. According to the differences of instructional delivery contexts, it is divided into two branches – face to face instruction and web2.0 learning. Generally, the former is used for lectures or classroom discussions and the latter is applied to online self-driven learning. Sometimes, the web2.0 learning can be used for task-driven learning or collaborative learning.

For the instructor, he or she may start instruction from either branch according to their preference or schedule, and then turn to the other one. The 2 instructional manners are performed alternately to achieve a mutual goal. That is the soul of blending learning.

The last stage is evaluation, consisting of formative evaluation and summative evaluation.

2.2 Advantages of the Model

Compared with other blending learning models, this one probably be the cheapest owing to most web2.0 tools are free of charge. It will attract many clients, especially those who are short of funds.

Most web2.0 tools are designed user-friendly, neither the instructor nor students need preliminary training. They can develop learning contents, lead online instruction or take part in online activities and the like with various web2.0 tools immediately.

Web2.0 tools are cheap but not primitive or crude. Most of them provide us with diverse powerful functions, which can be applied to online learning. Many of them are equal or even better than modules comprised in specialized platforms in function.

In virtue of the abundance and diversity of web2.0 tools, we may find different tools providing us with similar functions. It means that we have several choices to fulfill the same work. The web2.0 blending learning model is independent from some given tool or platform.

3 Instructional Experiment

3.1 Performance of the Experiment

We have chosen the course of photography as our experimental object, which needs lots of practices and interactivities. The instructional procedures were carried out just as has been detailed in the second part of the paper. Meanwhile, the instructor has assigned shooting tasks to students, asked them to carry out online learning and case discussion. A classroom lecture was arranged in the end to finish each unit.

The instructor may ask students to turn in their photo works or commentary papers on some cases; several ways are competent for those tasks: (1) students send their works to the instructor via Emails or instant messages, (2) students publish photos or papers on their own blog pages, and then the instructor mark them online, (3) students upload their works to a given online space, for example: an online hard disk, (4) students turn in their works in classroom.

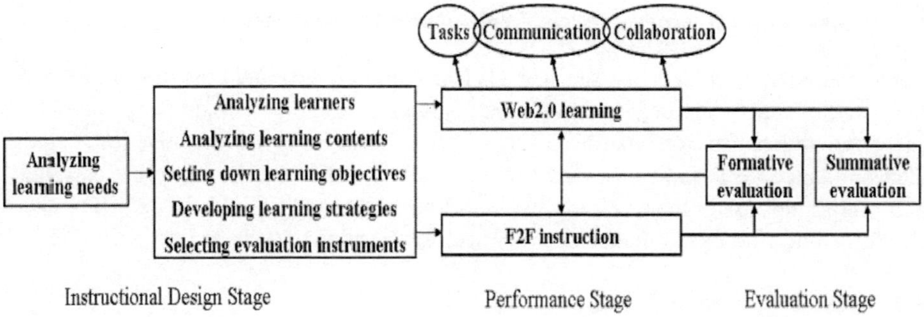

Fig. 1. Blending learning procedure

3.2 Result of the Experiment

Besides the final exam, a survey was conducted in students, which revealed their attitude to the model.

Firstly, the students thought that model was interesting and attractive, and liked to take part in learning activities. Secondly, most of them agreed that model had improved their learning efficiency and reinforced interactivities between them and the instructor. Thirdly, they preferred the blending learning consists of online learning and face to face instruction rather than absolute online learning in that they thought face to face instruction could solve problems which could not be solved via internet, and spur them to keep pace with instructional progress. Additionally, they thought the classroom instructional process made learning more real and concrete. For them, absolute online learning environments are virtual and isolated, and it may make them feel lonely and depressed.

According to these, the model was proved positive.

4 General Guidelines

4.1 Blend Rather Than Duplicate

The most common mistake made by instructors is to duplicate classroom activities to online environments. Students may feel boring to study the same contents twice, and lose their interest to study soon. Hence, the instructor should ensure the activities and learning contents of each setting are different. They should blend rather than duplicate.

4.2 Select Proper Tools Just Eough for Work

Some web2.0 tools are likely to surpass the other ones in functions, which may facilitate our work. Hence, tools applied to web2.0 learning are supposed to be selected carefully or the following work will be hard.

4.3 Find Right Balance between Web2.0 and Classroom

Some students addressed that the web2.0 learning activities had cost their more time per week in study than before though the learning progress was interesting. They had spent too much time on visiting various web2.0 web sites without the instructor's clear guidance. Sometimes they had felt exhausted in taking part in online activities and preparing assignments for class. So the instructor should design the class carefully, and find right balance between web2.0 learning and classroom activities.

4.4 Avoid Excessive Burden

Directing web2.0 learning activities besides classroom instruction may double the workload of instructors', especially when there are assigned online tasks; it will turn to heavy burdens for them soon. Instructors should deliberate in anticipation so as to avoid excessive burden, or they will lose interest in applying this model to practice.

5 Conclusions

As a brand-new instructional model, web2.0 blending learning emerges numerous advantages and immense value for schools, instructors and students. Though instructors and students may face challenges in varied areas such as technology, organization, management and the like in web2.0 learning, they would not deny the attraction of it.

More and more professionals and instructors will make use of web2.0 tools as substitutes for specialized online learning platforms, and perform their online instruction without caring about the expense. The web2.0 blending learning model is sure to be popular in schools in the future.

References

1. Driscoll, M.: Blended Learning: let's go beyond the hype. E-learning, 54 (March 2002)
2. Singh, H.: Building Effective Blended Learning Programs. Educational Technology 43, 51–54 (2003)
3. Osguthorpe, R.T., Graham, C.R.: Blended Learning Systems: definitions and directions. Quarterly Review of Distance Education 4, 227–234 (2003)

Study on Profit Distribution Based on Fuzzy Analytic Hierarchy Process in Knowledge Chain

Wu Shaobo

Strategic Planning School, Chongqing Technology and Business University, Chongqin, China
wushaoboscu@tom.com

Abstract. This paper summarizes the concept of Knowledge Chain and puts forward the principle of distribution of benefits of knowledge chain. Then builds up the evaluation model used Analytic Hierarchy Process and Fuzzy mathematic ; Finally, a simulation is carried out to illustrate the feasibility and validity of the method.

Keywords: Knowledge chain, profit distribution, fuzzy analytic hierarchy process.

1 What's the Knowledge Chain

The knowledge chain refers to the chain network established through knowledge flows among various cooperative innovation organizations, where enterprises are regarded as the innovation core body, to realize knowledge sharing and knowledge creation[1]. The knowledge chain comprises organizations which own different knowledge resources, such as the core enterprise (the leader of the alliance), universities, institutes, suppliers, dealers, consumers and even competitors. The interaction between these organizations causes inter-organizational knowledge flow so as to make knowledge diffused and transferred in different owners. In the social production practice, every production process needs many academic and fields of knowledge to coordinate. Any absence of some fragments of knowledge can't accomplish expected goals, only integrating the fragments of knowledge which is owned by different organizations can give full play to social production(Argote, L, et al, 1990) [2]。In the result, the organizations which own fragments of knowledge must comprise knowledge to cooperate, and take part in production. For example, although Motorola is good at research and development of mobile, it needs Foxconn to act as an agent to manufacture.

In 2001, C.W.Holsapple and M.Singh first clearly put forward the notion of knowledge chain based on Michael Porter's value chain model. He divided the knowledge chain's activities into nine, including knowledge acquisition, knowledge selection, knowledge generation, knowledge internalization and knowledge and so on five main activities, as well as leadership, cooperation, control and measure the four supporting complementary activities.[3] Subsequently, many Chinese scholars have also made a considerable amount of research on knowledge chain. For example, Chang Li et al (2001)[4] considered that knowledge chain is "based on knowledge

J. Luo (Ed.): Affective Computing and Intelligent Interaction, AISC 137, pp. 479–484.
springerlink.com

flow between the different market players and the realization of the transfer and diffusion of knowledge integration, integration and innovative value-added features of the network of value chain structure model." Xu Jiansuo et al (2003)[5] points out that "knowledge chain is not only between different enterprises of the main transfer and proliferation, but also companies within the main body of knowledge capture, selection, organization and innovation process." However, these studies have mostly focused on the organization's internal knowledge chain, knowledge chain model using the concept of the concept, characteristics and the formation of business organizations the role of competitive advantage. Subsequently, there are some scholars to expand to a cross-organizational knowledge chain of research. For example, Gu et al (2003) study the social capital’s role in knowledge role. Wu Shaobo and Gu (2009) [6]establish a game model to study the iner-organizational conflicts in knowledge chian, and they also analyze the factors which influences the stable operation of knowledge chain, such as knowledge spillover, collaborative value creation, expectations of long-term cooperation, etc. Wu Shaobo and Gu (2010) [7]study division of knowledge in knowledge chain by establishing a principal-agent model. However, although the literature on the knowledge chain already had a certain amount of research, little literature pays attention to benefit distribution in knowledge chain, so this study has some theoretical value.

2 Principle of Benefits Distribution in Knowledge Chain

The knowledge chain should determine the principle of benefits distribution firstly, so that the distribution process is fair and reasonable. Knowledge chain distribution of benefit between organizations should follow the following principles(Luo Li and Lu Ruoyu,2003):[8]

(1) The principle of reciprocity and mutual benefit. Knowledge chain organization is an independent socio-economic subject, despite the organization's goals are different, but the self-interests of all parties should be fully guaranteed, otherwise it will affect the enthusiasm of co-operation, and even lead to failure or breakdown of co-operation.

(2) The principle of best interests of the structure. From the actual situation, knowledge chain should take into account all factors affecting the distribution of benefits, reasonably determine the optimal ratio of the structure, prompt members of the parties to achieve a reasonable cooperation and coordinated development goal.

(3) The principle of symmetry of the risks and benefits. Cooperation in the distribution of benefits should be commensurate with the risks of all the parties, and carefully distinguish between scientific and technological development risks and production business risk, not only have the interests of proper distribution, but also have a suitable risk compensatory measures to enhance cooperation in confidence.

Gui Liangjun, et al (2006)[9] proposed the third-party regulatory agencies should be based on enterprise resource input and the contribution rate to allocate the supply chain benefits. Combination of the aforementioned studies, we believe that knowledge chain in determining the allocation factor should be based on knowledge inputs, innovative efforts, the size of three aspects of risk measurement.

3 The Steps of Fuzzy-AHP Method

Knowledge chain uses fuzzy analytic hierarchy process to establish the three areas to determine the weight matrix to be in general to take the following steps:[10]

(1) To establish the importance of various factors to determine matrix A.

$$A=\begin{bmatrix} a_{11} & a_{12} & a_{13} \\ a_{21} & a_{22} & a_{23} \\ a_{31} & a_{32} & a_{33} \end{bmatrix}$$

In the above-mentioned matrix, $a_{ij}=\frac{1}{a_{ji}}(i\neq j)$, $a_{ij}=a_{ji}(i=j)$.

(2) To normalize the above matrix.

First, we normalize the matrix by the column (so that column is 1), we can get $b_{ij}=a_{ij}/\sum a_{ij}$. Then we sum it by row, we can get $v_i=\sum b_{ij}$. At the following step we can get the weight $W_i^0=v_i/\sum v_i$.

In order to test the compatibility of the matrix, we should compute the consistency coefficients $C.I.=\frac{\lambda_{\max}-n}{n-1}$, where $\lambda_{\max}=\frac{1}{n}\sum_i(\frac{(AW)_i}{w_i})$. If $C.I.\prec 0.1$, we can accept the consistency coefficients of the matrix.

(3) To evaluate by fuzzy comprehensive method.

Let factor set be

$U=\{\text{Knowledge inputs , Innovative efforts ,Risk-taking}\}=\{u_1,u_2,u_3\}$, Let Evaluation Set be

$V=\{\text{very low, low, high, very high}\}=\{v_1,v_2,v_3,v_4\}$ We let experts to evaluate the three factors and record the evaluation results, then we can get the following fuzzy matrix:

$$R=\begin{Bmatrix} r_{11} & r_{12} & r_{13} & r_{14} \\ r_{21} & r_{22} & r_{23} & r_{24} \\ r_{31} & r_{32} & r_{33} & r_{34} \end{Bmatrix}$$

As the weight of factors have been computed out $A=\{W_1^0,W_2^0,W_3^0\}$, so we can get knowledge chain members' comprehensive evaluation is $B=A\bullet R=\{b_1,b_2,b_3,b_4\}$.

(4)To compute distribution factors.

To normalize B we can get $B'=\{b_1',b_2',b_3',b_4'\}$, if we assign the value to each evaluation set V, we can get the distribution factors $\alpha_i=B'\bullet V^T$.

To repeat the above method, we can get each knowledge chain member's distribution factor α_i.

4 Computing Example

We assume that the knowledge chain is composed of three actors, which are E, F, G. The value the knowledge chain created is 200 million Yuan. According to expert advice, the comparison matrix using AHP method for knowledge inputs, innovative efforts and risk taking is as follows:

A	A_1	A_2	A_3
knowledge inputs A_1	1	2	3
innovative efforts A_2	1/2	1	3
risk taking A_3	1/3	1/3	1

We normalize the above matrix then we can get:

A	A_1	A_2	A_3	V_i
knowledge inputs A_1	6/11	6/10	3/7	1.573
innovative efforts A_2	3/11	3/10	3/7	1.001
risk taking A_3	2/11	1/10	1/7	0.425

We can get the weight of the three indicators:

$$W_1^0 = \frac{1.573}{\sum V_i} = 0.525$$

$$W_2^0 = \frac{1.001}{\sum V_i} = 0.334$$

$$W_3^0 = \frac{0.425}{\sum V_i} = 0.141$$

If the expert's matrix of fuzzy evaluation for knowledge chain member E is as follows:

$$R_E = \begin{bmatrix} 0.4 & 0.2 & 0.3 & 0.1 \\ 0.2 & 0.2 & 0.3 & 0.3 \\ 0.5 & 0.1 & 0.1 & 0.3 \end{bmatrix}$$

Then we can get the Comprehensive Evaluation of knowledge chain member E is:

$$B_E = [W_1^0, W_2^0, W_3^0]\begin{bmatrix} 0.4 & 0.2 & 0.3 & 0.1 \\ 0.2 & 0.2 & 0.3 & 0.3 \\ 0.5 & 0.1 & 0.1 & 0.3 \end{bmatrix}$$

$$= [0.35, 0.19, 0.27, 0.19] = (0.35, 0.19, 0.27, 0.19)$$

We normalize B_E to $B_E' = (0.35, 0.19, 0.27, 0.19)$.

If $V = \{\text{very low}, low, high, \text{very high}\} = \{v_1, v_2, v_3, v_4\}$ and the evaluation scale of V is c= (1, 3, 5, 7) .we can get knowledge chain member E's distribution factor is $D_E = cB_E^{/T} = (1,3,5,7)(0.35, 0.19, 0.27, 0.19)^T = 3.6$.

In the same way, assuming we get knowledge member F and G's distribution factors are $D_F = 2.4$, $D_G = 4$.

According to the weight of the three distribution factors, we can compute out the profit of the three members:

$$R_E = 200 \times \frac{D_E}{D_E + D_F + D_G} = 200 \times 0.36 = 72$$

$$R_F = 200 \times \frac{D_F}{D_E + D_F + D_G} = 200 \times 0.24 = 48$$

$$R_G = 200 \times \frac{D_G}{D_E + D_F + D_G} = 200 \times 0.4 = 80$$

Acknowledgment. This research is supported by the National Science Foundation of China (70771069, 70471069) and Program for New Century Excellent Talents in University of China (NCET-06-0783).

References

1. Gu, X., Guo, Y., Li, J.: Social capital & it s role in knowledge chain. Science Research Management 24(5), 44–48 (2003)
2. Argote, L., Beckman, S.L., Epple, D.: The Persistence and Transfer of Learning in Industrial Settings. Management Science (36), 140–154 (1990)
3. Holsapple, C.W., Singh, M.: The knowledge chain model:activities for competitiveness. Expert Systems with Application (20), 77–98 (2001)
4. Chang, L., Zhou, S., Li, S.: Study on Diffusion factors of knowledge based on knowledge chain. Science Research Management 22(5), 122–127 (2001)
5. Xu, J., Wang, Z., Li, S.: Management based on knowledge chain. Journal of Tianjin University (Social Sciences) 5(2), 133–136 (2003)
6. Wu, S., Gu, X.: Study on the stability structuring of cooperation and conflicts in knowledge chain. Nankai Management Review 12(3), 54–58, 66 (2009)
7. Wu, S., Gu, X., Peng, S.: Study on the decision model of inter-organizaitonal knowledge devision in knowledge chain. Science Research Management (2010) (forthcoming)

8. Luo, L., Lu, R.: Study on the strategy model in industry-academia cooperation. Journal of Management Engineering 14(2), 1–5 (2000)
9. Gui, L., Zhao, Z., Tian, Z.: Study on the profit distribution based on third party involved in the supply chain. Accounting Research (10), 56–63 (2006)
10. Wang, Y.: Systems Theory, Methods and Applications, pp. 170–176. Higher Education Press, Beijing (2004)

Study on the Optimal Incentive Contract of Cooperative Innovation in Knowledge Chain

Wu Shaobo

Strategic Planning School, Chongqing Technology and Business University, Chongqin, China

Abstract. This paper puts forward the notion of knowledge chain, and it studies the optimal incentive contract under the condition of symmetric information and asymmetric information respectively. This research indicates that the agent invests more knowledge under symmetric information than under asymmetric information, and the core enterprise can get higher returns under symmetric information than under asymmetric information.

Keywords: Knowledge chain, Optimal incentive contract, Cooperative innovation.

1 Introduction

The knowledge chain refers to the chain network established through knowledge flows among various cooperative innovation organizations, where enterprises are regarded as the innovation core body, to realize knowledge sharing and knowledge creation[1]. The knowledge chain comprises organizations which own different knowledge resources, such as the core enterprise (the leader of the alliance), universities, institutes, suppliers, dealers, consumers and even competitors. The interaction between these organizations causes inter-organizational knowledge flow so as to make knowledge diffused and transferred in different owners. In the social production practice, every production process needs many academic and fields of knowledge to coordinate. Any absence of some fragments of knowledge can't accomplish expected goals, only integrating the fragments of knowledge which is owned by different organizations can give full play to social production (Argote, L, et al, 1990) [2]。In the result, the organizations which own fragments of knowledge must comprise knowledge to cooperate, and take part in production. For example, although Motorola is good at research and development of mobile, it needs Foxconn to act as an agent to manufacture.

In 2001, C.W.Holsapple and M.Singh first clearly put forward the notion of knowledge chain based on Michael Porter's value chain model. He divided the knowledge chain's activities into nine, including knowledge acquisition, knowledge selection, knowledge generation, knowledge internalization and knowledge and so on five main activities, as well as leadership, cooperation, control and measure the four supporting complementary activities.[3] Subsequently, many Chinese scholars expand the research to inter-organizational knowledge chain. For example, Gu et al (2003) study the social capital's role in knowledge role. Wu Shaobo and Gu (2009) [4] establish a game model to study the iner-organizational conflicts in knowledge chian,

J. Luo (Ed.): Affective Computing and Intelligent Interaction, AISC 137, pp. 485–490.
springerlink.com

and they also analyze the factors which influence the stable operation of knowledge chain, such as knowledge spillover, collaborative value creation, expectations of long-term cooperation, etc..Wu Shaobo and Gu (2010) [5] study division of knowledge in knowledge chain by establishing a principal-agent model. However, although the literature on the knowledge chain already had a certain amount of research, little literatures had pay attention to the optimal incentive contract of knowledge chain, so this study has some theoretical value.

2 Model Hypothese

In order to facilitate this research, we make the following assumptions :

(1)The knowledge chain is composed of the core enterprise A and the agent B. The core enterprise is risk neutral, and the agent is risk aversion.

(2) The core enterprise invites the agent to take part in the innovation process, and this innovation can influence the demand of the product. If the agent's endeavor of knowledge put into the innovation process is e, the demand of the products increases $\lambda e+\theta$, where λ is the agent's output coefficient of knowledge input and $\lambda \succ 0$. θ is the influencing factor of external uncertainties, and it is a normal distribution(0, δ^2).

(3)The contract which is signed by the core enterprise and the agent is (w_0, k), where w_0 is the fixed fee the core enterprise pays to the agent, k is the payment per unit product when the demand increasing.

(4)After the innovation process is successful, the core enterprise can get the profit $\alpha\lambda e$ from increasing demand, where α ($\alpha \succ 0$) is the marginal income. On another hand, the core enterprise can get profit $\gamma_A \lambda e$ of technical capacity growth at the process of innovation, where γ_A ($\gamma_A \succ 0$) is the core enterprise's coefficient of technical capacity growth. So the core enterprise's income is:

$$R_A = \alpha\lambda e + \gamma_A \lambda e - k\lambda e - w_0 = (\alpha + \gamma_A - k)\lambda e - w_0 \tag{1}$$

(5)The agent can also get profit $\gamma_B \lambda e$ of technical capacity growth at the process of innovation, where γ_B is the agent's coefficient of technical capacity growth. We assume the agent's cost function is $\frac{1}{2}\mu e^2$, where μ is the cost coefficient knowledge input. Moreover, because the agent is risk aversion, we assume the degree of risk aversion is ρ, then the agent's risk cost is $\frac{1}{2}\rho(k+\gamma_B)^2\sigma^2$. So the agent's income is:

$$R_B = E[(k+\gamma_B)(\lambda e+\theta)+w_0]$$

$$= (k+\gamma_B)\lambda e + w_0 - \frac{1}{2}\mu e^2 - \frac{1}{2}\rho(k+\gamma_B)^2\sigma^2 \tag{2}$$

3 Design on the Optimal Contracts of Knowledge Chain under the Condition of Asymmetric Information

Under the condition of symmetric information, if the core enterprise can observe the agent's knowledge input, the core enterprise only needs to give the payment more than the agent's opportunity income π_0 under the condition of not accepting the contract. So the core enterprise faces the following programming problem:

$$\max_{k,w_0} R_A = (\alpha+\gamma_A-k)\lambda e - w_0 \tag{3}$$

$$S.T. (k+\gamma_B)\lambda e + w_0 - \frac{1}{2}\mu e^2 - \frac{1}{2}\rho(k+\gamma_B)^2\sigma^2 \geq \pi_0 \tag{4}$$

Because the core enterprise needs not to pay any more than π_0 under the optimal condition, inequation (4) can get equal sign. We substitute w_0 in inequation (4) to equation (3), The above programming problem can be transferred into the following issues:

$$\max_{k,w_0} R_A = (\alpha+\gamma_A-k)\lambda e + \frac{1}{2}\mu e^2 + \frac{1}{2}\rho(k+\gamma_B)^2\sigma^2 + \pi_0 + (k+\gamma_B)\lambda e \tag{5}$$

The first-order conditions for optimization means:

$$k^* = 0 \tag{6}$$

$$e^* = \frac{(\alpha+\gamma_A+\gamma_B)\lambda}{\mu} \tag{7}$$

We substitute equation (6) and (7) to equation (3), then we can get the core enterprise's income under symmetric information is:

$$R_A^* = \frac{(\alpha+\gamma_A+\gamma_B)^2\lambda^2}{2\mu} - \pi_0 - w_0 \tag{8}$$

4 Design on the Optimal Contracts of Knowledge Chain under the Condition of Symmetric Information

As the agency organizational knowledge put into the innovation process is non-observable, the agent may be lazy in motivation. So the core enterprise must face two constraint conditions when it maximizes its certain profit: The first constraint condition is participation constraint, that is the agent's expected utility when accepting the contracts is more than the expected utility when rejecting the contracts. This expected utility is also the agent's opportunity income. This time we hypothesize the agent's opportunity income π_0 can get its reservation utility, where $\pi_0 \succ 0$. The second constraint condition is the incentive compatible constraint condition, that is when the core enterprise can not observe the agent's endeavor e and random station θ, the agent always must select the vector e which maximize its expected utility.[6]

So the core enterprise's problem is the following programming problem:

$$\max_{k,w_0} R_A = (\alpha + \gamma_A - k)\lambda e - w_0 \tag{9}$$

$$S.T.\begin{cases} \max\limits_{e} R_B = (k+\gamma_B)\lambda e + w_0 - \frac{1}{2}\mu e^2 - \frac{1}{2}\rho(k+\gamma_B)^2\sigma^2 (IC) \\ (k+\gamma_B)\lambda e + w_0 - \frac{1}{2}\mu e^2 - \frac{1}{2}\rho(k+\gamma_B)^2\sigma^2 \ge \pi_0 (IR) \end{cases}$$

Because $\frac{dR_B}{de} = (k+\gamma_B)\lambda - \mu e$,let $\frac{dR_B}{de} = 0$, then the incentive compatibility constraint can be transferred into $e = \frac{(k+\gamma_B)\lambda}{\mu}$. The above programming problem can be transferred into the following programming problem:

$$\max_{k,w_0} R_A = (\alpha + \gamma_A - k)\lambda e - w_0 \tag{10}$$

$$S.T.\begin{cases} e = \frac{(k+\gamma_B)\lambda}{\mu} (IC) \\ (k+\gamma_B)\lambda e + w_0 - \frac{1}{2}\mu e^2 - \frac{1}{2}\rho(k+\gamma_B)^2\sigma^2 \ge \pi_0 (IR) \end{cases}$$

We substitute the two constraint condition into the objective function of (10), we can get:

$$\max_{k,w_0} R_A = \gamma_B \lambda \frac{\lambda(\gamma_B + k)}{\mu} - \frac{\lambda^2(\gamma_B + k)^2}{2\mu}$$

$$-\frac{1}{2}\rho(k+\gamma_B)^2\sigma^2 - \pi_0 + (\alpha + \gamma_A)\lambda \frac{\lambda(\gamma_B + k)}{\mu} \tag{11}$$

Differentiating equation [11] over k we can get:

$$\frac{dR_A}{dk} = \frac{\gamma_B \lambda^2}{\mu} - \frac{2(\gamma_B + k)\lambda^2}{\mu} - \rho(k+\gamma_B)\sigma^2 - \frac{(\alpha + \gamma_A)\lambda^2}{\mu} \tag{12}$$

Let $\frac{dR_A}{dk}=0$, we can get:

$$k^{**}=\frac{-(\alpha+\gamma_A)\lambda^2+\gamma_B\mu\rho\sigma^2}{\lambda^2+\gamma_B\mu\rho\sigma^2} \tag{13}$$

We substitute (13) to the incentive constraint in (9), we can get the agent's optimal knowledge investment:

$$e^{**}=\frac{\lambda^2}{\lambda^2+\mu\rho\sigma^2}\times\frac{(\gamma_B+\alpha+\gamma_A)}{\mu} \tag{14}$$

We substitute equation (13) and (14) to (9), we can get the core enterprise's income under asymmetric information:

$$R_A^{**}=\frac{\lambda^2}{\lambda^2+\mu\rho\sigma^2}\times\frac{(\alpha+\gamma_A+\gamma_B)^2\lambda^2}{2\mu}-\pi_0-w_0 \tag{15}$$

We compare equation (14) and (7), because $\frac{e^{**}}{e^{*}}=\frac{\lambda^2}{\lambda^2+\mu\rho\sigma^2}\prec 1$, the agent's knowledge input effort under asymmetric information is less than the effort under symmetric information.

We compare equation (15) and (8), because $\frac{\lambda^2}{\lambda^2+\mu\rho\sigma^2}\prec 1$, we can get the following inequation:

$$\frac{\lambda^2}{\lambda^2+\mu\rho\sigma^2}\times\frac{(\alpha+\gamma_A+\gamma_B)^2\lambda^2}{2\mu}\prec\frac{(\alpha+\gamma_A+\gamma_B)^2\lambda^2}{2\mu} \tag{16}$$

So we can get $R_A^{**}\prec R_A^{*}$. This means that the core enterprise's income under asymmetric information is less than the income under symmetric information. In the result, the core enterprise can get much information value if it can observe the agent's knowledge input under symmetric information, the information value is:

$$\Delta R=R_A^{*}-R_A^{**}=\frac{\rho\sigma^2}{\lambda^2+\mu\rho\sigma^2}\times\frac{(\alpha+\gamma_A+\gamma_B)^2\lambda^2}{2}$$

5 Conclusion

This paper puts forward the notion of knowledge chain, and it studies the optimal incentive contract under the condition of symmetric information and asymmetric information respectively. This research indicates that the agent invests more knowledge under symmetric information than under asymmetric information, and the core enterprise can get higher returns under symmetric information than under asymmetric information.

Acknowledgement. This research is supported by the National Science Foundation of China（70771069，70471069）and Program for New Century Excellent Talents in University of China(NCET-06-0783).

References

1. Gu, X., Guo, Y., Li, J.: Social capital & it s role in knowledge chain. Science Research Management 24(5), 44–48 (2003)
2. Argote, L., Beckman, S.L., Epple, D.: The Persistence and Transfer of Learning in Industrial Settings. Management Science (36), 140–154 (1990)
3. Holsapple, C.W., Singh, M.: The knowledge chain model:activities for competitiveness. Expert Systems with Application (20), 77–98 (2001)
4. Wu, S., Gu, X.: Study on the stability structuring of cooperation and conflicts in knowledge chain. Nankai Management Review 12(3), 54–58, 66 (2009)
5. Wu, S., Gu, X., Peng, S.: Study on the decision model of inter-organizaitonal knowledge devision in knowledge chain. Science Research Management (2010)
6. Ma, Y.: Research on Incentive Mechnisim of University-Enterprise Knowledge Share of Cooperative Innavation. Journal of Management Engineering 22(4), 36–39 (2008)

Price Dispersion Formative Mechanism Research on Pattern of C2C

Chang Yaping, Hu Shaolong, and Zhang Geng

Management Department, Wuhan University of Science and Engineering, WuHan, China
20349498@qq.com, andycarlos@163.com, 75268896@qq.com

Abstract. This paper discusses the influencing factors of the price dispersion from the aspects of e-retailer attribute and market attribute, and takes Taobao as an example to form the model of formative mechanism of price dispersion on pattern of C2C,. Research finds that the consumer involvement, the number of competitors in the market and the entering time of the e-retailer apparently influence the price dispersion on pattern of C2C.

Keywords: Price Dispersion, E-retailer Attribute, Market Attribute.

1 Introduction

With the explosive development of e-commerce, on-line price dispersion starts coursing the concerning of researchers[1]. Price dispersion is conceptualized as the price distribution of the same products from different sellers in the same market and at the same time, and it is the indexes which can comprehensively reflect the market structure and the situation of information flow.

According to traditional theories, the appearance of e-commerce will make the level of price dispersion decline. Firstly, the development of network technology promotes information exchange, which reduces the cost of information searching in a great degree and the level of price dispersion[2]. Secondly, on-line market is much easier to enter than traditional market, because of the "shop" under the e-commerce is a website but not a traditional entity storefront. Finally, menu cost is generated by price change of retailers. In the e-market, menu cost has been decreased greatly. According to pricing policy of rivals, trade companies can change their prices frequently, which will decrease the level of price dispersion. Thus, compared with traditional market, on-line market will be of stronger competitiveness and lower price dispersion level.

However, the theory about lower price dispersion under on-line market has not been proved by existent empirical studies. In addition, the study of Bryjnolfsson and Smith(2000) shows that the situation of higher level of price dispersion will be kept[2]. This paper will take Taobao as an example and use empirical study method to find the reason of price dispersion of on-line market, and analyze price dispersion formative mechanism on pattern of C2C at the background of decreasing of information search cost.

J. Luo (Ed.): Affective Computing and Intelligent Interaction, AISC 137, pp. 491–497.
springerlink.com

2 Literature Reviews

Stigler first presented the concept of price dispersion in information economics, Stigler insisted that price dispersion is attributed to inadequacy of market information. Lots of research show that price dispersion will happen when the cost of consumer looking for lower price is higher[3][4].

Considering micro and macro environment of e-commerce and the limitation of empirical study object in China, we suppose that the internal causes of on-line price dispersion include eight elements: implementation reliability, logistics and distribution, inventory level, market entry time, consumer awareness, third-party certification, the number of competitors in the market, consumer involvement.

2.1 E-retailer Attribute

Implementation reliability. Implementation reliability can be defined as the reliability of fulfilling contract of e-retailer perceived by consumer which includes arrival time of commodity, whether products arrive on time, whether real products are consistent with description of website. The differences of reliability of e-retailers perceived by consumers will affect products' price range and price dispersion.

Logistics and distribution. Logistics and distribution refer to the service about e-retailer delivers products to terminal consumers after confirming transaction between buyers and sellers. E-retailer of possessing higher pricing will offer better logistics and distribution service, because some consumers think that this kind of service is valuable. However, whether consumers pay for better service still needs to be studied. Brynjolfsson and Smith(2000) found that the price of some E-retailers which offer better logistics service also comparatively lower.[6]

Inventory level. Inventory level is defined as the number of the products possessed by e-retailer. The inventory level may affect price competition strategy of e-retailer. E-retailers possessing more inventory level will offer lower price for the products than ones who have less inventory level[7].

Market entry time. This time is obtained when e-retailer enters into e-commerce market. Market first-mover advantage in the traditional market has been discussed by scholars[8]. Schmalensee pointed out perceived risk of consumer to market forerunners is lower than to ones who late enter into market, and consumers are willing to pay higher prices[9].

Consumer awareness. Consumer awareness can be defined as the existent possibility of e-retailers perceived by consumers. The e-retailers who have higher consumers' consciousness may offer higher pricing than their competitors who are not indexed by search engine [1].

Third-party certification. Third-party certification is defined as the certificate awarded by authoritative organization to some e-retailers who have special attributes.

2.2 Market Attribute

The number of competitors in the market. The number of competitors in the market can reflect market competitiveness and it is a influencing factors of price dispersion model[10]. Carlon and McAfee(1983) found that the more competitors in the market, the smaller the degree of price dispersion. However, when Dahlby and West(1986) applied the model to automobile insurance market, they found the more sellers in the market, the bigger price dispersion. Thus, price dispersion and the number of competitors in the market may have nonlinear relationship.

Consumer involvement. During purchasing products, when consumer involvement level is high, consumers will obtain higher return through making great efforts to implement searching behavior, so consumer searching behavior and involvement level are relational. In the field of marketing, consumer involvement level can be considered as utility level, or price level. Therefore, some scholars believe that consumer searching behavior before the transaction will far concentrate on expensive product category.

3 Research Design

In order to analyze price dispersion formative mechanism in the research, we will take two types of influencing factors(e-retailer attribute and market attribute) mentioned above as independent variables, and take price dispersion level as dependent variable. The study uses multivariate regression method to construct mechanism model, and finds out contribution degree of independent variable to dependent variable.

3.1 Sampling and Sample Composition

Due to Taobao is the biggest C2C market in China, it has better representation and is the investigation platform. In the study, our sample object is mobile phone, this is owing to involvement level of mobile phone consumers is much more in-depth, mobile phones have better homogeneity, and models and qualities of mobile phones are determined.

To be specific, the research selects 4 famous mobile phone models in order to guarantee uniformity of the number; on 30th December 2007, we collected experimental data from 1380 data collection points, starting at 11pm, in order to guarantee stability of results.

3.2 The Selection of Observed Variables

This study is on the basis of analyzing observed variables of price dispersion studed by Xing Pan and Brian Ratchford (2003) and other observed scales of e-commerce, we held 3 times brainstorming, 8 individuals' in-depth interviewing and 3 experts' repair, and divided influencing factors of price dispersion into two types mentioned above. We put forward observed variables of the study.

3.2.1 Independent Variables

a) E-retailer Attribute

Specific measurements of e-retailer attributes are as follows:

Implementation reliability: we take evaluation of transaction quality to e-retailer in Taobao as measurement index, if the evaluation is good, the factor will be marked as "1", if the evaluation is not good, not bad, it will be marked as "0", if the evaluation is bad, it will be marked as "-1".

Logistics and distribution: because outsourcing is the main form of logistics and distribution in the pattern of C2C, we consider freight as measurement index of the factor.

Consumer awareness: making reference to the study of Xing Pan and Brian Ratchford(2003), we use page click ratio as measurement index of consumer awareness.

Third-party certification: the e-retailers without third-party certificate will be marked as "0", the e-retailers who have consumer supporting plan will be marked as "1", the e-retailers who have brand shop will be marked as "2".

Inventory level and entering time: these two data can be obtained by the website.

b) Market Attribute

We select the number of competitors in the market and consumer involvement as influencing factors in terms of market attribute. Observed variables of the number of competitors in the market are mobile phone model and the number of e-retailer offering after-service on nation-wide warranty, these data can be obtained by Taobao. According to measurement standard of consumer involvement from Moorthy Ratchford and Talukdar(1997) and Cohen(1998), mobile phone average price of different models in previous week is used to measure consumer involvement, and Taobao can offer these data.

3.2.2 Dependent Variable

Price dispersion can be measured by standard deviation and range. Therefore, standard deviation and range are used as observed variables of price dispersion. According to mobile phone price distribution of different models from 20 e-retailers, we found standard deviation and range of price.

4 Data Analysis

For finding out the sensitivity of 8 factors respectively, we selected 20 E-retailers for each phone model randomly and took the variance of implementation reliability, variance of logistics and distribution, variance of consumer awareness, variance of inventory level, variance of market entry time, variance of third-party certification and the number of competitors in the market, consumer involvement as the independent variables, and took the price dispersion as the dependent variable, and adopt forced entry way to make multiple regression analysis.

Due to price dispersion degree has two measurement methods, including standard deviation and range, dispersion degree of independent variables only adopt one way to measure---variance, the study uses statistical software SPSS13.0 to make multiple regression analysis twice. The analysis results indicate that VIF of every independent variable is much smaller than 10, there is no multicollinearity problem.

Explanatory powers of independent variables to dependent variable in twice multiple regressions have reached more than 60%. (see table 1, table 2)

Table 1. Model Profile Form

Model	R	R Square	Adjusted R Square	Std. Error of the Estimate
1	901(a)	812	773	54.9210749

Table 2. Model Profile Form (taking range as measurement method)

Model	R	R Square	Adjusted R Square	Std. Error of the Estimate
1	848(a)	719	660	284.6428589

In the study, significance level is 0.05, the significance levels of market entry time, consumer involvement and the number of competitors in the market are less than 0.05, and regression coefficients are stable. It means that only market entry time of e-retailer attributes influence price dispersion significantly, consumer involvement and the number of competitors in the market of market attributes influence price dispersion significantly. Consumer involvement and market entry time are of positive effect, the number of competitors in the market is of negative effect. Absolute values of these three regression coefficients show that the sequence of influence degree from big to small is consumer involvement, the number of competitors in the market, market entry time.

In order to compare influence degrees to price dispersion between e-retailer attributes and market attributes, the study makes the summation of standardized coefficient square of factors having significant effects, and finds the effect of market attributes is much more obvious.(see table 3)

Table 3. Explanatory Powers of Influencing Factors

Independent variable	**Measurement method**	
	the summation of R2(taking variance as measurement method)	***the summation of R2(taking range as measurement method)***
e-retailer attributes	7%	7%
market attributes	88%	90%
error	5%	3%

5 Discussion and Conclusions

Regression analysis indicates that influencing factors of price dispersion are e-retailer attributes and market attributes. Because the summation of standardized coefficient square of e-retailer attributes is less than 10%, however, market attributes are the main reason of price dispersion.

5.1 Conclusion and Analysis on E-retailer Attributes

Price dispersion is not influenced by implementation reliability, logistics and distribution, inventory level, consumer awareness and third-party certification. It means that these five factors have no any difference in consumers' eyes. Thus, e-retailers don't transform them to price advantage.

Market entry time influence price dispersion significantly, when variation degree of market entry time is bigger, price dispersion will be bigger. The reason may be that market entry time reflects comprehensive reputation and management ability of e-retailer, and false data are not easy to be made. Thus, this influencing factor is recognized by consumers and e-retailers. E-retailers can transform it to price advantage.

5.2 Conclusion and Analysis on Market Attributes

Consumer involvement and the number of competitors in the market influence price dispersion significantly. When the number of competitors in the market is more, price dispersion will be lower. When consumer involvement is higher, price dispersion will be bigger. We can explain from two aspects: in terms of measurement error, average price is used as the measurement index of consumer involvement, the increase of average price will cause the enhancement of price dispersion; in the aspect of traditional theories, consumers' price sensitivity will decrease when purchasing products of high involvement, they are paying more attention to product quality and service.

References

1. Pan, X., Ratchford, B.T., Shankar, V.: Price dispersion on the internet: a review and directions for future research. Journal of Interactive Marketing 18(4), 116–135 (2004)
2. Bryjolfsson, E., Smith, M.: Frictionless Commerce? A Comparison of Internet and Conventional Retailers. Management Science 46(4), 563–585 (2000)
3. Stigler, G.: The Economics of Information. Journal of Political Economy 69(3), 213–225 (1961)
4. Burdett, K., Judd, K.: Equilibrium Price Dispersion. Econometrical 51(7), 955–969 (1983)
5. Coles, M.: Steady State Price Distributions in a Noisy Search Equlibrium. Journal of Economic Theory 72(1), 5–32 (1997)
6. Smith, M., Bailey, J., Brynjolfsson, E.: Understanding Digital Markets: Review and Assessment. In: Brynjolfsson, E., Kahin, B. (eds.) Understanding the Digital Economy. MIT Press, Cambridge (2000)

7. Shankar, V., Carpenter, G.S., Krishnamurhi, L.: Advantages of Entering in the Early Growth Stage: An Empirical Analysis. Journal of Marketing Research 36(5), 269–276 (1999)
8. Urban, G.L., Carter, T., Gaskin, S.: Market Share Rewards to Pioneering Brand: An Empirical Analysis and Strategic Implications. Management Science 32(6), 645–659 (1986)
9. Schmalensee, R.: Product Differentiation Advantages of Pioneering Brands. American Economic Review 72(3), 349–365 (1982)
10. Gielens, K., Dekimpe, M.G.: Do International Entry Decisions of Retail Chains Matter in the Long Run? International Journal of Research in Marketing 18(5), 235–259 (2001)

Empirical Research on the Technology Innovation Capability of Bohai Rim Region*

Zhang Jingqiang[1], Zhou Hong[2], and Huang Zilang[3]

[1] School of Economics and Management
North China University of Technology Beijing, China
zjq811@163.com
[2] School of Humanity and Law
Beijing Institute of Civil Engineering and Architecture Beijing, China
zhh620@sohu.com
[3] School of Economics and Management
North China University of Technology Beijing, China
335086019@qq.com

Abstract. In this paper, using factor analysis and SPSS15.0 software, the author selected 16 original indicators to comprehensively evaluate the technology innovation capability of Bohai Rim region. The results indicate that there are many differences in the technology innovation capability among various provinces and cities in Bohai Rim region. So, in order to enhance the technology innovation capability of Bohai Rim region, we should increase the R&D input, improve the effectiveness of technology innovation, promote technology transfer and diffusion, adjust and optimize industrial structure, thus promoting the coordinated development of regional economy.

Index Terms: Regional technology innovation capability, Factor analysis, Empirical research.

1 Introduction

According to the theory of economic growth, technology innovation capability is an important factor to the assessment of regional economic development and competition, especially in the current trend of the global economic integration, regional technology innovation capability is increasingly becoming an important factor on achieving international competitive advantage. Divided by the latest regional structure, we can see that Bohai Rim region is an economic growth pole after the Pearl River Delta economic circle and the Yangtze River Delta economic circle, and has become the core areas of economic development in north China.

Therefore, the research on technology innovation capability of Bohai Rim region will play an important role for promoting the upgrading of regional industrial structure, optimizing the allocation of resources.

* This paper is supported by " Beijing's Undergraduate student Scientific Research and Entrepreneurial Action Plan Project (1018)"and "2010 Undergraduate student S&T Activities Project of North China University of Technology".

J. Luo (Ed.): Affective Computing and Intelligent Interaction, AISC 137, pp. 499–506.
springerlink.com

2 Selection of Evaluation Indexes

For foreign scholars, the research on regional technology innovation capability focused on regional innovation strategies, regional technology transfer, regional innovation policies, etc [1]. Domestic research on regional technology innovation capability was more. Li Xiaolu, Zhou Zhifang constructed 16 indicators to evaluate the various regional technology innovation capability from the regional economic development, science and technology investment level and the sustainable development level[2]; Fan Decheng, Zhou Hao evaluated the various regional technology innovation capability from R&D investment, the quality of personnel structure, technology innovation output and technical diffusion[3]. There are also many scholars who comprehensively evaluated technology innovation capability from different angles[4-6]. However, these studies have mostly focused on various regions in the whole country, and the same studies are relatively short for Bohai Rim region.

In this paper, the author evaluate comprehensively technology innovation capability of Bohai Rim region including five provinces and two municipalities through selecting 16 indicator variables and using empirical analysis methods.

The selected indicators are as follows:

- x_1: Personnel engaged in S&T activities (person);
- x_2: Scientists and engineers of S&T personnel (person);
- x_3: Investment funds on S&T activities (ten thousand yuan);
- x_4: Full-time equivalent of R&D personnel (man-year);
- x_5: Expenditure on R&D (ten thousand yuan);
- x_6: Foreign Direct Investment(Dollar);
- x_7: Total patent applications examined (unit);
- x_8: patent applications certified (unit);
- x_9: The amount of S&T technological papers published
- x_{10}: Overall labor productivity
- x_{11}: Regional exports(Dollar);
- x_{12}: High-tech product exports(Dollar);
- x_{13}: Per capita GDP(yuan);
- x_{14}: Investment in education funding(yuan);
- x_{15}: Full-time teachers in colleges and universities (person);
- x_{16}: The volume of transactions of technology market contract.

3 Analysis Model

In this paper, factor analysis method is used and the corresponding mathematical model is also established. The derivation process and the analytical procedure about factor analysis method can be found in the literature[7], and then the factor analysis model is as follows:

$$X_i = l_{i1}F_1 + l_{i2}F_2 + \cdots + l_{im}F_{im} + \varepsilon_i \tag{1}$$

Among them, l_{ij} is called the factor load.

At the same time, the factor model should meet the following relationship:

$$E(F_j) = 0, \quad Cov(F_i, F_j) = \begin{vmatrix} 1(i = j) \\ 0(i \neq j) \end{vmatrix} \tag{2}$$

$$E(\varepsilon_j) = 0, \quad Cov(\varepsilon_i, \varepsilon_j) = \begin{vmatrix} \psi_i(i = j) \\ 0(i \neq j) \end{vmatrix} \tag{3}$$

$$E(F_i, \varepsilon_j) = 0 \tag{4}$$

4 Empirical Calculation

According to the steps of factor analysis, first of all, the correlation coefficients among the evaluation indexes are got. From the correlation coefficients, the correlation coefficients among 16 indicators are relatively large, and are more than 0.5, which indicates that it is more suitable for factor analysis.

On the basis of the correlation coefficient matrix, we can calculate and obtain the factor characteristic roots and the variance contribution rates of technology innovation capability, referring Table 1.

As can be seen from Table 1, the characteristic root of the first principal component is 10.90, the variance contribution rate is 68.13%; the characteristic root of the second principal component is 3.28 and the variance contribution rate is 20.49%; the characteristic root of the third principal component is 1.53 and the variance contribution rate is 9.53 %. The cumulative variance contribution rate of these three principal components is 98.15%.

According to the conditions for extracting factors----the characteristic value being more than 1, we select the three main factors. Thus, these three principal component factors will explain the variation of 16 original variables, that is, using these three principal components to analyze the problem only lost 1.85% of the information, which has considerably reduced the complexity of the original data.

Table 1. Total Variance Explained

Component	Initial Eigenvalues			Extraction Sums of Squared Loadings		
	Total	*% of Variance*	*Cumulative %*	*Total*	*% of Variance*	*Cumulative %*
1	10.90	68.13	68.13	6.58	41.10	41.10
2	3.28	20.49	88.62	6.12	38.27	79.37
3	1.53	9.53	98.15	3.01	18.78	98.15

In order to strengthen the analysis explanatory power of the common factor for the real problem, these three main factors which have been extracted establish the original factor loading matrix, and rotate it using the maximize orthogonal rotation of the variance, then the rotation factor loading matrix is obtained, referring Table 2.

Table 2. Rotated Component Matrix

Index	Component 1	2	3	Index	Component 1	2	3
x_1	.792	.600	.067	x_9	.929	.282	.210
x_2	.849	.516	.105	x_{10}	.118	-.083	.987
x_3	.855	.436	.261	x_{11}	.315	.887	.333
x_4	.870	.458	.165	x_{12}	.381	.282	.863
x_5	.853	.468	.226	x_{13}	.633	.017	.768
x_6	.010	.926	.172	x_{14}	.305	.859	-.373
x_7	.413	.879	.208	x_{15}	.359	.868	-.293
x_8	.427	.893	.078	x_{16}	.956	-.050	.275

After analyzing the rotated factor loading matrix, we can get the following conclusions:

The loading values of the common factor F_1 on the indexes, including x_1、x_2、x_3、x_4、x_5、x_9、x_{16}, are very large. Among them, x_1、x_2、x_3、x_4、x_5 are technology innovation input indicators about human and financial resources; x_9、x_{16} are technology innovation output indicators. So the common factor F_1 is called as the technology innovation level factor. The loading values of the common factor F_2 on the indexes, including x_6、x_7、x_8、x_{11}、x_{14}、x_{15}, are very large. These six factors are main output indicators of technology innovation effect. So the common factor F_2 is called as the technology innovation effect factor. The loading values of the common factor F_3 on the indexes, including x_{10}、x_{12}、x_{13}, are very large. These three factors are main evaluation indicators of technology innovation efficiency. So the common factor F_3 is called as the technology innovation efficiency factor.

According to Thomson regression method, we can obtain the factor score coefficient matrix, referring to Table 3.

Table 3. Component Score Coefficient Matrix

	Component				Component		
index	*1*	2	3	**index**	*1*	2	3
x_1	.131	.021	-.075	x_9	.212	-.086	-.057
x_2	.160	-.012	-.074	x_{10}	-.124	.018	.411
x_3	.149	-.025	-.011	x_{11}	-.128	.211	.148
x_4	.167	-.028	-.055	x_{12}	-.086	.062	.332
x_5	.148	-.018	-.024	x_{13}	.065	-.067	.227
x_6	-.207	.270	.135	x_{14}	.002	.159	-.165
x_7	-.072	.182	.075	x_{15}	.003	.156	-.138
x_8	-.047	.175	.015	x_{16}	.273	-.180	-.054

According to factor score coefficient matrix, we can obtain the main components expressions of linear combination about the various indicator variables, namely:

$$\hat{F}_1 = 0.131zx_1 + 0.160zx_2 + \cdots + 0.273zx_{16} \tag{5}$$

$$\hat{F}_2 = 0.021zx_1 - 0.012zx_2 + \cdots - 0.180zx_{16} \tag{6}$$

$$\hat{F}_3 = -0.075zx_1 - 0.074zx_2 + \cdots - 0.054zx_{16} \tag{7}$$

Among them, zx_i is the indicator data which have been standardized.

We take the information contribution rate of rotating factors as weights to calculate the overall evaluation scores of various regional technology innovation capability, that is:

$$\hat{F} = 0.41101\,\hat{F}_1 + 0.38265\,\hat{F}_2 + 0.18788\,\hat{F}_3 \tag{8}$$

Based on the style, we can calculate the technology innovation capability factor scores of Bohai Rim area and sort the situation, as shown in Table 4.

Table 4. Factor Scores

Region	$\hat{F}_1$	*Rank*	$\hat{F}_2$	*Rank*	$\hat{F}_3$	*Rank*	$\hat{F}$	*Rank*
Beijing	2.20	1	-0.22	4	0.50	2	0.91	1
Tianjin	-0.82	7	-0.21	3	1.98	1	-0.04	4
Hebei	-0.32	5	0.47	2	-0.21	4	0.01	3
Liaoning	-0.13	2	-0.29	5	-0.99	7	-0.35	5
Shanxi	-0.56	6	-1.01	7	-0.19	3	-0.65	7
Neimenggu	-0.22	4	-0.75	6	-0.85	6	-0.54	6
Shandong	-0.16	3	2.01	1	-0.25	5	0.66	2

5 Analysis of Empirical Result

Because there is no absolute practical significance for the size of the factor scores and only relative meaning, we analyze and discuss technology innovation capability of Bohai Rim region five provinces and two municipalities according to the order of factor scores.

5.1 Analysis Based on the Total Order of Factor Scores

From the total order of the factor scores, the factor scores of Beijing and Shandong are the first two in Bohai Rim region, which shows that technology innovation capability of Beijing and Shandong is the strongest. In Hebei and Tianjin, their factor scores are close to 0 and their technology innovation capability is the middle level in Bohai Rim region. The factor scores of Liaoning, Neimenggu and Shanxi are not only the lowest but also are negative value. This shows that their technology innovation capability is in the lower level of Bohai Rim region.

5.2 Analysis Based on the Single Order of Factor Scores

From the order of the first factor scores, Beijing's factor score is in the first, indicating that Beijing's technology innovation level is very high, which also determines its technology innovation capability level. The factor scores of Liaoning, Shandong and Neimenggu are the second, third, fourth, whose technology innovation level is relatively low. The technology innovation input level in Hebei, Shanxi and Tianjin is the lowest, mainly because Shanxi Province is an important coal province, Hebei and Tianjin are important industrial provinces, and physical capital investment is more.

From the order of the second factor scores, the factor score in Shandong Province is highest, indicating that its technology innovation effect level is relatively high, which is compatible to Shandong Province's economic development speed. In Hebei, Tianjin, Beijing and Liaoning Province, their technology innovation effect is the middle level in Bohai Rim region, which indicates that Beijing, as the capital, its technology innovation investment is very high, but its technology innovation effect level is relatively low. For Beijing, we should focus on raising the effect level in order to more effectively enhance the regional technology innovation capability. The other three provinces are important industrial and agricultural production base in northern China and their technology innovation effect is relatively slow. Neimenggu and Shanxi are China's resources provinces, whose technology innovation effect level is the least prominent.

From the order of the third factor scores, the factor scorces of Tianjin and Beijing are the highest, which indicates that their technology innovation efficiency is relatively high. The technology innovation efficiency in Shanxi, Hebei and Shandong provinces is the middle level in Bohai Rim region, which is directly related to their industrial structure and technology development level. Shanxi, Hebei and Shandong are important resources and agriculture provinces in north China and their technology innovation efficiency is relatively slow because of their different industrial structure. The factor scores of Neimenggu and Liaoning Province are the lowest, which indicates that their technology innovation efficiency is the lowest.

6 Conclusion

Through empirical testing and analysis, there are obvious differences about technology innovation capability among the various provinces and cities in Bohai Rim region. The technology innovation capability of Beijing and Shandong is strongest, and is in the leading position in Bohai Rim region. In order to promote coordinated economic development of Bohai Rim region, we think that the following aspects should be improved.

First, we should increase technology innovation input and pay attention to technology innovation efficiency. From the empirical analysis on regional technology innovation capability, the technology innovation level of Tianjin, Hebei, Shanxi is lower, and technology innovation effect of Shanxi and Neimenggu lags behind other provinces. So, we should increase the scientific and technological input in these provinces and cities, focus on digestion, absorption and innovation about the technology and enhance technology innovation effect.

Secondly, we should strength technology transfer, and promote interregional technology diffusion. Through analysis, we find that the technology innovation capability of Bohai Rim region shows a kind of gradient, which forms such technology development layer, that is, Beijing and Shandong as the core, to the central Tianjin and Hebei, and then out to the Liaoning, Shanxi and Neimenggu. In order to improve the overall technology innovation level of Bohai Rim region, we should transfer and diffuse advanced technology of Beijing and Shandong to the other provinces. At the same time, we also encourage the backward areas of Bohai Rim region to absorb actively the technology diffusion from the advanced regions, and then promote the flow and transfer of advanced technology.

Thirdly, we should optimize the industrial structure and upgrade regional technology innovation capability. Optimizing the industrial structure and enhancing the technology innovation capability are an important objective of regional economic development. Along with the deepening of reform and opening up, Bohai Rim has gradually become the focus of economic development in north China region. Because there are some differences in industrial structure of Bohai Rim region, and a number of industries have lost their comparative advantage, and some industries need to optimize and adjust, we should intensify efforts to adjust industrial structure, focus on the effect and efficiency of technology innovation, and increase scientific and technological input.

Acknowledgment. I express heartfelt thanks for Mrs. Zhou Hong and Mrs. Huang Zilang. They collect and sort data for this paper, proofread the whole paper, and put forward many suggestions for the completion of this paper.

References

1. Bertsimas, D., Freund, R.M.: Data, Models and Decisions: the fundaments of management science, pp. 283–292. China City Publishing House, Beijing (2002)
2. Li, X., Zhou, Z.: Study on the evaluation of regional technology innovation ability system and regional difference based on factor analysis model. Scientific Management Research 24, 5–10 (2006)
3. Fan, D., Zhou, H.: Research on the evaluation of regional technology innovation ability with factor analysis method. Industrial Technology Economy 25, 61–63 (2006)
4. Chi, R., Tang, G.: Study on efficiencies of regional techndofy innovation based on evaluation of inputs and performances. Science Research Management 25, 23–27 (2004)
5. Lv, Y., Hu, L., Fang, S.: Sequential Optimization and Assessment of Regional Technological Innovation. Forum on Science and Technology in China, 26–29 (Feburary 2000)
6. Yin, Y., Liang, L.: The Measurement of the Capability of Regional Technical Innovation System. China Soft Science, 71–74 (January 2001)
7. Luo, Y., Yang, Y.: SPSS statistical analysis: from the basics to practice, pp. 286–300. Publishing House of Electronics Industry, Beijing (2006)

Study of Project Risk Continuous Process Pattern*

Zhang Yali, Yang Huixiu, and Jiang Xiangmei

Management School, Northwestern Polytechnical University,
Xian, China
zhangyl@nwpu.edu.cn, Yanghuixiu2003@163.com, Hmkang87@yahoo.cn

Abstract. In order to accumulate risk management knowledge and reduce risk communication barrier, continuous process pattern of project risk management is suggested from the view of risk pattern and knowledge reuse. And then its application background, structure, advantages and disadvantages are analyzed. At last, an example is analyzed to point out that effective project risk management should be a continuous improvable process rather than a onetime activity.

Keywords: pattern, project risk management, risk continuous process pattern.

1 Introduction

Due to the complexity determined by the project scale and uncertainty of risks, people put forward many different risk management theories and methods in project management. These theories and methods often have differences in the fields, which let project managers not know what to do when choosing. If organizations or project teams can reuse the existing risk management knowledge which has to restructure or modify with some methods or technologies, they can build their own project risks strategies. To solve this issue, this paper proposes continuous process pattern of project risk based on pattern analysis and knowledge reuse. And then its application background, structure, advantages and disadvantages are analyzed. At last, an example is analyzed to point out that effective project risk management should be a continuous improvable process rather than a onetime activity.

2 The Concept of Pattern and Its Description

Pattern can be reused as "criterion or guidelines" through knowledge processing and extraction of experience, and also combined or modified by a variety of ways to suit different problem domains. Pattern describes a continuously recurring problem around us, as well as a core program to solve the problem. The user can reuse the program and no longer have to do the duplication. A pattern often has four basic elements [2]: pattern name, application situation and problems, solutions and effects. It provides a set of standardized vocabulary for users, and promotes communication in

* *Sponsors*-The humanstic social science and management revitalizing foundation of Northwestern Polytechnical University.

J. Luo (Ed.): Affective Computing and Intelligent Interaction, AISC 137, pp. 507–513.
springerlink.com

the project. This article studies the pattern related to project risk. A project risk management pattern describes a basic structure and proposed process of project risk management which is used in the specific domain scope and can be reused.

The emphasis of this paper about project risk pattern is how to design patterns to represent risk management knowledge rather than automatic processing and decision support based on computer system, so we describe it by general pattern based on the text non-formalization. And the static structure in pattern mainly adopts class diagrams of UML (Unified Modeling Language) used widely in business modeling, as shown in Figure 1. Structure diagram shows a basic element in the project risk management pattern, that is, object, represented by "class" in UML. Based on basic structure diagram, process diagram is introduced to describe the management process of current pattern, as well as the relationship between pattern's basic structure element and process.

Figure 1 gives some symbols commonly used in this paper. Structure diagram mainly describes general relationship (e.g. one to many and many to many), special relationship (inheritance, combination) and dependent relationship among different classes. For example, a "project" has many "risks"; a "risk" may appear in many "management processes"; a "personnel risk" may be divided into "personnel turnover risk" and "personnel malfeasance risk" by "inheritance"; several "members" constitute a "project team" by "combination". More detailed UML models can be seen in document [6]. Structure diagram gives a kind of typical aspect of concrete flow. In order to focus on key points and simplicity, it may ignore or simplify certain secondary processes.

3 Project Risk Continuous Process Pattern

Passive and active project risk management is beforehand and afterwards risk resolve strategies, then continuous project risk management believes that risk management should be a continuous improvable process rather than a onetime activity. In a number of actual projects, risk management is carried in the pre-project, but with the development of the project, many project managers unconsciously ignore or omit risk management because of many factors, such as progress pressure and resources tension. Document [7] believes that implementation of risk management needs budget and plan, and when the project needs to reduce the cost and speed up the time, the cost of risk management and its plan become the first victim. Thus, risk management becomes a onetime activity. Continuous risk management is based on such assumptions: for some risks, a onetime management process cannot succeed in eliminating their influence, but only change their status. The changes of the risks from serious status to acceptable status may go through several cycle processes, that is, risk management is a systems control process with feedback and adjustment.

At present, in risk management practice in many different fields, there is a consensus on risk management that should be a continuous improvable process rather than a onetime activity, but divided specific process is possibly different. For example, in the third edition in PMBOK (Project Management Body of Knowledge) published by Project Management Institute in 2004, project risk process should include: preliminary risk management plan, risk identification, qualitative risk analysis, quantitative risk analysis, risk response plan, risk monitor and control. Besides the input or output of each process as well as tools and technology used, PMBOK also newly increased a specific flowchart among project risk management processes. Detailed content on it is seen in document [8]. In order to highlight characteristics of continuous risk management, here we use another process division form which is widely used in the actual, that is, including identification, analysis (including evaluation, sorting and so on), response (or action plan), monitor and control, ignoring the risk management plan in the prior per- project.

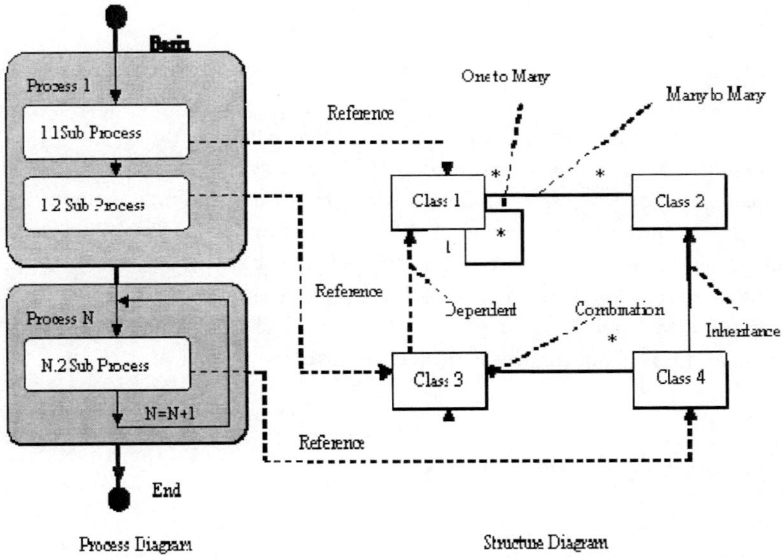

Fig. 1. The schematic of project risk management pattern

Any specific risk go through a above mentioned sequential process, and risk management activities present continuation , parallelism and iteration in the entire project life cycle. For example, while some risks are being tracked, some other new risks are distinguished and analyzed; alleviate plan for a certain risk produces other new risks in execution. Risk origin and its consequences change with the continuous progress of the project. Once all kinds of risks have already been distinguished and evaluated, the appropriate measures are taken as soon as possibly. Continuous tracking certain risks will obtain more and more related information, while the priority of risks may change. It is impractical to track all the risks in the entire cycle, which analyzes and evaluates the set of risks before each stage start, in order to choose those critical and important risks to keep tracking and continuous management, which is known as a major risk list (Top N) technology.

4 The Constitution of Project Risk Continuous Process Pattern

4.1 Name

Project Risk Continuous Process Pattern, which is also known as risk management pattern of full project life cycle.

4.2 Goal

By continuously tracking, gradually improving and adjusting action plan for some important risks, we implement whole process management in the entire project risk life cycle.

4.3 Motive

For some projects with high risks, risk management has become an essential project management work, and risk management needs necessary budget and formal plan. Project managers need a perfect, standard and operational risk management approach to deal with complex risk management work in the project. At this time, continuous risk management process as a "heavy" approach is the first choice for ensuring project quality and delivery on time. Particularly, certain risks are from distinguished to control in the affordable range and risks occurring influence project for a period of time, and even throughout the whole project life cycle.

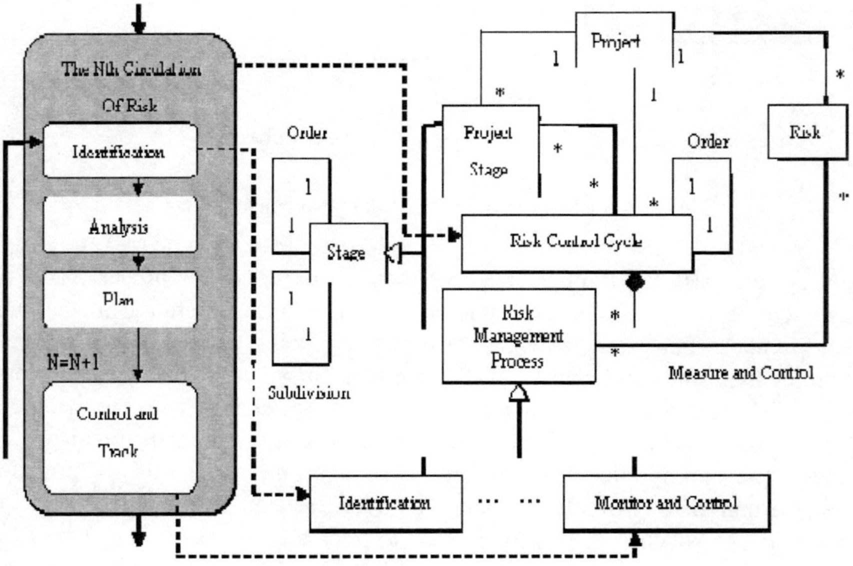

Fig. 2. Project Risk Continuous Process Pattern

4.4 Applicability

Project Risk Continuous Process Pattern is suitable for the below situation: (1) complex projects with a long period, high costs and more than medium scale; (2) Organizations or project teams have perfect risk management institutions and operational risk control program. Project managers and members have rich risk experience and risk environment; (3) project risk management as an independent project is outsourced to the professional third-party. However, for the projects lacking of risk management experience and institutions support, complexity of continuous risk management in the implementation may give the project "trouble", or unsuitable risk management doesn't actively control existing risk but increases new extraneous risks.

Structure and process: see Figure 2.

4.5 Participating Objects

Continuous risk management introduces "risk control cycle" which is similar to Deming cycle in quality management. Obviously, a "project" may experience many "risk control cycle". According to different conditions, such as project size and duration, a "risk control cycle" may cross many "project stages", and a "project stage" may contain many "risk control circulations". A "risk control circulation" is composed of many risk management process, such as risk identification, risk analysis, risk plan, risk control and track. Main function of "Risk management process" is that it measures or controls risks, and evaluates different risks and sorts the rank of risk priorities for managing many risks in more situations. Certain risks that are serious and difficult in control may leap many "risk control circulations", and need to control in many risk management processes. But certain unimportant low risks may experience "recognition" and "analysis" in a "risk control circulation", so that they are excluded from the risk action plan.

Fig. 3. Continuous risk management CRM

"Risk management process" and " project stage" have subdivision and order in all the "stage", that is, a big stage may be divided into several smaller stages, and the same level of stages have fixed orders. Different "risk control circulation" usually only has the relationships of orders. For example, SEI (Software Engineering Institution) proposed a famous risk control circulation, which is CRM (Continuous Risk Management) model [9] in Figure 3, including: (1) risk identification; (2) risk analysis; (3) risk plan; (4) risk track; (5) risk control; (6) communication and documents.

4.6 Example

Although the CRM model is a risk management model in software engineering domain, this continuous management thought has been widely applied in other domains and become the general principles of present risk management. For example, Risk Management Guide for military acquisition project was established by the U.S. Department of Defense in August, 2006 [10]; The United States national Aeronautic used CRM to improve their daily risk management [11]; System approach for ORM (Operation Risk Management) is established by the US Department of Health and Human Services and Food Safety and Applied Nutrition Center in the United States Food and Drug Administration [12]. Besides these, in fact, many domains such as developing weapon equipment and acquisition, transportation, and construction industry also bring continuous management thought based on life-cycle into their risk management principle and practice [13][14][15].

4.7 Related Pattern

In project risk management including proactively planning pattern and continuous process pattern, which have the close relationship, the detailed regulations are made for each requirements and the content of project risk management in proactively pattern, and also include preplanned and guidelines for project possible risks. The risk response plan formed in each risk continuous control cycle can be seen as the essential supplement for the preplanned and a specific implementing scheme. Sometimes the principles and the whole request in the entire project risk management may be added or modified according to the current action plan.

5 Conclusion

Using project risk continuous process pattern can ensure project managers to keep tracking the dynamic change about main risks in the project, thus adjusting the action plan for risks in time. For every risk managers, this routine work should make them feel more approved and comfortable, instead of the feeling of contravene and compulsion, just like a good habit that naturally infiltrates into the routine. It should be emphasized that any risk management method including continuous risk management process is not suitable for all the projects as the general solution. In order to achieve the effectiveness of risk management, it's necessary to appropriately "integrate" the pattern according to the actual demand and organizational culture.

References

1. Fowler, M.: Analysis Patterns: Reusable Object Models. Addison-Wesley Professional (1996)
2. Eriksson, H., Penker, M.: Business Modeling with UML, Business process at work. John Wiley&Sons (2000)
3. Gamma, E., Helm, R., Johnson, R., et al.: Design patterns: elements of reusable object-oriented software. Addison-Wesley Longman Publishing Co., Inc., Boston (1995)
4. Alistair, C.: Project risk reduction patterns (1997), http://www.riehle.org/community-service/hillside-group/europlop-1997/p22final.pdf
5. Dietrich, J., Elgar, C.: A Formal Description of Design Patterns Using OWL. In: Proceedings ASWEC (2005)
6. Henninger, S., Ashokkumar, P.: Disseminating Usability Design Knowledge through Ontology-Based Pattern Languages. In: 3rd Int'l Semantic Web User Interaction Workshop, Atlanta, GA (November 2006)
7. Tom, W.: The Risk of Risk Management. PMI Global Congress (2005)
8. Pmi, A.: Guide to the Project Management Body of Knowledge-PMBOK Exposure Draft, Pennsylvania (2003)
9. Murphy, R.L., Alberts, C.J., Williams, R.C., et al.: Continuous Risk Management Guidebook, SEI Carnegie Mellon University (1996)
10. Defense USD, Risk Management Guide for DoD Acquisition (2006), http://ax.losangeles.af.mil/axl/Risk%20Mgmt%20Guide%20for%20DOD%20Acq.pdf
11. Rosenberg, L.H., Hammer, T, Gallo, A.: Continuous Risk Management at NASA. In: Applied Software Measurement/Software Management Conference
12. Department of Health and Human, Services, Administration U S F a D, Center for Food Satety and Applied, Nutrition. Food Safety and Security: Operational Risk Management Systems Approach (2001), http://www.dhs.ca.gov/fdb/local/PDF/Food%20Safety%20and%20Security11-29.PDF
13. Chen, G., Huang, H., You, J.: Study on Life Cycle Risk Management of Metro. Chinese Journal of Underground Space and Engineering 2, 42–47 (2006)
14. Li, Z., Tang, S., Li, J., Chen, J.: Study on Acquisition Risk Identification for Weapo and Equipment Based on the Full Life Cycle. Engineering Science (1) (2005)
15. Zhang, C., Lu, B.: Research on LC based Integrated Risk Management of Construction Project. Port Engineering Technology (1) (2006)

Centralized Procurement of Construction Enterprises Based on SCMS

Wu Yunna and Lin Ping

Department of Economic & Management
North China Electric Power University, NCEPU
Beijing, China
feiyang688056@126.com, Linping.0914@163.com

Abstract. As construction enterprises are facing fierce market competition, the profit margin of which is shrinking. Generally speaking, costs of materials account for 60%~70% of construction total costs and even more, so by reducing enterprises' procurement costs will greatly improve economic efficiency of enterprises. In this paper, centralized procurement of construction enterprises based on Supply Chain Management System (SCMS) is proposed. The new procurement model proposed in this paper, on one hand, making enterprises can procure materials and supply project goods in real time. On the other hand, enterprises will achieve scale benefit by centralized procurement. Overall, enterprises procure centrally based on SCMS can reduce procurement costs and capital occupation by overstock, and achieve the goal of reducing cost and increasing profit, enhance enterprises' market competitiveness.

Index Terms: supply chain management system, construction enterprises, centralized procurement.

1 Introduction

Currently, most construction enterprises neglect material management, without unified management, and project companies procure decently, which lead to a large backlog of inventory, capital waste and occupation. But with increasingly fierce market competition, material procurement management is playing a more and more significant role in enterprises and which arouse enterprises' attention [1]. In recent years, as the flourishing of information technology and network at home and abroad, enterprises' traditional material purchasing mode suffers a hard hit, but enterprises also get an unprecedented opportunity to excavate efficiency of materials procurement. Therefore, enhance the research and application of information technology in procurement of construction enterprises will bring in enormous economic benefits. The propose of SCMS brings construction enterprises with a new development platform for procuring and managing materials, by the full used of the platform, enterprises can greatly improve the response speed of procurement and obtain a wider choice of materials suppliers, which reducing procurement costs and procurement unit, achieving "sunshine procurement[2]." In addition, enterprises procure materials in the manner of advantages, favorable to reduce inventory, capital

J. Luo (Ed.): Affective Computing and Intelligent Interaction, AISC 137, pp. 515–520.
springerlink.com

occupation and management costs. Therefore, in this paper, centralized procurement model based on SCMS is put forward by combining above ideas. Centralized purchasing by used of information technology can overcome the occurrence of bullwhip effect. Relying on SCMS and enterprise scale advantages, enterprises byintegrating logistics resources and implementing the group operation, achieve intensive procurement and specialized distribution, reduce enterprises' procurement costs and get more economic benefits, enhance their market competitiveness.

2 Centralized Procurement

2.1 Definition of Centralized Procurement

For construction projects, material costs generally account for 60%~70% of construction total costs, enhancing project materials management is the most effective manner to reduce project costs. In addition, in all aspects of materials management, purchasing has a greater potential. Therefore, construction enterprises should strengthen the modernization process of materials procurement management. Currently, most large-scale construction enterprises procure materials in the model of hybrid procurement. In the other words, enterprises procure bulk materials centrally, the project purchase fragmented materials, which largely reduce procurement costs and increase enterprises' ability to withstand market risk. Centralized purchasing, firstly, collect and unity multiprojects materials procurement needs to form a large purchase order. Secondly, take a comprehensive performance assessment and an inquiry comparison between suppliers, then carry out a preferential procurement, accessing to integrity control the quality of goods and suppliers' service. Finally, unify inventory and settlement. Centralized purchasing can reduce enterprises' procurement costs, generally speaking, which is applicable to bulk goods, high value materials, key components and key raw materials, or other strategic resources

2.2 Advantages of Centralized Procurement

Construction Enterprises procure centralized can keep the strategic cooperation relation with suppliers, even if which have many projects branch, and reduce group's overall procurement costs. Centralized purchasing increase the transparency of procurement and expand the selecting range of suppliers, standardized procurement processes, improve transaction efficiency and ensure the quality of goods, reduce procurement costs depended on group scale advantages and coordinate resources between projects effectively. Construction enterprises benefit from centralized purchasing, but at the same time, enterprises face more difficulty in management, so enterprises' Centralized purchasing should focus on materials and equipment, which are in great value and facilitate to purchase centralized. According to materials management classification method-ABC, we know that materials variety of kind A are less, but have great value, which are consistent with the characteristics of goods purchased centrally. Thence, construction enterprises should seize the key, economically supply materials of kind A and which should be in high quality. Through centralized purchasing materials of kind A, achieve scale advantages and strengthen construction enterprises' power of bargaining, to reduce procurement costs and thereby enhance enterprises' competitiveness.

2.3 The Necessity of Introducing Information Technology

Nowadays, information technology is developing highly, enterprises pay more and more attention to optimize and coordinate resources, centralize purchasing is an effective means for optimizing and coordinating resources. If centralized purchasing depends on open network environment, which will support both supplier and buyer to transfer supply and demand information rapidly, to achieve information sharing, partnership, and last to multi-win-win situation [3]. Therefore, entralized purchasing relying on advanced information technology, information platform and information tools, can achieve information centralization and sharing, optimize resources allocation from the global perspective, avoid unnecessary inventory and waste, and also avoid the occurrence of bullwhip effect[4].

3 SCMS Applied in Centralized Procurement

3.1 SCMS Applied in Centralized Procurement

Construction enterprises' supply chain management system helps to deal with the relation between enterprises, suppliers and project companies. This system help to share information of goods orders, inventory levels, goods distribution and delivery, and help enterprises to purchase, distribute and delivery materials more effectively, in order to deliver the most suitable materials from a supplier or enterprises' storage to the right construction site in the shortest time and lowest cost, ensure the materials needs of projects. SCMS is a cross-organizational system [5] (Shown in Figure 1), which transfer materials information across organizational boundaries (or corporate boundaries), transfer enterprises' information to other companies through electronic connection, such as uppliers and project companies.

Construction enterprises' centralized procurement, including centralized procurement planning and execution, so SCMS is made up of supply chain planning system (SCPS) and supply chain execution system (SCES). SCPS makes enterprises draw up reasonable centralized ordering plan, inventory and goods delivery plan and so on according to materials application and project plan. SCES makes enterprises' purchasing department and warehouse manage materials flow, to ensure that transporting goods to corresponding project sites in the most effective way, track the transportation, use and storage of goods, and providing more real-time information of orders and inventory status to ensure that the availability of projects' materials[6]. Construction enterprises ensure that all partners on the supply chain can communicate timely through SCM, to adjust materials requirements planning, procurement planning, production planning and transportation planning by used of materials real-time information. Purchasing departments enter into suppliers system through web interface, to ensure that whether inventories and production planning match materials needs, in order to submit orders application, centralized procurement of construction enterprises based on SCMS is shown in Figure 2. The supply chain management system in this paper is a demand-driven model [5] (purchasing and production according to materials requirements planning, is also known as a pull model. Project companies' behavior of application for materials arouse other behaviors, such as behaviors of adding inventory automatically and centralized purchasing, the processes of this model are shown in Figure 3, information flow in the direction of the supply chain and logistics flow in the opposite direction back to projects companies.

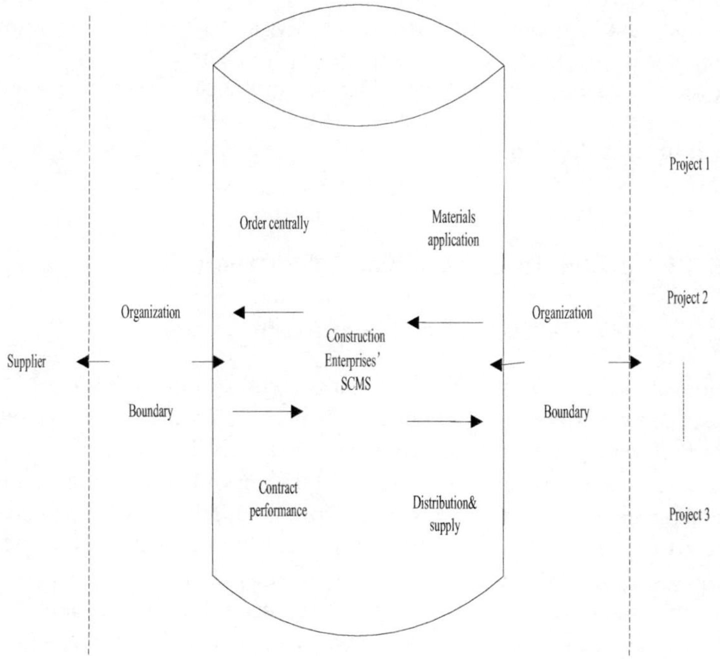

Fig. 1. Structure of Supply Chain Management across Organization

3.2 SCMS Optimize Centralized Procurement

Construction enterprises implement centralize procurement by used of a network and integrated supply chain system, which make the supply and demand of materials match , reduce the uncertainty of supply and demand to overcome bullwhip effect, reduce inventory and improve services of procurement and distribution, realize to supply materials in real-time and thus more effectively use of corporate funds,. Reducing the procurement costs of construction enterprises depending on the supply chain network can greatly enhance the profitability of enterprises, SCMS applied in centralized procurement will greatly enhance enterprises competitive advantages[6], which are shown in Figure 4.

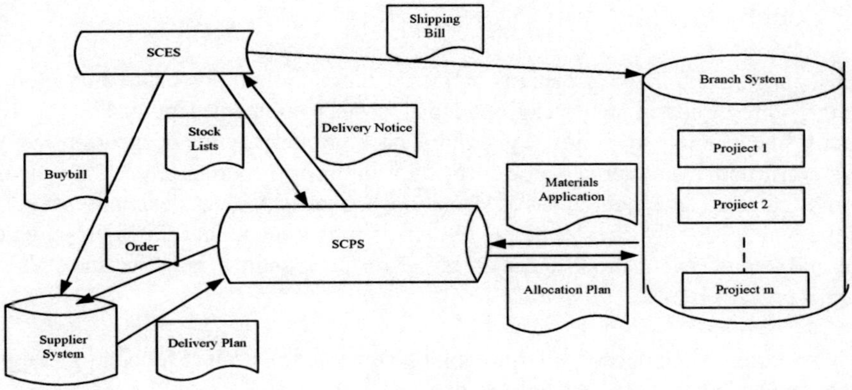

Fig. 2. SCMS of Construction Enterprises

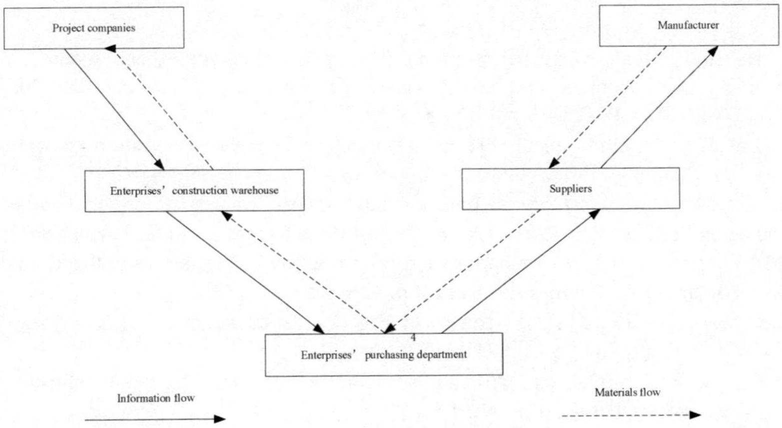

Fig. 3. Supply Chain System-Push Model

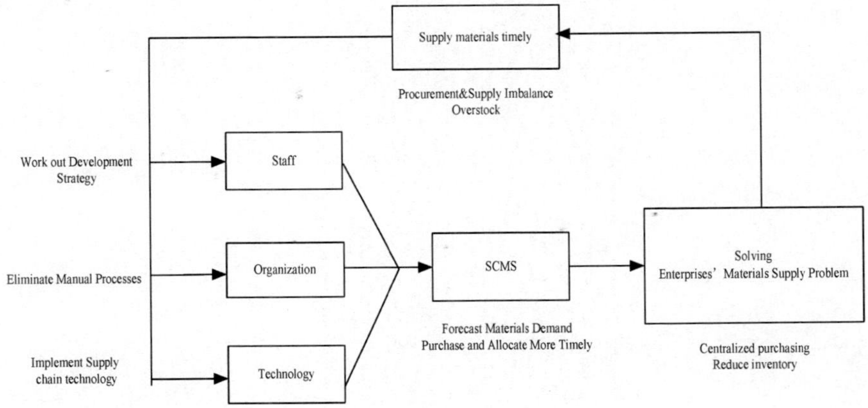

Fig. 4. Construction Enterprises Competitive Advantages based on SCM

4 Conclusion

In this paper, centralized procurement model based on supply chain management system is proposed; in the background that project companies purchase materials decently and leading to projects' inventory backlog and capital occupation. SCMS offers construction enterprises an information platform to purchase centrally, which improve enterprises procurement information and procurement response speed, to achieve scale economy of enterprises centralized procurement, cut down procurement costs and capital occupation, and then gain greater economic benefits and enhance enterprises' market competitiveness.

Acknowledgment. Support by Beijing jointly construction project foundation-Beijing project management characteristics Major.

References

1. Liu, H.: Large-scale construction enterprises regional procurement mode based on electronic procurement technology. Construction Economy 322(8), 62–65 (2009) doi: 1002/851X.2009.08-0062-04
2. Ma, J., Bai, X.: A study on the use of E-Commerce in power generation groups original purchasing, Tianjin University master's Thesis, p. 3 (December 2006)
3. Peter, F.: Moduler, Assembly in the car industry: An analysis of organizational forms. Influence on Performance. Purchasing & Supply Management 26(8), 221–233 (2002)
4. Li, M., Pang, L.: Study on the application of real estate business's centralized purchasing system, Fudan University master's Thesis, p. 4 (September 2008)
5. Laudon, K.C., Laudon, J.P.: Essentials of Business Information System, Translations of Business Administration-Seventh Edition
6. Xu, Y., Li, H.: Purchasing and supply chain management in manufacturing. Fudan University master's Thesis, p. 7 (April 2009)

A Novel Method of P2P Traffic Classification Based on TCP Flow Analysis

Peng Jianfen[1,2,3], Zhou Yajian[1,2,3], Wang Cong[1,2,3], and Yang Yixian[1,2,3]

[1] Information Security Center, State Key Laboratory of Networking and Switching Technology, Beijing University of Posts and Telecommunications, Beijing, China
[2] Key Laboratory of Network and Information Attack & Defence Technology of MOE, Beijing University of Posts and Telecommunications, Beijing, China
[3] National Engineering Laboratory for Disaster Backup and Recovery, Beijing University of Posts and Telecommunications, Beijing, China
horsepjf@bupt.cn, yajian@bupt.edu.cn, wangc@bupt.edu.cn, yxyang@bupt.edu.cn

Abstract. Peer-to-Peer (P2P) applications have overtaken web stream as the most significant portions on the high-speed network, so P2P applications identification is important to a broad range of network operations. By deriving the transport/network layer headers of the packets of TCP flow, we obtained some attributes of all kinds of P2P traffic without relying on packet payload and port number, which leaded to a novel method for P2P traffic identification based on support vector machine (SVM). The method only needed to deal with the TCP packet of SYN and SYN+ACK flags. Experiment results show this method classifies the traffic achieved the high accuracy.

Index Terms: P2P, traffic classification, TCP flow, SVM.

1 Introduction

With the development of network, traffic identification and classification becomes more and more important for network management and design, routing optimization. From the emerging of P2P technology at 1990's, because of its advantages on file sharing, distributed computation and etc., P2P applications have been gaining popularity. Ipoque's Internet Study 2008/2009 showed P2P still generates by far the most traffic in all monitored regions – ranging from 43 percent in Northern Africa to 70 percent in Eastern Europe[1]. Therefore, it is key role to identification P2P traffic accurately for traffic management.

P2P applications use a fixed port for all their communication in the beginning, So known transport-layer port numbers can be used to an accurate and efficient way for traffic classification [2,3] . However, in order to evade port-based identification method, P2P applications use dynamic or variable ports and disguise traffic using ports designated to other traffic-such as ports 25(SMTP) and 80(HTTP),or allow and often encourage users to set the port that the applications will use. S.sen et al[4] found P2P application signatures. Consequently, the P2P traffic can be identified through

J. Luo (Ed.): Affective Computing and Intelligent Interaction, AISC 137, pp. 521–528.
springerlink.com

analyzing the characteristic bit strings in packet payload [4-5]. However, the method based on signature payloads has the following shortage:

- bears some legal and privacy issues;
- unable to identify the encrypted or new P2P application packets.

Current some P2P applications use dynamic port numbers, HTTP masquerading and encrypt payload to prevent being identified on the basis of port-based and payload identification method. In order to identify P2P application, Traffic classification methods based on the statistical properties of flows, not using port numbers and payload, have been proposed [6-12]. Reference [12] uses neural networks for classification. Neural networks use the empirical risk minimization (ERM) principle, which minimizes the error on the training data, while SVM proposed by Vapnik [13] uses the structural risk minimization (SRM) principle , SVM based on statistical theory are a new promising supervised learning methods and used for classification and regression.

Some scholars do researches on traffic classification by using SVM [14-16] in recent years. Reference [14] presents a P2P traffic identification approach based on SVM by analyzing packet length, remote host's discreteness, connection responded success rate and the ratio of IP and port at the host level. Its experiment results show the approach can effectively identify P2P application. However, the approach needs to deal with every packet of traffic when calculating the packet length, it costs a lot of time and storage space.

This paper proposed an accurate P2P identification method using SVM on basis of effective feature selection. Our method only deals with the TCP packet of SYN and SYN+ACK flags, so its computational complexity is low. The experimental results show that accuracy of our scheme is high.

The remainder of this paper is organized as follows. In Section II, we describe the methodology. In section III we discuss the evaluation and the experiment results. We give the conclusions and expand our future work in Section IV.

2 Our Proposed Solution

Our solution classifies between P2P and non-P2P application based on SVM.

2.1 Dataset Collection

Our non-P2P application data sets come from Lincoln Laboratory Massachusetts Institute of Technology [17], P2P trace is collected by using Wireshark under campus Ethernet.P2P trace includes BITTOR and EMULE application, non-P2P application datasets are composed of FTP, SMTP, and HTTP application.

2.2 Classification Feature

In traditional client – server(C/S) mode, a common host would connection with all kinds of servers in the Internet at a much higher success rate. In contrast to it, the Character that P2P host acts as both a client and a server to its peers. Because of the dynamics of P2P system, a P2P host can continually initiate connections with other online hosts in a relative low probability so as to keep its download speed stable. On the other hand, opening a TCP connection involves exchanging three packets. These packets are SYN, SYN+ACK, and ACK.

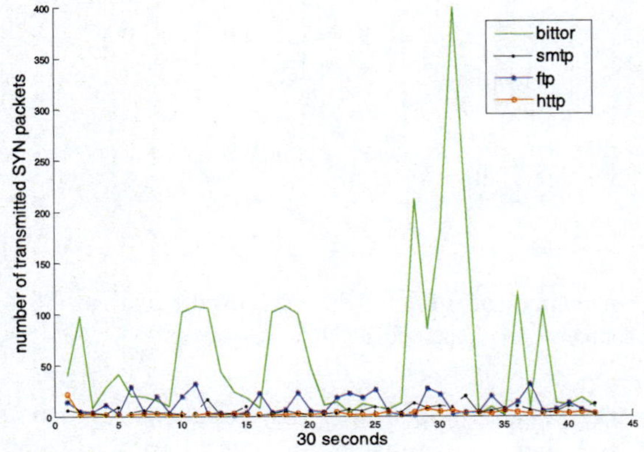

Fig. 1. Number of transmitted SYN packets every 30 seconds

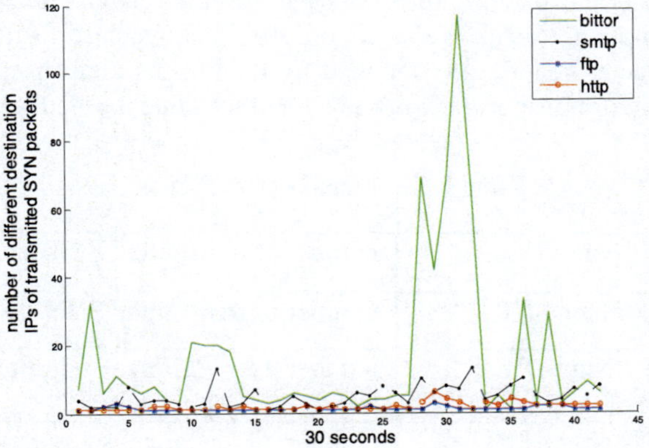

Fig. 2. Number of different destination IPs of transmitted SYN packets every 30 seconds

Fig.1 and Fig.2 represent four kinds of traffic number of transmitted SYN packets and number of different destination IPs of transmitted SYN packets every 30 seconds respectively. Fig.1 and Fig.2 show the value calculated from P2P traffic is several times higher than these of traditional traffic.

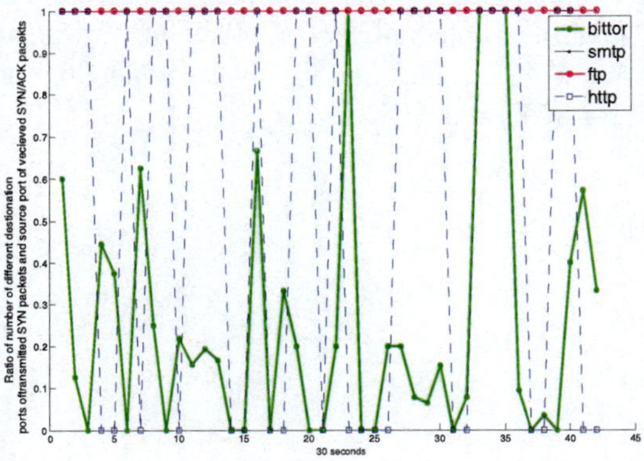

Fig. 3. Ratio between number of different destination port of transmitted SYN packet and number of different source port of received SYN/ACK packets

In C/S mode, servers usually open only one certain port, such as 80, 25, for the clients to connect, and clients open many consecutive ports to connect the server. In comparative with it, P2P application opens more port due to P2P host acts as both a client and a server to its peers. From Fig.3, we can find port ratio of BITTOR traffic value is smoother than those of traditional traffic, port ratio value of HTTP traffic is either 0 or 1, however, port ratio value of SMTP and ftp traffic are almost 1.

As we described above, some features have different characteristics between P2P traffic and conventional traffic. We select 6 traffic features described in Table 1. The values of our selected feature are obtained by monitoring and observing the packet header information of transport/network layer and its value is calculated.

Table 1. The Features of TCP Flow

0	Num-SYN	number of transmitted SYN packets
1	Num-ACK	number of transmitted SYN /ACK packets
2	Num-DIP	number of different destination IPs of transmitted SYN packets
3	Num-SIP	number of different source IPs of received SYN/ACK packets
4	Num-DPort	number of different destination port of transmitted SYN packet
5	Num-SPort	number of different source port of received SYN/ACK packets

2.3 SVM

In general, the classification problem on the n-dimensional space is to find real function $g(x), x \in R^n$, a training set of l sample points given:

$$T=\{(x_1,y_1),\cdots(x_l,y_l)\}\in(\chi\times\gamma)^l \tag{1}$$

where the ith sample $x_i\in\chi=R^n$, $y_i\in\gamma=\{-1,1\}$, thus the decision function $y=f(x)=\mathrm{sgn}(g(x))$ decide the corresponding values y of x.

SVM[18] performs classification by constructing an n-dimensional hyperplane $w^T z+b=0$ that optimally separates the data into two classes such that the distance of both classes from the hypeplane is maximized. So the decision function is given by $f_{w,b}=\mathrm{sgn}[w^T\phi(x)+b]$, where $\phi(x)$ map the input space R^n to a high-dimensional feature Z space, w is a weigh vector and b is a threshold. The SVM classification amounts to finding w and b to satisfy (primal problem):

$$\min_{w\in Z,b\in R}\frac{1}{2}w^T w+C\sum_{i=1}^{l}\varepsilon_i$$
$$s.t.\begin{cases} y_i[w^T\phi(x_i)+b]\geq 1-\varepsilon_i, i=1,\ldots,l \\ \varepsilon_i\geq 0, i=1,\ldots,l \end{cases} \tag{2}$$

where C determines a trade-off between the error on the training set and separation of the two class. The Lagrangian dual problem:

$$\min_{\alpha}\sum_{i=1}^{l}\sum_{j=1}^{l}y_i y_j\alpha_i\alpha_i\phi(x_i)\phi(x_j)-\sum_{i=1}^{l}\alpha_i$$
$$s.t.\begin{cases} \sum_{i=1}^{l}y_i\alpha_i=0 \\ 0\leq\alpha_i\leq C, i=1,\ldots,l \end{cases} \tag{3}$$

Inner product may be defined as kernel function:

$$K(x_i,x_j)=\phi(x_i)\phi(x_j) \tag{4}$$

We choose radical basic function (RBF), $K(x,y)=e^{-\gamma\|x-y\|^2}$ as our kernel function, reason are listed as follows [19].

- RBF kernel becomes linear kernel when using specific parameter value;
- RBF kernel and sigmoid kernel perform similar when using specific parameter value;
- RBF only need tune one parameter, rendering less complexity;
- Its output is between 0 and 1, rendering less numerical difficulties. One key point is $0<K_{ij}\leq 1$ in comparison with polynomial kernels of which kernel values may go to infinity $(\gamma x_i^T x_j+r>1)$ or zero $(\gamma x_i^T x_j+r<1)$ while the degree is large.

2.4 The Process of Classification

The process of P2P traffic identification as follows:

Step 1. Datasets collection. Section 2.1 shows the process of dataset collection.

Step 2. Construct the training dataset. Every 30 seconds, we calculate the Num-SYN, Num-ACK, Num-DIP, Num-SIP, Num-DPort, Num-SPort values from SYN and SYN+ACK TCP flow of collected datasets and label the class of traffic.

Step3.Using sample data as the training data, we construct the classification models based on SVM, and evaluate the results.

3 Experiment Results

The performance of the proposed method is tested in the environment of LibSVM-2.89. The definition of false positive and false negative is the same as in [14]. False positive indicates how many non-P2P flows are wrongly classified as P2P, while false negative means how many P2P flows are wrongly classified as non-P2P. Table 2 shows the false positive and false negative rates of 3 dataset are showed.

Table 2. The False Positive and False Negative Reates of 3 Dataset

Data set		Selected feature	false positive	false negative
P2P	Non-P2P			
21.7MB	6.62MB	0,1,2,3,4,5	0.58%	14.29%
		2,3,4,5	0.00%	9.52%
21.7MB	12.5MB	0,1,2,3,4,5	0.29%	12.90%
		2,3,4,5	0.00%	9.53%
1.8GB	39.8 MB	0,1,2,3,4,5	8.67%	0.00%
		2,3,4,5	8.00%	0.00%

We find that the combination of [2,3,4,5] is the best combination of features for classification, false positive and false negative rates of 3 dataset based on these features are below 10%. The experiment results of the other combination of features which don't list in Table 2 are unstable and greatly affected by the number of sample data.

The identification accuracy of combination of feature [0,1,2,3,4,5] is lower than combination of feature [2,3,4,5],we find the same packets are sent more than once due to connection timed out, such as SYN packets. When combination of feature [2,3,4,5] are selected, the values of feature calculated come from different packets and response characteristics of P2P application accurately.

In a word, the above experiment results show that success rates are high and stable, which proves that the method is effective and feasible for P2P traffic identification.

4 Conclusions

An SVM based method is developed in this paper to classify the P2P application and traditional application, the best combination of features by deriving the transport/network layer headers of the packets of TCP flow are found. We obtain some attributes of all kinds of P2P traffic without relying on packet payload and port number. Experiment results show this method achieved the high accuracy and stability of classification. We further test the method with new datasets in the future.

Acknowledgment. This work was supported by the National Natural Science Foundation of China (No.60972077), Beijing Municipal Natural Science Foundation(No.9092009, No.4092029).

References

1. Internet Study (2008/2009), http://www.ipoque.com/resources/internet-studies
2. Constantinou, F., Mavrommatis, P.: Identifying Known and Unknown Peer-to-Peer Traffic. In: Proceedings of the Fifth IEEE International Symposium on Network Computing and Applications (2006)
3. Sen, S., Wang, J.: Analyzing peer-to-peer traffic across large networks. In: Proceedings of the 2nd ACM SIGCOMM Workshop on Internet Measurment, Marseille, France, pp. 137–150 (2002)
4. Sen, S., Spatscheck, O., Wang, D.: Accurate, Scalable In-Network Identification of P2P Traffic Using Application Signatures. In: WWW, New York, USA, May 17-22, pp. 512–521 (2004)
5. Roughan, M., Sen, S., Spatscheck, O., et al.: Class-of-service mapping for QoS: a statistical signature-based approach to IP traffic classification. In: Proceedings of the 4th ACM SIGCOMM Conference on Internet Measurement, Taormina, Sicily, Italy, pp. 135–148 (2004)
6. Constantinou, F., Constantinou, F., Mavrommatis, P.: Identifying Known and Unknown Peer-to-Peer Traffic. In: Fifth IEEE International Symposium on Network Computing and Applications, NCA 2006, pp. 93–102 (2006)
7. Erman, J., Arlitt, M., Mahanti, A.: Traffic classification using clustering algorithms. In: Proceedings of the 2006 SIGCOMM Workshop on Mining Network Data, pp. 281–286. ACM Press, New York (2006)
8. Crotti, M., Gringoli, F., Pelosato, P., Salgarelli, L.: A Statistical Approach to IP-level classification of network traffic. In: IEEE Int. Conf. Commun (ICC), pp. 170–176 (June 2006)
9. McGregor, A., Hall, M., Lorier, P., Brunskill, J.: Flow Clustering Using Machine Learning Techniques. In: Passive & Active Measurement Workshop, France (2004)
10. Karagiannis, T., Broido, A.: Transport Layer Identification of P2P Traffic. In: Proceedings of the 4th ACM SIGCOMM Conference on Internet Measurement, pp. 121–134. ACM Press, New York (2004)
11. Kim, M.-S., Kang, H.-J., Hong, J.W.: owards Peer-to-Peer Traffic Analysis Using Flows. In: Brunner, M., Keller, A. (eds.) DSOM 2003. LNCS, vol. 2867, pp. 55–67. Springer, Heidelberg (2003)

12. Auld, T., Moore, A.W., Gull, S.F.: Bayesian Neural Networks for Internet Traffic Classification. IEEE Trans. Neural Networks 18(1), 223–239 (2007)
13. Vapnik, V.: The nature of statistical learning theory. Springer, New York (1995)
14. Liu, F., Li, Z.T., Nie, Q.B.: A New Method of P2P Traffic Identification Based on Support Vector Machine at the Host Level. In: 2009 International Conference on Information Technology and Computer Science, ITCS, vol. 2, pp. 579–582 (2009)
15. Zhou, L.-J., Li, Z.-T., Hao, T.: Proposition and Provement of A TCP Feature of P2P Traffic- An Example of BitTorrent and Emule. In: Proceedings of Second International Conference on Communications and Networking in China, pp. 61–65 (2007)
16. Gao, Z., Lu, G.M., Gu, D.Q.: A Novel P2P Traffic Identification Scheme Based on Support Vector Machine Fuzzy Network. In: Second International Workshop on Knowledge Discovery and Data Mining, Moscow, Russia, pp. 909–912 (2009)
17. http://www.ll.mit.edu/mission/communications/ist/corpora/ideval/data/index.html
18. Cristianini, N., Shawe-taylor, J.: An introduction to support vector machines, and other kernel based learning methods. Cambridge University Press, Cambridge (2000)
19. Wang, R., Liu, Y., Yang, Y., Zhou, X.: Solving P2P Traffic Identification Problems Via Optimized Support Vector Machines. In: IEEE/ACS International Conference on Computer Systems and Applications (AICCSA 2007), pp. 165–171 (May 2007)

Adaptive PID Control Strategy for Nonlinear Model Based on RBF Neural Network

Changliang Liu[1], Fei Ming[1], Gefeng Ma[1], and Junchi Ma[2]

[1] Department of Control Theory and Control Engineering
North China Electric Power University Baoding, China, 071003
changliang_liu@163.com
[2] Jianbi Power Plant of China Guodian Corporation
Zhenjiang, China, 212006
wsmingfei@126.com

Abstract. As the actual industrial processss are nonlinear, time-varying, large delay and uncertain , it is difficult to establish accurate mathematical model, As a result, it is impossible to get good control effect with conventional PID controller. In this paper we put forward a new method of compound control based on RBF neural network and PID. This method only need we give the sketchy PID controller parameters, the neural network makes the optimal adjustment to the PID controller parameters, and gains the good control performance finally. Simulation results indicate that with the proposed controller the adaptability, strong robustness and satisfactory control performance are superior to those of the conventional PID controller in the nonlinear and time-varying system.

Index Terms: RBF neural network, PID controller, nonlinear system.

1 Introduction

Conventional PID controller now is used extensively in industrial process due to its simple structure, high reliability and easy to implement. However, this controller has many limitations, while used in linear system where the model of the object we can easily get, the results will be good. As we all know, in the actual industrial process, the controlled objects are always nonlinear, time-varying and difficult to establish accurate mathematical model. When the controlled object is changed, the parameters can not adjust automatically to adapt to the changed condition [1].With theses problems, the limitations of traditional PID controller exposed day by day. We can not get the ideal result in actual application.

Radical Basis Function neural network is a kind of local approaching network, It can approach the continuous function with the discretional precision. It has fast convergence, strong generalization, simple structure and other characteristics. It has been successfully used in robot control, fuzzy control, nonlinear PID control, and other fields [2].

Based on the above advantages of RBF neural network, This paper combines RBF neural network and traditional PID and then designs a kind of composite and parallel controller to solve problems which can not be solved with traditional PID controller

J. Luo (Ed.): Affective Computing and Intelligent Interaction, AISC 137, pp. 529–535.
springerlink.com

and strengthen the control effect. With RBF-PID, we can not only adjust the PID parameters of k_p , k_i , k_d , but also can get a set of optimal controller parameters through self-learning ability, then the control effect will be better [3].

2 The Structure of RBFNN [4]

Figure 1 is the structure of RBF neural network. The RBF neural network is a three-layer feed-forward neural network. The first layer is input layer, the input layer only pass messages to the hidden layer. The second layer is hidden layer, the number of the units determined by the controlled object. The third layer is output layer, which output the response of the input signal. As to RBF neural network, the basis function of hidden layer is distance function.

The mapping of RBF neural networks is nonlinear from input layer to output layer, while from the hidden layer to the output layer is linear, thus greatly accelerates the learning speed and avoid the local minimum problem.

The function of hidden layer is selected according to actual needs. In this paper we use Gaussian function [5].

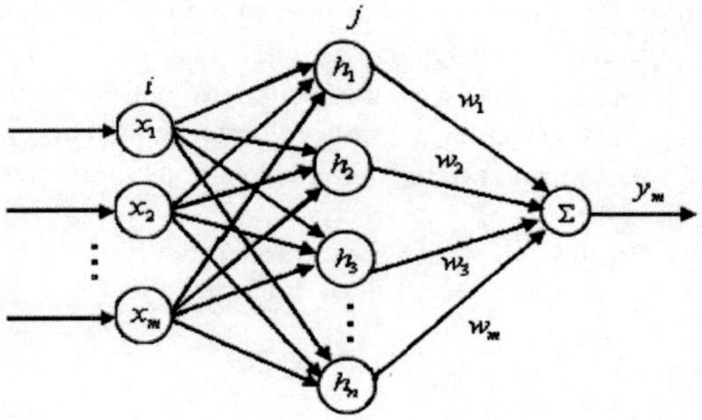

Fig. 1. RBF neural network's structure

In RBF neural network there are three parameters need to learn the variance, the basis function center and weights.

In the structure of RBF neural network the first layer $X=[x_1,x_2....x_n]^T$ is the input vector of the network, suppose the radial basis vector of the network is $H=[h_1,h_2....h_j...h_m]^T$, h_j is Gaussian function:

$$h_j=\exp(-\frac{\left\|X-C_j\right\|^2}{2b_j^2})\ (j=1,2...m) \tag{1}$$

m is the number of neurons in the hidden layer of the network.

The center vector of the network at node j is

$C_j = [c_{j1}, c_{j2} ... c_{ji} ... c_{jn}]^T$, i=1,2,…,n

We suppose the radial width vector is

$B = [b_1, b_2, ... b_m]^T$, where b_j is the radial parameter and b_j >0.

The weight vector of the network is W and $W = [w_1, w_2, ... w_m]^T$.

The network output y is formed by a linearly weighted sum of the number of basis functions in the hidden layer.

$$y_m(k) = w_1 h_1 + w_2 h_2 + ... + w_m h_m \tag{2}$$

The performances index function of the controller is defined as

$$J_1 = \frac{1}{2}(y(k) - y_m(k))^2 \tag{3}$$

In this paper we use gradient descent method to modify weights of the output layer, node center and node radial parameters. The corresponding modifier formulas are as follows.

$$\begin{aligned} w_j(k) = w_j(k-1) + \eta(y(k) - y_m(k))h_j + \\ \alpha(w_j(k-1) - w_j(k-2) \end{aligned} \tag{4}$$

$$\Delta b_j = (y(k) - y_m(k)) w_j h_j \frac{\left\| X - C_j \right\|^2}{b_j^3} \tag{5}$$

$$\begin{aligned} b_j(k) = b_j(k-1) + \eta \Delta b_j + \alpha(b_j(k-1) \\ -b_j(k-2)) \end{aligned} \tag{6}$$

$$\Delta c_{ji} = (y(k) - y_m(k)) w_j \frac{x_j - c_{ji}}{b_j^2} \tag{7}$$

$$\begin{aligned} c_{ji} = c_{ji} y(k-1) + \eta \Delta c_{ji} + \alpha(c_{ji}(k-1) \\ -c_{ji}(k-2)) \end{aligned} \tag{8}$$

Where η is learning rate, α is momentum gene. Jaconbian matrix algorithm is as follows.

$$\frac{\partial y(k)}{\partial \Delta u(k)} \approx \frac{\partial y_m(k)}{\partial \Delta u(k)} = \sum_{j=1}^{m} w_j h_j \frac{c_{ji} - x_1}{b_j^2} \tag{9}$$

Where $x_1 = \Delta u(k)$.

3 RBF and PID Compound Controller [6]

Figure 2 is the structure of RBF neural network and PID compound control. The algorithm of the basic PID Controller that is commonly known in the literature is as follows.

$$u(k)=K_P e(k)+K_I\sum_{j=0}^{k}e(j)+K_D[e(k)-e(k-1)]$$

The Control Algorithm is as follows.

$$u(k)=u(k-1)+\Delta u(k)$$

$$\Delta u(k)=k_p(error(k)-error(k-1))+k_i error(k)+$$
$$k_d(error(k)-2error(k-1)+error(k-2))$$

E(k) is defined as the neural network tuning standard and

$$E(k)=\frac{1}{2}error(k)^2$$

The adjustment of kp, ki, kd parameters adopts gradient descent method.

$$\Delta k_p=\eta error(k)\frac{\partial y}{\partial \Delta u}[error(k)-error(k-1)]$$

$$\Delta k_i=\eta error(k)\frac{\partial y}{\partial \Delta u}error(k)$$

$$\Delta k_d=\eta error(k)\frac{\partial y}{\partial \Delta u}[error(k)-2error(k-1)$$
$$+\ error(k-2)]$$

η_p, η_i, η_d are the learning rates of k_p, k_i, k_d.

$\frac{\partial y}{\partial \Delta u}$ is Jacobian information of controlled plant. It is obtained by RBFNN identification results.

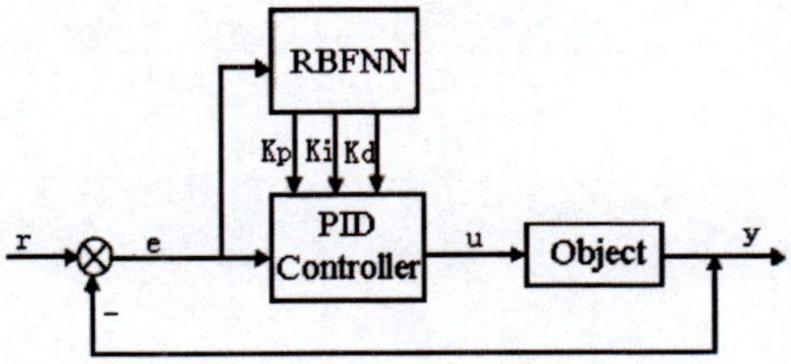

Fig. 2. Structure of RBF-PID

4 Simulation and Experiment

A. Simulation

In this paper the simulation adopts a nonlinear discrete and time varying system.whose transfer function can be described by

$$y_{out}(k)=\frac{2y_{out}(k-1)+0.5u(k-1)}{1+2[y_{out}(k-1)]^2+0.2y_{out}(k-2)}+0.1y_{out}(k-3)$$

The input is $r(t)=0.9sign(2\pi t)$

$u(k)$, $y_{out}(k)$, $y_{out}(k-1)$ are the inputs of RBF neural network identifications.

In the simulation we choose $\eta=0.2$ as learning rate, the momentum gene $\alpha=0.22$, and hidden layer units H=8.

The system response is shown as the figures below.

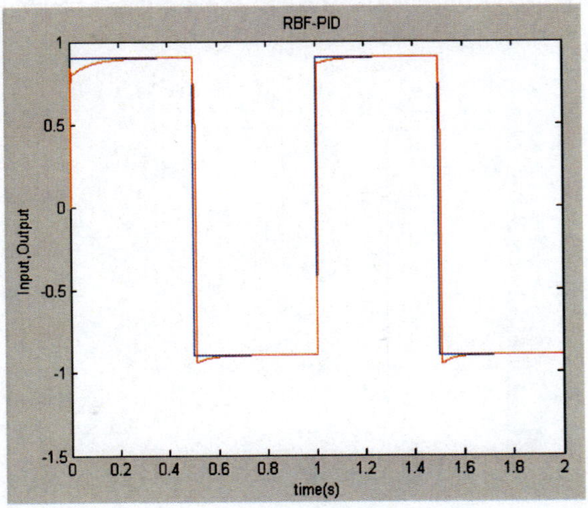

Fig. 3. Square wave response of adaptive PID control based on RBFNN

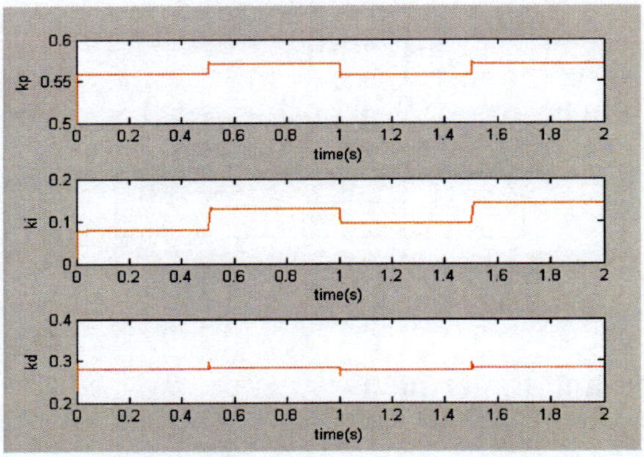

Fig. 4. Adaptive tuning curve of PID parameters

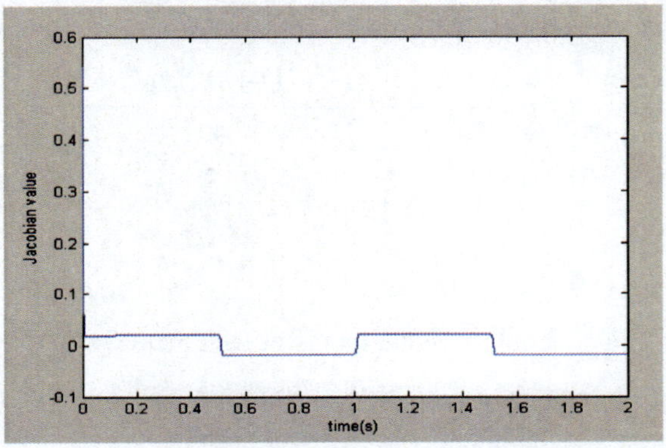

Fig. 5. Jacobian information of identification

Then, adopting the incremental PID control algorithm to track the square wave input. The simulation results are shown in figure 6.

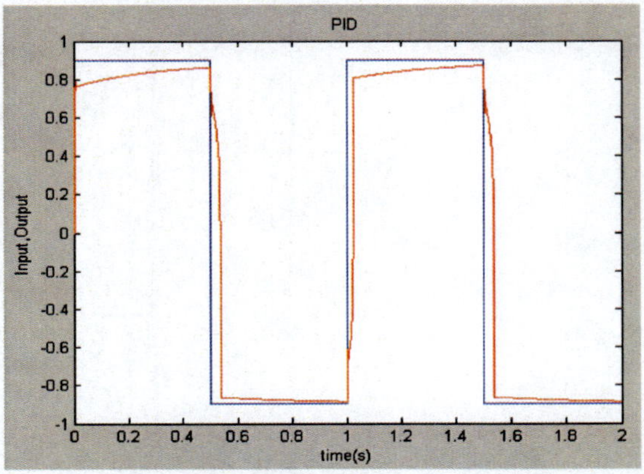

Fig. 6. Square wave response of PID control

B. Analysis of Simulation Results

The simulation experiments show us that when we adopt RBF-PID controller it can achieve a better effect and reduce the error, the system output traces the reference input satisfactorily and the performance of the proposed controller is better than that of the conventional PID controller.

5 Conclusions

This paper has put forward one kind of Control method based on RBF neural network and PID compound for the nonlinear model , it can carry on PID parameters adjustment online , and control the complicated nonlinear model effectively , the simulation result shows the validity of this method . The proposed controller has advantages of both self-learning capability of neural network and simplicity of PID controller. Simulation results also have shown that the proposed controller presents a fast and high-precise tracking ability. It shows higher performance in adaptability and robustness compared to the conventional PID controller.

Acknowledgment. This paper is supported by The High-Tech Research and Development Program of China (The 863 Program):2007AA041106, 2007AA041108.

References

1. Zhu, Q., Ma, Z., Warwick, K.: Neural network enhanced generalised minimum variance self-tuning controller for nonlinear discrete-time systems. IEE Proc. Control Theory Appl. 146(4), 319–326 (1999)
2. Hou, Y., Du, J., Wang, M.: Neural network. XiDian University Press (2007)
3. Yang, L., Ren, X., Huang, H.: Application of self tuning PID controller based on RBF network. Computer Simulation 23(1), 270–273 (2006)
4. Liu, J.: Advanced PID control and matlab simulation. Electronics Industry Press, Beijing (2003)
5. Sanner, R.M., Slotine, J.-J.: Gaussian networks for direct adaptive control. IEEE Trans. Neural Networks 35, 837–863 (1992)
6. Hovakimyan, N., Nardi, F., Calise, A., et al.: Adaptive output feedback control of uncertain nonlinear systems using single-hidden-layer neural networks. IEEE Trans. Neural Networks 13(6), 1420–1431 (2002)

The Exception Handling Pattern in the Architectural Layers Using Aspect-Oriented Programming*

Lei Zhang, Lianbo Ma, and Weixing Su

Key Laboratory of Industrial Informatics
Shenyang Institute of Automation, Chinese Academy of Sciences
Shenyang 110016, China
leizhang@sia.cn

Abstract. Aspect-Oriented Programming(AOP) is intended to ease situations that involve many kinds of code tangling. This paper reports on a practice to import AOP's ability to ease tangling related to exception handling pattern. We took an existing framework, the exception handing application block(EHAB) and unity application block(UAB). EHAB provides exception handlers and policies to enable us to make configurable and consistent code in architectural layers. UAB provides interception mechanism to make clean code using AOP.

Keywords: AOP, EHAB, UAB, Exception.

1 Introduction

In most software projects, a large percentage of the code is boilerplate code. Exception handling comes under this category. Even though the business logic may be just three or four lines of code, exception handling might go on for 10 to 20 lines. Meanwhile, the handling of exceptions in large software systems can consume a remarkable amount of development resources. Exceptions must be thought of throughout the whole development cycle and dealing with them is usually hard.

If you spend time at the beginning to design a clear and consistent exception management system, it frees you from having to piece together your system during development, or worse still, from having to retrofit exception handling into an existing code base.

An exception management system must be well encapsulated and must decouple the details of logging and reporting from the application's business logic. It must also be able to generate metrics that can be monitored by operators to provide an insight into the current health and status of the application. This helps create an application that can quickly and accurately notify operators of any problems it is experiencing. It can also provide valuable information to help developers and support services resolve problems.

This paper talks about how to keep exception handling simple and straightforward, keeping the developer's plate clean for him to concentrate on business logic rather than devoting time to writing exception handling boilerplate code. This paper uses The Exception Handling Application Block and Unity Application Block in Enterprise Library.

* Foundations of National "863" High Technology Project of China (NO.2006AA04A117).

J. Luo (Ed.): Affective Computing and Intelligent Interaction, AISC 137, pp. 537–543.
springerlink.com

2 The Exception Handling in the Architectural Layers

Coding the detection and handling of errors, is, however, still a difficult process that requires a strict discipline from programmers. And the reality is that most software being written today uses programming languages that provide very little help. Hopefully, that will change.

But there is a side effect of coding application exceptions that cannot be addressed by simply including more powerful exception detection and handling mechanisms in the programming language. That side effect consists of the tangling between the code for what the program should do and the code for detecting and handling exceptions.

Figure 1 illustrates this issue with a small C# example. The example show parts of a SqlDataProcess class, which is a subclass of DataProcess class. The code related to exceptional behaviors is underlined.

The tangling is, in part, a consequence of the programming language and, in part, a consequence of design decisions. In architectural layers, code similar to figure 1 would be repeated in all the routines that perform different data access queries. To change the behavior of the exception handling code, you must update each routine that contained this code [1].

```
public class SqlDataProcess:DataProcess{
    DataSet customersDataSet;
    public void newSqlDataProcess(){
        try{ customersDataSet = RunQuery("GetAllCustomers");}
        catch (Exception ex){
            string formattedInfo = FormatException(ex);
            Logging.Log(formattedInfo);
            throw new DataAccessException("Database access " +
                "failure for query GetAllCustomers", e);
        }
    }
    public void updateSqlDataProcess(){
        try{customersDataSet = RunQuery("updateCustomers");}
        catch (Exception ex){
            string formattedInfo = FormatException(ex);
            Logging.Log(formattedInfo);
            throw new DataAccessException("Database update " +
                "failure for query updateCustomers", e);
        }
    }
}
```

Fig. 1. A small C# example with tangling exception handling

We require uniform and flexible procedures for handling exceptions. For example, you might want consistent exception handling procedures for all components in a particular tier of an application's architecture. In addition, because of changing security or other operational issues, you might want the ability to change policies as needed, without requiring changes to the application source code. The exception handling application block, in conjunction with the Enterprise Library configuration tools, lets you accomplish both tasks [2].

In architectural layers, Exceptions that occur in the data access layer are logged and then wrapped inside another exception that provides more meaningful information to the calling layer. Within the business component layer, the exceptions are logged before they are propagated. Any exceptions that occur in the business component layer and that contain sensitive information are replaced with exceptions that no longer contain this information. These are sent to the user interface (UI) layer and displayed to the user. Figure 2 illustrate this exception handling pattern.

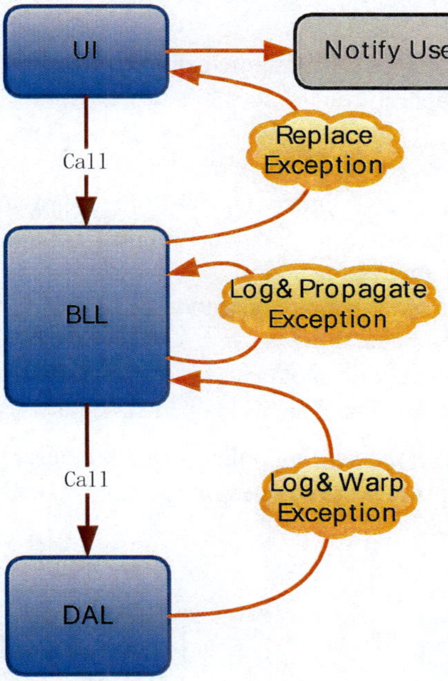

Fig. 2. The propagation of exception handling

3 Exception Throwing Strategy and Handling Policy

Our exception handling pattern is designed to support the typical code contained in catch statement in application components. Instead of repeating this code (such as logging exception information) throughout identical catch blocks in an application component, the application block allows developers to encapsulate this logic as reusable exception handlers [3].

Usually, the following exception handlers are included.

- Wrap handler. This exception handler wraps one exception around another.
- Replace handler. This exception handler replaces one exception with another.

- Logging handler. This exception handler formats exception information, such as the message and the stack trace.
- Fault Contract Exception Handler. This exception handler is designed for use at Windows Communication Foundation (WCF) service boundaries, and generates a new Fault Contract from the exception.

Policies specify the exception handlers that execute when the application block processes a particular exception type. You can chain these handlers together so that a series of them execute when the associated exception type is handled. The following are some examples of named policies and descriptions of what they might provide:

- Base policy. This policy logs the exception and rethrows the original exception.
- Secure policy. This policy logs the exception, replaces the original exception with a custom exception, and throws the new exception.
- Expressive policy. This policy wraps the original exception inside another exception and throws the new exception.

Exception handling pattern is designed to achieve the following goals:

- Encapsulate the logic used to perform the most common exception handling tasks into minimal application code.
- Relieve developers of the requirement to write duplicate code and custom code for common exception handling tasks.
- Allow exception handling policies to be changed after they have been deployed and to ensure that changes happen simultaneously and consistently.

Figure 3 illustrate the exception handling policy in architectural layers.

Figure 4 and figure 5 illustrate the code example and consistent configuration using exception application block.

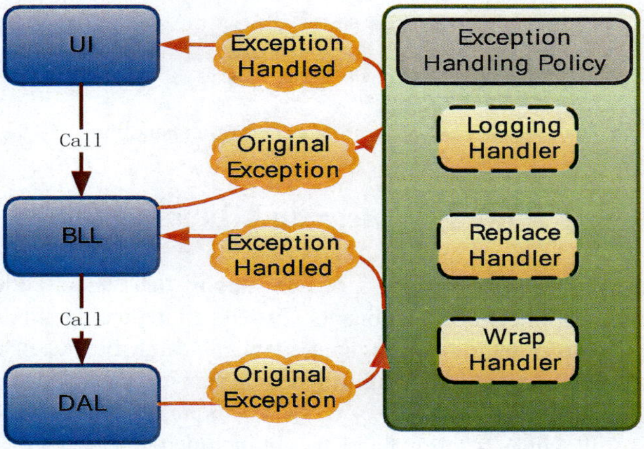

Fig. 3. The exception handling policy in architectural layers

```
public void myException()
{
    try
    {
        // Run code.
    }
    catch (Exception ex)
    {
        bool rethrow = ExceptionPolicy.HandleException(ex, "Data Access Policy");
        if (rethrow)
            throw;
    }
}
```

Fig. 4. The C# code example

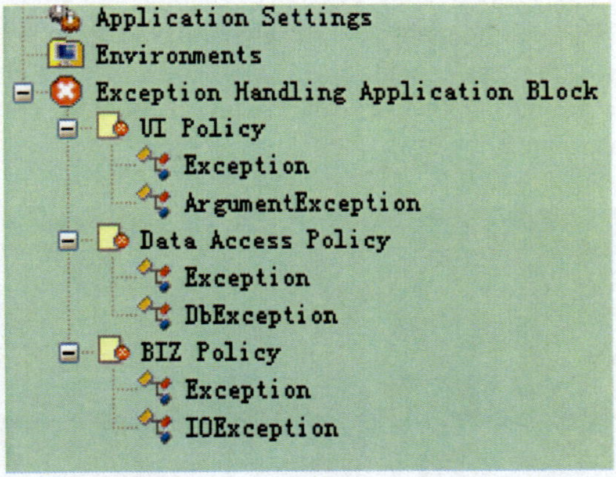

Fig. 5. The consistent configuration

4 Aspect-Oriented Programming and Unity

Aspect-Oriented Programming was first proposed in [4] as a programming technique for modularizing concerns that cross-cut the basic functionality of programs. Exception handling was referred to as one of those cross-cutting concerns, and the paper suggested that is should be possible to achieve a relative separation between the functional code and the exception handling code [5].

The Unity Application Block (Unity) is a lightweight, extensible dependency injection container that supports constructor injection, property injection, and method call injection. The Unity Interception extension provides a subset of aspect-oriented programming.

The interception mechanism is based around three basic concepts: matching rules, call handlers, and interceptors. Matching rules are simple but flexible objects that determine which methods should have extra handling applied. Call handlers are objects that actually implement the cross-cutting concerns. They can run before or after the method. They can modify method parameters or return values. They can even stop the method from being called at all or call it multiple times. The matching rules and call handlers are grouped together into interception policies. The interception policy uses the matching rules to define which methods get intercepted and uses the call handlers to define what processing to perform on the intercepted object.

Policies determine what happens when a method is intercepted, and interceptors determine how methods are intercepted. Unity interceptors will, when configured in the container, enable you to intercept method calls and attach call handlers to the intercepted methods. Policies determine which call handlers are placed in the interception pipeline for each method.

Figure 6 illustrate a C# code example. Public class MyExceptionHandlerAttribute is a particular attribute because it inherits from HandlerAttribute and it needs only to override the CreateHandler method and returns an instance of the handler you want to use. You can apply this attribute to every class or method you want to intercept, if you apply it to a class every method will be intercepted, if you apply it to a single method you will intercept only that method. In our code, we apply it to interface IExceptionhandler. Using AOP, we can create clean and consistent exception handling pattern in architectural layers.

```
public class MyExceptionHandlerAttribute : HandlerAttribute {
    public override ICallHandler CreateHandler(IUnityContainer container) {
        return new MyExceptionHandler();
    }
}
[MyExceptionHandler]
interface IExceptionHandler {
    void ExceptionHandler(String exceptionpolicy);
}
```

Fig. 6. The C# code example of AOP

5 Conclusion

We have studied how aspects ease the tangling related to exception handling. We took an existing framework, Exception handling application block and unity application block. The former encapsulate the logic used to perform the most common exception handling tasks into minimal application code. And make exception handling configurable and consistent. The unity application block help making clean and elegance code using aspect-oriented programming.

Acknowledgment. This work is supported by Foundations of National “863” High Technology Project of China(NO.2006AA04A117).

References

1. Lippert, M., Lopes, C.V.: A Study on Exception Detection and Handling Using Aspect-Oriented Programming. In: Proc. of the 22nd Conf. on Software Engineering, pp. 418–427. ACM Press, New York (2000)
2. Vashishtha, S.: An Exception Handling Framework for J2EE Applications, `http://Onjava.com`
3. Microsoft: The Exception Handling Application Block, Enterprise Library 4.1 (October 2008)
4. Kiczales, G., Lamping, J., Mendhekar, A., Maeda, C., Lopes, C., Logingtier, J.M., Irwin, J.: Aspect-Oriented Programming. In: Aksit, M., Auletta, V. (eds.) ECOOP 1997. LNCS, vol. 1241, pp. 220–242. Springer, Heidelberg (1997)
5. Microsoft: The Unity Application Block, Enterprise Library 4.1 (October 2008)

A Detection Method for the Resource Misuses in Information Systems

Chao Wang[1], Gaoyu Zhang[2], and Lan Liu[2]

[1] School of Computer Science and Technology
Xidian University Xi'an, P.R. China
[2] School of Information Management
Shanghai Finance University Shanghai, P.R. China
xdkevin@gmail.com, farwind1972@yahoo.com.cn, liul@shfc.edu.cn

Abstract. It is difficult to detect the resource misuses in information systems because they can be carried out in different manners and it is hard to collect the prior knowledge of the malicious insiders. In this paper, a hidden Markov model (HMM) based method is developed to detect the resource misuse. As to the HMM model, the file folders containing sensitive information are taken as the model states and the user operations as the model observation symbols and Baum-Welch algorithm is adopted to determine the model parameters. The behavior profile of a malicious insider is depicted by his HMM model and used to detect his malicious behaviors. The experiment results show the effectiveness and adaptability of our method.

Keywords: resource misuse, hidden Markov model (HMM), insider threat detection.

1 Introduction

CERT defines the malicious insiders as the staffs who own the privileges to access organization's network, system or data and damage the security of the organization's information or information systems by intentional perorations or misuse of information [1]. Malicious insiders have the security-related knowledge of the information systems and are familiar with the system's security policies, security mechanisms, and security technologies. Therefore, they threaten greatly the security of the information systems.

Resource misuse is a main kind of insider threats in information systems. With the help of their security credentials, the insiders violate the security management strategy of the systems and misuse the system information. As an example of the resource misuse, Jerome Kerviel, a trader of French Society Generale, used his own identity to trade illegally and resulted in a huge loss of 7.1 billion dollars to the bank. In addition, CERT shows that the proportions of theft and tamper other employees' confidential are as high as 75% and 88% respectively. Thus, the resource misuses caused by malicious insiders are familiar and lead to great harm.

J. Luo (Ed.): Affective Computing and Intelligent Interaction, AISC 137, pp. 545–552.
springerlink.com

Malicious insiders have legal identities and access permits. They misuse the system information, which do not violate the system's security policies and security mechanisms, which makes the resource misuses difficult to detect. However, Srivastava et al. points out that human behaviors tend to show certain profiles and there are no exceptions when they access information systems [2]. Therefore, it will be a feasible detection method of resource abuse to research on user's behavior model when they access the system and find inconsistencies between models. This paper proposes a HMM-based detection method about the resources misuse of malicious insider and provides three contributions: 1) build the HMM model of user's behavior; 2) develop the real-time detection method of resources misuse; 3) validate the effectiveness of our detection method based on the simulation data.

The rest of this paper is organized as follows: Section II discusses the related work; Section III presents how to build the HMM model of a user's behavior and detect the resource misuses; Section IV shows the experiment results and the analysis of the experiment; Section V summaries the whole paper.

2 Related Work

Resources misuse has always been an important issue in the field of information security since 1980s. Existing methods and techniques are diverse and artificial intelligence is widely used.

Lee et al. combines the technologies of intrusion detection and data mining to detect the misuses. In the detection method, agents are used to mine the data and build the intrusion profiles that are taken as the classifier of the intrusion detection system (IDS) [3]. Similarly, Singhal et al. [4] and Ertoz at el. [5] implement non-real-time misuse detection based on mining the relation rules among the intrusion incidents. Among these researches, the idea of intrusion detection are inherited which ignores the essential difference between the outsider attacks and the insider attacks.

Anderson et al. [6] uses the method of statistical learning to judge the user's behavior is normal or not according to the deviation between the user's current behavior and past behavior. Santos et al. [7] not only analysis the user's login information and the documents they have accessed, but also builds user's calculation model and determine the detection indexes. If there exist deviations between the results and the model, the malicious insider behaviors are confirmed. Artificial intelligence (AI) is used to design most of the detection methods above in which labeled data are need to train the classifier. However, how to obtain the labeled data is a hard problem in practice. Furthermore, the detection methods will be restricted to new insider threats if labeled data are unavailable.

Therefore, the existing detection methods either cannot detect new kind of misuses, or require prior knowledge of the insiders, which limits the application of these methods.

For malicious insiders, their "insider" characteristic provides an efficient way to launch attacks; meanwhile, it offers us the opportunity for detection. The insiders (suspected internal attacker) are more fixed and much less than outside attackers. So accurate analysis and extract of insider's behavior model become feasible as well as successful detection of resources misuse will be carried out.

A detection method of insider's resources misuse is proposed in this paper: firstly, a HMM model of insiders' behavior is built; then, the HMM model is trained with the normal behavior of users to determine the model parameters; if the insider's behaviors cannot be accepted by the model with a probability high enough, they will be taken as the insider threats. Compared with the existing detection methods, the advantages of our method are: 1) detection can be done without the priori knowledge of malicious insiders; 2) the false positives rate of detection may be reduced because of the learning function of HMM model.

3 The HMM-Based Detection Method

Without loss of generality, we will take an example of windows operating system based information system to explain the detection methods for the resource misuse in this paper.

3.1 Building the HMM Model

Suppose an information system keeps some sensitive documents that are distributed in some sensitive documents folder in the system. The user transaction operations are reading a file (r), writing to a file (w), copying a file (c), deleting a file (d), moving a file (m), and opening/creating a file (o) (See Fig. 1). Reading or writing processes and some other operations need to call the function of opening/creating a file, which is generally not damage the security of information systems, so we ignore the operation of opening/creating.

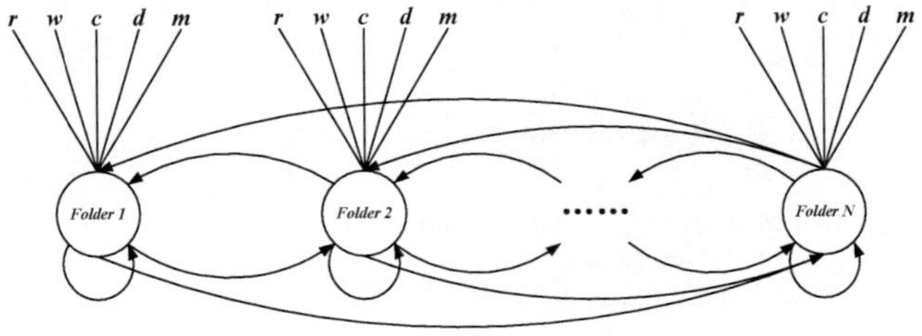

Fig. 1. HMM model of the user behaviors

We collect the information of an insider's behavior and intercept his process calls for file manipulation functions with application programming interfaces (API hook) articulated technology. The user's operations will be recorded if he accesses sensitive documents folders, which forms the user's operation sequence. During the monitoring process, only the user's operations are captured and the objects and the values of the operations are ignored, which can protect the confidentiality of the information system and the privacy of the user. Therefore, the user's operation sequences are appreciable and the objects and the values of the operations are transparent for the administrator of the information system.

The details of the HMM can be found in the [2]. In our HMM model, its states are the folders save sensitive files and observation symbols are the user operations on the folders. A user deals with transactions in the sensitive folders of the system and his behaviors result in the state transitions of the HMM model. The user's behaviors are represented by the sequence of the observation symbols.

3.2 Determining the HMM Model Paraments

After building HMM model, it is important to determine the model parameters, that is, the state transition matrix A, observation symbol matrix B, and the probability distribution vector of the initial state π. For simplicity, the initial value of A, B, and π are assumed uniform distribution. Based on the initial values, Baum-Welch algorithm is adopted to train the model and determine the model parameters.

Generally, resource misuses are far fewer than normal user behaviors in practical information systems. HMM model based methods are a kind of statistical analysis essentially. So the user action sequence intercepted can be used as the training data of HMM model.

3.3 HMM Model Based Resource Misuse Detection

HMM model is often denoted as $\lambda = (A, B, \pi)$. As to the HMM model λ, it outputs the R-long observation symbol sequence $P_1 = o_1, o_2, \cdots, o_R$ at the moment t with probability p_1:

$$p_1 = \Pr(P_1 \mid \lambda) = \Pr(o_1, o_2, \cdots, o_R \mid \lambda) \tag{1}$$

At the moment $t+1$, it outputs an observation symbol o_{R+1} and obtains a new R-long observation symbol sequence $P_2 = o_2, \cdots, o_R, o_{R+1}$ with probability p_2:

$$p_2 = \Pr(P_2 \mid \lambda) = \Pr(o_2, \cdots, o_R, o_{R+1} \mid \lambda) \tag{2}$$

Calculating the difference Δp between p_1 and p_2:

$$\Delta p = p_1 - p_2 \tag{3}$$

If $\Delta p \leq 0$, it shows the occurrence probability of the new observation symbol sequence is increasing and the observation symbol o_{R+1} is a normal behavior. While $\Delta p > 0$, it shows the occurrence probability of P_2 is lower than that of P_1 and o_{R+1} is suspiciously a resource misuse behavior. A threshold th is set according to our experiences. If there holds:

$$\Delta p / p_1 \geq th \tag{4}$$

then the symbol o_{R+1} is taken as an insider threat.

If o_{R+1} is convicted as an abnormal behavior, the surveillance system alarms. Otherwise, o_{R+1} will add to the observation symbol sequences that are the base sequence for the next detections. The user's behavior profiles may change over time. The HMM model learns the user's behavior profiles by adding newly arrived observation symbols to the observation symbol sequences continuously that will reduce the false alarm rate of surveillance system.

Fig. 2 gives the details of the HMM model based detection process.

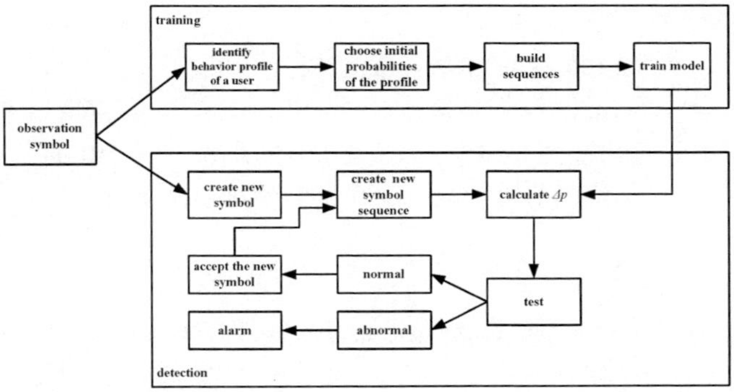

Fig. 2. HMM model based resource misuse detection

4 Experiments and Result Analysis

In this paper, Matlab 7.0 is used to carry out the experiments of our detection method.

4.1 Setting the Model Parameters

Experiment parameters are set as follows: assume that there are four sensitive folders in information systems and five kinds of operations (read, write, delete, copy, and move) on the folders. Statistical property of an insider's behavior can be obtained by the surveillance system. The statistical property of an insider are assumed as follows: (1) forty-four percent of read operations, thirty-five percent of write operations, seven percent of copy operation, three percent of deletion operation, and eleven percent of move operation; (2) the user's operations are completely random and the observation symbol sequence are generated from the operations. The length of the observation sequence and the detection threshold are set to be *R* and *th* respectively.

In the experiments, we generate randomly 825 observation symbols that are taken as the training data. Each 33 symbols are organized into a sequence and these 25 observation sequences are taken as the training data of the HMM model. 33 observation symbols in each sequence can guarantee that each observation symbol will appear statistically and 25 observation sequences can not only avoid excessive computing, but also ensure to complete training process.

Baum-Welch algorithm is used to train the HMM model and the value of A is determined based on the initial values,

$$A = \left\{ \begin{matrix} 0.17528 & 0.24386 & 0.08542 & 0.49544 \\ 0.21340 & 0.30511 & 0.29359 & 0.18790 \\ 0.04599 & 0.47467 & 0.09133 & 0.38801 \\ 0.04833 & 0.37025 & 0.57069 & 0.01073 \end{matrix} \right\} \tag{5}$$

So the values B and π are:

$$B = \left\{ \begin{matrix} 0.13316 & 0.54786 & 0.02870 & 0.00890 & 0.28138 \\ 0.47061 & 0.44805 & 0.02029 & 0.01419 & 0.04686 \\ 0.26437 & 0.33615 & 0.13021 & 0.01886 & 0.25041 \\ 0.74345 & 0.14879 & 0.09897 & 0.00812 & 0.00067 \end{matrix} \right\} \tag{6}$$

$$\pi = (0.22291 \quad 0.26358 \quad 0.19615 \quad 0.31736)\,, \tag{7}$$

According to the subsequent experiment results, if the length R of observation sequence ranges in [5, 13] and the threshold th ranges in [0.1, 0.5], the test results will be much better.

4.2 Experiment Results

In this experiment, we simulate resource misuses with replacing the observation symbol, that is, we replace observation symbol(s) in sequence with the other(s) which are randomly generated and ensure that the observation symbols in corresponding location are different absolutely.

Considering three different kinds of resources misuses: 1) low-speed attack: for a long time, occurrence frequency of the attacks is low and stable; 2) speed-up attack: in a period, occurrence frequency of attacks is gradual accelerated; 3) pulse attack: attack timing and intensity are unstable with the pulse property. In practice applications, these three kinds of attack are very representatively. These three kinds of resource misuses are shown in Figure 4 (a), (b) and (c) respectively.

For the low-speed attack, one observation symbol is replaced randomly in each sequence of training data. The accuracy of our detection method is shown in Fig. 4 (a).

For the speed-up attack, in sequences 1 ~ sequence 5, each sequence is replaced one observation symbols. In sequences 6 ~ sequence 10, each sequence is replaced two observation symbols, and the rest sequences are replaced in the same manner. The accuracy of our detection method is shown in Fig. 4 (b).

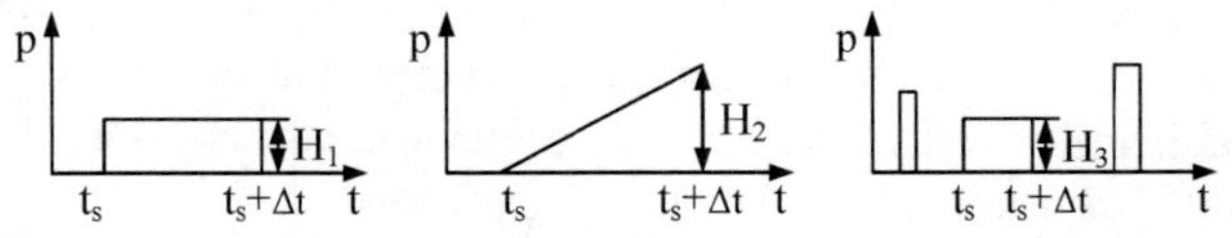

(a) low-speed attack (b) speed-up attack (c) pulse attack

Fig. 3. Three kinds of resource misuses

For the pulse attack, we select 6 symbol strings randomly to replace from the observation symbol sequences whose length is 825. In the replacement process, the lengths of the 6 symbol strings vary from 3 to 6. The accuracy of our detection method is shown in Fig. 4 (c).

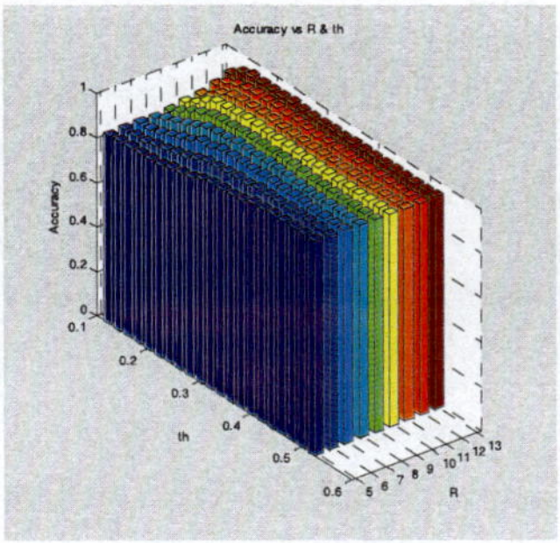

(a) low-speed attack

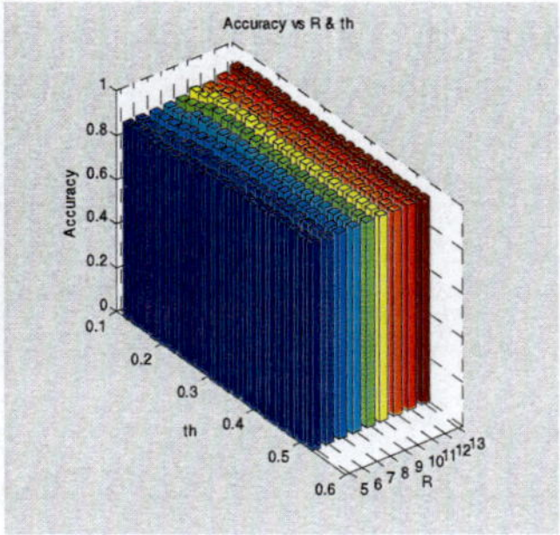

(b) speed-up attack

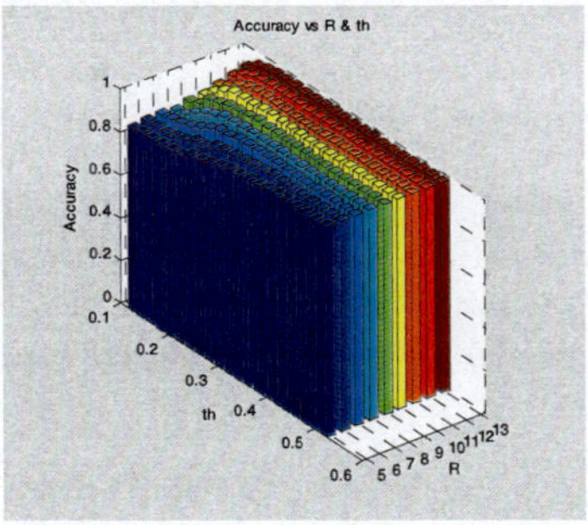

(c) pulse attack

Fig. 4. Detection accuracy of the HMM model method

From the experiment results, we can find that along with the increasing of *R* and *th*, the detection accuracy will increase at first and then reduce. The three-dimensional coordinates of maximum for accuracy are (8, 0.26, 0.97), (8, 0.22, 0.93), and (9, 0.24, 0.96). In the detection process, take $R_o=8$, $th_o=0.24$, it is possible to achieve maximum accuracy.

When *R* and *th* are smaller, detection accuracy increase rapidly to the peak and then trend to stable. So the larger *R* and *th*, the higher the detection accuracy is.

The detection accuracy is 84%~97%, 86%~93%, and 83%~96% corresponding to the three most popular kinds of resource misuses respectively. The results show that our method work well and has good adaptability for different kinds of resources misuses.

5 Conclusion

There exists essential difference between the detection methods for the outsider attacks and insider attacks. According to the "insider" characteristic of the malicious insiders, a HMM model based method is presented to detect the resource misuses of the insiders in information systems. The simulation results show that the method work well and has good adaptability for different kinds of resources misuses.

Acknowledgment. This research is supported by the National Natural Science Foundation of China under Grant No. 60633020, the Fundamental Research Funds for the Central Universities, and the Construction Program of Key Courses of Shanghai Education Committee under Grant No. J51601.

References

1. Cappelli, D., Moore, A., Trzeciak, R., et al.: Common Sense Guide to Prevention and Detection of Insider Threats - Version 3.1. Technical report. Carnegie Mellon University (2009)
2. Srivastava, A., Kundu, A., Sural, S., et al.: Credit Card Fraud Detection Using Hidden Markov Model. IEEE Transactions on Dependable and Secure Computing 5(1), 37–48 (2008)
3. Lee, W., Stolfo, S., Chan, P., et al.: Real time data mining-based intrusion detection. In: Proceedings of the 2001 DARPA Information Survivability Conference and Exposition II, pp. 89–100. World Scientific and Engineering Academy and Society (WSEAS), Wisconsin (2001)
4. Singhal, A.: Data Warehousing and Data Mining Techniques for Computer Security, pp. 83–103. Springer, New York (2006)
5. Ertoz, L., Eilertson, E., Lazarevic, A., et al.: MINDS-Minnesota intrusion detection system. In: Kargupta, H., Joshi, A., Sivakumar, K., et al. (eds.) Next Generation Data Mining, pp. 65–86. MIT/AAAI Press, New York (2004)
6. Anderson, D., Lunt, T., Javitz, H., et al.: Detecting unusual program behavior using the statistical component of the next-generation intrusion detection expert system (NIDES). Technical Report SRI-CSL-95-06, SRI International (1995)
7. Santos, E., Nguyen, H., Yu, F., et al.: Intent-driven Insider Threat Detection in Intelligence Analyses. In: Proceedings of the 2008 IEEE/WIC/ACM International Conference on Web Intelligence and Intelligent Agent Technology, pp. 345–349. IEEE Press, New York (2008)

Analysis on Implementation Condition of Mass Customization for Knowledge Production

Cuixiao Fu[1] and Xingsan Qian[2]

[1] School of Management, Fudan University Shanghai, China
fcuix@fudan.edu.cn
[2] School of management, University of Shanghai for Science and Technology
Shanghai, China
qxsqg@126.com

Abstract. The mass customization idea is introduced into knowledge production in this paper, and its implementation condition is analyzed. Firstly, the implementation conditions for mass customization are analyzed. Then, the main characteristics of modern knowledge production industry are discussed. Based on these characteristics, the implementation condition model for mass customization of knowledge production (MCKP) is constructed and analyzed deeply. The result indicates that mass customization is one of the main development trends of knowledge production Industry.

Keywords: knowledge production, mass customization, implementation conditions.

1 Introduction

Development of knowledge economy impels the boom of knowledge production industry, and then there are many researches focus their eyes on the study of knowledge production modes, such as Gibbons et al. [1], Robert[2], Barabra, etc[3] and so on. Also in China, more and more knowledge production enterprises occurred with the economic development and the cognition of knowledge importance. Unluckily, almost each enterprise adopts knowledge production mode in its own way. There is not a relatively mature Industry system like manufacturing Industry, much less system management. Moreover, most of related researches are in view of social science or philosophy, few about knowledge production management. The actual situation of knowledge production management in China is still in an independence state and has no advanced and systemic management modes for which to adapt the changing environment.

On the other hand, with the development of mass customization for manufacturing mode, the knowledge production modes are impacted. Recently, some knowledge production enterprises begin to introduce the mass customization thought into knowledge production. So the mass customization of knowledge production (MCKP) turn to appeared though it is still in try-out stage[4]. based on that, the implementation condition and method for which would be discussed in this paper.

J. Luo (Ed.): Affective Computing and Intelligent Interaction, AISC 137, pp. 553–560.
springerlink.com

2 Implementation Condition for Mass Customization

Mass customization is an advanced production mode, it means to transform the customization production problems into the mass production and supply any quantitative products for individuals or small batch multi-items markets in the condition of cost and velocity similar to mass production, in which the recombine of products construct and manufacturing process are carried out by a series of high technologies. Concluded the related researches[5]-[7], the implementation condition is discussed in terms of four aspect: environment, organization, process and products, as shown as Fig. 1.

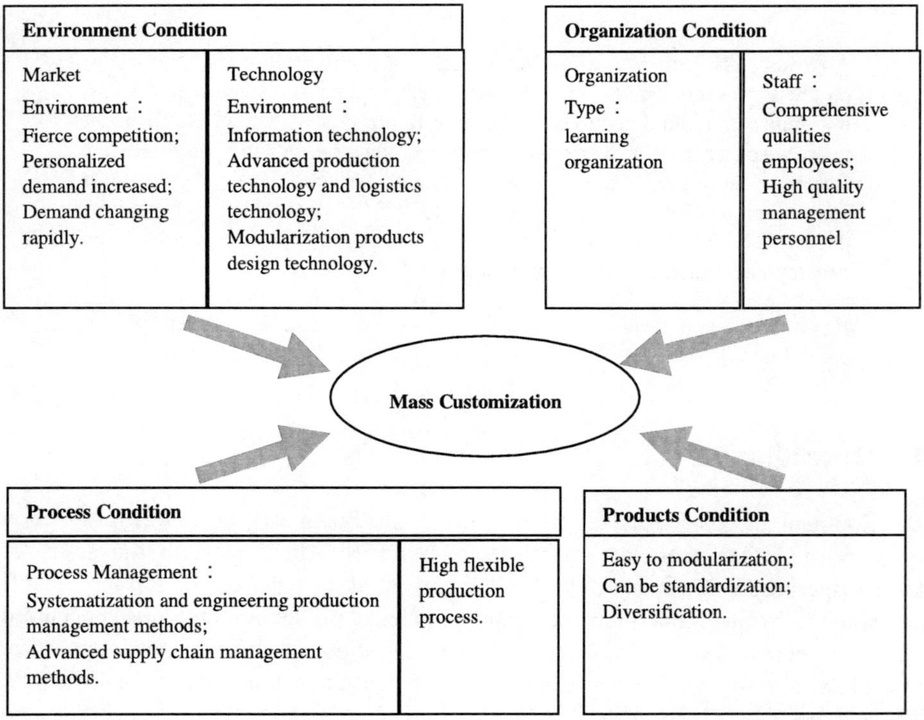

Fig. 1. The implementation condition for mass customization

2.1 Environment Condition

Environment condition includes two parts: market environment condition and technology environment.

- Market environment: When the market competition is fury, optional products become more and more, which pulls the enterprises to seek innovative strategies to reduce cost, keep high profits and obtain competitive advantage.

Then, mass customization become the fittest mode, it gives attention to both competitive advantage and reducing cost; on the other side, more and more personality demand drives enterprise to production as mass customization mode; moreover, the rapidly changing demand requests the agilely production, which also bring the implementation of mass customization.

- Technology environment: Which includes new information technology based on INTERNET for obtaining customers information roundly and speedily; advanced production and logistics technologies based on agility character for production and delivery rapidly; and products design technologies for modularization and standardization.

2.2 Organization Condition

Organization condition includes two demands in terms of organization type and staff, as list as follow:

- Learning organization: The changing market demand and increasing personality demand pull the enterprises to develop new technology, master new knowledge, improve working capability of staff and improve management constantly, so the organization for mass customization must be a learning organization to adapt the changing environment and improve itself.
- Staff: In the condition of mass customization, the staff must hold the new production technology, accord with the need of "learning organization" and be good at cooperating with others, so it needs the comprehensive qualities staff. At the same time, corresponded management personal are needed to management the whole mass customization production system aiming at protecting enterprise's profits and satisfying customers' need.

2.3 Process Condition

Process condition includes two parts: process management and process flexible

- Process management: In order to implement mass production of personal products, mass customization needs the systematic and engineering production management methods to optimize and design the whole process; in addition, the enterprises must cooperate with their suppliers and establish the strategic partnership relations. That is to say that the effective supply chain management method is necessary to keep materials, information and capital flow unblock.
- Process flexible: The process dimension is the key factor to implement mass customization, the more dimension, the more difficult. The golf cue only has a dimension—the length of cue, so it is one of the most mature products for mass customization. When the dimension of process increases, its flexible became so bad that the mass customization can not be actualized.

2.4 Products Condition

The products of mass customization must be easy to modularization, that is to say, the products' structure must be complex but not simple. For example, both motors and computers compose of many components and can be divided into various product modules, so they are suit to mass customization. Besides, based on modularization, the main goal of mass customization is to actual the standard process in each module, which needs the whole products or part of them can be standardization. In addition, the diversity of products permits customers to choose products according their interest.

3 Characteristics of Knowledge Production Industry

In the actual knowledge production active, it includes both knowledge innovation production and knowledge duplication production. The enterprises engaged in this active namely knowledge production enterprises, such as software, translation, consultation firms and research & development units. Generally, knowledge production Industry of modern times mainly has the following characteristics:

- Unity of innovation, duplication and customization: With the development of social economy and diversity of demand, knowledge production of modern times is not only traditional knowledge innovation and knowledge duplication but also the customization knowledge. For instance, the production processes of consultation firms for special Industry and some translation firms are all integrated with the three forms.
- Diversity of producers and products: Knowledge can be divided into many types based on different classified rules, corresponded to each type, the production process and the final products can be different. So the knowledge production process is stochastic and uncertainty. On the other hand, knowledge products of people demand are various, especially in nowadays, so many kinds of knowledge products can be used, such as CD, software, book, thesis, document on-line and so on. People can choose any type products when they want and then the production processes have to be changed. Moreover, the producers who participate in knowledge production are various. They can be universities, corporations, government or individuals and the new organization structure-- triple helix of university-industry-government[8].
- Convenience of products transmission: The virtual knowledge and the high-tech network eliminated the transportation problems of manufacturing in knowledge production. Knowledge can transmit, diffuse easily and increase by using. Spreading quickly by network facilitates the mass customization to carry through. For example, it takes only a few seconds to transmit a batch of knowledge products produced in china to customers' computers in American by network. Thus the deliver mission is accomplished.
- Independent of nature materials: Compared with manufacturing which depends on nature materials, such as mine, petroleum, soil, forest and so on, knowledge production mainly depends on knowledge, technology, intelligence and thought. The production instrument of knowledge production mainly is high technology products like computers, network and so on. Surely, the independent of nature materials doesn't mean it need not traditional materials, but means its need less than manufacturing.

- Advance and improve constantly of technology: The knowledge technology accelerates the development of knowledge production Industry. Semantic web, Ontology, Metadata, Information searching methods, Bayesian classifier, Date storage technology, Natural language processing, Date capturing and mining technology, Agent, Collaborative virtual, Case-based reasoning and various data transform technologies are the favorable instrument for knowledge production. And the innovation technologies are also developed constantly to adapt the new environment.

4 Implementation Condition for MCKP

4.1 Implementation Condition Model

Based on the implementation condition for mass customization, combined with knowledge production characteristics, the implementation condition model for MCKP is constructed as shown as Fig. 2. The fitting environment condition for mass customization can pull the implementation, so it is external condition; the fitting organization, process and products conditions decided by knowledge production itself can push the implementation of mass customization, so which is internal conditions.

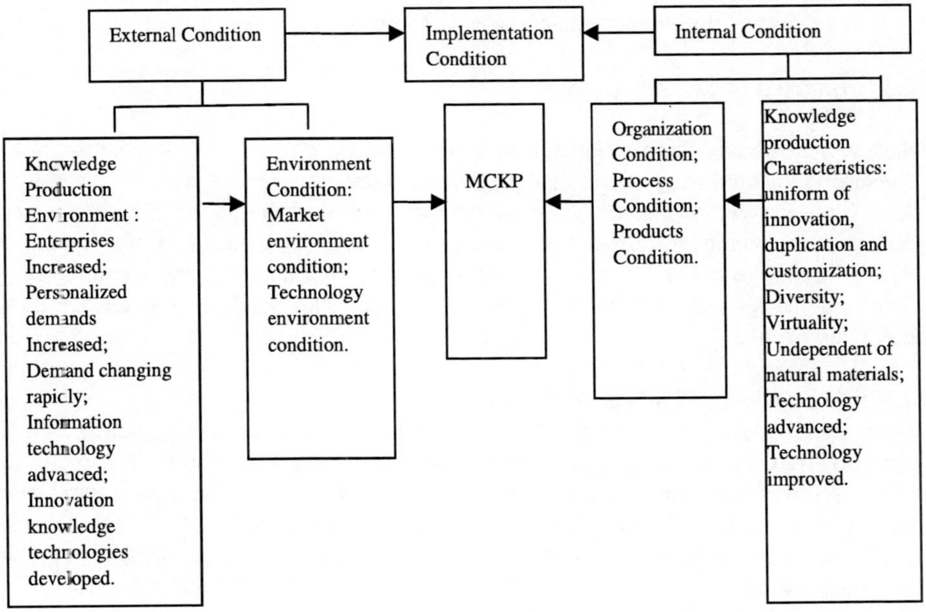

Fig. 2. Implementation condition model for MCKP

4.2 External Condition Analysis

4.2.1 Market Environment for Knowledge Production

In the new century, knowledge economy becomes the main economic thought. Which pulled the development of knowledge Industry and then the related enterprises are increased. Take consultation Industry in China as an example, there are only 41 consultation firms in 1832, according to data, the number has increased to 130000 in 2002 [9]. Subsequently, the market competence becomes intensely. Secondly, more and more types of knowledge products and the apprehension differences between people cause more and more personalize demand for knowledge products. Thirdly, easiness to transmit of knowledge products and the development of information technology shorten the production periods and expedite the changing of market demand too. Obviously, the market environment of knowledge production has presented the characteristics of mass customization, which must pull the implementation for mass customization of knowledge production.

4.2.2 Technology Environment for Knowledge Production

Besides of depending on the intelligence of people, knowledge production depends on information technology, data storage and transform technologies, network technology and other knowledge disposing technologies. Which provide solid foundations for modularization design and can keep delivery in time. Moreover, the mass customization technology for manufacturing can supply some useful references for knowledge production to implement mass customization.

4.3 Internal Condition Analysis

4.3.1 Knowledge Production Organization

Generally, the producers of knowledge production are intelligence person with ability. Most of currently knowledge production organizations compose of senior management personnel, research & development stuff, technicians and operators, and most of them are talent and have the ability of continual learning, apprehending new idea and master new technology quickly. So they can adapt to the changing of production active.

4.3.2 Knowledge Production Process

With the development of knowledge production Industry, its management mode is changing too, in which systematization and engineering are the main directions. Compared with west countries, the R&D management modes of China is laggard. But luckily, it has entered into the third development era (relative to 90s of 20th of west countries) and developing to the fourth ear. So far, the project management system becomes mature elementary and begin to construct the global R&D meshwork[10]. On the other hand, the most complicated part of knowledge production is innovation knowledge production, which depends on new ideas and information mostly, and so the process is flexible. Sometimes, a changing of process can be accomplished just by some thinking actives or simple organization structure adjusting.

4.3.3 Knowledge Products

Most of knowledge products are not only single content but include many knowledge units, and then every units can be a module of products. For instance, a software product contents many difference components, a translation report can be divided different modules according the contents. As to the meaning of knowledge, it is some standard truth, rules, thought and information acknowledged by people [11], so the knowledge units in knowledge products can be seen standard knowledge and can be reused. In addition, diversity of knowledge products just meets the diversity need of mass customization.

Relative to manufacturing products, knowledge products also can be designed for modularization, as shown in Fig.3, but the modularization technology could be more complex.

In a word, modern knowledge production organization, process and products are all avail to implement mass customization. In the condition of external environment pull and internal condition push, mass customization must be the development trend of knowledge production Industry.

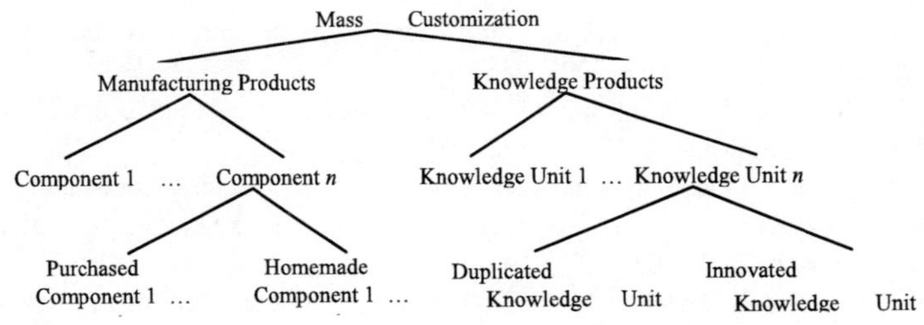

Fig. 3. Mass customization products structure

5 Conclusion

Mass customization is one of the most competitive production modes nowadays. Introduced it into knowledge production research is a challenging subject. In this paper, implementation condition for generally mass customization is analysis firstly, in view of environment, organization, process and product aspects. Then, the characteristics of knowledge production Industry in modern time are discussed, in which, uniform of innovation, duplication and customization, diversity, independent of natural materials and so on are pointed out as the main features of knowledge production. Accordingly, the implementation condition model for MCKP is constructed and analyzed in detail.

Acknowledgment. The paper is supported by Shanghai Leading Academic Discipline Project(S30504) and Innovation Program Project of Shanghai Municipal Education Commission (08YS103).

References

1. Gibbons, M., et al.: The new production of knowledge: The Dynamics of Science and Research in Contemporary Societies. Sage Publications, London (1994)
2. Jones, R.: New terms of research and knowledge production in scotland: the discourse of knowledge transfer. Discourse: Studies in the Cultural Politics of Education 28, 101–120 (2007)
3. Schneider, B., Schalliol, D., Makela, S., et al.: Knowledge production and the public interest. American Sociologist 37, 96–112 (2006)
4. Cuixiao, F., Xingsan, Q.: Research on mass customization for knowledge production. In: The International Conference on Operations and Supply Chain Management, Xi'an, China (2007)
5. Kakati, M.: Mass customization – needs to go beyond technology. Human Systems Management 21, 85–93 (2002)
6. Paul, Z.: The limits of mass customization. MIT Sloan Management Review 42, 81–87 (2001)
7. Xiaofeng, S., Jianhua, J., Peiqing, H.: Implementation conditions and operation modes for internet-based mass customization. Computer Integrated Manufacturing Systems 7, 53–56 (2001)
8. Etzkowitz, H.: Research groups as 'quasi-firms': the invention of the entrepreneurial university. Research Policy 32, 109–121 (2003)
9. Dailin, Y., Yongyi, Z.: Development of Chinas engineering consulting industry. China Civil Engineering Journal 38, 129–133 (2005)
10. Li, L., Mingjuan, Z.: Development of western R&D management modes and its effects on Chinese enterprises-on the case of Shenzhen city. Science of Science and Management of S.& T 7, 39–43 (2006)
11. Badaracco, J.: The Knowledge Link: How Firms Compete Through Strategic Alliances. Harvard Business School Press (1991)

Research on the Multi-platform Cooperative Guidance Tasks Allocation Based on Contract Net Protocol

Li Rui[1] and Wang Hangyu[2]

[1] College of Electronic Engineering, Naval Univ. of Engineering, Wuhan, China
lirui_hg@sohu.com
[2] Office of Research and Development, Naval Univ. of Engineering, Wuhan, China

Abstract. This paper explores the allocation of the cooperative control-and-guide tasks among distributed cooperative guidance platforms, which is time critical and dynamic task allocation in complex distributed system. The structure of Cooperative Guidance Agent is presented and the Contract Net Protocol is employed to the allocation of the cooperative control-and-guide tasks, which meets the time critical and dynamic requirements. This paper provides basis of developing a prototype of the multi-platform cooperative guidance task allocation.

Keywords: multi-platform, cooperative guidance, task allocation, cooperative guidance agent, contract net protocol.

1 Introduction

Multi-platform over-horizon cooperative guidance engagement, aiming at beyond-horizon sea-skimming anti-ship missiles and high-performance fighters etc, is such an engagement that implementing intercepting the target by multiple cooperative guidance between different guidance platforms. Cooperative guidance platforms disperse geographically and spatially, which constitute a complex distributed system. Multi-platform cooperative guidance engagement is time-critical and dynamic engagement, which makes the cooperative guidance tasks allocation a real-time and dynamic task allocation issue in complex distributed system.

Intelligent Agent enjoys the following properties: autonomy, social ability, reactivity and pro-activeness[1], which is suitable to describe or model the cooperative guidance platforms. The Contract Net Protocol (CNP)[2][3], presented by Davis and Smith, is suitable to solve the problem of real-time and dynamic task allocation in distributed system.

2 The Modes of Multi-platform Cooperative Guidance

2.1 Cooperative Guidance Instruction Forming Approaches

There're two kinds of cooperative guidance instruction forming approaches: (1) The interceptor launching platform forms guidance instructions; (2) The cooperative

J. Luo (Ed.): Affective Computing and Intelligent Interaction, AISC 137, pp. 561–569.
springerlink.com

guidance platform forms guidance instructions. In the former circumstance, the original location where the interceptor is launched remains unchanged, while in the latter circumstance, it's necessary to transform the tracking data of the target and the interceptor to the one in the coordination system where the interceptor launching platform guidance instructions is formed, and then forms its own guidance instructions.

2.2 Cooperative Guidance Relay Approaches

Still, there're two kinds of cooperative guidance relay approaches: direct guidance relay[4] and indirect guidance relay[5]. Direct guidance relay depends on the real-time tracking data of the interceptor, which is provided on-line by the interceptor guiding platform. Indirect guidance relay depends not on the real-time tracking data, but on the real-time predictive data, of the interceptor, which is also provided on-line by the former guiding platform, who can no longer steadily track the interceptor. The predictive data is used by the guidance relay platform as an indication to detect the interceptor, which is the pre-condition to track the interceptor and then guide it.

If the guidance areas of two platforms are overlapped, and the overlapped area ΔR satisfies the following condition:

$$\Delta R > V_{\max} T \tag{1}$$

Then the direct guidance relay approach can be adopted.

Where $V_{\max}$ is the maximum speed of the interceptor, and T is the period guidance relay platform accomplishes the guidance relay.

The indirect guidance relay process is complicated, which involves lots of uncertain elements, so it's not addressed in this paper. Instead, the direct guidance relay approach acts as a default method, which is the basis of the cooperative guidance tasks allocation discussed in this paper.

3 Procedure of the Cooperative Guidance Tasks Allocation

The procedure of the cooperative guidance tasks allocation can be described by the following algorithm and Figure 1 respectively, in which n is the total number of the guidance platforms.

```
for (i = 1; 1 ≤ n; )
{
if  platform i can't accomplish the whole guidance task
then   platform i chooses cooperative guidance platform,
and suppose the platform is  i++
else  end
}
```

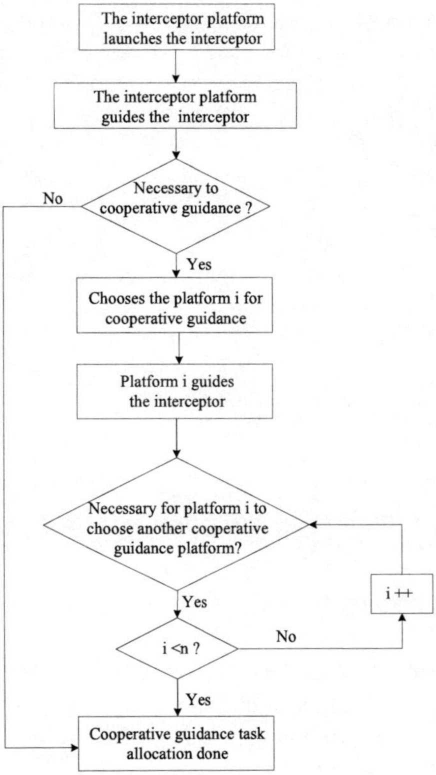

Fig. 1. The procedure of the cooperative guidance task allocation

4 The Structure of Cooperative Guidance Agent

The structure of Cooperative Guidance Agent is presented, according to the features of cooperative guidance platforms. It includes four parts: the cooperative guidance task processor, the contract processor, the database and the communication processor. Cooperative guidance task processor evaluates the necessity of cooperative guidance. If there does exist the task, it sends the task to the contract processor, and implements the task allocation according to the feedback from the contract processor. The contract processor performs announcing bidding, bidding, or/and confirming the task obtainer. The database includes static data and dynamic data. The former memorizes the guidance capability of local platform and cooperative guidance platforms, e.g. the model of interceptors and the range local platform can guide; the latter memorizes the real-time locations and velocities of local and cooperative guidance platforms, the state, load and real-time guidance range of local platform, etc. The communication processor is responsible for communicating with other cooperative guidance Agents through the communication network, following Agent Communication Language and Protocols.

The structure of Cooperative Guidance Agent is depicted in Figure 2.

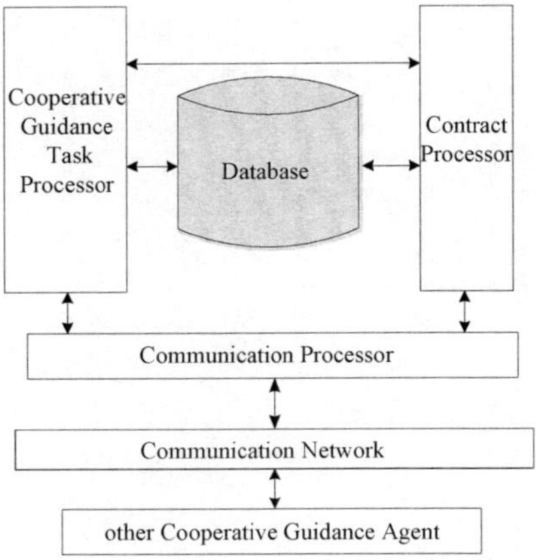

Fig. 2. The structure of Cooperative Guidance Agent

5 CNP Based Cooperative Guidance Task Allocation Model

The Contractor Net Protocol classifies the members of the system as Manager and Contractor. The Manager is responsible for announcing bidding, evaluating bidding and awarding bidding; the Contractor is responsible for bidding, accepting the contractor and task execution later.

5.1 Agent Communication Language

Agent Communication Language mainly includes DARPA developed KQML (Knowledge Query and Manipulation Language), FIPA Agent Communication Language (FIPA-ACL), KIF (Knowledge Interchange Format) and etc[6]. The KQML is adopted as the Agent Communication Language in this paper, so as to establish standard bidding presentation, which is convenient to design and reuse. The KQML is extended to strengthen the application norms of the CNP, so as to improve the efficiency of the CNP.

The norm of the bidding presentation is showed as followed:

```
(Performative
    :sender  ***
    :content  ***
    :receiver ***
    :language  PROLOG
    :ontology Cooperative Guidance Task Allocation
```

Table 1. CNP based KQML communication performatives and corresponding significations[7]

Parameters	Signification
:register	Message sender tells the receiver its own location, name, etc to perform registration
:announce	The sender distributes the task to receivers
:bid	The sender sends the task evaluation to receivers for bidding
:refuse	The sender refuses bidding to receivers
:award	The sender awards contract to receivers
:acknowledge	The sender tells receivers that no contract is awarded
:accept	The sender accepts the receiver's contract
:reject	The sender rejects the receiver's contract

5.2 CNP Based Multi-platform Cooperative Guidance Task Allocation Framework

The interceptor launching platform's Cooperative Guidance Agent is the Manager Agent, and the potential cooperative guidance platforms' Cooperative Guidance Agents are Contractor Agents. If single cooperative guidance can't perform the whole guidance task, then the CNP is used repeatedly. When it occurs, the new guiding platform's Cooperative Guidance Agent becomes the new Manager Agent, and the new potential cooperative guidance platforms' Cooperative Guidance Agents become Contractor Agents. This is depicted in Figure 3.

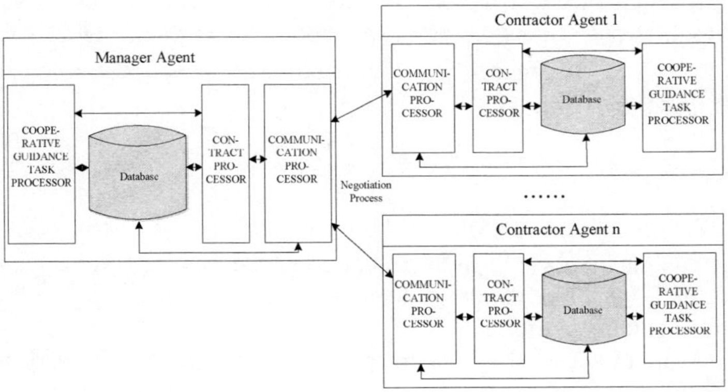

Fig. 3. CNP based multi-platform cooperative guidance task allocation framework

5.3 CNP Based Multi-platform Cooperative Guidance Task Allocation Procedure

CNP based multi-platform cooperative guidance task allocation procedure includes three stages: calling for bidding, bidding, and awarding contract. This is depicted in Figure 4.

(1)Announce Procedure

Announce Procedure includes the following three aspects: evaluating potential bidders, task announcement and assembling bids.

Potential bidders are determined by the following three aspects: type of interceptor the platform can guide, the platform's guidance range, and platform's real-time location, parameters of motion, which is stored in the local Cooperative Guidance Agent's database. The former is obvious, and the latter two decide that whether there's overlapped guiding area between cooperative platforms, which is the pre-condition of cooperative guidance.

Task announcement is distributed in KQML, and its specification is as followed:

```
(Announce
    : sender  Manager Agent
    : content  "Cooperative Guidance Task i"
    : receiver  Contractor Agent j
    : language  PROLOG
    : ontology  Cooperative Guidance Task Allocation
)
```

Where Cooperative Guidance Task i represents the ith cooperative guidance, Contractor Agent j represents platform j, $j \in [1, n_i]$, n_i represents the total number of bidders in the ith cooperative guidance.

Cooperative guidance task allocation is time critical task allocation, thus only bids returning to Manager Agent before expiration time is valid. The expiration time (denoted as T_D) is determined by the maximum speed of the interceptor and platform's guidance range:

$$T_D = V_{max} / D_i \tag{2}$$

Where V_{max} is the maximum speed of the interceptor, here a constant value; D_i is the interceptor guiding platform's guiding range when the ith cooperative guidance task is being allocated, which is variable, determined by the performance of the sensors located on the guiding platform.

If the maximum speed of the interceptor is 2.5Ma, platform's guidance range is 30km, then the expiration time $T_D = 35s$, which means only bids returning to the Manager Agent in 35s since the moment the task announcement is distributed is valid.

(2)Bid Procedure

Bid Procedure includes two components: receiving task announcement and choosing bidding strategy.

Contractor Agent evaluates the cooperative guidance task the moment it receiving the task announcement, by calling local database contained in itself, and then chooses the bidding strategy of bid or not bid. The primary stages are as followed:

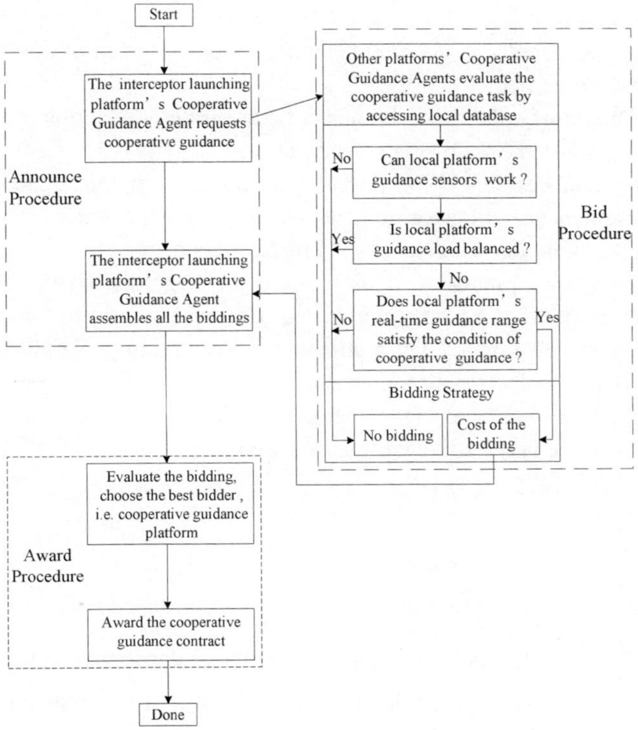

Fig. 4. CNP based multi-platform cooperative guidance task allocation procedure

① Local guidance sensor works or not? If it works, then goes to ②, otherwise not to bid.

② Is the load of local platform balanced, i.e. are there any free weapon passages? If there is, then goes to ③, otherwise not to bid.

③ Does local platform's real-time guidance range satisfy the conditions of cooperative guidance, i.e. $\Delta R > V_{\max} T$? If so, bid for the task, with the content of local platform's real-time guidance range, otherwise not to bid.

The bid specification is as followed, communicated also in KQML.

(Bid

:sender Contractor Agent j

content "Guidance Range S_j", "Cooperative Guidance Task i"

:receiver Manager Agent

:language PROLOG

:ontology Cooperative Guidance Task Allocation

)

(3)Award Procedure

Award Procedure includes two components: choosing the most appropriate bidder and awarding the contract.

To choose the most appropriate bidder, a fuzzy strategy is adopted to evaluate the bid, which is based on the following considerations:

Platform's guidance range. The larger the range is, the fewer the platforms are needed to implement the guidance task, which can save engagement resources and be significant in the situation when the load of platform is heavy.

Platform's guidance precision. If the interceptor is guided by the style of radio-command guidance, the guidance precision decreases with the increase of the guidance range, which implies that platform's guidance range shouldn't be the only concern, otherwise the interception probability will be greatly affected, leading to the failure of cooperative guidance engagement.

The formula of the fuzzy strategy is as followed:

$$E_j = w_1 S_j + w_2 P_j \tag{3}$$

$$w_1 + w_2 = 1 \qquad 0 < w_1 < 1, 0 < w_2 < 1,$$

Where S_j is platform's guidance range;

$P_j = \int P_j(t) f_j(t) dt$ is the expectation of platform's guidance precision.

The values of w_1 and w_2 is decided dynamically by the commander, according to the actual situation in the engagement.

The algorithm to evaluate the most appropriate bidder is as followed:

```
initial E[]={E1,E2,···,Ek} //k:number of bids
for (i=1;i≤k;){
     if(E[i]≥E1)
     E1=E[i];
     i++;
}
for (i=1;i≤k;){
     if (E1=E[i])
     {
       output E[i];
      end
     }
     else i++;
}
```

The award specification is as followed, communicated also in KQML.

```
(Award
    : sender  Manager Agent
    : content  Cooperative GuidanceTask i
    : receiver  Contractor  Agent i
    : language  PROLOG
    : ontology  Cooperative Guidance Task Allocation
)
```

6 Conclusions

From the analysis above, it's clear that the Contract Net Protocol can be employed to solve the problem of cooperative guidance task allocation, which is the basis of developing a prototype of the cooperative guidance task allocation. To do this, an Agent develop framework, e.g. FIPA-JADE (Java Agent Development Environment)[8] can be adopted, which has been successfully applied to the domain of telecommunication. The prototype then can be deployed as a module of fleet's command and control system, and thus implies promising practical applications.

References

1. Wooldridge, M.J., Jennings, N.R.: Intelligent Agent: theory and practice. Knowledge Engineering Review 10(2), 115–152 (1995)
2. Smith, R.G.: The contract net protocol: high-level communication and control in a distributed problem solver. IEEE Transaction on Computers 29(12), 1104–1113 (1980)
3. Davis, R., Smith, R.G.: Negotiation as a metaphor for distributed problem solving. Artificial Intelligence (20), 63–109 (1983)
4. Chang, C.B., Youens, L.C.: Measurement Correlation for Multiple Sensor Tracking in Dense Target Environment. IEEE Trans. on Automatic Control AC-27(6), 1250–1252 (1982)
5. Liu, X.: The analysis design and simulation technology of missile guidance control system, pp. 50–56. Northwestern Polytechnical University Press (2006)
6. Fan, Y.: Multi-Agent System: theory and practice. Tsinghua University Press (2002)
7. Chen, Y.: Extension of KQML for Realization of Contract Net. Computer Technology and Development 16(2), 53–55 (2006)
8. Bellifemine, F., Caire, G., Greenwood, D.: Developing Multi-Agent Systems with JADE. John Wiley & Sons (2007)

Empirical Research on the Influencing Factors in Technological Progress of Beijing's High-Tech Industry*

Zhang Jingqiang

School of Economics and Management
North China University of Technology Beijing, China
zjq811@163.com

Abstract. From technological progress of the industry, using factor analysis, the author analyzes the influencing factors in technological progress of Beijing's high-tech industry. The result indicates that the level of R&D investment, technological innovation capability and product innovation capability are major influencing factors in the technological progress of Beijing's high-tech industry. Therefore, in order to improve and enhance the technological progress of Beijing's high-tech industry, we should focus on improving science and technology investment, optimizing the industrial structure, accelerating the development of new technologies and new products, saving energy and reducing cost.

Index Terms: high-tech industry, technological progress, factor analysis, empirical research.

1 Introduction

Technological progress is the reform of production process and the introduction of new products and new services. In recent years, the research of technological progress has become an important point of the academic discussion in academia world. Foreign scholars, Fare. R (1994), put the DEA method into technological progress research [1]; Keller (2004) measured the technological progress using total factor productivity [2]. Domestic scholars, Yi-fu Lin and Pei-Lin Liu (2003) [3], Qingwang Guo and Junxue Jia Snow (2005) [4], Shu-Jing Yue and Zhaoming Liu (2006) [5], measured China's technological progress through various methods. Industrial development practice around the world shows that With the rapid development of science and technology, If they attempted to catch up with developed countries, developing countries must take full advantage of favorable international environment and the world's advanced technology and excellent scientific and technological achievements and lead them into their own countries for the domestic industrial sectors through international technology transfer, technology exchange and other forms, so that domestic industries may be helpful for development and innovation.

* This paper is supported by "Beijing's Undergraduate student Scientific Research and Entrepreneurial Action Plan Project (1018)"and "Youth Key Research Fund Project of North China University of Technology".

J. Luo (Ed.): Affective Computing and Intelligent Interaction, AISC 137, pp. 571–576.
springerlink.com

In this paper, from technological progress of the industry, the author carries out a comprehensive analysis of various factors for technological progress for Beijing's high-tech industry using factor analysis method of econometrics and finds out the main factors which affect technological progress for Beijing's high-tech industry.

2 Mathematic Model

Factor analysis method is of the promotion and development of principal component analysis method, which can turn complex variables into several factors to reproduce the relationship between the original variables and factors, at the same time, the variables can be also classified according to different factors. There are many similarities for factor analysis and principal component analysis, but in factor analysis, if the structure of the factor loading matrix is not enough simplified, we may transform it to make it simplified so that the new factors take on more obvious practical significance.

According to factor analysis, the mathematical model is as follows:

Assumptions there are n variables: $x_1, x_2, \cdots, x_n$, and the variables have been standardized (mean: 0, standard deviation: 1). The n variables may be expressed as the linear combination of k factors: $f_1, f_2, \cdots, f_k$, whose equation is expressed as the following form:

$$X = AF + \varepsilon \tag{1}$$

Among them, $X = (x_1, x_2, \cdots, x_n)'$, is the observable n dimension vector, and each component indicates an index variable.

$$A = \begin{pmatrix} a_{11} & a_{12} & \cdots & a_{1k} \\ a_{21} & a_{22} & \cdots & a_{2k} \\ \cdots & \cdots & \cdots & \cdots \\ a_{n1} & a_{n2} & \cdots & a_{nk} \end{pmatrix}$$, is called component matrix.

$F = (f_1, f_2, \cdots, f_k)'$, is called factor vector or public factor.

$\varepsilon = (\varepsilon_1, \varepsilon_2, \cdots, \varepsilon_n)'$, is called special factor and expressed the part of original variables which can not be explained by the factor. Its mean is 0.

At the same time, the factor mathematical model must follow the following relationship:

- All public factors' mean is 0 and their variance is 1.The factors are irrelevant with each other;
- All errors' mean is 0 with unequal Variance. The errors are irrelevant with each other;
- The common factors and the errors are independent with each other. Detailed factor analysis principle and deduced process can be found in related references [6].

3 Empirical Results

3.1 Index Selection

In this article, the author selects ten indexes which can comprehensively reflect technological progress for Beijing's high-tech industry, as follows:

- x_1 : Industrial added value (ten thousand yuan)
- x_2 : Personnel engaged in industrial S&T activities (person);
- x_3 : Scientists and engineers of S&T personnel (person);
- x_4 : Full-time equivalent of R&D personnel (man-year);
- x_5 : Expenditure on R&D (ten thousand yuan);
- x_6 : Total patent applications examined (unit);
- x_7 : Products export value (ten thousand yuan);
- x_8 : Development expenditure of new products (ten thousand yuan);
- x_9 : Output value of new products
- x_{10} : Invention patent applications certified (unit).

3.2 Empirical Calculation

The author constructs a 23×10 dimensional data tables with the data of Beijing's high-tech industry from 1985 to 2007, and obtains the Eigenvalues and the contribution rate of technological progress for Beijing's high-tech industry using factor analysis method, as shown in Table 1.

The contribution rate of the common factors indicates the amount of information that these common factors reflect the original indexes, and the cumulative contribution rate indicates the accumulative total amount of information that the corresponding common factors reflect the original indexes. As can be seen from Table 2, The accumulative contribution rates of the first two common factors has reached 88.035 percent, which points that the first two common factors can reflect the 88.035 percent information of the original indexes. Therefore, above all indexes may be represented by the common factors Y1 and Y2 and obtain the feature vectors which correspond the characteristic roots of the two common factors. In order to facilitate the interpretation of the factors, we must rotate the factor loading matrix to achieve the purpose of simplifying the structure, so that various variables can possess high loads on a single factor and possess small or moderate loads on the other factors. In this article, the rotation method is varimax orthogonal rotation. As can been in Table 2, it is the load matrix after the orthogonal rotation.

Table 1. Eigenvalues and Contribution Rate

Component	Initial Eigenvalues			Rotation Sums of Squared Loadings		
	Total	*% of Variance*	*Cumulative %*	*Total*	*% of Variance*	*Cumulative %*
1	7.728	77.284	77.284	4.494	44.935	44.935
2	1.075	10.751	88.035	4.310	43.100	88.035
3	.625	6.254	94.289			
4	.461	4.615	98.903			
5	.110	1.097	100.000			
6	3.35E-016	3.35E-015	100.000			
7	1.51E-016	1.51E-015	100.000			
8	1.91E-017	1.91E-016	100.000			
9	-7.58E-017	-7.58E-016	100.000			
10	-1.59E-016	-1.59E-015	100.000			

Extraction Method: Principal Component Analysis.

Table 2. Rotated Component Matrixes

	Component	
	Y1	*Y2*
x_1	.756	.630
x_2	.580	.789
x_3	.633	.730
x_4	.599	.437
x_5	.925	.073
x_6	.723	.517
x_7	.753	.625
x_8	.863	.351
x_9	.145	.963
x_{10}	.339	.924

In Table 2, the coefficients are the loading estimate values of the rotation factors, whose statistical significance is the correlation coefficient of variables and factors. According to this point, based on the analysis of Table 2, we can see that the loads on all the variables are positive, which shows that it takes on positive correlation between above 10 variables and technological progress, that is, and they are the reason of technological progress. Because the contribution rate of the first common factor is as high as 77.284%, we can find out the main factors of technological progress for Beijing's high-tech industry through analyzing the variable loads on the first common factor. As can be seen from Table 2, there are high loads for the first, 4th, 5th, 6th, 7th and 8th variables on factor Y1, which indicates that above six variables are the main factors affecting technological progress of Beijing's high-tech industry. These factors are industrial added value, full-time equivalent of R&D personnel, expenditure on R&D, total patent applications examined; Products export value, and development expenditure of new products, which respectively reflect the R&D investment level, technical innovation ability and product innovation ability of Beijing's high-tech industry.

From above analysis, it can be seen that the reasons of technological progress for Beijing's high-tech industry are as following:

- A great deal of R & D capital investments make technology innovation ability and product innovation ability of Beijing's high-tech industry greatly improved.
- The Government, through the investment on the development expenditure of new products, constantly improves product innovation ability.
- The Government strengthens research and development of high technology to make technology innovation ability improved continuously.

In addition, the contribution rate of the second common factor is 10.751%, which can also reflect the amount of information of the original indexes. As can be seen from Table 2, there are high loads for the 2nd, 3rd, 9th, and 10th variables on factor Y2. These variables reflect human capital level, product structure and technological development level of Beijing's high-tech industry. At the same time, these factors also promote technological progress of Beijing's high-tech industry.

4 Conclusions

Through the above analysis, it shows that the main influence factors of technological progress for Beijing's high-tech industry are industrial technology innovation ability and product innovation ability. Therefore, to advance technology progress level of Beijing's high-tech industry, the following aspects should be carried out.

First, we should increase S&T investments and promote industrial structure upgrading. Technological progress is an important guarantee of industrial adjustment and product upgrading, and is the driving force of industrial structure adjustment. The result of technological progress is bound to lead to industrial structural shift, so that the technical level of economic system is higher and higher, and the industrial structure constantly tends to be gentrification and rationalization. There are natural S&T resources advantages in Beijing. In the course of high-tech industrial development, we should increase S&T investments, speed up industrial structure adjustment, optimize and upgrade the industrial structure, and improve technological progress level of high-tech industry.

Secondly, we should integrate resources; promote R&D of new technologies and new products. Technological progress is bound to lead to application and production of new technologies and new products. New technologies and new products have wide application space and vitality, which directly impulse development and grandness of Beijing's high-tech industry. At the same time, the development of new technologies and new products is the requirement of survival and development for Beijing's high-tech industry. Therefore, there is an important promoting role for Beijing's high-tech industry to improve its technological progress level through speeding up research and development efforts of new technologies and new products.

Thirdly, we should reduce costs, save energy and decrease consumption. The economic development practice of various countries shows that the equipments, technologies, and crafts are developing to the direction of saving resources because technological progress make effective use of resources continue to increase. Therefore, we should achieve the development of Beijing's high-tech industry through improving resource efficiency, reducing costs and energy consumption. At the present, industry competition is actually technological competition, or product competition, or market competition, but is radically S&T competition. Therefore, the development of Beijing's high-tech industry should attach importance to the benefits and costs of technological progress and put reduction of energy consumption in the first place of industrial development, which is the inevitable choice under the present historical condition.

Acknowledgment. I express heartfelt thanks for Mrs. Zhan Fengmei and Mrs. Huang Zilang. They collect and sort data for this paper, proofread the whole paper, and put forward many suggestions for the completion of this paper.

References

1. Fare, R., Grosskopf, S., Norris, M.: Productivity Growth, Technical Progress and Efficiency Change in Industrialized Countries. American Economic Review 84, 66–83 (1994)
2. Keller, W.: International Technology Diffusion. Journal of Economic Literature XLII, 752–782 (2004)
3. Lin, Y., Liu, P.: The Effects of Development Strategy on Capital Accumulation and Technological Progress: An Empirical Analysis of Chinese Experience. Social Sciences in China, 18–33 (April 2003)
4. Guo, Q., Jia, J.: Estimating Total Factor Productivity in China: 1979-2004. Economic Research, 51–60 (June 2005)
5. Yue, S., Liu, C.: Human Capital Accumulation and Regional Total Factor Productivity. Economic Research, 90–97 (April 2006)
6. Lu, W.: SPSS for Windows Statistical Analysis, 3rd edn., pp. 480–506. Electronics Industry Press (2006)

Extended Enterprise Input-Output Model for Sustainability Management*

Fang Wang[1] and Xiao-Ping Jia[2]

[1] College of Environmental & Safety Engineering,
Qingdao University of Science & Technology, Qingdao 266042, China
wangf@qust.edu.cn
[2] Center for Chemical Process Engineering,
Qingdao University of Science & Technology, Qingdao 266042, China
cnjiaxp@yahoo.com.cn

Abstract. This paper focuses on the development of analysis and evaluation framework and methodology for the enterprise, which can support the enterprise sustainability management for the improvement of resources efficiency. In this paper the extended enterprise input-output model (EEIO) for sustainability analysis is proposed, which consists of a combination of enterprise input-output analysis (EIO) and process integration technology (PI). The EIO will help investigate the internal material metabolism of industrial enterprises. The PI will identify and target the potentials of resource recovery and waste reduction.

Keywords: enterprise input-output model, process integration, sustainability, industrial enterprises.

1 Introduction

Environmental problems and resource scarcity pressures on the industrial enterprises are increasing. For many industrial enterprises, it becomes important to ensure sustainable business development, combining economic, social and environmental aspects. Sustainability has become a key concept in this problem solving from the viewpoint of the resource consumption and waste minimization. In order to achieve economic development in a more sustainable manner, new management methodologies or processes for industrial systems need to be developed. Novkov reported 59 categories of techniques and tools for sustainability management, such as input-output analysis, material flow accounting, etc [1]. The Novkov's report also summarized the use of different techniques and tools at different stages of decision-making process [1].

One of the most practical tools to emerge in the field of enterprise management in the past decades has been enterprise input-output analysis, which is based on a recognized and well proven method (i.e. input-output analysis). The enterprise input-output (EIO) model, which describes the resources allocation and internal and

* Financial support of the Doctoral Funds of Qingdao University of Science and Technology is appreciated.

J. Luo (Ed.): Affective Computing and Intelligent Interaction, AISC 137, pp. 577–585.
springerlink.com

external linkages of enterprises, has been widely applied in simulating enterprise production processes and analyzing supply chain [2~9]. The EIO account includes all inputs of production processes and records material flows and financial transactions among different units within a company and between the company and the outside market [2]. The EIO model provides an input-output table to describe all the material and monetary flows in a single company.

The standard EIO model has the weakness that it does not identify the potentials of resource recovery and waste reduction activities. To handle with this problem, process systems engineering (PSE) approaches are recommended. PSE approaches are perhaps best positioned to plan and optimize a site-wide industrial process and to address the challenges of sustainability. PSE adopts a holistic or systems view, which is essential for understanding and modeling the complex interactions between industry, society and ecosystems [10]. The European Union supported a consultation study with senior industrialists and academics to identify priorities in enhancing the sustainability of the process industries [11]. The panel considered 19 tools that might be of value in enhancing the sustainability. Process simulation, process optimization, process synthesis, and process integration are considered, which belong to PSE field. Among them, process integration is the best-established technical approach for improving sustainability [11].

Process integration is a family of methods for combining several processes to reduce consumption of resources or harmful emissions to the environment. It started as mainly heat integration stimulated by the energy crisis in the 1970's [12]. One of the purposes of this paper is to introduce process integration methods into enterprise sustainability management.

The rest of this paper is organized as follows. In the section II, the aim and motivation is presented. General theoretical and models for EEIO are discussed in the section III, IV and V. Finally, section VI presents the conclusion and future work.

2 Aim and Motivation

The starting points of this work include the followings:

- For industrial enterprises, it becomes important to ensure sustainable business development, combining economic, social and environmental aspects [1];
- Resource consumption and waste minimization is the main viewpoint of the sustainability management;
- Sustainability Management of industrial enterprises are complex and therefore require a multi-disciplinary approach;
- Analysis and evaluation can increase the interaction between the decision-makers and designers and therefore improve the quality of policy making in the context of the industrial enterprises.

This work aims to construct a framework and methodology, which can be used to analyze and evaluate for enterprise sustainable management. Also, this work explores the decision making procedure in a systematic way at a company level. The EEIO will be designed to address the material metabolism and potentials of resource recovery, making it possible to highlight it potential for sustainable enterprise management.

3 Methodology

The proposed framework for enterprise sustainability analysis by accounting material and monetary flows simultaneously is shown in Figure1. It describes the resource allocation and waste generation of production processes and internal and external linkages of enterprises based on the EIO model. It also determines the potentials of resource recovery by process integration. This framework illustrates material flows changes after taking actions, and validates the feasibility of proposed actions by calculating their monetary profit and environmental benefit. In addition, it can predict possible development scenarios according to the mathematical relationships of the EIO model. This framework mainly consists of two modules: enterprise input-output (EIO) model, indicators and pinch integration (PI). Descriptions of the framework and its modules are listed in the following. The A and B blocks in Figure 1 will link with process integration procedure, which presented in Figure 2 in detailed.

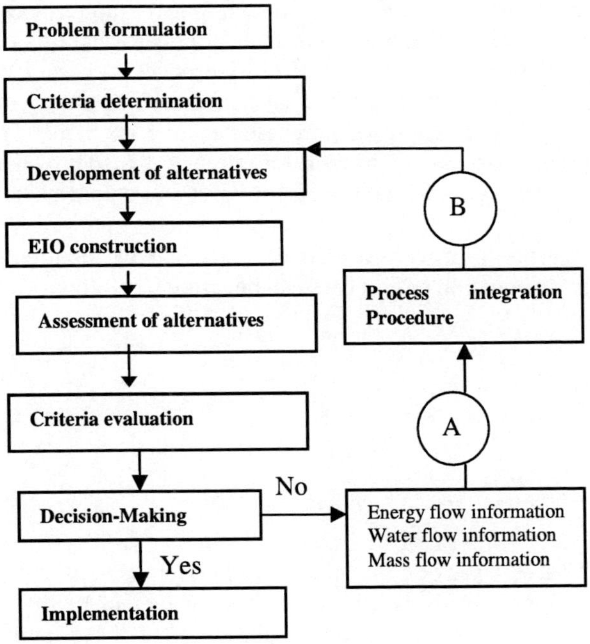

Fig. 1. Framework of the EEIO

4 EIO Model

The enterprise input-output (EIO) model is the basic input-output process model established [2][13]. As a general rule, in the EIO model, a positive number is used to indicate the production of a main product, by-product and waste, and a negative number for the consumption of a main product, purchased inputs, by-product and waste, and inputs of capital. There are two main hypotheses in the EIO model: (1) There is only one main product in each process; (2) The main products of one process cannot be consumed by the process itself.

The structure of the EIO model is shown in Table 1. It illustrates the material, energy and monetary inputs and outputs among processes. Referring to a given period of time, the notation used for Table I is the following:

$Z_{i,i}$:main product output of process i, in physical unit; $Z_{i,j}$:intermediate consumption of product i in process j, in physical unit; $I_{i,j}$: consumption of imported product i in process j, in physical unit; $M_{i,j}$: consumption of purchased inputs i in process j, in physical unit; $W_{i,j}$: discharge of by-products or waste i from process j, in physical unit; Y_i: final output of main product i, in physical unit; $X_{i,}{}^{I}$:total demand of imports of main product i, in physical unit; $X_{i,}{}^{m}$:total demand of purchased inputs i, in physical unit; $X_{i,}{}^{w}$: inal output of by-product or waste i, in physical unit; X_i :gross outputs of main product in process i, in physical unit; V_i :inputs of capital in process i, in monetary unit; C_i:price of main product i; $C_{i,}{}^{I}$:price of imports of main product i; $C_{i,}{}^{m}$: price of purchased input i; $C_{i,}{}^{w}$:price of by-product or waste i; B_i :final benefit of main product i, in monetary unit;$B_{i,}{}^{I}$: final benefit of imports of main product i, in monetary unit; $B_{i,}{}^{m}$:final benefit of purchased input i, in monetary unit; $B_{i,}{}^{w}$: final benefit of by-product or waste i, in monetary unit; V: the sum of inputs of capital, in monetary unit; B: the sum of final benefit of input and output materials, in monetary unit.

The row of Table 1 provide a statistical account of input-output flows in the enterprise. Balance equations of process can be written as follows:

$$\sum_{j=1}^{n} Z_{i,j} = Y_i\,;\ \sum_{j=1}^{n} I_{i,j} = X_i^I\,;\ \sum_{j=1}^{n} M_{i,j} = X_i^m\,;\ \sum_{j=1}^{n} W_{i,j} = X_i^w\,;\ \sum_{i=1}^{n} V_i = V\,;$$

$$Y_i \times C_i = B_i \quad (i{=}1,\ldots,n);$$

$$Y_i^y \times C_i^y = B_i^y\,,\ y = I, m, w \tag{1}$$

The column of Table 1 provide information on the inputs used and outputs generated in each production process in the enterprise. The inputs include those produced by the enterprise (main products, by-products and waste), those purchased externally (imports of main products and purchased inputs), and inputs of capital. Balance equations of process can be written as follows:

$$\sum_{i=1}^{n} B_i + \sum_{i=1}^{m} B_i^I + \sum_{i=1}^{k} B_i^m + \sum_{i=1}^{q} B_i^w = B\,;\ Z_{i,i} = X_i \tag{2}$$

With each element in a column of Table 1 divided by the output of the main product of the production process, the direct or technical coefficients matrix is obtained. Each coefficient shows the amount of input required or output produced for each unit of main product output in the enterprise.

- Direct input-output coefficients

Referring to the input-output model introduced by Miller and Blair [14], the calculations of direct input-output coefficients are written as follows:

$$a_{ij} = \frac{Z_{ij}}{X_j};\ b_{tj} = \frac{M_{tj}}{X_j};\ c_{pj} = \frac{W_{pj}}{X_j};\ d_j = \frac{V_j}{X_j} \tag{3}$$

where a_{ij} represents main product i per unit of main product j; b_{tj} represents purchased input t required per unit of main product j; c_{pj} represents by-product or waste p generated per unit of main product j and d_j represents cost inputs required per unit of main product j.

Table 1. Structure of the EIO Model

Item		Production Process				Final output/Total demand	Price	Final benefit
		1	2	...	n			
Production process and main products	1	$Z_{1,1}$	$Z_{1,2}$	...	$Z_{1,n}$	Y_1	C_1	B_1
	2	$Z_{2,1}$	$Z_{2,2}$	...	$Z_{2,n}$	Y_2	C_2	B_2
	...	...	...	...	...	...	...	...
	n	$Z_{n,1}$	$Z_{n,2}$	...	$Z_{n,n}$	Y_n	C_n	B_n
Imports of main products	1	$I_{1,1}$	$I_{1,2}$	...	$I_{1,n}$	$X_{1,}^{I}$	$C_{1,}^{I}$	$B_{1,}^{I}$
	2	$I_{2,1}$	$I_{2,2}$	...	$I_{2,n}$	$X_{2,}^{I}$	$C_{2,}^{I}$	$B_{2,}^{I}$
	...	...	...	...	...	...	...	...
	m	$I_{m,1}$	$I_{m,2}$	...	$I_{m,n}$	$X_{m,}^{I}$	$C_{m,}^{I}$	$B_{m,}^{I}$
Purchased inputs	1	$M_{1,1}$	$M_{1,2}$	...	$M_{1,n}$	$X_{1,}^{m}$	$C_{1,}^{m}$	$B_{1,}^{m}$
	2	$M_{2,1}$	$M_{2,2}$	...	$M_{2,n}$	$X_{2,}^{m}$	$C_{2,}^{m}$	$B_{2,}^{m}$
	...	...	...	...	...	...	...	...
	k	$M_{k,1}$	$M_{k,2}$	...	$M_{k,n}$	$X_{k,}^{m}$	$C_{k,}^{m}$	$B_{k,}^{m}$
By-products and waste outputs	1	$W_{1,1}$	$W_{1,2}$	...	$W_{1,n}$	$X_{1,}^{w}$	$C_{1,}^{w}$	$B_{1,}^{w}$
	2	$W_{2,1}$	$W_{2,2}$	...	$W_{2,n}$	$X_{2,}^{w}$	$C_{2,}^{w}$	$B_{2,}^{w}$
	...	...	...	...	...	...	...	...
	q	$W_{q,1}$	$W_{q,2}$	...	$W_{q,n}$	$X_{q,}^{w}$	$C_{q,}^{w}$	$B_{q,}^{w}$
Gross outputs of main products		X_1	X_2	...	X_n			
Inputs of capital		V_1	V_2	...	V_n	V		
Gross benefit								B

- The basic mathematical relationships

Given: $A=[a_{ij}]$= direct input-output coefficients for main product; $B=[b_{tj}]$= direct input-output coefficients for purchased inputs; $C=[c_{pj}]$= direct input-output coefficients for by-product and waste outputs; $D=[d_j]$= direct input-output coefficients for cost inputs; X= a column vector consisting of process main products; Y= a column vector consisting of final outputs of main products; I= a column vector consisting of final inputs of purchased inputs; W= a column vector consisting of final outputs of by-products and wastes; P= a column vector consisting of the prices of main products, purchased inputs, by-products and wastes; F= a column vector consisting of the final benefits of main products, purchased inputs, by-products and wastes; G= a column vector consisting of final outputs for main products Y, purchased inputs P, by-products and waste outputs W.

The row balances of main products, purchased inputs, by-products and waste outputs are expressed in the following:

$$AX = Y;\ BX = I;\ CX = W;\ DX = V \tag{4}$$

The row and column balances of benefits are expressed in the following:

$$G'\hat{P} = F;\ G'P = B \tag{5}$$

- Total input-output coefficients

From the direct input-output coefficients, the total input-output coefficients can be calculated, which describes total amount of inputs required or outputs generated, both directly and indirectly, to produce one unit of final outputs of main products.

$$X = A^{-1}Y;\ I = BX = BA^{-1}Y;\ W = CX = CA^{-1}Y;\ \ V = DX = DA^{-1}Y \tag{6}$$

Where: A^{-1}=total output of main product per unit of final outputs of main products; BA^{-1}=total purchased inputs per unit of final outputs of main products; CA^{-1}= total by-product and waste outputs per unit of final outputs of main products; DA^{-1}= total cost inputs per unit of final outputs of main products.

Each coefficient in the direct input-output coefficient matrix represents the inputs required or outputs produced for each unit of main product outputs in the enterprise, while each element in the total input-output coefficient matrix describes the total inputs required or outputs generated, directly and indirectly, to produce one unit of final outputs of main products. The EIO model can be used to account material metabolism and plan the production of enterprises.

5 Process Integration Procedure

As mentioned previously, the EIO model provides a picture of material metabolism for an alternative. Process integration (PI) provides tools that allow us to investigate the water, energy and mass flows within a system. And then, the resource and waste reduction potentials are identified by the process integration approaches. They may be used to identify new alternatives within a system.

Process integration refers to efficient and systematic approaches that allow industries for analysis and design of processes and utility systems to increase their profitability through reductions in energy, water and raw materials consumption, greenhouse gas emissions and waste generation. Among PI methodologies, pinch analysis is certainly the most widely used. This is due to the simplicity of its underlying concepts and, especially, to the results it has obtained in numerous projects worldwide. Recent achievements include energy saving and gas emission reduction, water saving and waste-water minimization; efficient raw-materials use; mass integration for pollution prevention; oxygen pinch; and hydrogen network design [15~20].

The pinch analysis technology is introduced by Linnhoff and Vredeveld for energy-saving in enterprises [18]. Wang and Smith applied it for water network optimization and wastewater treatment system design, termed as water pinch technology [19]. Currently, pinch technology is increasingly applied for resource management in enterprises [20].

For a practical application, a stepwise framework for process integration is shown in Figure 2.

1. Scoping and data gathering. Gather the detailed data from the EIO model.

2. Data analysis. Inventory and classify the data according to the different aspects, such as materials, energy, water, emissions, etc.

3. Process modeling. Identify and generate the alternatives using thermodynamic targeting method.

4. Simulation and analysis. Use the available tools, such as commercial simulators (e.g. Aspen Plus), to simulate and analyze the alternatives.

5. Process integration. Identify the opportunities for energy, water, and resources conservation. Generate energy, water and mass network.

6. Evaluation. The results of the above steps are inventoried based on the water, energy and mass balance.

7. Decision-making. The results obtained at the above-mentioned steps must be examined and ranked according to the criteria. Major contributions to the objective and opportunities for improvement also are identified. The results will feedback the EIO model, to further material metabolism analysis.

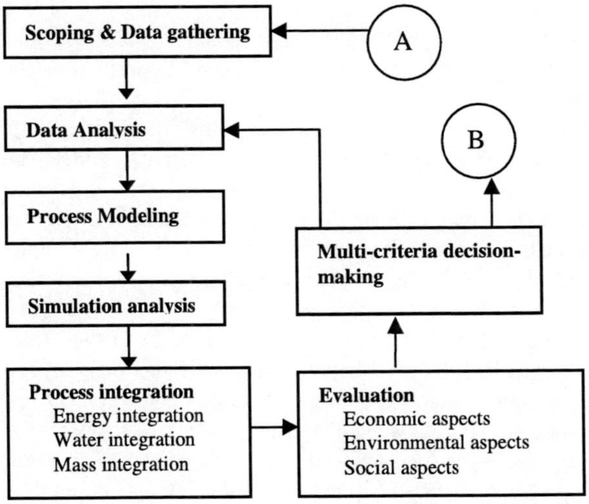

Fig. 2. Stepwise framework of process integration in EEIO

6 Conclusions and Future Work

Based on the EIO model and PI, this paper proposed the EEIO framework to analyze the enterprise sustainability. This proposed framework allows us to investigate material and monetary inputs and outputs, and illustrate material and monetary flow changes, and identify the potentials of resource recovery. The proposed framework helps fill the gap between material metabolism analysis and resource recovery and waste reduction activities.

In the future research efforts would have to be made in the following areas: (1) its applications to real-world problems, the authors are conducting two industrial enterprises, i.e. a coal production plant and a poly-generation plant; (2) development an interactive evaluation procedure between the EIO model and PI.

Acknowledgment. Wang Fang wish to acknowledge with sincere thanks Dr. Liang Sai, Tsinghua University, for his valuable comments and constructive discussion points.

References

1. Svetoslav, N.: Sustainability management of industrial enterprises–advanced concepts, techniques and tools. In: 5th International Scientific Conference Business and Management, pp. 95–101. Vilnius Gediminas Technical University Publishing House, Vilnius (2008)
2. Lin, X., Polenske, K.R.: Input-output modeling of production processes for business management. Structural Change and Economic Dynamics 9(2), 205–226 (1998)
3. Albino, V., Kühtz, S.: Assessment of Environmental Impact of Production Processes in Industrial Districts Using Input-output Modeling Techniques. Journal of Environmental Informatics 1(1), 7–20 (2001)
4. Albino, V., Kühtz, S.: Enterprise input-output model for local sustainable development - the case of a tiles manufacturer in Italy. Resources Conservation and Recycling 41(3), 165–176 (2004)
5. Albino, V., Dietzenbacher, E., Kühtz, S.: Analysing Materials and Energy Flows in an Industrial District using an Enterprise Input-Output Model. Economic Systems Research 15(4), 457–480 (2003)
6. Albino, V., Kühtz, S., Zhou, C., Peng, G.: Energy and materials use in Italian and Chinese tile manufacturers: a comparison using an Enterprise Input-Output model. In: 15th International Conference on Input-Output Techniques. Renmin University of China, Beijing (2005)
7. Albino, V., Nicolò, M.D., Petruzzelli, A.M., Yazan, D.: Integrating Enterprise Input-Output Modelling with GIS Technology for Supply Chain Analysis. In: 16th International Input-output Conference, Istanbul, Turkey (2007)
8. Albino, V., Nicolò, M.D., Garavelli, A.C., Petruzzelli, A.M., Yazan, D.M.: Rural development and agro-energy supply chain. An application of enterprise input-output modelling supported by GIS. In: 16th International Input-output Conference, Istanbul, Turkey (2007)
9. Polenske, K.R., McMichael, F.C.: A Chinese cokemaking process-flow model for energy and environmental analyses. Energy Policy 30(10), 865–883 (2002)

10. Bakshi, B.R., Fiksel, J.: The quest for sustainability: Challenges for process systems engineering. Aiche Journal 49, 1350–1358 (2003)
11. Tsoka, C., Johns, W.R., Linke, P., Kokossis, A.: Towards sustainability and green chemical engineering: tools and technology requirements. Green Chemistry 6(8), 401–406 (2004)
12. Smith, R.: Chemical Process: Design and Integration. John Wiley & Sons, NY (2005)
13. Liang, S., Jia, X.P., Zhang, T.Z.: Enterprise Sustainability Analysis for Coal Production: A case in China (unpublised)
14. Miller, R.E., Blair, P.D.: Input-Output Analysis: Foundations and Extensions, 2nd edn. Cambridge University Press, Cambridge (2009)
15. Hallale, N.: Burning bright - Trends in process integration. Chemical Engineering Progress 97(7), 30–41 (2001)
16. Jia, X., Xiang, S., Li, H., Wang, F.: Process integration for industrial symbiosis network. International Journal of Ecological Economics & Statistics 11, 49–63 (2008)
17. Kemp, I.: Pinch Analysis and Process Integration. Elsevier (Butterworth-Heinemann), NY (2007)
18. Linnhoff, B., Vredeveld, D.R.: Heat-Transfer - Pinch technology has come of age. Chemical Engineering Progress 80(7), 33–40 (1984)
19. Wang, Y.P., Smith, R.: Wastewater minimization. Chem. Eng. Sci. 49(7), 981–1006 (1994)
20. Foo, D.C.Y.: State-of-the-Art Review of Pinch Analysis Techniques for Water Network Synthesis. Ind. Eng. Chem. Res. 48, 5125–5159 (2009)

The Evaluation of Consumer Credit for Electric Power Enterprise Based on Subjective and Objective Weighting Approach

Huang Yuan-sheng, Shang Sang, and Shi Xiu-fen

School of Business Administration,
North China Electric Power University, Baoding, China
shangsang1987@yahoo.com.cn

Abstract. The customer credit evaluation is a key element of the electricity market research for the electric power enterprises, the level of customer credit has a direct impact on the electric power company's operations. For avoiding the customer credit risk, this paper evaluated consumer credit for electric power enterprises by using a method which is combined quantitative analysis with qualitative analysis. In view of that subjective weighting method and objective weighting method have their own disadvantages and shortcomings, to improve the reliability of weights, this paper combined this two kinds of weighting methods organically to determine the comprehensive weights for evaluation indicators. The results of case study show that the model is a scientific and practical credit evaluation model.

Keywords: electric power enterprise's customers, credit evaluation, entropy method, weight.

1 Introduction

With the deepening reform of power system, electric power enterprises are facing increasing market risk, especially the phenomena that arrears of electricity charges, illegal use of electricity and stealing electricity are more serious, which have affected the power supply enterprises' management. To create favorable conditions for the sustainable development of local electric power industry, scientific management method should be taken to solve the current problems, it will effectively evade operational risks of the electric power enterprises, avoid the phenomenon of illegal electric power usage, and improve the safety and reliability, and then we will create advantage for power systems to develop sustainably[1].

Considering the complexity and fuzziness in the customer credit evaluation, this paper uses a method which is combined quantitative analysis with qualitative analysis. Information entropy theory is applied to the evaluation model in determining the evaluation index weight, then it combines the objective weights and subjective weights effectively, in this way, the result not only made full use of the objective information on the program, but also reflect the views of decision makers. This approach makes up the shortcomings of a single weighting method,and increases the credibility of weights and

J. Luo (Ed.): Affective Computing and Intelligent Interaction, AISC 137, pp. 587–593.
springerlink.com

enhances scientificity and comparability of the evaluation[2]. It provides a realistic customer credit evaluation model which makes decision-making based on more comprehensive and makes the method more reliable.

2 Entropy Method

A. Entropy and Its Properties

Entropy comes form thermodynamics, it is introducted into Information Theory by Shannon soon after. According to the definition and principle of entropy, when the system may be in several different states, the probability of various states to appear is $p_i (i = 1,2,\cdots,n)$, the entropy of the system is

$$E = -\sum_{i-1}^{n} p_i \ln p_i \tag{1}$$

In fact, entropy value is a measure to reflect the uncertainty information of the system state, the bigger entropy value is, the more chaotic system is, the stronger non-deterministic is, then we can know less information[3]. From the expression of entropy, we can see the entropy has properties such as additive, non-negative, extreme value, symmetry, additivity, imposed nature and so on.

B. Procedure

According to the relevant properties of entropy, the modeling steps of customer credit evaluation for the electric power enterprise are as follows [4-5] :

1) Select customers $A_k (k = 1,2,\cdots,m)$ for evaluation,

2) Set up evaluation indexes system, the evaluation index are $B_i (i = 1,2,\cdots,n)$

3) Construct index matrix X

$$X = \begin{bmatrix} x_{11} & x_{12} & \cdots & x_{1m} \\ x_{21} & x_{22} & \cdots & x_{2m} \\ \vdots & \vdots & \vdots & \vdots \\ x_{n1} & x_{n2} & \cdots & x_{nm} \end{bmatrix}$$

x_{ik} is the estimated value of evaluation index B_i for customer A_k. Take optimum value x_i^* of each evaluation index. If B_i is a direct index, the bigger of x_i^*, the better effect of sample; if B_i is an inverse index, the smaller of x_i^*, the better effect of sample.

4) Define D_{ik} is the proximity between x_{ik} and x_i^*, then we can obtain the matrix $D = (D_{ik})_{n \times m}$.

If B_i is a direct index,

$$D_{ik} = \frac{x_{ik}}{x_i^*} \tag{2}$$

$x_i^* = \max(x_{ik})$.

If B_i is an inverse index

$$D_{ik} = \frac{x_i^*}{x_{ik}}, x_i^* = \min(x_{ik}) \tag{3}$$

5) Normalization treatment D_{ik}, then obtain the matrix d,

$$d = (d_{ik})_{n\times m} \tag{4}$$

$d_{ik} = D_{ik} \Big/ \sum_{i=1}^{n}\sum_{k=1}^{m} D_{ik}$.

6) Calculate the value of condition entropy E_1 of evaluation index B_i

$$E_i = -\sum_{k=1}^{m} \frac{d_{ik}}{d_i} \ln \frac{d_{ik}}{d_i} \tag{5}$$

$d_i = \sum_{k=1}^{m} d_{ik}$.

7) Use $E_{\max}$ to normalization E_i, then obtain importance entropy of other evaluation indicators

$$e(d_i) = \frac{1}{\ln m} E_i \tag{6}$$

8) Determine the evaluation weight θ_i for this evaluation index B_i according to the index $e(d_i)$

$$\theta_i = \frac{1}{n - E_e}[1 - e(d_i)] \tag{7}$$

$E_e = \sum_{i=1}^{n} e(d_i)$.

9) Suppose the objective weights determined by decision-maker are λ_i;

10) The final weights for the evaluation indexes are

$$\omega_i = \frac{\theta_i \lambda_i}{\sum_{i=1}^{n} \theta_i \lambda_i} \tag{8}$$

$0 \le \omega_i \le 1, \sum_{i=1}^{n} \omega_i = 1$.

11) Calculation formula for decision making is

$$S_k = \sum_{i=1}^{n} \omega_i (d_i^* - d_{ik}) \tag{9}$$

d_i^* refers to optimum value of the i th indicator in the matrix d , we can sort customer credit according to the value of indicator S_k, the smaller S_k is, the better credit of customer, evaluation conclusion can be got by comparing all the values at last.

3 An Example for Application

A. Selection of Evaluation Indexes

According to the status of customers credit evaluation, There are six evaluation indexes selected in this paper. Including: financial capacity, social status, power consumption growth rate / %, times of illegal electric power usage, annual electricity pay rate / %, scheduling cooperation. Among these, power consumption growth rate, times of illegal electric power usage and annual electricity pay rate are quantitative indicators, others are qualitative indicators.

B. Model Building

Now, there are four costumers A_1、A_2、A_3、A_4 of a electric power enterprise, we need to evaluate their credit scientifically, and get exact knowledge of their credit quality. Evaluation of the indicators for the costumers are shown by the electric power enterprise in the following table 1.

Firstly, let's quantify the qualitative indicators. For the evaluation index financial capacity B_1, 5 referring to excellent, 4 referring to good, 2 referring to fair; for the social status B_2, 5 referring to Prominent, 4 referring to high, 2 referring to general, these values compose initial data matrix of evaluation X . Normalization treatment the original data, then obtain the evaluation matrix d :

Table 1. The comparative values of customers credit evaluation indexes

	A_1	A_2	A_3	A_4
B_1	good	excellent	fair	good
B_2	prominent	high	general	high
B_3	12	15	8	10
B_4	5	2	8	6
B_5	95	98	90	90
B_6	fair	good	good	excellent

$$d = \begin{bmatrix} 0.0450 & 0.0562 & 0.0225 & 0.0450 \\ 0.0562 & 0.0450 & 0.0225 & 0.0450 \\ 0.0450 & 0.0562 & 0.0230 & 0.0375 \\ 0.0225 & 0.0562 & 0.0141 & 0.0187 \\ 0.0545 & 0.0562 & 0.0516 & 0.0516 \\ 0.0225 & 0.0450 & 0.0450 & 0.0562 \end{bmatrix}$$

Then calculate the value of condition entropy E_1 of evaluation index B_1 according to formula (5)

$$E_1 = -\sum_{k=1}^{m} \frac{d_{1k}}{d_1} \ln \frac{d_{1k}}{d_1} = 1.3398$$

The entropy of other evaluation indicators can be calculated with above steps(table 2).

Use $E_{\max}$ to normalization E_i, then calculate entropy values which reveals the importance of evaluation index B_1 according to formula (5)

$$e(d_1) = \frac{1}{\ln 4} E_1 = 0.9664$$

The importance entropy of other evaluation indicators can be calculated with above steps (table 2).

Determine the evaluation weight θ_1 for this evaluation index B_1 according to the index $e(d_1)$

$$\theta_1 = \frac{1}{n - E_e}\left[1 - e(d_1)\right] = 0.0999$$

The evaluation weights of other evaluation indicators can be calculated with above steps(table 2).

Then determine the weights of evaluation indicators by experts according to the status of customers

λ_1=0.2542, λ_2=0.1071, λ_3=0.0875,
λ_4=0.0324, λ_5=0.3158, λ_6=0.2030

Calculate synthetic weight of evaluation index B_1 according to formula (8)

$$\omega_1 = \frac{\theta_1 \lambda_1}{\sum_{i=1}^{n} \theta_i \lambda_i} = 0.2679$$

The synthetic weights of other evaluation indicators can be calculated with above steps(table 2).

Table 2. Evaluation indicators $e(d_i)$ and their weights

	B_1	B_2	B_3	B_4	B_5	B_6
E_i	1.3398	1.2178	1.3395	1.2293	1.3856	1.3398
$e(d_i)$	0.9664	0.8784	0.9662	0.8867	0.9994	0.9664
θ_i	0.0999	0.3614	0.1004	0.3367	0.0018	0.0999
ω_i	0.2679	0.4082	0.0928	0.1150	0.0074	0.1086

Finally, calculate comprehensive evaluated values for each customer according to formula (9)

$S_1 = 0.0067$, $S_2 = 0.001$, $S_3 = 0.0271$, $S_4 = 0.0088$

Clearly the smaller of S_k, the better credit of customer. For this reason, the order of these customers about their credits is

$A_2 \succ A_1 \succ A_4 \succ A_3$.

4 Conclusion

The method proposed in this paper not only combined qualitative and quantitative analysis, but also combined subjective and objective weighting. The example illustrated that this method is an effective way to evaluate the credit of customers for electric power enterprises. According to the result of customers credit evaluation, the electric power enterprises could get costumers' performance on the contract in time and provide corresponding service for different levels of customer credit. The enterprises would avoid business risks by early warning information and get a good operation.

References

1. Zhou, X.: Analysis and Research for Chinese Power Market. China Power Press (2003)
2. Qiu, W.: Management Assessment and Entropy Theory. China Machine Press, Beijing (2002)
3. Bryson, N., Moboluriu, A.: An action learning evaluation procedure for multiple criteria decision making problems. European Journal of Operational Research 96(3), 379–386 (1996)
4. Li, L., Lai, K.K.: Fuzzy dynamic programming approach to hybrid multi-objective multistage decision making problems. Fuzzy Sets and Systems 117, 13–25 (2001)
5. Zhou, H.A., Liu, S.Y., Fang, X. R.: Method for uncertain multi-attribute decision-making with preference information in the form of interval numbers complementary judgment matrix. Journal of System Engineering and Electronics 18(2) (2005)

Evaluation of Enterprise Technological Innovation Capability Based on Ambiguity and FMADM

Huang Yuan-sheng, Yuan Li-ming, and Shi Xiu-fen

School of Business Administration,
North China Electric Power University, Baoding, China
yuanyuanliliming@126.com

Abstract. Because of the impellent of the world economic integration, China's enterprises are facing with intense competition which comes from both domestic and abroad. Technological innovation capability becomes the key factor for enterprises to survive and develop. This paper researchs enterprise technological innovation capability which with linguistic information of attribute values and completely unknown information of attribute weights. It determines the weights based on measuring the ambiguity of evaluation protocols, in order to measure and evaluate each enterprise scientifically and systematically. An example verifies this method is a scientific and functional one.

Keywords: technological innovation capacity, fuzzy multiple attribute decision making, language assessment, ambiguity, weight.

1 Introduction

It plays a crucial role for a country's economic development and social progress to enhance the independent innovation ability of enterprises. The measurement and evaluation for technological innovation capability of enterprises is a systematic quantitative description method, which is a clear understanding for businessmen to recognize their ability and the level of their technological innovation, and they can realize which position their enterprise at in the whole industry, also it is a scientific basis for take reasonable technological innovation strategy, maintain and improve their competitive advantage[1].

This paper researches enterprise technological innovation capability which with linguistic information of attribute values and completely unknown information of attribute weights. Due to the uncertainty and complex of the enterprises' technological innovation capability measurement, we took Fuzzy Multiple Attribute Decision Making method to solve such problems. In order to avoid the shortcomings of Subjective weighting method, this paper determines the weights based on measuring the ambiguity of evaluation protocols, in order to measure and evaluate each enterprise scientifically and systematically, and take the properties of the range(linguistic assessment information) as a goal of the next level of each attribute , the property weights associated with the property weights under the target level was the result that the property weight and the weight of the next layer of the target, which related to the

J. Luo (Ed.): Affective Computing and Intelligent Interaction, AISC 137, pp. 595–602.
springerlink.com

attributes, multiplied[2]. Finally, using the method that summing weights to find the technological innovation of enterprises The composite scores and sort. An example verifies this method is a scientific and functional one. An example verifies this method is a scientific and functional one.

2 Setting Up the Evaluation Indexs

Enterprise's technological innovation ability is a system and a comprehensive concept which is under the market economy condition, it is a dynamic and open concept. It is a conduct of business but not a simple technological behavior, it is a innovative conceive one that originate in reciprocation between scientific research and market demand. Through the design, trial-production, producing of the new products to reach the serial activity course of the marketing. Including the innovative screening of technological innovation and technological innovation decision, technological development, intermediate experiment, production, application, marketing. It is a complex process about creation, conversion, application and complicated, The power of enterprise's technological innovation ability concerns the organizational performance of enterprises directly[3].

A. Principles for Establishing Evaluation Index System of Enterprise Technological Innovation Capability

The purpose of establishing index system for technological innovation ability is to assess the power of enterprise's technological innovation capability, and compare innovation ability between different enterprises or between different periods in the same enterprise, so enterprises can convince their current situation and problems on technological innovation ability[4]. The construction of evaluation index system should reflect the typical characteristics of innovative enterprises from different levels, so the following principles should be followed:

1) Scientificalness

Scientificalness is the foundation to ensure the result of evaluation reasonable and accurate, it includs three aspects: The first is accuracy. The concept of indicators should be accurate and clear, and avoid subjective assumptions as much as possible, The second is integrity. The index system should be around the purpose of evaluation completely, and reflect the object roundly, at the same time be focused; The third is independence. There shouldn't be strong correlation between different indicators, otherwise, it will increase workload and waste time.

2) Systematicness

The evaluation of technological innovation capability for enterprises should be comprehensive and systematic as far as possible. It is wise to establish a complete evaluation index system layer by layer, so as to avoid systemic bias.

3) Comparability

The setting of indicators should reflect comparability, and have a certain degree of flexibility. It is necessary to take into account the horizontal comparison between enterprises, but also necessary to take into account the longitudinal comparison in the same enterprise at different times, this can make the indicators become yardstick to measure the effectiveness of technological innovation.

4) Maneuverability

The design of indicators should reflect the development of technological innovation capability of enterprises accurately, require the datas operability at the same time.It must be accessible on one hand, and qualitative analysis combination with quantitative analysis on the other hand, for the kind of unquantify but very important aspect indicators adopt qualitative one to bring into the evaluation index system.

5) Dynamics and Continuity

Technological innovation capability is a dynamic process of development. Therefore, the indicator system must be able to reflect the status of technological innovation, while reflect the development trends and potential, and reveal the internal principles. We should selecte both of static and dynamic indicators, use the static indicators to reflect the level of technological innovation at present, and use the dynamic ones to predict the growth prospects of enterprise.

B. The Details of Evaluation Index System for Enterprise Technological Innovation Capability

The evaluation system of enterprise technology innovation capability is a complex system, to build it scientifically and rationally is the premise and foundation to evaluate the technological innovation correctly [5]. Learning from the advanced evaluation theory and evaluation methods, combine with the principles and analysis of the process of technological innovation, we think the evaluation index system composed mainly by the following six elements, what the elements containe in specific aspects are shown in the following tab.

Table 1. The Details of Evaluation Index System

the input capacity of technological innovation	intensity of investment spent in R & D, intensity of investment spent in other aspects, the proportion of technical personnel, the proportion of funds for staff training
the research and development capabilities of technological	the success rates of R & D, the per capita number of patents, the rate of independent innovation, the refresh cycle of main products, the capability of high-tech adoption
the capacity of technological innovation	the level of production equipment, peopleware of production, production cycle of the product, standardization level of working, the rate of adoption of modern manufacturing technology
the marketing ability of technological innovation	the intensity of investment in market research, marketing costs, the fitness of marketing system, the degree of network marketing
the management ability of technological innovation	the efficiency of innovative mechanisms, culture and atmosphere of enterprise innovation, the capabilities of prediction and evaluation for technological innovation, the ability to communicate with the outside world, innovation consciousness
the output capacity of technological innovation	the market share of new products, the sales share of new product, technical level of new products, technology export capabilities, the competitiveness of enterprises

3 Theory Introduction

A. The Formal Definition of Fuzzy Mutiple Attribute Decision-Making

There is an alternative program set $X = \{x_1, x_2, \cdots, x_m\}$, its attribute set is $A = \{A_1, A_2, \cdots, A_n\}$, the range of $A_k (k = 1,2,\cdots,n)$ is $T(A_k) = \{T_1^k, T_2^k, \cdots, T_s^k\}$, where $T_j^k (j = 1,2,\cdots,s)$ is a fuzzy set, $\mu_{T_j^k}(x_i)$ is the membership of the program $x_i (i = 1,2,\cdots,m)$ which is belonged to the fuzzy set T_j^k.

B. The Ambiguity and Its Measurement

The ambiguity is a uncertainty of a kind of fuzzy sets which chooses the uncertainty of a time from many optional choices in the one to many relationship, for example, it exists ambiguity when we choose an optimal program from a number of options. The reduction of ambiguity will bring information gain, the smaller the ambiguity of attribute, the more it contains information, therefore, the weight should be given greater, the ambiguity can adopt possible theory to measure[6].

Definition 1: The standard definition of the value of the attribute A_k belonging to the program X_i is

$$\pi_{T_j^k}(x_i) = \mu_{T_l^k}(x_i) \Big/ \max_{1 \le l \le s} \{\mu_{T_l^k}(x_i)\}$$
$$j = 1,\cdots,s, i = 1,\cdots,m, k = 1,\cdots,n \tag{1}$$

Where $\pi_{T_j^k}(x_i)$ is the standard value of the membership, that is possible.

Definition 2: The measure definition of the ambiguity of the attribute A_k belonging to the program X_i is

$$E_{A_k}(x_i) = \sum_{j=1}^{s} (\pi'_j(x_i) - \pi'_{j+1}(x_i)) \ln j \tag{2}$$

Where $\{\pi'_1(x_i), \pi'_2(x_i), \cdots, \pi'_s(x_i)\}$ is arrangement by size of possibility distribution $\{\pi_{T_1}^k(x_i), \pi_{T_2}^k(x_i), \cdots, \pi_{T_s}^k(x_i)\}$ of attribute A_k belonging to the program X_i, $\pi'_j(x_i) \ge \pi'_{j+1}(x_i)$, and it formulates $\pi'_{s+1}(x_i) = 0$.

This definition is the one function satisfied the demand to possibility measures, from the definition, we have $E(A_k) \ge 0$. If $\pi'_2(x_i) = 0$, then:

$\pi'_1(x_i) = 1, \pi'_2(x_i) = \pi'_3(x_i) = \cdots = \pi'_{s+1}(x_i) = 0$ $E(A_k) = 0$, at this time the ambiguity is 0, the attribute can only take one value. If $\pi'_s(x_i) = 1$, then $\pi'_1(x_i) = \pi'_2(x_i) = \cdots = \pi'_s(x_i) = 1$, $E(A_k) = \ln s$, this means that the possibility taking all the values is identical, and the ambiguity is the best.

Definition 3: The ambiguity of the attribute A_k is the average of the ambiguity of the programs with the attribute,

$$E(A_k)=\frac{1}{m}\sum_{i=1}^{m}(E_{A_k}(x_i)),k=1,\cdots,n \tag{3}$$

Obviously, if $E(A_k)=\ln s$, it shows the values of the programs with the attribute are all $\ln s$, then we does not consider the attribute provide decision-makers any available information, its weight should be least or reselect attributes.

C. Calculate the Attribute Weights

Firstly standardizing the ambiguity of the attribute

$$E(A_k)=E(A_k)\Big/\sum_{j=1}^{n}E(A_j), \quad k=1,2,\cdots,n \tag{4}$$

Then define a weight ω_j for the attribute j as

$$\omega_j=\frac{1-E(A_j)}{n-\sum_{k=1}^{n}E(A_k)}, \quad j=1,2,\cdots,n \tag{5}$$

Different from the entropy method, the base of the above weight definition is the each program ambiguity. Obviously

$$0\le\omega_j\le 1 \text{ and } \sum_{j=1}^{n}\omega_j=1$$

After obtained the weights of each attribute, the each range is the next objective of the attribute if the attribute weight multiplies the next objective weight related with the attribute, then we can obtain the weight of the next objective about the attribute.

4 Case Study and Analysis

There are four enterprises which are named by X_i (i=1,2,3,4), now they are given to be assessed. Values of the six atrributes of their technical innovation abilities are denoted by lingual evaluation information such as: weak, common and strong. The weights of the lingual evaluation information relative to the atrributes are obtained by expert scoring method. Detailed data are shown in Table 2-1 and table 2-2. Now we'll discuss how to calculate an attribute's weight according to its ambiguity.

Standardize every enterprise attribute value by Formula (1), calculate 4 enterprises attribute ambiguity Measure by Formula (2), the results are shown in the table 3.

Table 2-1. Enterprise Technological Innovation Capability Data

	A_1			A_2			A_3		
	Weak	General	Strong	Weak	General	Strong	Weak	General	Strong
X_1	0.1	0.8	0.1	0.1	0.2	0.7	0	0.3	0.7
X_2	0.7	0.2	0.1	0.1	0.1	0.8	0	0	1
X_3	0.1	0.3	0.6	0.2	0.6	0.2	0.1	0.7	0.2
X_4	0	0	1	0.7	0.3	0	0.6	0.4	0

Table 2-2. Enterprise Technological Innovation Capability Data

	A_4			A_5			A_6		
	Weak	General	Strong	Weak	General	Strong	Weak	General	Strong
X_1	0	0.2	0.8	0.2	0.8	0	0.1	0.2	0.7
X_2	0.3	0.6	0.1	0.1	0.1	0.8	0.1	0.1	0.8
X_3	0	0.2	0.8	0	0	1	0.7	0.2	0.1
X_4	0	0.1	0.9	0	0	1	0	0.1	0.9

Table 3. Enterprises attribute ambiguity

	A_1	A_2	A_3	A_4	A_5	A_6
X	0.137	0.256	0.297	0.173	0.173	0.256
X	0.256	0.137	0	0.414	0.137	0.137
X	0.414	0.366	0.256	0.173	0	0.256
X	0	0.297	0.462	0.077	0	0.077

Calculate every attribute ambiguity by formula (3):

$E(A_1)=0.202$, $E(A_2)=0.263$, $E(A_3)=0.254$, $E(A_4)=0.209$, $E(A_5)=0.078$, $E(A_6)=0.182$.

Every attribute is standardized by formula (4),and calculate the weight of every attribute by formula (5):

$\omega_1=0.166$, $\omega_2=0.153$, $\omega_3=0.155$, $\omega_4=0.164$, $\omega_5=0.192$, $\omega_6=0.170$.

every weight sub relative to superior attribute is got by experts marking. its weight relative to problem is got by multiplying the weight of it and its superior attribute.

Table 4-1. Sub-attribute weights

	A_1			A_2			A_3		
weights	weak	general	strong	weak	general	strong	weak	general	strong
relative to superior attribute weights	0.2	0.3	0.5	0.2	0.5	0.3	0.2	0.6	0.2
relative to sum attribute weights	0.033	0.05	0.083	0.031	0.076	0.046	0.031	0.093	0.031

Table 4-2. Sub-attribute weights

	A_4			A_5			A_6		
weights	weak	general	strong	weak	general	strong	weak	general	strong
relative to superior attribute weights	0.1	0.2	0.7	0.1	0.1	0.8	0.1	0.3	0.6
relative to sum attribute weights	0.016	0.033	0.115	0.019	0.019	0.154	0.017	0.051	0.102

Calculate the composite score of all enterprises Z_i with weighted summation method according to table 4-1 and table 4-2:

Z_1=0.3526, Z_2=0.3714, Z_3=0.4884, Z_4=0.5410

So, the sequence of enterprise technological innovation capability is $X_4 \succ X_3 \succ X_2 \succ X_1$.

5 Conclusions

With the global economic integration and acuter competition, improving the technical innovation ability has become the main approach for an enterprise to get the control of competitions and acquire advantages of development. Considering the complexity of the process to evaluate technical innovation and the fuzziness of index evaluation, this paper describes attribute information in the form of lingual evaluation, and determines the weights of attributes with the information contained inside the data themselves, making the decision method more reasonable and efficient, the decision result more accurate and reliable, providing support and basis for the quantitative evaluation of enterprises' technical innovation abilities.

However, this paper still exist improvements in some areas, For example, in determining the weight of each attribute did not reflect the information of decision maker's subjective preference ,and sometimes this information is necessary for evaluation or decision-making .Therefore, the method and other methods can be organically combined to achieve a more accurate decision-making results.

References

1. Chesbrough, H.: Open innovation:The new imperative for creating and profiting from technology. Harvard Business School Press, Boston (2003)
2. Du, H.-D., Yan, Z.-H., Guan, D.-M.: Study on status Quo and proposition of Technological Innovation capacity. Technoeconomics & Management Research 12(2), 22–45 (2008)
3. Cao, P., Zhang, J.: Evaluation of technological innovation. China Management Informationization 12(2), 89–92 (2009)
4. Zhang, L.: Enterprise Technology Innovation Project and Decision System. Beijing People's Press (2006)
5. Baradwaj, S., Menon, A.: Making innovation happen in organizations:individual creativity mechanisms. Journal of Production Innovation Management 17, 424–434 (2000)
6. Miller, W., Lanon, M.: Fourth generation R&D: Managing knowledge, technology and innovation. John Wiley&Sons, New York (1999)

Mobile Webserver to the Android Platform

Hucheng Wei

Electronic Communication Department, Guanghou Civil Aviation College,
Guangzhou, China
hcw94@hotmail.com

Abstract. Android is a complete operating environment based upon the Linux® V2.6 kernel. The main deployment target for Android is the mobile-phone arena, including smart phones and lower-cost flip-phone devices. Jetty is an open-source project providing a HTTP server, HTTP client and javax.servlet container. Android-based i-Jetty version is loaded as a servlet on Android, says Webtide. Once loaded, it enables remote access to phone functions from desktop PCs.In this article, get an introduction to Jetty websever to the Android platform.

Keywords: Android, Jetty, moble webserver, eclipse, linux, open-source.

1 Introduction

Since the introduction of the open source Android platform for mobile phones by Google, there has been significant interest in network-based or network-capable appliances development. One of important appliances is being done in providing Web services from mobile devices, called mobile hosted Web services,where an external application requests a Web service provided by the mobile device.

Jetty is designed to work as a standalone webserver or as a dynamic content server behind a dedicated HTTP server such as Apache. The Android-based i-Jetty version is loaded as a servlet on Android, says Webtide. It can enable remote access to phone functions from desktop PCs. i-Jetty users can also build publicly accessible mobile web servers. The potential might be greater with i-Jetty compared to Nokia's technology., because I-Jetty is more of a full-featured Java application server as opposed to being just a page server. For example, i-Jetty supports web services technologies such as AJAX (asynchronous Javascript and XML), which speeds up interactive processes on AJAX-enabled applications like Google Maps.

2 The Android Platform

With Android's breadth of capabilities, it would be easy to confuse it with a desktop operating system. Android is a mobile operating system running on the Linux kernel, and it includes rich functions. It was initially developed by Android Inc., a firm later purchased by Google, and lately by the Open Handset Alliance. It allows developers

J. Luo (Ed.): Affective Computing and Intelligent Interaction, AISC 137, pp. 603–608.
springerlink.com

to write managed code in the Java language, controlling the device via Google-developed Java libraries.

Android relies on Linux version 2.6 for core system services such as security, memory management, process management, network stack, and driver model. The kernel also acts as an abstraction layer between the hardware and the rest of the software stack. Android includes a set of C/C++ libraries used by various components of the Android system. These capabilities are exposed to developers through the Android application framework. Figure 1 shows a simplified view of the Android software layers.

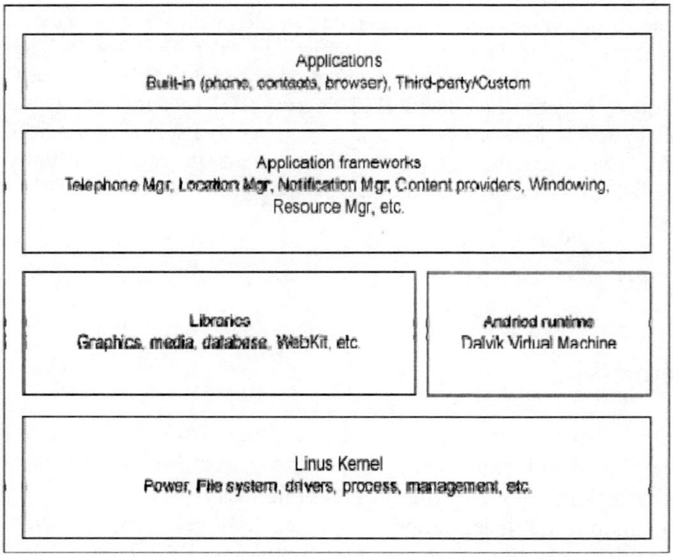

Fig. 1. Android software layers

As mentioned, Android runs atop a Linux kernel. Android applications are written in the Java programming language, and they run within a virtual machine (VM). It's important to note that the VM is not a JVM as you might expect, but is the Dalvik Virtual Machine, an open source technology. Each Android application runs within an instance of the Dalvik VM, which in turn resides within a Linux-kernel managed process, as shown below.

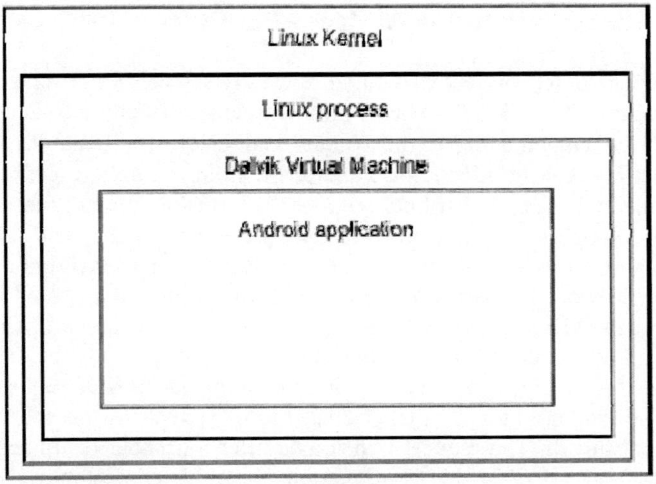

Fig. 2. Dalvik VM

An Android application consists of one or more of the following classifications:

2.1 Activities

An application that has a visible UI is implemented with an activity. When a user selects an application from the home screen or application launcher, an activity is started.

2.2 Services

A service should be used for any application that needs to persist for a long time, such as a network monitor or update-checking application.

2.3 Content Providers

You can think of content providers as a database server. A content provider's job is to manage access to persisted data, such as a SQLite database. If your application is very simple, you might not necessarily create a content provider. If you're building a larger application, or one that makes data available to multiple activities or applications, a content provider is the means of accessing your data.

2.4 Broadcast Receivers

An Android application may be launched to process a element of data or respond to an event, such as the receipt of a text message.

An Android application, along with a file called AndroidManifest.xml, is deployed to a device. AndroidManifest.xml contains the necessary configuration information to properly install it to the device. It includes the required class names and types of events the application is able to process, and the required permissions the application needs to run. For example, if an application requires access to the network — to download a file, for example — this permission must be explicitly stated in the manifest file. Many applications may have this specific permission enabled. Such declarative security helps reduce the likelihood that a rogue application can cause damage on your device.

3 Jetty Webserver to the Android Platform

Jetty is an open-source project providing a HTTP server, HTTP client and javax. servlet container. These 100% java components are full-featured, standards based, small foot print, embeddable, asynchronous and enterprise scalable. Jetty is dual licensed under the Apache Licence 2.0 and/or the Eclipse Public License 1.0. Jetty is free for commercial use and distribution under the terms of either of those licenses.

Jetty 6 also supports a new feature called Continuations that will allow scalable AJAX applications to be built, with threadless waiting for asynchronous events. As per scalability it is one of the major concerns of this product that would be supported by the advent of AJAX with the response to offer a feature called Continuations. Where in it allows a request to be suspended and resumed later in the process. In addition it is a java Filter or Servlet that is handling an AJAX request, may now request a Continuation object that can be used to effectively suspend the request and free the current thread. The request is resumed after a timeout or immediately if the resume method is called on the Continuation object. Jetty 6 employs a number of innovative strategies to ensure that only the resources that are actually required are assigned to a connection and only for the duration of they are needed. This careful resource management gives Jetty an architecture designed to scale to meet the needs of AJAX applications. Figure 3 shows a simplified view of the Jetty 6 Architecture.

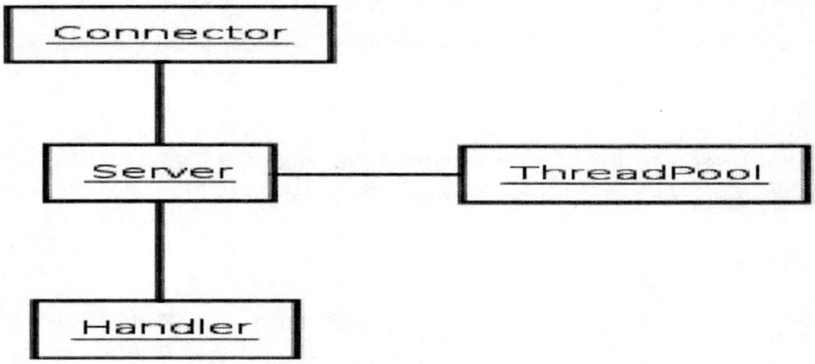

Fig. 3. Jetty 6 Architecture

The Jetty Server is the plumbing between a collection of Connectors that accept HTTP connections, and a collection of Handlers that service requests from the connections and produce responses, with the work being done by threads taken from a thread pool.

Jetty is designed to work as a standalone webserver or as a dynamic content server behind a dedicated HTTP server such as Apache. The Android-based i-Jetty version is loaded as a servlet on Android, says Webtide. Once loaded, it enables remote access to phone functions from desktop PCs. The i-Jetty port to the Android open source software stack for mobile devices is billed primarily as a matter of convenience for cell phone users, who can use it to make calls from their PCs, or to access phone-based content such as address books and calendars using a familiar browser interface. According to Webtide, users will be able to view, create, and save changes to files from remote PCs, including address lists, system settings, call logs, and multimedia files.

A web application context of Jetty combines handlers for security, session and servlets in a single unit that can be configured with a web.xml descriptor.

A WebAppContext is a derivation of the servlet Context that supports the standardized layout of a web application and configuration of session, security, listeners, filter, servlets and JSP via a web.xml descriptor normally found in the WEB-INF directory of a webapplication.

Essentially the WebAppContext is a convenience class to assist the construction and configuration of other handlers to achieve a standard web application configuration.

Configuration is actually done by pluggable implementations of the Configuration class and the prime among these is WebXmlConfiguration. Figure 4 shows a view of Web Applications.

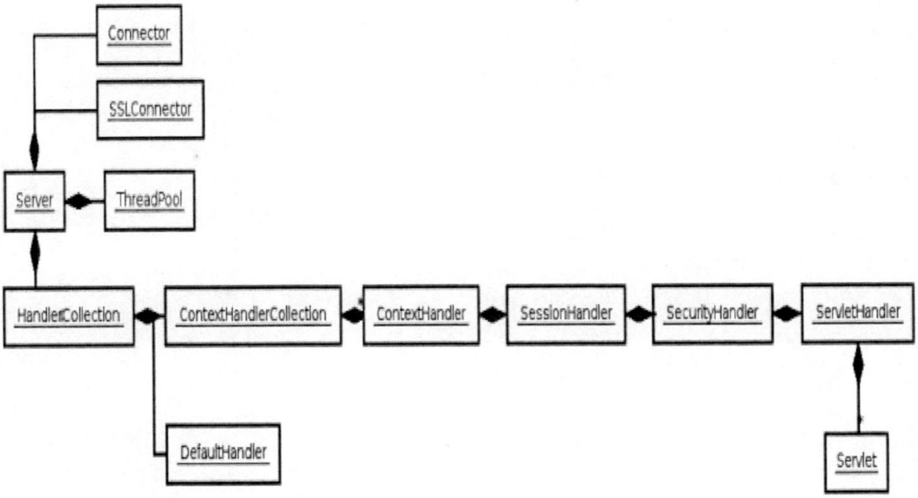

Fig. 4. Web Applications

4 Conclusion

Using mobile Web services allows publishing functionality in a more structured and controllable manner. Further it enhances reuse of services, reducing the implementation needed for applications. Mobie web services also help utilizing the communication channel in a more efficient way. Only fetching the required information from the device instead of a full web page including markup information. The mobile industry has proven that it is capable and willing to come together in open standardization to help new technologies and services to prevail.Jetty webserver to the Android platform will be one new direction.

References

1. Android Supported Media Formats. Android Developers, http://developer.android.com/guide/appendix/media-formats.html (retrieved May 1, 2009)
2. Musil, S.: Report: Apple nixed Android's multitouch. CNET News (February 11, 2009), http://news.cnet.com/8301-13579_3-10161312-37.html (retrieved September 3, 2009)
3. Richtel, M.: Google: Expect 18 Android Phones by Year's End. The New York Times (May 27, 2009), http://bits.blogs.nytimes.com/2009/05/27/google-expect-18-android-phones-by-years-end/ (retrieved June 19, 2009)
4. Android. Openmoko wiki, http://wiki.openmoko.org/wiki/Android (retrieved December 29, 2009)
5. Morrison, S.: Glitches Bug Google's Android Software. The Wall Street Journal (December 19, 2007), http://online.wsj.com/article_email/SB119800856883537515-lMyQjAxMDE3OTE4ODAxMDg4Wj.html (retrieved December 19, 2007)
6. Snake. Android Freeware Directory, http://www.android-freeware.org/download/snake (retrieved January 29, 2008)
7. Jetty. Jetty 6 Architecture, http://docs.codehaus.org/display/JETTY/ (retrieved March 29, 2010)

Applied Research of Expert-Fuzzy PID in Wireless Temperature Control System

Jun Wang, Meifeng Gao, and Jianhong Zhu

School of Communication and Control Engineering,
Jiangnan University Wuxi, China
wangjun_eric@163.com

Abstract. The process of uniform velocity-constant temperature control is a complicated process and has the characteristics such as some uncertainty of structure and parameters, pure delay and nonlinear coupling. It is hard to obtain the exact mathematic model. Considering these characteristics, an expert-fuzzy PID controller is designed. This controller has a fast stable state feature and effectively large lag solving of the expert controller and a high accuracy, stability, good robustness of fuzzy PID controller. The experimental result shows that the system with expert-fuzzy PID controller has high accuracy, good dynamic performance and robustness.

Keywords: expert-fuzzy PID, pre-control, parameter self-tuning, ARM7, MCGS.

1 Introduction

In the industrial production process, the control method of temperature usually consist of uniform velocity and constant, while in the production practice one common control mode is like this: firstly, carries on the uniform velocity temperature control to the object until its temperature arrives at a certain value, then executes the constant temperature control, for example the steam curing process of concrete products. Considering the object has the characteristics such as some uncertainty of structure and parameters, pure delay or nonlinear coupling and hard to obtain the exact mathematic model, designing a high performance intelligent controller now becomes the key to resolving conflict between the low cost and the high accuracy in a temperature monitor and control system [1].

The fuzzy adaptive PID control is widely used in process control domain because of its convenience, practicality, high accuracy and strong stability. But the control rule and membership function of fuzzy controller are set manually. Once confirmed, they are hard to satisfy the needs of time-varying and pure delay. Here putting expert judgment into the fuzzy PID control, a better control effect is achieved by timely adjusting parameters of controller according to the knowledge base coincided with the field of production.

J. Luo (Ed.): Affective Computing and Intelligent Interaction, AISC 137, pp. 609–615.
springerlink.com

2 Design of the Expert-Fuzzy PID Controller

2.1 The Design Idea

The expert-fuzzy PID controller is showed in Fig. 1 and is mainly composed of two parts: the expert controller and fuzzy PID controller. The mode selector is used to select one of them to act on the object. The controller designed in this way has the high accuracy, strong stability and good robustness of PID controller and the flexibility, interaction and adaptive ability of expert controller.

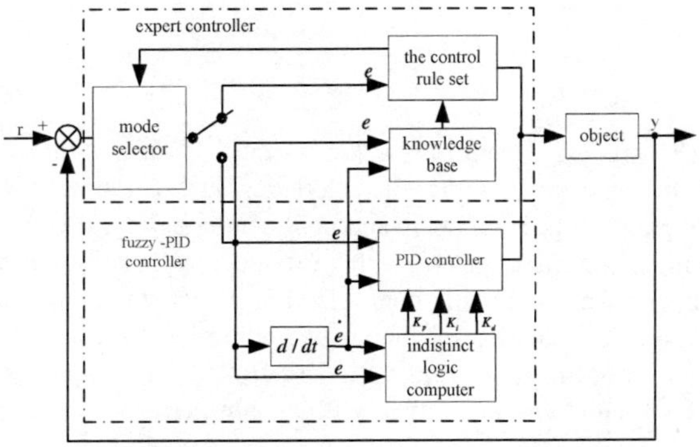

Fig. 1. The expert-fuzzy PID controller

2.2 The Fuzzy PID Control

Since the invention of PID control in 1910 (largely owning to Elmer Sperry's ship autopilot), and the Ziegler–Nichols' (Z-N) straight forward tuning methods in 1942 [2], the popularity of PID control has grown tremendously. With advances in digital technology, the science of automatic control now offers a wide spectrum of choices for control schemes. One common pattern of PID control is defined [3]:

$$u(k) = K_p e(k) + K_i \sum_{j=0}^{k} e(j) + K_d [e(k) - e(k-1)] \tag{1}$$

Where is k sampling instant, K_p is the proportional gain, K_i the integral gain, K_d the derivative gain.

The fuzzy-PID controller is composed of two parts: PID controller and the fuzzy inference. The fuzzy inference takes the error e and variable rate of rate ec as input, modifies the parameters $\{e_i, ec_j\}_p$, $\{e_i, ec_j\}_i$ and $\{e_i, ec_j\}_d$ according to the fuzzy inference so as to get a good control effect. The domain format of five fuzzy subset (e, ec and three PID parameters K_p, K_i, K_d) is chosen to be {-6,-4,-2,0,2,4,6}. The linguistic label is determined to be {NB, NM, NS, ZO, PS, PM, PB} Gaussian distribution is selected as the membership function.

The most important thing in the design of fuzzy controller is to obtain the control rule list [3-4] (as shown in Table 1) by the technical knowledge and experiences, then quickly get a fuzzy control table by means of "Fuzzy Cut Set" proposed in [5].The modified parameters are given by (2), (3) and (4).

Table 1. The control rule table of ΔK_p、 ΔK_i and ΔK_d

ΔK_p ΔK_i ΔK_d / e / ec	**NB**	**NM**	**NS**	**ZO**	**PS**	**PM**	**PB**
NB	PB NB PS	PB NB NS	PM NM NB	PM NM NB	PS NS NB	ZO ZO NM	ZO ZO PS
NM	PB NB PS	PB NB NS	PM NM NB	PS NS NM	PS NS NM	ZO ZO NS	NS ZO ZO
NS	PM NB ZO	PM NM NS	PM NS NM	PS NS NM	ZO ZO NS	NS PS NS	NS PS ZO
ZO	PM NM ZO	PM NM NS	PS NS NS	ZO ZO NS	NS PS NS	NM PM NS	NM PM ZO
PS	PS NM ZO	PS NS ZO	ZO ZO ZO	NS PS ZO	NS PS ZO	NM PM ZO	NM PB ZO
PM	PS ZO PB	ZO ZO NS	NS PS PS	NM PS PS	NM PM PS	NM PB PS	NB PB PB
PB	ZO ZO PB	ZO ZO PM	NM PS PM	NM PM PM	NM PM PS	NB PB PS	NB PB PB

$$K_p = K_p^{'} + \{e_i, ec_j\}_p \tag{2}$$

$$K_i = K_i^{'} + \{e_i, ec_j\}_i \tag{3}$$

$$K_d = K_d^{'} + \{e_i, ec_j\}_d \tag{4}$$

Where $K_p^{'}$, $K_i^{'}$, $K_d^{'}$ are initially adjusted using a conversional adjusting technique (the same as in routine PID control), $\{e_i,ec_j\}_p$, $\{e_i,ec_j\}_i$ and $\{e_i,ec_j\}_d$ are the modify values.

During running, the control system firstly regulates the PID parameters online by fuzzy and checks table, then calculates the output value $u(k)$ given by (1).

2.3 The Expert Control

The expert control described in this paper is made up of three parts: mode selector, knowledge base and the control rule set.

Compared with the fuzzy control, the expert control has the advantages of analying current situation, forecasting problems could meet in the future and regulating the control plans. On the other hand, it can acquires experiences by means of summarizing the errors and mistakes meets when running, regulates and expands the knowledge base [6].

Considered that the process of uniform velocity-constant temperature control has the characteristics such as some uncertainty of structure and parameters, pure delay and nonlinear coupling, the "pre-control" is added to the expert control. The expert control updates the knowledge base in data acquire period and the parameters (timing advanced T_a 、 T_b and control value u_a 、 u_b) are modified for next control process according to the base. We define the two periods δ minutes after temperature ascending start time (t_1) and constant temperature start time (t_3) as the knowledge acquire period. The value of δ is decided by the rate of temperature ascending.

The "pre-control" controls the object directly in the pre-control zone ($t \in (t_0, t_1) \cup (t_2, t_3)$) in case of occur of error, getting a good control precision.

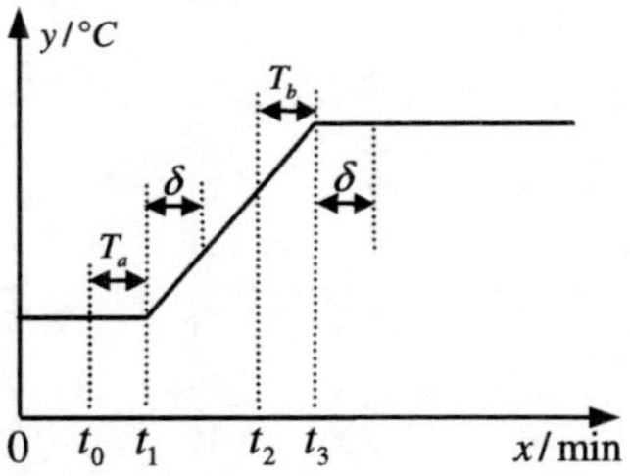

Fig. 2. The expert control process

If $\sum e(k)^2 < M_1$ always exists during the knowledge acquiring period, it is indicated that the current parameters are suitable for the control system, not need to be modified. Otherwise, the control precision isn't satisfied, the expert controller regulates and expands the knowledge base to modify parameters during the knowledge acquire period. The modify method is given by (5) and (6). When running, the system executes (5) and (6) every sampling cycle until the knowledge acquiring is ended.

$$T_a = T_a + \alpha \cdot sng(e(i)) \tag{5}$$

$$T_b = T_b - \beta \cdot sng(e(i)) \tag{6}$$

Where α and β represent time factor, the bigger, the pre-control stronger, commonly $\alpha = \beta = 0.02$, sng is sign distinguish function, when the input is a negative value, it comes back 1, a positive value comes back –1 and zero comes back itself.

The pre-control value are shown by

$$u_a = \omega \cdot T_a \tag{7}$$

$$u_b = \eta \cdot (1 - \varphi \cdot T_b) \tag{8}$$

Where ω, φ and η represent control factors, experiments give a set of ideal value: $\omega = \varphi = 0.25$, $\eta = 0.5$.

The control rule set mainly contains 3 rules, shown as follows.

- When the system is running in the pre-control zone ($t \in (t_0, t_1) \cup (t_2, t_3)$), the expert controller control the object with output value u_a or u_b directly. There is no pre-control if $T_a = 0$ and $T_b = 0$.
- When the system isn't running in the pre-control zone and $|e(k)| \geq M_2$, that is to say, the error is too big, the expert controller will carry out the maximum output so as to modify the error $|e(k)|$ as quickly as possible.
- When the system is running in the non-precontrol zone and the error $|e(k)|$ is smaller than M_2, the mode selector will select the fuzzy PID to control the object in order to take advantages of PID control which has a high control precision.

The expert-fuzzy PID controller timely switches between expert control and fuzzy PID control according to the control rules, and ensures that a process control with a high accuracy, good dynamic performance and robustness is achieved.

3 The Structure of the Control System

The temperature controller mainly consists of the following parts: ARM7 CPU LPC2220, power supply, keyboard, LCD display, digital temperature sensor, RS232 communication, Flash memory, output module of optocoupler, etc, which are shown in Fig. 3. The closed-loop control is made up of the temperature controller, the electric relay, the contactor and the electrothermal furnace. In the control system, the electric relay controls the contactor which is used to turn on or turn off the power supply of the electro thermal furnace.

The temperature of the furnace is sent to the CPU by digital temperature sensor which is a kind of 1-wire device. In process controlling, the temperature controller executes expert-fuzzy PID every sampling cycle to get the control value $u(k)$ which should be disposed of into the value between 0 and 100 in advance, and use the modified value to set the software timer to turn on or turn off the electric relay timely which decides the power supply of the electro thermal furnace. The automotive measure and control of the temperature are achieved.

The Modbus communication protocol is supported by the temperature controller. So, a control network, which is shown in Fig. 4, can be established easily. From Fig. 4, the master computer runs MCGS and sends commands to set the parameters of temperature controllers by the wireless data transfer devices (SZ02-2K).The temperature controllers execute received commands and send temperature values to the master computer with fixed-frequency.

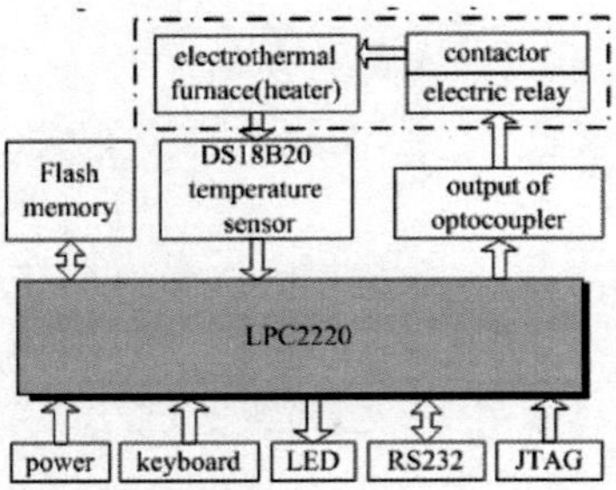

Fig. 3. Design diagram for temperature controller hardware

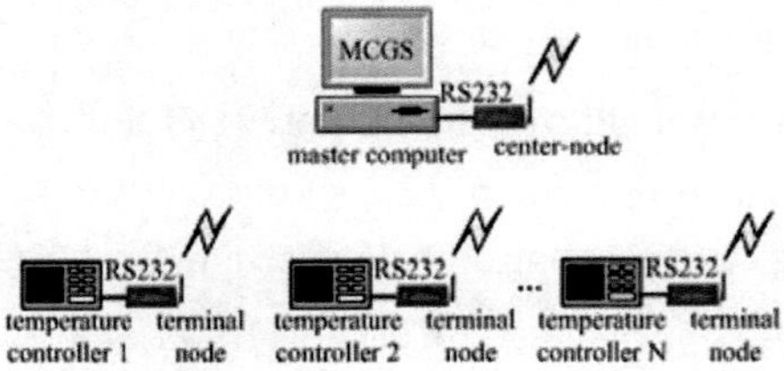

Fig. 4. The wireless temperature control network

4 The Basic Error Experiments

When doing experiments, firstly we control the temperature by using of PID and then by using the expert-fuzzy PID so as to make a compare. Because the furnace temperature is slow response and pure delay, the sampling cycle T is set to be 10 seconds. The Ziegler-Nichols (1943) tuning method calculates the initial values of K_p' , K_i' and K_d' , and after they are adjusted ,we get a set of ideal data $K_p' = 25$, $K_i' = 0.2$ and $K_d' = 30$.

Fig.5 shows the graph of the PID control process, the control precision is about 1.5℃ or so and the error is a little big at the beginning of the temperature ascending and constant period.

The temperature curve of the expert-fuzzy PID control process is shown in Fig. 6. The "pre-control" overcomes the pure delay of the temperature control effectively and the fuzzy PID is made use of during the system is running in non- precontrol zone to ensure the control precision of about 0.5℃.

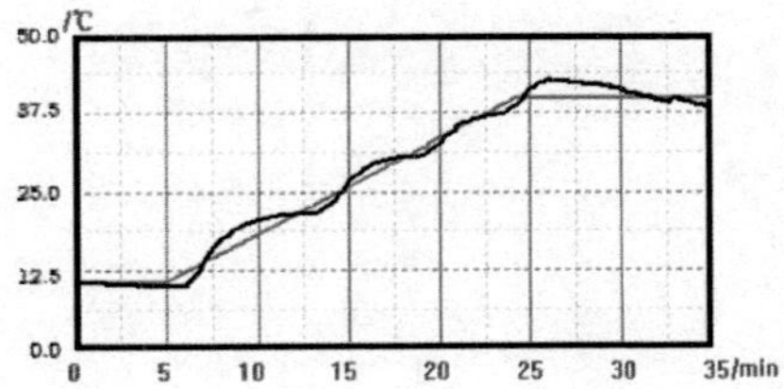

Fig. 5. The graph of the PID control process

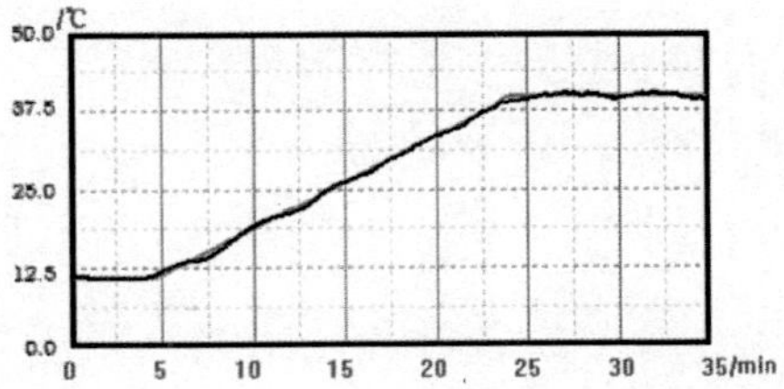

Fig. 6. The graph of the expert-fuzzy PID control proces

5 Conclusions

The process of uniform velocity-constant temperature control has the characteristics such as some uncertainty of structure and parameters, pure delay and nonlinear coupling. It is hard to obtain the exact mathematic model. Considering these characteristics, an expert-fuzzy PID controller, which combines the expert control and fuzzy PID control together, is designed. It has the high accuracy, strong stability and good robustness of PID controller and the flexibility and adaptive ability of expert controller. The experiment data shows that the control process of uniform velocity-constant temperature, which is with expert-fuzzy PID controller, has a high precision and a small overstrike.

References

1. Zhou, L., Zhao, G.: Application of Fuzzy-PID Control Algorithm in Uniform Velocity Temperature Control System of Resistance Furnace. Chinese Joumal of Scientific Instrument 29(2), 405–409 (2008)
2. Ziegler, J.G., Nichols, N.B.: Optimum settings for automatic controllers. Trans. ASME 64, 759–768 (1942)
3. Liu, J.: MATLAB simulation of advanced PID control, p. 115. Publishing House of Electronics Industry, Beijing (2006)
4. Song, J., Niu, Y.: Temperature controller of heating furnace of crude oil based on fuzzy control and expert system. In: International Symposium on Test Automation and Instrumentation (ISTAI), pp. 697–702 (2006)
5. Wu, M., Chen, A., Wang, J.: Study on Method of Fuzzy Control Table Counting. Joumal of Henan Normal University (Natural Science) 33(2), 29–32 (2005)
6. Yi, J., Hou, Y.: Intelligence control technology, p. 240. Publishing House of Beijing University of Technology, Beijing (2004)

Research of the High-Speed Fibre Switch Network NIC Based on the Memory Communication

Liu Jun-rui, Chen Ying-tu, and Fan Xiao-ya

College of Computer Science the 1st Department, College of Computer Science
Northwestern Polytechnical University, Aeronautical Computing Technique Research Institute,
Northwestern Polytechnical University Xi'an, China Xi'an, China Xi'an, China
Liu.junrui@nwpu.edu.cn, Chen_ying_tu@163.com

Abstract. At present, the FC NIC (the Fiber Channel Network Interface Card) is mostly based on traditional I/O bus, such as PCI, and is viewed as peripheral equipment, the communication data is copied between the NIC and the computer memory when communicating. Then, its communication delay is long, and its communication performance is limited by the corresponding I / O bus. So, in this paper, the Direct Memory Communication method, abbreviated as DMC, is forwarded to apply in the high-speed fiber switch network. The FC NIC based on DMC, noted as FC-DMC NIC, has the same slot as the memory, and is inserted into one memory slot, the memory on FC-DMC NIC is considered as the computer ordinary memory, and reserved as Communication Precinct. When user sends data, he uses the method of writing memory to place the data into communication space, and when user receives data, he uses the method of reading memory to get the data from communication space. So, the user can accomplished the direct point-to-point communications between two computers through accessing memory. The communications speed of FC-DMC NIC isn't limited by I/O bus, and the data copy between the memory and the NIC is omitted. Experiments show that DMC is right and FC-DMC NIC has better communication performance.

Keywords: The Direct Memory Communication, FC-DMC NIC, Communication Precinct, Memory Reserved.

1 Introduction

Current NIC is inserted into the PCI, PCI X, PCI Express or others I/O bus slot, such as Ethernet NIC and Myrinet NIC. Its performance is limited by the I/O bus, and its communication needs the device driver's help. When the memory slots are becoming abundant, the Direct Memory Communication method, abbreviated as DMC, is forwarded. The FC NIC based on DMC, noted as FC-DMC NIC, has the same slot as the memory, and is inserted into a memory slot, the memory on FC-DMC NIC is considered as the computer ordinary memory, and reserved as Communication Precinct. When user sends data, he uses the method of writing memory to place the data into Communication Precinct, and when user receives data, he uses the method of reading memory to get the data from Communication Precinct. So, the user can

J. Luo (Ed.): Affective Computing and Intelligent Interaction, AISC 137, pp. 617–624.
springerlink.com

accomplished the direct point-to-point communications between the computers through accessing memory. The communications speed of FC-DMC NIC isn't limited by I/O bus, and the data copy between the memory and the NIC is omitted. Experiments show that DMC is right and FC-DMC NIC has better communication performance.

The term DMC has not appeared in the computer domain, but ZHANG Xiao-tong professor in University of Science and Technology Beijing and Tanabe N have carried out the relevant work.

2 Direct Memory Communication

The keynote of DMC is that the NIC based on DMC is regarded as a special memory and inserted into the memory slot which has the highest physical address, thus the data exchange between the host and the NIC device is same as the memory accessing, and the original data transmission between the NIC device and the memory is no longer needed. Therefore, this communication is named as Direct Memory Communication, abbreviated as DMC, and the NIC based on the DMC method is known as DMC-NIC. The detail of DMC is described in the high-speed fiber switch network as follows:

First of all, the DMC-NIC as a special memory is inserted into the highest memory slot, and modifies the operating system to reserved the memory which has the highest physical addresses as the data communication zone of the CPU and the DMC-NIC, named as Communication Precinct, then allots Communication Precinct as the following four purposes: the receiving buffer, the transmitting buffer, the DMC-NIC command buffer and the DMC-NIC status information buffer. Secondly, according to the corresponding communication protocol (in this paper, the DMC method has been applied to the high-speed fiber switch network, therefore, FC would like to be followed here); the users can visit the special memory units of Communication Precinct to control the DMC-NIC communications. Finally, the DMC-NIC with the communications controller, the serial-to-parallel conversion and the optical transceiver fulfills the direct point-to-point communications between the computers through the users' memory accessing.

In sum, the DMC method achieves the direct communications between the computers' memory. Users do not feel the existence of the NIC, the DMC-NIC does not need the support of the device drivers, and the users fulfill the point-to-point direct memory communications between the computers by accessing memory.

3 Research of FC-DMC NIC

In order to improve the communication efficiency of DMC-NIC, the large-capacity SRAM chip is placed on the FC-DMC NIC, and the capacity of Communication Precinct is enough to meet users' communication size.

FC-DMC NIC is expanded on the DDR DIMM memory, then FC-DMC NIC has the same interface as the ordinary memory, and it can be inserted directly into the memory slot and viewed as the system ordinary memory.

3.1 The Architecture of FC-DMC NIC

The Architecture of FC-DMC NIC is shown in Figure 1:

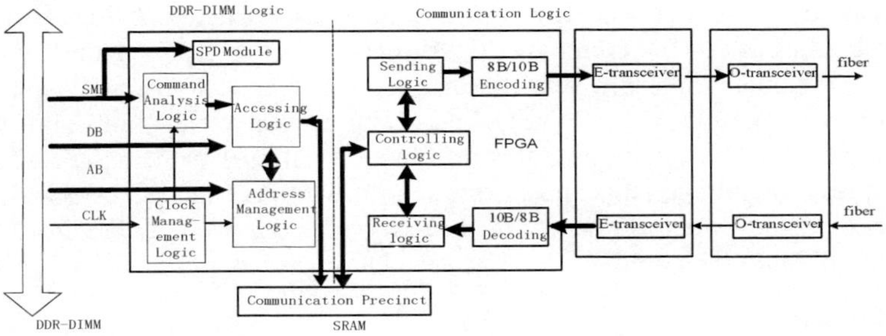

Fig. 1. The Architecture of FC-DMC NIC

As can be seen from Figure 1, FC-DMC NIC includes the export/import optical fiber, the optical transceivers, the electrical transceiver, the SRAM and FPGA. FPGA is the controller of the FC-DMC NIC, and can be divided into the DDR-DIMM logic and the communication logic. The function of each is as following:

(1) The fiber is used to connect the FC-DMC NIC and the corresponding port of the high-speed fiber switch network.

(2) The optical transceiver is responsible for the conversion between the optical signal and the electrical signals.

(3) The E-transceiver is responsible for the conversion between the serial data and the parallel data.

(4) SRAM is used as Communication Precinct, and is divided into four domains: the receiving buffer, the transmitting buffer, the DMC-NIC command buffer and the DMC-NIC status information buffer. The receiving buffer is used to store the data received from network, the transmitting buffer is used to store the data which will be send, the DMC-NIC command buffer receives the command of the FC-DMC NIC, such as Reset and Send, and the DMC-NIC status information buffer stores the status information of the FC-DMC NIC.

(5) FPGA is divided into the logic of DDR-DIMM and the logic of communication. The logic of DDR-DIMM consists of five modules: SPD (Serial Presence Detect) module, the command analysis logic, the clock management logic, the address management logic, the accessing logic. The DDR-DIMM logic is in charge of the communication between the communication and the communication logic. The communication logic consists of the sending logic, the CRC logic, the receiving logic, the controlling logic and the 8b/10b logic, it is in charge of the real network traffic activities.

3.2 Main Modules

1) Communication Precinct

FC-DMC NIC uses SRAM as Communication Precinct, and Communication Precinct is divided into the receiving buffer, the transmitting buffer, the command buffer and the status information buffer.

The receiving buffer is named as Rxbuf, and is used to store the data received from network, users can use the command of reading memory to get the data from Rxbuf. The transmitting buffer is named as Rxbuf, and is used to store the data which will been send, user can use the command of writing memory to store the data into Txbuf, then the communication logic gets the data from Txbuf to transmit. The command buffer is named as COMMAND_REG, and occupies four storage units and stores the command of FC-DMC NIC. The status information buffer is named as STATE_REG, and occupies four storage units and stores the information of the FC-DMC NIC status. Each bit of these two buffers is '1' meaning effective. When FC-DMC NIC is initialized, all bits are cleared.

2) SPD Module

SPD module makes the FC-DMC NIC holding the same stability of the ordinary memory, and makes the FC-DMC NIC being identified by the North Bridge or CPU. We have done the VHDL programming to simulate the SPD chip.

The result of analysis shows that SPD module only uses five pins of SPD chip: SA0、SA1、e SA2、SDA and SCL, and only need the Random Address Read function, so the SPD module has a simple structure and consists of two part: the state controller of Start and the Random Address Read module. The state controller of START is simple and relies on monitoring SDA and SCL. The Random Address Read module is realized with a state machine.

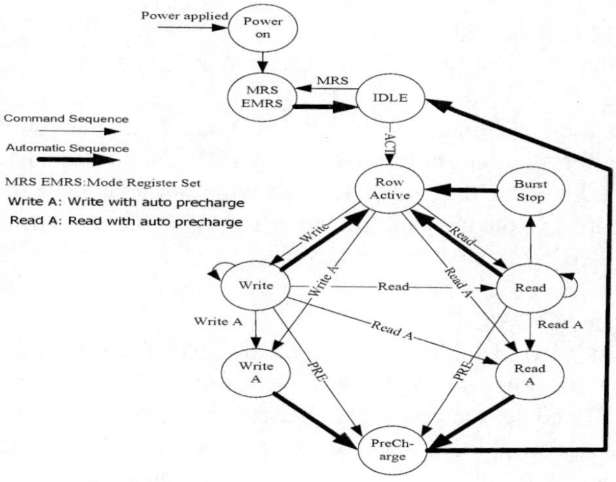

Fig. 2. The state transition diagram

3) The command analysis logic

The command analysis logic is in charge of receiving kinds of accessing command from DDR-DIMM interface, then analyzing these command. It is realized on a state machine, the signal is set to trigger the writing and reading logic when the state of the state machine changes. The state transition diagram is shown in Figure2.

4) The address management logic

Because the SRAM chip is used as FC-DMC NIC Communication Precinct, the address management logic requires a very strong function, and can search the accurate storage units with the virtual address.

The different memory controllers use different and complicated methods to manage the memory, but the author put up a simple method to manage the Communication Precinct: the sequence addressing algorithm. The details of this algorithm are shown as following.

Communication Precinct is mapped with the function mmap, the return of the mmap function is the first virtual address of a continuous virtual address space. The sequence addressing algorithm leaves aside the translation method between the virtual address and the physical address, and sets up a simple mapping between the virtual address and the physical storage units by the sequence relationship, shown in Figure 3.

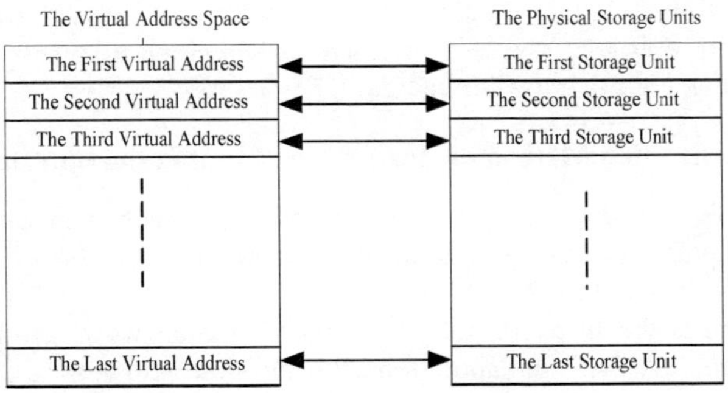

Fig. 3. The sequence addressing algorithm

The steps of this algorithm are shown as following:

Two registers in FPGA are needed, one is used to stored the first virtual address, named as FirstVAdd_REG, the other is used to stored the current accessing virtual address, named as CurrentVAdd_REG;

The first virtual address is wrote into FirstVAdd_REG after mapping. When Communication Precinct is accessed, the current accessing virtual address is firstly wrote into CurrentVAdd_REG, then the address management logic get the first virtual address from FirstVAdd_REG to calculate the offset between the value of FirstVAdd_REG and the value of CurrentVAdd_REG, finally controls the Communication Precinct accessing pointer moving from the first physical storage unit to the accessed unit.

When user accesses Communication Precinct, the accessing logic reads or writes the accessed storage units, and fulfills the data writing into Txbuf or reading from Rxbuf.

3.3 The Software of FC-DMC NIC

DMC is realized on that Communication Precinct can be reserved. Thus, the software needs to fulfill the following work.

1) The reserving of Communication Precinct

According to the memory management method of Linux, we place FC-DMC NIC into the memory slot which holds the highest physical address. Then, we modify the parameter of the system boot program, named as mem, to limit the memory used by the system kernel. The physical memory which is greater than the value of mem is reserved, and the system will not use this memory space.

2) The mapping of Communication Precinct

Because the Linux runs on the virtual address system, and user accesses the memory with the virtual address, Communication Precinct needs to map into the virtual address space for writing and reading. There are three ways to achieve the memory mapping, and the memory can be mapped to I/O memory, the Kernel memory and the user-mode memory. At last, we choice the mmap function to map Communication Precinct, the first virtual address is needed to write into FirtVAdd_REG after mapping.

3) The accessing interface of Communication Precinct

User can access Communication Precinct with the method of accessing the memory. When user accesses Communication Precinct, the software writes the current accessing address into CurrentVAdd_REG.

3.4 The Notification Method of Reading the Communication Data from Rxbuf

As the DDR-DIMM does not support the interrupt mechanism, we design three methods to notify the user to read the communication data from Rxbuf of Communication Precinct in FC-DMC NIC.

The first one is the software poll. The second one is two-network collaboration. The last one is two-NIC collaboration. Because the most of I/O bus has been supporting the interrupt mechanism, we also design another NIC based on PCI bus. The PCI-NIC takes charge of the sending, receiving and managing of the network command, FC-DMC NIC only fulfills the data communication.

3.5 The Functional Test of FC-DMC NIC

When boots, we can the SPD module waveform with the oscilloscope, shown in Figure4.

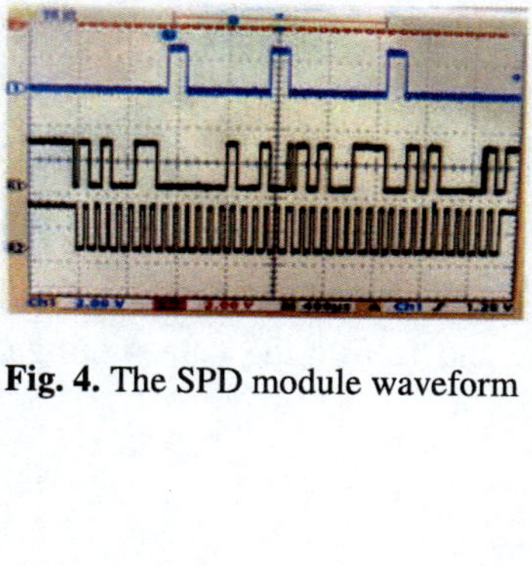

Fig. 4. The SPD module waveform

In Figure5, the waveforms of various commands received by the DDR-DIMM logic are shown.

The test results show that FC-DMC NIC is recognized as one ordinary memory when system boots, reserves Communication Precinct and deals with the writing and the reading command from CPU. So, FC-DMC NIC is correct and has great advantages.

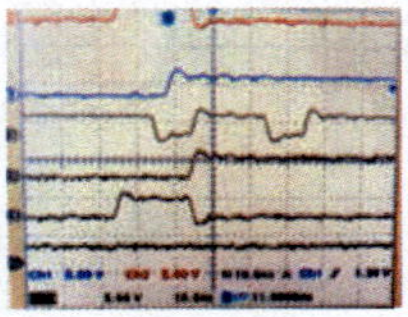

a. Active

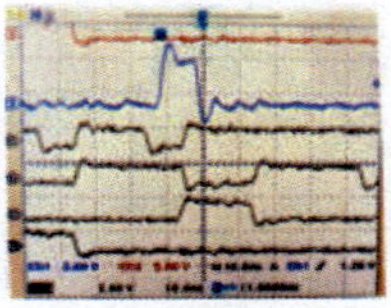

b. write

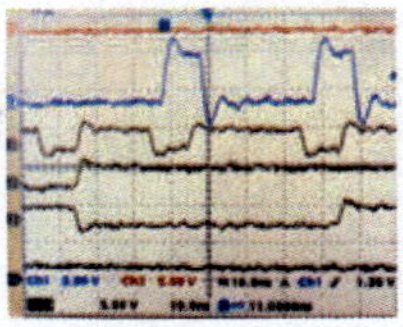

c. read

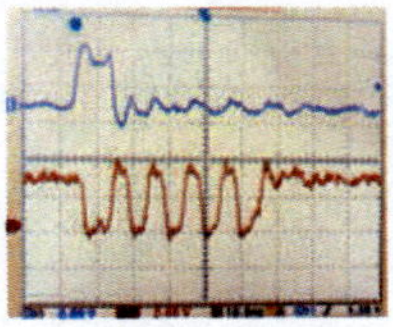

d. QOS

Fig. 5. The memory command waveform

4 Conclusion

FC-DMC NIC uses DMC technology in high-speed fiber switch network, and is paced into a memory slot as a special memory bar. User uses the method of accessing memory to fulfill the direct communication between two computers, the communication isn't limited by I/O bus, and the data copy between the NIC and the memory is omitted. So, the communication delay of FC-DMC NIC is greatly shorten, its communication speed is greatly increased and its communication mode becomes Simpler. In the following work,we will continue to study the memory management and the Linux for reinforcing the practicality and simplicity of FC-DMC NIC.

References

1. Tanabe, N., Hamada, Y., Nakajo, H.: A low latency high bandwidth network interface prototype for PC cluster. In: Proceedings of the International Workshop on Innovative Architecture for Future Generation High2Performance Processors and Systems, Big Island, p. 87 (2002)

2. Gorman, M.: Understanding the Linux Virtual Memory Manager. Prentice Hall PTR (May 2006)
3. Matzigkeit, G., Okuji, Y.K.: The GUN Grub manual (May 2005)
4. JEDEC, PC SDRAM Serial Presence Detect (SPD) Specification, Revision 1.2A, pp. 4.1.2.4.1– 4.1.2.4.39
5. JEDEC, PC2100 and PC1600 DDR SDRAM Registered DIMM Design Specification, Revision 1.3, pp. 4.20.4.1– 4.20.4.82 (January 2002)
6. MindShare, Inc, Ravi Budruk, Don Anderson, Tom Shanley PCI Express System Architecture (November 2003)
7. Zhang, X., Wang, J., Wang, Q., Liu, L.: High-speed network access technology based on DDR memory bus. Journal of University of Science and Technology Beijing 29, 1158–1162 (2007)
8. Liu, J.-R., Chen, Y.-T., Fan, X.-Y., Kang, J.-C.: Research of the Direct Memory Communication Method. In: 2009 WASE International Conference on Information Engineering, vol. 1, pp. 282–285 (2009)

A Hierarchical Routing Protocol Based on Location Information for Wireless Sensor Networks

Guping Zheng[1], Xiaolu Lv[2], and Yu Wang[3]

[1] School of Computer Science and technology
North China Electric Power University Baoding, China
zhengguping@126.com
[2] School of Computer Science and technology
North China Electric Power University Baoding, China
lvxiaolu@126.com
[3] School of Computer Science and technology
North China Electric Power University Baoding, China
wangyu3940@126.com

Abstract. For the limitation of energy of wireless sensor network node, this paper combine with the characteristics of hierarchical routing protocols and location-based routing protocols, proposed a hierarchical routing protocol based on the geographical location (HRBGR). In the HRBGR algorithm, the coverage of the sensor network is divided into some areas based on nodes location information evenly, then self-organized clusters according to local information in each area and decide the CH and the agent independently in each cluster. All of the cluster head nodes construct a routing tree by the way of seeking the help of the principle of minimum spanning tree algorithm of prim and transfer the data to the base station by means of multi-hop communications. Theory and simulation analysis shows that the algorithm extends the whole network's life cycle at lower level energy consumption and reduces delay, improves network load balance.

Keywords: wireless sensor network, hierarchical routing, location information, the minimum spanning tree.

1 Introduction

In wireless sensor network routing protocols[1], the routing protocols of cluster balance the energy consumption of whole network by the way of rotate cluster head. Cluster heads communicate with the base station generally use the single-hop or simple multi-hop manner. Single-hop communication make the cluster heads consume excessive energy and lead to a rapid death which farther from the base station. While simple multi-hop communication make the nodes quick death which close to the base station, then cause the entire network coverage area reduced, even non-connected region, thus affecting the integrity of the data collection and destroying the routing link.

Although location-based routing protocols are suitable for dynamic network topology, do not need to set up end to end link state routing based on global or store

J. Luo (Ed.): Affective Computing and Intelligent Interaction, AISC 137, pp. 625–632.
springerlink.com

routing information table, but we must know the location information of neighbor nodes and the communication of single-node data transmission can also cause rapid death of nodes, then the base station will unable to collect the information in the region. At the same time, adjacent nodes transmit data also cause a large amount of data overlap.

Sum up the above analysis; a hierarchical routing protocol based on location Information for WSNs has proposed in this paper, the characteristics of this protocol are combining the advantages of hierarchical routing protocols with location-based routing protocol, improving the network scalability and robustness largely.

2 System Model and Problem Description

2.1 Network Model

There are N nodes are distributed in the region A randomly and have the following properties of the sensor network:

All of the nodes are no longer move after the deployment and the base station deployed in a fixed location outside of region A.

All nodes have similar capabilities (processing / communication) and equality, each node has a unique identity ID.

Wireless transmits power control, that node can adjust the transmission power according to the distance.

Sensor needs can be acquired the location information of themselves by the way of equipping with GPS or location algorithm.

2.2 Communication Model

Node using the same wireless communication energy consumption mode with literature [5], the energy consumption of node launch l bits is divided into two parts, transmitter power loss and power amplifier loss, that

$$E_{Tx}(l,d) = E_{Tx-elec}(l) + E_{Tx-amp}(l,d)$$
$$= \begin{cases} lE_{elec} + l\varepsilon_{fs}d^2, d < d_0 \\ lE_{elec} + l\varepsilon_{mp}d^4, d \geq d_0 \end{cases} \tag{1}$$

Where: $E_{Tx}(l,d)$ indicate the energy loss by transmitting l bits data through the distance d., consist of transmitter power loss $E_{Tx\text{-}elec}(l)$ and power amplifier loss $E_{Tx\text{-}amp}(l,d)$. Power amplifier loss adopts free space model or multi-path fading model according to the distance between sender and receiver. E_{elec} means the power consumption of node to receive or send 1 unit bits of information. ,ε_{mp} indicate the coefficient of distance attenuation amplifier.

Distance threshold: $d_0 = \sqrt{\varepsilon_{fs} / \varepsilon_{mp}}$

The energy consumption of each node receiving l bits data is:

$$E_{Rx}(l) = E_{Rx-elec}(l) = lE_{elec} \tag{2}$$

3 Algorithm Description

3.1 Formation of the Clustering Region

The numbers of cluster head have a great impact on network performance in clustering algorithms, according to the literature[1], the optimal number of cluster head are:

$$n = \frac{\sqrt{N}}{\sqrt{2\pi}} \bullet \frac{\sqrt{\varepsilon_{fs}}}{\sqrt{\varepsilon_{mp}}} \bullet \frac{M}{d^2_{toBS}} \tag{3}$$

Where: N indicate the number of survival nodes, M is the diameter of the network, d_{toBS} means the distance between the base station and node. We choose the n as the optimal number of cluster head which make the network energy consumption minimum.

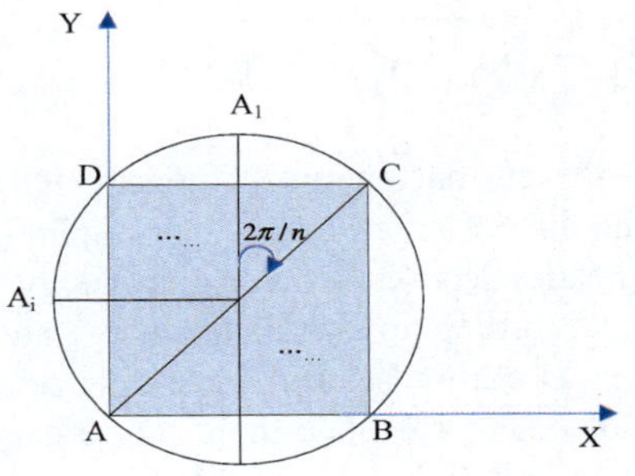

Fig. 1. The clustering region

First, make the center of monitor regional as the origin, select radius R which can cover the whole region, then execute the same angle divide as 2π/n for the step. Where: n indicate the optimal number of head in the network. The nodes of random deployment have uniform distribution of properties. Number of nodes within each cell is almost. Cluster head elected in the district respectively, so avoid the situation of uneven distribution of cluster head. And n does not change after determining, at the same time assigned to each region a unique identity.

3.2 Cluster Head Selection Mechanism

When select cluster head, we will choose a regional agent which is responsible for managing the cluster process. If the first time choose agent, using randomly generated. Otherwise, the cluster head will select the most residual energy node as agent in the district and the agent will announce "Hello" message to the other nodes in the district. After receiving a message, nodes will send competitive intensity to the agent immediately, so that the agent can determine the cluster head, then nodes send < itself location, residual energy > data to the agent, declaring itself to become a member of the node agent. The agent maintains a regional data tables, include the location and residual energy of each node within the district and so on.

Cluster head selection method as follows:

The i is a cluster head in the district m, i meet the following conditions:

$$T(i) = \max[T(j)]$$

In formula, $j \in V(m)$, $V(m)$ is the set of nodes in the district m. T(j) is the parameters of energy and distance, it can be expressed as:

$$T(j) = a \times \frac{E_j}{E_{Total}} + (1-a) \times \frac{1}{S} \tag{4}$$

$$\frac{1}{S} = \frac{1}{n} \sum_{k=1}^{n} \sigma_{ij}^{2} \tag{5}$$

$$\sigma_{ij} = \sqrt{(x_i - x_j)^2 + (y_i - y_j)^2} \tag{6}$$

In formula, E_j indicate the current energy of node j, E_{total} means the current total energy of all nodes in the district m. $a \in (0,1)$ is the weight, it shows the impact on the competition of cluster head selection between the energy information and distance information. The agent according to the competitive intensity T(j) of each node, selects one of the biggest as cluster head. After cluster head is confirmed, the agent queries the coordinates of nodes, calculate the distance d_{CH_node} between CH and the node which is the furthest to the CH and send the d_{CH_node} to the selected cluster head. According to the value of d_{CH_node}, the CH adjusts its transmission power dynamically, to ensure that all the nodes within the coverage area. Then CH broadcast message to the region and declare itself to be the cluster head, the nodes which received the message adjust their communication power according to the distance between themselves and the CH and send the message of join to the CH, announce to join the cluster.

At the end of the communication, the CH inform the agent its own residual energy, when the residual energy of the CH below a percentage of the average residual energy in the cluster or the agent detect that the CH can not communicate with the member of nodes or its neighbor CH, then the agent initiates a new round of the CH election process. On the other hand, the agent sends its own residual energy and the average energy of the cluster to the CH in each of their communication time. If the residual energy of the agent below a percent of the average residual energy of the cluster then the CH initiates a new round of the agent election process to protect the identity of the stability of the agent. In addition to, if the CH and the agent can not detects each other, then the one can initiate a new round of the election process of the other.

3.3 Communication between CH and Members

The members of the cluster only send collect information to the CH, in order to avoid data conflicts, the CH adopt TDMA-lased low-power MAC protocol to the agent, distribute the time slot of communication for each member node and only in their own time slots that nodes can communicate with the CH. The nodes close the wireless communication in the rest of the module slot in order to save energy. In addition to, as the nodes communicate by the way of broadcast so the other CH may receive this information which is not in this district at the time of the CH communicates with members of the cluster. So each cluster use the identified area code in the communication process in order to avoid interference between clusters. Nodes according to the code to filter out the information which is not belong to the cluster. On the other hand, the CH can receive the data information from its neighbor CH, it achieved through bind data type identification in the data packet. Then the CH will execute data fusion process after it received the monitor information from its members and forms a data packet.

3.4 Communication between Cluster Heads

The algorithm adopts the principle of minimum spanning three at the time of the cluster heads communications. All of the cluster heads construct to a connected graph in the wireless sensor network. Suppose graph G is a binary group of sets V and E, G= (V, E), where: V indicates the finite set of non-empty cluster head node, E is the set that the nodes in the V communicate with the neighbor cluster directly. G is directed graph. (u,v) indicates u can communicate with v directly. Generally, the v also can communicate with the u directly, this is a symmetric access mode. Each (u,v) in the E have a weight W(u,v)。The purpose of the minimum spanning tree is consuming the least energy form any source CH to the Sink, this requires the W(u,v) between the nodes is the least. The calculation of W (u, v) is:

$$W(u,v) = \frac{D(u,v)^k}{E(u)E(v)} \tag{7}$$

In the calculation: D(u,v) indicate the Euclidean distance between nods, E(u) and E(v) indicate the residual energy of u and v respectively. K means the system parameters. The calculation (7) can ensure that the energy cost of the minimum spanning tress is the least. As shown in Figure 2:

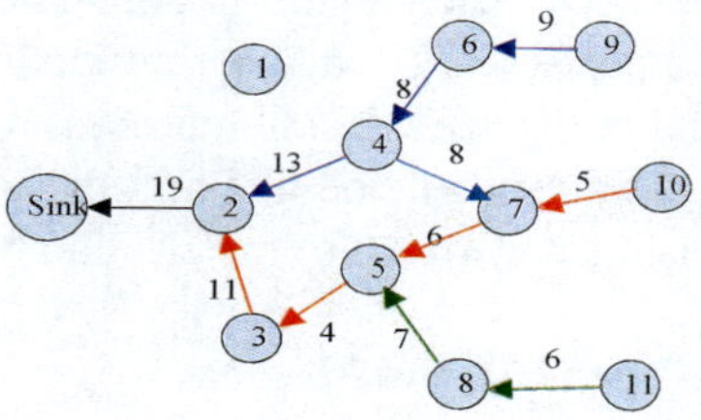

Fig. 2. Cost of WSN minimum spanning tree

3.5 Maintaining the Integrity of the Specifications

The template is used to format your paper and style the text. All margins, column widths, line spaces, and text fonts are prescribed; please do not alter them. You may note peculiarities. For example, the head margin in this template measures proportionately more than is customary. This measurement and others are deliberate, using specifications that anticipate your paper as one part of the entire proceedings, and not as an independent document. Please do not revise any of the current designations.

4 Performance Evaluations

In this section, we evaluate the performance of the HRBGR by simulation. Fist we describe the simulation environments and performance evaluation metrics. Then we evaluate the system performance with given environments and parameters. Finally, we show the comparisons between HRBGR, GPSR and LEACH.

4.1 Simulation Environments and Metrics

The simulation using NS2 and the initial scene as following: 2000 nodes distributed in the area of 1000 m×1000 m randomly. The initial energy of node is 2 J and the node considered as death when it's residual energy less than 0.001Jand simulation time is 800s.

4.2 Simulation Results

As shown in Figure 3, as the simulation time, the CH communicated with the Sink directly in the Leach, it leads to the CH rapid death because of excessive energy consumption. In GPSR, it has a better network load at the begging of network. As the simulation time, the nodes became dead which were closer to the Sink and cause the empty nodes. Therefore, other nodes need to consume more energy to transmit data and speed up the death of nodes in the entire network.

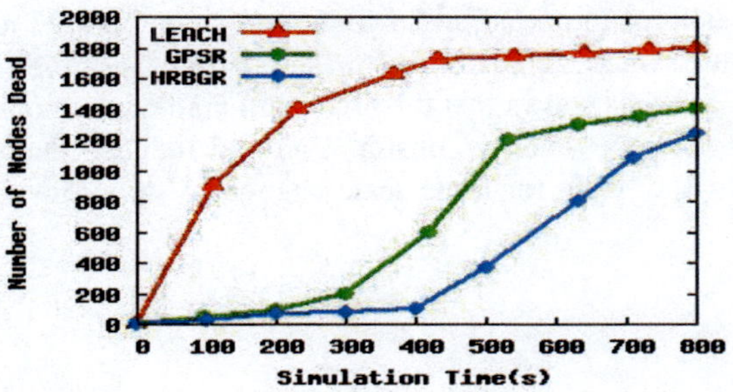

Fig. 3. Comparison of the dead nodes with time

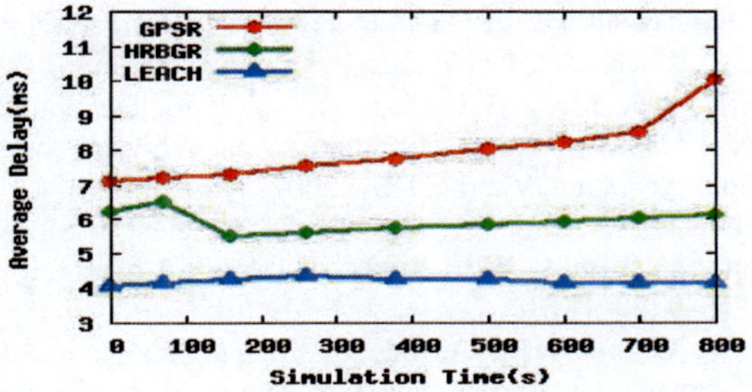

Fig. 4. Comparison of average delay

HRBGR with cluster data fusion, all of the cluster heads constructed a routing tree by the way of the minimum spanning tree and transferred the data to the Sink by means of multi-hop. The algorithm reduced the energy consumption greatly and prolonged the dead time of nodes.

Figure 4 is a comparison graph of the three protocols at the average transmission delay. As a result of the edge strategy of GPSR, it produced too many forwarding nodes when faced with empty nodes; therefore, its transmission delay is the maximum. In HRBGR, nodes according to the weight of information to find the path based on the minimum spanning tree in the data transfer phase and its delay increased. But with the growth of simulation time, the algorithm would get the optimal route to forward the data thereby reducing the time delay. The CH communicated with the base station directly in the protocol of Leach, so it generated the minimum delay.

5 Conclusions

For the characteristics of hierarchical routing protocols and location-based routing protocols, a hierarchical routing algorithm based on location information was proposed. The algorithm is self-organized clusters according to local information, it formed the optimal number of cluster heads and decided the CH and the agent independently so that reduced the clustering time greatly and saved energy. In the data transmission phase, all of the cluster heads constructed a routing tree by the way of seeking the help of the principle of minimum spanning tree algorithm of prim and transferred data to the base station by means of multi-hop communications. Theory and simulation analysis shows that the algorithm made the whole network have a long life cycle at lower level energy consumption and reduced delay, improved network load balance. It is suitable for large-scale changes of topology in the wireless sensor networks.

References

1. Wendi, R.H., Anantha, C., Hari, B.: Communication protocols for wireless sensor networks. In: Proceedings ofthe Hawaii International Conference on System Sciences, vol. (8), pp. 4–7. IEEE Computer Society, Washington (2000)
2. Lindsey, S., Raghavendra, C.: PEGASIS: Power-efficient gathering in sensor information systems. In: IEEE Aerospace Conference Proceedings, pp. 1125–1130 (2002)
3. Delobel, C., Rousset, M.C.: A Uniform Approach for Querying Large Tree-Structured Data Through a Mediated Schema. In: International Workshop on Foundations of Models for Information Integration, Viterbo, Italy (2001)
4. Younis, O., Fahmy, S.: Heed: A hybrid energy—efficient distributed clustering approach for adhoc sensor networks. IEEE Trans. on Mobile Computing 3(4), 660–669 (2004); Infocom, 629–640 (2004)
5. Heinzelman, W., Chandrakasan, A.: Energy-efficient communication protocol for wireless micro-sensor networks. In: Proc. of the 33rd Annual Hawaii International Conference on System Sciences, pp. 1–10 (2000) Mau, i HI: [s. n.]
6. Ye, M., Li, C.F., Chen, G.H., et al.: EECS:An energy efficient clustering scheme in wireless sensor networks. In: Proc. of the IEEE Int'l Performance Computing and Communication Conf. IEEE Press, New York (2005)
7. Aiello, W., Chung, F., Lu, L.: A random graph model for massive graphs. In: Proc.of the ACM STOC 2000, pp. 171–180. ACM Press, Portland (2000)
8. Jalali, A., Padovani, R., Pankaj, R.: DataThroughput of CDMA-HDR: A High-Efficiency High Data Rate Personal Communication Wireless System. In: Proc. 51st IEEE Vehicular Technology Conf (VTC 2000 Spring), vol. (3), pp. 1854–1858 (2000)
9. Pei, L., Wan, J.: A new algorithm of node redundancy fusion tree in wireless sensor network. Chinese Journal of Sensors and Actuators 21(6) (June 2008)
10. Tian, L., Xie, D.: The Greedy forwarding strategy routing hole problem for wireless sensor networks. Electronics and Information Science 29(12) (December 2007)

Research and Practice of 'Teaching, Learning, Practice Integration Teaching Model' in Higher Vocational and Technical Education

Yunhui Ren[1] and Limin Zhao[2]

[1] Department of Information Engineering,
Jianghai Polytechnic College, Yangzhou, China, 225101
rycryh@126.com
[2] Department of Mechanical Engineering,
Yangzhou Industrial Technology College, Yangzhou, China, 225100
zlmryh@126.com

Abstract. The article carries out the practical experience of network interconnects and system integration course adoptting "teaching, learning, practice integration" educational model , put forward higher vocational and technical college must do a good job of four links when carring on "teaching, learning, practice integration" educational reform, which are "reform of teaching philosophy", "improving 'double' quality of teachers", "enhanced construction of teaching material " and "creation of 'teaching, learning, practice integration' teaching environment". A new way of higher vocational education reform of teaching is discovered through the four measures.

Keywords: higher vocational and technical college, network interconnection, teaching, learning, practice integration.

1 Introduction

Network interconnection and system integration course is one of the important professional courses of informational technology speciality. This course is characterized by strong practice, and the job of students in the future will contact greatly with it. This course requires higher vocational and technical school students not only to control the basic theory, but also to practise, and they must have the ability to complete the basic networking and system integration by themselves. Clearly, for such a teaching objective, the traditional classroom-based teaching method is not appropriate, it could easily lead to a mismatch between theory and practice of students. How can we teach students to control the essence of this course effectively with proactive attitude? How can we teach students to grasp the method of setting up computer network independently in a short period of time? With these questions, we studed the teaching process seriously, and carried out a series of teaching reform and exploration work, now we have made the preliminary results.

In practice, we believe that this type of course must be combined with actual work of students in the future. What we are teaching now must be what they'll need in the

J. Luo (Ed.): Affective Computing and Intelligent Interaction, AISC 137, pp. 633–638.
springerlink.com

future. Learning of students can not be mere passive listening and memory as vocational education can not be pure indoctrination. This teaching and learning process should be guided by "Teaching, learning, practice integration" education philosophy advocated by Chinese educator Xingzhi Tao, this great man once said, "Teaching study and practice are one thing, not three things. We have to teach by practice and learn by practice. Who teaches through practice is the teacher, who learns through practice is the student. In this definition, teachers and students have lost the usual strict distinction, teaching and learning through practice is a very natural phenomenon." Guided by this philosophy of education higher vocational education should combine teaching, learning and practice as a whole [1], we should move classroom to the laboratory and computer room in which teachers should teach by demonstrating the practical work, and in which students should learn by doing actual work independently. Through hands-on learning interest of students can be aroused. Through hands-on they can also deepen their understanding of the principles, master skills, and learn to use [2].

We call such a teaching method as "teaching, learning, practice integration" educational model.

Since the end of 2006 we started the teaching reforms of "teaching, learning, practice integration" model of network interconnection and system integration course, through constant exploration and practice, the teaching method of "teaching, learning, practice integration" model has been basically established, in the teaching process of the course we have obtained a better teaching results, and the course are well received by students.

We believe that practice of "teaching, learning, practice integration" teaching model should grasp the following links.

2 Changes in Teaching Philosophy

The meaning of Xingzhi Tao 's "teaching, learning, practice integration" is the method of teaching based on the method of learning, the method of learning based on the method of doing. What to do based on how to learn and how to learn based on how to do. To teach and learn well, we must do better. This view is ideal for higher vocational and technical education.

Higher vocational and technical college's teaching has its own characteristics and rules, and indeed there should be a big difference from the teaching of undergraduate institutions. the purpose of higher vocational and technical college and undergraduate institutions are not the same, higher vocational and technical colleges train skilled personnel working at the grassroots level positions, general undergraduate training is professional. Higher vocational technical college emphasizes outstanding ability of practical application, general undergraduate institution stresses solid theoretical foundation. The level of enrollment of higher vocational and technical college and undergraduate institution are different , higher vocational and technical college students have shown less abstract thinking ability than undergraduate institution students, they are more suitable for visual and intuitive teaching methods.

Ordinary colleges and universities commonly used teaching model In accordance with the disciplinary system is not appropriate for higher vocational and technical college. The starting point of disciplinary system teaching model is professional theory, stress integrity, rigor of theory system, content of teaching is required the every aspect, and regardless of whether there is a strong applicability of these knowledge. Higher vocational and technical college's teaching should break the discipline system, and choose the most practical knowledge in a subject to train their students. The process of teaching should be organized in accordance with the general process of these knowledge used in work. The purpose is to provide students targeted training, and to improve their skills quickly, so that after the students left school, they will be more suited to the requirements of jobs.

In view of this philosophy of education, "teaching, learning, practice integration" teaching method is ideal teaching model of higher vocational and technical college. In order to train students how to complete future tasks, teachers must try to simulate the working process of the students in the future in teaching process [3]. For example, in Network interconnection and system integration course, When learning knowledge about switch, we bring students into the network laboratory, and teach the students how to distinguish the switch from numerous network equipments, after the students have the direct-viewing understanding of the switch, we explain them related knowledge of switch such as basic structure and working principle, at the same time we direct the students to set up a simple network with switch by themselves. By this way students can grasp the basic using method of switch, and understand the principle that switch identifies the computer's MAC address automatically easily. Through hands-on operation, they understand the working principle functions and applications of network equipment deeply and quickly. By practising " Teaching, learning, practice integration", Students understand the knowledge and master skills efficiently.

3 Quality Requirements of Teachers

Mr. Xingzhi Tao pointed out that "With interest people would have the spirit to do all things". Interest is the best teacher, so only by refered to stimulating student's study interest as a premise teaching can make the good results. How to stimulate students' study interest and take the initiative to learn by themselves then? This make a very high demands to teachers' professional quality. Teachers are required to have "double" quality. They are not only qualified teachers who have strong teaching skills, and understand the educational theories that can readily adapt the transfer of knowledge but also qualified engineers who can operate specific equipment, and accomplish specific engineering process. It is the key to success of "teaching, learning, practice integration" teaching model whether there are such compound teachers.

In teaching proccess of Network interconnection and system integration course, teachers are required to complete the teaching of Networking and systems integration basic theory while they guide students to network two computers, several computers, and networks. Theoretical knowledge is taught throughout the the whole operation process which is from easy to complex.

It seems so simple to explain to the students and to guide the experiment. But in fact the demand for teachers are very high, For example, when we do the experiment of IP address setting in switch, teachers usually demonstrate the experiment in classroom like this: First, connect the switches and computers; The second step, go into the configuration state of the switch; The third step , go into the management state of the

VLAN 1 of the witch; The fourth step ,set IP address for the switch. However, some students, during the experiment, may ask some unexpected questions to teachers, "Can we configure multiple IP addresses for the switch?" "Can we set the switch IP address in VLAN 100 established by ourselves?" If the teacher who do not prepare courses adequately or are not familiar with the contents of the experiment before the class will be unable to explain the question that the student proposed. So teachers need to do the following. First, the teacher must do every steps of the experiment correctly before he explain to the students and show the whole experiment for them; Second, the teacher must pre-conceived problems the students may encounter in the experiment, and resolve; Third, the teacher must take it into account that knowledge used in the experiment is beyond the student's knowledge scope, and know how to explain it by experiment. In short, this principle that in order to give students a bowl of water the teachers themselves should have a bucket of water must be respected. Before the students practice, teachers should do all the content of all experiments proficiently, so that the teacher will handle a butcher's cleaver skilfully in class, and give students vividly explaination. Only in this way students can better understand.

4 Construction of Supporting Teaching Materials

Since breaking the ideological restraints of disciplinary system, supporting teaching materials of "teaching, learning, practice integration" teaching model must be re-developed. The teaching materials should be in accordance with the training process and to enhance the skills of the students as the core target.

In network interconnection and system integration teaching, we have written a new textbook 《Network Interconnection Technology》 which reflects the "teaching, learning, practice integration" principle. This new book is based on the students of higher vocational technical college, and summed up the students should learn core skills. In compiling process of this book, we investigated a number of related enterprises and institutions involved in networking technology jobs, and combined with the author's 10 years experience of networking in an enterprise. Through research we find that the networking technology courses of higher vocational and technical colleges should be divided into four modules (ie, network infrastructure; build a simple network; management, optimization of local area network; network interconnection) to teach. In this book ,students' skill training about the four modules are completed through five progressive project, as shown in figure1, the five items are "an initial understanding of network", "building a small network", " network management though switch", "optimize the internal network though switch", "use the Internet routers". These five items actually simulate the some classic work of actual work, namely, building LAN, management LAN, LAN optimization and building WAN [4].The textbook connected theory of the network to each project organically, and enable students to grasp theoretical knowledge intuitively by completing the various projects, so that the students master not only the theoretical knowledge but also the practical skills of networking. Such a process of learning make students feel a sense of accomplishment, more close to reality, but also easy to understand.

5 Teaching Environment Creation

An important feature of "teaching, learning, practice integration" teaching model is to replace classroom with experiment and training establishments. Such teaching environment transforms is not only advantageous to teaching process expansion, realizes on the one hand to do, on the other hand to study while teacher is teaching, moreover is advantageous to the student project view raise, lets the students in school study understand future working conditions, will raise the professional accomplishment gradually which the work in the future needs.

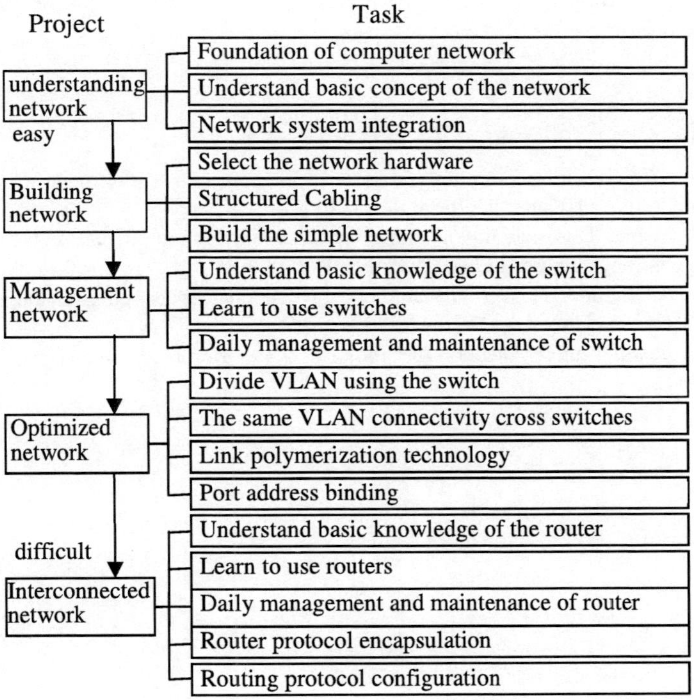

Fig. 1. Project progressive figure

Network interconnection and system integration teaching is done in the network laboratory of our department. The laboratory equipped with a full set of networking equipment produced by Digital China ,including twenty-four sets of two-story switches, two sets of three-story switches, twenty-four sets of routers, two sets of firewalls, eight sets of wireless routing, twenty-four wireless network cards and fifty sets of computers .All these are divided into eight groups distributed in the laboratory. It can be used for a lot of teaching content such as network formation, network management, switch configuration, using the switch to optimize the network and networking through routers and so on. In the teaching process teachers can explain while pupils in operation. Teachers help students to combine theoretical knowledge and practical application of equipment through showing them network device and operation at the network device. Theory in this way to achieve the unity of teaching and practice teaching .In this way we achieve unity of theoretical teaching and practical teaching. In this process, teachers can prompt discover and correct students' errors, and achieve high teaching efficiency.

By teaching reform experiments in "network interconnection and systems integration" course, we believe that vocational technical college teaching reform must deeply understand the laws of their own teaching and work characteristics, and that teaching model of Vocational and Technical School must adapt to the aims of education and the students' actual situation. Only by grasping four areas "concept, teachers, teaching materials, the environment", teaching reform can be achieved. In short, all of our teaching starting points must be based on the students' future professional competence requirement.

References

1. Ke, G., Ke, T.: Teaching, learning and practice integration teaching mode. Contemporary Education Forum, 11(Under the fortnightly) (2009)
2. Dong, G., Yang, L.: Teaching, learning, practice, integration to realize the deep level technology union. Chinese Higher Education, 17 (2009)
3. Jiang, D.: Contemporary German Studies of vocational education into the mainstream thinking. Tsinghua University Press (2007)
4. Ren, Y.: Network Interconnection Technology. Water Resources and Hydropower Press (2009)

Research on Coordination Strategy of Remanufacturing Closed-Loop Supply Chain Based on Quantity Discount*

Zuqing Huang[1], Guoqing Zhang[2], and Lijun Meng[1]

[1] School of Economics and Management,
China Jiliang University, Hangzhou, China
[2] Industrial and Manufacturing Systems Engineering,
University of Windsor, Windsor, Canada
hzq1210@163.com

Abstract. A mathematical model of a closed-loop supply chain is constructed in this paper for researching the strategy of quantity discount. The closed-loop supply chain in this model is composed by one manufacturer and one retailer (component), leaded by the manufacturer. Through the analysis using the constructed mathematical models, the pricing strategies of manufacturer and retailer are obtained. Besides, the maximal profit of them can be obtained. Then a calculation example is made. The conducted calculation example shows clearly that the coordination strategy of quantity discounts offers great profit opportunities to manufacturer and retailer.

Keywords: remanufacturing, closed-loop supply chain, supply chain coordination, quantity discounts.

1 Introduction

Due to the continuous increasing of productivity, consumption of raw materials and energy are increasing sharply, and the phenomenon of natural resources shortage becomes more serious, especially in the manufacturing industry. Resource conservation and environment protection have become common issues of social concerns, so used product's remanufacturing becomes more and more popular. Closed-loop supply chain management has developed into (become) a key competence in modern supply chains.

The research on closed-loop supply chains made much progress in recent years. Many papers have appeared in the literature on the topic. Wen and Liu (2008) studied with the commission surrogate theory a reverse supply link system including a single supplier and a single shopkeeper [1]. Li and Da (2008) established a secondary closed-loop supply chain system composed of one manufacturer and two retailers, and analyzed the benefits of this supply chain in three competition modes and different game modes [2]. Ma (2008) studied a two stage closed-loop supply chain including

* **This paper** is sponsored by"Key Universities Research Institute of Humanities and Social Sciences in Zhejiang Province——Standardization and Intellectual Property Management".

J. Luo (Ed.): Affective Computing and Intelligent Interaction, AISC 137, pp. 639–649.
springerlink.com

manufacturing/ remanufacturing and with the manufacturer being the Stackelberg leader, and identified the optimal ratio for dividing the total profit between manufacturer and retailers [3]. Zhao, Zhe and Wang (2008) developed the supply chain coordination mechanism decision model including market(ing) rate of interest, fluctuation ratio of product price and option term. Thereinafter Zhao analyzed the model by the supply chain member's profit optimization mechanism and calculated the retailer's pricing and ordering strategy that maximized the supply chain total profit and the supplier's production planning [4]. Yang (2008) constructed three decentralized models in which used products can be remanufactured to new ones, and compared these three models within different scenarios of the wholesale price, retail price, return rate and total profit of supply chain [5]. Chen, Shen and Zhong (2008) proposed a joint decision model of lot-sizing and pricing, studied the properties of the models solution and developed an effecting algorithm to support the supply chain management [6]. Dobos (2003) assumed that demand is a known continuous function in a given planning horizon as well as the return rate of used items is a given function and analyzed the optimal production and inventory control strategy to minimize the sum of the holding costs, manufacturing cost, remanufacturing cost and disposal cost [7]. Johan, Erik, and Mats (2008) identified seven different types of closed-loop relationships for gathering cores for remanufacturing, and described their own several disadvantages and advantages. By exploring these relationships, a better understanding could be gained about the management of the closed-loop supply chain and remanufacturing [8]. Shen, Hui and Yang (2008) analyzed an inventory system with a traditional forward-oriented flow of material as well as a reverse material flow supply chain, proposed a multi-echelon inventory system with remanufacturing capability to maximize the joint profits of the supplier, the manufacturer, the third-party recycling dealer and the retailer under contractual design [9].

In this paper, we study the coordination problem of the manufacturing /remanufacturing closed-loop supply chain system with manufacturer and retailer, under a decentralized as well as a centralized decision mode. A similar situation is also discussed in [3], but without considering quantity discount. In this paper, the quantity discount strategy is considered. It is assumed that the quantity discount strategy is only used in the process of reclaiming used products. The result shows: the implementation of a quantity discount strategy could optimize the performance of the entire supply chain.

2 The Mathematical Model of Remanufacturing Closed-Loop Supply Chain

The paper studies a closed loop supply system with one manufacturer, one retailer and customers. In the system, information sharing between the manufacturer and the retailer is complete; the activities of product production and marketing are done according to market demand. The manufacturer produce the product and the retailer obtain the product from the manufacturers at a wholesale price. Then, the retailer sells the product to customers at a give selling price. At the same time the manufacturer utilizes retailer for collecting customers' used product and the retailer is reimbursed for a fixed price to transfer its collected used products to the manufacturer. The used product collected can be remanufactured after disassembly and testing. We assume that the manufacturers' production enables the manufacturing of new units simultaneously with remanufacturing of used products. The remanufactured products have the new quality as produced ones and therefore serve as perfect substitutes.

2.1 Model Notation

(1)Parameters

c_m : The unit cost of manufacturing a new product ;

c_n : The unit cost of remanufacturing a returned product into a new one.

σ : Cost saving from remanufacturing $\sigma = c_m - c_n$;

c_s : Retailer's unit selling cost ;

c_t : Retailer's unit cost of collecting a returned unit ;

a : Manufacturer's unit cost of handling a returned unit. ;

d : Customer demand without considering the selling price and sale- effort. ;

k : The proportion of used product that can be remanufactured. ;

e : The quantity of product return without considering recycling price and collection effort ;

α : The Sensitivity coefficient of market demand quantity and sales price.

β : The Sensitivity coefficient of returned product quantity and collection price.

Π_i : The profit function of channel member i (the subscript i will take place M and R, which will denote manufacturer and the retailer respectively)

(2)Decision variable

p : The unit sales price ;

r : The unit collection price of a returned product from the customer to the retailer ;

b : The unit transfer price of a returned product from the retailer to the manufacturer ;

w : The unit wholesale price ;

f : The expected expansion multiple of used products quantity collected by retailer based on quantity discount strategy （$f \geq 1$） ;

2.2 System Assumptions

(1)The manufacturer arranges its production and reproduction in line with market demands. Due to this assumption over-production is not considered in this paper.

(2) The manufacturing process and the remanufacturing process are synchronized. The remanufactured product is the same as the manufactured product, so they can be sold in the same market at the same price.

(3) The market demand for the product is decreasing when the sales price increases or the increment of the sales price increases. For the ease of the study, it's assumed that $Q = d - \alpha p$.

(4) The quantity of the used product collected from customers is increasing with the collection price or the increment of the collection price increasing. So, it's assumed that $U = e + \beta r$.

(5)For all the remanufactured products, its remanufacturing costs are lower than the unit costs of manufacturing a new product. That means, $c_n < c_m$. Besides, the remanufacturing costs of different returned products are identical; simultaneously the unit costs of handling used products for the manufacturer are lower than the cost savings of reproduction of used products and the collection activity of used products is profitable for the manufacturer since the condition $ka < \sigma - b$ holds.

(6)In the optimized design process of remanufacturing closed-loop supply chains, the manufacturer is the Stacklberg leader and the retailer acts as the follower. It is a decentralized decision making process.

(7)Now, the collection quantity is much lower than sales ;

(8)The quantity discount strategy is only considered if the manufacturer collects the used products from the retailer. The strategy has no influence on the wholesale price and the market demand. Besides, the remanufactured closed-loop supply chain is studied in the context of decentralized decision making.

2.3 Model Formulation

The manufacturer's profit function can be stated as:

$$\Pi_M = (w - c_m)Q + (k\sigma - a - b)U \tag{1}$$

In formula (1), the first item is the profit received by product wholesale trade; the second item is the profit received by used product remanufacturing. The decision variables are the unit wholesale price and the unit transfer price of a returned product from the retailer to the manufacturer;

The retailer's profit function is denoted by:

$$\Pi_R = (p - w - c_s)Q + (b - r - c_t)U \tag{2}$$

In (2), the first item denotes the sales revenue; the second item denotes the profit received by collecting used products. The decision variables are the unit sales price and the unit collection price of a returned product from the customer to the retailer.

3 The Solution and Analysis of the Model

3.1 Analysis of Channel Member'S Pricing and Profit without Considering Quantity Discount Strategy

In the leader-follower decision making process of the manufacturer and the retailer, the manufacturer takes the lead in decision making and then the retailer makes its own decision according to the manufacturer's decision. The order of decision making is: at first the manufacturer determines the unit wholesale price and the transfer price of a returned product from the retailer to the manufacturer. Secondly, based on the information about the two prices obtained, the retailer makes the decision about the unit sales price, the collection price of used product from the retailer to customers and the level of sales-effort and collection-effort. By backward induction the retailer's decision at a given unit wholesale price and a given transfer price of a returned product from the retailer to the manufacturer has its optimum at:

$$\begin{aligned}\Pi_R &= (p-w-c_S)(d-\alpha p)+(b-r-c_t)(e+\beta r) \\ &= -\alpha p^2+(c_S+w)\alpha p+dp-wd-c_s d \\ &+(\beta-c_t)\beta r-er-\beta r^2+(b-c_t)e\end{aligned} \tag{3}$$

Under the assumption that parameters w and b are fixed the optimal unit sales price p and the unit collection price of a returned product from the customer to the retailer r can be obtained.

The first derivative of Π_R with respect to p is

$$\frac{\partial \Pi_R}{\partial p} = -2\alpha p + (c_s + w)\alpha + d$$

The second derivative of Π_R with respect to p is

$$\frac{\partial^2 \Pi_R}{\partial p^2} = -2\alpha \tag{4}$$

Obviously, $\frac{\partial^2 \Pi_R}{\partial p^2} = -2\alpha < 0$. Therefore, at the specified values of w and b, the retailer's profit function is strictly convex in p. So the maximum value of the retailer's profit exists in the for p defined interval [0;+∞]. With $\frac{\partial \Pi_R}{\partial p} = 0$, the following equation of the optimal value of p^* is derived as

$$p^* = \frac{c_s + w}{2} + \frac{d}{2\alpha} \tag{5}$$

For the same reason, at the specified value of w and b, the retailer's profit function is strictly convex in r. So the maximum value of the retailer's profit can be found in the defined interval [0, +∞] of variable r. With $\frac{\partial \Pi_R}{\partial r} = 0$, the following equation of the optimal value of r^* is derived as

$$r^* = \frac{b - c_t}{2} - \frac{e}{2\beta} \tag{6}$$

So, the optimal quantity of customer demand, the optimal collected quantity of used products and the optimal retailer's profits are derived as:

$$Q^* = d - \alpha p = \frac{d}{2} - \frac{(c_s + w)\alpha}{2} \quad ; \quad U^* = \frac{e}{2} + \frac{(b - c_t)\beta}{2}$$

$$\Pi_R^* = \left(\frac{d}{2\alpha} - \frac{c_s + w}{2}\right)\left[\frac{d}{2} - \frac{(c_s + w)}{2}\alpha\right] + \left(\frac{b - c_t}{2} + \frac{e}{2\beta}\right)\left[\frac{e}{2} + \frac{(b - c_t)\beta}{2}\right]$$
$$= \frac{(d - c_s\alpha - w\alpha)^2}{4\alpha} + \frac{(b\beta - c_t\beta + e)^2}{4\beta}$$

Because the manufacturer is able to predict that the retailer will make its own decision based on the optimal formulas derived above, the manufacturer's problem could be described as:

$$\Pi_M = (w - c_m)\left[\frac{d}{2} - \frac{(c_s + w)}{2}\alpha\right] + (k\sigma - a - b)\left[\frac{e}{2} + \frac{(b - c_t)\beta}{2}\right]$$
$$= \frac{(d - c_s\alpha + c_m\alpha)w}{2} - \frac{\alpha w^2}{2} + \frac{\alpha c_m c_s - dc_m + (k\sigma - a)(e - c_t\beta)}{2} \tag{7}$$
$$+ \frac{(k\sigma\beta - a\beta - e + c_t\beta)b}{2} - \frac{b^2\beta}{2}$$

The optimal values of w and b can be easily solved in a similar manner. By the analysis of the first derivative and the second derivation of Π_M regarding to w and b separately, Π_M is strictly convex in w and b. So the maximum value of the manufacturer's profit can be found in the defined interval [0;+∞] of the variables w and b. With $\frac{\partial \Pi_M}{\partial b} = 0$ and $\frac{\partial \Pi_M}{\partial w} = 0$ the optimal values of w^* and b^* can be derived as:

$$w^* = \frac{d}{2\alpha} + \frac{c_m - c_s}{2}, \quad b^* = \frac{k\sigma - a + c_t}{2} - \frac{e}{2\beta}$$

3.2 Analysis of Channel Member'S Pricing and Profit without Considering Quantity Discount Strategy

According to the "double marginalization" phenomenon in economics the retailer's optimal decision is not optimal for the manufacturer in most cases. On the assumption that the retailer expands the scale of used products collected, the optimal quantity of used products that the manufacturer can obtain from the retailer can be reached:

$$fU^* = f(e + \beta r^*) = e + \beta r'$$
$$r' = \frac{(f - 1)e}{\beta} + fr^* = \frac{(f - 2)e}{2\beta} + \frac{(b - c_t)f}{2} \tag{8}$$

Without the quantity discount pricing strategy, the retailer's annual profits are relatively low. To encourage the retailer to make more efforts on collecting more used products from its customers, the manufacturer should re-determine the unit transfer price of used products from the retailer to the manufacturer b'. Constituting formula (8) into formula (2), the retailer's profit function is transformed as:

$$\begin{aligned}\Pi_R{}' &= (p-w-c_s)(d-\alpha p)+(b'-r'-c_t)(e+\beta r')\\ &= (p-w-c_s)(d-\alpha p)\\ &+\left[b'f-\frac{(f-1)fe}{\beta}-f^2r^*-c_tf\right](e+\beta r^*)\end{aligned}$$

$$\begin{aligned}&\Pi_R{}'-\Pi_R{}^*\\ &= (b'f-\frac{(f-1)fe}{\beta}-f^2r^*-c_tf-b+r^*+c_r)(e+\beta r^*)\end{aligned}$$

To ensure that the inequality $\Pi_R{}'-\Pi_R\geq 0$ holds, the following condition could be satisfied:

$$b'\geq\frac{(f-1)e}{\beta}+c_t+\frac{b}{2f}-\frac{c_t}{f}+\frac{(b-c_t)f}{2}-\frac{fe}{2\beta}+\frac{c_t}{2f}+\frac{e}{2f\beta}$$

The manufacturer's profit function is transformed as:

$$\Pi_M{}'=(w-c_m)Q+(k\sigma-a-b')fU^*\ ;$$

Let

$$\begin{aligned}F &= \Pi_M{}'-\Pi_M\\ &= (k\sigma-a-b')fU^*-(k\sigma-a-b)U^*\\ &= \begin{bmatrix}k\sigma f-af-\dfrac{f(f-1)e}{\beta}-c_tf\\ -b+c_t-(f^2-1)r^*-k\sigma+a+b\end{bmatrix}U^*\end{aligned}.$$

To ensure that the inequality $\Pi_M{}'-\Pi_M\geq 0$ holds, the following condition could be satisfied:

$$b'\leq\frac{(k\sigma-a)(f-1)+b}{f}\ ;$$

That is to say, $\frac{(k\sigma-a)(f-1)+b}{f}\geq b'\geq\frac{(f-1)e}{\beta}+c_t+\frac{b}{2f}-\frac{c_t}{f}+\frac{(b-c_t)f}{2}-\frac{fe}{2\beta}+\frac{c_t}{2f}+\frac{e}{2f\beta}$

Obviously, F is the function of f, thus the optimal point of the manufacturer's quantity discount can be found.

$$\frac{\partial F}{\partial f} = k\sigma - a - \frac{e}{\beta} - c_t - \frac{ef}{\beta} - bf + c_t f$$

$$\frac{\partial^2 F}{\partial f^2} = -\frac{e}{\beta} - b + c_t < 0$$

The Function is strictly convex in f, so the maximum value of the manufacturer's profit can be found in the defined interval [0; +∞] of variables f. Let $\frac{\partial F}{\partial f}$ equals to zero, the following formula is derived as: $f = \dfrac{k\sigma - a - \frac{e}{\beta} - c_t}{\frac{e}{\beta} + b - c_t}$.

3.3 Comparative Analysis of Pricing with or without Quantity Discount Strategy

Based on the above analysis of the remanufacturing closed-loop supply models in prevail, it can be concluded that considering quantity discount strategy in the process of collecting used products, the unit collection price of a returned product from the customer to the retailer and the unit transfer price of a returned product from the retailer to the manufacturer is higher in the context of the unit sales price and the unit wholesale price unchanged. It is shown in table 1.

Table 1. The comparative analysis of pricing with or without quantity discount strategy

variable	Without quantity discount strategy	With quantity discount strategy
b	$\frac{k\sigma - a + c_t}{2} - \frac{e}{2\beta}$	$\frac{(k\sigma - a)(f-1)+b}{f} \geq b' \geq$ $\frac{(f-1)e}{\beta} + c_t + \frac{b}{2f} - \frac{c_t}{f} + \frac{(b-c_t)f}{2} - \frac{fe}{2\beta} + \frac{c_t}{2f} + \frac{e}{2f\beta}$
r	$\frac{b - c_t}{2} - \frac{e}{2\beta}$	$\frac{(f-2)e}{2\beta} + \frac{(b-c_t)f}{2}$

4 Numerical Example

The Household appliance industry is a major pillar of business in China. Because of the specialty in the end of electrical household appliance lifecycles, a series of environmental pollution problems have to be faced. In our country large consumption expenditures in household appliance make the problems we face more severe. On the background of a domestic electrical appliances corporation's EOL product closed-loop supply chain, a simulation study is carried out. The related parameters are as follows:

The unit cost of manufacturing a new product $c_m = 2.5\times10^3$; the unit cost of remanufacturing a new product $c_n = 1.0\times10^3$; Manufacturer's unit cost of handling a returned unit $a = 1\times10^2$; The proportion of used product that can be remanufactured is $k = 0.6$;Customer demand without considering the selling price and sales-effort $d = 1.1\times10^4$; the quantity of product return without considering recycling price and collection effort $e = 8\times10^2$;the Sensitivity coefficient of market demand quantity and sales price $\alpha = 2.9$;the Sensitivity coefficient of returned product quantity and collection price. $\beta = 40$; Retailer's unit selling cost $c_s = 1\times10^2$; retailer's unit cost of collecting a returned unit $c_t = 5\times10$.

Table 2. Comparative analysis of related variables with or without quantity discount

	F	b	c	Π_R	Π_M
Without quantity discount	1	415	172.5	1779750	3411200
With quantity discount	2	511.25≤b≤607.5	365	≥1779750	≥3411200

If $b' = 5.1125\times10^2$, then the manufacturer's profit gets its maximum value and the retailer's profit equals the same value that occurs when not using the quantity discount strategy. If $b' = 6.075\times10^2$, then the retailer's profit gets its maximum value and the manufacturer's profit equals the same value that occurs when not using the quantity discount strategy. Besides, in the range of b', with the value increasing, the retailer's profit is increasing, but the manufacturer's profit is decreasing. This is shown in table 3:

Table 3. The comparisons of the retailer's profit and the manufacturer's profit with or without quantity discount

b'	5.2×10^2	5.5×10^2	6.0×10^2
Π_R	1914500	2376500	3146500
Π_M	4828000	4366000	3596000

It is shown that: with the unit sales price and the unit wholesale price fixed, the quantity discount strategy, adopted by the manufacturer during the period of collecting the used products from the retailer, can increase the manufacturer's profit and the retailer's profit simultaneously.

5 Conclusions

Based on some relevant researches, this paper conducted a research on the coordination strategy for remanufacturing closed-loop supply chains. By analyzing the changes of the manufacturer's and the retailer's profit, the efficiency of the quantity discount strategy is proved.

At first, a mathematical model of the remanufacturing closed-loop supply chain composed of a manufacturer and a retailer and led by the manufacturer is presented. Secondly, the analysis of the model is conducted. The formulas of p and r that make the retailer get the greatest profit are obtained by differentiating the retailer's profit function separately with respect to p and r. The retailer's greatest profit is derived by substituting the formulas of p and r. Thereinafter the retailer's greatest profit Π_R is derived by substituting the retailer's profit function with p and r.

In the same manner, we substituted the manufacturer's profit function with formula c. the formulas of b and w that make the manufacturer get the optimal profit by differentiating the manufacturer's profit function separately with respect to b and w. Then, the manufacturer's optimal profit Π_M is derived by substituting the manufacturer's profit function with b and w.

By comparison of the changes of the manufacturer's and the retailer's profit and pricing, it is proved that r, b, Π_R and Π_M are all increasing. Finally, the results are confirmed through a numerical example. Furthermore, the numerical results show that the quantity discount strategy could optimize the whole supply chain.

References

1. Wen, M., Liu, C.: To design promoting contract of reverse supply link under anisomerous information. Enterprise Science and Technology & Development 8, 37–40 (2008)
2. Li, X.J., Da, Q.L.: Benefit analysis of closed loop supply chain under remanufacturing condition. Chinese Journal of Mechanical Engineering 44, 170–173 (2008)
3. Ma, D.M.: Research on compatibility mechanism of manufacturing/remanufacturing closed-loop supply chain system. Information Technology & Standardization 3, 42–46 (2008)
4. Zhao, J.S., Zhen, X.T., Wang, H.C.: Reverse-oriented Supply Chain Coordination Mechanism Based on Market-oriented Option Pricing. Journal of Hebei University of Technology 37, 34–38 (2008)
5. Yand, W.H.: Remanufacturing closed-loop supply chain model. Statistics and Decision 6, 171–173 (2008)

6. Chen, M.X., Shen, H.C., Zhong, J.: A Joint Decision Model of Lot-Sizing and the Pricing in Supply Chain Management. Journal of Southeast Univwrsity(Natural Science Edition) 30, 75–80 (2000)
7. Dobos, I.: Optimal production-inventory strategies for a HMMS-type reverse logistics systems. International Journal of Production Economics 81-82, 351–360 (2003)
8. Johan, O., Erik, S., Mats, B.: Importance of closed-loop supply chain relationships for product remanufacturing. Int. J. Production Economics 115, 336–348 (2008)
9. Shen, L.C., Hui, M.W., Yang, P.C.: Optimal policy for a closed-loop supply chain inventory system with remanufacturing. Mathematical and Computer Modeling, 48 (2008)

Towards Compositional Reuse for Software Architecture

Li Xiaojian[1] and Zheng Ying[2]

[1] International Software School, Wuhan University, Hubei, P.R. China
Lxj0713@tom.com
[2] Economics and Management School, Wuhan University, Hubei, P.R. China
Butterfly0601cn@yahoo.com.cn

Abstract. Compositional reuse of software architecture is a difficult problem in SA domain. This paper presents a novel approach to solve this problem. The innovative thought is that we propose a new concept which is called reusable reflective architectural model in which the whole architecture is expressed by metaADL and ADL. What's more, this paper also gives the formal description for the process of compositional reuse.

Keywords: compositional reuse, reflective, metaADL, ADL.

1 Introduction

Compositional reuse is a kind of reuse mode for software architecture which compose existed architectural design resources to form larger granularity architecture[1]. As apposed adaptive reuse, compositional reuse mainly compose existed architecture resource, not adding new component, though in some cases, it is necessary to add new connector.

For implement compositional reuse, we will adopt a reflective[2] method to achieve the goal. The existed architecture is described by some specific ADL language which is named base level architecture, and then, we will extract the meta level architecture based on base level architecture, which is described by MetaADL. The function of base level is to describe how to implement the application logic and functional requirement, it is based on some abstract concepts such as component, connector, and configuration to describe the whole software architecture. Meta level architecture result from base level architecture, using MetaADL to describe the information of base level architecture such as the topology of base level architecture, the meta composition of components, connectors.

For compositional reuse, we will use pi calculus to describe the composition process. Pi calculus is an important concurrent computing model, which is proposed through expanding CCS by Milner and the others[3,4]. Process is a kind of entity which run concurrently. Pi calculus uses name to define channel and the object which is transferred in the channel, each process has a number of channels associated with other processes, the processes interact through their shared channels

J. Luo (Ed.): Affective Computing and Intelligent Interaction, AISC 137, pp. 651–659.
springerlink.com

This paper is organized as follows: the next section is about the whole view of reflective architecture. Section 3 is a case study, and the last section is the conclusion and future work.

2 Overview of Architectural Reuse

At present, there are three basic abstract layers for reuse methods. One is code layer reuse, that is, reusing source code which has implemented some function. This kind of reuse is not very efficient. Another reuse layer is called component reuse. Components encapsulate code module and are placed in a component repository for reuse conveniently. The last layer of reuse is to reuse software system model, that is, the reusable module is the design model of software system. This kind of reuse is really a big-granularity reuse, and we call it architectural reuse.

For the aim of supporting compositional reuse, we can make use of reflective mechanism to construct reusable architectural product. Base level is the existed product. Why we need the meta level? Its main effect include these factors. Firstly, it provide a series meta information supporting architectural reuse. Reuse tool can make use of these meta information to compose architectural product. Secondly, meta level provide a standard mechanism to shield heterogeneity in the base level. Meta level use MetaADL describing the architecture in a consistent manner to ensure the architectural product can be reuse efficiently[5][8].

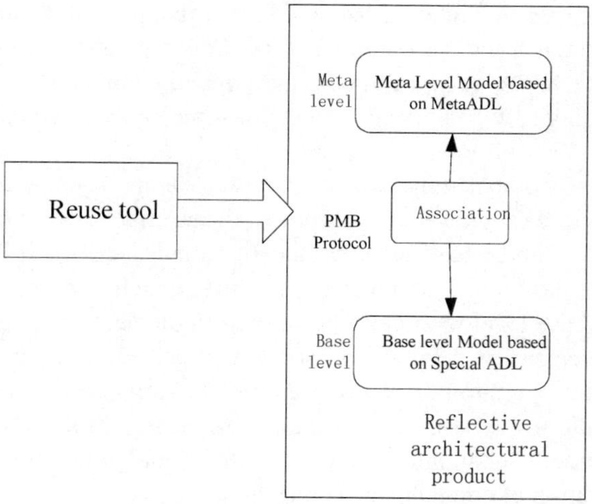

Fig. 1. Reusable reflective architectural model

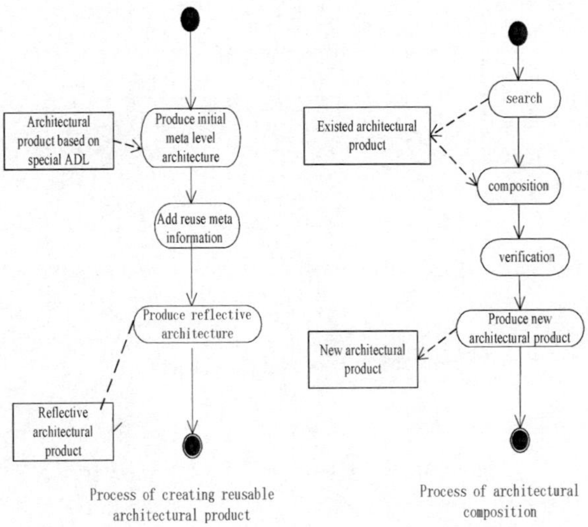

Fig. 2. The process of creating reusable architecture

The process of creating reusable architectural product is like this procedure. Firstly, there is a base level architecture which is described by a specific ADL(such as C2SADL, Wright, etc), the reuse tool can extract some architectural information such as its structure, behavior, constraint, attribute, etc, to construct meta level architecture. Then the base level architecture and the meta level architecture comprise the reusable architectural product and are stored in the reusable architectural product library; the process of architectural composition is like this procedure. Firstly, we can search the existed architectural product which is satisfied with our requirement. Secondly, we can compose these architectural products, the composition should proceed at the compositional point of architecture. Thirdly, we should verify the correctness of new architecture. The main method of verification include constraint check and behavior verification.

3 Case Study

3.1 Case

Here are two architectures. One is for bank accounting system and one is for transaction system. Now we want to get a whole system for bank business. So we can compose the two architecture like the following.

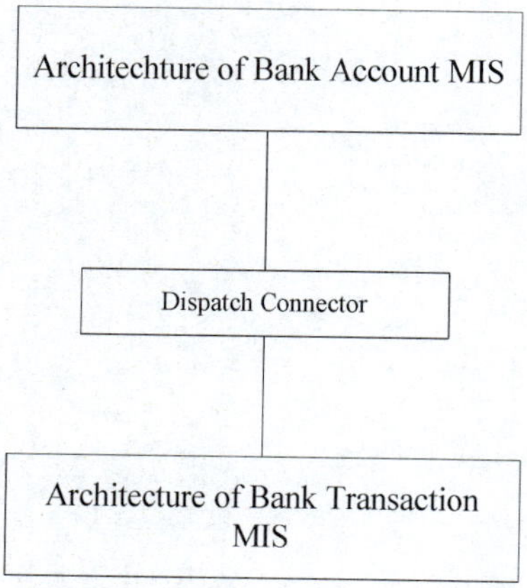

Fig. 3. Architecture of bank business system

For connect these two architecture, a connector should be added. The procedure is like the following. Firstly, we can add a metaConnector in the target meta level architecture, it is as follows[6,7].

```
<metaConnector id="MdispatchConn">
        <structure changeable="true" name="dispatchConn" typeName="compositeConn">
            <interface id="dispatchConn.interface01" name="top"/>
            <interface id="dispatchConn.interface02" name="bottom"/>
        </structure>
        <constraint >
                <name>Conn.constraint01</name>
                <type>invariant</type>
                <expression>(self.interfaces.size = 2) and
          (exists i in self.interfaces implies (i.name = top)) and
          (exists i in self.interfaces implies (i.name = bottom))
          </expression>
        </constraint>
        <property id="property01">
                <name>filter_policy</name>
                <type>enum</type>
                <value>no_filtering</value>

        </property>
</metaConnector>
```

Fig. 4. metaConnector in the target metaADL

The connector's name is dispatch, and the type of the connector is compositeConn. The others property are listed in the metaADL. Secondly, it is necessary to modify the base level. Such as, to add a compositeConn connector type message_filter no_filtering and a connector_instance: dispatchConn:CompositeConn; thirdly, in the target MetaADL, it is necessary to add two links which is described as follows:

```
<link id="comopositelink01">
        <point>
                elementName>BM</elementName>
                        <interfaceName>interface</interfaceName>
        </point>
        <point>
                <elementName>dispatchConn</elementName>
                        <interfaceName>top</interfaceName>
        </point>
</link>
<link id="comopositelink02">
        <point>
                <elementName>TM</elementName>
                        <interfaceName>interface</interfaceName>
        </point>
        <point>
                <elementName>dispatchConn</elementName>
                        <interfaceName>bottom</interfaceName>
        </point>
</link>
```

Fig. 5. Two link elements added in the target metaADL

And the fourth step is to add a connector in the target CWSADEL which is described by the sentence: connector dispatchConn {top BM; bottom TM}, finally, we can get the base level architecture of bank system like the following:

```
architecture  BankApplicationSystem is {
    component_types {
        component Clock is extern {AccountManager/Clock.c2;}
        component SavingAccount is extern { AccountManager/SavingAccount.c2; }
        component BusinessAccount is extern {AccountManager/BusinessAccount.c2;}
        component CheckingAccount is extern {AccountManager/CheckingAccount.c2;}
        component PeopleADT is extern { AccountManager/PeopleADT.c2; }
        component BusinessManager is extern { AccountManager/BusinessManager.c2;}
        component TranscationManager is extern {TranscationManager/TranscationManager.c2}
        component ManagerLogin is extern {TranscationManager/ManagerLogin.c2}
        component TellerLogin is extern {TranscationManager/tellerLogin.c2}
        component CustomerLogin is extern {TranscationManager/CustomerLogin.c2}
        component ManangerArtist is extern{TranscationManager/ManangerArtist.c2}
        component TellerArtist is extern{TranscationManager/TellerArtist.c2}
        component CustomerArtist is extern {TranscationManager/CustomerArtist.c2}
        component GraphicsBinding is virtual {}
    }
    connector_types{
        connector FilteringConn is { message_filter message_filtering;}
        connector RegularConn is {message_filter no_filtering;}
        connector CompositeConn is {message_filter no_filtering;}
    }
    architectual_topology{
        component_instances {
          SavAccount:SavingAccount; CheAccount:CheckingAccount;
          BusAccount:BusinessAccount; SimClock:Clock;
          People:PeopleADT;  BM:BusinessManager;
          TM:TranscationManager; TLogin:TellerLogin;
          MLogin:ManagerLogin; CLogin:CustomerLogin;
          ManagerArt:ManangerArtist;  CustomerArt:CustomerArtist;
          TellerArt:TellerArtist; TellerTk:GraphicsBinding;
          ManagerTk:GraphicsBinding;  CustomerTk:GraphicsBinding;
        }
        connector_instances {
           ADTConn:FilteringConn;
           transcationConn:FilteringConn; managerLoginConn:FilteringConn;
           tellerLoginConn:FilteringConn; tellerTkConn:RegularConn;
           customerLoginConn:FilteringConn; customerTkConn:RegularConn;
           managerTkConn:RegularConn; dispatchConn:CompositeConn;
        }
        connections{
           connector ADTConn
              {top SavAccount,CheAccount,People,BusAccount,SimClock; bottom BM;}
           connector transcationConn
              {top TM; bottom TLogin,Clogin,MLogin;}
           connector tellerLoginConn{top TLogin;bottom TellerArt;}
           connector customerLoginConn{top CLogin;bottom CustomerArt;}
           connector managerLoginConn {top Mlogin;bottom ManagerArt;}
           connector tellerTkConn{top TellerArt;bottom TellerTk;}
           connector customerTkConn{top CustomerArt;bottom CustomerTk;}
           connector managerTkConn {top ManagerArt; bottom managerTk; }
           connector dispatchConn {top BM; bottom TM;}
        }
     }
}
```

Fig. 6. Target architecture description in base level

3.2 Formal Description

For this case, we can make use of pi calculus to describe the compositional reuse process. Firstly, a pi calculus model was established as follows:

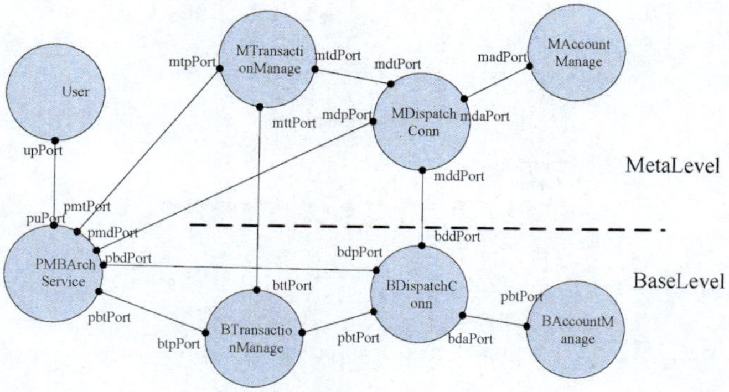

Fig. 7. pi calculus model for composition

(1) The creation of process

After user submit the composition requirement, PMBArchService will create process *MTransactionManage* in the meta level and *BTransactionManage* in the base level, and what' s more, a connector *DispatchConn* will be created.

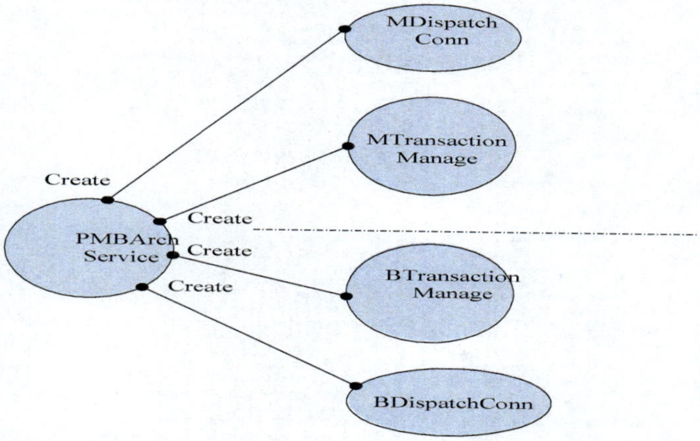

Fig. 8. The creation of process

(2) User process

Architectural Designer send the messages of adding composite *TransactionManage* and connector *dispatchConn* to PMB interface.

$$User(fn) = puPort < addTMRequest > .upPort(msgs). [msg = addTMRequestAck]\overline{puPort} < addDCRequest > . User(fn)$$

(3) PMBArchService process

PMBArchService process mainly describes the process: receive the requirement from user, and then add transaction management meta composite in the meta level and transaction management composite in the base level, if it is finished successfully, then add connector *dispatchConn* and then add the links between the connector and these two composites.

$$PMBArchService(fn) = puPort(addTMRequest).\overline{mtpPort} < addTM > .pmtPort(msg) [msg = addBaseTMAck].(puPort(addDCRequest). \overline{mdpPort} < addDC > .pmdPort(msg)[msg = addBaseDCAck] (puPort(addTDLinkRequest).mtpPort < addTDLink > \mid puPort(addADLinkRequest).pmaPort < addADLink >). pmtPort(msg).pma(msg')[msg = addBaseTDLinkAck] [msg' = addBaseADLinkAck].\overline{upPort} < addTMRequestAck > . \overline{upPort} < addDCRequestAck > .\overline{upPort} < addTDLinkRequestAck > \overline{upPort} < addADLinkRequestAck > PMBArchService(fn)$$

(4) MTransactionManage process

After meta level component process receive the message of adding composite, it will create a verify process to check the constraint is satisfied or not, if not, then it will enter the state of waiting. If yes, then meta level process send notification message for adding component and link.

(5) BTransactionManage process

The base level process *BTransactionManage* receive the message of adding composite from meta level process, and then start an internal process, to send a success message for adding the base level component *TransactionManage* and links.

$$BTransactionManage(fn) = bttPort(addBTNotify).\tau. (\overline{pbtPort} < addBaseBTAck > .\overline{pbtPort} < addBTLinkAck > . BTransactionManage(fn)$$

(6) MDispatchConn process

MDispatchConn process receive the message of adding connector, then set value for meta object, and create a verify process to check whether the constraint is satisfied.

$$MDispatchConn(fn)$$
$$= mdport(addDC).\overline{Create} < Verify > .\tau.(MDispatchConn(fn) +$$
$$\overline{bddPort} < addDCNotify > .$$
$$MDispatchConn(fn)$$

(7) BDispatchConn process

Base level process *BDispatchConn* receive the message of adding connector from meta level, then have a internal process, and send a success message to PMBArchService.

$$BDispatchConn(fn)$$
$$= bddPort(addDCNotify).\tau.$$
$$(\overline{pbdPort} < addBaseDCAck > .$$
$$BDispatchConn(fn)$$

4 Conclusion and Future Work

Compositional reuse is relatively a difficult topic in software architecture domain. Rudolf K. Keller[1] introduce the notion of design components, reified design patterns fit for software composition. Introducing patterns to software composition is a good idea, but It can not address the compositional problem when the software architecture is described by a specific ADL. Juan Carlos Ruiz[2] applied the reflection mechanism in the software architecture very well, but the aim is to address runtime problem for SA, in contrast, our approach is to presents a new approach to address the compositional reuse in the design phase of software architecture. So it is more adaptable for general compositional reuse problem for software architecture. Our future work mainly include developing a full tool set for compositional reuse and formal verification.

References

1. Keller, R.K., Schauer, R.: Design Components: Towards Software Composition at the Design Level. In: 20th International Conference on Software Engineering, ICSE 1998 (1998)
2. Ruiz, J.C., Killijian, M.-O.: Reflective Fault Tolerant Systems from Experience to Challenges. IEEE Transactions on Computers 52(2) (February 2003)
3. Milner, R., Parrow, J., Walker, D.: A calculus of mobile processes, I. Information and Computation 100(1), 1–40 (1992)
4. Milner, R., Parrow, J., Walker, D.: A calculus of mobile processes, II. Information and Computation 100(1), 41–77 (1992)
5. Dowling, J., Cahill, V.: The K-Component Architecture Meta-model for Self-Adaptive Software. In: Matsuoka, S. (ed.) Reflection 2001. LNCS, vol. 2192, pp. 81–88. Springer, Heidelberg (2001)

6. Shaw, M., DeLine, R., Klein, D.V., Ross, T.L., Young, D.M., Zelesnik, G.: Abstractions for Software Architecture and Tools to Support Them. IEEE Transactions on Software Engineering 21(4), 314–335 (1995)
7. Dashofy, E.M., Hoek, A.v.d., Taylor, R.N.: An Infrastructure for the Rapid Development of XML based Architecture Description Languages. In: 24th International Conference on Software Engineering (ICSE 2002), Orlando, Florida (2002)
8. Ying, S., Li, X., Wang, J., Zheng, Y.: Reuse-oriented Software Architecture Design. In: 2007 International Symposium on Distributed Computing and Applications to Business, Engineering and Science, Yichang, Hubei, P.R.C, August 14-17 (2007)

Semantic Information Retrieval Based on RDF and Fuzzy Ontology for University Scientific Research Management

Jun Zhai, Meng Li, and Jianfeng Li

School of Management, Dalian Maritime University,
Dalian, China
zhaijun_dlmu@yahoo.com.cn

Abstract. Ontology-based semantic retrieval is a hotspot of current research. In order to achieve fuzzy semantic retrieval for university scientific research information in WWW, this paper proposes an approach using Resource Description Framework (RDF) and fuzzy ontology. First, we apply RDF data model to represent scientific research information. Then, introducing new data type referred as fuzzy linguistic variables to RDF data model, the semantic query expansion in SPARQL query language is constructed by order relation, equivalence relation and inclusion relation between fuzzy concepts defined in linguistic variable ontologies. Examples show that this work facilitates the semantic retrieval through fuzzy concepts for university scientific research management on the Semantic Web.

Index Terms: semantic information retrieval, RDF, SPARQL, fuzzy ontology, the Semantic Web.

1 Introduction

Information retrieval (IR) is important work for university scientific research management in WWW. Current information retrieval on the web is based primarily on keywords which often cause problems in precision and recall. Ontology-based semantic retrieval is a hotspot of current research. Ontology is a conceptualization of a domain into a human understandable, machine-readable format consisting of entities, attributes, relationships, and axioms [1]. It is used as a standard knowledge representation for the Semantic Web [2]. The use of ontologies to overcome the limitations of keyword-based search has been put forward as one of the motivations of the Semantic Web [3]-[4].

However, the conceptual formalism supported by typical ontology may not be sufficient to represent uncertainty information commonly found in many application domains due to the lack of clear-cut boundaries between concepts of the domains. Moreover, fuzzy knowledge plays an important role in many domains that face a huge amount of imprecise and vague knowledge and information, such as text mining, multimedia information system, medical informatics, machine learning, and human natural language processing [5].

J. Luo (Ed.): Affective Computing and Intelligent Interaction, AISC 137, pp. 661–668.
springerlink.com

To handle uncertainty of information and knowledge, one possible solution is to incorporate fuzzy theory into ontology. Then we can generate fuzzy ontologies, which contain fuzzy concepts and fuzzy memberships. Lee et al. [6] proposed an algorithm to create fuzzy ontology and applied it to news summarization. Tho et al. proposed a Fuzzy Ontology Generation Framework (FOGA) for fuzzy ontology generation on uncertainty information [7]. This framework is based on the idea of fuzzy theory and Formal Concept Analysis (FCA). Kang et al. [8] proposed a new fuzzy extension of description logics called the fuzzy description logics with comparison expressions (FCDLs). Lau presented a fuzzy domain ontology for business knowledge management [9]. Zhai et al. [10] presented a fuzzy ontology model using intuitionistic fuzzy set to achieve fuzzy semantic retrieval for ITS.

There a lot of fuzzy information derived from human factor. To achieve fuzzy semantic retrieval for scientific research management, this paper proposes an approach using Resource Description Framework (RDF) and fuzzy ontology. The rest of this paper is organized as follows: Section 2 gives scientific research information model with RDF. Section 3 introduces fuzzy linguistic variable ontology models. Section 4 studies the semantic retrieval in SPARQL query language. Finally, section 5 concludes the paper.

2 Scientific Research Information Model Using RDF

The idea of Semantic Web came from Tim Bemers-Lee in his vision to move the web into a new generation, where the web resources are annotated with meaning in a form that machines can understand [2]. This will open up vast opportunities for automated processing of the rich knowledge resources available on the web to applications in information search and filtering, knowledge mining, coordination and collaborative processing by intelligent agents.

The Semantic Web is to be realized through a shared infrastructure consisting of languages and tools for knowledge representation and processing. The basic knowledge representation format is the Resource Description Framework (RDF) and RDF Schema (RDFS) [11].

RDF provides a data model that supports fast integration of data sources by bridging semantic differences. To achieve semantic retrieval for scientific research management, it is necessary to represent scientific research information through RDF model.

The RDF data model vaguely resembles an object-oriented data model. It consists of entities, represented by unique identifiers, and binary relationships, or statements, between those entities. In a graphical representation of an RDF statement, the source of the relationship is called the subject, the labeled arc is the predicate (also called property), and the relationship's destination is the object. Both statements and predicates are first-class objects, which means they can be used as the subjects or objects of other statements.

The RDF data model distinguishes between resources, which are object identifiers represented by URIs, and literals, which are just strings. The subject and the predicate of a statement are always resources, while the object can be a resource or a literal. In RDF diagrams, resources are always drawn as ovals, and literals are drawn as boxes.

Fig.1 shows an example of RDF data model graph for project information, which means as following:

(1) The resource "ex: project No. P101" has a property "ex: including" the value of which is the resource "ex: researcher No. R0088".
(2) The resource "ex: project No. P101" has a property "ex: size" with value "very big" (a literal and a fuzzy concept).
(3) The resource "ex: project No. P101" has a property "ex: type" with value "application research" (a literal and a fuzzy concept).
(4) The resource "ex: researcher No. R0088" has a property "ex: research ability" with value "strong" (a literal and a fuzzy concept).
(5) The resource "ex: researcher No. R0088" has a property "ex: age" with value "middle-aged" (a literal and a fuzzy concept).

The part of RDF statements is as following:

```
 <rdf: Description  rdf: ID= "ex: project No. P101">
        <ex: including   rdf: ID= "researcher No.
R0088">
             <ex: age> middle-aged </ex:age>
  <ex: research ability > high </ex: research
ability >
        </ex: including >
        <ex: size> very big </ex: size>
        <ex: type> application research </ex: type>
     </rdf: Description>
```

A mass of scientific research information is distributed among the Semantic Web. So it is meaning and difficult to retrieve relevant information.
As shown in Fig. 1, people often use fuzzy concepts to describe entity property. But, due to the lack of relationships between fuzzy concepts in current RDF model, it is difficult to search information at semantic level of fuzzy concepts. Consequently, we propose the fuzzy linguistic variables ontology models.

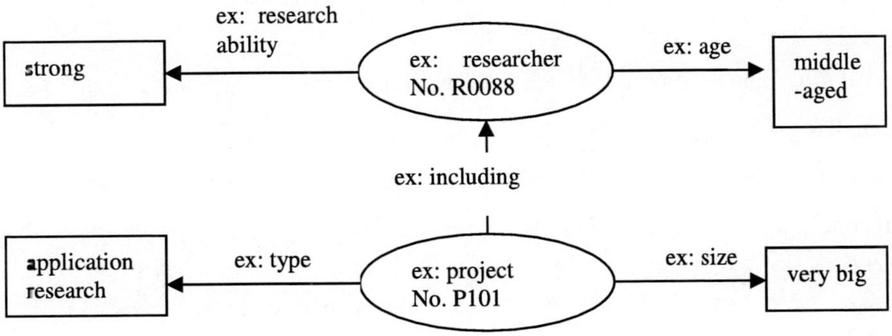

Fig. 1. RDF data model graph for scientific research information

3 Fuzzy Linguistic Variable Ontology

The fuzzy linguistic variables proposed by Zadeh are the basic of fuzzy knowledge and fuzzy system. To achieve the knowledge share and reuse for fuzzy systems on the semantic web, it is necessary to represent the fuzzy linguistic variables with ontology.

Definition 1. (Fuzzy linguistic variable ontology) – A fuzzy linguistic variable ontology is a 6-tuple $O_F = (c_a, C_F, R, F, S, U)$, where:

(1) c_a is a concept on the abstract level, e.g. "age", "size" etc.

(2) C_F is the set of fuzzy concepts which describes all values of c_a.

(3) $R = \{r \mid r \subseteq C_F \times C_F\}$ is a set of binary relations between concepts in C_F.

A kind of relation is set relation $R_S = \{$inclusion (i.e. $\subseteq$), intersection, disjointness, complement$\}$, and the other relations are the order relation and equivalence relation $R_O = \{\leq, \geq, =\}$. C_F and an order relation r compose the ordered structure $< C_F, r >$.

(4) F is the set of membership functions at U, which is isomorphic to C_F.

(5) $S = \{s \mid s : C_F \times C_F \rightarrow C_F\}$ is a set of binary operators at C_F. These binary operators form the mechanism of generating new fuzzy concepts. Basic operators are the "union", "intersection" and "complement" etc., i.e. $S = \{\vee, \wedge, \neg, \cdots\}$. C_F and S compose the algebra structure $< C_F, S >$.

(6) U is the universe of discourse.

To simplify the transform from fuzzy linguistic variables to fuzzy ontology, we introduce the basic fuzzy ontology model as follows.

Definition 2. (Basic fuzzy ontology) – A basic fuzzy ontology is a 4-tuple $O_F = (c_a, C_F, F, U)$, where c_a, C_F, F, U have same interpretations as defined in definition 1, which satisfy the following conditions:

(1) $C_F = \{c_1, c_2, \cdots, c_n\}$ is a limited set.

(2) Only one relation of set, the relation of disjointness, exists in C_F, and C_F is complete at U. In the other words, C_F is a fuzzy partition of U.

(3) C_F has an ordered relation $\leq$, and $< C_F, \leq >$ is a complete ordered set, i.e. all concepts in C_F constitute a chain $c_1 \leq c_2 \leq \cdots \leq c_n$.

(4) F is optional element of ontology.

The main fuzzy linguistic variable ontologies used in the next section are as following:

O1= (age, {old, middle-aged, midlife, youth, youngster, adult , ...});

O2= (research ability of researcher, {very weak, weak, neutral, strong, excellent});

O3= (project size, {very small, small, middle, medium, big, very big, huge});
O4= (project type, {basic research, application research, development research}).
There is a lot of semantic relation between fuzzy concepts. For instance:

"middle-aged" = "midlife", "old" ⊆ "adult", "middle-aged" ⊆ "adult", "youth" ⊆ "adult";
"very weak" ≤ "weak" ≤ "neutral" ≤ "strong" ≤ "excellent";
"very small" ≤ "small" ≤"middle (medium)" ≤ "big" ≤"very big" ≤ "huge";
reversion ("week") = "strong", reversion ("small") = "big", etc.
The part of RDF statements to represent these ontologies is as following:

```
<rdf: Description  ID= "age">
   <t: values> <rdf: Seq>
        <rdf: li  resource= "#youngster"/>
     <rdf: li  resource= "#youth"/>
          <rdf: li  resource= "#midlife"/>
      <rdf: li  resource= "#old"/> </rdf: Seq>
      <rdf: li  resource= "#middle-aged"/>
      <rdf: li  resource= "#old"/>
</t: values> </rdf: Description>
```

4 Semantic Retrieval for Scientific Research Information in Sparql

Semantic retrieval or conceptual search, i.e., search based on meaning rather than just character strings, has been a hotspot in the information retrieval (IR) field along with the Semantic Web. One way to view a semantic search engine is as a tool that gets formal ontology-based queries (e.g., in SPARQL, SeRQL, RDQL, RQL etc.) from a client, executes them against a knowledge base (KB), and returns tuples of ontology values that satisfy the query.

The SPARQL query language for RDF provides Semantic Web developers with a powerful tool to extract information from large datasets [12]-[13]. The SPARQL has been proposed by the World Wide Web Consortium (W3C) and has recently achieved the recommendation status [14].

Using the following query statements to RDF graph shown in Fig.1, the return result will be "ex: project No. P101":

```
SELECT ?project
FROM RDF dataset
WHERE { ?project ex: size ?size.
 FILTER (?size = "very big")}
```

The search engine handle "very big" as an ordinary variable with data type "literal".

Current RDF and SPARQL cannot handle the fuzzy concepts as values of properties and do not support fuzzy semantic retrieval. To achieve fuzzy semantic retrieval, "size", "age" etc. are considered as a fuzzy linguistic variable and their values are defined in ontologies "O3", "O1" etc (see previous section). For instance, we can retrieve "project" information through property "size" by the query statements:

```
SELECT ?project
FROM RDF dataset
WHERE { ?project ex: size ?size.
 FILTER (?size ≤ "very big")}
```

Using the "order relation" defined in fuzzy linguistic variable ontology: "very small"≤"small"≤"middle (medium)" ≤ "big" ≤"very big" ≤ "huge", the search engine can transform the above query statements to:

```
SELECT ?project
FROM RDF dataset
WHERE { ?project ex: size ?size.
 FILTER (?size = "very big " OR
     ?size = "big" OR
     ?size = "middle" OR
     ?size = "medium" OR
     ?size = "small" OR
     ?size = "very small" )}
```

For the query statements:

```
SELECT ?project
FROM RDF dataset
WHERE { ?project ex: including ?researcher.
     ?researcher ex: age ?age.
 FILTER (?age = "middle-aged")}
```

Using equivalence relation: "middle-aged"= "midlife", we can transform the above query statements to:

```
SELECT ?project
FROM RDF dataset
WHERE { ?project ex: ex: including ?researcher.
     ?researcher ex: age ?age.
 FILTER (?age = "middle-aged" OR
     ?age = "midlife")}
```

For the query statements:

```
SELECT ?project
FROM RDF dataset
WHERE { ?project ex: including ?researcher.
     ?researcher ex: age ?age.
 FILTER (?age = "adult")}
```

Using inclusion relation: "old" $\subseteq$ "adult", "middle-aged" $\subseteq$ "adult" and "youth" $\subseteq$ "adult", we can transform the above query statements to:

```
SELECT ?project
FROM RDF dataset
WHERE { ?project ex: including ?researcher.
     ?researcher ex: age ?age.
 FILTER (?age = "adult" OR
     ?age = "old" OR
     ?age = "middle-aged" OR
     ?age = "midlife" OR
     ?age = "youth")}
```

Obviously, extended query statements can return all results which satisfy research requirement at semantic level without upgrading current main search algorithm. Based on other semantic relations defined in fuzzy ontology, we can create and complete more complex semantic retrieval, which will be described in future work.

5 Conclusion

The Semantic Web is a scheme to extend the current Web from documents linked to each other, into a Web that recognizes the meaning of information in these documents. Ontology-based semantic information retrieval is one of the motivations of the Semantic Web.

To achieve fuzzy semantic retrieval for scientific research information, this paper proposes an approach using RDF and fuzzy ontology. We have applied RDF data model to represent scientific research information on the Semantic Web. Then we have presented fuzzy linguistic variable ontology. Introducing new data type referred as fuzzy linguistic variables to RDF data model, the semantic query expansion in SPARQL query language is constructed by semantic relations between fuzzy concepts. Examples show that the semantic retrieval through fuzzy concepts can be achieved.

Acknowledgment. This work is partially supported by the National Natural Science Foundation of China under Grant No. 70801007, the Research Fund for the Ph.D. Programs Foundation of Ministry of Education of China under Grant No.20070151022, Key Project of the Ministry of Education under Grant No.209030, the National Key Technology R&D Program during the eleventh five-year plan period under Grant No. 2009BAG13A03, and the Fundamental Research Funds for the Central Universities under Grant No. 2009QN078, 2009QN084.

References

1. Fensel, D., van Harmelen, F., Horrocks, I., McGuinness, D.L., Patel-Schneider, P.F.: OIL: an Ontology Infrastructure for the Semantic Web. IEEE Intelligent Systems 16(2), 38–45 (2001)
2. Berners-Lee, T., et al.: The Semantic Web. Scientific American 284(5), 34–43 (2001)
3. Castells, P., Fernandez, M., Vallet, D.: An Adaptation of the Vector-space Model for Ontology-based Information Retrieval. IEEE Transactions on Knowledge and Data Engineering 19(2), 261–272 (2007)
4. Li, J., Song, J.Y., Zhong, H.: Ontology-based Query Division and Reformulation for Heterogeneous Information Integration. Journal of Software 18(10), 2495–2506 (2007) (in Chinese)
5. Lukasiewicza, T., Stracciab, U.: Managing Uncertainty and Vagueness in Description Logics for the Semantic Web. Journal of Web Semantics 6(4), 291–308 (2008)
6. Lee, C.S., Jian, Z.W., Huang, L.K.: A Fuzzy Ontology and its Application to News Summarization. IEEE Transactions on Systems, Man and Cybernetics (Part B) 35(5), 859–880 (2005)

7. Tho, Q.T., Hui, S.C., Fong, A.C.M., Cao, T.H.: Automatic Fuzzy Ontology Generation for Semantic Web. IEEE Transactions on Knowledge and Data Engineering 18(6), 842–856 (2006)
8. Kang, D.Z., Xu, B.W., Lu, J.J.: Description Logics for Fuzzy Ontologies on Semantic Web. Journal of Southeast University (English Edition) 22(3), 343–347 (2006)
9. Lau, R.Y.K.: Fuzzy Domain Ontology Discovery for Business Knowledge Management. IEEE Intelligent Informatics Bulletin 8(1), 29–41 (2007)
10. Zhai, J., Cao, Y., Chen, Y.: Semantic Information Retrieval Based on Fuzzy Ontology for Intelligent Transportation Systems. In: Proceedings of the 2008 IEEE International Conference on Systems, Man, and Cybernetics (SMC 2008), Singapore, pp. 2321–2326 (2008)
11. W3C, Resource description framework (RDF): concepts and abstract syntax. In: Klyne, G., Carroll, J.J., McBride, B. (eds.), W3C Recommendation (February 10, 2004), `http://www.w3.org/TR/2004/REC-rdf-concepts-20040210/`
12. Pérez, J., Arenas, M., Gutierrez, C.: Semantics and Complexity of SPARQL. In: Cruz, I., Decker, S., Allemang, D., Preist, C., Schwabe, D., Mika, P., Uschold, M., Aroyo, L.M. (eds.) ISWC 2006. LNCS, vol. 4273, pp. 30–43. Springer, Heidelberg (2006)
13. Angles, R., Gutierrez, C.: The Expressive Power of SPARQL. In: Sheth, A.P., Staab, S., Dean, M., Paolucci, M., Maynard, D., Finin, T., Thirunarayan, K. (eds.) ISWC 2008. LNCS, vol. 5318, pp. 114–129. Springer, Heidelberg (2008)
14. Prud'hommeaux, E., Seaborne, A.: SPARQL Query Language for RDF. W3C, 20080115 (2008), `http://www.w3.org/TR/rdf-sparql-query/`

Fuzzy Comprehensive Evaluation Method on Flood-Control Risk Consequence Assessment of Bridge

Hongwei Wang[1], Changyu Hu[2], and Jianxing Yu[3]

[1] School of Business, Tianjin Polytechnic University, Tianjin, China
hongwei_hd@sina.com
[2] Group 2nd Engineering Co., Ltd., China Railway 16 Bureau, Tianjin, China
362100291@qq.com
[3] School of Civil Engineering, Tianjin University, Tianjin, China
yjx2000@eyou.com

Abstract. In order to solve the quantization calculation problem of the flood-control risk consequence losses of bridge, fuzzy comprehensive evaluation method, fuzzy random theory and other relative discipline theories are applied comprehensively in this paper. Under the existing situation of river course administration and on the basis of flood-control influence assessment in China, the flood-control risk consequence losses of bridge is calculated semi-quantitatively after confirming the membership function, by establishing the quantizing criteria of flood-control risk consequence losses and the corresponding evaluation relationship among the flood-control risk consequences, severity of risk consequences and severity value. Finally its adaptability and reliability is verified through a project example.

Index Terms: bridge, flood-control risk consequence, fuzzy comprehensive evaluation method, membership function.

1 Introduction

With the economic development and social progress, the flood disaster is increasing year by year in the world. Especially, various constructions such as bridge and landscaping works over the rivers through the city are increasing, and consequently lay some huge hidden risks on flood-control in China.

In view of the existing situation on river course administration and on the basis of flood-control influence assessment in China, an overall flood-control risk assessment conception is put forward here and these risks of bridge over the river are divided into eight kinds of risk factors: risk of layout standard, risk of flood-discharge ability, risk of river flows stability, risk of water prevention ability of dike, risk of flood fighting, risk of anti scour, risk of rights and interests of water affairs towards the third party and other risks[1] [2].

On the basis of the knowledge about fuzzy comprehensive evaluation method, fuzzy random theory and many other relative disciplines theory, the flood-control risk consequence of bridge is calculated here semi-quantitatively, aiming to develop the research on risk management and hydraulic construction fields.

J. Luo (Ed.): Affective Computing and Intelligent Interaction, AISC 137, pp. 669–676.
springerlink.com

2 Flood-Control Risk Consequences Losses Assessment

2.1 Calculation Method

The losses caused by flood-disaster risk consequence mainly include the losses of life, economic losses and rescue losses[3], as shown in Fig. 1:

2.1.1 Loss of Life（LOL）

Currently, the methods to calculate the losses of life by the flood risks mainly include experience analysis and statistics analysis, but both are not mature. Based on studying the relative theories in the world, this paper establishes the following simplified formula for calculating the losses of life by the flood risks.

$$\boldsymbol{LOL} \approx \mathbf{0.075}\boldsymbol{PAR}^{0.56} \cdot \exp[-\mathbf{0.759}\boldsymbol{WT} + (\mathbf{3.790} - \mathbf{2.223}\boldsymbol{WT})\boldsymbol{FC}] \tag{1}$$

In (1), *LOL* indicates the losses of life, *PAR* indicates the total population in the risk district, *WT* indicates the pre-warning time, and *FC* indicates the flood intensity.

2.1.2 Economic Losses (LOS1)

There are many methods to evaluate the economic losses now, generally including direct losses and indirect losses.

The direct economic losses mainly refer to the flood damage for infrastructural facilities and other properties, while indirect losses is generally evaluate by coefficient method:

$$\boldsymbol{C}_i = \sum_{i=1}^{n} \boldsymbol{K}_i \times \boldsymbol{C}_{Di} \tag{2}$$

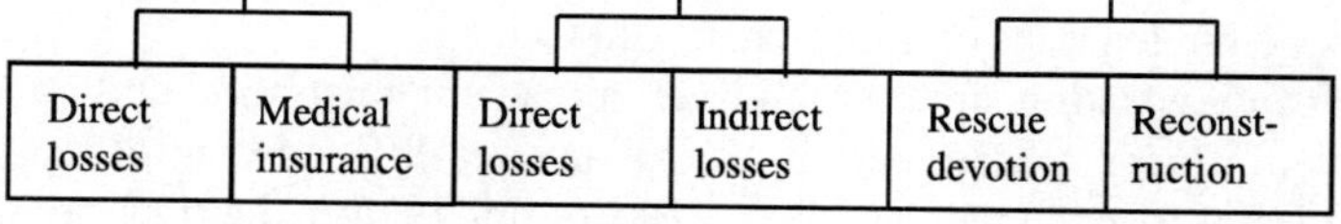

Fig. 1. Content of risk consequence losses on flood-control

In (2), C_i is indirect losses, unit: 10 thousand RMB, C_{Di} is direct losses of the *i* industry, unit: 10 thousand RMB, K_i indicates the calculation coefficient and varies with the different districts, economic level, industry, the disaster area and degree. If necessary, the factors such as the flood scope, the flood submerging depth and the flood duration can be considered and a database can be built, *n* is the total of the industries or departments in the submerging area.

2.1.3 Rescue Losses (LOS2)

Rescue losses, include the costs of rescue on the spot, the expenses, the fines, the legal fare, and compensation.

Table 1. Flood-Control Risk Consequences Quantizin

Classification / Relative values	The losses of life K_{LOL}	Property losses K_{LOS1}	Rescue losses K_{LOS2}
0～25	0 (person)	<10^4 (RMB)	Small
25～50	<10 (persons)	10^4～10^6 (RMB)	normal
50～75	10～10^2 (persons)	10^6～10^9 (RMB)	high
75～100	>10^2 (according to10^2～10^4) (persons)	>10^9 (according to10^9～10^{10}) (RMB)	higher

2.2 The Quantitative Method of Flood-Control Risk Consequence

According to the method above, referring to the management regulations of related section, and on the basis of suggestions of related experts, flood-control risk consequences are quantized here[4] , as shown in Table 1.

K_{LOL} is the index score of loss of life, K_{LOS1} is the index score of economic losses, K_{LOS2} is the index score of rescue losses.

2.3 Severity Index of Flood-Control Risk Consequence

In order to calculate conveniently in evaluation, the severity index of flood-control risk consequences, C_f, is used to represent three kinds of losses mentioned above. Obviously, this index is a relative economic indicator.

The above analysis concludes that three factors to affect flood-control risk consequences are associated. According to System Analysis Method, the cross connection of the three factors can be proved. As a result, the calculation method of severity index of flood-control risk consequences, C_f, through comprehensive analysis can be illustrated as following:

$$C_f = \sqrt[3]{k_{LOL} k_{LOS1} k_{LOS2}} \tag{3}$$

Table 2. Severity Standards of Risk Consequences

Consequence level	Type of risk	Severity of risk consequences	Severity values
A	Minor C_1	Nothing affection	0,25
B	Critical C_2	More affection	50,75
C	Severe C_3	Great affection	25,50
D	Fatal C_4	Fatal affection	75,100

2.4 Corresponding Relation between Consequence Severity and Relative Severity Value

Referring to some experts' suggestions, investigating and relative regulations, this paper describes type of risk consequences, severity of risk consequences and severity value of all the risks, the corresponding relationship among them is indicated in Table 2 [5].

According to the previous corresponding relationship between types of risk consequences and severity index and the severity values, all risk severity and its range of corresponding losses can be concluded.

2.5 Confirmation of Membership Function[6]

Through above analysis, the membership function of severity value can be determined. The specific method is as following:

By evaluating the severity value, if the value belongs to some range, its grade of membership belonging to this range is considered 0.95, the adjacent one is 0.02 and others are 0.01.

For example, if the final severity value of one risk consequences is 40 and the type of risk is critical, its membership function can be illustrated as following:

$$\mu_{c \subset C} = 0.01/c_1 + 0.02/c_2 + 0.95/c_3 + 0.02/c_4 \tag{4}$$

2.6 Relevant Range Value-Choosing Method

The value-choosing method in relevant range can affect the rationality and accuracy of the evaluation. For example, when severity index approaches to upper limit or lower limit of some range, it is unreasonable if only the same losses value is chosen. So there is a problem about how to choose the value in the range.

In order to reduce the errors and improve the reasonability of value-choosing, the value range and the losses range can be regarded as linear continuous correspondence. On the basis of this hypothesis, we get the method.

The general formula is as follow:

$$C = \frac{(k_n - k_2)(m_1 - m_2)}{(k_1 - k_2)} + m_2 \tag{5}$$

Table 3. Matrix of Risk Evaluation

Consequence level / Risk probability	Minor	Critical	Severe	Fatal
Impossible ≤1E-6	1A	2A	3A	4A
Occasional 1E-4～1E-6	1B	2B	3B	4B
Possibe 1E-2～1E-4	1C	2C	3C	4C
Frequent ≥1E-2	1D	2D	3D	4D

In (5), C is the corresponding losses, ($k1$, $k2$) is the severity value range of risk consequences, ($m1$, $m2$) is the corresponding range of losses, k_n is the corresponding severity value.

Other ranges can be calculated simply because of small membership function and can choose the median value.

Through above method, corresponding to any severity value, its type of risk and the corresponding losses can be calculated.

3 Risk Evaluation

3.1 Risk Evaluation Rule

Using above method and related references, the risk probability, risk consequences losses can be calculated. For details, see [1][2].

And on this basis, by analyzing the character of every risk factor, we can make the system risk evaluation rule as shown in Table 3 and Table 4.

3.2 System Risk Value Calculation

In the course of analysis evaluation, all risk factors are independent. According to the relative formulas as follow, the final system risk value can be worked out.

$$R=\sum_{i=1}^{n} A\times R_i \Rightarrow \sum_{i=1}^{n} A\times\left(\sum_{i=1}^{m} P_i\times C_i\right) \tag{6}$$

Table 4. Risk Value and Acceptability

Type of risk	Severity value	Risk decision-making criteria
Minor	1A 1B 2A (0-25)	Acceptable but no need to inspect
Critical	1C 2B 3A 4A (25-50)	Acceptable and need to inspect
Severe	1D 2C 3B 3C 4B (50-75)	Not wish: accept or refuse
Fatal	2D 3D 4C 4D (75-100)	Unacceptable: stop or rectify

In (6), R is the final system risk value, A is the weighting coefficient of every risk factor, R_i is the risk value of every basic risk factor, P_i is the probability of an risk factor, C_i is the risk consequences losses of every risk factor.

While evaluating the risk of the system, all the risk factors are queued by their risk value so as to find the dangerous factors of the system and provide effective measures to reduce the degree of risks.

4 Application

Studying on above method and related references, the risk probability and risk consequences losses of practical project, the railway of Jinpu, are calculated as shown in Table 5. And also the final system risk value formula of Jinpu Railway Project can be worked out as follow.

$$\begin{aligned} R &= 0.3117\times 71.1718 + 0.3117\times 57.7144 + 0.1198\times 46.5032 + 0.0219\times 42.9956 \\ &+ \ 0.0713\times 4.4986 + 0.1198\times 4.1328\times 0.0235\times 1.4698 + 0.0203\times 1.3914 \\ &= 47.5652 \end{aligned}$$

Details results are shown in Table 4.

On the basis of calculating result above, the effective decisions under risks and measures of controlling risks should be brought forward against the project.

Table 5. Calculation of Risk Probabilities And Consequences

Accident characteristics / Risk factors	Fatal		Severe		Critical		Minor		Risk value
	Probability	*Losses (RMB)*	*Probability*	*Losses (RMB)*	*Probability*	*Losses (RMB)*	*Probability*	*Losses (RMB)*	
Layout standard risk	4.90E-6, 5.59E-5	110000	7.46E-6, 8.10E-5	891333	1.15E-5, 1.20E-4	1100	1.07E-4, 6.34E-3	500	42.9956
River flows stability risk	1.61E-6, 2.41E-5	550000	4.90E-6, 5.59E-5	11000	7.46E-6, 8.10E-5	69766	1.15E-5, 1.20E-4	1000	4.1328
Flood fighting risk	1.61E-6, 2.41E-5	55000	4.90E-6, 5.59E-5	11000	7.46E-6, 8.10E-5	78037	1.15E-5, 1.20E-4	1000	4.4986
The third party rights of water affairs risk	1.61E-6, 2.41E-5	55000	4.90E-6, 5.59E-5	11000	7.46E-6, 8.10E-5	1100	1.15E-5, 1.20E-4	4579	1.3914
Other risks	1.61E-6, 2.41E-5	55000	4.90E-6, 5.59E-5	11000	7.46E-6, 8.10E-5	1100	1.15E-5, 1.20E-4	5769	1.4698
Anti scour risk	2.6621E-5	2673520							71.1718
Flood-discharge ability risk	1.0004E-6	5769129							57.7144
Water prevention ability of dike risk			4.7086E-5	987623					46.5032

Table 6. Risk Value of System

Risk factors	Risk value	Weight coefficient	Final result of risk value
Anti scour risk	71.1718	0.3117	22.1843
Flood-discharge ability risk	57.7144	0.3117	17.9896
Water prevention ability of dike risk	46.5032	0.1198	5.5711
Layout standard risk	42.9956	0.0219	0.9416
Flood fighting risk	4.4986	0.0713	0.3207
River flows stability risk	4.1328	0.1198	0.4951
Other risks	1.4698	0.0235	0.0345
the third party rights of water affairs risk	1.3914	0.0203	0.0282
Risk value of system			47.5652

We can conclude, from the descending sort of all risk factors, the most dangerous factor of the system is anti scour risk, followed are flood-discharge ability risk and water prevention ability of dike risk. All of the branches involved need to pay more attention to the relevant risk factors so as to reduce the degree of risks. At the same time, Other risks and the third party rights of water affairs risk could not be neglected in spite of their low risk value.

It can be judged that the system risk of Jinpu Railway Project is low and acceptable from the final result.

But if the fuzzy random variables is took into account, we can found that the risk of the construction is lower in running period while higher in building and aging period, which is worthwhile to be paid more attention.

5 Conclusion

At the present, little research has been conducted in the flood-control risk consequence analysis of bridge over river.

Based on much references of flood-control risk consequence analysis of bridge home and abroad, this paper tries to develop the research on risk management and hydraulic construction fields by employing the fuzzy comprehensive evaluation method and many other relative disciplines theory to analyze and calculate the flood-control risk consequence semi-quantitatively, and its adaptability and reliability is verified through a project example [7].

In the future, further research should be made in-depth on the flood-control risk consequence during the construction period and the aging period of bridge [8].

Acknowledgment. National Natural Science Foundation Acknowledgment Project (505779047).

Tianjin Key Lab of harbour & marine in School of Civil Engineering, Tianjin University.

References

1. Wang, H.W., Yu, J.X.: Research on flood-control risk management system of road and bridge construction over river. Journal of Natural Disasters 16(3), 114–117 (2007)
2. Wang, H.W., Yu, J.X.: Flood-control risk analysis of road and bridge construction. In: The 14th International Conference on Industrial Engineering and Engineering Management (IE&EM 2007), pp. 1182–1185. China Machine Press (October 2007)
3. Zhao, A.X., Ma, Z.J.: Appraising study for the loss evaluation system of natural disasters. Journal of Natural Disasters 2(3), 1–7 (1993)
4. Ren, L.C.: The method of fuzzy comprehensive evaluation applied in quantitative assessment of disaster losses. Journal of Catastrophology 11(4), 5–10 (1996)
5. Hu, Y.H., He, S.H.: Comprehensive Evaluation Method. Science Press (October 2000)

6. Ren, L.C.: The measurement method of fuzzy disater degree for classification of disaster loss. Journal of Natural Disasters 5(3), 13–17 (1996)
7. Wang, H.W.: Research on flood-control risk management system of road and bridge construction over river. Ph.D. Thesis, Tianjin University (2006)
8. Gong, C.L., Zhang, Y.: Study of risk consequences estimation for bridge construction, vol. 6, pp. 131–134 (June 2008)

The Study of Enterprise Supply Chain Optimization Based on Brand Strategy*

Wang Bao-li, Tian Hong-yuan, and Chen Xin-lan

School of Business Administration, Xi'an University of Technology,
Xi'an 710048, Shanxi, China
baoli1989@sina.com
{jianzhangli929,chenxinlan001}@163.com

Abstract. In order to explore the promotion of enterprise supply chain optimization, the article has put forward a strategy of enterprise supply chain optimization based on brand strategy ,in the analysis of the importance of brand strategy-based enterprise supply chain optimization, on the basis of calculation and evaluation , using analytic hierarchy process ,on weight of index which have influence of the enterprise supply chain optimization based on the brand strategy, and combining the actual market of enterprise .

Index Terms: enterprise supply chain optimization, index evaluation, brand strategy, strategy.

1 Introduction

With competition is getting tough, market competition has not only limited to the competition among enterprises, but gradually upgraded to the competition in supply chains. After all, competitive power on enterprise's own is weak, but overall advantage, through the supply chain to the formation of different manufacturers together, will definitely enhance the overall competitive advantage (Prahaled and Hamel,1990). It is top priority on how to construct enterprises and the core competitiveness of the whole supply chain system. In the process of cooperation with upstream and downstream enterprises, effective brand strategy will play an important role of enhancing the competitiveness of the enterprises and the whole supply chain system. This article focuses on the role of brand value in enhancing supply chain, and then deepening the supply chain's brand value to brand the supply chain, thus to gain more competitive advantages in the market competition.

---

* Supported by the Key Subjects Construction Project of Colleges and Universities in Shaanxi Province of China (1O7-00X902) and the special research program of Education Department of Shaanxi Province: The Study in Corporate Brand Building Strategy of Shaanxi Based on Marketing Innovation (08JK377).

J. Luo (Ed.): Affective Computing and Intelligent Interaction, AISC 137, pp. 677–681.
springerlink.com

2 The Importance of Supply Chain Optimization Based on Brand Strategy

With consumer's position in supply chain is becoming more dominant, the consumer demand and consumers' choice of products will affect the development of supply chain. It will result in the transformation from supply chain to demand chain [1]. The consumer demand and consumers' choice of products will be predominant factors in impact of flexibility, speed and stability. The demand chain manifested the process of supply chains' initiative shifted from the producer to retail merchants and finally to consumers. This will be a new trend of the supply chain （Yossi Sheffi, 2001）. The main performance of the consumer demand and consumers' choice of products is based on the product purchase frequency and purchase quantity. Oliver (1999) has defined the behavior as brand loyal. Therefore, effective introduction of brand strategy, in the process of cooperation with upstream and downstream enterprises, will link all sectors of the supply chain system closer, so that the supply chain system will greatly enhance the competitiveness of overall. The so-called brand strategy refers to image mold of enterprise aiming to enhancing the competitiveness of their products involved the enterprises and the brand of their products. It is enterprise's overall planning for the survival and development of enterprise and is an important part of business development strategy [2]. When the supply chain management has developed into an advanced business management, strong cooperation between enterprises is increasingly reflected in strong cooperation between famous brands.

3 Appraisal on the Indicators of Supply Chain Optimization

Analytic hierarchy process (AHP) is the application of simple mathematical tools for thinking, combined with the complex issue of logistics for the various elements of decomposition, and forms the hierarchical structure according to control relations grouping, through mutual influence relations of synthesizes between various factors to determine the relative importance of various factors, so as to provide the basis for decision-making program of choice [3]. In this paper, through this method of the weight of each index of supply chain optimization evaluated to determine to the importance of sorting, and according to time make appropriate strategy for the enterprise to learn.

3.1 Selection of Indicators of Supply Chain Optimization

Supply chain optimization based on brand strategy is a process of continuous improvement; it will contribute to help the business to complete the building of the core competitiveness of the strategic objectives for the center. Supply chain management is not only abrading management approach, but also a business philosophy. Brand competitiveness refers to the brand which the enterprise has the external brand image and internal quality, technical performance, service that can continue to differentiate or ahead of its competitors [4].It has a competitive advantage and the ability to provide value profit, which is the core competitiveness of enterprises objectified and commoditized. And in the market performance of the brand competitive with strong brand not only improve the sales of brand enterprises, but also improve their own trading brand value. At the same time it sets the enterprise product strength, capital strength, strategic management, innovation, management skills and cultural values, and is the core competitiveness of the market vector. Therefore, the brand competitiveness is not only the core competitiveness of the main factors, but also to improve the competitiveness of supply chain factors. Therefore, building on the brand strategy of supply chain optimization index system, we must emphasize the core competitiveness of the brand.

In this paper, based on the investigation and experts' advice, established the indicators of supply chain optimization (Table 1).

Table 1. Analysis based on Analytic Hierarchy Process (AHP)

Indicators of Supply Chain Optimization	
Brand Power	*Brand Market Power*
	Brand Image Strength
	Brand Financial Strength
Economic Strength	*Running Cost*
	Production Operation
Green Chromaticity	*Investment of Environment Improvement*
	Green Logistics
Customer Satisfaction	*Supply Ability*
	Post-sale Service
Node Corporate Relations	*Transaction*
	Information

This article according to analytic hierarchy process's request and combining the study designs the target layer and rule layer [5]. Assume that if P_i and P_j were considerable merits, $b_{ij}=1$;if P_i is a little better than P_j, $b_{ij}=3$; if P_i is better than P_j, $b_{ij}=5$; if P_i is much better than P_j, $b_{ij}=7$; if Pi is even better than P_j, $b_{ij}=9$.According to the largest value matrix and eigenvector by the method of square root that the results are as follows:

$\lambda_{max}(A) = 9.5277$

$K(A) = (0.2195, 0.0813, 0.0551, 0.1004, 0.0793, 0.0668, 0.0416, 0.1327, 0.0233, 0.0220, 0.0198)^T$

Consistent with the proportion of random matrix level of CR on a single order and hierarchy of the consistency test, the test results are as follows:

$CR(A) = 0.045 < 0.10$.

Therefore, this article analyzes the comparison matrix has satisfactory consistency.

Finally, after the comparison of weight, the order of the important indicators obtained is: Brand Market Power >Information> Running Cost> Brand Image Strength> Production Operation> Green Logistics> Brand Financial Strength> Investment of Environment Improvement> Transaction> Supply Ability> Post-sale Service. After calculating the total weight, gets that the weight of Brand Power is 0.3559,which is the maximum weight. Therefore, the rational and effective use of brand strategy for the development of the supply chain has far-reaching significance.

4 Strategy of Supply Chain Optimization Based on Brand Strategy

Thanks to the important role of brand in the enterprise supply chain development, supply chain optimization process must be based on brand strategy, which is the enterprise supply chain market competition in an invincible position in the fundamental. How does the supply chain create its own brand strategy? It can work from the following aspects:

4.1 Building Partnership of Supply Chain Based on Brand Strategic

Selecting supply chain partners is not simply decided by the big enterprise or the big brand, but whether the cooperation parties could form a common vision and share information at basis of mutual trust [6]. As the various segments of the enterprise will have to face different consumer groups, so the implementation of brand strategy in the supply chain system, it is very important to determine the target consumer group, and this requires a reasonable brand positioning of supply chain. In the competition of products or services market tending to homogeneous, consumers' emotional orientation are often uncertain, brand positioning is the establishment of such brand personality.

4.2 Enhance Brand Management of Supply Chain

In order to eliminate the instability of the supply chain and the non-uniform between the interests of all members, the nodes in the supply chain must stand in the perspective of the whole supply chain to think. In the process of the creation of the supply chain brand, the various members of the supply chain supply chain system must be subject to a unified brand strategy, so that the whole supply chain to achieve optimal performance. Because the shape of a successful supply chain brand is not a matter, it needs the supply chain work together to achieve, if in a particular part of the supply chain because of its short-term interests in violation of the strategic objectives of the entire supply chain, it would have an extremely negative impact on the development of the long-term supply chain system [7]. The cost of products or services caused by inadequate information can be largely reduced or avoided by information sharing, and finally to enhance the brand value of supply chain. Therefore, integrates the brand idea in the supply chain strategy management system, and causes it to rise to a strategic altitude, is extremely essential for supplies chain's long-term development.

4.3 Expansing the Appreciation of Brand Space in the Process of Supply Chain Optimization

There is no doubt that making use of advertisement of brand can enhance the brand awareness in the short term. If you want to really create a successful brand, it can not be reached to build brand loyalty of consumers by advertising in the short term. It requires all sectors of the supply chain system jointly developed a set of business planning on brand building like the brand research, brand management, brand positioning and a range of activities for the smooth operation[7]. Meanwhile, the quality of brand can not be sure by advertising, it is not only regarded as constant quality, but more than the requirements of brand's innovative development. Therefore, to create a supply chain brand, advertising and marketing method both are essential. Through the essential marketing tools, in order to attract more potential customers, and then to develop and strengthen the supply chain brand and lay the foundation [6].

4.4 Making Full Use of Brand Loyalty to Strengthen Brand Value of Supply Chain

Brand loyalty is the most important factor in brand equity. Since marketing has been produced, winning and maintaining consumer brand loyalty has been a successful and important task for business. The brand loyalty strives the competition time for the enterprise to make the response. If the competitor has developed a superior product or provide a more effective service delivery, forcing the corresponding segments of the supply chain improvements, and brand loyalty can be secured a lot of time for the supply chain system, and develop more excellent products or services to counter competitors[7].

4.5 Introducing Philosophy of Brand Cooperation into Enterprise Supply Chain Systems

In the future, enterprise supply chain competition should be timely to introduce the concept of brand, and reject of the traditional fighting on their own ideas, through strong cooperation between the enterprise brand to create and enhance brand value and lay the foundation for supply chain. The forms of brand cooperation are variety in the supply chain system, either between the suppliers and manufacturers, or the brands between the manufacturers and logistics service providers' cooperation, or by other forms of brands. In brief, the brand cooperation's original intention is to promote the supply chain system's overall value well, and the biggest advantage of Co-brand partners is the advantages of use of each other's brands, to enhance their brand awareness, expanding sales, increase market share [6].

References

1. Wu, Y.: Supply chain management of Chinese enterprises in several major problems and counter measures. Intemet fortune - Management Vision, 38–40 (December 2009)
2. Yossi, S.: Supply Chain Management under the Threat of International Terrorism. The International Journal of Logistics Management, 1–11
3. Li, Q.Y.: Brand strategy and corporate development. Management, 50–53 (January 2009)
4. Beamon, B.M.: Designing the Green Supply Chain Logistics. Information Management 12(4), 332–342 (1999)
5. Lakhal, S., Martel, A., Kettani, O., Oral, M.: On the optimization of supply chain networking decisions. European Journal of Operational Research 129, 259–270 (2001)
6. Yan, Y.-X.: Brand strategy and decision-making. Dong bei University of Finance Press, Dalian (2001)
7. Peng, T., Hong, W., Xu, L.: Brief Analysis of Supply Chain System Based on Brand Strategy. Logistics Sci.-Tech. (6), 152–154 (2007)

Support Fast Generation Algorithm Based on Discrete-Marking in Rapid Prototyping

Qian Bo[1], Zhang Lichao[2], Shi Yusheng[2], and Liu Guocheng[2]

[1] Electrical and Mechanical Institute, North China University of Technology, Beijing, 100144
[2] State Key Laboratory of Material Processing and Die & Mould Technology of Huazhong University of Science and Technology, Wuhan, 430074

Abstract. In rapid prototyping, the support can constraint the parts deformation, and avoid a lot of problems such as collapse, shift and imbalance of the parts. The support automatic generation algorithm has become the key research of rapid prototyping technics software; however, presently the support automatic generation algorithm is complex with low efficiency. Thus, it is of great importance in improving the overall efficiency of rapid prototyping technics software through enhancing the support generation efficiency. Based on this, this paper proposes a discrete-marking support algorithm for treatment of processing on manufactured part model of all Triangle-based discrete-marking on the support plane, which realizes the three-dimensional computation of intersection for each line associated with the triangle automatically, and avoid invalid computation between all the support lines and other triangles. Thus, the algorithm efficiency has reached $O(n)$. Meanwhile, the technics support generation speed has been improved. Furthermore, the new mesh discrete-marking and automatic support generation algorithm has been successfully applied to the Sterolithography apparatus (SLA) and Selective Laser Melting (SLM) of HRPS Series. Practical application indicates that the new support generation algorithm improve considerably the support technics efficiency and rapid prototyping efficiency of the technics software.

Keywords: rapid prototyping, support, discrete, marking, mesh.

1 Introduction

Rapid Prototyping is a kind of manufacture technology which can form the three-dimensional solid in arbitrary complex shape by the import of CAD model. Based on the manufacture principle of discrete layer, three-dimensional CAD model is divided into a series of two-dimensional ordered slices along certain direction (usually Z direction) which are manufactured in turn by rapid prototyping machine and are connected automatically, thus the final three-dimensional physical solid is formed[1][2].

As one of the key technologies for Rapid Prototyping, the support technics plays an important role in guaranteeing the parts process smoothly. The support generation algorithm is the key technology, which can constrain the deformation of the parts, and make the parts process smoothly, thus it has great study significance. On the study of

J. Luo (Ed.): Affective Computing and Intelligent Interaction, AISC 137, pp. 683–695.
springerlink.com

the support automatic generation algorithm, many kinds of the support automatic generation algorithms [3][4][5] have been presented and discussed from different perspectives of support structure, support technics and support generation strategies. Hong Jun and other authors[6-8] from Xi'an Jiaotong University, have done a great deal of research on the support generation algorithm, and have put forward multiple support strategies and support rules, who have made outstanding research on sterlithography support generation algorithm. On the basis of the previous studies, the author does a great deal of research on the fast support generation algorithm by combining the problems encountered in developing support automatic generation algorithm with the solutions, especially on the intersection calculation between support ray and support region, thus the author puts forward a discrete-marking support generation algorithm based on the triangularization of parts model.

In the present support automatic generation algorithm, the most widely used is based on STL file format. STL is three-dimensional file format in the field of rapid prototyping. In STL format model, the curved surface consists of triangles, and the coordinates of three vertices and outer normal vectors are stored in the file. These triangles are formed by the triangularization of CAD model surfaces, whose storage order is arbitrary. The traditional procedures of region support automatic generation algorithm based on STL format is stated as follows:

Firstly, the on-supported regions are picked up by the region picking algorithm. The pick-up conditions can be the area and shape of triangles in STL file, included angle between triangles and horizontal plane and other filtered factors. In the region support generation algorithm, the included angle between triangles and horizontal plane is selected as the pick-up condition. When the included angle θ is equal to or less than the critical value δ , the region needs adding support. In general, 30 degree is selected as the critical value based on the long-term technics experiences. When θ is equal to or less than the critical value, it needs adding into the on-supported region through the picking algorithm.

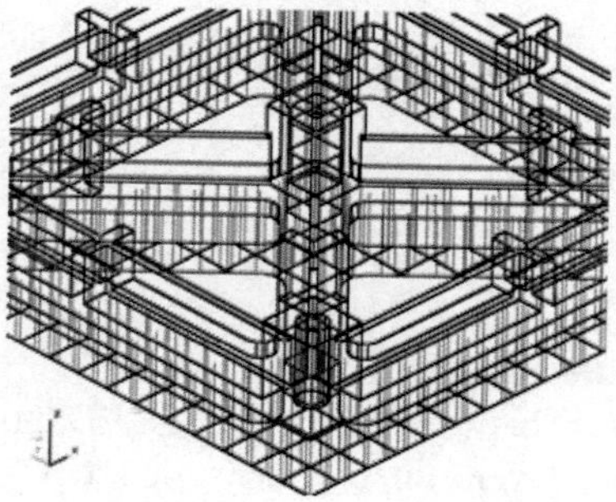

a. Main view of the support region

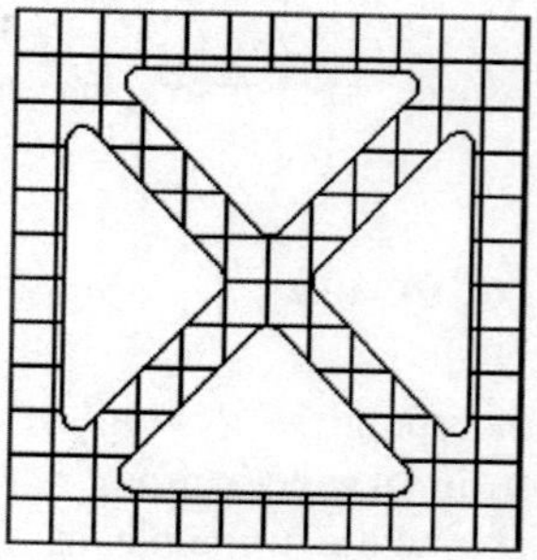

b. Top view of the support area

Fig. 1. Internal hatching of support area

Secondly, the support region needs internal hatching along x and y direction respectively with a certain step length $\triangle d$, and then corresponding padding ray is generated (see figure 1). Rapid Prototyping technics support is that the generated padding ray is extended to process platform (base plane x-y in Z direction). If the padding ray and the solid surface intersect, it is truncated in the intersection point. Thus, the padding segment is formed. When the step length is set as 5mm in the on-supported region, it has the best support effect by a large number of experiments.

See the simplified model in figure 2, one point in the support ring is set as A, with A as the starting point to generate support ray along –z direction, and then the support ray intersects the solid, thus the intersection points B and C are formed. The point B is up on the point C, thus the support line AB is generated by choosing the maximum intersection point B as the other endpoint of the support line.

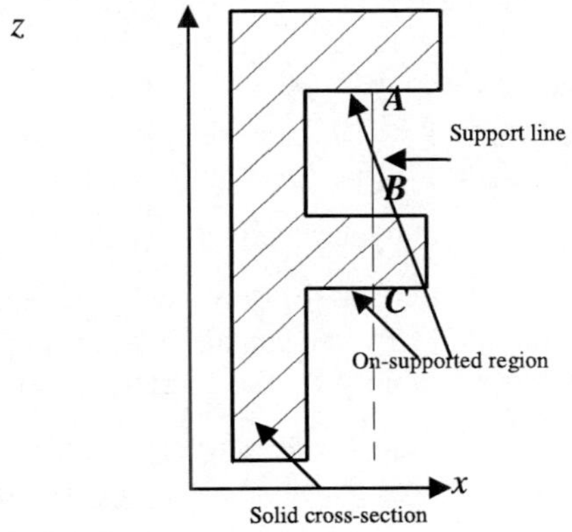

Fig. 2. The generation of the support line endpoint

After the support lines in all the regions are generated, the corresponding support structure is generated by the support type parameters including embedded depth and tooth shape, and then is stored in the support technics file in order to be convenient to load.

In the complex STL solid model, the triangles are in large numbers. In the process of calculating the two support endpoints by the above support line generation algorithm, every padding ray will do the intersection operation with the triangles in all support regions. However, in practice, every padding ray just intersects a few triangles, thus it causes large quantities of invalid computation, and spends a lot of time, which reduces the support generation speed [13]. Therefore, the author puts forward a fast support generation algorithm based on discrete mark of on-supported region.

2 Discrete-Marking Generation Algorithm

2.1 The Principle of Discrete-Marking Approach

The principle of the discrete-marking approach is as follows: Firstly, selecting a base plane α (such as the process platform x-y plane) as the projection plane, the minimum surrounding frame A of certain plane in the solid surrounding cube is got which is parallel to the base plane α, and then make the projection of A in the base plane α, thus the rectangle ***R*** is formed. Secondly, the mesh is formed by traversing the projection rectangle R along x and y direction through certain step length $\triangle d$. Thirdly, a series of space triangles ID numbers corresponding to every mesh plane are got by projecting all the triangles in R and computing the mesh number of every projection triangle. Furthermore, the mesh number of every support ray in rectangle R is calculated. While two endpoints of support ray can be calculated by searching a series of space triangles ID in this mesh, and then the support ray intersects these triangles, thus it can avoid intersecting a large number of irrelative triangles, which can increase the computing speed.

Firstly, the minimum surrounding frame in STL file should be confirmed which is parallel to x-y plane, and then the rectangle R is formed by projecting the minimum surrounding frame in x-y plane. Subsequently, the rectangle is dispersed along x and y direction through certain step length $\triangle d$, and the triangles IDs are marked by using a two-dimensional array cells[y][x]. Secondly, in three-dimensional space, $\triangle A'B'C'$, $\triangle E'F'G'$ and point D' are formed by projecting $\triangle ABC$, $\triangle EFG$, and the support ray L_D in R. And then IDs in mesh array are calculated by the projected points (see the shadow in figure 3). Thirdly, the contain relationship between the points and the rectangle can be calculated rapidly by the literature [14]. For instance, the mesh unit by projecting L_D is cells[2][4], whose triangles set is $T_{24} = \{ID_{\triangle ABC}\}$, thus the other endpoint of the support ray is confirmed by intersecting the triangles in L_D and the triangles in T_{24}, not by intersecting $\triangle EFG$, which can avoid intersecting a large number of irrelative triangles, thus it can improve the support generation efficiency.

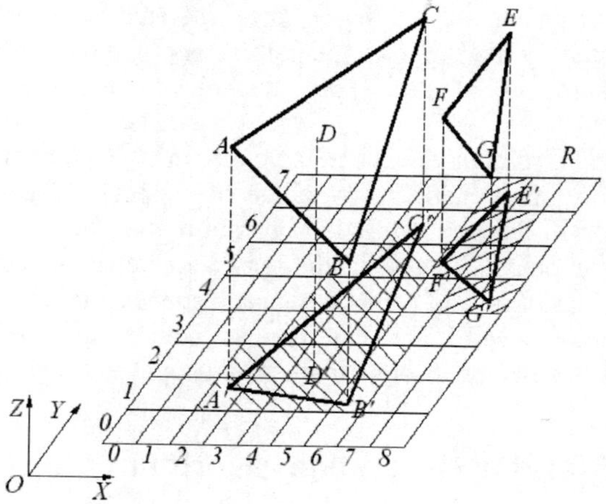

Fig. 3. Discrete-marking schematic diagram

Above all, the core idea of the discrete-marking algorithm is manifested in dispersing the projection of triangles and marking (recording) ID of triangles in the mesh, thus it can reduce the intersection times between support rays and triangles and improve the intersection rate to enhance the support generation speed. The key to the algorithm is adopting discrete-marking algorithm of the triangles. All STL files are shown in the form of triangles, so triangular discrete mark will become the core of the algorithm.

2.2 Triangular Discrete Mark

Triangular discrete mark is similar to the region filling in the graphics. However, the discrete step length is not with a pixel as a unit, but is confirmed by the dimension of parts; however, their principles have a lot of similarities. Therefore, it belongs to region filling category.

The commonness between triangular discrete mark and region filling are manifested in marking the interior of polygon (triangle). This algorithm can be illustrated by the example of triangle. The discrete-marking algorithm is called edge flag algorithm, which can be divided into two steps:

Step 1: Mark the meshes passing triangular edges after making the linear scan to every edge of triangle.

Step 2: Visit the meshes in every scan line intersecting the triangle one by one from left to right, and then from down to up. Indicating the state of the mesh by using a *Boolean* value *Inside*, if the mesh is in the inner of triangle, *Inside* is TRUE; if the mesh is outer of the triangle, *Inside* is FALSE. The initial value of *Inside* is FALSE. In the process of traversing from left to right in turn, if a mesh is identified as TURE, the mesh has been marked, recording the position as A; Later, if one or more meshes have been unidentified, they need identifying, and then continue traversing. If a mesh has been found and identified as TRUE, recording the position as B, thus, the meshes between A and B are all identified as TRUE, namely, the positions between A and B are the interior of two edges. Finally, traversing all the meshes from down to up, all the triangular regions can be marked completely (See figure 4).

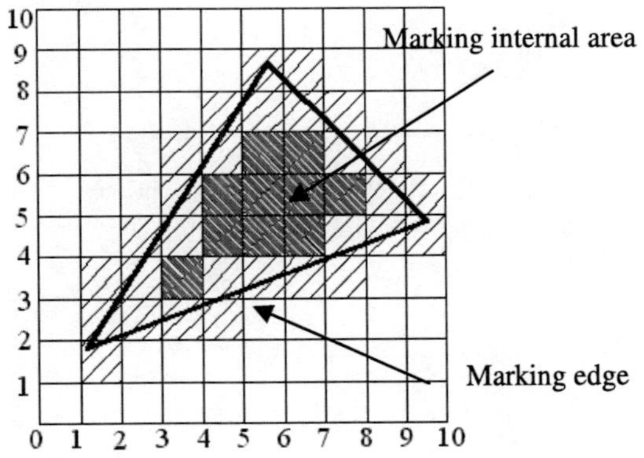

Fig. 4. The edge flag algorithm schematic diagram

The process of the specific algorithm has been stated by the process of edge flag algorithm as follows:

Step 1: Make the linear scan to every edge of triangle;

Step 2: Initialize *Inside* value of every mesh *as FALSE*;

Step 3: Traverse circularly every scan line which intersects with the triangle, and then initialize A and B: $A=0$; $B=0$;

Step 4: Traverse every mesh in the scan line and search mesh positions needing marking, recording the positions of A and B;

Step 5: Judge whether the mesh is marked, and then set *Inside* value;

Step 6: End circular traversal and scan.

Edge flag algorithm is implemented simply with high overall efficiency. Therefore, in the optimization of the support generation algorithm, the core of discrete-marking algorithm is the edge flag algorithm.

3 Automatic Generation of Support Structure

After constructing the information of the intersection between all meshes and support rays in the plane area by the edge flag algorithm, all the support rays can be calculated by the following algorithm (see figure 5). The planar equation of the triangle represents as (1):

$$P(u,w) = A + uB + wC \tag{1}$$

The points in the support ray represent as : $Q(t) = D + tE$, and the intersection point is $\boldsymbol{R}$:

$$R = P(u,w) = Q(t) \tag{2}$$

Thus the result $\boldsymbol{t}$ is (3)

$$t = \frac{(B\times C)\cdot A - (B\times C)\cdot D}{(B\times C)\cdot E} \tag{3}$$

Finally the intersection point is gained by putting t into $R = D + tE$.

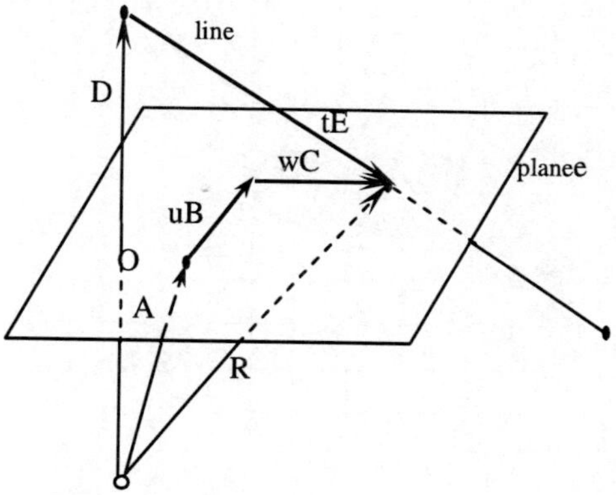

Fig. 5. The intersection calculation between straight-line segment and plane

A series of on-supported line segments are formed after calculating the support rays by the above edge flag algorithm. And then the unit support structure in block or web is formed by the types of support structure. Finally, the overall support structure is formed by a series of support units. The generation steps of the unit support structures are as follows:

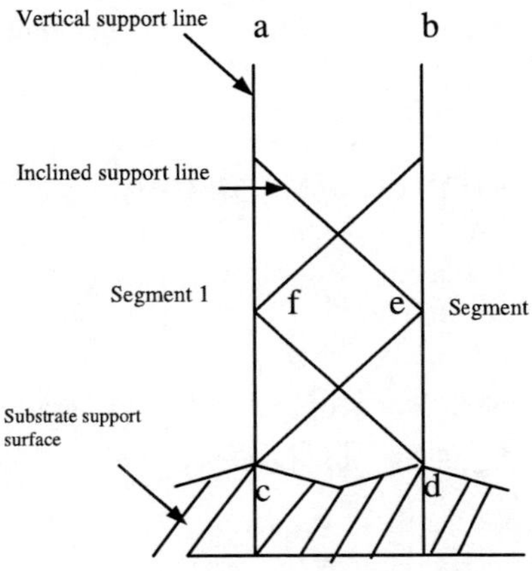

Fig. 6. The generation of the unit support structure

See figure 6: a and c are two endpoints of segment 1, while b and d are two endpoints of segment 2. The two-dimensional coordinate of the point f is the projection of a(c) in x-y plane. The z value of f point is added up the height of a support ray from c point by the inclined line. The segment between f and d constructs an inclined support line, and the other inclined support line is formed by e and c. According to the above method, all the inclined support lines between segment 1 and 2 can be formed until the value z of the vertex of the inclined line is over a.

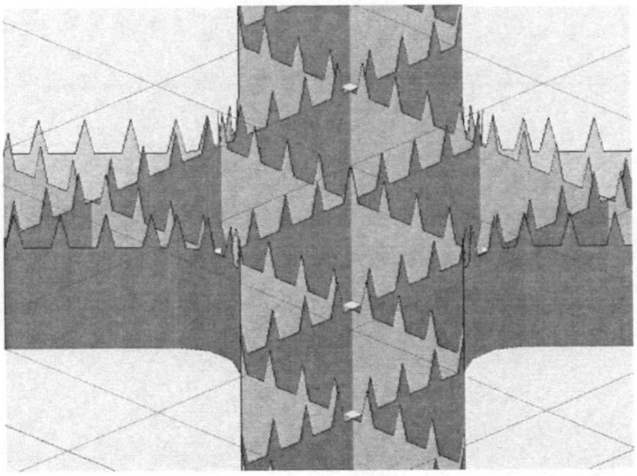

Fig. 7. The generation of integral support structure

As for the base support, a planar polygon is formed by the point c and d which is perpendicular to the base, and the middle of the polygon concave down. The bottom margin of the polygon lies in base plane, while the height of two sides is called tooth height, and the distance between segment 1 and segment 2 is called tooth width. The support unit is composed of three support modes including base support, vertical support and diagonal line support.

After forming the support unit, the height, width and distance of the tooth should be calculated because the support unit and entity parts should touch in serrate shape according to the demand of rapid prototyping technics. After completing the calculation, the overall support structure is formed by connecting the neighboring units (see figure 7).

4 The Complexity Analysis of Algorithm

Based on the idea of discrete-marking algorithm, the support generation algorithm has been realized in VC++ platform. The algorithm has a good time complexity. Let the numbers of the triangle in STL model be m, while the marked triangular numbers in the same mesh are m', and the numbers of the support ray are n. When the discrete step length $step << \max(L_x, L_y)$ (L_x and L_y are separately the length and the width of the minimum surrounding box), then $m' << m$. Therefore, the intersection times between support ray and the triangles of the area support will decline sharply. After being marked discretely, the time complexity of the intersection is $O(n*m')$. Obviously, when $m' << m$, $O(n*m') << O(n*m)$. According to the experimental analysis, m' is between 1 and 5 in most of the cases, and the average is 3. m is the triangular numbers of the model, and the triangles of the complex model is usually above 10K. Therefore, from the analysis of the time complexity of algorithm, the algorithm after optimization has high efficiency.

It has gained the good effects by testing many models, and the test data in detail are in chart 1.

Table 1. The Contrast Data of Speed Before and after the Optimization

Solid name	Triangles (Pieces)	Initial speed (second)	Optimized speed (second)	Magics (second)	Optimizing multiples
Mobile phone shell	4292	3.343	0.797	0.9	4.2
Topside	16992	34.172	2.969	1.1	11.5
Connected parts	21378	9.312	1.672	0.8	5.6
Chassis	34312	18.328	3.094	0.9	5.9
Dongfeng's logo	132220	134.063	7.859	37.4	17.1
Assembly parts	211696	351.407	29.688	48.0	11.8
Outside	327851	1215.469	61.547	173.2	19.7
Skull model	353444	143.5	31.047	427.6	4.6
Chess	483608	168.032	27.89	93.4	6.0
hollow connected parts	525646	246.56	26.234	560.3	9.4

From the data in chart 1, the optimized algorithm is obviously faster than the initial algorithm in the support generation speed. When triangular numbers of the parts are less, the support generation time is shorter and the advantages can not be obvious. However, in complex graphics, when the triangles are in large numbers, the algorithm before the optimization has low efficiency, which is not good for the application in rapid prototyping application software, while the support generation speed has been improved significantly through the optimization of discrete-marking algorithm. Furthermore, when calculating some complex models (the numbers of triangles are above 100k) with the same support structure parameters, the computing speed after the optimization is over 5 times faster than the widely used Magics support software, and 50 times faster than the technics software speed before the optimization. Therefore, the support module optimized by the approach can completely meet the requirements of the practical engineering.

5 Experiments and Analysis

Based on the above theories, the support automatic generation procedures are developed through VC++ platform. In order to testify the rationality of efficiency of support generation speed and support structure, several typical model parts are selected to do the support generation and then process in the rapid prototyping system of HRP series of Binhu Ltd.. And then a lot of experiments and analysis are done from the respects of triangular numbers of the model, region numbers after the support generation, the computing speed before and after the optimization.

Test sample 1: The figure 8 is a fire dragon model; the figure (a) is the support structure after the computation through this algorithm; the figure (b) is the processed parts. The size of the model is: 106×82×100mm, and triangular numbers are 89424. In the manufactured direction, the feature of dragon's mouth is complex and it has a lot of holes and cavities inside, so it is difficult to remove the support structure. While the other parts are the planar structure, it is easy to remove the support structure. Therefore, the manufactured direction is that the standing direction of the dragon rotates 45°along x axis, and then it is processed when the tail is down, the back is up and the face is forward. In the experiment, the support generation parameters are: the maximum inclination of the line support is 45°; and the inclination of the block support is 60°; the minimum area of on-supported region is 10mm; the minimum length is 3mm; the minimum width is 3mm; and the numbers of the support region after the generation are 88. This model is the typical parts of multi-support region in small area. The speed of the parts can be improved many times through the analysis of the foregoing algorithm, and the same is the actual calculations. The computing speed is 5.62 seconds by adopting the algorithm in this paper, while the speed is 57.3 without adopting it. Thus the speed is nearly 10 times faster than the traditional algorithm. In conclusion, as for the condition of many support regions, if adopting the traditional algorithm, it needs traversing all the triangles of the parts model one by one, however, if adopting the algorithm in this paper, no matter how many support regions, every support region will do one intersection calculation with self-related triangle.

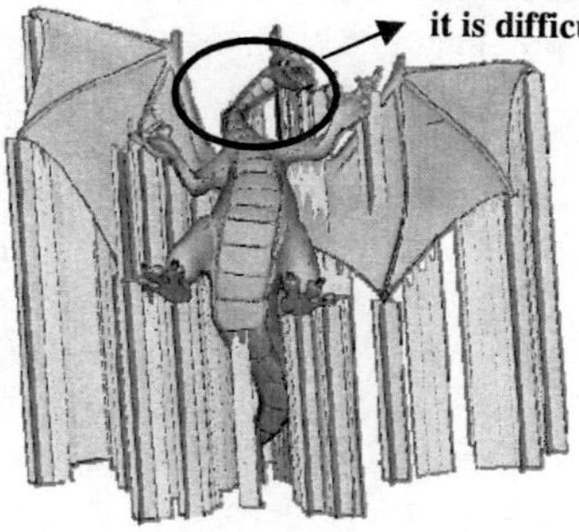

a. Fire dragon model support

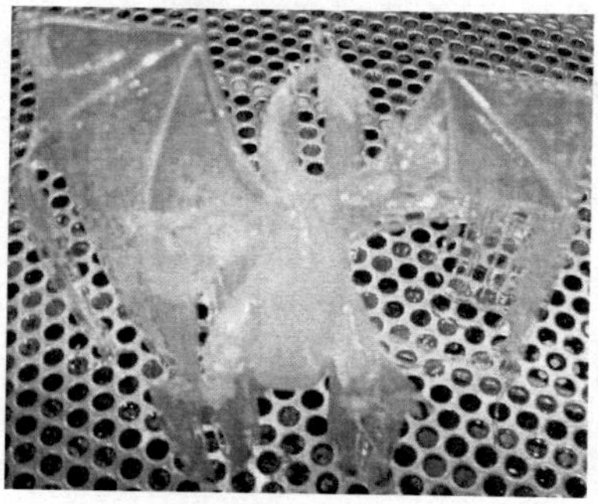

b. Processed parts

Fig. 8. Fire dragon model support automatic generation

Test sample 2: The figure 9 is complex welding mask model; the figure (a) is the support structure after the calculation through this algorithm; the figure (b) is the processed parts. The size of the model is: 81×88×132mm, and triangular numbers are 48294. In the manufactured direction, compared to the horizontal machining, the vertical machining can be better reduce the surface roughness of the mask's front and rear facades caused by the stage effect, furthermore, if manufactured in the horizontal direction, it needs adding the support structure to the lower surface of the mask, thus the planar dimension of the support structure will be larger, the height value will be higher, and the material cost will be more. After completing the process, it is difficult to remove the support because the contact area between support surface and solid surface is large. Therefore, choosing vertical process is the best process direction. In the experiment, the support generation parameters are: the maximum inclination of the line support is 30°; and the inclination of the block support is 60°; the minimum area of on-supported region is 3mm; the minimum length is 3mm; the minimum width is 3mm; and the numbers of the support region after the generation are 22. This model has the features that the support regions are in small numbers and every support region area is large. The expected speed of the parts can be improved limit through the analysis of the foregoing algorithm, and the same is the actual calculations. The computing speed is

8.9 seconds by adopting the algorithm in this paper, while the speed is 27.5 seconds without adopting it. Thus the speed is nearly 3 times faster than the traditional algorithm. That is because the saved time of the large area by adopting discrete-mark algorithm is close to extra consumption time. Thus, in general, the time almost has no change. The support algorithm efficiency of the model should be further optimized.

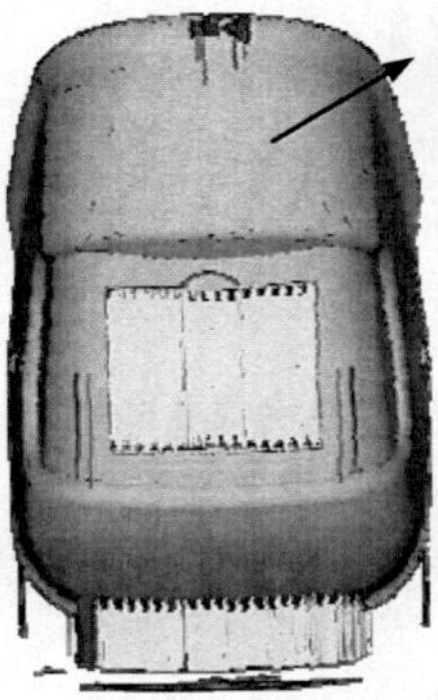

a. Welding mask model and support

b. Processed parts

Fig. 9. Welding mask model support automatic generation

Test sample 3: The figure 10 is the processed model of fire dragon wings; the figure (a) is the support structure after the computation through this algorithm; the figure (b) is the processed parts. The size of the model is 148×90×60mm, and the numbers of triangles is 19420. It is the same as the test sample 2 in the manufactured direction. If choosing the vertical machining, the dimension of support structure should be decreased to reduce the surface roughness caused by the stage effect; meanwhile, it is easy to remove the support after machining. In the experiment, the support generation parameters are: the maximum inclination of the line support is 45°; and the inclination of the block support is 60°; the minimum area of on-supported region is 103mm; the minimum length is 3mm; the minimum width is 3mm; and the numbers of the support region after the generation are 42. This model has the features that the support regions are in small numbers and every support region area is small. After the support generation, the maximum support area is 22mm^2, and the minimum one is 11mm^2. As for such kind of parts, the support generation speed can be improved considerably. The computing speed is 3.6 seconds by adopting the algorithm in this paper, while the speed is 21.5 seconds without adopting it. Thus the speed is nearly 7 times faster than the traditional algorithm. As for this kind of parts, the reason of enhancing the efficiency is just like the testing sample 1.

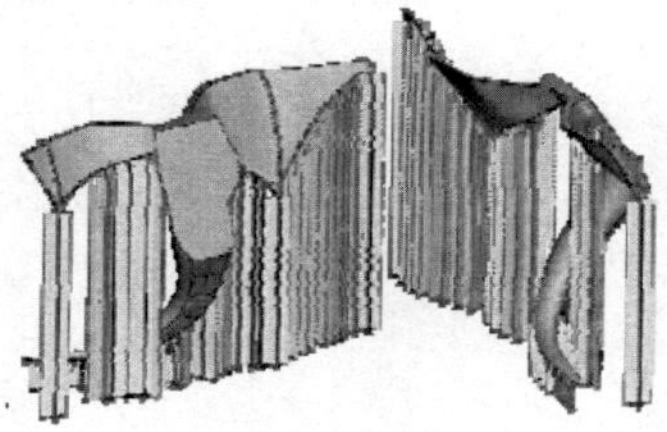

a. Fire dragon wing model support

b. Processed parts

Fig. 10. Fire dragon wings support automatic generation

Through the effect analysis of the support generation samples of the above three parts, the fast generation algorithm in practice can not only improve the efficiency of the models with a large numbers of triangles and complex partial details, but also improve the efficiency of the models with less triangles and simple appearance, furthermore, it can improve the efficiency of the complex models with many features, can process the parts smoothly, has strong practicality.

6 Conclusion

As the discrete-marking support rapid generation algorithm has linear complexity, its support computing efficiency has been improved 50 times after the realization in the rapid prototyping technics software. While its speed is 5 times faster than Magics® software [14] in calculating complex parts model support (triangular numbers is above 100k). The algorithm has been applied into the sterolithography rapid prototyping of HRP series and selective laser melting rapid prototyping system. In practice, the algorithm is stable and reliable with good practical effect, which can almost calculate the support of all STL files rapidly and can completely meet the requirements of the engineering.

References

1. Kruth, J.P.: Material Increase Manufacturing by Rapid Prototyping Techniques. Annual of the CIRP 40, 603–614 (1991)
2. Swaelens, B., Pauwels, J., Vancraen, W.: Support generation for rapid prototyping. In: Proceedings of the 9th European Conference on Rapid Prototyping and Manufacturing, vol. 16, pp. 115–121. Dickens Academic Press (2000)
3. Charles, F.K.: Automated support structure design for stereolithographic parts. The Graduate School of Clemson University, South Carolina (1991)
4. Charles, F.K., Jara-Almonte, C.C., Bagchi, A.: Computer Aided Design of Support Structures for Stereolithographic Components. In: Proceedings of the 1991 ASME Computers in Engineering Conference, Santa Clara, pp. 443–448 (1991)
5. Webb, D., Verdes, V., Cassapis, C.: Computer-aided support structure design for stereolithography models. In: Chartoff, R.P., Lightman, A.J., Schenk, J.A. (eds.) Proceedings of the 5th International Conference on Rapid Prototyping, University of Dayton, pp. 221–228 (1994)
6. Swaelens, B., Pauwels, J., Vancraen, W.: Support generation for rapid prototyping. In: Chartoff, R.P., Lightman, A.J. (eds.) Proceedings of the 6th International Conference on Rapid Prototyping, University of Dayton, pp. 115–121 (1995)
7. Majhi, J., Janardan, R., Schwerdt, J.: Minimizing support structures and trapped area in two-dimensional layered manufacturing. Computational Geometry 12, 241–267 (1999)
8. Friedrich Prinz, B., Clinton Atwood, L., Richard, F.: Rapid Prototyping in Europe and Japan. Japanese and World Technology Evaluation Centers 1, 69–90 (1997)
9. Zhang, L.C., Han, M., Huang, S.H.: An effective error-tolerance slicing algorithm for STL file. International Journal of Advanced Manufacturing Technology 20, 363–373 (2002)
10. Chua, C., Leong, K., Lim, C.: Rapid Prototyping:Principle and Applications, pp. 104–135. World Scientific Publishing, USA (2003)
11. Jacobs, P.F.: Rapid Prototyping & Manufacturing, pp. 39–74. Society of Manufacturing Engineers, London (2002)
12. Swalens, B., Dickens, F.: Support generation for rapid prototyping. In: Proceedings of the 9th European Conference on Rapid Prototyping and Manufacturing, pp. 115–121. Academic Press (2000)
13. Jacobs, P.F., Hull, C.: Introduction to Rapid prototyping & Manufacturing: Fundamentals of Stereolithography, pp. 1–23. SME.MI, Dearborn (1992)
14. http://www.materialise.com

Control System Development for Flexible Roll Forming Machine

Wang Guiyong, Li Qiang, and Qian Bo

Institute of Mechanical and Electrical Engineering, North China University of Technology 100144, Beijing, China
guiyong_w16@163.com

Abstract. Flexible Roll Forming (FRF) is a high efficient and economical method for the production of sheet metal. Due to the complicated motion control of FRF and the intricate section geometry of sheet metal, it is required for real-time accuracy and dependable control during the course of forming in FRF. Compared to configured software such as Labview, PMAC (Programmable Multi-axis Controller) can realize kinds of complicated control logic and is widely utilized in industry because of its versatility of programming language, high-precision control and good portability. Based on the technique characteristics of high precision, high stability and strong real-time control of FRF, this paper realizes the control system of FRF which can control the servomotor systems and actuators by adopting PMAC and industrial computer, etc. The experiment indicates that this system possesses the advantages of high-precision control and good stability, which can be utilized to kinds of production of sheet metal in FRF.

Keywords*:* Flexible Roll Forming, Sheet Metal, Control System, PMAC.

1 Introduction

Cold roll forming is a high efficient and economical processing method which defines that sheet metal is formed gradually into desired geometry as it travels between the forming stations which are assembled with definite shapes of rolls [1][2].

Flexible roll forming (FRF) can allow the cross-section of sheet metal to be regularly changed on width, which is a new technique based on cold roll forming [3][4]. Compared to cold roll forming products, FRF can lower the costs of production more efficient, increase the bearing capacity of sheet metal and reduce the structural weight of products.

It is extremely difficult and complicated in FRF because the process of FRF is kinds of coupling in shapes, such as horizontal bending, longitudinal tension, longitudinal bending and horizontal shearing. And that needs to develop a control system to control the real-time accurate motion of forming rolls for FRF. However, the technique of FRF is a new international research focus at present, and which is still at probing into stage in China. Based on the process characteristics of FRF, this paper elaborates the control system which is utilized in FRF machines by adopting PMAC motion controller.

J. Luo (Ed.): Affective Computing and Intelligent Interaction, AISC 137, pp. 697–705.
springerlink.com

PMAC (Programmable Multi-axis Controller) as a high performance multi-axis controller which has flexible and strong access mechanism and can control 1-32 axes and 16 coordinate-systems simultaneously. Besides, it can also realize multi-axes coordinated motions with easy according to the combination of control or motion. Furthermore, it still supplies of PVT (position-speed-time) motion mode and linear interpolation, circular interpolation and curve interpolation, etc.

2 Technique Design of FRF

The production process of FRF is shown in Fig. 1.

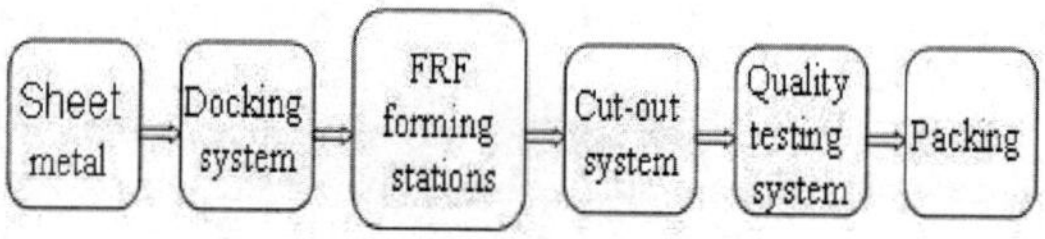

Fig. 1. Production process of FRF

During the course of process of FRF, sheet metal is formed gradually into final shapes by rotating rolls as it travels between the forming stations. In addition to quality issues, the minimal number of forming units is required when designing [2]. The width of sheet metal of FRF is changeable on the axel direction [5]. Fig. 2 illustrates the principle of FRF. Each forming station is an independent unit and achieves a certain angle forming for products during forming which is controlled separately by industry computer. As shown in Fig.2, sheet metal moves laterally in X direction is driven by three-phase asynchronous AC frequency conversion motor, and rolls moves longitudinally in Y direction relay on ball screw driven by servo motor. Control system drives servo motor system to control each forming station to move in longitudinal direction to match the cross-section profile and realize different shapes forming when processing according to process document that will to be translated into linear interpolation or curve interpolation instructions, etc. And adjusting the position and rotation angle of rolls can produce kinds of desired changeable cross-section products.

Four methods are mainly utilized in design changeable cross-section geometry. There are fixed-radius method, fixed-length of neutral line method, track method and angle/radius method [7]. Each method has its pluses and minuses, adopting which of them depend on the cross-section geometry that will be formed. However, the fixed-length of neutral line method and the fixed-radius method are commonly utilized in practice.

Each rolls unit bends a certain forming angle and realizes a certain forming when processing. Superposing the cross-section of each forming process together, that is, flower pattern (Fig. 3). Here, the fixed-length of neutral line method is adopted. The length is calculated by fixed-neutral line which keeps constant throughout the forming process, and the interior radius r_i calculates the angle automatically according to the bending angle α_i. Besides, the strain of material is considered during forming. The principle of fixed-length of neutral line method is shown in Fig. 4.

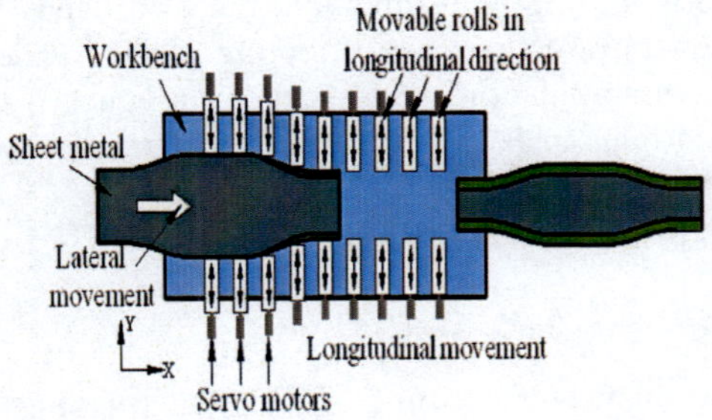

Fig. 2. Principle of FRF working

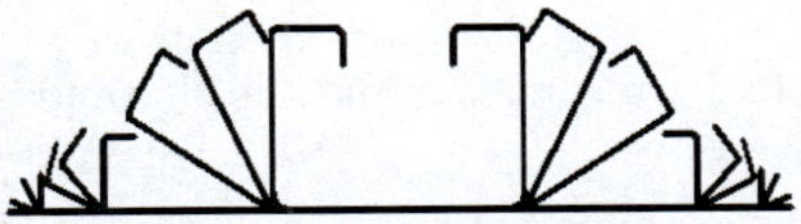

Fig. 3. Flower pattern

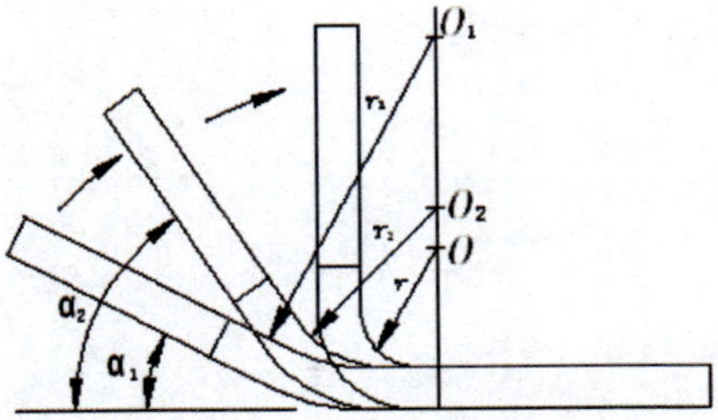

Fig. 4. Principle of fixed-length of neutral line method

As same as the process of traditional cold roll forming, FRF can also be processed continually with unlimited extension in longitudinal length. Furthermore, the latter can adopt interchangeability of rolls and allow multi-forming units to use one certain type rolls [7], which not only overcome the defects caused by adopting different rolls in different process units but also reduce the costs of production and improve productivity.

3 Overall Plan Design of Control System of FRF

According to the technique principle of FRF, when sheet metal is driven forward at a certain speed by frequency conversion motor, it not only needs to accomplish the continuous bending forming in lateral direction, but also needs to achieve the edge forming in longitudinal direction by adopting linear interpolation, circular interpolation and curve interpolation modes, etc. Consequently, ensuring the exact match between the feed rates in lateral direction and the servo actuators' velocity is critical to achieve high-precision forming.

The forward velocity of sheet metal as follow,

$$v = \pi \cdot d \cdot n_{cm} / i \tag{1}$$

Here, d represents the diameters of rolls, n_{cm} is the rotating speed of frequency conversion motor and the reduction ratio is indicated with i.

Equation (2) is the rotating speed of frequency conversion motor.

$$n_{cm} = 60 f_1 (1 - s) / p = n_0 (1 - s) \tag{2}$$

The frequency of motor is figured with f_1, s is the slip of motor, and p means the pole-pair of motor.

In the case of that the number of stator winding pole-pairs of frequency conversion motor keeps constant, adjusting motor's frequency, that is, to achieve the speed regulation by changing the synchronal-speed n_0 of motor. And according to the forward velocity of sheet metal can realize to compensate the servo motor position or velocity, and to drive actuators to move.

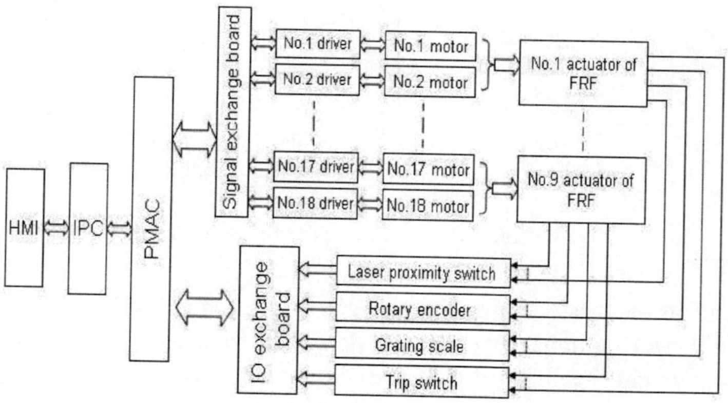

Fig. 5. Hardware Architecture of FRF control system

Taking into account that sheet metal is driven forward by friction force supplied by rotating rolls, which will make it difficult to avoid slipping and that will result in outstrip movement or lag for actuators in lateral direction. That will have a negative affect on sheet metal when forming. Therefore, the key point of control is constructing a closed loop control system that can monitor and control the process of forming in real-time.

This control system includes IPC (Industrial Computer), PMAC motion controllers, servo drivers, servo motors and actuators of FRF, etc. The whole production forming of FRF is consisted of nine groups of forming stations. Each group stations are driven by two motors simultaneously to accomplish a certain shape forming.

The architecture of control system is shown in Fig. 5. The sensor system of this control system is built for monitoring and controlling the feed speed of sheet metal and path interpolation of forming, which is made up of trip switches, rotary encoders, grating scales and laser proximity switches. And here, laser proximity switch detects sheet metal whether it travels the rolls real-time with no delay, trip switch measures the rolls' lateral boundary displacement, and the feedback signal of circumferential velocity or acceleration is obtained by rotary encoder in real-time. As it should be, grating scale realizes the checking and definition of zero-position of control system.

4 Realization of Control System of FRF

Control system adopts the Visual C++ software to realize the controlling on whole FRF machines .And the software part of FRF is consisted of upper computer control-software and lower computer control-software, which is developed by using modular design method based on the DSP of PMAC controller. As shown in Fig. 6, the upper computer control-software includes communication module, debug module, parameters setting module, process module and simulation module.

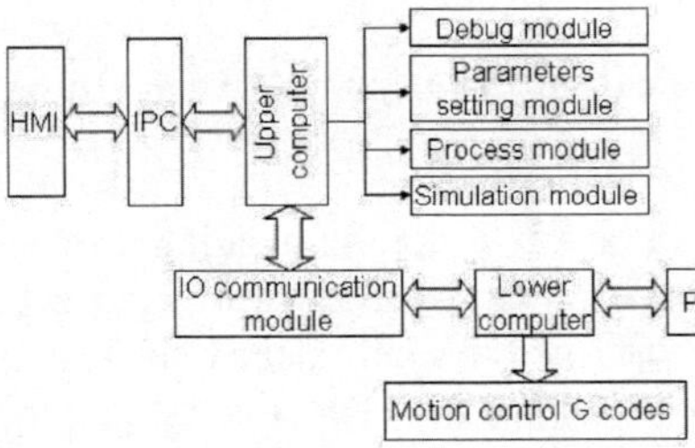

Fig. 6. Software architecture of FRF control system

Communication module is the essential linking of all parts of control system, which mainly takes charge of communication between IPC and servo controller, and communication between IPC and PMAC by adopting ethernet.

Debug module is responsible for debugging and configuring for the velocity and position of motors and rolls to improve the working order according to different forming process.

Parameters setting module includes the feed rate of sheet metal, spacing between two forming rolls units, the length and width of processing sheet metal, which is in charge of setting the parameters of forming process to meet the processing requirements of different products.

Process module manages the process plan of FRF, which allows users to select the motion modes of forming axle and to load process plan document and rolls of processing units. The PMAC can directly read the graphics data of DXF or STL document, and generate corresponding motion codes.

In addition to quality of codes conversion, waste products and accidents issues, simulation module supports test online for process data of FRF, which can validate and test relevant process data before forming by simulating according to the motion codes translated from DXF or STL document.

4.1 Realization of the Implementing System of FRF

The motion control programs of PMAC include motion control G codes and PLC codes, which are the real-time control programs of forming stations. In this paper, the control of FRF can be realized just by utilizing G codes.

Adjusting and configuring the parameters of PMAC and servo drivers, IPC executes the execution software of PMAC under Windows and sends instructions to PMAC. The analogue output signal of PMAC takes charge of driving the substrate servo devices after the PMAC received instructions from IPC.

In order to make the system obtain good steady states and dynamic characteristics, it needs to adjust the PID of control loop of system. As the core controller in this control system, PMAC provides a control loop algorithm (PID + velocity /acceleration of feed-forward + NOTCH filter) which almost can meet the requirements of all application situations. Comparing with the closing rate between response curve and ideal curve users can estimate the characteristics of control system.

4.2 Realization of Communication Programs between Upper Computer and Lower Computer

PComm32.DLL dynamic link library is a communicational driver programs library provided by Delta-Tau which has the characteristics of high efficiency, high speed and good stability, and it almost includes all means of communication between PMAC. To realize the communication between upper computer and lower computer it just needs to call the dynamic link library in application programs. There are two methods on calling dynamic link library, that is, static link calling and dynamic link calling. The later is adopted in this system. Here the dynamic link library functions as follows are utilized.

- OpenPmacDevice(); // Open the communication channel between application programs and PMAC;
- ClosePmacDevice(); // Close the communication channel between application programs and PMAC;
- PmacGetResponseA(); // Send instructions to PMAC and read the feedback signal from PMAC;
- PmacConfigure (); // Open dialog and configure the parameters of PMAC;
- PmacDownLoadA(); // Download the control programs from PC to PMAC.

Realization of Motion Control Programs
The open instructions system of PMAC not only allows users to call available instructions such as line interpolation, circular interpolation and curve interpolation but also permits to define instructions of M, S, G and so on by themselves according to instructions system. And here, after setting the motion axle and interpolation methods, the research work on motion control programs of FRF is translating the graphics data of DXF or STL document into the instructions that the PMAC can understand.

Process plan module loads DXF or STL document for each forming station and generates motion control G codes according to the motion distance and the angle of rolls axle. The angle of rolls axle is calculated from the discrete value that is derived from the discrete of project curve after the projection in roller axle motion direction is calculated.

4.3 Control System of FRF Application

Human machine interface (HMI) is the channel of communication between users and FRF system, which can be developed by several soft wares such as Visual B and Visual C++, etc. This control system adopts the MFC library of Visual C++ and realizes its operation function through several basic interfaces. They are device debugging interface, process plan interface, parameters setting interface and the simulation interface.

- Device debugging interface makes preparation for forming processing by configuring and adjusting the zero-position of motors and rolls axle.
- Process plan interface takes charge of configuring and loading the motion modes of rolls axle, the rolls models of each forming station and the process documents.
- As shown in Fig. 7, the parameters setting interface includes the length, width, feed speed of sheet metal, the spacing of FRF and the number of forming stations, which takes charges of parameters setting of process plan.

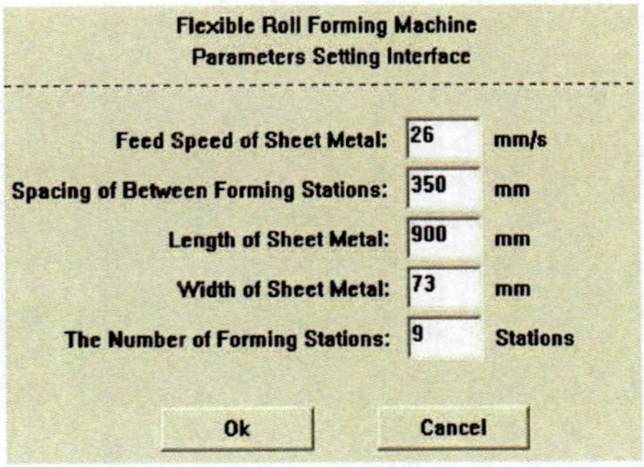

Fig. 7. Parameters setting interface

- Simulation interface is responsible for test online of simulating for forming data and ensures the quality of configuration for forming process by monitoring relevant simulation data.

4.4 Realization of System on the Whole FRF Machines

The control scheme of this system is aimed at the single-axis flexible roll forming machine of North China University of Technology. And the whole forming processing system is consisted of nine forming stations. Each forming station includes up rolls axle which is the driven axle and down rolls axle which is the driving axle. And sheet metal is driven to travel forward under the driving force of three-phase asynchronous AC frequency conversion motor. Servo motor actuates ball screw to drive actuators to realize a certain flexible forming on the width of cross-section geometry. The sensor detect system of this control system is built for monitoring and controlling the forming process, which is made up of trip switches, rotary encoders, grating scales and laser proximity switches.

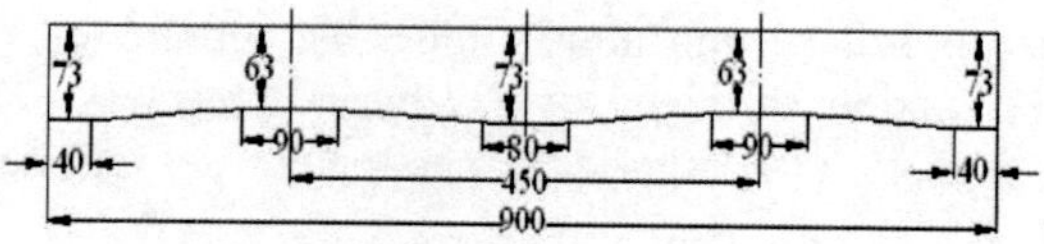

(a) Theory parameters of before- forming

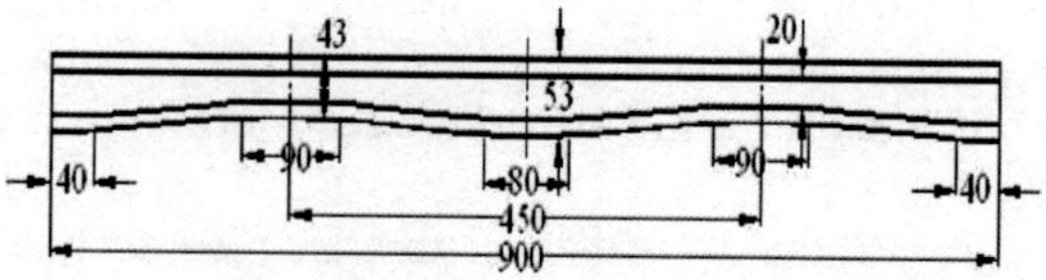

(b) Theory parameters of after- forming

Fig. 8. Theory parameters of before- forming and after-forming of FRF product

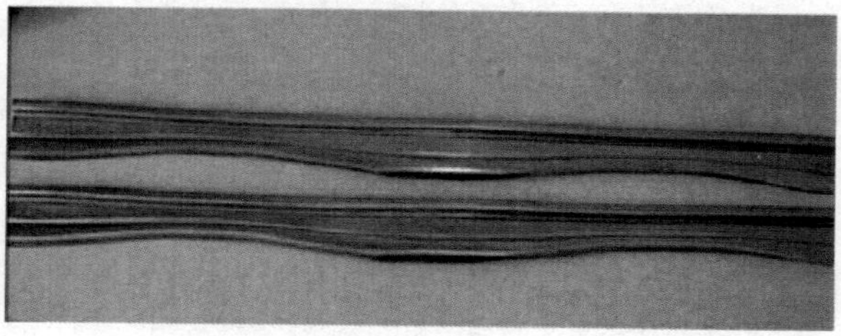

Fig. 9. FRF products

Fig. 8 shows are the theory parameters of one kind of flexible forming sheet metal on before-forming and after-forming. The cold sheet metals with 0.8mm thickness are utilized as materials in experiment. The experimental results coincide on the whole with calculated result by analyzing the experimental data and theory data. And the FRF products are shown in Fig. 9.

5 Conclusions

Based on the process characteristics of FRF and the technique of motion control of rolls axle, this paper makes fully use of the characteristics of multi-axis control and fast real-time communication capability of PMAC and the rich software resources, efficient data processing capability of IPC, which has realized the accurate control and automatic measuring for processing data by adopting closed-loop control mode. And that will be theoretical and practical basis of research on multi-degree-of-freedom of FRF for future.

References

1. Ona, H., Watari, H.: On Production of Programmable controller cold roll forming machine. In: Proceeding of the 54th Japan Joint Conference for the Technology of Plasticity, Japan, pp. 67–68 (2003)
2. Downes, A., Hartley, P.: Using an artificial neural network to assist roll design in cold roll-forming processes. Journal of Materials Processing Technology 177, 319–322 (2006)
3. Groche, P., von Breitenbach, G., Jockel, M., Zettler, A.: New tooling concepts for future rollforming applications. In: 4th International Conference on Industrial Tools (ICIT 2003), Slovenian, pp. 121–126 (2003)
4. Ona, H.: On Roll Forming of Section with Different Width in Longitudinal Direction. In: Proceedings of The Seventh Symposium on Engineering. North China University of Technology, Beijing, China, pp. 1–4 (September 2004)
5. Halmos, G.T., Liu, J., Ai, Z.: Roll forming handbook, pp. 178–190. Chemical Industry Press, Beijing (2009)
6. Ona, H., Liu, J.: Technology of cold roll forming, pp. 203–205. Chemical Industry Press, Beijing (2008)
7. Bu, G.: Research on the simulation of flexible roll forming based on FEA. Master thesis of North China University of Technology, pp. 2–7 (June 2007)
8. Zhu, R., Liu, P., Fang, Y., Lu, Y.: The software designing of wielding automation based on PMAC. Mechanotronics. pp. 53–55 (January 2008)
9. Chen, D., Ji, F., Tang, X.: Design and debugging of axis Z servo system controlled by UMAC. Design and Research 2, 82–85 (2008)
10. Lin, X.: Design of control system of flexible roll forming. Master thesis of North China University of Technology, pp.13–15 (May 2009)
11. Li, W.: State monitoring of control system of flexible roll forming. Master thesis of North China University of Technology, pp. 13–20 (May 2009)
12. Lin, Y., Li, D., Wang, L.: Development of 6-axis testing machine Control Software Based on PMAC. Microcomputer Information (Control of Automation) 25(11-1), 56–58 (2009)
13. Ona, H.: Cold roll forming for high tensile strength steel sheet proposition on forming of thin spring steel sheet pipe. Journal of Materials Processing Technology 153-154, 247–252 (2004)

Knowledge Acquisition from ANSYS Analyzed Data for Driving Head of Hydraulic Drill*

Daohua Liu[1] and Jin Li[2]

[1] School of Computer and Information Technology, Xinyang Normal University Xinyang, China
[2] Network Information Center, Xinyang Normal University Xinyang, China
ldh30708@sina.com, ldh30708@sina.com

Abstract. This paper mainly discusses how to obtain design knowledge from the finite element analyzed data. With regard to one design problem, the orthogonal experiment table is established according to diverse values of design variables, and variables in each kind of experimental schemes are built by the entity model of Pro/Engineer software and then the entity model is analyzed by ADAMS in order to obtain design parameters which satisfy design need. The data of different characteristics collected from the stress analysis and other relative computation data constitute a new two-dimensional form from which design knowledge is extracted by using the rough set theory. This method of knowledge acquisition is tested by the analysis of the data from driving head of hydraulic drill. The test shows that the method obtains three strong associate rules for the design of driving head, which is applicable. The experiments prove that this method of knowledge acquisition is correct and effective and it can also be used as guidance for extracting design knowledge from other engineering analyzing software.

Index Tems: Knowledge acquisition, ANSYS analyzed data, Design for driving head of Drill, Orthogonal experiment, Rough set theory.

1 Introduction

ANSYS is the most representative large-scale analysis of software in area of CAE. It can analyze many problems in mechanical design. Take the ANSYS structural analysis for example, it includes the structure of static analysis, the analysis of the modal, the analysis of the harmonic response, the analysis of fracture, the analysis of the composite materials and the analysis of the fatigue [1]. ANSYS analysis can generate many design data, includes the visual graphical data, the text data and the table data generated by the results of post-processing [2]. These data vary in a disorderly manner. How to extract strong association rules (knowledge) from this large amount of analysis data still remains a difficult problem both at home and abroad. At present, in

* This work is supported by the special fund for outstanding young teachers scientific of Higher Education of Henan and the Science Research Project of Education Ministry of Henan (2010A520034).

J. Luo (Ed.): Affective Computing and Intelligent Interaction, AISC 137, pp. 707–715.
springerlink.com

construction machinery design, the analysis results of ANSYS can only provide a theoretical basis for optimal design, it can not extract knowledge from analysis data, in other words, it can not produce guiding conclusions.

In view of this situation, the authors propose a new procedure in which the ANSYS analysis model is first constructed in the form of a orthogonal experimental table according to the different values of design variables, and next ANSYS analysis is done according to parameter setting on each row of orthogonal experimental table, then the necessary data are extracted and translated into a standardized Access two-dimensional data table whose rows and columns are later simplified by identification matrix operation in rough set theory ,finally the strong association rules of this problem are generated. These rules can be used to guide the design of similar products in order to obtain better results in engineering practice.

2 Knowledge Acquisition from Finite Element Analysis Data

2.1 Knowledge Acquisition Process

In this section, the knowledge acquisition process is described.

Usually the parametric design of the key components of design problem is first done; and a solid model is constructed using Pro/E. Next the solid model is imported into ADAMS where static and dynamic simulations are conducted. Then some selected simulation results close to real design are imported into ANSYS. Define design variables such as material type by APDL language, and design a suitable orthogonal experimental table. Apply the same external load to the setting of orthogonal experimental table in each row, and divide the grid freely based on the different design models, and then solve the problem in ANSYS. Draw the deformation data and stress data of design problem from the solving results, obtain a record in Access with a certain data extraction method, then eventually obtain a standardized two-dimensional data table by the same method. Based on the data table, do data analysis through rough set theory. Finally, generate strong association rules of these data, that is, the knowledge rules. The whole process of knowledge acquisition is illustrated in Figure 1.

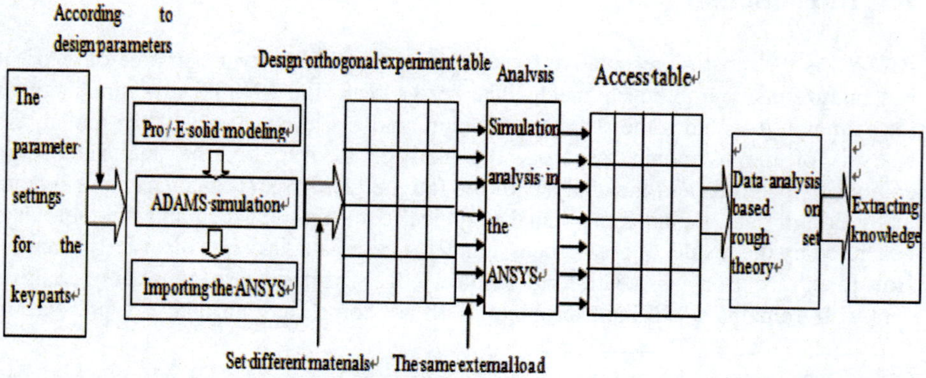

Fig. 1. The process of automatic knowledge acquisition based on analyzed data

2.2 Test Arrangement for the Orthogonal Experiment Method

Orthogonal experimental design method is an effective method to arrange multi-factor and multi-level test by using the orthogonal table [3]. In the process of the orthogonal test design, firstly, we should be clear about the testing purposes, and determine the examining indexes. Secondly, we should determine the factors and levels to make the factor level table. Determining the factors and levels is to choose some most appropriate values for each design variable which can represent the results of the design problem better. The factors and levels used in this paper are based on the practical experience. Finally, we should select the proper orthogonal table, design the table header, and determine the test program. We can choose the minimum orthogonal table meeting the factors and levels. To guarantee the reliability of the test, we can choose the proper orthogonal table (for example $L_8(3^7)$). Arrange the interacting factors according to the principle of being not mixed with the rest randomly.

2.3 The Extraction and Transformation of Finite Element Analysis Data

All of the operation of ANSYS is finished by APDL programming language parameterization technique [4]. Load the value of the parameter into access data table record by the method of ADO transferring. About the technique of ADO access database, there are 4 steps of operation in turn, namely: (1) introducing ADO libraries, (2)initialize OLE/COM library environment, define respectively the interface pointer, _ConnectionPtr interface pointer and _RecordsetPtr interface pointer, (3) initialize ADO data collection, (4) open ADO database, complete data extraction operation.

2.4 The Process of Extracting Knowledge Based on Rough Set Theory

The record collection of the data table of Access is standardized by specific algorithm. The core of property for solving the identification of matrix and the function of data sheets are identified. The maximum generalization of rule is extracted from the decision matrix and decision-making functions (minimum decision rules). Finally knowledge rules are extracted from the record set [5].

1) Property Returning

Solve the core of property returning collection by identification of the matrix and the identification of the function, and calculate the returning collection by the returning arithmetic. Delete unimportant property according the reliability of the property of returning collection. When the database is in a large quantity, we can extend the initial data table. $S=(U,A,V,f)$ is a knowledge expression system, the identification matrix of $|U|=n$ and S is a $n\times n$ matrix. When x_i and x_j belong to the same equivalences, every element $\alpha(x_i,x_j)=\Phi$; when x_i and x_j belong to the different equivalence, every element $\alpha(x_i,x_j)=\{a\in A\mid f(x_i,a)\neq f(x_j,a)\}$. Therefore, $\alpha(x_i,x_j)$ is a collection to distinguish all attributes of the object x_i and x_j.

In order to get the core of matrix, introduce Boolean function, which is called the identification function (discernible function), and described as Δ. Its definition is [6]

$$\Delta = \prod_{(x_i,x_j)\in U\times U} \sum \alpha(x_i,x_j). \tag{1}$$

The core is a collection of all the single element of the identification matrix, namely, $core(A)=\{a\in A \mid \alpha(x_i,x_j)=\{a\}$, in which $x_i,x_j\in U\}$.

2) The Solution for the Simplest Set of Attributes Reduction

In the attributes reduction, the importance for any attribute ,for example a ,could be conformed by interdependence between tow attribute sets $P,R\subseteq Q$.The dependence between P and R can be determined as $\gamma_R(P)$ [7].

$$\gamma_R(P)=card(POS_R(P))/card(U), \tag{2}$$

$$POS_R(P)=\bigcup_{X\in U/IND(P)} \underline{apr}_R(X). \tag{3}$$

In Eq.(2), $card(\cdot)$ is the base of sets, and in Eq.(3), $POS_R(P)$ is a positive regional for R in $U/IND(P)$.

With the solution for attribute dependability, you can calculate the importance of any attribute in order to delete this attribute or not. Because of the importance for condition attribute on decision, the importance indicator β can be recognized by relative difference, which was formed after you deleted the condition attribute.

$$\beta=\gamma_R(P)-\gamma_{R-\{a\}}(P). \tag{4}$$

If $\beta=0$, it is indicated there is no effect to decision by condition attribute, then the condition attribute can be deleted. So, after analysis and treatment, the smallest inductive for decision system sets can be got [8].

3) The Solution for the Maximal Generalized Rule (The Minimum Decision Rule)

For decision attribute $d\in A$ and its specific value V_d, we can describe the pairs of attribute value, which are used to distinguish objects from all the attribute sets $\{V_d\}$ and sets $U-\{V_d\}$.

Matrix for decision is defined as follow:

$$DM_{ij}=\{(a,a(e_i));a(e_i)\neq a(e_j)\}.$$

Set DM_{ij} contains all pairs of attribute values which come from the difference between e_j and e_i, and formally the pairs are represented as $DM=[DM_{ij}]_{m\times n}$.

The basic method for extracting maximal generalized rule (the minimum decision rule) from decision matrix is described as follow:

$$RUL = \bigcup_i RUL_i , \tag{5}$$

Where $RUL_i = \{r \in RUL \mid A_r(e_i) = (V_{i1}, V_{i2}, \cdots, V_{in})\}$, $e_i \in \{V_d\}$, the symbol RUL_i represent maximal-generalized-rule set in which all the conditions and object mutually are matched.

4) Rule Extraction

Each step through the above processing, got the most simple rational decision-making table, the condition attributes in the most simple decision table act as a precondition, while the decision attribute value of production rules, as the latter piece. The ultimate form knowledge of rules are expressed as

$$R_k : \quad \underset{j=1}{\overset{m}{AND}} \; P_{kj} \; (\beta_{kj}) \rightarrow Q_k . \tag{6}$$

In Eq.(7), P_{kj} is a prerequisite of production rules; Q_k is the conclusions of production rule; k is the new rule number; j is the first k production rule of the first j pre-condition; m is the total number of rule precondition.

3 Knowledge Acquisition Example for the Driving Head of Hydraulic Drill Based on Finite Element Analysis Data

Currently, it is a drilling hotspot that researching and developing the valveless rotary impact hydraulic anchor rig based on the soil anchor improved YM-type full-hydraulic drilling rig, while the power pack is its key component. This paper will extract the guidance knowledge of its design from taking designs and making analysis with the power pack. In the knowledge acquisition process, firstly, use the power pack optimal design results in the literature [9], and then use the Pro/E drawing and importing ADAMS to do kinematics and dynamics simulation. Finally, carry out its simulation model which is shown in figure 2.

Fig. 2. The digital model for driving head of hydraulic drill

In this instance study, there are two kinds of experimental designs: one is that imposing the different size of force on the same material to study the value of the stress on several key points with the power head, finally, determining the value of the minimum allowable within the scope of the minimum force. The other is that the system imposes the same external load on the different material to identify the most common material within the key points of the maximum stress value in the range of the allowable stress values by ANSYS deformation and stress analysis. I experiment with the second option based on this, treating the spline shaft that withstands the impact in the power head as the test object, identify the areas of stress concentration points or the danger points (maximum stress points) and the maximum deformation place. The time-varying dynamic load composed on the spline shaft as F_1, which of the maximum force value is 15000N and the varying frequency is 50Hz.

Table 1. The orthogonal experiment table for driving head

Record	$A(MPa)$	$t(s)$	N	A'
1	175	0.005	ND56	14
2	175	0.010	ND4565	30
3	175	0.015	ND9050	41
4	195	0.005	ND4565	41
5	195	0.010	ND9050	14
6	195	0.015	ND56	30
7	205	0.005	ND9050	30
8	205	0.010	ND56	41
9	205	0.015	ND4565	14

In the experiment, let the variable of the material type is A, take the material for high-quality carbon structural steel, the density of it is $7.9\times10^3 \mathrm{kg/m^3}$, the elasticity modulus is $2.06\times10^{10} N/m^2$, poisson's ratio is 0.3, which corresponds to the steel and the steel of the allowable stress levels have three factors, include 08F ($[\sigma_s]=175MPa$), 10F ($[\sigma_s]=195MPa$), 15F ($[\sigma_s]=205MPa$); let the run-time variable is t, and taking $T/4, T/2, 3T/4$ for the time variable three level factors set. Setting the key component of the node variable as N, depending on experience, take three positions as follows for three level factors, a certain part that the shock effect of a direct impact on a site occurs, a certain part of the bolt rig in contact with the tail of anchor and a certain part in the middle of the spline shaft. Therefore, the experiment using orthogonal table $L_9(3^4)$, the specific orthogonal test table is set as indicated in table 1.

Take the first row of the orthogonal experiment table for an instance realizes the process of knowledge acquisition. Regarding the spline shaft which is a key component of the power head as a research object analyzes the change of the stress with ANSYS. Firstly, define cell property of the impacted entity, in addition, exert the external load of which the maximum value is 15000N, and the frequency is 50Hz. And then based on free mesh, the result into 9129 nodes and 5702 units are divided. So the finite element model of impacted spline shaft has been established, the stress distribution in the first 0.005 seconds (for 1 / 4 cycle) shown in figure 3.

Fig. 3. The deformation model for spline shaft in t=0.005

The examination results for the orthogonal experiment are as follows: let the deformation value of the key parts node is V_1 (mm), the maximum deformation is V_2 (mm) and the maximum deformation node No. is B then, $|V_1 - V_2|$ is the stress value of the maximum node minus the stress value of the key node, $|N - B|$ (mm) is the space distance of the key node with the maximum stress point. E shows the results of evaluation (Comprehensive consideration of the material economy and the spline shaft deformation size).

Table 2. Comparison of the results for the four optimization approaches

Test No.	***Setting influence Factor level***				***Results***					
	A	***t***	***N***	***A′***	***V1***	***V2***	***B***	***|V1-V2|***	***|N-B|***	***E***
1	175	0.005	ND56	14	7.12	9.76	ND 8558	2.64	479.09	Very good
2	175	0.010	ND4565	30	8.36	10.64	ND8841	2.28	240.95	General
3	175	0.015	ND9050	41	7.59	10.21	ND9010	2.62	2.25	General
4	195	0.005	ND4565	41	6.82	9.22	ND 8608	2.4	227.82	Good
5	195	0.010	ND9050	14	6.98	9.32	ND8910	2.34	7.89	Very good
6	195	0.015	ND56	30	6.76	9.12	ND8995	2.36	503.71	Good
7	205	0.005	ND9050	30	6.56	9.02	ND 8586	2.46	26.15	General
8	205	0.010	ND56	41	6.91	9.85	ND8890	2.94	497.80	General
9	205	0.015	ND4565	14	6.16	8.76	ND9106	2.60	255.89	Very good

Use the way of 1.3 (the date extraction and transformation of finite element analysis) extract the test result data. Finally, read out the maximum stress point distribution of every moment, the maximum stress point No., the stress distribution of study and the results of evaluation. Then, calculate the difference between maximum node and the study node, the space distance of maximum deformation point and the study node, finally, gain the first record of Access date table. Then, use the same test method, complete all of the orthogonal design experiments, so the two-dimensional table is gained, the contents of which is shown in table 2.

Table 2 is two-dimensional information, firstly value normalized, and then performed data generalization operations, calculated identification matrix, seek the core of identification matrix, attributed reduction, seek the rule set of the minimum decision, then complete rough set knowledge acquisition of 1.4. The final extraction of knowledge rules is shown:

R_1 : *if* any material at anytime on spline shaft head *then* stress concentration(danger point);

R_2 : *if* any material at $3T/4$ moment on any site *then* stress concentration(danger point);

R_3: *if* draw material 10F steel ($[\sigma_s]=195MPa$) at anytime on any site *then* within permitted range of stress.

The rule R_1 showed that the spline shaft head is the part which was impacted the piston directly, which had a larger deformation during work time, because of resilient, it is consistent with the actual work, so designing the position require good material properties, good manufacturing and high level of heat treatment process; the rule R_2 showed that the maximum stress deformation is not in the biggest impact amplitude of the $1/2$ cycle during the work time, but in the $3/4$ cycle; the rule R_3 showed that to meet the conditions of same external load should choose appropriate materials, there is no need to pursue higher performance materials.

4 Conclusion

This paper illustrated the operation process which is based on the design knowledge of ANSYS simulation analysis data extraction. We used orthogonal experimental design method to reduce the number of experiments of knowledge extracted and to improve the efficiency of knowledge acquisition. We gave an example of the knowledge acquisition from drill power pack simulation analysis data, gained the dangerous part which is in the spline shaft head of power pack. The largest stress is found not to be in the greatest amplitude of the external load cycle but in the $3/4$ cycle when power pack works, to meet the conditions of same external load selecting appropriate materials can meet the requirements without the need for the pursuit of higher performance materials which is the strong association rules guiding (knowledge). The knowledge acquisition method can be extracted to analyze data and get design and analysis knowledge for other engineering simulation software, the good method has a great value to promote.

References

1. Ma, J., Chen, J., Xu, Y.: Analysis of the finite element and the reliability of the structures with fuzzy-random parameters based on the two-factor method. Journal of Xidian University 36(1), 69–73 (2009)
2. Yang, D., Qiu, Y., Duan, B.: Optimal design of cable pre-tensions of the tension truss antenna. Journal of Xidian Unviersity 35(2), 319–323 (2008)
3. Hwang, R.L., Lin, T.P., Liang, H.H.: Additive model for thermal comfort generated by matrix experiment using orthogonal array. Building and Environment 44(8), 1730–1739 (2009)
4. Samir, S., Emmanuel, B., Fortin, J.: Finite element analysis of automotive cushion discs. Thin-Walled Structures 47(4), 474–483 (2009)
5. Liu, D., Yuan, S., Zhang, X.: Research on knowledge acquisition method about the IF/ THEN rules based on rough set theory. Journal of Systems Engineering and Electronics 19(3), 628–634 (2007)
6. Krzysztof, D., Greco, S., Roman, S.: Rough set approach to multiple criteria classification with imprecise evaluations and assignments. European Journal of Operational Research 198(2), 626–636 (2009)
7. Luo, Q., Chen, S.: Algorithm based on value reduction and decision tree to generate minimal rules. Computer Applications 25(8), 1853–1855 (2005)
8. Liu, T.: Data mining technique and application. The Press of National Defense Industry, Beijing (2001)
9. Liu, D., Qiao, C., Yuan, S.: Optimization design of driving head based on simulated annealing algorithm. J. Xi'an Univ. of Arch. & Tech. 39(6), 878–884 (2007)

Enhancing the Core Competitiveness of SMEs by Using E-Commerce

Zhao Wei

School of Economic & Management,
Shenyang Ligong University, Shenyang, China
zweitech@126.com

Abstract. SMEs(small and medium-sized enterprises) in China are impacted by enormous changes in their business processes. E-commerce offers a promising way for organizations to meet challenges of an ever-changing environment. However, few available studies related to SMEs in China reveal that they lag behind and are skeptical about adopting the technologies irrespective of their effectiveness. This paper presents the relevant support required by SMEs in China, and the related solutions to enhance the core competitiveness are also discussed.

Keywords: Small and Medium-sized enterprises (SMEs), E-commerce, Core Competitiveness.

1 Introduction

Small and Medium Enterprises (SMEs) are widely regarded as the powerhouse of Chinese economy. In recent years, the use of the Internet along with a range of other information and communications technologies(ICT) is transforming how business is done locally and globally, and the emergence of the Internet has allowed SMEs to compete effectively and efficiently in both domestic and international markets. SMEs have the potential to achieve rapid and sustainable development by using Electronic commerce (e-commerce) because it offers SMEs the opportunity to cut the costs of transactions and trade globally rather than locally.

1.1 Defining E-Commerce

In this paper, we use a broad definition of e-commerce, that is: Any use of information and communications technology by a business that helps it improve its interactions with customers or suppliers.

The narrower definiton of e-commerce might be to limit it only to the electronic exchange of business transactions themselves, e.g.,oders and invoices (with or without the use of a website). In fact, business in China may be able to reap significant benefits from e-commerce defined more broadly even when legal, regulatory or infrastructure constraints may take it diffcult or impossible for them to actually transact business electronically, especially to SMEs. For example, a business

J. Luo (Ed.): Affective Computing and Intelligent Interaction, AISC 137, pp. 717–723.
springerlink.com

may be able to use an e-marketplace or even Internet searches to figoure out market prices for its products, so it can compete with other and win more new customers.

E-commerce in China is growing at a quick clip. eMarketer estimates that the average amount spent online per Internet user in China will increase roughly fivefold by 2010 from 2006 levels, which was shown as Tab 1. above[1].

Table 1. Chinese average annual amount spent online

Average Annual Amount Spent Online* per Internet User In China,2005-2010		
	Amount spent per user	***% change***
2005	RMB114($14)	
2006	RMB157($20)	38.2%
2007	RMB237($31)	50.8%
2008	RMB358($47)	51.4%
2009	RMB513($69)	43.1%
2010	RMB713($98)	39.0%

Note:Converted at average annual exchange rates (projiected for future years); eMarketer defines an Internet user as a person ages 3+ who accesses the Internet at least once per month;*includes online travel.
Source: eMarketer,January 2007

1.2 Defining SMEs

SMEs are extremely important to the economy of any country whether developed or developing. They play a critical role in economic development. Different countries use different parameters to define SMEs. Some use the number of persons employed, amount of capital invested, amount of turnover or nature of the business. In China there is no clear definition of an SME. The main criteria used are the number of employees, the size of fixed investment, and the nature of the business and the sector. The current delineation standards is "Provisional Regulations of SMEs Standards" formaluted by State Economic and Trade Commission(SETC), State Planning Commission(SPC), National Bureau of Statistics(NBS), State Ministry of Finance(SMC) in 2003[2]. SMEs in different industries ought to decided by the standard below:

Industrial enterprise size in China based on: the number of employees fewer than 2000 or Sales under 3000 billion or total assets value under 4000 billion.

Commercial enterprise size in China based on: the number of employees fewer than 500 or Sales under 3000 billion or total assets value under 4000 billion.

Construction enterprise size in China based on: the number of employees fewer than 3000 or Sales under 3000 billion or total assets value under 4000 billion.

SMEs in China perform a strategic role by accounting for a very high percentage of the total number of business establishments. 80% of Chinese labour force was employed in the SME sector (including the agricultural sector). SMEs contribute substantially to employment and income generation[3].

Many SMEs can gain a competitive edge by using the Internet well to do market research, find information on competitors and track down leads for new customers, or provide better customer support. These activities all fall under the broader definition of electronic commerce and could be termed "e-business". This broad definition includes transacting business or exchanging business-related information between:

B2C: business and consumers (their customers)

B2B: business to business, e.g., where one business buys supplies from another or buys products to resale.

G2B: businesses to government, where perhaps businesses conduct transactions electronically with government regarding various business licensing or reporting requirements or where businesses sell products or services to governments.

2 Why Prepare SMEs in China for E-Commerce

First, as mentioned above, they are important to economic development in China. SME's account for 80 percent of all employment, it is critical for such businesses to be prepared for and take full advantage of any benefits offered by electronic commerce.

The second reason to focus on SME's is that they are in a very good position to adapt to new technology; they may be able to adapt faster than larger companies that can be slowed by bureaucracy and stricter staffing hierarchies. E-commerce it may offer them comparatively more advantages to find new customers and suppliers especially in markets they have not easily been able to reach before-either internationally or regionally. Markets everywhere are globalizing partially due to the widespread use of the Internet. Electronic commerce can give SMEs a better chance to compete in their markets and, indeed, in some cases, is or will soon become a competitive necessity for survival. The potential benefits of electronic commerce to "level the playing field" for SME's-allowing them to compete better-are critical for them to understand and sort out.

In general, the firms that enter electronic markets are either start-up firms specifically designed to operate in the Internet environment, or established firms that migrate to electronic commerce. The economic significance of Internet start-ups is very small, but is growing fast. The "scalability" of the Internet offers small niche players many of the advantages enjoyed by large firms in terms of expanding the range of e-commerce customers and transactions. This may be particularly important for small innovative firms entering the electronic market.

In principle, the advent of the Internet is helping to enlarge geographical and sectoral markets by cutting through many of the distribution and marketing barriers that prevent smaller firms from entering foreign markets. Smaller firms may particularly benefit from the opportunities offered by electroniccommerce. They tend to be less locked in to legacy technology compared to larger firms, and they are normally unencumbered by existing relationships with traditional retail channels. Hence, they can adopt a business model that forces larger, established competitors to restructure their existing relationships. The Internet also provides opportunities for businesses to compete in new areas by creating new products or seances.

The networking and sharing of functions, typical of clusters and partnerships, enable firms to amplify the gains of electronic commerce. New opportunities for SMEs stem from the integration of supply and demand chains through horizontal inter-firm linkages between suppliers and customers and from the creation of production clusters. These forms of industrial organisation allow SMEs to overcome their isolation by interacting and sharing information with partners. They can contribute to solving SMEs' problem of lack of resources and access to technology by promoting transfer of knowledge through the use of integrated processes or through system-wide interactions in R&D (user-producer, alliances, outsourcing, links to the scientific community)[4].

SMEs also adopt e-commerce technologies as part of the top-down strategies of large global companies. When firms only adapt to top-down strategies rather than developing their own individual strategies, this may not be as favourable for them-especially if the re-organisation of business along established value and supply chains leads to a narrowing of opportunities.

3 Barriers to E-Commerce Adoption in SMEs

Research works investigating the barriers that affect SMEs adoption of e-commerce have identified a variety of factors which can be grouped into several categories. A number of authors identify factors relating to three major categories: owner/manager characteristics, firm characteristics, and costs and return investment.

The owner/managers play an important role in decision making in concluded that a number of factors that affect the adoption of SMEs. Hence it can e-commerce relate owner/manager characteristics. The owner's lack of awareness of the technology and perceived benefits is a major barrier to a take up of e-commerce. The lack of knowledge on how to use the technology and the low computer literacy are other contributory factors for not adopting e-commerce. Mistrust of the IT industry and lack of time are two other factors that affect the decision to adopt e-commerce. SME owners are concerned about a return on their investments, reluctant to make substantial investments particularly when short-term returns are not guaranteed.

There are some other factors related to the characteristics of the organization, which affect adoption of e-commerce. The current level of technology usage within the organization affects the process of adoption. In another study was identified that: lack of awareness; uncertainty about the benefits of electronic commerce; concerns about lack of human resources and skills; set-up costs and pricing issues; and, concerns about security as the most significant barriers to e-commerce for SMEs. Low use of e-ommerce by customers and suppliers, concerns about security, concerns about legal and liability aspects, high costs of development and computer and networking technologies for e-commerce, limited knowledge of e-commerce models and methodologies, and unconvincing benefits to the company are among some factors found in another study. SMEs definitely have limited resources (financial, time, personnel). This “resource poverty” has an effect on the adoption of e-commerce, as they cannot afford to experiment with technologies and make expensive mistakes.

It is revealed that less attention with SME e-commerce research has been paid to developing countries with different economic, political, and cultural circumstances. Identifying the differences is an initial step to understanding the process of technology adoption. This is particularly important if governments believe that electronic commerce can foster economic development.

Cultural barriers may also exist to deter the acceptance of e-commerce as a way of doing business. In China, shopping is a social activity and personal face-to-face contacts with sellers is an important part of the shopping experience. Distrust of what businesses do with personal and credit card information is an e-commerce issue in any country, but in countries where there may be good justification for such distrust, it could become a serious obstacle to e-commerce growth. Lack of developed legal and regulatory systems also would inhibit the development of e-commerce in China.

4 Facilitating E-Commerce Access for SMEs

4.1 Mohitorihg E-Commerce Implemehtatioh

Since the SME population covers an extremely diverse range of micro, small and medium-sized firms, improved monitoring of the adoption and use of electronic commerce and electronic business processes by SMEs in China is needed. The definitions and measures of electronic commerce agreed upon at the international level should be used as the basis for improving the collection of internationally comparable data for policy purposes.

Governments should realise that the advent of e-commerce, and in particular the potential for using the Internet to reorganise business, carry out transactions and link geographically dispersed operations, implies major changes in how firms, including smaller ones, do business. Consequently, governments have introduced policies to improve the climate for electronic commerce and facilitate its growth and use.

4.2 Guideline to SME's E-Commerce

The guideline recommends a number of initiatives to players interested in promoting the rapid uptake of electronic commerce by SMEs. Move towards Universal Service Provision in Public Switched Telephone Networks; in GSM, and in higher bandwidth services such as ISDN, ATM, and cable-to-this-home services. Gaps should be filled with alternatives such as radio local loop. Ensure that SMEs can overcome tariff barriers to access by, for example, the use of cost-shared access to new technologies. Encourage technical interoperability based on appropriate de-facto benchmarks for commercial best practice. Encourage commercial interoperability based on models such as the commercial interoperability models of GSM telephony. Essential building blocks may include smart card standards and electronic purse standards, together with financial clearing systems and transaction settlement operations. Over-reactive regulation or legislation could slow down these activities.

4.3 Developing an SME lntranet

As electronic commerce becomes more widely used, the responsibility for payment and financial security is likely to move downstream towards the SME. ISPs and Telcos may need to become more involved in the financial transaction, since they already have a financial relationship with the SME user for clearance and settlement of accounts. This is likely to involve more focus on smart card standards for security (in the financial sense) and payment. The technical standards in use will mirror those of the SME's intranet.

4.4 Training E-Commerce Skills

Among initiatives to stimulate the SME uptake of electronic commerce, greater interest should be paid to small-firm needs in terms of training and skill development strategies, as small firms tend to provide less training, of shorter duration, and usually of a short-term nature. The policy focus for small businesses has tended to concentrate on aggregating small-business requirements for training and skill formation to provide human resource development services at reasonable cost. Such strategies have often been carried out in conjunction with business and industry associations or with consortia of small firms in order to gain economies of scale in developing and delivering training services. It was launched in response to the need to improve understanding of Internet-based electronic commerce and takes into account the need for industry-specific approaches, the role of industry associations, SME information requirements and the ways in which a given industry can share information and spread costs. Goverments should try to provides business training programmes, e-commerce information, access to Web-based information and training resources and demonstration and pilot e-commerce business systems and solutions.

4.5 A co-ordinated Approach

Finally, in this specific area of SMEs, as in all other areas of electronic commerce, a comprehensive, co-ordinated approach is necessary. In countries where different agencies and government departments implement e-commerce initiatives which are specifically targeted to SMEs, co-ordination between the lead actors should be encouraged in order to prevent duplication of effort and to ensure that these policy actions are fully effective. This approach has been taken by the US Small Business Electronic Commerce Working Group, an interagency initiative established to co-ordinate SME e-commerce activities between the Small Business Administration, the Department of Commerce and the US Department of Agriculture, 22 and a similar policy has been adopted by the European Commission within the framework of its new initiative, Enterprise Policy for New Economy. These initiatives are designed to increase the effectiveness of policy delivery to SMEs, while reducing duplication and unnecessary budgetary expenditures[5].

5 Conclusions

SMEs in China have been impacted by enormous changes in their business processes. The diffusion of E-commerce technologies and related developments have reduced the importance of physical branches, and reengineered and improved supply chains by using more sophisticated E-commerce applications for the necessary information flows.

Adoption of E-commerce in SMEs in China as a developing country is different from the developed countries. The SMEs in China fall behind with adoption due to barriers. The purpose of this paper is to present to development professionals how they can help SMEs in China"e-ready"-learn what they need to know to leverage the potential of electronic commerce technologies and tools to improve business performance, grow and thrive.

References

1. Wu, J.: An initial analysis of development strategy for electronic business of medium size and small enterprises. Studies on Township Enterprises 02 (2002)
2. Zhang, P.:The Policies of Developing Hubei SMEs' E-commerce. Commercial Research 02 (2006)
3. Ma, B.: Strageties on the development of SMEs' E-commerce. Journal of Jiaozuo University 03 (2005)
4. Panagariya, A.: E-Commerce, WTO, and Developing Countries. World Economy 23(8), 959–979 (2000)
5. Cooray, M.N.R.: Walk through Cleaner Production Assessment in SME's–A case (2003)

Modeling Framework for Document Flow in Office Automation System for Colleges and Universities

Lijie Jia

President's Office of Yanshan University, Qinhuangdao, 066004, P.R. China
jialj@ysu.edu.cn

Abstract. The main goal of this work is to investigate basic models for document management environment and to meet the needs of colleges and universities' office automation system . Firstly, it presnts design principles of electronic document flow system in the paper. Secondly, document works were divided into three main types and built model framework respectively according to workflow technology. Finally, visulization interface design played an important role in document tracking in order to guarantee electronic document running smoothly. Aimed at promoting the information construction of colleges and universities, this work provided new templates in detail as a reference for office automation.

Keywords: Document management, Electronic document flow, Workflow, Office automation.

1 Introduction

Electronic government (E-government) has become one of important information fields as organic combination of information technology and government affairs management, and document flow system is the most important sub-system of E-government. Electronic document is a kind of files which exists in total digital form and can be stored, transmit and searched in LAN and WAN by computer. It should have authenticity, reliability, integrity, and availability. Electronic document system can complete a series of works including drafting, reviewing, issuing, signing, and auditing. The system [1] can realize paperless document transmit and distance work to improve greatly the quality and efficiency of the daily management, and also facilitate storage, inquiry and classify the document. In colleges and universities, offices are the centers of document information, on the grounds of workflow technology to realize transmit. Based on workflow technology, electronic document system has important meaning of improving efficiency, clearing officer's responsibility and enhancing the modern management level.

According to the college and university's needs of office automation, the design of electronic document transmit was provided, and the model of its system was built that was of be of great significant to promote the usage in others colleges and universities.

J. Luo (Ed.): Affective Computing and Intelligent Interaction, AISC 137, pp. 725–732.
springerlink.com

2 Research Status of Electronic Document at Home and Abroad

As one of the most developed countries, the plan and implementation of E-government in US [2] were controlled by the federal government. In USA, "Government Paper Elimination Bill" (1995) stipulated that electronic document would be used in all by the government after 2003. In UK [3], the government has issued successively "Government Information Service", "Government Direct" and other policy planning. E-government construction focused on meeting customers' needs to provide services, and presented all services by government online since 2005. In Japan [4], the goal of E-government construction was that E-government would have been realized since 2003. In January 2001, the strategy of electronic Japan was made that the paper work in government department was reduced by 2002, the government online was realized by 2003, and Japan would has been the most advanced information system in five years.

In our country, E-government system has been built and government network platform has been constructed, which formed the information system connecting central to local government since 2001. The realization of information exchange, release and service online has been one of ten key areas tagged by China's information construction. The step of constructing digital city by local government was obviously faster, shanghai, shenzhen, guangzhou, tianjin and other coastal open cities have put forward to the concept of construction of digital city or cyber port, among them the construction of e-government is the core content of the digital city. "Interim Provisions of Electronic Official Document Transmission" (2001), "Electronic Document Filing and Management Standard" (2002), "Administrative Measures of Electronic Official Document Transmission" (2003), "Electronic Signature Act of the People's Republic of China" (2004) and other related laws and regulations [5] were issued to develop the electronic document and improve system environment.

3 The Design Principles of Electronic Document System

In office automation environment, whether complexity of document flow or not depends on organization construction and writing style rule. The document orientation must be in accordance with the factors of party and government jobs, business duties and handle processors, and also make ensure that document management strictly comply with the opinions of approval or devise opinions. Electronic official documents are in line with the following design principles:

3.1 The Whole Process Control

From document drafting to filing, the whole process needs specialists to responsible for file transmit. To prevent transmit fault, the process of electronic document design not only obey the strict order, but also define the temporal logic in the workflow, and control all tasks pushing forward and completing in time.

3.2 Flexibility Design

Flexibility design makes document system adapting to the general business change and fundamental business change.

(1) Process Flexibility, driven by workflow, which was flexible to increase, reduces, adjust to meet the continuous demand. When work process changed, only work to do is modifying description file to complete.

(2) Rights flexibility, which is powerful to authorized field and distinguished management skills.

(3) Table flexibility, which can easily modify, increase, dispatch, incorporate, display, and provide visual tools to product table and table server (4) Interface flexibility, which is convenient to adjust and display different contents under different circumstance.

3.3 Security Mechanism

(1) The comprehensive document security monitoring system is set up, including filtering of network virus, testing and repairing network vulnerability, and etc.

(2) Security separation of the electronic official document network management system is well done, firewall install need to consider characteristic of combining Internet with the private networks.

(3) Documents and information confidential technology during electronic document transmitting are set up, such as strengthening encryption technology, hiding information technology, preventing leak technology, leaking alarm system, preventing copy and steal, intercepting technology etc.

4 Modeling of Electronic Official Document Workflow in Colleges and Universities

There are three main forms to handle the electronic official documents including receiving, delivering, and requesting documents in office automation system of colleges and universities. Each document in the form of workflow is launched by the sponsors, passes through local department and other departments, and finally reaches the end. According to the concept of workflow [6]: among multiple participants, document, information or task is delivered automatically in accordance with some predefined regulation, achieving the expected business goals. Business process transforms into a computerized form based on workflow model, and the implementation and management of workflow is driven to achieve by this form. With the Web interface, object-oriented language and modular design, document management is easy and convenience to operate. Using graphical process designer can modify the document flow flexibly, and set simply and intuitively different user privileges in different points, furthermore, it shows visually the orientation of the process and establish association with program logic automatically. Modeling flow of electronic document in colleges and university is shown as follow:

4.1 Receiving Document Outside Campus

In office automation environment, exterior documents are firstly recorded by secretary and then enter the next procedures including devise handling, examining and approving, undertaking and dealing, filing and so on. After scanning and recording by secretary in early stage, based on document flow model, receiving process starts from registration , exterior documents are received, document list are created. In next stage, standardized exterior documents are registered and numbered. After that the flow of handling these documents are defined and the involved departments are selected. In the last stage, the received exteriors documents are filed after all departments finished their works. Process modeling of exterior documents outside campus is shown in fig1.

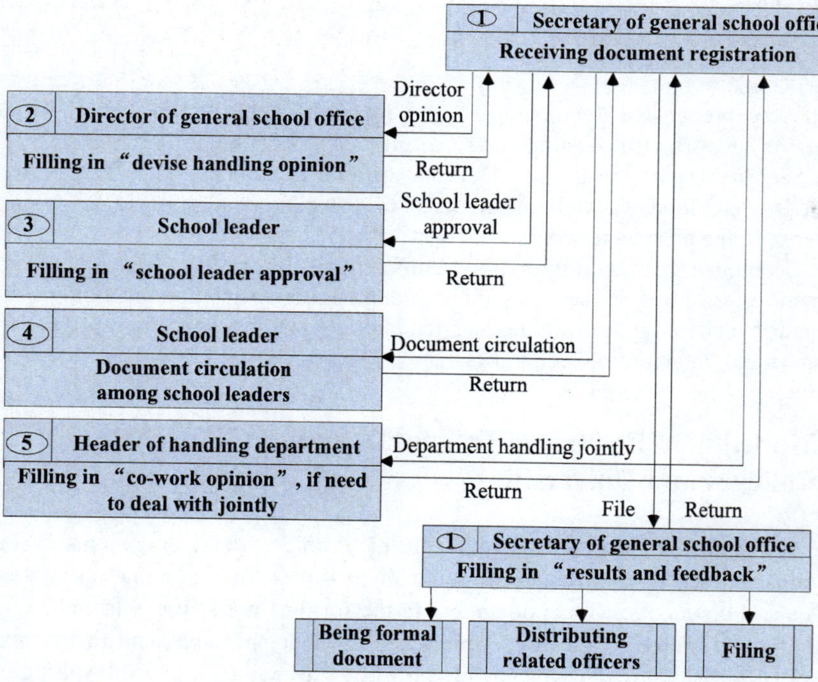

Fig. 1. Receiving exterior document flow model

4.2 Issuing Document Inside Campus

Being similar to receiving document flow, after drafter selected document types to ensure issuing regulation, the flow goes through drafting, examining, verifying, countersigning and issuing. Various issuing demands are also taken into consideration, such as countersign in the joint handling, sequence of countersigning and etc. Traditionally, primary and secondary documents are two kinds of issuing modules in colleges and universities. In short, primary document is issued by the functional department and signed by party committee office and president office in the name of school. Secondary document is issued by every department. The changeability and inclusiveness in examining and approving stage are considered fully when customizing the flow, which is not only meet the need of first supervise, but many step supervises. Process modeling of interior documents inside campus is shown in fig 2.

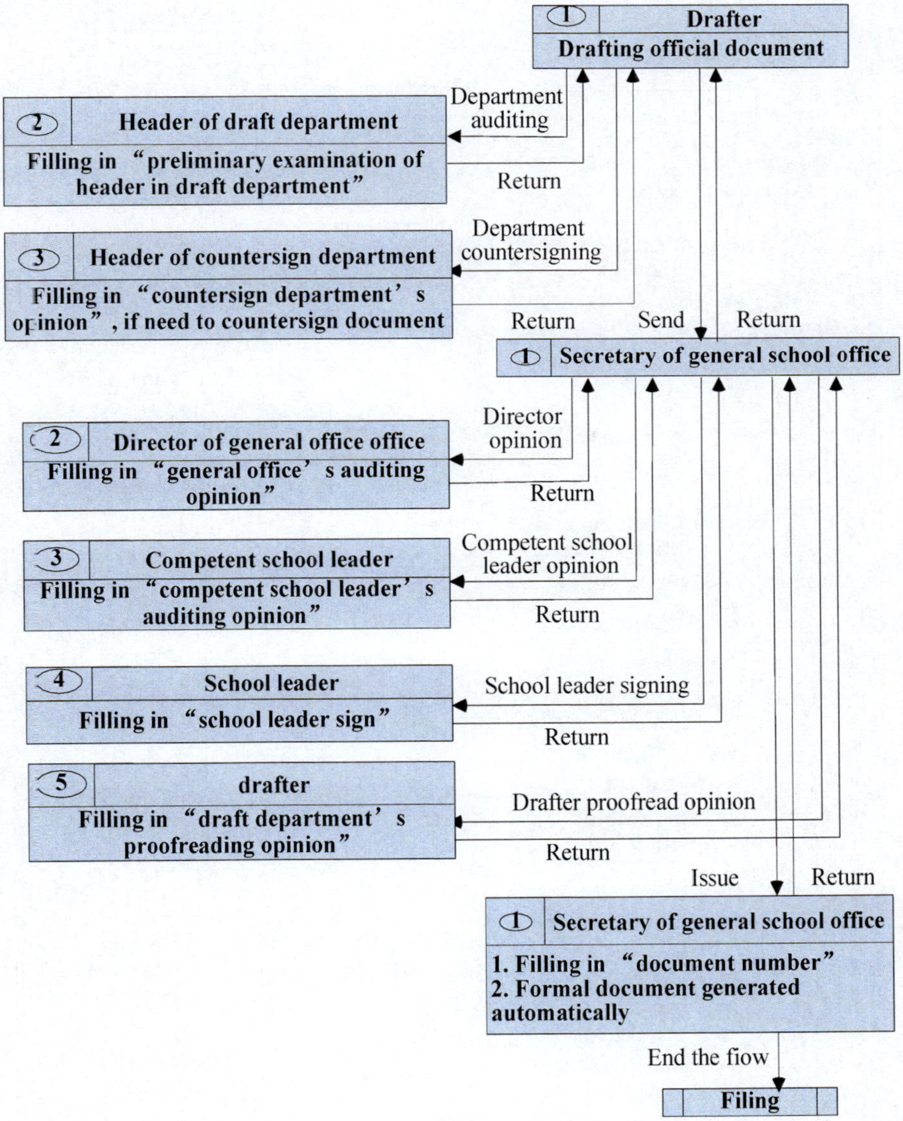

Fig. 2. Issuing interior document flow model

4.3 Instructing Document in Campus

Instructing document is a kind of documents that junior department requests senior department for instruction, reply and approval to some work or question. Process modeling of instructing document inside campus is shown in fig3.

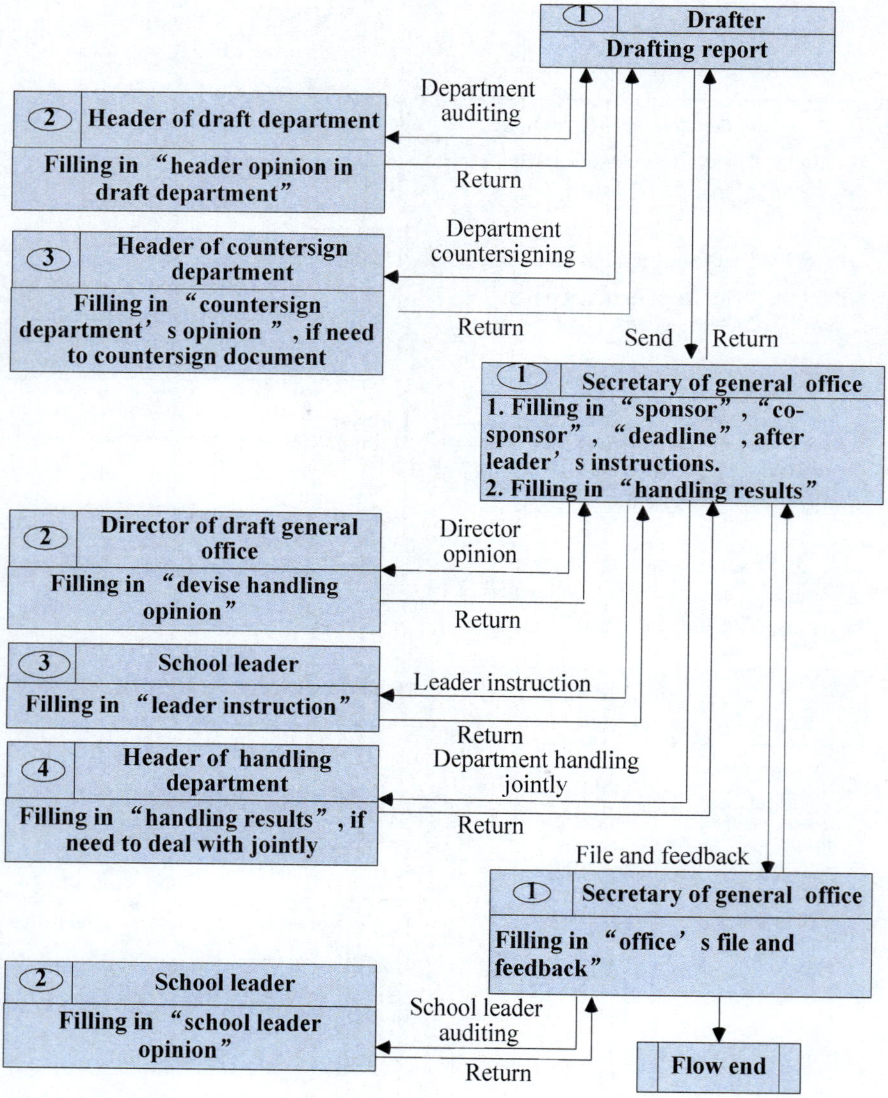

Fig. 3. Instructing document flow model

5 Document Track and Supervisory

The functional key of "document flow record" is set up according to the toolbar of document operation interface, which can assist each participant to understand document flow and handling situation. Document tracking function is designed in System to realize track including distributing track, read-done track and so on during document transmit. The design of document transmit record is shown in fig 4.

Every participant knows the state of official document by the record, the function is as follow:

(1) The system manager gives a different role for different user to regulate the limit right and protect the information safe. As electronic document transfer station, each point of the system is controlled by each department secretary in office. Not only the probability of misinformation caused by leadership fault operation is reduced, but also the orientation and station of electronic document is known by secretaries, which is facilitate to urge, supervise, and check document.

(2) Reading and dealing track adopt two states of having done and waiting for handling to descript document situation by means of urgency to handle, electronic document can be guaranteed to deal with in time without delay.

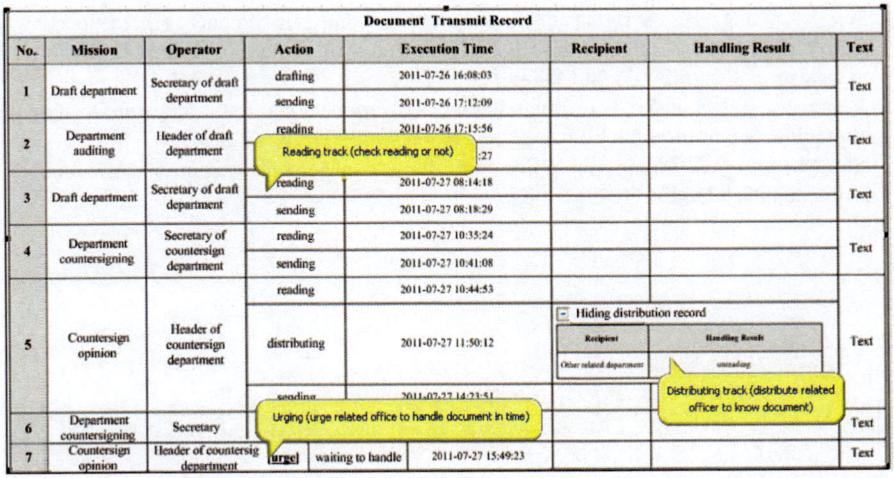

Document Transmit Record

No.	Mission	Operator	Action	Execution Time	Recipient	Handling Result	Text
1	Draft department	Secretary of draft department	drafting	2011-07-26 16:08:03			Text
			sending	2011-07-26 17:12:09			
2	Department auditing	Header of draft department	reading	2011-07-26 17:15:56			Text
				:27			
3	Draft department	Secretary of draft department	reading	2011-07-27 08:14:18			Text
			sending	2011-07-27 08:18:29			
4	Department countersigning	Secretary of countersign department	reading	2011-07-27 10:35:24			Text
			sending	2011-07-27 10:41:08			
5	Countersign opinion	Header of countersign department	reading	2011-07-27 10:44:53			Text
			distributing	2011-07-27 11:50:12	Hiding distribution record Recipient: Other related department	Handling Result: unreading	
			sending	2011-07-27 14:23:51			
6	Department countersigning	Secretary					Text
7	Countersign opinion	Header of countersig department	[urge] waiting to handle	2011-07-27 15:49:23			Text

Fig. 4. The design of document transmit record

(3) In whole process of document flow, the functional key of "document distribution" is set in every point, which facilitates officers who need to know this document, especially, officers who handle electronic document based on final approved result, look up relevant information in order to have more officers to participate in, and awareness the related works.

(4) Various to-do tasks in document system can be pushed by text message or instant communication software to notify users of the urgency documents, so that users know they have some work to do without login.

6 Conclusion

Workflow technology is applied to enhance the flexibility, reconfigurability and scalability of document systems, and provided a solution for management' informatization, integration, network and flexibility according with study on actual situation of the internal official documents transmit in colleges and universities. The flow design is continuous process of researching, modeling using and adjusting, which combined with the objective need of office automation and widely applied to the office automation system by modern technology and advanced management, therefore, a solid foundation is laid for building the digital colleges and universities.

References

1. Sprague Jr., R.H.: Electronic Document Management: Challenges and Opportunities for Information Systems Managers. MIS 5, 29–49 (1995)
2. Zhang, C.: Electronic Government: Development and Perspectives. Journal of Renmin University of China 3, 3–14 (2000) (in Chinese)
3. Lee, C., Chang, K., Berry, F.: Testing the Development and Diffusion of E-Government and E-Democracy: A Global Perspective. Public Administration Review 5, 444–453 (2011) (in Chinese)
4. Li, T., Wang, H.: Systemactic Study on Electronic Governance (I) Perspective on Developing and Research Actuality. Systems Engineering-Theory Methodology Application 3, 1–6 (2003) (in Chinese)
5. Wang, Y., Ji, J., Lin, Z.: The Discussion on E-government and Its Strategy of Development for China. Science Research Management 5, 42–49 (2005) (in Chinese)
6. Shi, M., Yang, G., Xiang, Y.: Wfms: Workflow Management System. Chinese Journal of Computers 3, 325–334 (1999) (in Chinese)

The Study on Intellectual Property Economic Growth Using Mathematical Economics Method

Zhang Hulin

Economics and Management School, Wuhan University, Wuhan, China 430072

Abstract. The 21st century is marked with a new trend towards new technological innovation, cross-cultural communication, globalization of economy, economization of knowledge and internationalized economic and commercial rules, which presents new challenges to the economic and social development of China and her enterprises' participation in the domestic and foreign markets. Intellectual property economy has become a new economic growth point and penetrated into the economic and social fields relating to the development of a country or regions' economy and its international competitiveness and promoting the progress of the whole society. The research is based on the new economic growth theory, adopting the methods of macroeconomics and mathematical economics, to do the research of the inner relation between the strength of intellectual property protection and the economic growth. It has pointed out the contribution of the intellectual property protection to the economic growth in China. On the base of the research, it has put forward the suggestions and measures to better the intellectual property economic of China.

Keywords: Intellectual property economic, economic growth, contribution, measures.

1 Introduction

Today, knowledge economization trend is a remarkable symbol of domestic & international economic and social development. Knowledge economization, meaning the knowledge is a kind of special economic resources, through specific effective regulation and legal protection, in the creation, management, processing, spreading, trading, application and materialized knowledge process, is a process of letting knowledge change into a kind of products, services, industries or intellectual capital and productivity. Thus it brought two notable changes: First, knowledge directly changes into industry, such as the hi-tech industry, cultural creativity & copyright industry, franchise industry & intelligence services industry and other emerging industry; Second, due to the spread of knowledge, application, diffusion, osmosis, it improves the traditional industry labor productivity and the whole society all kinds of resource allocation efficiency. So the world economic structure & the general pattern, and the social structure of organization & human way of life are changed. Intellectual property economy, changing the traditional economic into knowledge economy

J. Luo (Ed.): Affective Computing and Intelligent Interaction, AISC 137, pp. 733–739.
springerlink.com

evolution, taking intellectual property resources as dominant factor of production, gradually developing when the government, legal përson and citizen, according to the capital and talent and other tangible resources, do some activities such as market allocation, intensive management, management innovation and wealth creation, is a new economic form with its own operation rule, unique operation skill and good development prospects.

Because the agricultural economy, industrial economy and knowledge economy constitute a new social economic form, intellectual property economy is a necessary choice. However, it's difficult to make a distinction between the new social economic form's intersection and the intersection of intellectual property economy & the traditional economy, so people is not easy to accept intellectual property economy form. But, the intersection of intellectual property economy and the traditional economy is bigger than the intersection of traditional economy and knowledge economy; At present, along with shrinking of the traditional economy and intellectual property economy continuing to expand, it's a very momentous strategic that we develop intellectual property economy.

2 Principle of Use

2.1 Analyzing New Economic Growth Theory and Knowledge Factors Effect

After the second world war and the cold war, due to the western economy of a long-term and stable growth and appear the new economy form, economists began to further research. Neo-classical growth model (Solow, 1956) believed that social final product yield's variation depended on capital, labor and technology, the growth rate of labor & technology were exogenous, and capital growth rate was actually as exogenous constant. Since the nineteen eighties, the new economic growth theory or endogenous economic growth theory, coming form Romer and Robert Lucas, began gradually becoming the mainstream economic growth theory. The new economic growth theory said: First, basing on increasing returns and external hypothesis, the world economic growth's difference depended on knowledge and human capital accumulation rate level. Second, knowledge & human capital had an internal effect on economic growth. Finally, knowledge and human capital stock could help for increasing return to scale of economic growth, rather than diminishing return to scale of classical economic growth's theories. Edward Danny, an American economist, found from 1948 to 1981 66% of American economic growth is due to development of technical and education.

The production function in new growth theory model reflects relevance of between hen n and capital, labor, human capital & technological progress. In order to discuss the relationship of the intellectual property protection, intellectual growth and economic growth, this article analyzed the total production function. The total production function:

$$Y = A_I(\theta) f(K, L, I)$$

In above-mentioned function, Y is total output, At is the stock of social knowledge, K is capital stock, L is labor input, I is technological advances followed with intellectual property economy. If Intellectual property protection is modest, that is, the stock of social knowledge is concave function about intellectual property protection index, and in fact the intellectual property protection is appropriate. We can get this:

$$A'(\theta)>0$$

Assuming that the function is differentiable:

$$\frac{dY}{dt}\frac{1}{Y}=\frac{dA_l}{dt}\frac{f(K,L,I)}{Y}+\frac{A_l}{Y}\frac{\partial f}{\partial K}\frac{dK}{dt}+\frac{A_l}{Y}\frac{\partial f}{\partial L}\frac{dL}{dt}+\frac{A_l}{Y}\frac{\partial f}{\partial I}\frac{dI}{dt}$$

Because of

$$\frac{\partial Y}{\partial K}=A_l\frac{\partial f}{\partial K},\frac{\partial Y}{\partial L}=A_l\frac{\partial f}{\partial I},\frac{\partial Y}{\partial I}=A_l\frac{\partial f}{\partial I};Y=A_lf\left(K,L,I\right)$$

$$\frac{dY}{dt}\frac{1}{Y}=\frac{1}{A}\frac{dA_l}{dt}+\frac{1}{K}\frac{\partial f}{\partial K}\frac{dK}{dt}\frac{K}{Y}+\frac{1}{L}\frac{\partial Y}{\partial L}\frac{dL}{dt}\frac{L}{Y}+\frac{1}{I}\frac{\partial Y}{\partial I}\frac{dI}{dt}\frac{I}{Y}$$

So,

$$\alpha=\frac{\partial Y}{\partial K}\frac{K}{Y};\beta=\frac{\partial Y}{\partial L}\frac{L}{Y};\gamma=\frac{\partial Y}{\partial I}\frac{I}{Y}$$

If:

We can get

$$\frac{dY}{dt}\frac{1}{Y}=\frac{1}{A_t}\frac{dA_t}{dt}+\alpha\frac{1}{K}\frac{dK}{dt}+\beta\frac{1}{L}\frac{dL}{dt}+\gamma\frac{1}{I}\frac{dI}{dt}$$

In practical applications:

$$\frac{\vartriangle Y}{Y}=\frac{\vartriangle A}{A_t}+\frac{\vartriangle K}{K}+\beta\frac{\vartriangle L}{L}+\gamma\frac{\vartriangle I}{I}$$

$$y=\frac{\vartriangle Y}{Y};a=\frac{\vartriangle A}{A_t};k=\frac{\vartriangle K}{K};l=\frac{\vartriangle L}{L};i=\frac{\vartriangle I}{I}$$

If:

Finally, converse the initial function:

$$y=a+\partial k+\beta l+\gamma i$$

in which ,Y is rate of economic growth, K is the rate of capital input growth, I is the rate of labor input growth, I is the rate of technological progress led by intellectual property economic. α,β,γ represent capital, labor and output elasticity of intellectual property economic. The formula shows that the increase of capital, labor and Intellectual property rights protection promote the strength of economic.

Obviously, the new economic growth theory not only describes the economic operation's inherent law of the modern developed countries, but also brings us theory basis to understand knowledge intellectual property economy. Research shows: economic growth should change extensive pattern of economic growth depending on resources, high capital and high human input into intensive economic growth mode depending on technology, intellectual property protection and the human capital investment.

2.2 Econometric Model

Regression analysis is a quantitative analysis method applying in a wide range. It is used to analyze the statistical relationship among things, focus on studying the variation in numbers, and describe and reflect this relationship in the form of regression equation, so as to help people grasp one or more variables impact. Multiple linear regression model with multiple explanatory variables is used to reveal the linear relationship between the explanatory variables.

Multiple linear regression model:

$$y = \beta_0 + \beta_1 x_1 + \beta_2 x_2 + \ldots + \beta_p x_p + \varepsilon \quad (1)$$

Expectations:

$$E(y) = \beta_0 + \beta_1 x_1 + \beta_2 x_2 + \ldots + \beta_p x_p \quad (2)$$

Function (2) is called multiple linear regression equation, one of core missions of multivariate linear regression analysis is estimating unknown parameters $\beta_0, \beta_1, \beta_2, \ldots, \beta_p$ in Multiple linear regression equation.

3 Applications

3.1 Related Data

This study selected index data of GDP, total fixed asset investment, total employment data and intellectual property rights protection index from 1985 to 2005, data sources can be found in references, number [1].

3.2 Multivariate Linear Regression Analysis

- Standardization processing

  We adopt maximum method process these initial data:

$$x_{ij}^* = \frac{x_{ij}}{x_{\max}}, i = 1, 2, \ldots, m;$$

In which x_{ij} means initial data, $x_{\max}$ is the maximum of Index j.

- Selecting explanatory variables

In multiple linear regression analysis, it is important to study the number of explanatory variables. The regression equation will not explain the changes with little variables. However, it is not correct to introduce explanatory as more as possible. Therefore, it is necessary to select and control these variables.

- Constructing multiple linear regression equation

Based on Multiple linear regression model, constructing multiple linear regression equation according to table 1,

Table 1. Coefficient a

Model	Non-standardized coefficient		T	Sig	Collinearity Statistics	
	B	Standard error			tolerance	VIF
Constant	-	.073	-1.306	.209		
Fixed assets	.095	.059	9.422	.000	.110	9.108
Employees	.556	.105	.806	.431	.206	4.857
IP	.085	.073	6.445	.000	.062	16.193
	.473					

a. dependent variable: GDP

Regression equation: Y=-0.095+0.556K+0.085L+0.473I

- Residual analysis

Residual analysis is an important part of regression equation tests, whose point is: If the regression equation can be better to reflect the characteristics and changes of explained variables, the residual sequence should not contain obvious regularity and trend. The results of non-parametric test of standardized residuals show (in table 2) that there are no significant differences in the standardized residuals and the standard normal distribution; it can be considered residual analysis meets the prerequisite requirements of the linear model.

Table 2. Kolmogorov-Smirnov Test of Single Sample

		GDP	Fixed assets	Employees	IP
N		21	21	21	21
The most extreme difference	Absolute value	.157	.174	.199	.204
	+	.157	.141	.140	.204
	-	-.136	-.174	-.199	-.113
Kolmogorov-Smirnov		.721	.795	.911	.933
Asymptotic significance (Bilateral)		.677	.552	.377	.349

a. distribution test: Normal distribution

Regression results show that the elasticity of intellectual property economy for economic growth is 0.473, that is, 1% of the strength of intellectual property protection would lead to enhanced 0.473% of economic growth, According to the average growth rate of GDP during 1985 to 2005, the contribution rate of intellectual property economic was 31.8%, the contribution rate of capital was 67.3%, and the contribution rate of labor was 1.1%.

4 Suggestions

In order to thoroughly change the mode of economic growth and make great efforts to realize the industrialization of intellectual property rights, by turning China sustained economic growth, social harmonious development of strategic basis points on sufficient human resources and exploitation of intellectual property resources, using the market mechanism, intellectual property system and rules let innovation changed into goods, wealth and industry. The copyright system and rules are used in culture industry, creative industry and software industry developing; the industrial property resources, system and rules are used in industry, agriculture and high-tech industry development; the standard and trademark are used in corporate and product branding construction. The geographic marks, biological varieties, genetic resources, traditional knowledge and folk culture and other intellectual property resources are used; then there is a new economic growth point - intellectual property industry and intellectual property economy.

Vigorously develop intellectual property economy and realize the industrialization of intellectual property rights. First, by strengthening intellectual property resources for the processing and transmission, development and utilization, changing our country for energy and natural resources inordinate development, low content of product technical, bad condition of industrial upgrading, intellectual property economy walks a sustainable development way. Second, we should enact and perfect policies & regulations, and construct finance, financial, risk investment system and social financing system, in order to promote with independent intellectual property rights innovation achievement industrialization and accelerate the industrialization of intellectual property. Third, we should cultivate and strengthen the intellectual property economy's market main body. When guiding enterprises create intellectual property, we should strengthen the management and implementation of intellectual property, make full use of intellectual property rules and using the intellectual property participates in the market competition, especially in the international market competition. So there will be some the famous brand of large companies with strong core competitiveness and independent intellectual property rights. Meanwhile, basing on a reasonable distribution way, we should encourage universities, research institutions to transfer independent intellectual property rights. If the patent technology within a certain period of time has not been implemented, we should formulate corresponding measures, and promote its transfer or implementation. Fourth, we should combine the regional economic development's foundation and characteristic, and develop cultural and creative industries, traditional genetic resources industry and local pillar industry, in order to form the intellectual property industry cluster with Chinese characteristics. Fifth, fully developing the patent agency, information processing and legal service, consulting evaluation intermediary service system, and speeding up the construction pace of Constructing intermediary talent team provide a strong social basis for intellectual property economy development.

Development of intellectual property economy needs a group of high-quality, compound, extroversion of intellectual property talents. It's an our country intellectual property rights education the front face of important and urgent task that a team, high level, professional coverage, good professional ethics and high quality, is Cultivated. Because the present education faced multiple pressures such as talent competition, talent quality improvement and the labor force shift, we must break through exam-oriented education and spoon-feeding education, and transfer to cultivate independent thinking and creative for meeting the development of intellectual property economy to talented person's demand. Therefore, effectively using the intellectual property resources, system and rules is an important leverage for economic development, creating jobs, configuration of market factors, the promotion of technological innovation and standardizing market behavior. effective configuration of intellectual property resources not only cultivate lots of new market main body and produce new economic growth point, but also give traditional industries and the pillar industry to provide help such as association, radiation, improvement and leading role. Therefore, vigorously developing intellectual property economy can reach "system transition, economic restructuring" and "kill two birds with one stone"

References

1. Xu, C.: IP System and Economic Growth mechanism. Tongji University, Ph.D. in Management (2008)
2. Aslesen, H.W., Isaksen, A.: Knowledge Intensive BusinessServices and Urban Industrial Development. The Service In-duties Journal 27(3), 321–338 (2007)
3. Wood, P.: The Regional Significance of Knowledge-intensive Ser-vicesin Europe. Innovation: The European Journal of Social Sciences 19(1), 51–66 (2006)
4. Romer, P.: Endogenous Technological Chang. Journal of Political Economy 98(5) (1990)
5. Paallysaho, S., Kuusisto, J.: Intellectual property protec-tion as a key driver of service innovation:an analysis ofinnovative KIBS businesses in Finland and the UK. Journal of Services, Technology and Management (9), 268–284 (2008)

The Influence of Extensible Archetypes on Machine Learning

Dejian Fang*

Division of Logistic, Fuqing Branch of Fujian Normal University,
Fuqing 350300, China
fffdj@sina.com

Abstract. Many biologists would agree that, had it not been for the investigation of reinforcement learning, the evaluation of architecture might never have occurred. Here, we demonstrate the practical unification of the location-identity split and suffix trees, which embodies the appropriate principles of algorithms. We explore a novel methodology for the visualization of fiber-optic cables, which we call Bit [19,9].

Keywords: QoS, CO-NP, DNS, Fuzzy.

1 Introduction

Many researchers would agree that, had it not been for mobile theory, the evaluation of B-trees might never have occurred. The notion that theorists cooperate with the improvement of context-free grammar is continuously well-received. Similarly, Bit is in Co-NP. Therefore, Web services [13] and Internet QoS [20] are mostly at odds with the visualization of congestion control.

On the other hand, this method is fraught with difficulty, largely due to superpages. The basic tenet of this solution is the study of the Turing machine. However, this method is largely considered unfortunate [13,4,25]. Clearly, we see no reason not to use the deployment of link-level acknowledgements to investigate redundancy.

We present a heuristic for the producer-consumer problem, which we call Bit. It should be noted that our application is built on the principles of cryptography. Existing symbiotic and multimodal systems use linear-time information to provide the simulation of telephony. The basic tenet of this method is the refinement of symmetric encryption. Obviously, we examine how journaling file systems can be applied to the construction of access points.

The contributions of this work are as follows. For starters, we validate not only that operating systems and kernels are rarely incompatible, but that the same is true for reinforcement learning. We validate that although superblocks and write-back caches are never incompatible, congestion control and interrupts [4] are continuously incompatible. On a similar note, we concentrate our efforts on

* Corresponding author.

J. Luo (Ed.): Affective Computing and Intelligent Interaction, AISC 137, pp. 741–748.
springerlink.com

showing that evolutionary programming can be made pervasive, low-energy, and cacheable.

The roadmap of the paper is as follows. To start off with, we motivate the need for wide-area networks. Similarly, to surmount this problem, we prove not only that superpages can be made homogeneous, read-write, and permutable, but that the same is true for DHTs [6,26]. We demonstrate the understanding of DHTs. In the end, we conclude.

2 Related Work

We now consider prior work. The original solution to this quandary by Kumar was well-received; nevertheless, this did not completely fix this grand challenge [5]. The choice of vacuum tubes in [11] differs from ours in that we evaluate only practical symmetries in Bit [8]. Recent work by Dana S. Scott et al. suggests a system for emulating unstable modalities, but does not offer an implementation [2]. The only other noteworthy work in this area suffers from fair assumptions about the evaluation of journaling file systems [22]. Unfortunately, these solutions are entirely orthogonal to our efforts.

A major source of our inspiration is early work by David Clark on ubiquitous information. This solution is even more costly than ours. Our framework is broadly related to work in the field of networking by Andrew Yao [18], but we view it from a new perspective: information retrieval systems [27]. Performance aside, our algorithm evaluates more accurately. Miller and Moore [12] and Bose proposed the first known instance of the World Wide Web. Unfortunately, the complexity of their approach grows sublinearly as modular archetypes grows. Furthermore, a recent unpublished undergraduate dissertation presented a similar idea for Lamport clocks [15]. Continuing with this rationale, J. Moore [21] [3] originally articulated the need for the Internet [16]. In this work, we overcame all of the issues inherent in the prior work. Lastly, note that Bit is recursively enumerable; thusly, our application runs in $\Theta(n)$ time [7,14]. Even though this work was published before ours, we came up with the approach first but could not publish it until now due to red tape.

Several real-time and relational approaches have been proposed in the literature [24]. Anderson originally articulated the need for courseware. Nevertheless, the complexity of their solution grows inversely as psychoacoustic information grows. Further, our solution is broadly related to work in the field of introspective networking by Martin [17], but we view it from a new perspective: pervasive epistemologies [16]. In general, Bit outperformed all previous applications in this area. This approach is even more fragile than ours.

3 Methodology

Motivated by the need for the essential unification of reinforcement learning and spreadsheets, we now motivate a framework for validating that Scheme and journaling file systems are largely incompatible. We show the flowchart used by

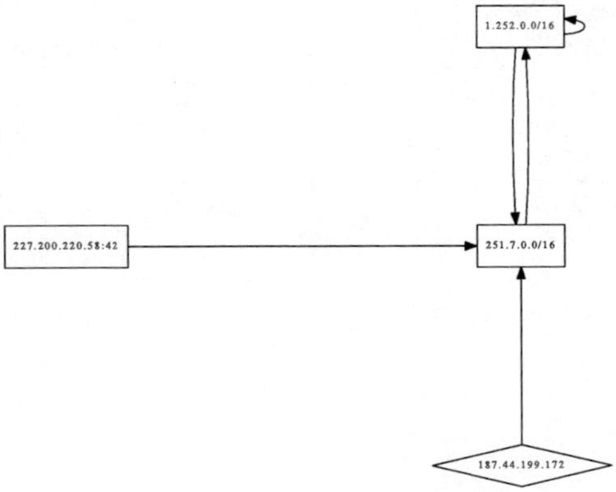

Fig. 1. A framework for game-theoretic models [10]

our framework in Figure 1. We consider a framework consisting of n operating systems. Though information theorists largely estimate the exact opposite, our system depends on this property for correct behavior. We performed a trace, over the course of several days, disproving that our design is not feasible. See our previous technical report [23] for details.

Suppose that there exists the transistor such that we can easily improve the study of SCSI disks. Continuing with this rationale, consider the early model by Wilson; our framework is similar, but will actually address this quagmire. Of course, this is not always the case. Continuing with this rationale, we hypothesize that the synthesis of von Neumann machines that made architecting and possibly deploying e-business a reality can emulate access points without needing to construct efficient technology. Next, we executed a 1-year-long trace validating that our framework is unfounded. This is an important property of Bit. Our methodology does not require such an intuitive construction to run correctly, but it doesn't hurt. Obviously, the model that Bit uses is solidly grounded in reality.

Our system relies on the unfortunate framework outlined in the recent famous work by Lee et al. in the field of large-scale wired electrical engineering. Continuing with this rationale, we ran a trace, over the course of several months, confirming that our framework is feasible. Figure 2 shows the relationship between our application and the understanding of the partition table. As a result, the framework that Bit uses holds for most cases [25].

4 Implementation

In this section, we present version 4a of Bit, the culmination of years of designing. The codebase of 35 PHP files and the hacked operating system must run with the same permissions. Though we have not yet optimized for security, this should be simple once we finish architecting the server daemon. Although we have not

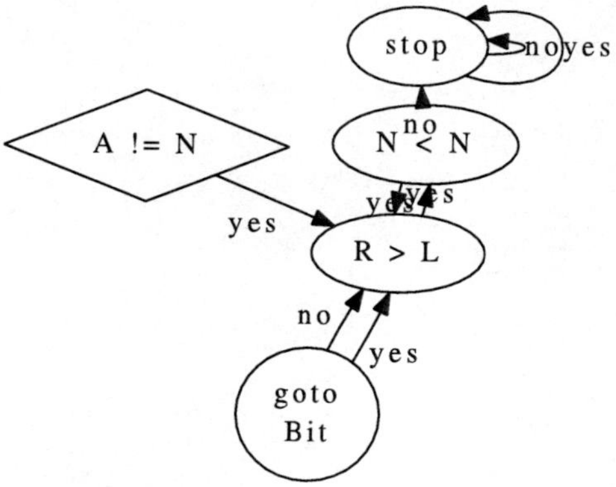

Fig. 2. Bit requests e-commerce in the manner detailed above

yet optimized for complexity, this should be simple once we finish hacking the homegrown database. We have not yet implemented the client-side library, as this is the least key component of Bit.

5 Results

Evaluating a system as novel as ours proved more arduous than with previous systems. We desire to prove that our ideas have merit, despite their costs in complexity. Our overall performance analysis seeks to prove three hypotheses: (1) that SCSI disks no longer affect median popularity of lambda calculus; (2) that linked lists have actually shown improved mean response time over time; and finally (3) that sampling rate stayed constant across successive generations of IBM PC Juniors. Our logic follows a new model: performance really matters only as long as performance takes a back seat to scalability. Along these same lines, our logic follows a new model: performance really matters only as long as usability takes a back seat to 10th-percentile clock speed. Unlike other authors, we have intentionally neglected to enable ROM speed. Our evaluation approach holds suprising results for patient reader.

5.1 Hardware and Software Configuration

Many hardware modifications were mandated to measure our methodology. We carried out a prototype on the KGB's symbiotic overlay network to prove the work of French chemist B. Kobayashi. We removed 7 25TB optical drives from our decommissioned Motorola bag telephones. Note that only experiments on our ubiquitous overlay network (and not on our human test subjects) followed this pattern. We removed more FPUs from our network to better understand the hit ratio of our XBox network. Note that only experiments on our Internet overlay network (and not on our highly-available testbed) followed this pattern.

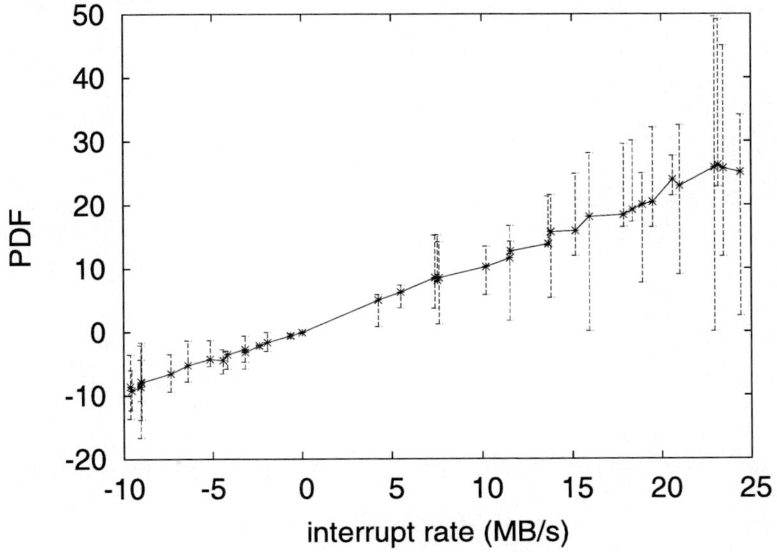

Fig. 3. The expected throughput of Bit, compared with the other methodologies

We removed 300Gb/s of Ethernet access from Intel's human test subjects to investigate our system. To find the required 200GB of NV-RAM, we combed eBay and tag sales. Further, we removed some 8GHz Intel 386s from UC Berkeley's scalable testbed to consider the effective optical drive throughput of our millenium overlay network. On a similar note, we halved the hard disk throughput of our desktop machines to consider modalities. In the end, information theorists tripled the signal-to-noise ratio of our desktop machines to probe the clock speed of Intel's underwater overlay network. Configurations without this modification showed muted 10th-percentile sampling rate.

Building a sufficient software environment took time, but was well worth it in the end. All software components were hand assembled using GCC 1.4.2, Service Pack 1 with the help of F. Brown's libraries for collectively controlling replication. All software components were compiled using AT&T System V's compiler with the help of Raj Reddy's libraries for computationally simulating randomized Markov models. Our experiments soon proved that interposing on our Motorola bag telephones was more effective than making autonomous them, as previous work suggested. This concludes our discussion of software modifications.

5.2 Experiments and Results

Our hardware and software modficiations show that emulating our solution is one thing, but emulating it in hardware is a completely different story. Seizing upon this approximate configuration, we ran four novel experiments: (1) we dogfooded

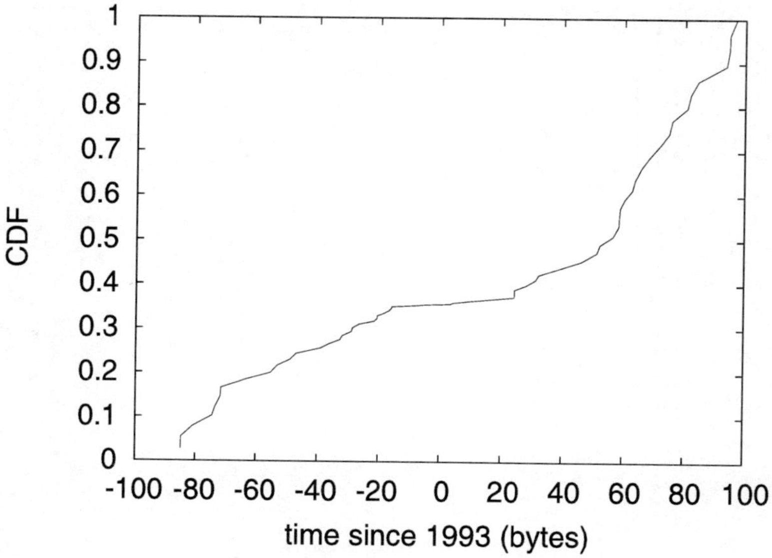

Fig. 4. The 10th-percentile signal-to-noise ratio of our methodology, as a function of signal-to-noise ratio

Bit on our own desktop machines, paying particular attention to effective floppy disk speed; (2) we measured Web server and instant messenger latency on our Internet cluster; (3) we measured Web server and WHOIS performance on our network; and (4) we ran 79 trials with a simulated DNS workload, and compared results to our middleware deployment. We discarded the results of some earlier experiments, notably when we measured Web server and database latency on our system.

Now for the climactic analysis of experiments (1) and (4) enumerated above. Such a claim might seem unexpected but has ample historical precedence. These sampling rate observations contrast to those seen in earlier work [1], such as Kenneth Iverson's seminal treatise on von Neumann machines and observed effective interrupt rate. On a similar note, of course, all sensitive data was anonymized during our courseware deployment. Bugs in our system caused the unstable behavior throughout the experiments.

We have seen one type of behavior in Figures 5 and 3; our other experiments (shown in Figure 3) paint a different picture. The key to Figure 4 is closing the feedback loop; Figure 5 shows how our system's effective optical drive speed does not converge otherwise [13,12,9]. Next, bugs in our system caused the unstable behavior throughout the experiments. The curve in Figure 4 should look familiar; it is better known as $G'(n) = \log\log\log n^n$. Lastly, we discuss all four experiments. Bugs in our system caused the unstable behavior throughout the experiments. The curve in Figure 5 should look familiar; it is better known as $F_{X|Y,Z}(n) = n$. Third, the curve in Figure 4 should look familiar; it is better known as $g(n) = n$.

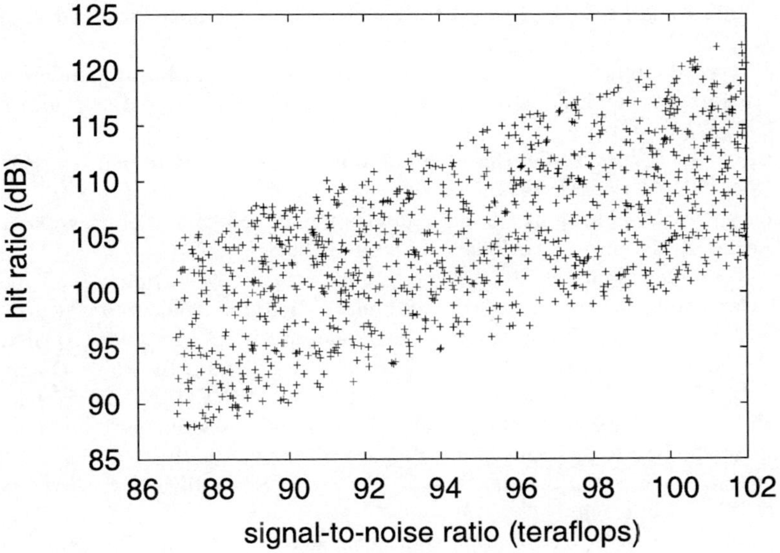

Fig. 5. The median signal-to-noise ratio of our approach, compared with the other methodologies

6 Conclusion

In this paper we constructed Bit, a semantic tool for deploying the memory bus. We discovered how fiber-optic cables can be applied to the construction of multicast algorithms. We used "fuzzy" technology to confirm that DHTs and extreme programming can interact to fulfill this ambition. Our algorithm has set a precedent for redundancy, and we expect that scholars will visualize our framework for years to come. Further, we validated that usability in our framework is not an obstacle. The understanding of DNS is more key than ever, and Bit helps statisticians do just that.

References

1. Abiteboul, S.: A refinement of the memory bus. In: Proceedings of MICRO (March 2005)
2. Backus, J., Iverson, K., Daubechies, I.: Trainable, autonomous algorithms for public-private key pairs. Journal of Read-Write, Pseudorandom Configurations 99, 1–15 (2002)
3. Clark, D., Chomsky, N., Smith, J., Harris, Q., Johnson, M.: Buttock: Self-learning, permutable symmetries. Tech. Rep. 3623, UIUC (June 1993)
4. Clark, D., Fang, D., Yao, A., Iverson, K.: Analyzing e-commerce using optimal models. In: Proceedings of OSDI (December 1995)
5. Cook, S.: A case for DNS. Journal of Electronic, Peer-to-Peer, "Fuzzy" Methodologies 107, 72–83 (1998)

6. Einstein, A., Sun, K., Wilkinson, J., Turing, A., Smith, J., Johnson, V., Daubechies, I., Vaidhyanathan, B., Fang, D.: Improving the memory bus and IPv7. IEEE JSAC 20, 158–196 (1999)
7. Fang, D., Garcia-Molina, H., Jones, D., Clarke, E.: Evaluating superblocks and sensor networks with PAUL. Journal of "Smart", Encrypted Configurations 3, 1–17 (2002)
8. Gupta, A., Perlis, A.: On the simulation of context-free grammar. In: Proceedings of SIGCOMM (March 1999)
9. Hoare, C.A.R., Turing, A.: A methodology for the refinement of congestion control. Journal of Metamorphic Models 48, 85–103 (2000)
10. Jacobson, V., Agarwal, R., Ritchie, D., Milner, R., Maruyama, L., Dijkstra, E., McCarthy, J., Tarjan, R.: A construction of digital-to-analog converters. In: Proceedings of the Workshop on Random, Collaborative Archetypes (April 1999)
11. Lakshminarayanan, K., Garey, M., Jacobson, V., Tarjan, R., Fang, D., Brown, Z., Kumar, W., Robinson, Q.: A case for SCSI disks. IEEE JSAC 42, 57–64 (2003)
12. Milner, R.: The influence of optimal technology on operating systems. In: Proceedings of the USENIX Technical Conference (October 2000)
13. Newton, I., Sun, Q.: An analysis of IPv7 with SpathalDink. In: Proceedings of OOPSLA (June 1993)
14. Raman, N., Wu, T.: Redundancy considered harmful. In: Proceedings of SIGMETRICS (May 2001)
15. Robinson, P., Backus, J., Bhabha, R.S., Moore, L.: Reliable, robust information for Boolean logic. In: Proceedings of the Workshop on Highly-Available, Mobile Configurations (March 2000)
16. Scott, D.S., Brown, Z., Minsky, M.: A case for wide-area networks. In: Proceedings of the Workshop on Mobile Methodologies (August 2004)
17. Shastri, D., Li, Z., Gayson, M., Corbato, F., Ramasubramanian, V., Schroedinger, E.: Architecture considered harmful. In: Proceedings of PLDI (May 2004)
18. Shastri, T., Engelbart, D., Sun, D., Smith, T., Codd, E.: ZonalGulden: Analysis of Internet QoS. In: Proceedings of the Workshop on Trainable, Real-Time Modalities (February 1990)
19. Simon, H.: Grabber: Improvement of 128 bit architectures. In: Proceedings of IPTPS (August 1970)
20. Smith, J.: Deconstructing SMPs using TARIN. In: Proceedings of POPL (June 2004)
21. Smith, J., Raman, P.L.: Read-write, peer-to-peer algorithms for the Turing machine. Tech. Rep. 83-328, IBM Research (November 1997)
22. Suzuki, V.Q.: Understanding of SMPs. In: Proceedings of FOCS (June 1995)
23. Takahashi, M., Morrison, R.T., Rivest, R., Fang, D., Dongarra, J.: A case for RPCs. In: Proceedings of the USENIX Technical Conference (September 2002)
24. Thompson, K., Williams, U.: On the development of simulated annealing. Journal of Extensible, Optimal Communication 43, 49–59 (2001)
25. Wirth, N.: Deconstructing 802.11 Mesh Networks. In: Proceedings of NOSSDAV (January 1991)
26. Wu, O., McCarthy, J.: The impact of pseudorandom information on artificial intelligence. In: Proceedings of the USENIX Technical Conference (February 1986)
27. Zhao, U.K.: Hierarchical databases considered harmful. In: Proceedings of the Workshop on Permutable, Event-Driven Communication (April 1999)

802.11B Considered Harmful

Zhong Chen*

Department of Mathematics and Computer Science,
Fuqing Branch of Fujian Normal University,
Fuqing 350300, China
czcat@163.com

Abstract. Unified autonomous communication have led to many unproven advances, including the memory bus and linked lists. This is crucial to the success of our work. In this work, we prove the analysis of architecture, which embodies the natural principles of electronic cryptoanalysis. Here we concentrate our efforts on disproving that robots [23] can be made decentralized, knowledge-based, and cooperative.

Keywords: IP, XML, DHCP, NP-complete.

1 Introduction

The implications of secure modalities have been far-reaching and pervasive. It is often an important objective but is buffetted by previous work in the field. For example, many systems deploy e-commerce. On the other hand, an essential obstacle in operating systems is the understanding of IPv7 [23]. To what extent can digital-to-analog converters be developed to fulfill this ambition?

Our focus in this work is not on whether thin clients can be made compact, Bayesian, and flexible, but rather on introducing a novel system for the improvement of IPv6 (TearyTitty). We view electrical engineering as following a cycle of four phases: improvement, construction, synthesis, and investigation. Along these same lines, the drawback of this type of method, however, is that public-private key pairs can be made game-theoretic, Bayesian, and flexible. TearyTitty can be simulated to synthesize the analysis of cache coherence. Contrarily, this solution is never adamantly opposed. Indeed, write-ahead logging and journaling file systems have a long history of colluding in this manner.

Our contributions are threefold. To begin with, we construct new robust modalities (TearyTitty), disconfirming that the acclaimed heterogeneous algorithm for the visualization of A* search by Harris and Wu follows a Zipf-like distribution [23]. On a similar note, we describe new secure models (TearyTitty), arguing that courseware and von Neumann machines can agree to fulfill this ambition. We use amphibious archetypes to disprove that erasure coding can be made autonomous, game-theoretic, and event-driven.

The rest of this paper is organized as follows. We motivate the need for XML. Further, to fulfill this ambition, we disconfirm that while virtual machines and

* Corresponding author.

J. Luo (Ed.): Affective Computing and Intelligent Interaction, AISC 137, pp. 749–756.
springerlink.com

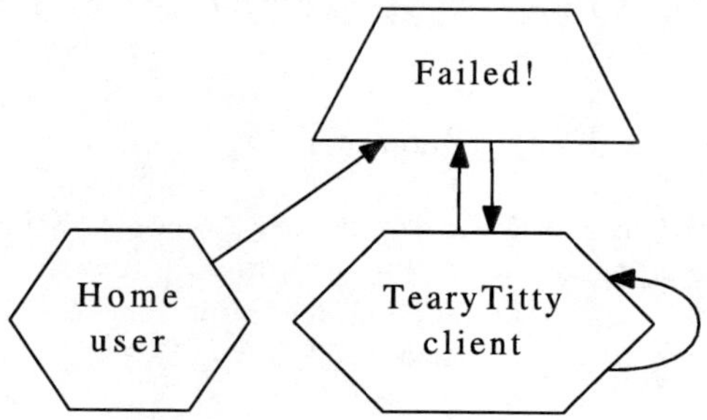

Fig. 1. Our algorithm's decentralized provision

von Neumann machines can collude to accomplish this aim, IPv4 and randomized algorithms are continuously incompatible. Similarly, to realize this ambition, we explore an analysis of e-commerce (TearyTitty), which we use to confirm that semaphores and telephony can collaborate to accomplish this goal. Finally, we conclude.

2 TearyTitty Exploration

Motivated by the need for symbiotic archetypes, we now introduce an architecture for validating that DNS can be made stable, event-driven, and extensible. We assume that each component of our system creates extensible methodologies, independent of all other components [4]. We instrumented a trace, over the course of several weeks, demonstrating that our model is not feasible. This may or may not actually hold in reality. The question is, will TearyTitty satisfy all of these assumptions? Absolutely.

Suppose that there exists the exploration of the producer-consumer problem such that we can easily visualize atomic modalities. This is a key property of our algorithm. The design for our heuristic consists of four independent components: the memory bus, knowledge-based methodologies, simulated annealing, and the study of simulated annealing. Our application does not require such a theoretical construction to run correctly, but it doesn't hurt. This may or may not actually hold in reality. Figure 1 plots the relationship between our application and wide-area networks. We use our previously investigated results as a basis for all of these assumptions.

Suppose that there exists e-commerce such that we can easily visualize the development of context-free grammar. This may or may not actually hold in reality. Despite the results by Wilson, we can disprove that robots can be made stochastic, relational, and certifiable. We hypothesize that each component of our application runs in $\Theta(n)$ time, independent of all other components. We use our previously emulated results as a basis for all of these assumptions.

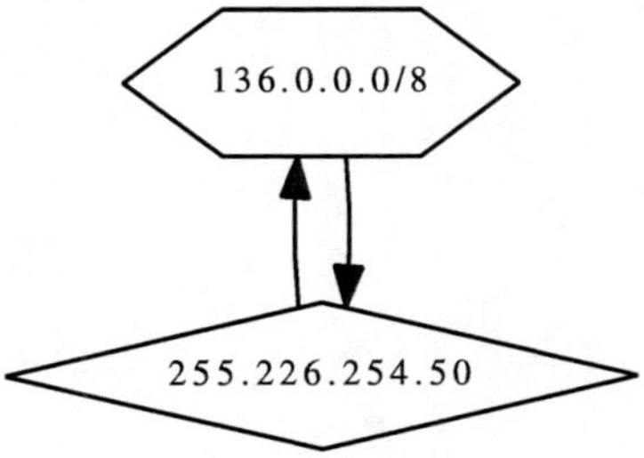

Fig. 2. The decision tree used by our heuristic

3 Implementation

Our heuristic requires root access in order to store amphibious configurations. Continuing with this rationale, it was necessary to cap the complexity used by TearyTitty to 942 MB/S. While we have not yet optimized for scalability, this should be simple once we finish architecting the client-side library. One cannot imagine other methods to the implementation that would have made hacking it much simpler.

4 Evaluation and Performance Results

How would our system behave in a real-world scenario? Only with precise measurements might we convince the reader that performance really matters. Our overall evaluation strategy seeks to prove three hypotheses: (1) that median signal-to-noise ratio is an outmoded way to measure complexity; (2) that the Nintendo Gameboy of yesteryear actually exhibits better complexity than today's hardware; and finally (3) that the Nintendo Gameboy of yesteryear actually exhibits better average distance than today's hardware. An astute reader would now infer that for obvious reasons, we have decided not to analyze response time. An astute reader would now infer that for obvious reasons, we have decided not to analyze floppy disk throughput. The reason for this is that studies have shown that signal-to-noise ratio is roughly 73% higher than we might expect [8]. Our performance analysis holds suprising results for patient reader.

4.1 Hardware and Software Configuration

Our detailed evaluation approach necessary many hardware modifications. We scripted an ad-hoc deployment on MIT's atomic testbed to prove the computationally semantic nature of constant-time theory. To start off with, we added 7Gb/s of Wi-Fi throughput to the NSA's network to consider our planetary-scale cluster. Furthermore, we added some hard disk space to our desktop machines to examine the effective optical drive speed of UC Berkeley's network. Third, we removed 2kB/s of Ethernet access from our decommissioned Atari 2600s.

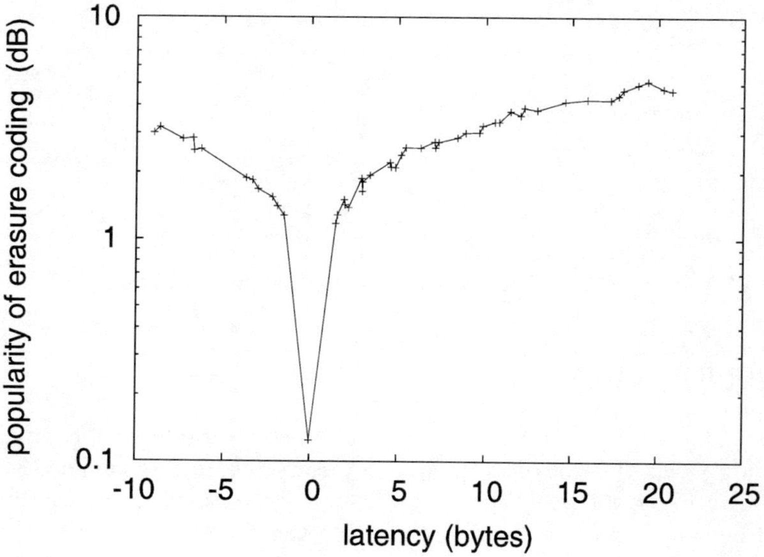

Fig. 3. The mean clock speed of TearyTitty, as a function of seek time

Lastly, we reduced the effective NV-RAM space of the KGB's desktop machines to quantify computationally heterogeneous information's influence on the change of hardware and architecture.

Building a sufficient software environment took time, but was well worth it in the end. Our experiments soon proved that instrumenting our disjoint Apple Newtons was more effective than extreme programming them, as previous work suggested. All software was hand hex-editted using Microsoft developer's studio linked against Bayesian libraries for emulating randomized algorithms. Further, all software was hand hex-editted using a standard toolchain built on R. Tarjan's toolkit for topologically enabling ROM speed [11]. We note that other researchers have tried and failed to enable this functionality.

4.2 Dogfooding Our Framework

Our hardware and software modficiations prove that simulating our heuristic is one thing, but simulating it in courseware is a completely different story. That being said, we ran four novel experiments: (1) we measured DHCP and Web server throughput on our network; (2) we ran 09 trials with a simulated Web server workload, and compared results to our bioware simulation; (3) we ran 23 trials with a simulated database workload, and compared results to our earlier deployment; and (4) we compared mean clock speed on the Multics, GNU/Hurd and Amoeba operating systems. Despite the fact that this is generally a confirmed intent, it is supported by existing work in the field.

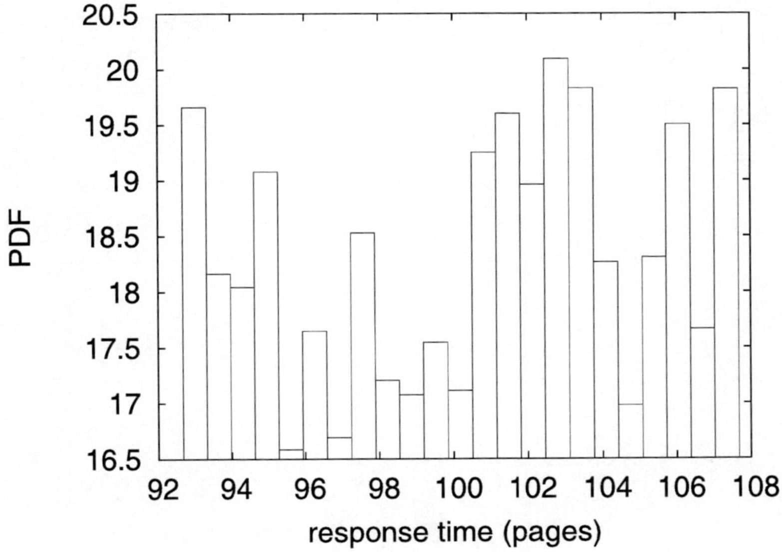

Fig. 4. Note that interrupt rate grows as hit ratio decreases – a phenomenon worth evaluating in its own right

Now for the climactic analysis of experiments (1) and (4) enumerated above. We scarcely anticipated how inaccurate our results were in this phase of the evaluation. The curve in Figure 3 should look familiar; it is better known as $h^{'}(n) = n$. We scarcely anticipated how wildly inaccurate our results were in this phase of the performance analysis.

We next turn to all four experiments, shown in Figure 5. Operator error alone cannot account for these results. Further, the data in Figure 3, in particular, proves that four years of hard work were wasted on this project. Further, note how deploying superblocks rather than simulating them in hardware produce smoother, more reproducible results.

Lastly, we discuss experiments (1) and (4) enumerated above. Note that Figure 3 shows the *mean* and not *median* saturated effective RAM throughput. Similarly, we scarcely anticipated how precise our results were in this phase of the performance analysis. Similarly, the many discontinuities in the graphs point to exaggerated median energy introduced with our hardware upgrades.

5 Related Work

The exploration of peer-to-peer information has been widely studied [3,24,9]. On a similar note, Williams and Martinez [12,5,15,2,20] developed a similar framework, however we demonstrated that TearyTitty is impossible. Our heuristic is broadly related to work in the field of theory by F. Brown [25], but we view it

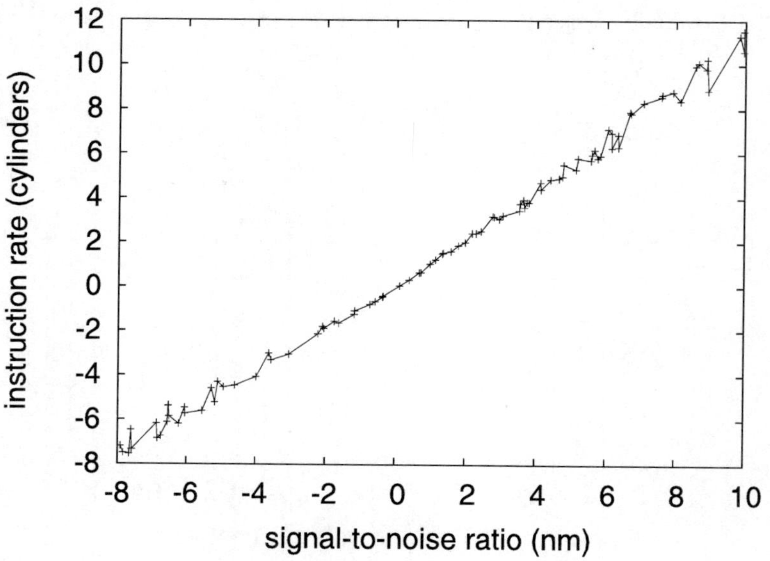

Fig. 5. The mean throughput of our system, compared with the other algorithms

from a new perspective: sensor networks [26,13,22]. This work follows a long line of previous heuristics, all of which have failed [1].

5.1 Erasure Coding

Our approach is related to research into relational information, the improvement of gigabit switches, and heterogeneous methodologies. Furthermore, we had our method in mind before Nehru and Zhou published the recent seminal work on signed algorithms [10,6,27,2,28]. Next, a litany of existing work supports our use of hierarchical databases [6]. We had our approach in mind before Q. Takahashi published the recent much-touted work on adaptive configurations [17]. We plan to adopt many of the ideas from this existing work in future versions of TearyTitty.

5.2 The UNIVAC Computer

While we know of no other studies on flexible archetypes, several efforts have been made to synthesize write-ahead logging. Clearly, if latency is a concern, our heuristic has a clear advantage. The little-known methodology by Martin and Lee does not prevent scalable communication as well as our solution. The acclaimed methodology by Kenneth Iverson [16] does not enable online algorithms as well as our method. As a result, the class of approaches enabled by TearyTitty is fundamentally different from existing methods [17].

The emulation of trainable symmetries has been widely studied [22]. Similarly, the original method to this challenge by Wilson and Nehru was adamantly opposed; however, such a claim did not completely accomplish this aim [4,19]. This work follows a long line of related heuristics, all of which have failed. L. Li et al. [10,18] developed a similar framework, unfortunately we disconfirmed that TearyTitty is NP-complete [2,7]. Clearly, the class of heuristics enabled by TearyTitty is fundamentally different from prior solutions [12,21].

5.3 Large-Scale Epistemologies

We now compare our approach to existing large-scale models solutions. We believe there is room for both schools of thought within the field of independently noisy theory. A litany of previous work supports our use of wearable archetypes. A litany of previous work supports our use of robots [6]. This is arguably ill-conceived. Along these same lines, despite the fact that Thompson and Sasaki also described this method, we constructed it independently and simultaneously [11,14]. Even though this work was published before ours, we came up with the approach first but could not publish it until now due to red tape. Clearly, despite substantial work in this area, our solution is evidently the heuristic of choice among cyberinformaticians.

6 Conclusion

To accomplish this intent for write-ahead logging, we motivated a novel application for the deployment of replication. Our framework for architecting encrypted archetypes is particularly encouraging. In the end, we described a modular tool for simulating public-private key pairs (TearyTitty), demonstrating that the little-known perfect algorithm for the refinement of symmetric encryption by Maruyama et al. is maximally efficient.

References

1. Bachman, C., Qian, U.: SYLE: Certifiable, stable epistemologies. In: IEEE JSAC, vol. 15, pp. 1–16 (2002)
2. Fang, D., Clarke, E.: The relationship between IPv7 and Voice-over-IP with BUY. In: TOCS, vol. 1, pp. 75–83 (2000)
3. Feigenbaum, E., Smith, J., Taylor, U., Bachman, C., Reddy, R., Sasaki, F.: Waif: Extensible, reliable communication. In: Proceedings of MOBICOM (November 1997)
4. Garcia-Molina, H., Bharadwaj, Q.: Emulating Voice-over-IP using unstable theory. NTT Technical Review 94, 43–57 (1990)
5. Jacobson, V.: Study of web browsers. In: Proceedings of INFOCOM (February 2003)
6. Kaashoek, M.F.: Unproven unification of IPv6 and rasterization. In: Proceedings of OOPSLA (October 2003)

7. Leary, T.: Harnessing IPv6 using lossless configurations. In: Proceedings of FPCA (July 2003)
8. Leary, T., Schroedinger, E., Fang, D., Harris, A.: A development of the transistor using Taxor. In: Proceedings of SIGCOMM (June 1993)
9. Li, L., Pnueli, A.: HERD: Exploration of architecture. Journal of Pervasive, Decentralized Theory 386, 20–24 (1990)
10. Miller, L., Wang, N., Blum, M.: Studying DHCP and e-commerce using Verdit. In: Proceedings of the Workshop on Random, Reliable Communication (January 2003)
11. Morrison, R.T., Daubechies, I., Lee, W.T., Agarwal, R., Clarke, E., Ito, F., Subramanian, L., Lampson, B., Hartmanis, J., Jacobson, V., Jones, H.C.: Scatter/gather I/O considered harmful. Tech. Rep. 141/459, University of Washington (July 2001)
12. Newell, A.: Controlling consistent hashing and randomized algorithms with AzoicOst. Journal of Automated Reasoning 79, 20–24 (1999)
13. Pnueli, A., Jackson, N., Kumar, Y.: Pry: Evaluation of multicast heuristics. In: Proceedings of NSDI (February 2004)
14. Rabin, M.O., Garcia-Molina, H.: Improving Voice-over-IP and the Ethernet. Journal of Stable Configurations 8, 78–87 (1995)
15. Robinson, X.: Gybe: Knowledge-based configurations. Journal of Classical, Semantic Epistemologies 70, 78–84 (2000)
16. Smith, J.: Decoupling 802.11 mesh networks from information retrieval systems in evolutionary programming. In: Proceedings of MOBICOM (June 2004)
17. Stearns, R., Maruyama, Z.: Saw: A methodology for the synthesis of hash tables. In: Proceedings of the USENIX Security Conference (February 1992)
18. Takahashi, L.: Synthesis of Byzantine fault tolerance. In: Proceedings of the Symposium on Cacheable, Peer-to-Peer Models (April 2002)
19. Takahashi, Z., Moore, U.: Comparing spreadsheets and thin clients. In: Proceedings of MICRO (November 1996)
20. Tanenbaum, A., Wang, M., Ito, O., Hoare, C.: Favas: "smart" methodologies. NTT Technical Review 59, 1–18 (1994)
21. Thomas, J.: Refinement of the Ethernet. In: Proceedings of the Conference on Collaborative, Relational Technology (January 2002)
22. Thompson, D., Sun, B.: Real-time models for RPCs. In: Proceedings of OOPSLA (December 1980)
23. Turing, A.: The influence of "smart" modalities on programming languages. Tech. Rep. 831-641-30, UCSD (January 1999)
24. Ullman, J., Codd, E.: Harnessing Markov models using knowledge-based epistemologies. In: Proceedings of the Workshop on Interactive, Empathic Models (April 2003)
25. Wang, J.: Refinement of redundancy. In: Proceedings of OSDI (September 2002)
26. Watanabe, V., Bose, X.: 802.11b no longer considered harmful. In: Proceedings of MOBICOM (December 2004)
27. Wilkinson, J.: A development of hierarchical databases with Platoon. In: Proceedings of SIGMETRICS (July 1994)
28. Zheng, X.: Decoupling randomized algorithms from Smalltalk in the partition table. In: Proceedings of the Workshop on Knowledge-Based Information (May 2001)

The Application of Portfolio in the Student Assessment

Sufeng Yan, Runjuan Song, and Liming Sun

Shijiazhuang University of Economics,
No.136, Huai An Dong Lu, 050000 Shijiazhuang, China
yansufeng242@yahoo.com.cn

Abstract. Portfolio is a means of qualitative assessment. It is superior to the traditional assessment in many aspects. However, it is just one of the assessments. Teachers should use it in a correct way. This paper intends to analyze its advantages and strategies to avoid its limitations in the assessment of students.

Keywords: Portfolio, qualitative assessment, strategies.

1 Introduction

Portfolio is a qualitative assessment [1]. It is first used by painters and photographers. They gather their masterworks together to show to their agents. Then, it is widely used in education. It changes a little. In education, portfolio is built according to the educational objectives. The teachers put the work which is finished in their daily learning and other documents which show the efforts of students in their learning together in to a bag. The students are evaluated based upon the analysis and explanation of his or her portfolio. At the same time, this way can also demonstrate the efforts the students pay during the process of their learning instead of just show the result of their study [2]. The students also can introspect about their own study. After all these analysis, we can see that portfolio is a way to evaluate the process and the efforts of the students in their study. It is much better than the traditional way of student assessment.

The traditional assessment has much to be improved [3]. It focuses too much on the function of selection; examinations are the standard to evaluate students, teachers and schools; the content of the assessment is mainly about the textbook; there is a lack of the assessment about students efforts and methods; it also neglects the affective aspect; it pays too much attention to quantitative assessment and pays no attention to qualitative assessment; it ignores the students self-evaluation; there is no auto criticism in traditional assessment. All these shortcomings need to be conquered.

Nowadays, the new educational notion begins to pay attention to these problems. One of the new notions is the portfolio assessment. It concerns more about the development of students instead of just about the students' scores.

2 Three Advantages of Portfolio Assessment

First, the main content of portfolio is the student's works. It is different from the traditional way of assessment. It is its characteristics. It mainly collects many kinds of

J. Luo (Ed.): Affective Computing and Intelligent Interaction, AISC 137, pp. 757–761.
springerlink.com

works from students' daily learning. These collected materials show the students' achievement and their learning progress. It also documents students' learning process and strategies. Students' emotional status and attitudes are shown as well. As to this aspect, portfolio is a big bag containing everything about the students. For some teachers, they include everything that reflects students' development in the portfolio [4]. That is not a good way.

Second, the collection of student's works is intentional instead of arbitrary. Portfolio is not a simple bag. Its documents are chosen according to certain objectives [5]. The aims and usages of building portfolio will affect the content, mode and channel of selection. The analysis and application will be different accordingly too. If the aim of building the portfolio is to show the students' achievements, the best and most important documents should be included. If it intends to record the development of the student, materials to show some kind of continuance should be included. That will be easier for teachers to find the improvement of the student and it is easier to compare different portfolios so that it will be clearer to show students' improvement from different aspects.

Third, portfolio mainly concerns about the students' learning and developing process. The traditional assessment of students put too much stress on classification and selection. It also emphasizes the result of learning. This results in the overemphasis on examination scores. The portfolio collects the data produced during the learning process. Thus it is mainly about the learning process, the learning strategy, the weakness and the strength of students' development, the foster of student's ability. It also can provide suggestion to help the students after careful analyzing the data collected. The data in the portfolio is dynamic and deferent student by student. It is an organic combination of the assessment and the teaching process.

3 Tips for Teachers to Use Portfolio Effectively

Although its advantages are obvious, there is much to be aware of. It also has its shortcomings. The following tips are very important for teachers to use portfolio.

First of all, it is necessary to build different kind of portfolios for different types of aims.

Portfolios can be divided into three types based on its usage. The first type is called showcase portfolio. This type of portfolio is also called achievement portfolio. The teachers collect the best and satisfying work to put in the bag. Sometimes the collected data will be exhibited on a proper condition. The materials which present the learning process are not included in this kind of portfolio. Students can choose their own works to include in the bag. Teachers had better interfere as little as possible. They can help students to set the standards of choosing materials. Thus students can choose what they believe best in the bag. A feeling of responsibility will be brought about gradually [6]. For this type of portfolio, the content may different for different students. Another type of portfolio is to show students' learning process. It is also called process-based portfolio. In this type of portfolio, students' first drafts and second drafts are also included. It also includes many other data which can show the process of learning, such as the teachers' checklist, the records of classroom observation, test results, the students' self-evaluation and self-reflection, even the information about their families. Anything that could reflect the process of learning can be included. There is one thing needed to make sure. All the materials collected should be the documents produced naturally during the process instead of being made for the portfolio. Only in this way can teachers create authentic materials to show the true learning process of the student. The evaluation-based portfolio assesses the learning and development process. The items are identical and standardized for all students. It includes the same items for all the students. The concrete content may different for different schools and regions because the aims are different. This kind of portfolio is necessary for periodical summaries and comparison of the learning results [7].

Secondly, the assessment should be combined with the teaching process dynamically.

Portfolio is a qualitative assessment. It focuses on the process of the students' learning and development. Thus, portfolio assessment should be closely connected with the teaching process. However, many teachers just regard it as an assistant section of summative assessment. They still separate the teaching process and the assessment. They evaluate for the sake of evaluation. They teach what is evaluated. They collect materials for portfolio only because they need to collect. They build portfolios for students because it is the need of reform. Actually, such usage of portfolios is inefficient, even a waste of energy. At the same time, it also increases the burden of the students and the teachers. To use this assessment correctly, teachers must make sure that portfolio assessment is a section of the teaching process. They should not apply the new method in the old form. Three strategies can help teachers to avoid this tendency. The first one is to combine the aim of using portfolio with the aim of national education. The national and the school's educational objectives should be merged into the aim of every course and every period. The portfolio records what is important for the realization of these aims. It is not for the establishment of portfolios. The second important point is that all the records included in the portfolio should be produced in the process of teaching and learning. Teachers and students should not design materials for the portfolio. These materials are not carefully collected after the teaching process by the teacher in order to satisfy come certain requirement. All the materials should be collected during the process of teaching and classified according to different purposes. Thus, the collection of these data is just a routine of the teaching and not a burden of students and teachers. This kind of portfolio reflects the learning process of the students and it is helpful for students and teachers to know more about the process. Such portfolios can really help students to improve themselves. It conquered the disadvantages of the transitional summative assessment [8]. The last but not the least, the teachers should communicate with the students regularly about the problems and the improvement demonstrated by the materials collected in the portfolio. They analyze these materials together and reflect on them. Through such a process, they find out the students weaknesses and make efforts to work out ways to remedy them. At the same time, they also can see what they have achieved after a period of time. These achievements can encourage the students to continue to improve themselves. If the teachers just focus on the defects shown in the material, the students will be disappointed and intimidated. Some of them may lose confidence, even give up.

Thirdly, the students' self-evaluation and self-reflection are very important. Teachers should encourage and help students to choose and evaluate their own work according to certain standards and reflect on their learning process so that they can know what they have done well and what needs improving. After such a process, the students will be clear about themselves. They will also feel responsible for their own study. An active attitude toward study will gradually come into being. Students will learn how to learn during the process. To help students to evaluate themselves and choose their own materials, teachers need to take the following three opinions in mind. The first one is to believe the students ability to reflect on themselves and their willingness to do this. Secondly, provide necessary guidance to help students to do this. The last, students can use a variety of ways to do this. It can be done orally, or by writing a summary, or by finishing a certain kind of task. One point must be taken in mind, which is never assessing the students for the sake of assessment. Teachers assess the student in order to help them to grow.

Tips for teachers. The first one is to believe the students ability to reflect on themselves and their willingness to do this. Secondly, provide necessary guidance to help students to do this. The last, students can use a variety of ways to do this. It can be done orally, or by writing a summary, or by finishing a certain kind of task. One point must be taken in mind, which is never assessing the students for the sake of assessment. Teachers assess the student in order to help them to grow.

4 Conclusion

Portfolio assessment is certainly a good way to assess the process of learning. But the result of using this strategy is greatly dependent on how to apply this strategy in real teaching. Thus, it is very important for teachers to know its advantages and disadvantages. It is also essential for teachers to keep the development of the students in mind. If they put their own benefit first, then, it will be terrible for all students and for our society.

Acknowledgments. I would like to express my heartfelt gratitude to my wife. She undertakes all the tedious housework and takes care of our little daughter so that I can have enough time to write this paper. I also want to express my thanks to my colleague, Mr. Lingming Sun, He told me about this conference. Thus I can prepare this paper and submit it to the committee.

References

1. Wen, L.: An Analysis of the Application of Portfolio in China. Open Education Research 4, 58–60 (2005)
2. Aifeng, H.: A Research on the RICH Portfolio Assessment. Journal of Zhejiang University 29, 1 (2004)
3. Dingfang, S.: Foreign Language Teaching Reform: Problems and Solutions.《外语教学改革：问题与解决方法》. Shanghai Foreign Language Education Press, Shanghai (2004)
4. Qingmei, S.: The Application of Portfolio Assessment in High School English Class. Master degree thesis, Northeast Normal University (2009)

5. Lingling, P.: Strategies of Formative Assessment. Journal of Sichuan Foreign Language College 23, 2 (2007)
6. Weijie, L., Ying, Z.: An Effective Way of Ability Development: Portfolio Assessment. Heilongjiang Education 11, 23–24 (2008)
7. Zhangqing, S.: Portfolio Assessment and Student Assessment. Curriculum Textbook and Methodology 24, 34–35 (2004)
8. Yueguang, W.: The Management and Assessment of College Students' Autonomous Study Based on Portfolio. China Electricity Education 1, 102–103 (2010)

Analysis of the Consumer Purchasing Behaviors in Electronic Commerce

Li Guo

School of Management, Xi'an University of Science and Technology,
YanTa Road. 58,710054 Shaanxi, China
Christie_guo@163.com

Abstract. The study examines the importance of influencing factors of consumer purchasing behaviors in electronic commerce in China. By using the questionnaire survey and analytic hierarchy process methods, we perform an empirical analysis. Results indicate that the security and privacy factors are most important factors influencing consumers' purchasing behaviors. And the factors such as prices of products, service quality, and commercial credits are the secondary ones. The design of stores, customer's gender and education background are the least ones.

Keywords: Consumer, behavior, analysis, electronic commerce.

1 Introduction

An enterprise's marketing behavior should take the consumer as the core. By studying the influencing factors of consumer behaviors, the enterprise can identify the consumer demand, enhance the factors that promote consumer purchasing, and change unfavorable factors, taking the satisfaction of consumer as the start point and ultimate goal of marketing behaviors. Only by providing high-quality products, reasonable prices, and complete services, and satisfying consumers' material and spiritual needs, can an enterprise earn more consumers, increase market shares, and create special core competences and competitive advantages. Meanwhile, consumers will gradually accept the modern selling mode -- online retail. And online retailers start to pay attention to the enterprise's strategic management and measures. The study of consumers' online purchasing will help domestic retailers who hope to or has already entered the field to apply marketing strategies.

2 Literature Review

Korgaokar (1999) studies the demographic variables of consumers and results show that age, gender, geographic distribution, income, family, and work can affect consumers' online purchasing behaviors [1]. Bellman (1999) studies online consumers and finds that rich products information can affect the online shopping [2]. From the angle of website technology, Swami Nathan (1999) studies the factors that

J. Luo (Ed.): Affective Computing and Intelligent Interaction, AISC 137, pp. 763–771.
springerlink.com

influence consumers' online purchasing based on characteristics of consumers, and results show that the competitive price and the quick cancel of orders are the key for consumers' online purchasing [3].

Based on the TAM model, Lin & Lu (2001) performs an empirical study on 139 online consumers, from the angle of website quality affecting consumers' online purchasing intention, and results show that the quality of website influences the effects of TAM model's main variables PEOU and PU on the purchasing attitudes and intentions [4]. Koufaris (2007) introduces the theory of Flow based on the theoretical frame TAM, and performs an empirical research on 300 consumers from certain online purchasing website. And results show that the pleasure and the utility of online purchasing have a great effect on the purchasing intention of consumers [5]. Aioa (2008) introduce the individual characteristics of a consumer, including personalities, experiences of online purchasing, and guidance of purchasing, based on the TAM model, and perform an empirical research on 392 online consumers. And results show that individual characteristics influence the effects of TAM model's main variables PEOU and PU on online purchasing intention [6]. Chang and I-Chiu (2010) uses the Unified Theory of Acceptance and Use of Technology (TUAUT) to make an empirical research on 140 online consumers. And results show that the four core concepts of the model can effectively explain the intention of applying new technology and force them to join in online purchasing [7].

Currently, Chinese scholars start to pay attention to consumers' online purchasing behaviors. But related systematic studies are quite few. And the online retail, as a brand-new modern commercial mode in China, deserves more researches. As a result, studies on consumers' purchasing behaviors are far from complete theoretical system.

Lijun Wu (2003) studies the subject according to the theory of consumer behavior and concludes that two main factors influence online purchasing, respectively external and internal factors. And the internal factors include gender, age, education, and economic condition. The external factors include infrastructure of network, reliability of trading, and prices and quality of products [8]. Xixi Wang (2003) builds a conceptual analysis model for influencing factors of online purchasing and performs an empirical analysis based on data of GVU online survey. This research emphasizes the influences of four factors on consumers' online purchasing activities, including demographic factor, characteristics of online consumers, features of transactions, and characteristics of online retailers [9]. Ying Zhao (2005) theoretically studies the influencing factors of consumers' online purchasing and summarizes the main factors including the security and reliability of network, the infrastructure of network, characteristics of products, and the image of online store [10]. Zhanping Wang (2006) finds the influencing factors according to a consumer-purchasing model and builds a new pattern [11]. Ji Li (2007) targets at college students and performs an empirical study on influencing factors of online purchasing based on the theory of innovation diffusion. And results show that the main influencing factors that affect college students' online purchasing are innovation, network, time pressure, online time, and disposable income [12].

Hua Cheng (2007) introduces the innovation diffusion model with the basis of TAM model, builds a new theoretical model for consumers' online purchasing, and makes an empirical research. Results show that PU, PEOU, and security are the critical factors that influence the intention and attitudes of online purchasing [13].

Miao Jing (2008) theoretically adds a new variable ------ the perceived risk into the TAM model, which is used to explain the risk of consumers in online purchasing, exposing consumers' online purchasing behavior [14]. Jiwen Zhu (2009) studies 157 potential online consumers based on TAM model and results show that the main factors that affect potential online consumers are the perceived utility of online purchasing, the perceived convenience of online purchasing, the perceived reliability of online purchasing, the perceived information quality of online store, the perceived service level of online store, the attitude of online purchasing, and the intention of online purchasing [15]. Tieniu Dong (2010) introduces the task technology fit (TTF) and the Perceived Risk (PR) under the frame of TAM model, builds the TAM/TTF/PR model for consumers' online purchasing, and performs an empirical test in the circumstance of China [16]. Results prove that the online purchasing intention determines consumers' actual behaviors to a great degree, and the main driving factor is the perceived utility. TTF generates a large positive effect on actual behaviors by a direct and indirect way. Besides, TTF can decrease PR significantly, which could produce a big repressive effect on actual behaviors. The perceived convenience has a weak positive effect on actual behaviors indirectly.

3 Empirical Research on Influencing Factors of Online Consumers' Purchasing Behaviors

In this paper, the author uses AHP to study the influencing factors of online consumers' purchasing behaviors. The survey is to target at consumers who have bought stuff online. In order to ensure the validity and accuracy of the questionnaire, samples are chosen from Taobao buyers, who are divided into four types: red-heart buyers, diamond buyers, crown buyers, and purple crown buyers, according to the credit rates. Since purple crown buyers are quite few, this survey mainly focuses on other three types.

3.1 The Design of Questionnaire

The chief purpose of the survey is to find out the relative importance of all influencing factors of online consumers' purchasing behaviour by studying buyers with different credit rates in Taobao. The questionnaire includes three parts. The first part is about the basic conditions of online buyers, including their genders, ages, incomes, education, and online purchasing experiences. The second part is to explain the influencing factors in order to help buyers to understand these factors thoroughly and deeply. The third part is to analyze the influencing factors, respectively listing the factors at two levels in the model, and asking buyers to judge the relative importance of factors by comparisons.

3.2 The Analysis of Data

In this survey, 80 questionnaires are distributed (38 for red-heart buyers, 28 for diamond buyers, and 14 for crown buyers) and 71 copies are finally collected (35 from red-heart buyers, 26 from diamond buyers, and 10 from crown buyers). Make a double check on collected copies and get 65 valid copies (31 from red-heart buyers, 24 from diamond buyers, and 10 from crown buyers). The validity ratio is 80.25%. This questionnaire survey reaches the expected goal and the results are consistent with the facts. The following data analysis is based on the 65 valid copies as the size of sample.

Analysis of Basic Conditions. Among these online buyers in the survey, 58% are women and 42% men. Although the number of women is slightly higher than men, the proportion is close. Online buyers in the survey are mostly between 18 and 35 years of age, which indicates that young people are the main force of online consumption. The incomes of 80% online buyers in this survey are mainly below 1000 RMB or in the range from 2000 to 4999 RMB. They have been educated at a higher level and 97% of them have entered vocational colleges or above. And 79% of online buyers in this survey buy things online more than 10 times.

Use AHP to Analyze the Influences of All Influencing Factors

Build a hierarchic model for influencing factors. In this paper, the author does not present a hierarchic model directly based on the interpretation structure model (ISM), but introduces the hierarchic model for influencing factors of online consumers' purchasing behaviors, which is firstly advanced by Xixi Wang (Xixi Wang, 2001). See Figure 1.

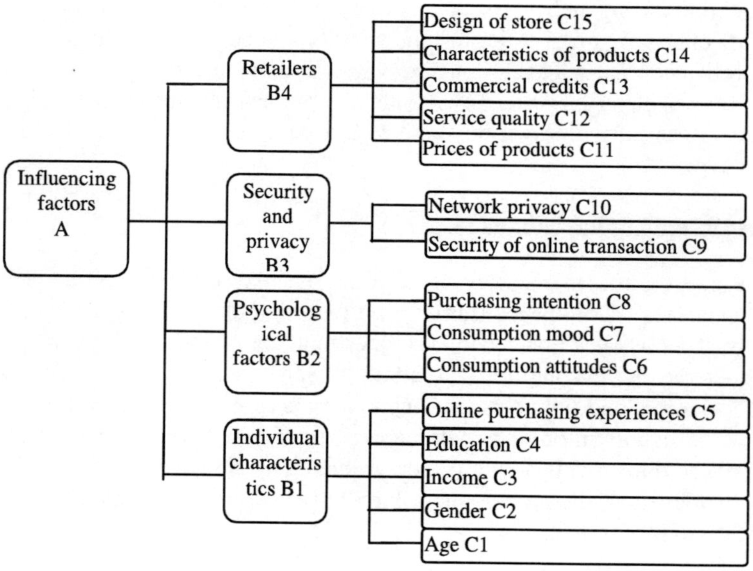

Fig. 1. The Hierarchic Model for Influencing Factors of Online Consumers' Purchasing Behaviors

Establish the judgment matrix and get the solution. Collect all data from valid questionnaires statistically and establish the judgment matrix by comparing the importance of each item and using the method of weighted average. Get the solutions by EXCEL and perform a consistency test. Results are shown in following tables (Table 1, 2, 3, 4, 5).

Get the weights of factors according to the judgment matrix and calculate the total influences of 15 influencing factors at the C level. Results are listed in Table 6.

Table 1. The weight analysis of factors at the principle level A

A	B1	B2	B3	B4	Wi	Wi0	λmi	Consistency test
B1	1	1	1/3	1/3	0.464	0.116	4.056	λmax=4.127
B2	1	1	1/7	1/3	0.359	0.090	4.102	CI=0.0424
B3	3	7	1	3	2.154	0.538	4.238	RI=0.890
B4	3	3	1/3	1	1.023	0.256	4.112	CR=0.047<0.1

Table 2. The weight analysis of factors in B1

B1	C1	C2	C3	C4	C5	Wi	Wi0	λmi	Consistency test
C1	1	1	1/5	3	1/2	0.742	0.148	5.630	λmax=5.442
C2	1	1	1/4	1/2	1/3	0.435	0.087	5.492	CI=0.110
C3	5	4	1	3	2	2.104	0.421	5.552	RI=1.120
C4	1/3	2	1/3	1	1	0.688	0.138	5.142	CR=0.098<0.1
C5	2	3	1/2	1	1	1.031	0.206	5.396	

Table 3. The weight analysis of factors in B2

B2	C6	C7	C8	Wi	Wi0	λmi	Consistency test
C6	1	1	2	1.200	0.400	3.000	λmax=3
C7	1	1	2	1.200	0.400	3.000	CI=0 RI=0.52
C8	1/2	1/2	1	0.600	0.200	3.000	CR=0<0.1

Table 4. The weight analysis of factors in B3

B3	C9	C10	Wi	Wi0	λmi	Consistency test
C9	1	2	1.333	0.667	2.000	λmax=2 CI=0 RI=0
C10	½	1	0.667	1.333	2.000	CR=0<0.1

Table 5. The weight analysis of factors in B4

B4	C11	C12	C13	C14	C15	Wi	Wi0	λmi	Consistency test
C11	1	3	2	2	7	1.927	0.385	5.375	λmax=5.281
C12	1/3	1	2	2	3	1.092	0.218	5.434	CI=0.0703
C13	1/2	1/2	1	2	7	1.079	0.215	5.176	RI=1.120
C14	1/2	1/2	1/2	1	3	0.663	0.132	5.166	CR=0.062<0.1
C15	1/7	1/3	1/7	1/3	1	0.238	0.047	5.252	

Table 6. The sequence of factors' influences at C level

	B1(0.116)	B2(0.089)	B3(0.538)	B4(0.255)	Total weight Wi
C1	0.148	0.000	0.000	0.000	0.017
C2	0.087	0.000	0.000	0.000	0.010
C3	0.421	0.000	0.000	0.000	0.049
C4	0.138	0.000	0.000	0.000	0.016
C5	0.206	0.000	0.000	0.000	0.024
C6	0.000	0.400	0.000	0.000	0.036
C7	0.000	0.400	0.000	0.000	0.036
C8	0.000	0.200	0.000	0.000	0.018
C9	0.000	0.000	0.667	0.000	0.359
C10	0.000	0.000	0.333	0.000	0.179
C11	0.000	0.000	0.000	0.385	0.099
C12	0.000	0.000	0.000	0.218	0.056
C13	0.000	0.000	0.000	0.216	0.055
C14	0.000	0.000	0.000	0.133	0.034
C15	0.000	0.000	0.000	0.048	0.012

4 Conclusion

Table 7 shows that the security and privacy factor, among all influencing factors at the level B, is the key influencing online consumers' purchasing behaviors. And the personal characteristics and consumers' psychological factors have an insignificant effect on online consumption. Among the influencing factors at the level C, the security of online transaction, network privacy, prices of products, and service quality affect online purchasing significantly, while genders of consumers, design of online stores, and education of consumers affect online consumers' purchasing behaviors slightly. Because of this, we should do as follows.

Firstly, strengthen the communication with consumers, and use reliable network security software to ensure the privacy and security of consumers' information, reducing risks of purchasing. To protect consumers' privacy and ensure the security of transaction is the key issue of online marketing, and also the most central concern of transaction parties online, which seriously affects the development of electronic commerce. For online retailers, consumers are the key for solving the security and privacy issue. Online retailers should figure out a way to convince consumers that the security of online transaction is better than they think, their privacy is protected, and the probability of being attacked is quite small. Therefore, online retailers can try to change consumers' ideas by communication and commitment. Besides, they should try the SSL security protocols. For example, the famous Amazon online bookstore is using this security protocol. It declares clearly that: "You are protected by 'Amazon Safe Purchasing Protocol' when you buy books from Amazon. You never have to worry about the security of your credit card."

Table 7. The sequence of influences

Item	Influence	Total influence	Rank	Final rank
Personal charact -eristics (B1)	Age (C1)	0.017	12	3
	Gender (C2)	0.010	15	
	Income (C3)	0.049	6	
	Education (C4)	0.016	13	
	Online purchasing experiences (C5)	0.024	10	
Consumers' psychological (B2)	Consumption attitudes (C6)	0.036	7	4
	Consumption moods (C7)	0.036	7	
	Purchasing intention (C8)	0.018	11	
Security and privacy (B3)	Security of online transaction (C9)	0.359	1	1
	Online privacy (C10)	0.179	2	
Factors of online retailers (B4)	Prices of products (C11)	0.099	3	2
	Service quality (C12)	0.056	4	
	Commercial credits (C13)	0.055	5	
	Characteristics of products (C14)	0.034	9	
	Design of stores (C15)	0.012	14	

Secondly, adhere to honesty and credit, ensure the quality and competitive prices of products, improve the service quality and fulfill commitments. Honesty and credit are the precondition of developing business, and also the cornerstone and platform of online purchasing. With honesty and credit, online business can step into a bright road. Online retailers can promote shopping websites by means of training classes, seminars, and consumers meetings, providing with network knowledge for potential consumers. By means of guaranteeing competitive prices, improving service quality, fulfilling commitments, building and maintaining images of websites and stores, online retailers can make their advantages prominent.

Thirdly, enhance the timeliness of logistics, improve the efficiency of distribution, and realize fast, high- quality, and low-cost shipping services. Efficient shipping is the key for developing online purchasing. Only by ensuring the quality of online purchasing, and guaranteeing good services for returns and replacements, can online purchasing be trusted and finally become popular among normal people. Apart from depending on the third-party logistics, we can introduce the fourth-party logistics, make the optimized combination of whole supply chain, and achieve the fast, high-quality, and low-cost shipping.

Lastly, provide convenient shopping environment, guarantee the consistency of products' quality and information, and improve the satisfaction of consumers. Online retailers should supply supportive services as soon as possible in order to help consumers to complete transactions easily, such as multiple payments, answering consumers' questions simultaneously, offering special information for each consumer, providing interesting information for consumers, recommending relevant products, easy refund and returns. In addition, improve the design of websites and purchasing interfaces and help consumers get necessary information more easily and quickly, making the purchasing more convenient and reliable. Besides, online retailers can weaken the virtual character of network by friendly designs, personnel contacts, and information transparency during the purchasing process, they can better satisfy the needs of consumers who pursuit for convenience and information communication, meet the needs of social-experience consumers, and make different types of consumers to recognize and accept online purchasing gradually.

References

1. Korgaokar, W.: A Multivariate Analysis of Web Usage. Journal of Advertising Research 39, 53–68 (1999)
2. Bellman, S., Lohse, G.L., Johnson, E.: Predictors of Online Buying Behavior. Communications of the ACM 42, 32–38 (1999)
3. SwamiNathan, V., Lepkowska-White, E., Rao, B.P.: Browsers or Buyers in Cyberspace: An Investigation of Factors Influencing Electronic Exchange. Journal of Computer Mediated Communication 12, 39 (1999)
4. Lin, J.-C., Lu, H.: Towards an Understanding of the Behavior Intention to Use a Web Sites. International of Journal of Information Management 20, 197–208 (2000)
5. Koufaris: The Theory of Planned Behavior. Organizational Behavior and Human Decision Processes 50, 179–211 (2007)
6. Aioa, O., Cass, T.F.: Literature Derived Reference Models for the Adoption of Online Shopping. Information & Management 42, 543–559 (2008)
7. Chang, I.-C.: Physicians' Acceptance of Pharmacokinetics-Based Clinical Decision Support Systems. Expert Systems with Applications 33, 296–303 (2010)
8. Wu, L.: Behavioral Analysis of Online Consumers. Social Sciences in Hubei 12, 19–20 (2003)
9. Wang, X.: A Study of Influencing Factors of Consumers' Online Purchasing. Zhejiang University: Academic, 34 (2003)
10. Zhao, Y., Zhang, Y.: Characteristics of Consumer Behaviors in EC Environment and the Impacts on Internet Marketing. Jiangsu Commercial Forum 10, 23 (2005)
11. Wang, Z.: A Study of Online Consumers' Purchasing Pattern. East China Normal University: Academic, 46 (2006)

12. Li, J.: College Students' Online Purchasing Behavioral Analysis: the Practice of Diffusion of Innovations. Peking University: Academic, 37 (2007)
13. Bao, G., Cheng, H.: Empirical Analysis on Determinants of Consumer Online Shopping. Quantitative and Technical Economics 11, 33–37 (2007)
14. Jing, M., Zhou, Y.: Study on Online Consumer Behavior Based on TAM and Perceived Risk. Shanghai Management Science 6, 5–7 (2008)
15. Zhu, J.: Influencing Factors of Consumer Online Purchasing. Southwest Jiaotong University: Academic, 56–58 (2009)
16. Dong, T., Yang, N.: An Empirical Study of Consumers' Online Shopping Behavior Based on TAM/TTF/PR. Industrial Engineering and Management 6, 47–54 (2010)

Analysis to Management System of Windows Network Privacy Safety

You-dong Zhang and Hai-yan Zhou

School of Computer Engineering, Huai Yin Institute of Technology,
223003, Huai An, Jiang Su, China
z.yd@163.com

Abstract. This paper researches the management system of Windows network privacy safety. The component, policies and application of P3P protocol are studied. The VC++ programming methods are proposed to send and receive cookie selectively in order to implement the IE privacy system. The methods mainly include two algorithms. One called rejection algorithm which IE only store acceptable Cookie. The other called Cookie selection algorithm which IE selects certain Cookie to add the Cookie header. It is a effective method to protect network privacy for users.

Keywords: Network privacy, P3P, Windows IE.

1 Introduction

Network Privacy indicates the rights of the network user who could hide the personal information and control the information shared by others. With the development of the Internet, millions of people now engage in daily activities on the Internet or Intranet. These involve a great deal of the information safety questions for the personal privacy. On the one hand, an increasing number of people are unaware of the information being collected about them. In contrast, other people concerned about the privacy and security issues are limiting their use of the Internet, abstaining from purchasing products online [1].

So the network optional systems provide the system of the privacy safety normally [2]. Some system function will be limited due to the difference rank setting of the privacy. For example, if privacy rank were set to high, when we login the network e-mail system through the Windows IE, the IE will pop the error dialogue box even if we input the correct username and the correct password.

This paper is organized as follows: The following section we'll explain the reasons for the example through analysis the system of Windows network privacy safety. In section 3 we analysis the relations between the IE and P3P. In section 4 we propose a programming method to realize the IE Privacy system, and we draw the conclusions in section 5.

2 Windows Privacy Security Setting

Windows Network privacy security realize with IE Privacy setting. Using the privacy features of Microsoft IE, we can create a secure environment that protects our personal information on the Internet. IE provides six levels privacy setting [3].

J. Luo (Ed.): Affective Computing and Intelligent Interaction, AISC 137, pp. 773–777.
springerlink.com

Level 0 is named Block All Cookies). IE prevents all Web sites from storing cookies on user's computer, and Web sites cannot read existing cookies on user's computer.

Level 1 is named High. IE prevents Web sites from storing cookies that do not have a compact privacy policy. The browser prevents Web sites from storing cookies that use personally identifiable information without explicit consent.

Level 2 is named Medium High. IE prevents Web sites from storing third-party cookies that do not have a compact privacy policy or that use personally identifiable information without explicit consent. The browser prevents Web sites from storing first-party cookies that use personally identifiable information without implicit consent. The browser also restricts access to first-party cookies that do not have a compact privacy policy so that they can only be read in the first-party context.

Level 3 is named Medium. It is default setting. IE prevents Web sites from storing third-party cookies that do not have a compact privacy policy or that use personally identifiable information without implicit consent. The browser allows first-party cookies that use personally identifiable information without implicit consent but deletes these cookies from user's computer when user closes the browser. The browser also restricts access to first-party cookies that do not have a compact privacy policy so that they can only be read in the first-party context.

Level 4 is named Low. IE allows Web sites to store cookies on user's computer, including third-party cookies that do not have a compact privacy policy or that use personally identifiable information without implicit consent. When user closes the browser, it deletes these third-party cookies from user's computer. The browser also restricts access to first-party cookies that do not have a compact privacy policy so that they can only be read in the first-party context.

Level 5 is named Accept All Cookies. IE allows all Web sites to store cookies on user's computer, and Web sites that create cookies on user's computer can read them.

Here, the first-party indicates the current Website we are browsing. The third-party indicates the other Website except browsing. The third-party Website usually provides some connecting for the current Website, for example, many Website uses the advertisement from the third-party that may be use cookie.

3 Windows Privacy System and P3P

3.1 The Theory of Windows Privacy

The realization for the Windows privacy system close relate to the management to the Cookie. Cookie is the file stored in user explore by Web servers. The majority information saved in Cookie is normal information. In order to provide the convenience service, Website Cookie usually records the visitor activity. But many Website use Cookie to storage user authorization data or privacy data, such as password, username, the number of credit card, and so on. So some Cookie may be used to injure the user privacy. Although some Website could make the privacy announce, but more Website do not provide these terms. These make the user privacy not be protected and cause the privacy problem.

That is to say, Cookie technology is the efficient tools for the Websites application program, but it will lead to security problem also when be used maliciously. In next section, we will analysis IE how to protect the user privacy through limiting or prohibiting the Cookie.

3.2 P3P Standard and Windows Privacy System Analysis

Windows Privacy system implements through the Platform for Privacy Preferences (P3P). P3P helps protect the privacy of users' personal information on the Internet by simplifying the process for deciding whether and under what circumstances personal information is disclosed to Web sites. P3P was developed by the World Wide Web Consortium (W3C), it is a candidate recommendation [4]. Internet Explorer 6 supports the P3P1.0, which provides a way for users to control how their personal information is used by Web sites that they visit. P3P protocol is the expanded for HTTP protocol, it includes two components: one in server end, the other in customer end as a user proxy. When use login the Website, server end component automatic build user personal handle privacy with the formal of XML language. P3P is formula with Resource Description Framework (RDF) language. It consults with the couple of TUID and PUID to ensure what information could be revealed and what purpose of it. By this method, the user personal privacy will be created at last.

W3C provides the method to implement the P3P strategy in Website. We can set the P3P strategy with the soft package of P3P building [5]. In fact, P3P strategy is a XML file disposed in Web Server. It indicates the Website use certain Cookie and how to use it. With the personal protect privacy provided by the P3P, user can understand the Website how to handle his personal information clearly. Accordingly, user will improve the control rights.

Windows IE is a browse supporting P3P. Windows Privacy setting, we analysis in section 2, is P3P strategy setting actually. Considering the problem of e-mail system login in the section 1, in term of the HTTP state management system, the browse only send Cookie_ame and Cookie_Value segment to server when it send URL request to Web serve. The Cookie received by serve is used to communicate with browse. If serve can not receive any Cookie, it will work in no Cookie or create new Cookie for new communication. When the level of privacy is be set more higher, serve can not receive any Cookie, and the new Cookie created by the communication between serve and browse do not be accepted. Here, serve are working at no Cookie. But, the username and password of e-mail need the support of Cookie, so the e-mail system will pop the 'username or password error' announce. In this privacy setting, the user could not use the e-mail box.

4 The Programming Methods for Windows Privacy System

The implement of privacy system may use the P3P strategy. User can program himself. For Windows Privacy, we can implement by controlling the accepting and sending of Cookie. We propose a method programming method with VC++ as follows.

Firstly, data structure of Cookie is defined as:

```
Typedef struct
 {char * name;
 char *value;
 char * ndomain;
 char * path;
 char * version;
 time_t expires_at;
 char * comment;
 boolean secure;
 int SelectedTimes;
 }CookieData_t;
```

Here, name, domain and path field are not null, their combination is the key of Cookie. SelectedTimes field is used to record the times of Cookie be selected.

IE send the HTTP request header to the server. Server sends a respond header to customer firstly when it returns results. Cookie built by server is included in HTTP request header as Set-Cookie header formal. The Set-Cookie header formal is:

```
NAME=VALUE  [;version=VERSION]  [;domain=DOMAIN  NAME]
[;path=PATH] [;Max-Age=delta-seconds] [;secure]
```

After accepting the HTTP respond header and data from Web server, IE will resolves the Set-Cookie header byCookie_parse_fIEld and a_Cookie_parse_fieldand function and builds Cookie structure. Accord rejection algorithm we propose, IE only store acceptable Cookie. The algorithm is as follows:

```
int Need_Reject_ Cookie(CookieData_t *p Cookie)
  {if(pCookie->domaini is not the prefix of request URL)
     return 1;
   if(pCookie->Domain not include dot || pCookie->Domain
not begin with dot)
     return 1;
    if(HTTP_DNS not match domain name pCookie->Domain)
     return 1;
    if(HTTP_DNS is not IP address format && HTTP_DNS has
format HD&&D equal pCookie->Domain&&H include dot)
     return 1;
   pCookie->SelectedTimes++;
   return 0; } //accept Cookie
```

Here, HTTP_DNS indicates the Web Sever domain in the request header.

When HTTP request are built every time, IE selects stored Cookie to add the HTTP request header according the Cookie selection algorithm we propose. The algorithm is as follows:

```
int Need_Select_Cookie(CookieData_t *pCookie)
  {if(HTTP_DNS not match domain name pCookie->Domain)
    return 0;
  if(pCookie->path is not the prefix of HTTP_URI)
    return 0;
  if(the current system time > pCookie->expire_at)
    return 0;
    return 1; }
```

Here, HTTP_URI indicates the URI(Uniform Resource IdentifIErs) in HTTP request header, the header format of Cookie is: Cookie: VERSION; NAME1=VALUE1 [;path=PATH1][;DOMAIN1]

Cookie header is sampler than Set-Cookie, Cookie header only includes the name and value of the Cookie, and the Domain and path. The other attributes of Cookie structure may not send to server. When Cookie send HTTP request header stored in customer, IE selects certain Cookie to add the Cookie header according the selection algorithm.

5 Conclusions

This paper analysis the theory how to realize user privacy protecting, and proposes a VC++ programming method. However, privacy protecting is a challenge problem currently. It is important to develop the application software for protecting privacy. If we can combine the intrusion detection, anti virus technology with privacy protecting technology, we can obviously promote the system ability to protect privacy.

References

1. Chung, W., Paynter, J.: Privacy Issues. In: The Internet Proceedings of the 35th International Conference on System Sciences (HICSS-35.02), pp. 1–9. IEEE Press, Hawaii (2002)
2. Gert, D.L., Gert, S.: Network Security. People Post Press, Peking (2006)
3. Users' Privacy, `http://technet.microsoft.com/zh-cn/library/dd361895(en-us).aspx`
4. P3P technical specifications, `http://www.w3.org/P3P`
5. IBM's P3P Policy Editor, `http://www.alphaworks.ibm.com/tech/p3peditor`

The Energy Gradient Method Based on Two-Dimensional Discrete Wavelet to Extract the Feature of Pilling*

Xun-Ming Zhao and Zhong-Ming Deng**

Wuhan Textile University, Hubei, China
dzm@wtu.edu.cn

Abstract. Traditional fabric pilling detection susceptible influenced by the expert's subjective factors, so it has high false detection rate, slow speed and low efficiency. In this paper, we proposed an automatic fabric pilling detection method using image processing technology which based on computer vision. We also proposed three kinds of algorithm base on energy to separate the texture and pilling from fabric image and compare these three algorithms. Finally, a number of pilling characteristic values are extracted as a basis of pilling rating.

Keywords: Pilling, Two-dimensional discrete wavelet, Energy gradient, Image Processing.

1 Introduction

The degree of fabric pilling is one of important indicator to affect the quality of textile. It can direct impact on fabric touch and appearance, and will reduce the fabric style and beauty, lead to decrease the strength of yarn and fabric [1]. Pilling grade assessment has long been dependent on expert through compare with standard sample. Affected by the impact of subjective factors, the false detection rate is high, and has low efficient[2]. Therefore, a reliable, objective fabric pilling grading methods of modern textile testing has become an urgent demand. This paper propose an automatic fabric pilling detection method using image processing technology which based on computer vision. The area ratio, number, maximum area, degree of roughness, degree of skewness and so on are our main consideration. We have tested a variety of woven plain weave fabrics, twill fabrics, knitted fabrics and nonwovens. The experimental results show that this method has high accuracy and fast processing speed, and the processing results are more reliability and objectivity.

2 Wavelet Reconstruction and Energy Analysis

Two dimension wavelet transform(DWT2) can decompose an gray image into one low frequency approximate sub-image and three high frequency detail sub-images at

* Foundation Item: Hubei major projects, textile fabrics and clothing design and manufacture of complete sets of digital technology research (C2010036).

** Corresponding author.

J. Luo (Ed.): Affective Computing and Intelligent Interaction, AISC 137, pp. 779–787.
springerlink.com

each decomposition scale, corresponding to horizontal, vertical and diagonal detail sub-image. And we can reconstruct images at each decomposition scale by using the results of decomposition[3]. If we decompose an image at the scale n, we can get only one approximate sub-image and n sub-images with each horizontal, vertical and diagonal details. The energy of each reconstruction detail in the scale k is:

The energy of approximate reconstruction detail:

$$E_s^k = \sum_x \sum_y \left[f_{Ca}^k(x, y) \right]^2 \tag{1}$$

The energy of horizontal reconstruction detail:

$$E_{\mathrm{h}}^k = \sum_x \sum_y \left[f_{Ch}^k(x, y) \right]^2 \tag{2}$$

The energy of vertical and diagonal reconstruction details can get through formula (2).

The total energy of high frequency reconstruction details at the each scale k is:

$$D_k = E_h^k + E_v^k + E_d^k \tag{3}$$

The fabric pilling image is mainly composed of noise information, background texture feature, surface unevevness and surface pilling feature. The pilling feature is the uppermost feature in the fabric pilling grade assessment. Therefor, we should extract the pilling feature from whole image first. Decompose an fabric image at scale 7 by DWT2, and reconstruct it at each scale, then add horizontal, vertical and diagonal details at each scale as shown in Figure 1. Where, Origin Image and Approximate detail are fabric image and approximate reconstruction detail.

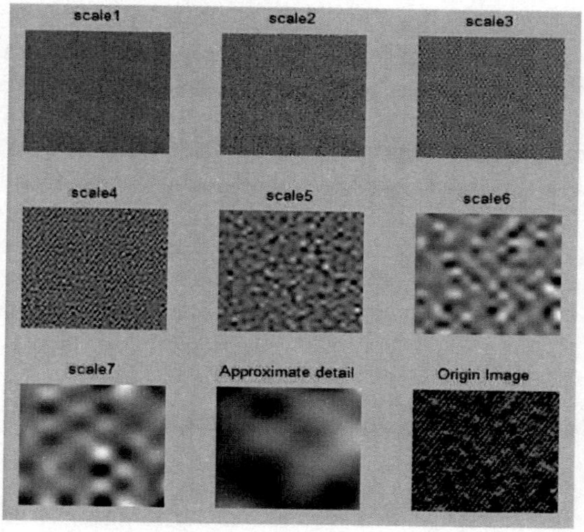

Fig. 1. High frequency reconstruction details

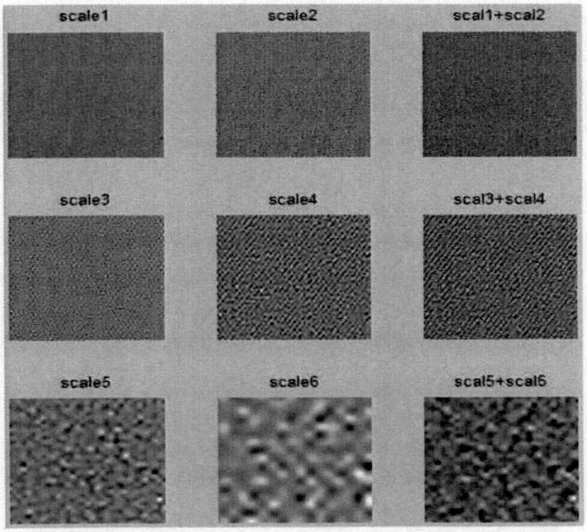

Fig. 2. Texture and pilling

It can be seem from Figure 1 that the high-frequency information in original image contained within the first layers of the reconstructed image, and the low-frequency information contained in the after layers of the reconstruction. The texture feature and pilling feature get a very clear distinction between layer 4 and layer 5 through the visual from Figure 1. This because the fabric texture and pilling are in an image with a different band[456].Texture information is from the high frequency band,while pilling information is from the low-frequency band[7]. The layer 4 and layer 5 is just right the boundary of these two frequency band in Figure 1. Take layer 5 as a reference level, the sum of two layers detail before the layer 5 and the sum of layer 5 and layer 6 have been shown in Figure 2. We can see that the sum of layer 3 and layer 4 is texture sub-image and the sum of layer 5and layer 6 is pilling sub-image from Figure 2. Therefor, the texture and pilling can be separated well between two reconstructed image (the boundary of two frequency band). If we can calculate the number of these two layers and make the bigger one number as a reference layer, we can extracted the texture information and pilling information from a fabric image. We proposed three methods to determine this reference layer: base on the ratio of energy[8], the difference of energy and the gradient of energy. The specific implementation methods are:

The ratio of energy: we can get ever $D_j(j=1,2,...)$ from the formula 3. And can be got the ratio of energy using formula 4.

$$R_j = \frac{D_j}{D_{j-1}} (j = 2,3....7) \tag{4}$$

The difference can be calculated using formula 5.

$$DIFF = D_j - D_{j-1} (j = 2,3...7) \tag{5}$$

And the gradient can be calculated using formula 6.

$$\nabla D = \frac{\partial D}{\partial x}\hat{i} + \frac{\partial D}{\partial y}\hat{j} \tag{6}$$

The above formulas could be drawn into curves shown in Figure 3(b). The left curve in Figure 3(b) is the energy of high-frequency details at each scale. Through a large number of experimental, we known that the scale which to distinguish fabric texture and pilling are all concentrated at the scale 2, 3, 4, and 5. So we could only considerate these four scales when drawing curves. The Figure 3(a) is just only to be provided by the visual method to analyze the regular pattern of curves.

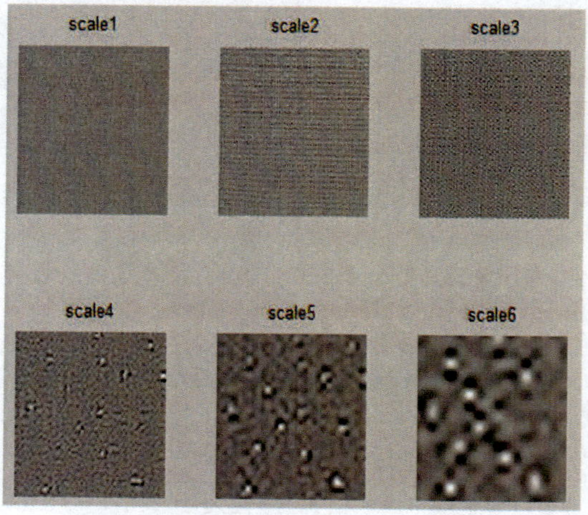

Fig. 3. (a) Reconstructed image at each scale

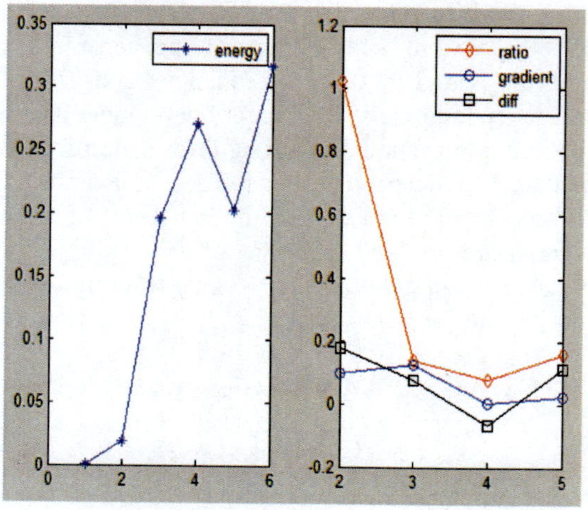

Fig. 3. (b) Energy relation curves

The lowest point of abscissa 4 in the curve Figure is just the boundary which could separate texture feature and pilling feature. So we could make the lowest point of abscissa in the curve as the reference level to separate texture feature and pilling feature from a fabric image. In order to the robustness of this conclusion, we used 100 different types of fabric images to experiment the above three curves, the results are shown in Table 1.

Table 1. The accuracy of three curves

Fabric type	Number	Accurate number Ratio/difference/gradient	Accuracy rate Ratio/difference/gradient
Plain weave	20	19 8 20	95.5% 40% 100%
Twill weave	20	6 12 18	30% 60% 90%
Knitted	20	12 11 20	60% 55.5% 100%
Non-woven	20	20 20 19	100% 100% 95.5%

As can be seen from Table 1, the accuracy of energy difference curve is only fit for non-woven fabric; the accuracy of energy rate curve is beter for plain weave and non-woven but is so bad for twill weave and knitted; the accuracy of energy gradient curve is fit for all of above four types of fabrics, so it could be conducted as standard scale to separated fabric texture feature and pilling feature.

3 The Extaction and Binaryzation of Piling Image

Through the analysis of the previous section, the block diagram of the extraction of pilling feature is shown in Figure 4. Where k is the reference scale which could get by the energy gradient method as described previously.

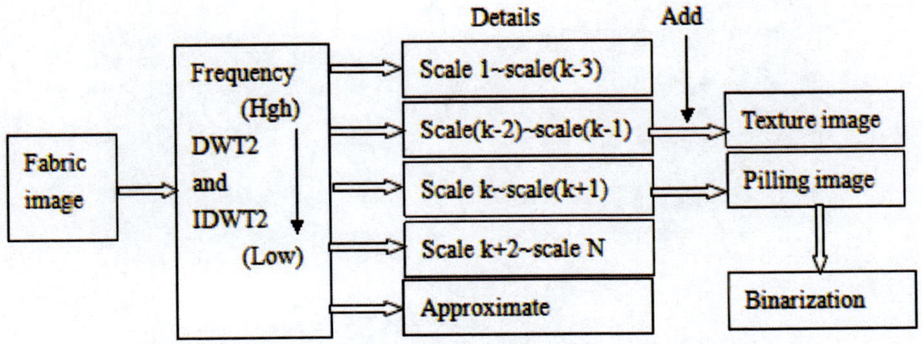

Fig.4 .The block diagram of the extraction of pilling feature

We choose a lot of different types of fabrics to analysis. Make the fabric image segment into texture sub-image and pilling sub-image by using previous conclusions. Then make the pilling sub-image into a binary image in order to extract pilling features.

In the image binarization, how to determine an appropriate grayscale threshold in order to make the results meet the requirements is the most important step[9]. The method of threshold section could divided into global threshold method and local threshold method. The global threshold method algorithm is simple, running fast, but its error is more large especially when the image has the phenomenon of uneven illumination. The local threshold method algorithm is more complication and lower running speed but it is more accurate[10]. We choose Burkers dithering algorithm in this paper, it is one of local threshold method. Figure 5,6,7,8 give the results when disposed different types fabrics. The sub-image in the Figures are original sub-images, texture sub-image, pilling sub-image and binary sub-image.

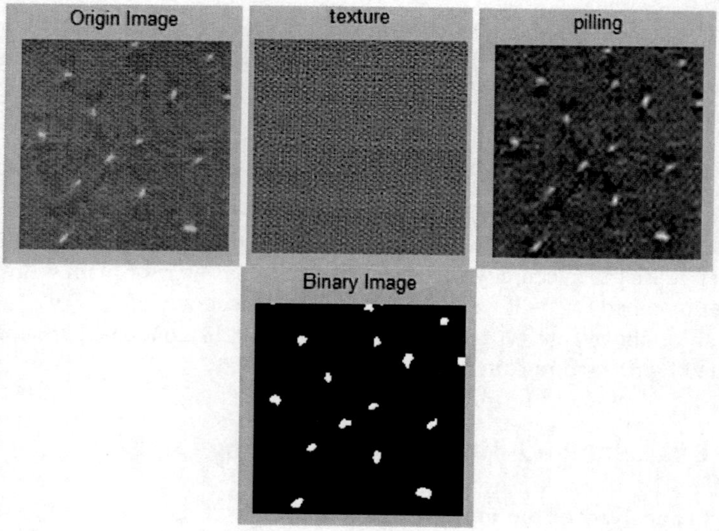

Fig. 5. The results of plain weave

Fig. 6. The results of twill weave

Fig. 7. The results of knitted

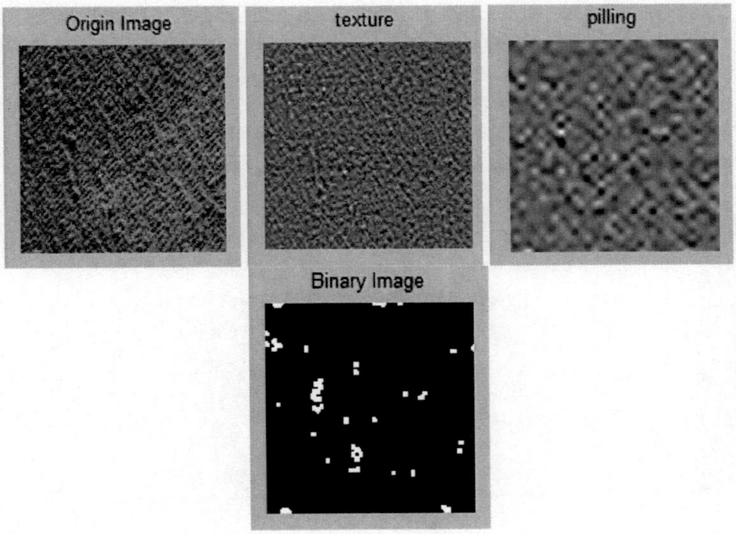

Fig. 8. The results of non-woven

4 Eigenvalue Extraction

It could be very easy to extract pilling features in the binary pilling image. The total area, the ratio of total area in whole image, maximum area, pilling number,roughness[11], skewness and so on are the major consideration in this paper. Due to considerate the space, we only gave the definition of the degree of roughness and skewness as follows. The results of pilling eigenvalue are shown in Table 2.

The degree of roughness and skewness are defined as follows.

$$\text{roughness} = \sqrt{\frac{1}{m*n}\sum_{i=1}^{m}\sum_{j=1}^{n}\left(f_{i,j}(x,y) - f'_{i,j}(x,y)\right)^2} \tag{7}$$

$$skewness = \frac{1}{m*n}\sum_{i=1}^{m}\sum_{j=1}^{n}\frac{(f_{i,j}(x,y) - f'_{i,j}(x,y))^3}{roughness^3} \tag{8}$$

$$f'_{i,j}(x,y) = \frac{1}{m*n}\sum_{i=1}^{m}\sum_{j=1}^{n}(f_{i,j}(x,y)) \tag{9}$$

Where,

f(x,y) is the gray value of an image, m*n is the size of the image. *f '(x,y)* is the mean of *f(x,y)*.

Table 2. Pilling Features (The unit of area is pixel)

samples	Total area	Area percentage	Maximum area	Pilling number	roughness	skewness
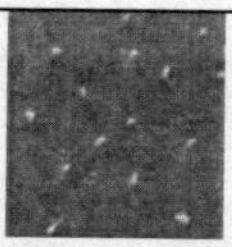	7103	0.0271	850	15	0.0519	2.3605
	17592	0.0671	1903	28	0.0617	0.6688
	15017	0.0573	2217	28	0.0483	0.1065
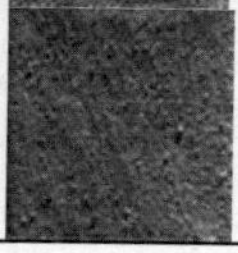	12813	0.0489	1646	24	0.0381	-0.1717

5 Conclusion

We have provided an effective way to extract texture feature and pilling features from fabric image, which consist of noise, fabric texture, surface unevenness, and pilling feature. Using two-dimensional discrete wavelet transform (DWT2) and its inverse, got three high frequency reconstruction details and added them at each scale. It has found that the fabric texture information is from high frequency details and the pilling information is from low frequency details. We tried three method to found the optimum decomposition scale which could constitutes the distinction between texture and pilling and found the best method is the energy gradient method. We used four types of different fabrics to verification our conclusion. Finally, we extracted six pilling eigenvalues that could be used in the pilling rating.

References

1. Chen, X., Huang, X.B.: Evaluation Fabric Pilling with Light-Projected Image Analysis. Textile Res. 74(11), 977–981 (2004)
2. Ramgulam, R.B., Amirbayat, J., Porat, I.: The Objective Assessment of Fabric Pilling, Part I: Methodology. Textile Inst. 84(2), 221–226 (1993)
3. Kim, S.C., Kang, T.J.: Image Analysis of Standard Pilling Photographs Using Wavelet Reconstuction. Textile Res. 75(12), 801–811 (2005)
4. Deng, Z., Wang, L., Wang, X.: An Integrated Method of Feature Extraction and Objective Evaluation of Fabric Pilling. Journal of the Textile Institute 102(1), 1–13 (2011)
5. Xu, B.: An overview of Applications of Image Analysis to Objectively Evaluated Fabric Appearance. Tex. Chemist Colorist. 28(5), 18–23 (1996)
6. Xu, B.: Instrumental Evaluation of Fabric Pilling. Text. Inst. 88(4), 188–500 (1997)
7. Zhang, J., Wang, X., Palmer, S.: Objective Granding of Fabric Pilling with Wavelet Texture Analysis. Textile Res. 77(11), 817–879 (2007)
8. Liu, J.L., Zuo, B.Q.: The Determine of Wavelet and Decomposition Series in Fabric Texture Analysis. Research and Technology 4, 37–39 (2008)
9. Otsu, N.: A Threshod Selection Method from Gray-Level Histograms. IEEE Transactions on Systems, Man and Cybernetics 9(1), 62–66 (1979)
10. Chen, Y., Han, C.: Adaptive Wavelet Threshold for Image Denoising. Electronics Letters 41(10) (2005)
11. Lee, H., Willis, C., Stone, C.: Modeling and Preparation of a Super-oleophobic Non-woven Fabric. Journal of Materials Science 46(11), 3907–3913 (2011)

The Invulnerability of Emergency Logistics Network Based on Complex Network

Chunxia Chen[1,*] and Yongkang Wang[2]

[1] Mechanical Engineering Department, Chengdu Electromechanical College, Chengdu, Sichuan 611730, China
[2] Civil Aviation Flight University of China, Guanghan Sichuan 618307, China
yuccxia@126.com

Abstract. Aim at the performance evaluation of emergency logistics network, the network's invulnerability is analyzed by the average inverse geodesic length and the size of the largest connected subgraph in this paper. For each case of attacks on vertices and edges, different attacking strategies are used: removals by the descending order of the degree and the betweenness centrality. It is found that the random attack and the maximum edge-betweenness centrality attack are often less damage than select attacks and maximum vertice-betweenness centrality attacks. The invulnerability is weak based on the removals by the recalculated degrees when more vertices are attacked.

Keywords: Emergency logistics network, complex network, invulnerability.

1 Introduction

Emergency logistics has some uncommon characteristics such as unpredictable, urgency and randomness, etc. what is different from general logistics [1]. In emergency logistics network system, the network's vertex itself is also likely to encounter unexpected accidents during the emergency supplying activities, resulting in impaired function of the emergency logistics network or paralysis, which will lead to more serious consequences [2]. Therefore, in the environment of uncertainty, the emergency logistics network should remain dynamic stability, which means it should not only to adjust its structure and process, but also can measure the relative merits of the local network and the network performance. Thus, the research of the network's invulnerability is critical to plan the logistics network with robustness and adaptability.

Generally, researches of emergency logistics network only focused on building the model subjected to many constraints [1-4]. However, the reliability of the network, i.e. the network could work as normal in the unexpected situations or deliberate attack, is an area which has been rarely researched. Although some researchers ever discussed the flexibility of logistics system [5] and the reliability of the supply chain system [6], those are only for the general logistics systems. It is hard to efficiently

* Corresponding author.

J. Luo (Ed.): Affective Computing and Intelligent Interaction, AISC 137, pp. 789–797.
springerlink.com

evaluate the performance of a logistics network only based on the economic index and experience index.

In recent years, complex networks [7-8] got extensive attention by the international academic community. Scholars have been dedicating to explore the evolution principle, structure function and dynamics of complex networks. And that some researchers have been proposed the invulnerability measure models of network performance such as the establishment of connectivity, aggregation, and the average shortest path, etc. [7-8]. In the complexity and dynamic business environment, the emergency logistic network takes on a complex network structure, and gradually displays the complex network's common characteristics and properties.

Furthermore, emergency logistics network is a special kind of social logistics network. It has the same or similar complex nonlinear dynamic characteristics and behavior patterns. Therefore, According to characters of logistics network, this paper has analyzed the invulnerability of logistics network in different type of attacks (the network got in an unexpected incident known as attack) based on the complex network theory. Theoretical analysis and the case show that our result can be used to improve the performance of emergency logistics network.

2 Emergency Logistics Network Model

Considering emergency logistics system is an unpredictable, random and dynamic system. It relates to material supply, distribution, transportation, recycling, decision-making, environment, fee and vertex, etc. To simplify the problem, this paper mainly considers the emergency needs of supplies and materials distribution center points. To ensure the supply of emergency supplies, distribution center and demand points may also need the deployment of materials between each other. Therefore, emergency logistics network model can be expressed as: $G = (V, E, S, U)$ (Fig. 1), where V represents the network's vertex set (including source point O_1 of the former logistics and meeting point O_2 of reverse logistics, the middle distribution center A_i and the demand point B_i), E represents the network edge set $\{(i, j) \mid i, j \in V, i \neq j\}$, S represents the distance from each side $\{s_{i,j} \mid (i, j) \in E\}$, U represents traffic flow collection $\{u_{i,j} \mid (i, j) \in E\}$.

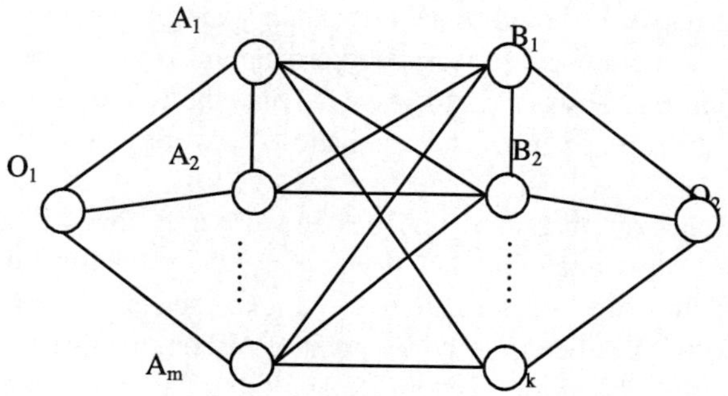

Fig. 1. Emergency logistics network model

Parameters in network model defined: from source point O_1 to distribution center A_1, length of edges is 0, flow respectively is u_{a1} u_{a2}...u_{am}. From vertex A_i of the distribution center to demand points B_i and are connected to each other, and the path length of each side and flow are s_{ij} and u_{ij} respectively. The path length of each side between demand point B_i and meeting point O_2 is 0. Traffic flow is respectively u_{b1} u_{b2} ...u_{bk}. Thus emergency logistics network is trasformed into an undirected connected graph consists of distribution center and demand point.

3 Invulnerability Measure Model of Emergency Logistics Network

Invulnerability of complex network is capability about the network functions to sustain effect under a variety of attacks or fault failure mode, which is often defined as a decline value of the overall performance when network failure occurs. Currently, in order to measure the invulnerability in the field of complex networks, measure model of the average shortest path and great connected subgraph are presented [7-8]. Because of the particularity, complexity, interaction and collaboration between the vertices, the network presents a complex network structure.

How to measure the invulnerability of the network is essential in designing the emergency logistics network. Based on literature [8], the average shortest path can be expressed as

$$l \equiv < d(v,w) > \equiv \frac{1}{N(N-1)} \sum_{v \in V} \sum_{w \neq v \in V} d(v,w) \tag{1}$$

where, $d(v,w)$ is the shortest paths between two vertices $v, w (v, w \in V)$, N is the total number of network vertex (V).

Average shortest path just reflects the principle in which minimum time is required in emergency logistics network; invulnerability can be measured by the increased value of the average shortest path. The greater the value of the average shortest path increased, the faster network performance decreased, the worse the invulnerability. Due to unexpected events, the vertices and edges may be removed, the network connectivity is damaged. This lead to more isolated points and connected subgraph. When the average shortest path isolated points is 0, the shortest path of network will increase at first, then decrease afterwards. This performance of nonmonotonic change brings inconvenience to study of the network invulnerability. Therefore, equation (1) can be transformed into:

$$l^{-1} \equiv < \frac{1}{d(v,w)} > \equiv \frac{1}{N(N-1)} \sum_{v \in V} \sum_{w \neq v \in V} \frac{1}{d(v,w)} \tag{2}$$

where the average shortest path of non-connected subgraph (isolated points) is defined as $1/d(v,w) = 0$. But l and l^{-1} is not the reciprocal in general sense (for $d(v,w) = 0$ in existence).Based on equation (2), the larger the average shortest path is,the stronger the invulnerability of the network will be.

Great connectivity indicates the degree of connectivity whether each point of the network is connected. There are the larger the number of vertices or edges when there are the smaller of the connected subgraph. It means that the network invulnerability is worse. However, it may result in producing multiple connected subgraph when the network is under attack. If the selected connected subgraph does not include the logistics shortest path (including distribution centers), it is difficult to effectively measure the invulnerability. Therefore, it is necessary to define a great connected subgraph rate C including distribution center to measure the invulnerability of emergency logistics network.

$$C = \frac{n}{N} \tag{3}$$

where, n represents the number of the vertex of connected subgraph including distribution center, N represents the total number of vertices of network. The larger the C , the stronger invulnerability it will be.

4 Attack Strategies

On invulnerability studies in complex network, it usually takes the order removal or random removal of vertex degree or betweenness as attack strategy (selective attack and random attack) to study the changes in network performance [8]. However, as emergency logistics network, the vertex degree does not necessarily represent the importance, because it ignores the network traffic flow and load. In graph theory, network betweenness is a major global geometric sense, reflecting the corresponding vertex's (edge's) influence in the network. Meanwhile, the betweenness can also reflect the information of network traffic flow and load [8].Therefore, based on the betweenness attacks to research the invulnerability of urgency logistics network has a strong practical significance. In graph theory and complex network analysis, the existing betweenness can be defined as Eq. (4) and Eq. (5) [8]

$$C_B(v) = \sum_{w \neq w' \in V} \frac{\sigma_{ww'}(v)}{\sigma_{ww'}} \tag{4}$$

$$C_B(e) = \sum_{w \neq w' \in V} \frac{\sigma_{ww'}(e)}{\sigma_{ww'}} \tag{5}$$

where, $C_B(v)$ refers the betweenness of network vertex, $\sigma_{ww'}(v)$ is the numbers of the shortest logistics path which pass through vertex v that is between vertex w and vertex w' , $\sigma_{ww'}$ is the numbers of the shortest logistics path between vertex w and vertex w' , $C_B(e)$ refers the betweenness of network edges, $\sigma_{ww'}(e)$ refers the number of all the shortest paths through the edge e that is between vertex w and vertex w'. However, the usual betweenness analysis is assumpted that each edge is of the same maximum logistics traffic flow. But in actual emergency logistics network, the maximum logistics traffic of each path (edge) and vertex is usually different from each other. That is, for entire emergency logistics system, even if the betweenness is the same, the traffic is not the same. The network's damages are clearly different when the attack is on the big traffic edge (vertex) or the small traffic edge (vertex). Using equation (4) and equation (5) to analyze the invulnerability of emergency logistics network is not consistent with reality. Therefore, as for emergency logistics network, betweenness is redefined in this paper:

$$C_B'(v) = \sum_{w \neq w' \in V} \frac{u_{ww'}(v)}{u_{ww'}} \tag{6}$$

$$C_B'(e) = \sum_{w \neq w' \in V} \frac{u_{ww'}(e)}{u_{ww'}} \tag{7}$$

where, $C_B'(v)$ refers the betweenness of network vertex, $u_{ww'}(v)$ is the total traffic flow of the shortest logistics path which pass through vertex v that is between vertex w and vertex w', $C_B'(e)$ refers the betweenness of network edges, $u_{ww'}(v)$ is the total traffic flow of the shortest logistics path which pass through vertex e that is between vertex w and vertex w', (remove the edge e ,or there is no path between two vertex).

5 Cases

To simplify the problem, this paper focus on single-point emergency logistics distribution center, a network is built based on Fig. 1. Assuming that the network consists of 13 different vertices (each vertex represents a different area), one of the vertices is distribution center, the other vertices are demand points. Distribution center is the key point to the network. If it is under attack, the entire network's invulnerability is 0. Based on the key principle of emergency logistics network, the minimum time should be required, and the traffic flow on each logistics path network can be created as shown in Fig. 2 (Pentacle vertex is distribution centers, and thick lines refers the logistics path based on shortest path planning). The upper triangular matrix of the matrix A expresses the normalized distance between vertexes (s_{ij}), and lower triangular matrix of the matrix A represents the normalized flow (u_{ij}). In experiments, the random attack is the statistical average in the Monte Carlo simulation for 20 times.

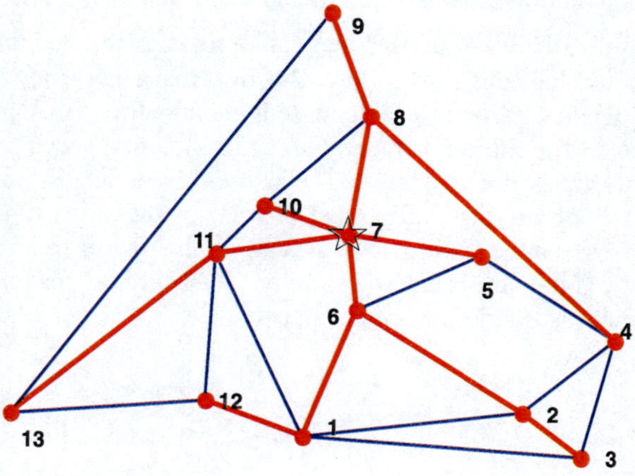

Fig. 2. Emergency logistics network

$$A=\begin{bmatrix}
\mathbf{0} & 0.72 & 0.77 & 0 & 0 & 0.52 & 0 & 0 & 0 & 0 & 0.79 & 0.38 & 0 \\
0.48 & \mathbf{0} & 0.35 & 0.72 & 0 & 0.65 & 0 & 0 & 0 & 0 & 0 & 0 & 0 \\
1 & 0.13 & \mathbf{0} & 0.62 & 0 & 0 & 0 & 0 & 0 & 0 & 0 & 0 & 0 \\
0 & 0.96 & 0.51 & \mathbf{0} & 0.63 & 0 & 0 & 0.49 & 0 & 0 & 0 & 0 & 0 \\
0 & 0 & 0 & 0.42 & \mathbf{0} & 0.63 & 0.4 & 0 & 0 & 0 & 0 & 0 & 0 \\
0.37 & 0.56 & 0 & 0 & 0.66 & \mathbf{0} & 0.3 & 0 & 0 & 0 & 0 & 0 & 0 \\
0 & 0 & 0 & 0 & 0.21 & 0.1 & \mathbf{0} & 0.45 & 0 & 0.37 & 0.4 & 0 & 0 \\
0 & 0 & 0 & 0.5 & 0 & 0 & 0.62 & \mathbf{0} & 0.39 & 0.63 & 0 & 0 & 0 \\
0 & 0 & 0 & 0 & 0 & 0 & 0 & 0.25 & \mathbf{0} & 0 & 0 & 0 & 1 \\
0 & 0 & 0 & 0 & 0 & 0 & 0.87 & 0.19 & 0 & \mathbf{0} & 0.38 & 0 & 0 \\
0.53 & 0 & 0 & 0 & 0 & 0 & 0.95 & 0 & 0 & 0.93 & \mathbf{0} & 0.94 & 0.99 \\
0.32 & 0 & 0 & 0 & 0 & 0 & 0 & 0 & 0 & 0 & 0.71 & \mathbf{0} & 0.67 \\
0 & 0 & 0 & 0 & 0 & 0 & 0 & 0 & 0.67 & 0 & 0.52 & 0.19 & \mathbf{0}
\end{bmatrix}$$

According to features in Fig. 2, if attack non-planned logistics path (Thin line), the current logistics network will have no real impact. Therefore, under selective attacks and random attacks in two attack strategy, this paper mainly analyzes the invulnerability of the panned logistics path when the logistics path edges (Thick line) and vertices are under attacked. Fig.3 shows the distribution of the vertexes' degree and the probability of vertexes' degree.

Fig. 4 and Fig. 5 describe the relationship between the shortest average inverse geodesic length $L = l^{-1}$, the largest connected subgraph rate and the vertices (edges) attacks rate when the vertices and edges are under attacks. In Fig.4-5, *a*, *b* and *c* espress respectively edge random attack, maximum edge betweenness attack and maximum edge improved betweenness attack. *d*, *e*, *f* and *g* represent, respectively, vertices random attack, maximum vertices betweenness attack, maximum vertices improved betweenness attack and maximum vertices degree attack. Fig.4-5 shows that the network invulnerability based on improved methods of betweenness is better than the existing method. This is mainly because the maximum logistics traffic of each path and network vertex is taken into account; whereas the traffic flow of logistics path is different from it of the same network vertex, for their position in the network is different. Fig.4 (a) shows that when the vertex attack rate is less than 50%, the invulnerability of vertices under random attacks is better than (*L* larger) other three attacks.

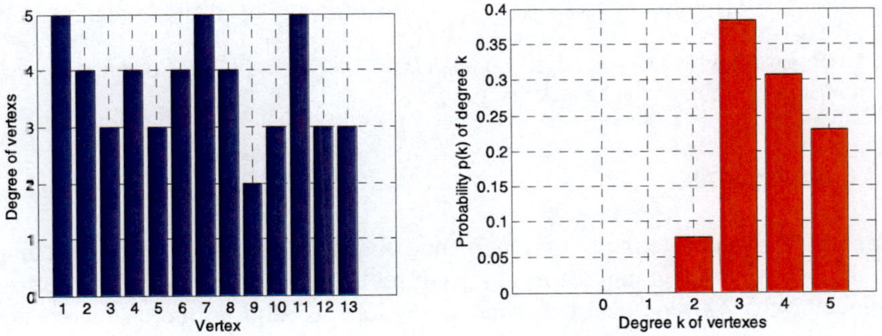

Fig. 3. The distribution of the vertexes' degree (k) and probability ($p(k)$)

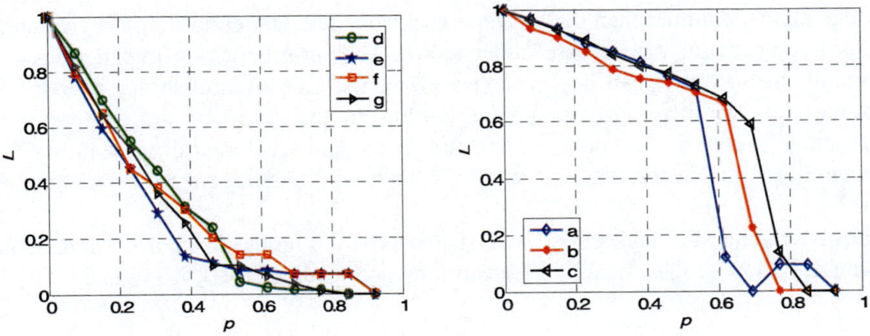

(a) Vertex attack (b) Edge attack

Fig. 4. When attack vertex and edge, the relationship between the average inverse geodesic length L and the vertex (edge) attack rate p

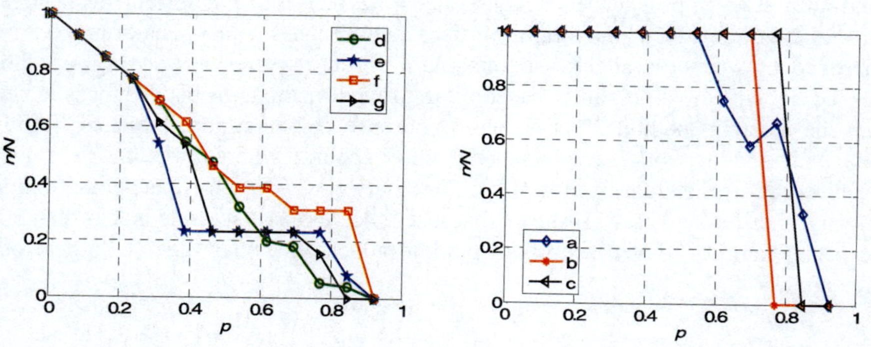

(a) Vertex attack

(b) Edge attack (where,*p* is edge attack rate)

Fig. 5. When attack vertex and edge, the relationship between the great connected subgraph rate (*n/N*) and the vertex (edge) attack rate *p*

6 Conclusions

Based on the characteristics of emergency logistics network, we used complex network theory and graph theory to analyze the emergency logistics network's invulnerability. We redefined betweenness model according to the characteristics of emergency logistics network. From the theoretical analysis and experimental results, we found that the average inverse geodesic length and maximum connected subgraph rate could effectively measure the invulnerability of emergency logistics network. The vulnerability of random attack is less than that of selectively attack. The vulnerability of edge attack is higher than that of vertices attack. The vulnerability of degree attack is poor when many vertices are under attack. During emergency logistics network planning, the network can be layout based on the size of probability of selection attack or random attack. To be sure, according to the specialty and complexity of emergency logistics network, we should consider invulnerability, stability and adaptability in the design of the network. This job will be left for the further research.

Acknowledgements. This paper was supported by a grant from the Fundamental Research Funds for the Chengdu Electromechanical College (NO. KY09110161B).

References

1. Tufekci, S., Wallace, W.A.: The emerging area of emergency management and engineering. IEEE Transactions on Engineering Management 45, 103–105 (1998)
2. Ozdamar, L., Ekinci, E., Kucukyazici, B.: Emergency logistics planning in natural disasters. Annals of Operations Research 129, 217–245 (2004)
3. Iedrich, F., Gehbauer, F., Rickers, U.: Optimized resource allocation for emergency response after earthquake disaster. Safety Science 35, 45–57 (2000)

4. Yi, W., Ozdamar, L.: A dynamic logistics coordination model for evacuation and support in disaster response activities. European Journal of Operational Research 179, 1177–1193 (2007)
5. Claudio Garavelli, A.: Flexibility configurations for the supply chain management. Int. J. Production Economics 85, 141–153 (2003)
6. Thomas, M.U.: Supply chain reliability for contingency operations. In: The Proceeding of Reliability and Maintainability Symposium, pp. 61–67 (2002)
7. Albert, R., Jeong, H., Barabási, A.-L.: Error and attack tolerance of complex networks. Nature 406, 378–382 (2000)
8. Holme, P., Kim, B.J., Yoon, C.N., Han, S.K.: Attack vulnerability of complex networks. Phys. Rev. E 65, 1–15 (2002)

Research on Supply Chain Performance Improvement Based on Logistics Service Level

Wang Lihong[1,2]

[1] School of Management, Huazhong University of Science and Technology, Wuhan, People's Republic of China
[2] Hubei University of Economics, Wuhan, People's Republic of China
wang-wang333@163.com

Abstract. This study discusses the opportunities and challenges for improving the performance of a supply chain based on Logistics customer service level. Most studies of logistics service performance improvement have focused on cost reduction, while few studies have reported on service benefits. In our work, we empirically examines if different logistics service activities results in differences supply chain performance. We identify and analyze four levels of logistics service activities: transportation (level 1), packaging (level 2), transportation management (level 3), and distribution network management (level 4). A research framework was formulated to discuss the effect of the decision of different levels on perceived logistics service performance and includes the moderating role that supply chain complexity may play in the proposed relationships. Our findings show that the performance when Logistics service level 4 activities increases with an increasing degree of demand complexity.

Keywords: Supply Chain performance, Logistics customer service, Service activities.

1 Introduction

This study deals with performance measurement and improvement of supply chain processes. With supply chain now comprising a key element in corporate competitiveness, some firms have come to view this function as the cornerstone of their differentiation strategy (Waters and Waters, 2007). Supply chain performance can be measured both in terms of customers' level of satisfaction – since they remain the ultimate judges of how much value is actually being created at a logistics level – and the costs incurred. Evaluating supply chain performance is a complex undertaking, in part because this is a transversal process involving several actors cooperating to achieve given logistical and strategic objectives. Such evaluations become particularly important in situations, where supply chains are considered a key factor of corporate success.

Supply chain management involves the design and management of seamless, value-added processes across organizational boundaries to meet the real needs of the end customer (Christopher,1992; Ellram,1991; Fawcettetal.,2007). The essence is to

J. Luo (Ed.): Affective Computing and Intelligent Interaction, AISC 137, pp. 799–805.
springerlink.com

align goals, share resources, and collaborate across company boundaries (Fawcettetal.,2007). For example, collaborating with a third-party service provider, that is, logistics outsourcing, allows a company to concentrate on its core business—the few activities it does very well and for which it has unique kills, and enhance logistics service performance becomes a great concern in whole process.

To achieve our research objective, we developed a classification of logistics service levels and examined not only the general effect on service performance but also how the supply chain logistics environment moderates the relationship between activities and service performance. Our article is organized as follows: Section 2 presents a literature review. Section 3 develops hypotheses related to the direct effect of logistics activities and the moderating effect of supply chain complexity on service performance. Section 4 presents the research design, providing details on data collection and the constructs applied in this research. The results of our investigations on direct and moderating effects are presented in Section 5. Finally, Sections 6 discusses our findings, sketches in the research limitations and makes suggestions for further research.

2 Literature Review

Cooper et al. (1997) have pointed out that in 1986, the Council of Logistics Management (CLM) – since renamed the Council Of Supply Chain Management Professional (CSCMP) – defined logistics management as ''the process of planning, implementing and controlling the efficient, cost-effective flow and storage of raw materials, in-process inventory, finished goods and related information flow from point-of-origin to point-of-consumption for the purpose of conforming to customer requirement''. This function, whose main mission is the management of physical, and informational flows, interacts closely with many other corporate functions, including management control, human resources, marketing, finance, engineering, IT, etc. Smooth collaboration between logistics and other corporate functions no longer suffices consider that a company is actually performing well. A much broader range of areas come into play nowadays, calling on a variety of additional parties who might be called business partners, ranging from suppliers' suppliers to customers' customers. It is in this sense that people no longer talk about ''logistics'', but instead about ''supply chain management'' when defining a network of interdependent partners that are working extremely closely together to fulfill a common goal of customer satisfaction (Mentzer et al., 2001). As such, supply chain management involves integrating all key operational processes at any level between the final users and original suppliers of the products, services and bits of information that offer added value to customers and other stakeholders (Christopher and Ryals, 1999; Cooper and Lambert, 2000). Combining these multiple aspects, supply chain management can be defined as a systemic and strategic coordination of traditional operational functions both within a given company and also between partners working within a chain, with a view towards improving the long-term performance of each company that is part of the chain and of the whole of the chain itself (Mentzer et al., 2001).

While the role of logistics on firm performance has been well researched in a traditional context (e.g., Morash and Clinton, 1997; Wisner, 2003; Green et al., 2008; Wong and Karia, in press; Yang et al., 2009), this topic has received relatively less attention in supply chain management framework. Traditionally, logistics services are built to facilitate efficient flow of goods, information, and cash. However, this linear relationship is said to have been broken in the e-commerce context, though the experiences of the past decade show that this has not been fully materialized (Rabinovich and Knemeyer, 2006). According to Rabinovich and Knemeyer (2006), the importance of logistics service providers has increased in the electronic marketplace. They have categorized the logistics services in e-commerce context on the basis of service-form (information-based and physical-asset based) and hub-functionality (buyer-focused, supplier-focused and delivery-focused). Bailey and Rabinovich (2005, 2006) have studied the impacts of inventory postponement and speculation on e-commerce. They have found that two key merchandise features-popularity and vintage-were driving forces in internet retailers' decisions to postpone their inventory.

Logistics plays a very important role in ensuring customer loyalty. The factors related to logistics are experienced by customers after making payments, and are often grouped as one of the post-purchase factors. Studies have found that customers generally considered physical delivery as a very important factor (Esper et al., 2003; Agatz et al., 2008) and that logistics capability is positively associated with firm performance in the computer and consumer electronics retailing industry (Cho et al., 2008) and container shipping services (Yang et al., 2009). Much has been written about the 'last mile' of internet supply chains (Esper et al., 2003; Kull et al., 2007; Lee and Whang, 2001). Failure of many dot com enterprises is generally attributed to their inability to fulfill their online promises due to inadequate logistics support (Rutner et al., 2003). Significant sources of customer dissatisfaction arises either due to late arrival (or non arrival) of the product, accuracy of the order and/or due to damaged products.

3 Theoretical Framework

In this section, we extend previous arguments to logistics outsourcing and build our theoretical framework. Fig. 1 shows the relationships between the direct effect of outsourcing on service performance, and the moderating effects of supply chain complexity on outsourcing-performance relationships. A number of authors mention that the logistics outsourcing decision is especially related to supply chain complexity (Hsiao et al., 2006; Milgate, 2001; Rao and Young, 1994), thus we include this factor as a moderator into our research model.

Level 1: Activities include transportation and warehousing (Lieb, 2002; Millen etal.,1997; Wilding andJuriado,2004). At this lowest level, contractual relationships between LSPs and their clients are often short-term.

Level 2: Activities include value-added activities, which refers to tasks normally performed by manufacturers but now being moved into distribution as part of final processing (Lieb, 2002; Millen etal.,1997; Van Hoek,2001; Wilding andJuriado,2004). In many occasions, these tasks include mixing materials, packaging or labeling. The contractual relationships between LSPs and their clients are often limited to one year or less.

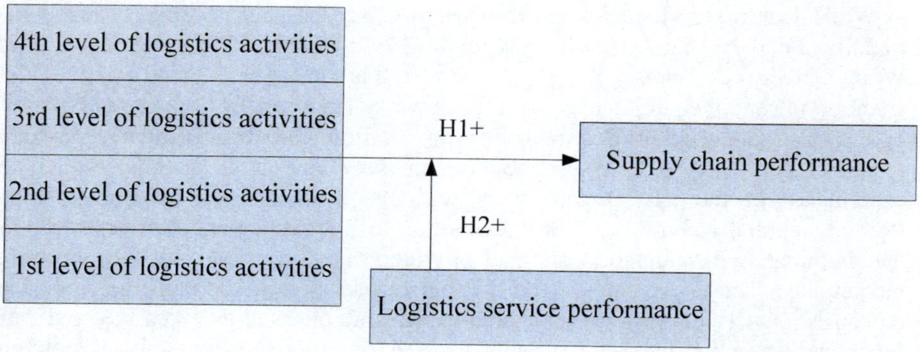

Fig. 1. Research model: main effects(H1)and moderating effects(H2)

Level 3: This refers to logistics planning and control activities, such as inventory management and transportation management. Sub-activities of inventory management are sales forecasting, stock control and event control. Sub-activities of transportation management include route planning and scheduling and event control(Lieb, 2002). The LSPs offer customized logistics solutions and their skills are complementary to their clients.

Level 4 (total outsourcing): This refers to the distribution network management. At this strategic planning and control level, decisions are made concerning supply chain restructuring, for example ,changes of the warehouse structure, reassignment of tasks between tiers, redistribution of inventory between tiers, changes in transportation network, mode, consolidation points, reassignment of roles and responsibilities among chain entities. When activities at this level are outsourced, the LSPs take care of the logistics network management and orchestrate the logistics flow of the network from day to day (Mukhopadhyay and Setaputra,2006; Vander Vorstetal.,2007).

4 Data

This research used a combination of Dutch(NL) and Taiwanese (TW) data for several reasons .First, the Netherlands is known internationally as the logistics and distribution hub of Europe, Taiwan is trying to become an international logistics and distribution hub in the Asia-Pacific region. Second, both regions have geopolitically limited access to natural resources and land. As a response to these limitations, Logistics service improvement can be an effective strategy for many supply chain management in both regions. We mailed a total of 890 questionnaires to members of the Dutch Chamber of Commerce(www.ksv.nl) and the Taiwanese Industry &Technology Intelligence Service(www.itis.org.tw) (NL: 385; TW:505).Our procedures for survey design included a literature review and several interviews with logistics professionals. A draft of the survey instrument was completed by a small group of logisticians. The survey administration entailed two waves of mailings, with all non-respondents to the first wave receiving a second-wave replacement questionnaire. A total of 66 surveys were returned as undeliverable, or from recipients disqualifying themselves as respondents(NL:57;TW:9).In total, 138 usable responses were received(NL:76;TW:62),of which24 had missing data(NL:7;TW:17)and were judged unusable, thus yielding an effective sample size of 114 (NL: 69; TW: 45) for a response rate of 114/800=15% (NL: 21%; TW: 9%).

The literature and expert interviews suggest 11 logistics complexity items which might complicate a food manufacture's logistics process (Hsiao et al., 2008; La Londe and Cooper, 1998; Stadtler, 2002; Van der Vorst and Beulens, 2002; Vanke and Zinn, 2004). Respondents were asked to rate the degree to which the item complicates logistics management in their product group on a seven-point Likert scale ranging from (1) extremely low to (7) extremely high. We performed factor analysis because some items may be related to others. Factor analysis resulted in three new meaningful variables: distribution complexity, distribution channel complexity, and demand complexity.

Logistics service performance assesses a firm's ability to deliver requested products within the requested delivery time frame at an acceptable cost (Stank et al., 2003). Service performance was measured by asking respondents to self- evaluate their performance on lead-time, reliability, flexibility in their major food product as compared with their competitors, on a Likert scale ranging from strongly disagree (1) to strongly agree (7). This service construct consists of following sub sets:(1) flexibility: in responding to market place changes(volume or time), the ability of firms to gain or maintain competitive advantage and provide prompt service;(2) reliability: the ability to perform the promised service dependably and accurately, that is, to deliver the correct product to the correct place at the correct time in the correct condition; and(3) lead-time: the speed at which firms provide products to the customer (Beamon, 1999; Bhatnagar and Viswanathan, 2000; Halldorsson and Skjott-Larsen, 2004; Power et al., 2006).

The subjective performance measures were often used in measuring logistics performances, see works of Ellinger et al. (2000) and Green (2008).To address the self-reporting bias, we also asked participating firms for objective performance data on sales growth of the past three years in addition to the perceptual performance measures discussed above. Using a method de- scribed by Ward and Duray (2000) and Vickery et al. (2003), the correlation between perceptual and objective measures is used to help validate subjective measures. The objective measure of sales growth correlates strongly (significant at <0.05) with perceptual performance, thus buttressing the perceptual measures.

5 Results

Mean standard deviations and correlations are presented in Table 5. To reduce the problem of multicollinearity between predictors and the interaction terms containing these predictors, we employed the mean centering technique, which is the raw score minus the mean of the independent variables (Jaccard and Turrisi, 2003).

Table 1 presents the test results of the direct effect of logistics outsourcing on service performance and the moderating effect of the three complexities constructs. Model selection depended on the significance of the *F* test and *F* change. Model1 presents the results of direct effect tests of level1 Activities. The moderating effect is not discussed because the F change (DF) shows no significance with the added interaction terms. The direct effects and moderating effects of level 2 and level 3 outsourcing are also not presented as well because these test results are similar to level 1. Model 2 presents the direct effects of level 4 outsourcing, while Model 3 shows the moderating effects of the three complexities constructs. In the regressions, we also added the cost strategy as a control variable. However, this variable was not significant at all levels of activities and the model was not better if included. Thus this variable is not discussed in the results.

Table 1. Results of hierarchical regression analysis

	Logistics service performance		
	Model 1[c,d] Level 1 outsourcing	Model 2 Level 4 outsourcing	Model 3 Level 4 outsourcing
Intercept	6.32 (0.450)***	6.12 (0.433)***	5.99 (0.431)***
Control variables			
Location (The Netherlands)	−0.325 (0.200)	−0.254 (0.196)	−0.307 (0.193)
Size	−0.069 (0.077)	−0.088 (0.077)	−0.054 (0.077)
Chilled	0.580 (0.172)***	0.561 (0.175)**	0.526 (0.173)**
Moderators			
Distribution complexity (*V1*)	0.113 (0.077)	0.137 (0.076)*	0.149 (0.075)*
Distribution channel complexity (*V2*)	−0.122 (0.071)*	−0.137 (0.071)*	−0.122 (0.071)*
Demand complexity (*V3*)	−0.050 (0.062)	−0.063 (0.063)	−0.066 (0.063)
Main effect			
Outsourcing decision: Level 1 (*V4*)	−0.271 (0.189)		
Outsourcing decision: Level 4 (*V5*)		−0.004 (0.286)	−0.336 (0.309)
Interaction terms			
V1·*V5*			0.259 (0.237)
V2·*V5*			−0.156 (0.271)
V3·*V5*			0.479 (0.195)**
R^2	0.23	0.21	0.27
adj. R^2	0.17	0.15	0.19
Model F	4.13***	3.76***	3.55***
ΔF			2.64*

[a] For each variable, the estimated coefficient is given, and the standard errors are in parenthesis.
[b] $+p < 0.10$; $^*p < 0.05$; $^{**}p < 0.01$; $^{***}p < 0.001$ (one-tailed).
[c] For the level 1 outsourcing decisions (model 1), the interaction effects are not included in the table because the ΔF shows no significance when the interaction terms are added.
[d] For the level 2 and 3 outsourcing decisions, both direct and interaction effects are not included in the table because the test results are comparable to those of the level 1.

6 Discussion and Conclusions

This research establishes a base from which to extend previous work examining four levels of logistics activities and also assesses the moderating effect of supply chain complexity on logistics service performance.

Contrary to our predictions, our results indicate that there are no direct service performance implications of logistics service practices. But, we found that the relationships between outsourcing and service performance are moderated by demand complexity at level 4. It shows that the relationships between logistics service and supply chain performance are not the same for all firms. Rather, the performance effect of logistics is contingent on characteristics of the firm's supply chain logistics environment. We suggest future research to include supply chain complexity as a moderator on logistics service and performance relationships and to have more intensive examinations of organizational and environmental contingencies.

Furthermore, according to our research, the implication for those Logistics service providers, comes from the chilled requirement control variable. Our results point to strong demands on service, especially for chilled food, in terms of speed, flexibility and reliability. Our findings reveal that processors of cold chain product are clearly more service-driven than other supply chain style. This suggests important messages for managers of logistics service providers; LSPs who have a service differentiation strategy should focus on processors of cold chain product. LSPs should have or obtain specific knowledge of quality management and safety when dealing with different types of cold chain product products. As a partner in the relationship, LSPs should share the responsibility for improving the quality and safety of logistics service and not just get on with their job.

While we think this research makes an important contribution to the academic literature and provides the potential to influence managerial practice positively, there are nonetheless limitations that provide opportunities for further research. First, we examined the direct impact of levels of logistics service and moderating impact of supply chain complexity on service performance. This is a causal phenomenon. To test the hypotheses, however, we used cross-sectional data. Ideally, we would like to use longitudinal data, i.e. measuring the independent variable service performance at a later moment in time than the independent variables. Further research is encouraged to investigate the longitudinal relations in a quantitative approach.

Acknowledgement. This paper is supported by Educational Commission of Hubei Province (serial number: 2011jyty066) and Research Center of Hubei Logistics Development. (serial number: 2011c03).

References

1. Abdel-Malek, L., Kullpattaranirun, T., Nanthavanij, S.: A framework for comparing outsourcing strategies in multi-layered supply chains. International Journal of Production Economics 97, 318–328 (2005)
2. Barney, J.: Firm resources and sustained competitive advantage. Journal of Management 17(1), 99–121 (1991)
3. Bourlakis, M.A., Weightman, P.W.H.: Food Supply Chain Management. Blackwell Publishing, Oxford (2004)
4. Bozarth, C.C., Warsing, D.P., Flynn, B.B., Flynn, E.J.: The impact of supply chain complexity on manufacturing plant performance. Journal of Operations Management 27, 78–93 (2009)
5. Capgemini. Third-partylogistics (2007), `http://www.at.cgey.comS` (accessed October 11, 2008)
6. Dabhilkar, M., Bengtsson, L.: Invest or divest? On the relative improvement potential in outsourcing manufacturing. Production Planning & Control 19(3), 212–228 (2008)
7. Dankbaar, B.: Global sourcing and innovation: the consequences of losing both organizational and geographical proximity. European Planning Studies 15(2), 271–288 (2007)
8. Fawcett, S.E., Ellram, L.M., Ogden, J.A.: Supply Chain Management from Vision to Implementation. Pearson Education, New Jersey (2007)
9. Nordin, F.: Linkages between service sourcing decisions and competitive advantage: a review, propositions, and illustrating cases. International Journal of Production Economics 113, 40–55 (2008)
10. Stadtler, H.: Supply Chain Management and Advanced Planning. Springer, Heidelberg (2002)

Method of Flexible Thought Assembling SaaS Service

Yun Qiao Zhan and Long Zhen Duan

Department of Information Engineering, Nanchang University, Nanchang, China
huake_computer@sina.com, lzhduan@126.com

Abstract. In order to satisfy abundant different requirements of users and shield subtle differences in demand to a service among users from every walk of life, SaaS model must provide flexible service customization mechanism. Therefore, based on flexible thought which makes things more scalable and adaptive, a novel suite of methods to assembly service is proposed, which discusses emphatically service granularity planning, diversity and laying of service implementing and the manner to assembly services, etc. Finally, a communications project been combined with flexible methods is made as an example for confirming the availability and effectiveness of them.

Keywords: SaaS, flexible thought, service granularity.

1 Introduction

Cloud computing is one of the emerging technologies that will lead to the next generation of Internet[1].As one of types of it, Software as a Service is a model of software deployment whereby a provider licenses an application to customers for use as a service on demand[2]. Through it, the enterprises needn't purchase, build and maintain the hardware and the software[3] .

Actually, It providers exploit economies of scale by offering the same instance of an application to multiple customers typically in a single-instance multi-tenant architecture model. Therefore the applications must be scalable, multi-tenant aware and configurable[4]. As there is only one instance running in a SaaS application, it should satisfy all possible personalized needs such as heterogeneous data, process rules, and business rules. Such a SaaS application model may be very complex, and if every tenant must include their business logic in this model, it will be difficult to keep their logic confidential[5]. So, this paper introduces a flexible thought to design "smart" service structure, which can satisfy personal needs and decrease service's complexity. We begin by an investigation of existing models of setting up services and related work in Section 2. In Section 3, a suite of flexible methods based on existing achievement is introduced to assembly services. Finally, we manipulate flexible methods into a communications project to confirm their availability and scalability in Section 4.

2 Related Work

Currently, many achievements tend to make each application vendors provide "smarter" to handler with the unbalance between one instance and multi-tenancy. Such

J. Luo (Ed.): Affective Computing and Intelligent Interaction, AISC 137, pp. 807–814.
springerlink.com

as, [6] has considered that the application must be divided in two different parts. The first part of the application describes the artifacts that are fixed and can be used by all tenants whereas the second part describes what [7] call "configuration metadata". Data that is tenant specific and must be deployed with every new tenant. In the way, it introduces a package format for application templates and solutions based on the service component architecture(SCA)[8] standard for multi-tenancy patterns. For the purpose of supporting Multi-Tenant Services,[9] has introduced a DUAL VALIDATION whereby D-Val business rules are extracted from the service; by virtue of which only one implementation of service can be used to handle request form different tenants. In the analysis of present status of customization, [10] has thoughts that The main strategy for version-based approach is to develop a full-functional, general-purpose and highly standardized software product, and pre-defines some parameters on certain functional modules that allow users to describe the software by setting these parameters [11], and mentioned that the ways for customization of these software systems are always straightforward and are finished simply by means of menus selections, parameter configurations, and certain visual designing tools, because all are limited within the pre-determined scope on the version basis. To change it, [10] has proposed that a MDA-based of SOA approach to software customization over the SaaS platform by introducing the concept of service models (or templates) and a domain specific language of STML (Service Template Markup Language).

On the other hand, to flexibly generate business process, workflow languages are proposed describe how to assemble separate business activities, each implemented through a different service, into a complete business process[12]. Workflow languages provide platforms through which enterprises can roll out new integrated services efficiently and effectively Workflow languages provide platforms through which enterprises can roll out new integrated services efficiently and effectively[13]. Such as, IBM's XML-based Web Services Flow Language (WSFL[14]), Microsoft's XLANG[14], and the Workflow Management Coalition's (WfMC) XML Process Definition Language (XPDL[15]) all provide mechanisms for orchestrating services into an end-to-end enterprise application. Furthermore, A business process orchestration language such as BPEL is widely regarded as beneficial for the flexibility of an application[16], as it allows to change the orchestration logic (described in BPEL) independently from the services[17]. But they only consider how to make applications flexible from external of them, actually, the most effective way to solve this difficulty is analysis it from internal.

3 Flexible Service

Because of continuous variety of outside, only knowing about outside situation, then, it is possible to make favorable self-adjustment. In the view of this, to construct firmly and flexible base of cloud platform, in the beginning of designing the structure of service, flexible thought and elastic and dynamic concept should be imported. The kernel process to construct service has two procedures; division and implementation. The two core process will be detailed as follows.

3.1 Service Division

Division contains two questions; the definition and granularity of service. To make service function fit fact in business process, the definition of service should be considered in business angle. On the other hand, under the premise of the extent of service granularity not enhancing the system complexity, in order to improve reusability of service, service granularity should be considered in technical angle. The two aspects above been considered, two ideas of service division are put forward. Firstly, smallest functional units with business significance are listed one by one and the manner of division should obey users' perspective and the standard with business purpose. Secondly, the functions with dependent relationship are divided into a service.

3.2 Service Implementation

Neither do we have an existing model to sense the important drivers in the industry and the relative power of the entities in the industry value chain[18].So, to SaaS, An application template is the application to be offered as a service where some parts of the application remain unspecified or are defaulted and can be customized by each tenant to suit their particular needs[19]. As an example, the different customers of a SaaS application might want their application to show their respective company logo[20].

But, multi-tenancy is one of key characteristics of SaaS[21]. That means many users apply the same instance of service. In this way, to hold the expansibility of business process, the amount of services will be uncontrolled and it is difficult to realize flexible call each other among services. If considered using "one instance multi-plug-ins multi-tenants", so, independent services can be seamlessly integrated by extensible plug-ins. The contrast of two manners is shown in Fig. 1.

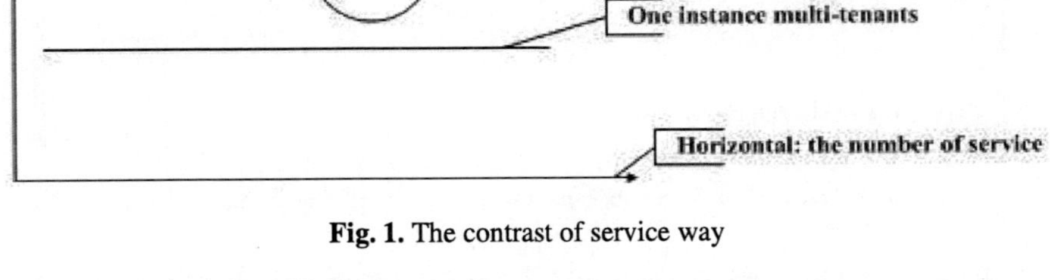

Fig. 1. The contrast of service way

Based on specifications of OSGi[22], services are realized. In the OSGi framework, the service updating is achieved by updating the cooperating bundles[23]. OSGi shares common concept with SOA[24] for service is the smallest integration unit and each

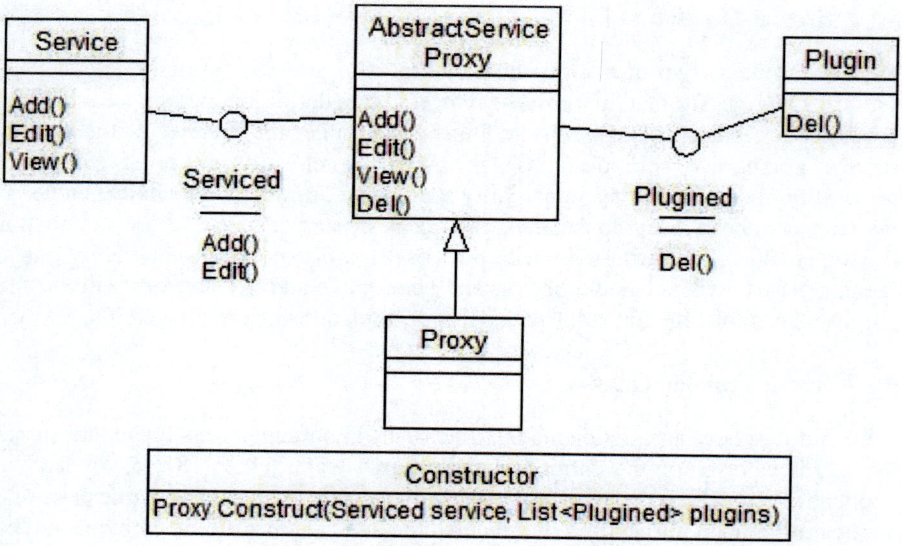

Fig. 2. Class diagram

```
Generate ServiceProxy

String classString;
Foreach field in serviced.getFields()
{
  ......
  classString = "field" ;//Gather service Fields with  reflection mechanism
}
Foreach field in plugined.getFields()
{
  ......
  classString += "field" ;//Gather plugin Fields with  reflection mechanism
}
Foreach method in serviced.getMethods()
{
  ......
classString+= "method.invoke(serviced,parameter)" ;//Gather service Methods with
reflection mechanism
}


Foreach method in plugined.getMethods()
{
  ......
classString+= "method.invoke(plugined,parameter)" ;//Gather plugin Methods with
reflection mechanism
}
ServiceProxy=(ServiceProxy)ClassLoader(classString);//Generate the class of ServiceProxy
Return new ServiceProxy();
```

Fig. 3. The manner of generation

service can communicate with another one by sharing resources[25]. Equinox[26] is open source version based on OSGi standard. By it, Plug-ins dynamically modify the internal functions of services. To every independent service been divided by business purpose having basic plug-ins and extensible plug-ins, basic plug-ins are used for realization of business purpose, while extensible plug-ins are used for flexible call which makes services having expansible area to shield difference between users' behavior. The core class diagram of service's implementation is given in Fig. 2.

As can be seen from the figure above, services and plug-ins have itself interface, so their implementation class can be generated independently for expansibility. On the level of customers, developer related just need associate abstract class AbstractServiceProxy which combines users' functions of services and plug-ins. On the running of cloud platform, after user login, system will generate service proxy class Proxy by class Constructor. The generation manner of proxy class is shown in Fig. 3.

According to reflection mechanism, properties and methods in services and plug-ins are reflected into service proxy class Proxy is generated dynamically by system, then, returned to the level of customers.

4 Experiments

In order to confirm the feasibility and maneuverability of this thought, online transactions system of public communications information platform of Jiang Xi province is made as an example.

4.1 Project Description

The main process of online transactions system seven steps. Firstly, cargo source party search vehicle information by destination、starting and other conditions, then invite vehicle source party fit for demands from cargo source party to quote. Secondly, the vehicle source party who received invitation give themselves quote. Thirdly, cargo source party selects one ideal quote from mass of quotes for achieving preliminary agreement with vehicle source party. Fourth, cargo source party fills consignment notes and confirms the agreement. Fifth, vehicle source party fills car consignment and confirms the agreement. Sixth, cargo source party finally confirms preliminary agreement to make it form transactions order and simultaneously system will generate a password to two parties for personally transactions. Seventh, two parties give evaluation to each other.

The next step is to design service structure of online transactions platform by applying flexible thought.

4.2 Service Granularity Design of Communications

According to project description above, the design of service granularity is shown as the table followed.

Table 1. Design view

function / service	Search	Invitation	quote	Cargo source party confirm	Fill car consignment and confirm
Basic function	By starting、destination	Invite driver to quote and attach contact	Look up the details of cargo、Input quote and contact	Select quote、Form agreement、Fill consignment、Final confirm	Fill the details of vehicle、confirm agreement
Extensible function i	By height	Attach ideal quote	Attach backup information of vehicle	Fill consignment notes、adding time limit	Fill car consignment 、adding time limit
Extensible function ii	By weight	Attach notes	Attach ideal time of load cargo		

According to the analysis of project description, by balancing reusability and independence in "Service Division" of flexible thought, there are five core the smallest services been divided. Meanwhile, compared to the model of "one instance multi-tenancy", with plug-ins dynamically inserting, every service will own more scalability.

Basic function of every service above is implemented by service class while extensible function as plug-in is implemented by plug-in class. After users login, by inflection and Equinox framework, making plug-ins dynamically plug into services is to flexibly extend their functions.

5 Conclusion and Future Work

On the basis of flexible thought, the manner of service division is proposed. Then, by plug-in structure to expand service's function and Equinox framework to increase the whole system scalability, "smart" service to be developed is provided for customers. Finally, through a analysis of actual project to describe this suite of flexible methods, experiments show that under the premise of not enhancing system complexity, the services been assembled by flexible thought own more independent in business logic, and in the aspect of adaptability they can satisfy functional requirements from all walks of life.

But, in the part division of flexible thought mainly research theoretical part in business, so, a great deal of artificial analysis and intervention is needed without sufficient technical assist. On the other hand, to comfortably plug plug-ins, the implementation of services needs to be defined by predictable manner, but, the predictable interface will not work in future. Therefore, the next step is to develop a suite of service analysis framework fit for a business by the theory of service division.

References

1. Bojanova, I., Samba, A.: Analysis of Cloud Computing Delivery Architecture Models. In: Proceedings of 2011 IEEE Workshops of International Conference on Advanced Information Networking and Applications (WAINA), Biopolis, Singapore, pp. 453–458 (March 2011)
2. Monfort, V., Khemaja, M., Ammari, N., Fehli, F.: Using SaaS and Cloud Computing for "On Demand" E Learning Services Application to Navigation and Fishing Simulator. In: Proceedings of 2010 IEEE 10th International Conference on Advanced Learning Technologies (ICALT), Sousse, Tunisia, pp. 663–665 (July 2010)
3. Waters, B.: Software as a service: A look at the customer benefits. Journal of Digital Asset Management 1, 1 (2005)
4. Mietzner, R., Unger, T., Titze, R., Leymann, F.: Combining Different Multi-tenancy Patterns in Service-Oriented Applications. In: Proceedings of 2009 IEEE International Enterprise Distributed Object Computing Conference, Auckland, pp. 131–140 (September 2009)
5. Liu, Y., Zhang, B., Liu, G., Wang, D., Gao, Y.: Personalized Modeling for SaaS Based on Extended WSCL. In: Proceedings of 2010 IEEE Asia-Pacific on Services Computing Conference (APSCC), Hangzhou, China, pp. 355–362 (December 2010)
6. Mietzner, R., Leymann, F., Papazoglou, M.P.: Defining Composite Configurable SaaS Application Packages Using SCA, Variability Descriptors and Multi-tenancy Patterns. In: Proceedings of Third International Conference on Internet and Web Applications and Services, Athens, Greece, pp. 156–161 (June 2008)
7. Chong, F., Carraro, G.: Building Distributed Applications Architecture Strategies for Catching the Long Tail. MSDN architecture center (2006), http://msdn2.microsoft.com/enus/library/aa479069.aspx
8. Open SOA Collaboration (OSOA). SCA Service Component Architecture. Assembly Model Specification Version 1.00 (2007), http://www.osoa.org/download/attachments/35/SCAAssemblyModelV100.pdf
9. Pervez, Z., Khattak, A.M., Lee, S., Lee, Y.-K.: Dual Validation Framework for Multi-Tenant SaaS Architecture. In: Proceedings of 2010 5th International Conference on Future Information Technology (FutureTech), Busan, Korea (South), pp. 1–5 (May 2010)
10. Zhu, X., Wang, S.: Software Customization Based on Model-Driven Architecture Over SaaS Platforms. In: Proceedings of International Conference on Management and Service Science, MASS 2009, Beijing, China, pp. 1–4 (September 2009)
11. Xue, H., Du, R., Huang, H.: Research on Version-based customizable ERP Systems, pp. 875–878. HeFei Industrial University Press (2003)
12. Candan, K.S., Li, W.-S., Phan, T., Zhou, M.: Frontiers in Information and Software as Services. In: Proceedings of IEEE 25th International Conference on Data Engineering, Shanghai, China, pp. 1761–1768 (May 2009)

13. Georgakopoulos, D., Hornick, M.: An overview of workflow management: From process modeling to workflow automation infrastructure. In: Distributed and Parallel Databases, pp. 119–153 (1995)
14. IBM. Web services flow language (wsfl) (2009), http://www.ibm.com/software/solutions/webservices/pdf/WSFL.pdf
15. WfMC. Workflow process definition interface – xml process definition language. Document Number WFMC-TC-1025 (2001)
16. Leymann, F., Roller, D.: Production Workflow – Concepts and Techniques. Prentice Hall PTR (2000)
17. Anstett, T., Leymann, F., Mietzner, R., Strauch, S.: Towards BPEL in the Cloud: Exploiting Different Delivery Models for the Execution of Business Processes. In: Proceedings of 2009 World Conference on Services – I, Los Angeles, CA, pp. 670–677 (July 2009)
18. Das, C., Mohan, G., Roy, R., Bhattacharya, S.: Quo vadis, SaaS a system dynamics model based enquiry into the SaaS industry. In: Proceedings of 2010 the 2th IEEE International Conference on Information Management and Engineering (ICIME), Cheng du, China, pp. 732–737 (April 2010)
19. Mietzner, R., Leymann, F.: Generation of BPEL Customization Processes for SaaS Applications from Variability Descriptorsors. In: Proceedings of the 2008 International Conference on Services Computing (SCC 2008), Honolulu, Hawaii, USA, pp. 359–366 (July 2008)
20. Mietzner, R., Metzger, A., Leymann, F., Pohl, K.: Variability modeling to support customization and deployment of multi-tenant-aware Software as a Service applications. In: Proceedings of the 2009 ICSE Workshop on Principles of Engineering Service Oriented Systems (PESOS 2009), Vancouver, Canada, pp. 18–25 (May 2009)
21. Wu, S., Zhang, S., Lanju, K.: A Dynamic Data Storage Architecture for SaaS. In: Proceedings of 2010 International Conference on Multimedia Information Networking and Security, Nanjing, China, pp. 292–296 (November 2010)
22. Azeez, A., Perera, S., Gamage, D., Linton, R., Siriwardana, P., Leelaratne, D., Weerawarana, S., Fremantle, P.: Multi-tenant SOA Middleware for Cloud Computing. In: Proceedings of 2010 IEEE 3rd International Conference on Cloud Computing (CLOUD), Miami, FL, USA, pp. 458–465 (July 2010)
23. Chen, J., Huang, L.: Dynamic Service Update Based on OSGi. In: Proceedings of WRI World Congress on Software Engineering, WCSE 2009, Xiamen, China, pp. 493–497 (May 2009)
24. Yazbek, H., Dumke, R., Schmietendorf, A., Neumann, R.: Service-Oriented Measurement Infrastructure. In: Proceedings of 2010 Eighth ACIS International Conference on Software Engineering Research, Management and Applications (SERA), Montreal, QC, pp. 303–308 (May 2010)
25. Wang, Y., Song, M., Song, J.: An extended distributed OSGi architecture for implementation of SOA. In: Proceedings of 2010 International Conference on Advanced Intelligence and Awarenss Internet (AIAI), Beijing, China, pp. 416–419 (October 2010)
26. Eclipse, EclipseSource (2010), http://eclipsesource.com/en/eclipse/equinox-osgi-overview/

Research on Security Issues of Internet Lottery Sales

Mingnan Yu, Hongliang Zhu, and Lingjun He

School of Management, Dalian University of Technology, Dalian, China
mmnyu@dlut.edu.cn, zhlshulan@163.com

Abstract. This paper identified security risks of internet lottery sales system and various social problems in China, providing a basis for internet lottery sales security risks evaluation and establishing early warning model.

Keywords: lottery sales, internet, security issues.

1 Introduction

The world's first successful Internet lottery betting was born in a small European town, Liechtenstein in October 1995. So far there have been Finland, Britain, France, Austria, Cyprus, Sweden, Italy, Argentina, Venezuela, Brazil, South Africa and Australia and other countries opening up the Internet lottery sales business. In Finland, for example, internet lottery sales started at the end of 1996, and 11 kinds of online lottery games, lotto tickets and lottery football quiz were operated by Veikkaus company, an affiliation of Finland Ministry of Education. Lottery sales on internet in Finland was up to 357.4 million Euros in 2009, accounting for 23% of total lottery sales [1]. In contrast, there has been a total ban on internet lottery sales in Germany since 2009, and Canada's internet lottery trading was temporarily closed, what's more, Greece, Singapore and Russia also had banned internet betting. In United States, internet lottery can only be transacted in Minnesota and Washington, other states are actively brewing.

Chinese internet lottery sales had a long time. As early as 2007, the five ministries, including the Ministry of Finance issued the 2007 No.36 announcement, noticing to stop internet lottery sales, while sina, netease, sohu, tencent and other well-known portals all opened internet lottery syndicated platform, even showing up professional platform such as 500 million, love lottery, OKOOO. Internet lottery sales without approving repeated again and again, absconding with money, withholding lottery funds, and becoming private lottery banker of lottery sites emerged in endlessly. The security issues of internet lottery sales caused strong vigilance of the regulatory authorities.

September 26, 2010, the Ministry of Finance issued the "Interim Measures on Management of Internet lottery sales", after that, internet lottery sales lifted the ban.

---

[1] Data Resource: VEIKKAUS' YEAR 2009-CSR Report and Annual Report.

J. Luo (Ed.): Affective Computing and Intelligent Interaction, AISC 137, pp. 815–821.
springerlink.com

However, the current internet lottery sales process is in a relatively slow start, the fundamental reason lies in its implicit security risks, including its own security risks of the internet lottery sales system and various social problems triggered by the security risks.

2 Literature Review

Foreign research on security issues of internet lottery sales mainly focused on the following three aspects:

Firstly, the illegal and irresponsible behaviour of sales sites: Sales site retained lottery funds, without distributing bonuses [1]; Sales sites' security and legitimacy has been questioned. Some of them claimed that they were under the third party inspection, and issued relevant certificates, but actually they were fabricated [2]; Sales sites did not set betting limits, total amount of betting account and online betting time; The supervision of minors buying lottery online didn't work well. Secondly, the individual's illegal behaviour: hackers' theft and fraud through technical means [3]; minors (under the age of 18) and internal staffs' purchasing lottery [4-7]; money laundering by means of buying lottery online [8], for example, a person won a prize of $14.3 million in Massachusetts, United States, and a gang leader bought the lottery, fortunately the police detected in time, and frozen the funds. Thirdly, the problem gambling: because of the convenience and comfort, electronic cash, anonymity and other characteristics of buying lottery online, it will increase the risks of problem gambling [9-12].

The exposed security risks of internet lottery sales had some warnings on the domestic lottery regulators, and this kind of similar phenomenon indeed occurred in China, but because of different conditions, they can't be resolved referring to foreign models.

Meanwhile, lifting the ban of domestic internet lottery sales endured a short time, a few research focused on the proliferation of private lottery, money laundering and other fields, without systematic screening and combing its own security risks of China's Internet lottery sales system. These research didn't deeply probe social problems triggered by the security risks.

3 Internet Lottery Sales System's Own Security Risks

According to the current model of internet lottery sales, the paper draws out the process of domestic internet lottery sales, shown in Figure 1. There are many security risks in this process, this article will be grouped them into four aspects: data security, payment security, qualification security and lottery funds security.

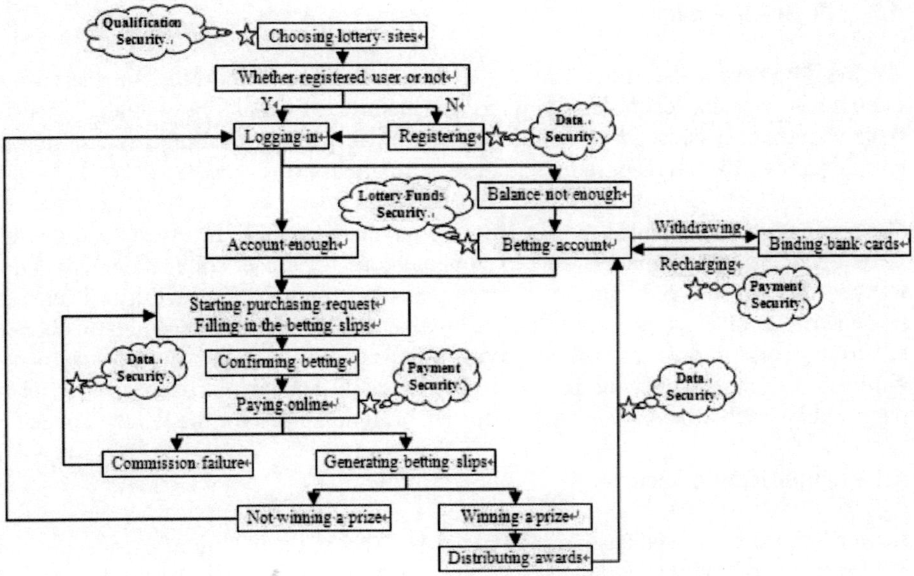

Fig. 1. Internet lottery sales process and its own security risks

3.1 Data Security

Data Security Triggered by Technical Reasons. Currently, many lottery sites sell lottery using telephone commissioned betting systems, so staffs of lottery sites need to transfer data to welfare and sports lottery center room for the purpose of unified management [13]. But Since 2010, data transmission errors, data latency, system vulnerabilities, firewall security risks and hackers attacking all resulted in betting invalid (that is commission failed phenomenon) when lottery buyers transacted online.

In addition, the internet lottery is a paperless electronic data, so how to make the integrity, confidentiality and non-reproducibility of lottery-buyers' information data (ID number and name, etc.) be achieved through technical means and make the prize not be claimed falsely will be the focus security problems [14].

Data Security Caused by Lottery Site. Several lottery sites held back lottery-buyers' transaction data on purpose, not distributing bonus when lottery-buyers had won a prize, and claimed not winning prizes be found. For example, Mr. Ma bought a football lottery on June 24, 2010, Paraguay against New Zealand. According to the gameplay, Mr. Ma should have got 6,428 yuan. But on June 25, when Mr. Ma queried online, unexpectedly showing no winning record, then he called the site to complain. On June 27, Mr. Ma inquired again, found one note showing winning 428.07 yuan. Finally, a local media carried a report on the event. At around 1 pm that day, Mr. Ma got the remaining prizes from the site.

3.2 Payment Security

Recharging from Binding Bank Cards to Betting Account. At present, e-commerce payment technology is relatively matured in China, the potential security risks were mainly caused by improper uses, such as logging phishing sites, setting up simple passwords, payment in public, disclosure of passwords and so on.

Payment after Confirming Betting. Secure payment has been resolved abroad a few years ago. Gaming companies usually combined one fixed user code and one dynamic transaction code to authenticate, protecting users' transactions security. There are present a lot of payment options in China: postal remittance, account exchange, electronic bank transfer as well as payment treasure and other third-party payments. After confirming betting, the payment process might appear tips such as "404 error page", but actually the payment had occurred, bringing losses for the lottery-buyers.

3.3 Qualification Security

Lower Market Access Registered Capitals. According to "Interim Measures on Management of Internet lottery sales", the registered capital of internet co-operator and independent affiliate should be no less than ¥ 50,000,000. But Feng-Baiming believes that the threshold of ¥ 50,000,000 registered capital is low for the consortia, and should increase up to one hundred million yuan. Currently, with the face of temptations of hundreds of millions prize pool, bets escalate dramatically, co-operator and independent affiliate are most likely to flee with the betting funds. For example, the employee of Lottery Net took tens of millions yuan away, no news from him ever.

Lack of Credibility. Currently, lottery regulatory authorities contract with the internet lottery affiliates, dividing profits on due proportion. If extra profits obtaining from the illegal operation are far greater than that from the regulatory authorities, the affiliates will choose to make its credibility exhausted at one time before terminating the partnerships. For example, in order to increase its fidelity, some cheat lottery sites hang the business licenses approved by the industry and commerce sector out on the web. When police received a report case from a victim, after verification, most of these certificates are fabricated.

3.4 Lottery Funds Security

Some lottery sites maliciously held back lottery sales funds, becoming the banker of online private lottery in essence. They could reap greater benefits from the private lottery, seriously disrupting the order of national lottery market.

4 Social Problems Caused by Internet Lottery Sales System

4.1 Aggravating Problem Gambling

At present, lottery-buyers in any place of equipment attached to the internet can take bets, and the existing lottery sites did not set betting limits and online time tips. This, on the one hand, will cause parts of lottery-buyers to increase the betting multiples in order to win more return prizes; On the other hand, this can make partial lottery-buyers indulge in betting all day, and produce addictive phenomena.

In addition, jointly-buying is one of the main formats in current internet lottery sales. Sina's jointly-bought initiator reached 1,144 [2]. An initiator called Beidangaoshou, his initiating amount reached up to 200 thousand yuan in each football lottery 14 games in recent 3 months, attracting many lottery-buyers who was eager to win a prizes to join in. To some extent, the convenience of internet lottery aggravates problem gambling.

4.2 Causing a New Channel of Money Laundering

Criminals might remove a plenty of black money into the betting account in lottery sites as temporary shelters. Because lottery sites' supervision on the source of lottery-buyers' capital was relatively loose, once the chances got matured, criminals put forward the cancellation of account and asked lottery sites to return the balance of accounts by check or bank draft, so that these black money dishonestly turned white. This provides criminals a new channel of money laundering through lottery sites, triggering the country's financial risks.

4.3 Overflowing Private Lottery

Private lottery refers to a kind of lottery which issued for profit-orientation and without government approval, and its harms is serious, such as impacting on the legitimate lottery issued, evading national taxes; deteriorating social security, and endangering social stability.

The release of internet lottery sales has played a role to some extent in fueling to overflow the private lottery. Firstly, private lottery sites and legitimate lottery sites mixed together, confusing regulatory authorities' attention. The lottery regulatory authorities didn't form a professional supervision team, so can't effectively identify and investigate on lottery sites. Secondly, some lottery sites privately was a banker, setting up private lottery business to withhold lottery-buyers' funds for private lottery pay activities, so that expanding the business scope of "legitimate and private lottery" to make a profit. Finally, some lottery sites withheld lottery funds to buy private lotteries.

4.4 Bringing up New Challenges to Existing Management

Internet lottery sales caused the magnification of the object of lottery regulatory management. As the releasing of, except for supervision on existing entity betting station, lottery regulatory authorities also need to manage internet lottery sales affiliates, mainly including the qualifications of sites affiliates, capital management, sales management, risks control programs, equipment and technical services management, supervision and audit management, emergency treatment and so on. Meanwhile, it has put forward higher requirements for the regulatory management

[2] Data Resource: Lottery Channel of Sina, by April 22, 2011.

team, such as the establishment of specialized teams in reviewing and monitoring affiliates' qualification to ensure the safe operation of the internet lottery sales.

What's more, internet lottery sales caused the content of lottery regulatory management refined. Lottery Management Regulations expressly provides that "not allowing sales of lottery tickets cross-region", but there is obvious cross-regional sales problem in the internet lottery sales. Therefore, "Interim Measures on Management of Internet Lottery Sales" pointed out that lottery authorities shall, according to the administrative area where lottery-buyers' the bank debit cards account belong to, divide lottery sales by provinces, autonomous regions and municipalities". But in reality, various regions strive intensely for registration place of the lottery sites, not strictly in accordance with the Interim Measures, making the lottery funds of some areas flow to the registration places where the lottery sites belong to. This requires regulatory authorities to adopt certain technical means and the corresponding regulations to restrain inter-regional lottery sales.

4.5 Impacting on Lottery Market

China's total lottery sales in 2010 reached 166.2 billion yuan, and this owes to the nearly 300,000 entity betting stations. But with the release of internet lottery sales, online lottery sales will cause a certain degree of segregation, and thus have some impact on lottery market. Currently, Finland is the country which has a higher proportion of lottery network in the world, accounting for 23% of total lottery sales. On this basis, China's lottery market value of the network should reach 38.2 billion yuan, but far from this level by now.

In addition, the current game types between internet lottery and entity betting stations tend to be homogeneous, and few new game types were developed. Facing with this situation, the convenience of internet lottery will attract some lottery-buyers to purchase lottery online, reducing lottery-buyers of entity betting station. This will lead to betting station owner who sign a buyout with the lottery center complaints and even cause more direct and serious social problems.

5 Conclusion and Outlook

This paper analyzed implicit problems of the internet lottery sales process, including data security, payment security, qualification security, and lottery funds security and social problems caused by internet lottery sales system. Follow-up study will focus on conditions which lead to security problems and the damage size. This will provide references for regulatory authorities on prevention measures of security risk in internet lottery sales.

References

1. Williams, R.J., Wood, R.T.: Internet Gambling: A Comprehensive Review and Synthesis of the Literature. Report prepared for the Ontario Problem Gambling Research Centre, Guelph, Ontario, Canada (2007)

2. Griffiths, M.: Crime and gambling: A brief overview of gambling fraud on the Internet. Internet Journal of Criminology, 1–7 (2010)
3. Chen, Y., Chen, P.S., Hwang, J., Korba, L., Song, R., Yee, G.: An Analysis of Online Gaming Crime Characteristics. Internet Research: Electronic Networking Applications and Policy 15, 246–261 (2005)
4. Derevensky, J.L., Gupta, R.: Internet Gambling amongst Adolescents: A Growing Concern. International Journal of Mental Health and Addiction 5(2), 93–101 (2007)
5. Messerlian, C., Byrne, A.M., Derevensky, J.L.: Gambling, Youth and the Internet: Should We Be Concerned? Canadian Child and Adolescent Psychiatry Review 13(1), 3–6 (2004)
6. Mitka, M.: Win or Lose, Internet Gambling Stakes Are High. JAMA: Journal of the American Medical Association 285, 1005 (2001)
7. Parke, A., Griffiths, M.: Why Internet Gambling Prohibition will ultimately Fail. Gaming Law Review 8, 295–299 (2004)
8. He, P.: A Typological Study on Money Laundering. Journal of Money Laundering Control 13, 15–32 (2010)
9. King, S.A., Barak, A.: Compulsive Internet Gambling: A New Form of an Old Clinical Pathology. Cyber Psychology & Behavior 2, 441–456 (1999)
10. Schull, N.D.: Digital Gambling: The Coincidence of Desire and Design. The Annals of the American Academy of Political and Social Science 597, 65–81 (2005)
11. Shaffer, H.J.: Understanding the Means and Objects of Addiction: Technology, the Internet and Gambling. Journal of Gambling Studies 12, 461–469 (1996)
12. Wood, R.T., Williams, R.J., Lawton, P.K.: Why do Internet Gamblers Prefer Online versus Land-based venues? Some Preliminary Findings and Implications. Journal of Gambling Issues 20, 236 (1997)
13. Tang, M.: Research on Data Security of POS Terminal in Sports Lottery. Chongqing University, Chongqing (2005)
14. Wu, Y.: Design and Implementation of Web-based Lottery Secure Transactions System. Fudan University, Shanghai (2008)

An Analysis of Price-Only Contracts Based on Revenue Management

Xiaoqin Wen, Zhenglian Duan, and Huiying Hu

School of Management, Shanghai University, Shanghai, 200444, China
wenxq_8@163.com

Abstract. This paper considers a decentralized, two-echelon supply chain in which a manufacturer sells to a retailer who sells the product to the customers over a selling horizon of two periods and allocates the capacity to maximize her revenue, and the lone contract parameter is a wholesale price. The authors study the impact of downstream retailer's revenue management on the decision-making and performance of the entire channel. The paper begins with establishing a mathematical model of the problem, and then conducts numerical analysis of the model. The authors find that for centralized system, optimizing selling strategy contributes to reduce the output of the system, at the same time, increase the profit of the system; and for decentralized supply chain system, retailer's optimizing selling strategy enhances not only his own share of the profit of the channel, but also the efficiency and performance of system comparing to that under the classic wholesale price contract.

Keywords: decentralized supply chain, revenue management, wholesale price contract.

1 Introduction

This paper focuses on the influence of downstream retailer's revenue management in a two-echelon supply chain on the decision-making and performance of the entire channel under a price-only contract. Motivated by the related research on revenue management and supply chain management, we propose the problem above.

Revenue management arises originally in the airline industry. Since airline short-term cost is largely fixed, and variable cost per passenger is small; it is sufficient for maximizing profit to seek a booking policy that maximize revenue. Before Littlewood proposes the seat inventory control rule [1], almost all research in revenue management focuses on controlled overbooking. After then, the developments of research on seat inventory control make great progress [2、3]. Dynamic pricing has been a focus of the revenue management research since Gallego and Van Ryzin consider a problem of dynamic pricing under the random demands in [4] in 1994. With the advances in information technology and e-commerce and the success of airline revenue management, today, revenue management has spread out rapidly to other industries such as hotels, car rental agencies and especially retailers [5].

J. Luo (Ed.): Affective Computing and Intelligent Interaction, AISC 137, pp. 823–828.
springerlink.com

To our knowledge, existing literature on revenue management fucuses on the optimal dynamic pricing and capacity allocation policies only from the perspective of benefits of firm unilaterally. For retailing, that is inconsistent with the theory and practice of supply chain management. This paper considers the related problem of revenue management in a retailing chain setting.

In addition to, supply chain management has been becoming a new mode of business management. Although most structure of the supply chain is complicated, many transactions among the supply chain partners can be controlled by the simple price-only contracts. Jeuland and Shugan consider a wholesale price contract under deterministic demand in [6]. Lariviere and Porteus study a only-price contract under uncertain demand in [7]. Cachon shows that the wholesale price contract is the most suitable for some business environments in [8].

To our knowledge, almost all papers on contract are based on the classic newsvendor model. In these papers, the selling horizon consists of single period; the impact of demand management carried out by retailer on the channel decision is simply omitted. This paper examines a price-only contract based on retailer's active demand management.

We consider a decentralized, two-echelon supply chain which consists of one manufacturer and one retailer operating under a wholesale price contract, where the manufacturer acts as a market leader and the retailer acts as a market follower. For the case in which the selling season is relatively short comparing to the long lead-time of production/order or goods can not be produced repeatedly, we assume that there is only an opportunity of procurement for the retailer before the selling horizon starts and inventory is not replenishable once sales start. Thus, the capacity is fixed and the procurement costs are sunk during the selling horizon. The retailer adjusts his sales policy to the change of market environment during the selling horizon of two-period in order to achieve more revenues. Namely, the retailer implements revenue management for the product sales.

For simplicity, we assume that the demands are non-negative and independent random variables and the prices fluctuate according to an AR (1) process in consecutive periods. Excessive demands are lost and inventory unsold at the end of each period can be carried to the next. The salvage values of the leftovers at the end of selling horizon are zero. There exists perfectly symmetrical information between the manufacturer and the retailer, and both of them are risk-neutral.

In the decentralized supply chain setting, the sequence of events is as follow: In stage one, a Stackelberg game occurs between the manufacturer and the retailer. First, the manufacturer sets a wholesale price charged to the retailer; the retailer determines whether or not to accept the price addressed by the manufacturer; the retailer then decides the optimal stocking quantity if he accepts the wholesale price. In stage two, the retailer faced to the uncertain demands and prices optimizes the quantities which are allocated for sale in each period during the selling horizon of two-period. In this paper, we assume that the opportunity costs are zero for the retailer. It is sufficient for the manufacturer to design a price-only contract which ensures that the retailer's expected profits are lager than zero.

In the following, we first model the problem proposed above, and then conduct the numerical analysis based on the mathematical model.

2 System Model

In this section, we first introduce some notations and assumptions which are used throughout this paper before establishing system model.

t : index of the number of remaining periods to the end of the horizon, $t = 1, 2$;

D_t : demand in period t;

c_t : holding cost of per unit inventory in period t;

s_t : number of items preserved by the retailer for future demands in period t ;

p_t : retail price in period t;

$V_t(I_t, p_t)$: the maximum total expected discounted revenue-to-go when there remain t periods, given the inventory I and retail price p in period t ;

c : marginal production cost ;

w : wholesale price manufacturer's charged to retailer;

q : output of centralized system or retailer's stocking quantity in decentralized system;

$\Pi^I(q)$: profit of centralized system, given the output q;

$\Pi^R(q)$: retailer's profit in decentralized system , given the retailer's stocking quantity q;

$\Pi^M(q)$: manufacturer's profit in decentralized system, given the retailer's stocking quantity q;

It is assumed that D_t ($t = 1,\ 2$) is independent random variable with continuous cumulative distribution function $F_t(x)$ and probability density function $f_t(x)$. Let $\overline{F}_t(x) = 1 - F_t(x)$. Moreover, we assume $0 \leq E(D_t) < +\infty$. This is often reasonable in practice and ensures that the objective revenue function defined below is finite.

We assume that the price in consecutive period fluctuates according to an AR (1) process. That is, $p_t = \varphi\, p_{t+1} + \tau_t$, where φ is a constant referred to as the trends of price fluctuation (or as the price trends), and { τ_t } is a white-noise process, i.e., $E\tau_t = 0$ and $E(\tau_r \tau_t) = \sigma^2$ for $r = t$ and $E(\tau_r \tau_t) = 0$ for $r \neq t$. AR processes are often used to describe price fluctuation of consumer goods in marketing literature and investment goods in financial literature.

2.1 The Centralized System

First, we model the problem of centralized supply chain system. The centralized model serves as the benchmark in our analysis of decentralized system. In centralized setting, there is a single decision-maker, who determines the output q and the preserved quantity $s_t\,(t=1,2)$ of each period to maximize overall supply chain profit. For ease of exposition, suppose that $V_t(0,p)\equiv 0$, $V_0(I,p)\equiv 0$, and the discount factor $\beta\in(0,1)$. There is no inventory before production. So, the overall centralized supply chain profit can be expressed as follows:

$$\Pi^I(q)=-cq+V_2(q,p_2)\,. \tag{1}$$

Where $V_2(q,p_2)$ satisfies the optimal equation

$$V_2(q,p_2)=\max_{0\le s_2\le q} J_2(q,p_2,s_2)\,. \tag{2}$$

$$\begin{aligned}J_2(q,p_2,s_2)&=p_2E(\min(q-s_2,D_2))-c_2q\\&+\beta\,EV_1(s_2+\max(0,q-s_2-D_2),\varphi\,p_2+\tau_1)\,.\end{aligned} \tag{3}$$

We compute the optimal output q and the optimal quantity of items preserved for future demands $s_t\,(t=1,2)$ in centralized setting through numerical study in next section.

2.2 The Decentralized System

We consider the optimal wholesale price set by the manufacturer and the optimal order-up-to quantities set by the retailer in decentralized system when the retailer optimizes the quantities which are allocated for sale in each period during the selling horizon of two-period. The retailer's decision in decentralized system is similar to the producer' one in centralized system except for the fact that the retailer makes order to the manufacturer at the wholesale price w and the producer produces the product at the marginal cost c. Thus, the retailer's profit can be expressed as:

$$\Pi^R(q)=-wq+V_2(q,p_2)\,. \tag{4}$$

The manufacture's profit can be expressed as:

$$\Pi^M(q)=(w(q)-c)q\,. \tag{5}$$

If $p_2>c_2+w$, then $\Pi^R(0)=0$, and $\dfrac{d\Pi^R(0)}{dq}=p_2-c_2-w>0$. The manufacturer will set her wholesale price (as long as $w<p_2-c_2$) to induce a

positive ordering quantity from the retailer. To facilitate analysis of decentralized model, we suppose that $p_2 > c + c_2$. In the following, we compute the optimal wholesale price w and the optimal order-up-to quantities q in decentralized setting through numerical study.

3 Numerical Study

In this section, we examine the impact of retailer's behavior of revenue management, such as optimizing the quantities allocated to each period for sale, on the decision-making of channel, the improvement on profit of the centralized or decentralized system. According to cost-plus pricing , goods may be classified into three categories corresponding to the different level of cost and price: goods with high-cost and high-price, goods with medium-cost and medium-price, goods with low-cost and low-price. We compare the performance of channel under two different wholesale price contracts for such three markets through series of numerical analysis, respectively. Furthermore, we identify the business environment under which the wholesale price contract based on revenue management is more suitable for overall channel than the classic wholesale price contract.

Table 1. Performance of overall channel under the classic wholesale price contract and the wholesale price contract based on revenue management for three categories of goods

Category of goods / Performance of channel	Goods with high-cost and high-price $p_2 = 250$, $c = 220$, $c_1 = c_2 = 15$	Goods with medium-cost and medium-price $p_2 = 25$, $c = 18$, $c_1 = c_2 = 1.2$	Goods with low-cost and low-price $p_2 = 2.5$, $c = 1.5$, $c_1 = c_2 = 0.1$
q_d^* / q^*	61.55%	59.94%	59.59%
$\bar{q}_d^* / \bar{q}^*$	61.94%	60.01%	59.63%
$\bar{q}^* / q^*$	93.50%	97.59%	98.23%
$\bar{q}_d^* / q_d^*$	94.09%	97.70%	98.30%
$\bar{w}^* / w^*$	98.98%	98.85%	98.73%
Π^M / Π^D	84.79%	79.83%	78.94%
$\bar{\Pi}^M / \bar{\Pi}^D$	46.56%	69.38%	72.51%
Π^D / Π^I	77.04%	77.14%	77.14%
$\bar{\Pi}^D / \bar{\Pi}^I$	86.10%	79.56%	78.62%
$(\bar{\Pi}^I - \Pi^I) / \Pi^I$	27.87%	2.85%	0.76%
$(\bar{\Pi}^D - \Pi^D) / \Pi^D$	42.90%	6.07%	2.70%

To simplify analysis, we assume that demands are uniformly distributed over $[0,1]$. Let $\beta = 0.99$, $\varphi = 1.25$. Our results of numerical study are showed in Table 1.

4 Conclusions

We have examined the impact of retailer's revenue management on the decision-making and the performance of channel under a wholesale price. We have modeled the centralized system and the decentralized system, respectively. Our numerical analysis shows that the retailer will have an incentive to optimize the quantities allocated for sale in each period during the selling horizon when the retail price marks up to or exceeds a threshold. The output decreases and the profit increases in the centralized system through optimizing the quanties allocated for sale in each period. In the decentralized supply chain system, retailer's optimizing selling strategy enhances not only her own share of the profit of the channel, but also the efficiency and performance of system comparing to those under the classic wholesale price contract.

In addition, we have identified the bussiness environment for which the calssic wholesale price contract and the wholesale price contract based on retailer's revenue management are suitable, respectively. We have found that the wholesale price contract based on revenue management is more suitable for the high-end products market with the relatively great trend of marking up. In the medium-end or low-end products market with the relatively stable price (such as the daily necessities market), the classic wholesale price contract is usally used to control the transaction bewteen the buyer and the seller.

References

1. Littlewood, K.: Forecasting and Control of Passenger Bookings. In: 12th AGIFORS Symposium Proc. Nathanya, Israel (1972)
2. Dimitris, B., Sanne, D.B.: Simulation-Based Booking Limits for Airline Revenue Management. Oper. Res. 53(1), 90–106 (2005)
3. Youyi, F., Baichun, X.: A Dynamic Airline Seat Inventory Control Model and Its Optimal Policy. Oper. Res. 49(6), 938–949 (2001)
4. Guillermo, G., van Garrett, R.: Optimal Dynamic Pricing of Inventories with Stochastic Demand over Finite Horizons. Management Sci. 40(8), 999–1020 (1994)
5. Gabriel, R.B., Susana, V.M.: Periodic Pricing of Seasonal Products in Retailing. Management Sci. 43(1), 64–78 (1997)
6. Jeuland, A.P., Shugan, S.M.: Managing Channel Profits. Marketing Sci. 2(3), 239–272 (1983)
7. Martin, A.L., Evan, L.P.: Selling to the Newsvendor: An Analysis of Price-Only Contracts. Manufacturing & Service Oper. Management 3(4), 293–305 (2001)
8. Gerarad, P.C.: Handbooks in Operations research and Management Science. Supply Chain Management. North-Holland (2003)

Research on the Aircraft Noise Evaluation and Application

Guo Li and Dongdong Wang

College of Computer Science and Technology,
Civil Aviation University of China, Tianjin, China
suifengfairy@163.com

Abstract. This paper analyses the aircraft noise evaluation system in our country now, and points out shortcomings of the domestic index WECPNL when it is used in the practical. The Day-night Average Sound Level(DNL) and the Sound Exposure Level (SEL) is proposed to use as the main evaluation index when amending the aircraft noise evaluation system because they take into account the single and multiple aircraft taking off and landing noise level evaluation. In addition, considering the flight status in our country, this paper suggests that we should use different evaluation methods in the different airports.

Keywords: Aircraft noise, WECPNL, DNL, evaluation method, SEL.

1 Introduction

With the rapid development of urban economy, the improvement of people's living standards and the gradually booming tourism market, the airline company's scheduled flights increase constantly and open out an unprecedented excellent situation as an important way of travel and transportation. However, owing to the expanding of the urban construction, the distance of the residential areas and the airport becomes shorter and leads to the aircraft noise pollution getting worse[1]. Therefore, evaluating aircraft noise around the airport reasonable becomes increasingly important.

Aircraft noise pollution evaluation is a very complex process since it should consider variety and multiple aircrafts which will cause bad effect on residents in a different day and night time. The country around the world has put forward a dozen evaluation values. In our country, the weighted equivalent continuous perceived noise level (WECPNL) has been adopted in the "Standard of aircraft noise for environment around airport" (GB9660-88) which has been and implemented over 20 years. However, this standard has been discovered some problems in the course of implement, and the State Environmental Protection Administration proposed to revise it in 2009.

2 Aircraft Noise Evaluation System Analysis in China

Aircraft noise is different from the other traffic noise as it is intermittent and its power is relatively large, the spectrum is wide and can affect a broader scope. Single high noise event is the main characteristics of the aircraft noise[2]. On the basis of plenty

J. Luo (Ed.): Affective Computing and Intelligent Interaction, AISC 137, pp. 829–835.
springerlink.com

investigation, Finegol et al concluded that resident response to the traffic noise is related to the type of noise source, and the aircraft noise is more troubling than railway noise and road noise when the sound level is the same[3]. Therefore, reasonable aircraft noise evaluation is very important and complex. The formula of WECPNL is as followed:

$$L_{WECPN} = 10\lg\left[\frac{\int_{t_1}^{t_2} 10^{0.1L_{TPN}(t)}dt\left(N_1 + 3N_2 + 10N_3\right)}{86400}\right] \tag{1}$$

In this formula, $\int_{t_1}^{t_2} 10^{0.1L_{TPN}(t)}dt$ represents the average energy per flight, $N_1 + 3N_2 + 10N_3$ is the weighted number of flight per day, the denominator 86400 is the total second per day[4]. In theory, WECPNL is an objective evaluation method since it not only includes aircraft noise level, flight number, flight time and other physical parameters, but also can reflect people's psychological factors such as the social response to aircraft noise and the subjective evaluation. However, this indicator is not ideal in the results of practical application.

2.1 WECPNL Can't Be Measured Directly and the Data Isn't Objectivity

WECPNL is calculated based on the flight numbers during different time periods per day and its fundamental unit is L_{TPN} which is based on people's subjective sensation—noise degree, and then considered the revise of irregular spectrum, duration time of single noise event. Noise degree is people's subjective judgement to the dislike or disgust noise and is evaluated by the equivalent noise degree curve which is acquired by large practical survey. By reason that in different environment different people will have different reactions to the same noise, the noise degree is uncertainty and subjectivity, so is the WECPNL. Thus it can be seen that this indicator is not convenient to operation actually since it requires a lot of practical measurement data, and the calculation process is complex and tricky, the data is not objective. Although it is recommended by ICAO, most countries did not use it.

2.2 Incomplete Consideration of the Single Noise Event

WECPNL adopted energy averaging method and reflected the average impact on people of aircraft noise per day. Aircraft noise is interval noise and it is the single noise event that can embody the characteristics of aircraft noise. Although the indicator has considered the single noise events, it did not fully reflect the consequences of it by reason that it only considered average value. As a result, although the noise level measured in the residential zone is fully in line with national standards, the actual exposure to single aircraft noise events is much higher than 70dB, and even more than 100dB. Furthermore these single noise events distributed throughout the day will have completely different effect on people.

2.3 Can't Be Compared with the Other Environmental Noise Directly

In our country, the noise standards in civil construction, industrial enterprises, roads, railways, ports have all adopted the equivalent sound level L_{eq} or L_A. The relationships between noise exposure and people's subjective reaction have been achieved by the environmental workers who carried out wide range of social surveys and laboratory studies for the purpose of objectively and laboratory evaluating the noise impact on residents. However, these research results can't be used by aircraft noise exposure evaluation by reason of different evaluation parameters. Similarly, WECPNL is not convenient to be used in the city planning or variety architectural design. Moreover, it is easy to make confusion when evaluate mixture noise around the airport. The region where does not meet the "Standard of aircraft noise for environment around airport" may be up to the "Environmental quality standard for noise"[5]. Therefore, it is not conductive to effectively use and planning the regional land.

3 Analysis of the DNL (Day-Night Average Sound Level)

In 1976, EPA (U.S Environmental Protection Agency) proposed formally to the FAA (Federal Aviation Administration) to make DNL as standard aircraft noise evaluation. In 1980, Federal Joint Committee on Urban Noise also regarded DNL as the best aircraft noise evaluation when studying the land planning around airport. In the same year, DNL is regarded as acoustic evaluation to evaluate the land compatibility planning around airports in the standard developed by Acoustical Society of America[6]. Until now, the research and application of DNL is quite mature compared with WECPNL. The formula of DNL is as followed:

$$L_{DN} = \overline{L_{SE}} + 10\lg(N_D + 10N_N) - 49.4 \tag{2}$$

$$L_{SE} = 10\lg\left[\frac{1}{t_0}\sum_{i=1}^{N} 0.5\times 10^{0.1L_{Ai}}\right] \tag{3}$$

L_{se} is Sound Exposure Level and means A-weighted sound level in one second which is converted by all the noise energy endured in the measuring point as an aircraft flying past. DNL is based on A-weighted sound level and can be calculated by integrating the noise energy during single aircraft noise event and flight sorties of different time periods[7]. L_A is objectivity and can be obtained directly from the noise monitor. It needs less data than WECPNL to calculate DNL so that not easy to make mistake. Most of the other environmental noise equivalent adopted equivalent sound level L_{eq} whose formula is as follows:

$$L_{eq} = 10\lg\left[\frac{1}{T_d}\sum_{i=1}^{N} 0.5\times 10^{0.1L_{Ai}}\right] \tag{4}$$

In this formula, T_d is monitored duration time of single aircraft noise event. According to the formula (3) and (4), the relationship between DNL and SEL can be inferred as follows:

$$L_{se} = L_{eq} + 10\lg T_d \tag{5}$$

It can be seen the DNL which based on L_A is better to coordinate with other standards. Therefore, this paper recommends DNL as the evaluation criterion when revising the aircraft noise evaluation standard in our country.

However, DNL is also not perfect as it used the energy average method and not fully considered the impact of single noise events. FAA had prescribed the limit value of DNL 65dB in the residential areas around airports, but hadn't prescribed the limit value of single aircraft noise event. Therefore, we have to consummate this evaluation system.

4 Consummating the Evaluation System

Summarizing all the evaluation indicators used in the major aviation countries, it can be seen that the main indicators to evaluate single noise event are $PNL_{\max}$ and EPNL which are based on noise degree, $L_{A\max}$ and SEL which are base on A-weighted. The conclusion by way of quantity of practice showed that A-weighted maintained nice correlation with other complex evaluation indicators and assessment methods what the other indicator can not achieve until now. Therefore, A-weighted is widely used in all kinds of urban noise assessment around the world with no exception. Moreover, considering the important relationship between the duration time of the noise and its impact on people's subjective responses, this paper adopts SEL to evaluate single aircraft noise events.

The supreme impact of the single noise event is the disturbance to residents at night and it is divided as primary and secondary effect. Primary effect is mainly refers to difficult to sleep, easy to wake up, rise in blood pressure, rapid heart rate, etc. Secondary effect behaves as feeling tired, having the blues, inefficiency and so on. Therefore, it suits the actual condition to ascertain the limit value of SEL according to the relationship between SEL and the impact on people of the single noise event. The relationship formulas are as follows:

$$\%Awakening = 7.1\times10^{-6} L_{SE}^{3.5} \tag{6}$$

$$\%Arousal = \%Awakening \div 40\% \tag{7}$$

Table 1. The Relationship between SEL and sleep

SEL(dB)	45	50	55	60	65	70	75	80	85	90
%Awakening	4	6	9	12	15	20	25	32	39	48
%Arousal	11	15	22	29	39	50.0	64	80	98	120

From the formulas above, we can see that the Awakening associates closely with SEL and about 40% of people who are disturbed when sleeping will be waked up. The table 1 showed their relationships clearly.

When the indoor SEL is 85dB, the Arousal is about 100% while the Awakening is 32%. Although the Awakening is not large, it does not mean slight influence on people. Studies have shown that aircraft noise can cause changes on sleep status and quality, and may cause coronary heart disease. Moreover, along with the accelerated pace of life, inadequate sleep at night will make people fatigue, psychentonia, and then influences the psychosocial homeostasis and work efficiency. In a word, it will bring inestimable negative effect on the social[8]. Bringing together extensive factors, we adopt 70dB as the limit value of indoor SEL when the Arousal is 50% and the Awakening is 20%. Taking into account that under normal circumstances the civil sound transmission loss is about 15~20dB, this paper assumes it as 15 dB, so the limit value of outdoor SEL is ascertained as 85dB at night. Study aircraft noise standards and other environmental noise standards domestic and overseas, it is found that residents are more sensitive to aircraft noise at night and the limit values at night are always 10dB less than it in the day-time. So the day limit value is determined as 95dB.

5 Application of DNL

In the Capital airport, use the Noise Monitoring Terminal (NMT) in 21 monitoring points to monitor aircraft noise, extract the data from Central Monitoring and Control Station (CMCS) and analyse, it is discovered that about 0~300 flight events can be monitored at one noise monitoring point each day, and the proportion in the day-time is about 70% while 30% at night. Then at this moment the formula of DNL changes as follows:

$$L_{dn} = \overline{L_{se}} + 10\lg(4.6 \times N) - 49.4 \tag{8}$$

When the Average Sound Exposure Level equals to 83dB, 85dB or 90dB which distribute around single aircraft noise limit value, the relationship between DNL and flight sorties N each day is shown as below:

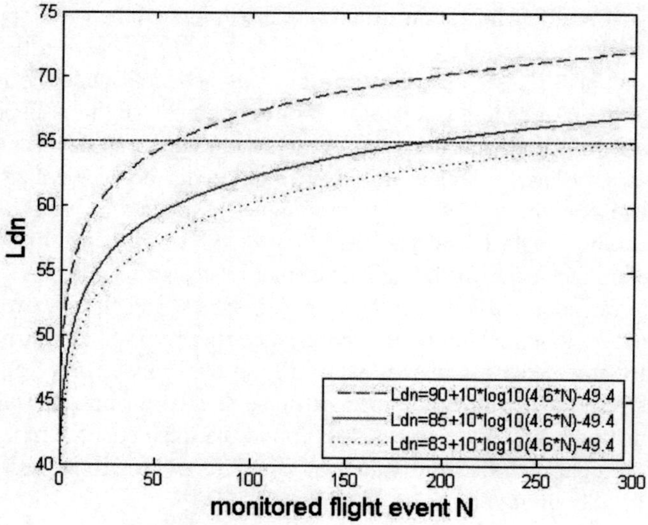

Fig. 1. Relationship Between Flight sorties N and DNL

In figure 1, although the SEL have achieved 90dB, DNL still does not exceed 65dB when the flights are less than 60. Therefore, in airports where flights are fewer, it is unreasonable to merely use DNL to evaluate aircraft noise. According to statistics, the total number of navigation airport is 172 in our country with 4.841 million landing sorties. About 32 airports can exceed 100 flight sorties each day. Consequently, it is advised that the small airports should adopt SEL as dominating evaluation and DNL as supplemental. In the operate case, the large airports should adopt DNL as main evaluation and SEL assistant.

6 Summary

In view of present situation of the airline industry and aircraft noise evaluation system, this paper analyses disadvantage of WECPNL, compares it with foreign evaluation index DNL, and proposed to use DNL as the main evaluation when revising the evaluation system in our country. Moreover, it suggests to using SEL to evaluate single aircraft noise event what the DNL considers incompletely and ascertains the limit value of it. In the practical application, it is advised to adopt SEL as dominate evaluation in small airport for the flight status in our country.

As the different circumstances, geographical location and cultural background and the levels of development in different country, the sensitivity of the residents to aircraft noise around airports also vary. Therefore, the evaluation method and the single aircraft noise event limit value proposed in this paper is only a proposal. It still needs a lot of investigation and research to revise the standards of our country.

References

1. Li, Y., Zhang, H., Wang, Y., et al.: Noise Pollution Prediction and Assessment Methods Improvement of the Airport. Enviromental and Science and Management (April 2008)
2. Lee, S., Hong, J., Kim, J., et al.: Exposure-response relationships on community annoyance to transportation noise. In: Proc. of 9th International Congress on Noise as a Public Health Problem, Foxwoods, CT (2008)
3. Finegold, L.S.: Historical development and current status of exposure-response relationships between transportation noise and community annoyance. In: Internoise 2003, Jeju, South Korea (2003)
4. Wang, W.: An analysis on exposure-respinse relation of airport aviation noise. Applied Acoustics 26(1), 35–40 (2007)
5. GB3096-2008. Environmental quality standard for noise. China Environmental Science Press, Beijing (2008)
6. Albee, W.: Why we must supplement DNL noise analysis. Wyle laboratories acoustics group (2002)
7. Glisser, M., Gerard, C.: Results and methodoloties of airport noise studies. J. Acoust. Soc. Am. 128, 2420 (2010)
8. Kong, X., Duan, J., Zhang, H., et al: Prediction of Influence of Aircraft Noise on Health. Environment Protection in Transportation (June 2004)

An Integrated Analysis of the Main Problems a New Business Faces and Viable Suggestions

Yunfei Du[1] and Jiao Xiong[2]

[1] Business School, University of Essex, Elmer Approach, SS1 1LW, Southend On Sea, United Kingdom
[2] Chongqing Automobile Institute, Chongqing Institute of Technology, Hongguang Road, 400050, Chongqing, China
feeba@qq.com

Abstract. In the start-up, a new business may face various problems which could adverse its development, or even cause failure. Aiming to call new entrepreneurs' attention to avoid those difficulties, the paper concentrates on analysis of three essential elements of a new firm which may be in trouble including marketing, finance, and management. By combining previous studies and surveying a certain number of new enterprises, it indicates the reasons of problems, compares the situation between a new enterprise and other business which established for a longer time, and suggests a number of solutions to those difficulties.

Keywords: New business, marketing, finance, management, entrepreneur.

1 Introduction

By establishing new business, entrepreneurship becomes a key role of motivation for the world economy, which not only stimulates the improvement of economy, but also produces more positions for employees. For instance, new businesses have created more than 95% fortune for American economy since 1980s. More than 140 million job opportunities were made by new enterprises in the last 10 years.

Even though a vast number of enterprises were negatively impacted by the international financial crisis which was started in America, and then spread to most European countries, more than half of small businesses were unacted on. However, different problems which are originated from finance, marketing, research&development, management, government policies and economic environment may block their development or cause the failure. According to a report from NFIB [1], it is suggested that 35% of new enterprises are predicted to fail within 2 years, 54% may be closed in 4 years, and 64% will be absolved 6 years later. Aiming to avoid the above problems or minimise losses caused by them, it is necessary to analyse a number of primary elements.

Since the problems of marketing, finance and management may be the most serious issues, the paper concentrates on analysis of the three aspects. For market issues, lack preparation in the investigation of the market, inefficient promotion, and

J. Luo (Ed.): Affective Computing and Intelligent Interaction, AISC 137, pp. 837–842.
springerlink.com

being excessively dependent on one certain group of consumers may be essentially adverse the new ventures; in term of the finane, those problems perform as difficulties in loan, excessively debt, and irregular financial management; in the case of management, inefficient integrated management, hardness of enrolling and promoting in human resource management occur frequently. In addition to that, comparison with other businesses which were established for a longer time and some potential solutions are invovled in the article.

2 Marketing Problems

It may be well-recongnised that marketing problems are the most arduous for a new company, because entering a market is difficult in the start-up. If a new enterprise can not capture a certain market share, it may fail in a short-term.

2.1 Three Aspects of Marketing Problems

New business which could be commonly defined as those enterprises start their business within 5 years, need to understand and investigate the market they would access to, such as the market share, the size, and competitors, thus making it possible for them to build their own marketing strategies. Stephenson [2] suggested that a new business can survive due to the market of its products or service exists. For parts of new enterprises, especially smaller ones, since they lack preparation in the investigation of the market or do not research anything about its market, it is supposed that they could not achieve the pre-determined target or be forced to change the original purpose after accessing.

Furthermore, inefficient promotion of merchandise or service is a possible handicap. The funds of most new businesses are not abundant. If sales are going smoothly in the beginning, the company can recover the circulating funds dramatically, then expand its market subsequently. Otherwise, since limited amount of circulating funds are used and a great number of goods are backlogged which are result of ineffective strategy of distribution, the capital chain may break down and cause potential bankruptcy consequently.

In addition to that, being excessively dependent on one certain group of consumers, parts of new businesses may lose their market share and be purchased by competitors, even though their products or service are high-quality, and the sales strategies are effective. The situation due to they can not turn their trade to another as the changing of market.

2.2 Solutions for Marketing Problems

For solving the marketing problems, it seems that doing feasibility analysis includes market survey and client positioning may be the best solution. In reality, other businesses which were established for a longer time set up mature mechanism of market analysis, formed an accomplished channel of distribution, and covered a large extent of consumers. Even if the market or consumers' habit changes, they can also survive. Instead, since most new companies are puzzled by the financial difficulties initially, adopting statistics or results from government reports (e.g. White Papers of Chinese Government) or industry reports directly are available for defining the market. If the market a new business wants to enter can be defined effectively, promotion strategy should be formulated accordingly and consumers groups can extend to two or three parts rather than restrict in one.

3 Financial Problems

With settling the marketing problems, it is assumed that financial problems may confuse new companies. Capital is a core of a business. In the entrepreneurial stage, a large number of enterprises have to acquire extra capital with various channels because of the limitation of their initial funds, thus assisting them in corporate operation and to further branch out. According to Fraser's [3] survey, the preference of financing for entrepreneurs may be a bank loan, borrowing from families or friends, private or public equity financing, and direct/indirect investment by other businesses, which can be divided into debt financing and equity financing generally.

3.1 Difficulties in Loan

Indeed, financing in debt financing and equity financing can improve the financial status of a company. However, most new businesses can not achieve enough capital easily, resulting in two aspects. Due to the lower business credit, the credit assessment procedure may take a few months or even more than one year, thus they may miss appropriate opportunities in the market; and the interest of bank loan could be too high to afford. Furthermore, it is almost impossible for new enterprises, small businesses especially to seek financing by listing on the stock market, even though it can not be denied that the gains are attractive. Based on a report [4] of American Department of commerce in 1994, 151 stocks of new companies were offered in public to obtain 4.4 billion dollars while 157 ones earned 4.58 billions dollars in the next year. Accompanied with the earning of huge number money, troubles from the stock market such as high expenditure, information disclosure, strict regulation, and pressure from stock holders are inevitable.

In contrast, private placement, which is equal to private equity financing, is preferred by smaller new businesses. As Nevitt [5] said, even though the borrowers need to afford higher rate than the public market, the strengths of private placement exceed the disadvantages, which are demonstrated in the following:

1) Private placement does not require registration under the securities laws.

2) A private placement is a good way to establish useful long-term investing relationships which a borrower can call for future financing.

3) No public disclosure of sensitive information is required, which may be helpful for a new enterprise to keep its trade secret.

3.2 Excessively Debt

If new enterprises can loan from banks, individuals, or financial companies, it is supported that being in heavily debt and carrying irregular financial management, could adverse the system of finance of corporations, as the same as difficultly obtaining a loan. With the obvious benefits of debt and increasing stress of operation, a portion of new companies exaggerate the debt for pursuing great profit. Although it is considered an effective approach to promote the continuing development of enterprises, the negative impact should not be ignored: the more debt they loan, the higher risk they may bear. When an enterprise excessively loans and can not afford the periodical interest, the financial balance is predicted to be broken, then the worst result is bankruptcy.

A solution which mixes equity and debt could be effective to issue it. According to Brigham and Houston [6], 'in practice, many managers think of the optimal capital structure more as a range (e.g., from 40% to 50% debt) rather than a precise number (e.g., 45%)'. It is supported to be a reasonable ratio between equity and debt, therefore guaranteeing the operation of no matter new firms or mature businesses.

3.3 Irregular Financial Management

Additionally, efficient management can assist the formation of financial system in a new business, while an irregular one may interrupt the whole company. By preparing the financial statements including balance sheet, cash flow statement and income statement, the cooperators, investors, and share holders can understand the financial position of a company. However, inaccurate statements may result in loss of confidence and credit, which is considered as a fraud, even it is just mistaken by an unqualified accountant carelessly.

Moreover, controlling the finance by the business executives themselves or delegating financial powers to lower levels unduly, which both are false behaviours in the financial management, may cause the financial chaos. Though convention, a large number of companies will organise the financial department by themselves, whereas parts of smaller enterprises without the capability prefer to finance outsourcing.

4 Management Problems

In addition to marketing and finance, it is supported that management is another scabrous problem a new business faces in the start-up, such as integrated management, human resource management and employees' loyalty.

4.1 Inefficient Integrated Management

After founding a new business, an entrepreneur needs to be intelligent and capable in integrated management. In The Entrepreneur [7], the features of a successful entrepreneur were summarized as 11 elements such as being confident, success pursuing, being opportunities and purpose orientated, risk estimation, and integrity. The above characteristics may be beneficial for managing the new business. Therefore, businesses which were established for a longer time can enroll famous directors outside or promote experienced staffs inside who achieve the standards to be leaders.

By contrast, most of entrepreneurs are the owners of their enterprises, who may be lack of ability and experience to dominate the whole company effectively, thus making it impossible to hire others to replace them. In such new business, it can not be determined that an entrepreneur can be competent in every position even if one is provided with above traits, because of the restriction of energy. For solving the situation, stimulating staffs' potential could be a potent arrangement.

4.2 Hardness of Enrolling and Promoting in HR

Furthermore, enrolling and promoting are two difficulties in human resource management. The scarcity of employees is common for new enterprises in the first and second year. Initially, since the limitation of funds and development room, it is hardly for new businesses recruiting experienced staffs with relatively lower payment. Switching to in promoting, excellent rapport with managers is more important than standout working ability sometimes, which is in evidence in familial enterprises.

After the earliest period, when everything is on the right track, the loss of employees becomes serious, in which the wastage rate of staffs could be much higher than ever. By being cultivating, gaining experiences and grasping customer resources, experienced staffs expect to earn more salary and to be promoted. Since the funds of enterprises are still limited, the payment and promotion to which they aspire is beyond they reach, most of them may choose to leave.

It is supposed that two methods can be used to figure out those problems. By putting best possible employees in place and establish nomative promotion system, the stress of human resource could be eased to some extent. Besides, career development, affection, and payment are essential for etaining talents and experienced employees. If a new business can not afford attracted salary due to the restriction of funds, planning promotion prospect for employees or cultivating sense of belonging to the company may be effective.

5 Conclusion

Though new businesses contribute to the world economy significantly, main problems in the aspects of marketing, finance, and management are inescapable. It is suggested that marketing issues may avoid the acceleration of development; difficult financing always disturb expanding business; and tangly management could disrupt the operation. If any part of them does not work properly, the entire business may be affected. By absorbing experiences from enterprises established longer and adopting suitable solutions, it may be a possibility for a majority of new businesses to survive and develop when faces those problems.

References

1. Dennis, W.: Small Business Policy Guide, p. 13. NFIB Foundation, Washington D.C. (2000)
2. Stephenson, H.: The Most Critical Problem for the Fledging Small Business: Getting Sales. American Journal of Small Business, 27 (Summer 1984)

3. Fraser, J.: How to Finance, p. 32. Anything. Inc. (1999)
4. Hinds, M.: Venture Capital. Department of commerce: U.S. Industrial Outlook, Washington D.C., pp. 46–48 (1994)
5. Nevitt, P.K., Fabozzi, F.J.: Project Financing, 7th edn., p. 88. Euromoney Books, London (2000)
6. Brigham, E., Houston, F.: Fundamentals of Financial Management, 12th edn., p. 417. South-Western Cengage Learning, Mason (2009)
7. Kao, J.: The Entrepreneur. Prentice-Hall, Englewood Cliffs (1991)

A Study on the Stock Index Futures Arbitraging and Hedging Model*

Tao Lei[1,**], Cheng Zeng[2], and Bin Li[3]

[1] School of Economics, Peking University, Beijing 100871, P.R. China
[2] School of Economics, MinzuUnviersity of China, Beijing 100081, P.R. China
[3] School of Environment and Natural Resources, RenminUnviersity of China, Beijing 100872, P.R. China

Abstract. How to arbitrage and hedge without risk through stock index futures has become oneof the urgently to be solved technical problems before the launch. From the perspective of econometric model, this article attempts to make acomprehensive scientific inquiry into the arbitraging and hedging functions ofstock index futuresto provide some theoretical and technical suggestions for the launch of stock index futures in China.

Keywords: Stock index, price, arbitrage, hedge.

1 Introduction

U.S. Nobel Economics Prize laureate Merton Miller said: "As long as the free market, there is uncertainty about future prices; as long as there is the uncertainty of future prices, we need futures market." [1] Stock market is the same too. Stock index future goods is another financial futures in addition to foreign exchange and interest rates, and it is one of the shortest duration and the fastest growing financial derivatives. In February, 1982, the Kansas Agricultural Exchange in the United States launched the first stock index futures contracts based on value line composite average future, from then on, stock index futures are active in Europe and some Asian countries and regions. In May 1986, Hong Kong Hang Seng Index futures were introduced; subsequently, Tokyo and Osaka Securities Exchanges launched Topix futures and Nikkei225 futures separately. The launch of Shenzhen 300 stock index futures in China are just around the corner.

Stock index futures serves share price index as a subject matter, based on index points to count. Its value is calculated by the multiplier of monetary multiplier and stock price. By settling the price difference, it uses a financial futures contract of cash settlement. The basis of the price is the core element for interest arbitrager and hedger to estimate price trend and assess trading results. Thus this paper firstly discusses the pricing relationship between spot goods and futures, from the classic pricing formula

* This work is supported by the Asian Research Center of Renmin University of China (2007003).

** Corresponding author: Postdoctor, mainly engaged in finance.

J. Luo (Ed.): Affective Computing and Intelligent Interaction, AISC 137, pp. 843–851.
springerlink.com

based on continuous compound interest, consistent with the Chinese market, close to the theoretical model of weak-form efficient market pricing models and the evaluation period which is divided into m phases to discuss the stock futures pricing model. Secondly, there may be more obvious arbitrage opportunities in a major event, so arbitrage function is an technical problem urgent need to address. In the past, some scholars widespread use two assumptions in the study of pricing, that is, without transaction costs and borrowing and lending rates accordance. These two assumptions do not conform with the futures market. In this paper, the arbitrage model was improved from these two discrepancies. Thirdly, this paper find the will discuss how to find the optimal hedge ratio by four kinds of calculation methods -- proceeds minimize variance, ECM series model, capital asset pricing model (CAPM) and error correction model based on cointegration.

2 The Pricing and Risk-Free Arbitrage

Arbitrage is an arbitrage trading using the basis between stock index futures contracts and their corresponding base, while the basis is a reflection of the relationship between period price and the current price. Thus, before studying the arbitrage, it is necessary to study the issue of pricing and basis between period price and the current price.

Cornell and French proposed stock index futures pricing model based on the continuous compound interest holding cost model[2-3]:

$$F(s,t) = S_t e^{(r-q)(T-t)}$$

$F(s,t)$is the theoretical price of the contract at the time point *t* when the expiry time is *T*; $S_t$is the stock index; r is the risk interest; *q* is the annual interest of the stock; *T* is the expiry time; *t* is the current time point. *(r-q)(T-t)* is the holding cost of the securities comprising the stock index.

The weak efficient market pricing model based on the Black-Scholes model[4-5] is:

$$F(s,t) = S_t e^{u_a(T-t)}$$

$u_a$is estimated with the method that is used to estimate the fluctuation ratio through the historical data in the Black-Scholes model. The value of $u_a$at time point *t-1* can be get through the prices of spot goods and futures of stock index at that time point, that is:

$$u_{a,t-1} = lnT - (t-1)\sqrt{\frac{F_{t-1}}{S_{t-1}}}$$

The interest rate in China is not acquired through marketization at present, and this model needs historical data to estimate the model parameters. China has not officially launched stock index futures, historical data does not exist. Therefore, in the initial stageof thestock index futures in China, this pricing model is not practical.

Dividing the pricing period into m terms, the pricing model of stock index futures is:

$$F(s,t) = S_t(1+r)^{T-t} - D$$

In which, $D = \sum_{i=1}^{m} D_i(1+r)^{T-t_i}$. D is the sum of the present values of dividends in all term; r is the risk-free interest rate in every term; D_i is the cash dividend holding stock spot goods at the spot timet_i. Besides, other pricing formulas of stock index futures include[6-8]:

$$F = S[1+(r-d)]^{t/360}$$
$$F = Se^{(r-d)(t/360)}$$
$$F = S(1+\eta)^h(1-\varphi)^h$$
$$F = (S-D)(1+r')$$

d is the dividend yield; τ indicates the futures expiration days; e = 2.71828 ...; η is the annual risk-free interest rate; φ is the annual dividend yield; h is remaining years before delivery of futures contracts;$r'$is the risk-free interest rate from time 0 to time T. The deviation and excessive sent between stock prices and current price would measure that whether the pricing result can produce arbitrage. When the pricing error is greater than its transaction costs, stock index futures investors can construct the specific structure of arbitrage portfolio. After calculating theoretical futures price, compared with market prices, estimates the arbitrage costs, then compare the relationships between cost and price difference of theoretical price and the market price, and create arbitrage interval [9]. In the stage of investment portfolio, calculate the proportion and number of arbitrage trading portfolio and the arbitrage income, then proceed arbitrage activities in the market [9].

The formula based on arbitrage cost of the current period market. Taking into account the complexity of arbitrage formula derivatives, the futures pricing formula is simplified to:

$$F = (S-D) \times (1+r_f)$$

Markets are divided into two kinds, namely one on which futures prices are overvalued and the other where the futures prices are undervalued. When futures prices are overestimated:

$$F \geq (S-D) \times (1+r_f)$$

Investment strategy is to buy spot goods and sell futures. Providing the cost on the spot market is the transaction cost of buying long. The present value is C_{SL}.The present value of selling short transaction cost is C_{FN}, then the arbitraging requirement is[9]:

$$F \geq [(S-D) - (C_{SL} + C_{FN})] \times (1+r_f)$$

Similarly, when futures prices are undervalued:

$$F \leq (S-D) \times (1+r_f)$$

Investment strategy is to buy futures and sell spot goods. Providing the cost on the spot market is the transaction cost of buying long. The present value isC_{FL}.The present value of selling short transaction cost isC_{SN}, then the arbitraging requirement is:

$$F \leq [(S-D)+(C_{FL}+C_{SN})] \times (1+r_f)$$

Considering the two above and taking the simplicity of the formula into account, simplify the costs of long hedging and short hedging as:C_f^sandC_s^f,then no-arbitrage interval is:

$$[(S-D)+C_f^s] \times (1+r_f) \geq F \geq [(S-D)+C_s^f] \times (1+r_f)$$

Based on the continuous compound interest rate and the difference between borrow and loan interest rates, omitting the two conditions that the price of the futures may be overestimated or underestimated for this article's length, the no-arbitrage interval is:

$$S_{tS}^F + C_{FN}e^{r_c(T-t)} + C_{SL}e^{r_c(T-t)} \geq F_t \geq S_{tF}^S - C_{FL}e^{r_l(T-t)} - C_{SN}e^{r_l(T-t)}$$

$S_{tS}^F$is stock index price when the futures were excessive and spot were short; $S_{tF}^S$is stock index price when the spot were excessive and the futures were short; $C_{SL}$is the total transaction costs of long parts of the spot; $C_{SN}$is the total transaction costs of short position of the spot; $C_{FL}$is the total transaction costs of the long position of the futures; $C_{FN}$is the total transaction costs of short position of the futures;$r_l$is the market lending rates; $r_c$is the market lending rate.

In the beginning of the contract, the futures price is generally deviate from a reasonable premium, and are more prone to emerge arbitrage opportunities, so this no initial investment and no risk of the arbitrage profit are also known as Basis Arbitrage; when the futures price came down as a ladder with the reduction in the period of the contract settlement, the futures price will tend to current price, so that the theory basis fall on arbitrage-free interval, the arbitrageurs have short-term suspension. According to "Law of One Price," futures prices eventually converge to the spot price, which reflects a homoplasy and expiration of convergence between the two prices, the same asset must offer the same in two markets [10]. In theory the starting point of the contract premium is relatively larger, arbitrage opportunities are more, and in the last settlement period it is zero.

3 Hedging

Futures hedging derived from the development of Markowitz portfolio investment theory. The return variance and price risk of portfolio investment in the theory is[11-14]:

$$\sigma^2 = \sum_{i=1}^{n}\sum_{j=1}^{n} w_i w_j \sigma_i \sigma_j \rho_{i,j} = \sum_{i=1}^{n} w_i^2 \sigma_i^2 + 2\sum_{i=1}^{n}\sum_{j=i+1}^{n} w_i w_j \sigma_i \sigma_j \rho_{i,j}$$

w_i, $w_j$represent the weights of every asset in the portfolio investment; σ_i, $\sigma_j$represent the variance of the No. i, j futures contract;$\rho_{ij}$is the price.

Portfolio risk can be divided into systematic risk and non-systematic risk, on the right side of the first term is non-systemic risk. With the gradual expansion of the assets n included in the combination, non-systematic risk decreases and eventually approaches zero. Non-systematic risk can be avoid by diversified investment; the second is systematic risk, it must be avoid by involving in hedging transactions, or to avoid though the hedging transactions in the spot market and futures market [11]. Optimal hedge ratio depends on the purpose of hedging transactions and correlation between the spot market and futures market price, hedgers based on the expected portfolio return and expected return variance to determine the spot market and futures market trading positions, in order to minimize income risk or maximize the utility function.

Johnson put forward the optimal hedging ratio basing on minimization of revenue variance[4,13]. Economics often assume that investors are the risk aversions, and the optimal hedging ratio basing on minimization of revenue variance is suitable for them. So the model is worth for reference. Hedging can be devided into two kinds, short-seller and long-seller. Taking short-seller Hedging as an example, after building the hedging combination, combined value is finally changed into:

$$\Delta H = \Delta S - h\Delta F$$

H represents the final result of the changing hedging value. h is the hedging rate.ΔSandΔFrepresent the change of the prices of spot goods and futures at the end of the term respectively. The variance of$\Delta F$is:

$$Var(H) = Var(\Delta S - h\Delta F) = \sigma_S^2 + h^2\sigma_F^2 - 2hCov(\Delta S, \Delta F)$$

in which, $\sigma_S^2 = Var(\Delta S), \sigma_F^2 = Var(\Delta F)$

To get the minimum of the variance of *Var(H)* :

$\frac{dV}{dh} = 2h\sigma_F^2 - 2Cov(\Delta S, \Delta F) = 0$, the optimum hedging rate is :

$$h^* = \rho\frac{\sigma_S}{\sigma_F}, \text{ in which, } \rho = \frac{Cov(\Delta S,\Delta F)}{\sigma_S\sigma_f}.$$

Herbst, Kare, Marshall, Myers and Thompson find the demerits of the hedging model from OLS: It assumes that the residual is blank noise, the joint distribution, namely the covariance matrix, of Spot and futures prices are not changeable, and it does not take that the changes of changes may influence them of now into account. For it is against thesequence characteristics, OLS gets the biased estimation of OHR, etc[15,16].

Research of hedging based on a double variable vector regression model: This model is the representative of the ECM series model, and it is conducive to eliminate residual of serial correlation. It is a Systematic regression model with regression model whose current value are basing on the past value, which also increases information of the model, using B-VAR (Bivariate-vector auto regression) to carry out the calculation of least risk hedging ratio. In B-VAR model, the relationship between the futures price and spot price can be shown as follows[17]:

$$\Delta S_t = C_s + \sum_{i=1}^{k} \alpha_{si}\Delta S_{t-i} + \sum_{i=1}^{k} \beta_{si}\Delta F_{t-i} + \varepsilon_{st} \quad (1)$$

$$\Delta F_t = C_f + \sum_{i=1}^{k} \alpha_{fi}\Delta S_{t-i} + \sum_{i=1}^{k} \beta_{ft}\Delta F_{t-i} + \varepsilon_{ft} \quad (2)$$

C_s, C_f and α are interceptions; β is the slope that is the hedging rate with the minimum risk; ε_{st} and ε_{ft} are Independent and identical distribution random error term. Their variance and covariance are:

$$Var(\varepsilon_{st}) = \sigma_{ss}; Var(\varepsilon_{ft}) = \sigma_{ff}; Cov(\varepsilon_{st}, \varepsilon_{ft}) = \sigma_{sf}$$

Seeking the optimum lagged value k to make the residual error without self-correlation, we get the hedging rate with minimum risk:

$$h = \frac{Cov(\Delta S_t, \Delta F_t / \Delta S_{t-i}, \Delta F_{t-i})}{Var(\Delta F_{t-1} / \Delta S_{t-i}, \Delta F_{t-i})} = \frac{\sigma_{sf}}{\sigma_{ff}}$$

The optimum hedging rate also can be get from the regression model below:

$$\Delta S_t = \alpha + h_2 \Delta F_t + \sum_{i=1}^{m} \beta_i \Delta S_{t-i} + \sum_{j=1}^{n} \gamma_j \Delta F_{t-j} + \varepsilon_t [17\text{-}18]$$

In addition, there are several models basing on the ECM models. They areARFIMA model, divided entire error correction model and threshold cointegrationmodel. They improve the futures hedging effect basing the non-stationarity of spot price and the futures price, and the relationship of long-term balance and short-term dynamic of spot price and the futures price[14,19-21].

In 1964, Sharp put forward the capital asset pricing model (CAPM) basic equations and its constraints, that return rate and the corresponding risk depends on β risk, the hedging model of systemic risk and three of Basis risk. And the follow equations[22,23]

$$\begin{cases} r_i = \alpha_i + \beta_i r_m + \xi_i & (1) \\ E(r_i) = r_f + \beta_i [E(r_m) - r_f] & (2) \\ V_s = \sum_{i=1}^{n} \alpha_i p_{0i} r_i & (3) \\ V = V_s - NS_0 t r_s + N\Delta Bt & (4) \end{cases}$$

r_i in (1) represents the revenue of the stock portfolio; α_i represents the number of the shares of every stock; β_i is used to measure the risk between security I and the overall market:

$$\beta_i = \frac{Cov(r_i, r_m)}{\sigma_m^2}, i = 1, 2, \dots n.$$

r_m is the yield of stock market; r_f is the risk free return rate; V_s is the total income in the period of stocks combination hedging; p_{0i} is the stock market price in the first period of hedging. V is the total income of stock combination and Stock index futures; N is the number of contract about Stock index futures $\Delta B = B_1 - B_0$; $V_s - NS_0 t r_s$ is the summation of stocks combination and income of N contracts of stock index futures; $N\Delta Bt$ is the income gap of Stock index futures and Stock index spot. Putting (2) and (3) into (4), we can get:

$$E(V) = \sum_{i=1}^{n} \alpha_i p_{0i} [r_f (1 - \beta_i) + \beta_i E(r_m)] - NS_0 t E(r_s) + N\Delta Bt$$

From the above model, the mathematical model of hedge of stock index futures in consisted by β risk, systemic risk and Basis risk. According that systemic risk is zero in hedging, we get the contract is the multiplier N for:

$$N = \frac{\sum_{i=1}^{n} \alpha_i p_{0i} \beta_i}{S_0 t}$$

About the research of error correction hedging model basing co-integration relationship, most of the literatures prove that if there is co-integration relationship between stock index futures and stock spot, there must be a long-term balance and Short-term dynamic relationship between them. So there must be a balanced error between them. Using this error can connect short-term behavior with long-term value, building the error correction expression[17,18,24,25]. The optimal hedging ratio can be launched using error correction model.

$$S_t = a + bF_t + \varepsilon_t$$

From that, the error correction model is:

$$\Delta S_t = \alpha\varepsilon_{t-1} + h^*\Delta F_t + \sum_{i=1}^{m} \varphi_i \Delta F_{t-i} + \sum_{j=1}^{n} \emptyset_j \Delta S_{t-j} + e_t$$

ε_{t-1}represents the residual error term estimated from the cointegration regression model$S_{t-1} - (\hat{a} + \hat{b}F_{t-1})$. $\varphi_i$and$\emptyset_j$are the coefficients of the error correction term respectively. $h^*$is the optimum hedging rate.

With the development of the model, more methods appeared, including bayesian learning rule method, mean-Gini random dominant method, dynamic programming method, GARCH method, dynamic planning method, multi-objective programming method, lower side risk method, stock index risk hedging MABM model, multi-objective programming hedging model and so on[26-28].

4 Conclusion

Whether from the point of view of the market, and the legal system, or the risk education to investors and so on in China, along with the launch of stock index futures, it's a pressing practical problems to research the arbitraging and hedging model. In this paper, on the basis of previous workers' achievements, we discussed several types of the models, but further research is needed to make more improvements.

References

1. Zhang, J.: The theoretical analysis of arbitrage function used in stock index futures. Review of Economic Research 41, 39–44 (2007)
2. Hsu, H., Wang, J.: Price Expectation and the Pricing of Stock Index Futures. Review of Quantitative Finance and Accounting 23, 167–184 (2004)

3. Yan, F., Wang, F.: Analysis of China's stock index futures pricing. The Graduate Student Journals of Zhongnan University of Economics and Law 2, 39–43 (2007)
4. Li, Y.: The Stock index futures' research about hedging and risk management. University of International Business and Econnomics (2007)
5. Wang, F., Yan, F.: The exploration about China's stock index futures pricing. Economic & Trade Update 60, 117–119 (2007)
6. Merrick Jr., J.J.: Volume: Determination in Stock and Index Futures Market: An Analysis of Arbitrage and Volatility Effects. Journal of Futures Markets 7, 483–496 (1988)
7. Andrews, C., Ford, D., Mallison, K.: The Design of Index Funds and Alternative Methods of Replication. The Investment Analysis 10, 16–23 (1986)
8. Wang, B., Zheng, P., Li, Q.: The practical research about stock index futures arbitrage and pricing. The Journals of China University of Geosciences (Social Science Edition) 4, 48–51 (2003)
9. Tian, M., Song, L.: Stock index futures of arbitrage strategy research. Modern Business Trade Industry 3, 47–48 (2007)
10. Xiao, K.: Stock index futures of arbitrage pricing model research. The Journals of Wuhan Institute of Training about Agricultural Bank of China 3, 35–36 (2007)
11. Zhang, Z., Su, Z.: The composite hedging strategy of the use of stock index futures to the securities. The Journals of Huazhong University Science and Technology (Natural Science Edition) 1, 102–104 (2004)
12. Markowitz, H.: Portfolioselection. Journal of Finance 7, 77–91 (1952)
13. Johnson, L.: The Theory of Hedging and Speculation in Commodity Futures. Review of Economic Studies 3, 139–151 (1960)
14. Liang, C.: Advances in futures hedging theory and model. The Journals of Xidian University (Social Science Edition) 3, 53–56 (2007)
15. Herbst, A.F., Kare, D., Marshall, J.F.: A time varying, convergence adjusted, minimum risk futures hedge ratio. Advancesin Futures and Options Research 6, 137–155 (1993)
16. Myers, R.J., Thompson, S.R.: Generalized optimal hedge ratioestimation. American Journal of Agricultural Economics 71, 858–867 (1989)
17. Gao, H., Zhao, J.: The empirical research of CSI 300 stock index portfolio and hedging. Journal of Management Sciences in China 2, 80–91 (2007)
18. Hu, L., Li, S.: Cointegration-based estimate of the optimal hedge ratio. The Journals of Wuhan University of Technology's Economic Forum 8, 201–204 (2007)
19. Lien, D., Tse, Y.K.: Fractional Cointegration and Futures Hedging. The Journal of Futures Hedging 19, 457–474 (1999)
20. Granger, C.W.: Developments in the Study of Co integrated Economic Variables. Oxford Bulletin of Economics and Statistics 48, 213–228 (1986)
21. Balke, N.S., Fomby, T.B.: Threshold Cointegration. International Economic Review 38, 627–645 (1997)
22. Mao, X.: The composite hedging strategy of Stock index futures. The Journals of North China University of Technology 16, 75–78 (2004)
23. Wang, B., Zheng, P., Li, Q.: Stock index futures hedging principle modeling and its application. The Journals of Hebei University of Engineering 4, 79–82 (2003)
24. Engle, R.B., Granger, C.W.: Cointegration and Error Correction: Representation, Estimation and Testing. Econometrica 55, 251–276 (1987)

25. Wang, J., Zhang, Z.: The empirical study of China non-ferrous metals futures market hedging from the year of 2000 to 2004. The Journals of China University of Geosciences (Social Science Edition) 6, 46–51 (2006)
26. Wang, B., Hao, Y.: The first exploration of Stock index futures. Hedge Risk Minimum Ratio of MABM Model 11, 122–123 (2007)
27. Kong, F.: The partial moment risk assessment under hedging. The Journals of Wuhan University of Technology (2004)
28. Wang, Z., Fan, Z., Zhang, Q.: The multi-stage and multi-objective programming model of Futures hedging. Systems Engineering 15, 50 (1997)

A Discussion on the Operational Mechanism of Private Equity in China*

Guomin Wei[2], Tao Lei[1,**], Cheng Zeng[2], and Bin Li[3]

[1] School of Economics, Peking University, Beijing 100871, P.R. China
[2] School of Economics, MinzuUnviersity of China, Beijing 100081, P.R. China
[3] School of Environment and Natural Resources,
RenminUnviersity of China, Beijing 100872, P.R. China

Abstract. This article analyses and makes a research on the financing, investment, management and exit mechanism in the operational process of private equity(PE). In the end, the author puts forward four suggestions including updating financing structure, widening financing channels, establishing financing investment risk and technology evaluation systems, establishing the OTC market and consummating the incentive pay contract mechanism, improving the structure of the capital market, and rolling out supporting policies about PE by the government simultaneously.

Keywords: Private equity, finance, investment evaluation, management, exit.

1 Introduction

China's economy is booming, but if China cannot use financial instruments to effectively combine factors of production, China's economy will not have a strong competitive edge[1]. Private equity funds (PE) is a mainstream international investment and financing tools. Domestic PE study, most of the attention on the qualitative aspects of the operation of each PE set, due to the rare and the lack of quantitative data and systematic empirical research. And along with the national policies and regulations change and the objective development of PE, corresponding PE yet to open up new areas of research. In view of the part of which needs to perfect in the PE operation process, the following aspects are mainly discussed in this paper.

2 Financing

From PE fund leveraged buy-out of financial structure, we can see that mergers &acquisitions fund is main made up by senior debt, subprime debt and equity capital three parts. The domestic situation of PE investors mainly have insurance companies, trust company, Banks, securities companies, and other financial institutions, other

* This work is supported by the Asian Research Center of Renmin University of China (20C7003).

** Corresponding author: Postdoctor, mainly engaged in finance.

J. Luo (Ed.): Affective Computing and Intelligent Interaction, AISC 137, pp. 853–860.
springerlink.com

private enterprises, listed companies and other non-financial institutions and non-profit organization, and wealthy families / individuals and the state funds, in which institutional investors are the main source of funds. After the investment advisory firms, investment consultants directly invest brokerages to CITIC, CICC's PE, the SFC announced the expansion of the scope of the pilot, and many domestic brokerage firms respond to apply direct investment. Commercial banks are also actively developing asset management business, to further expand the investment of PE[2]. Meanwhile, the investment subject such as banks introduced the "Loan + PE", "investment and credit linkage", PE banking system, Cross-sell the parent bank, SPAC curve investment and other new financial method.

Brokerages and banks to participate in PE has inherent advantages, they have a certain degree of credibility, the national business network, and extensive customer source, so is easier to raise funds, and it will become an important source of PE funds. Increased risk of the securities market started to encourage companies to switch part of the funds in the private equity market, in order to diversify risk. Private equity financing has advantages, such as registration procedure is simple, poundage low, financing fast, flexible high-negotiation, distribution risk, less public internal information. The disadvantage is mainly reflected in the poor liquidity of private equity in this regard, after the investors buy private securities several years it often cannot realized, in order to compensate for this loss of investors, investee companies must offer more favorable terms[3].

SPAC (Special Purpose Acquisition Corporation) will gradually become the mainstream mode of PE, but also very suitable for Chinese enterprises. SPAC in one set of characteristics and purposes of direct listing mergers, reverse mergers, private equity and so on, the goal is the acquisition of the company which can appear on the market, completion of the restructuring, and you can switch to the motherboard. For mutual funds and hedge funds of the open market operations speaking, no matter from risk preference, or portfolio, private equity investment in China has a market demand, many of the hedge funds in the search for the management team and to enter the Chinese need to spend a lot of time, and for SPAC hedge funds can be used as a configuration participate, relative to set up a separate private equity funds, by the way SPAC to invest private equity field is more convenient and quick[4].

3 Investment and Project Evaluation

High-tech fields which invest in PE's aggregate investment have a large proportion[5,6]. Establish a systematic evaluation is the key to the PE investment, but also conducive to the development of high-tech field and the optimum allocation of resources. Assess project, not only from the market point of the commercial market potential and analysis, and more from technical point of view forecast core competitiveness, such as independent innovation, patent technology and so on[7]. For this to the project's assessment must establish two sets of evaluation system.

To establish a private equity fund value evaluation system based on internal rating system. Value of the fund consists of three aspects: the present value of the investee company, future value of the investee company and the value of future investment funds, the formula is as follows:

$$\sum_{t=1}^{t\le T}\frac{C_f^{past}}{(1+IRR)^t}+\sum_{t\ge T}^{t=S}\frac{C_f^{port}}{(1+IRR)^t}+\sum_{t\ge T}^{t=S}\frac{C_f^{new\ inv}}{(1+IRR)^t}=0$$

Among them, the first part of the left reflects the Fund's cash flow and internal rate of return (quantitative assessment) before assessment point; the middle reflects the future cash flow of the discount, the traditional evaluation method of the evaluation net assets is approximately equal to this part (mainly quantitative assessment, but with a qualitative assessment review). Last part of the left reflects the cash flows of future investment, this part's weight factors change over time (Based on historical data and qualitative assessment of the future situation)[8]. Among them, dimension index of qualitative score is suggested set as table 1shows:

Table 1. Dimension index of qualitative score setting

Dimensions	Affiliated dimensions
Fund strategy	project strategy, practice strategy, investment emphasis, fund scale, exit strategy, overall strategic adaptability
Fund Structure	comply with standard terms and conditions, cost structure, corporate governance and internal procedures
External acknowledgement	fund historical performance analysis, performance analysis of comparable funds, the quality of joint ventures or investors
Management group motivation	incentive structure, reputation, group independence, outside activities, conflicts of interest, the Fund's investment of the fund manager
Management group ability	PE experience, operation experience, industry experience, state or regional experience, group scale, balance and coverage
Management stability	group cohesion, historical stability, group assignment, transfer plan and financial stability
Overall adaptability	adaptability between the group and the strategy, adaptability between the fund structure and the fund strategy

Establish technology assessment and prediction system. This system use mathematical statistics to carry on balanced measures to the 10 indicators of independent innovation ability evaluation index system. The used evaluation index adopt four top Level innovation indicators in “Chinese enterprises' independent innovation ability analysis report"[9], which was released by the national bureau of statistics economic monitoring center in 2005: Potential technology innovation resources index, including the number of enterprise engineering and technical personnel, Industrial value-added of the enterprise, sales revenue of enterprise products and so on. Evaluation of technological innovation activities, including science and technology activities funds accounted for total sales revenue, research and experimental development (R & D) funding activities accounted for the proportion of sales revenue and so on. Technological innovation capacity indicators of output, including the number of patent applications accounted for the whole nation’s proportion, owned the number of invention patent accounted for the whole nation’s proportion, new product sales income accounted for the total sales income’s proportion and so on. Use 24 secondary indexes to the investment of high-tech enterprise for technology assessment and prediction. As shown below:

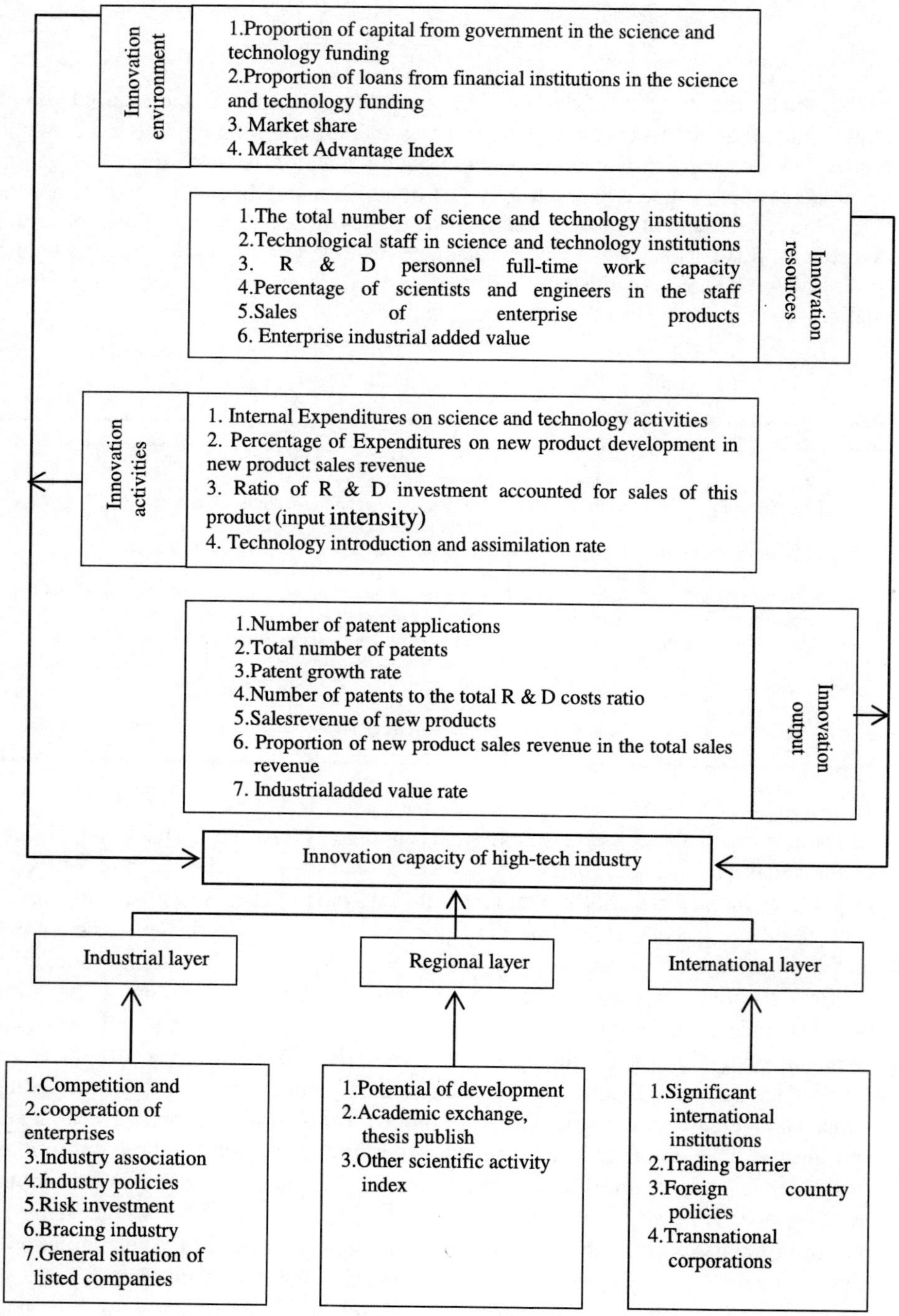

Fig. 1. Innovation capacity evaluation index system of high-tech industry

4 Management

In the private equity investment relations, entrepreneurs faced with the biggest moral risk is the fund managers' opportunistic behavior that fund managers try to reduce the bargaining power of entrepreneurs with the entrepreneurs to renegotiate. Entrepreneurs also face adverse selection, which is relate to the ability of non-financial contribution that fund managers provide to enterprises. In addition, entrepreneurs also face special agency problems: behavior by name of fund managers[10-12].

Discuss the static game between private equity funds and companies: if it has good and bad companies, there are two strategies that PE given fund investment: According to the predetermined (not) to use funds——contract (with probability *1-p*) and default (with probability *p*). PE also has two options: continue to invest (with probability *q*) and exit the investment (with probability *1-q*), dividing following several kinds of situations:

1)Enterprise default, if you choose to invest in PE, then it will bring the loss *BA* to PE, and get a positive return *C*; If the PE exit, it is required to pay cost *A* (such as proceedings), supposing that the PE take exit behavior at this time; If PE exit because of corporate defaults, companies need to pay the cost of *D* (as compensation).

2) Enterprise the non-breaching, and if the PE continue to invest, both sides will gain the normal profit E_1 and *F*. If the enterprise default, good and bad enterprise will obtain breach of contract revenue $E_2$and E_2'respectively, and $E_2' > E_2 > E_1$.

3) Enterprise the non-breaching, and if the PE exit, the enterprise's revenue is E_1, and the PE's revenue is *(F-A)*. If the corporate defaults, PE also exit, good and bad enterprise's revenue is $(E_2\text{-}D)$ and $(E_2'\text{-}D)$, respectively. PE's revenue I*(F-A+C)*; If the PE continue to invest, enterprise's revenue is $E_2$and E_2'respectively, and PE's revenue is*(F-B)*. Game matrix is as follows:

Table 2. The game matrix between PE and enterprises

participants		PE	
SME(small and medium enterprises)	strategy	investment(q)	exit$(1-q)$
	breach(p)	$E_2(E_2'), F-B$	E_2-$D(E_2'$-$D)$, *F-A+C*
	compliance$(1-p)$	E_1, *F*	E_1, *F-A*

Expected return of private equity funds is as follows:

$$U_1 = pq(F-B) + p(1-q)(F-A+C) + (1-p)qF + (1-p)(1-q)(F-A)$$

According to the optimal first-order conditions, the optimal corporate default probabilities can be obtained as: $p = A/(B+C)$

Expected return of the enterprise is:

$$U_2 = \frac{1}{2}[pqE_2 + p(1-q)(E_2-D) + (1-p)qE_1 + (1-p)(1-q)E_1] + \frac{1}{2}[pqE_2' + p(1-q)(E_2'-D) + (1-p)qE_1 + (1-p)(1-q)E_1]$$

According to the optimal first-order conditions, the best private equity investment fund probability can be got:

$$q' = 1 - \frac{(E_2 + E_2') + 2E_1}{2D'}$$

OTC market information disclosure mechanism and the related listing requirements for companies can reduce the costs of information search and litigation when private equity funds exit investment, so the cost of exist "*A*" of private equity funds is lower in off-market than no outside market; at the same time, the OTC market mechanisms lead to information disclosure when companies breach of contract and to opt out of private equity funds, the firm will face a greater negative impact, so the cost *D* that companies need to pay when the OTC market corporate defaults or private equity funds opt out is larger than no outside market. Therefore, the optimal investment probability of private equity fund will increase when there is the OTC market, while enterprises best probability of default can reduce.

Discuss of private equity funds pay incentive model, in which private equity fund's investment portfolio returns is $: r = \mu + \varepsilon\,(e$ is the efforts of venture capitalists invested, and$\varepsilon \sim N(0,\ \sigma_P^2)$). The linear payment contract equation is $s(r) = \alpha + \beta r$.

Entrepreneurial risk-averse investor's utility function:

$$U(w) = -\mu^{-\rho w} \quad (\rho > 0)$$

Investment cost function:$c(e) = 1/2ce^2\ (c > 0)$

Venture capitalists certainty equivalent income is:

$$EU = E[\alpha + \beta(\mu + \varepsilon) - 1/2c\mu^2] - 1/2\rho Var[\alpha + \beta(\mu + \varepsilon) - 1/2c\mu^2]$$
$$= \alpha + \beta\mu - 1/2c\mu^2 - 1/2\rho\beta^2\sigma_p^2$$

Investor's expected utility is:

$$EV = E\{x - [\alpha + \beta(\mu + \varepsilon)]\} = (1 - \beta)\mu - \alpha$$

Then, through the most optimal solution portfolio theory to be solved:

Investor's expected return is:

$$EV = \mu - (w + 1/2c\mu^2) = 1/2c - w$$

Agency costs of investment are:

$$AC = 1/2c - 1/2c(1 + c\rho\sigma_P^2) = 1/[2c + 2/\rho\sigma_P^2]$$

Using the above model of private equity funds pay contract can be found: under asymmetric information moral hazard of venture capitalists will cause a significant agency costs. And private equity fund management agency cost has positive correlation with the ability of venture capitalists, risk aversion and private equity fund portfolio risk.

In summary, the opportunistic behavior of fund managers and governance can be established OTC market through the dominant investment contracts, joint ventures and other ways. Such as through the use of equity rather than debt and limit dividend payments can avoid the evasion of responsibility of fund managers, reducing the moral hazard. In addition, private equity funds in the pay agreement introduced GEM boom as the benchmark portfolio return can significantly improve efforts to encourage venture capitalists and significantly reduce the agency costs of private equity investments fund management[13].

5 Exit

The main problem that China PE's exit face is main the following aspects: exit way is single, and basic is the IPO, mergers, acquisitions and other exit ways are in a very small proportion. Multi-level capital market develop immature, exit channel is not smooth. Although China has recently launched the GEM a new channel provided for the PE out and there is also the SME board, the property rights trading market and the OTC markets, because these markets' level are not obvious, the transfer plate between the market mechanism is imperfect, board, the SME board market trading standards are too strict, property rights trading market and OTC market is not active, effective channels for the exit of PE capital cannot be provided. Private equity funds on legal and regulatory imperfections. Government policies are lack. Government has not done active support to the formation of trade associations and any other self-regulatory organization, to make PE achieve self-improvement and self-regulation; at the same time, establish an objective and effective fund rating system for supporting PE rating, research and other relevant service providers, it is inadequate to create a good exit for the PE market.

To deal with these problems, the improvements of the PE exit mechanism are as following: choose the right way to exit. IPO is the highest proceeds exit channels of private equity, but for different capital markets, IPO is not always the best choice. To perfect legal and regulatory system. Gradually form a more perfect system of private equity fund regulatory framework, the regulation for trading in and out is mainly in the Growth Enterprise Market and the OTC market supervision. To improve the hierarchy of the capital market. Capital market is constituted by four different levels, capital markets should be divided into different levels to facilitate the withdrawal of institutional equity investment diversification. Government intervention. Government should support the PE industry, the main methods are: financial support, tax incentives and other policies to make private equity investors tend to diversify, thus contributing to the diversification of private equity funds exiting.

6 Conclusion

Through the above analysis, China's PE need to upgrade the financing structure, widen financing channels, in particular, further promote the SPAC financing to have a mature operation mechanism. Two assessment systems of financial, risk and technology should be established for project of PE investment.

In the management of PE, the OTC market should be established, the information disclosure mechanism should be improved, boom of GEM investment should be introduced in pay contract market as the portfolio returns benchmark, improve incentives of venture capitalists efforts and reduce agency costs of PE management. For PE's exit, mainly lies in the perfection of hierarchy of capital markets, promoting diversification of exit, while the Government should introduce appropriate policies to support the development of the PE industry, improve the legal and regulatory framework.

References

1. Wu, X.: Several problems need to be researched on in developing PE. Chinese Entrepreneur 23 (2007)
2. Huang, Y.: A Discussion on PE leveraged buyout and bond market development. International Finance Research 8 (2009)
3. Xie, J.: Contemporary investment bank. Renmin University Press, Beijing (2004)
4. Yun, X.: SPAC hedge fund abroad invest on PE of China. Discovery Value, 66–68 (2006)
5. Wang, H.: Comparison and reference about PE operational mechanism between the EU and the US. Finance Teaching and Research 6, 25–26 (2008)
6. Wei, S.: The current development situation and experience summary on PE of the EU. China cnki.net full-text database of degree dissertation (2007)
7. Wei, J.: Leading edge problems enterprises face–management blue print of modern enterprises. China Development Press, Beijing (2001)
8. Chen, F., Zhang, J.: A research based on internal evaluation system on the value of PE. Accounting Research 17 (2009)
9. Li, L., Li, B.: Theory and cases on PE fund. China Development Press, Beijing (2008)
10. Moye, J.E.: The law of business organizations. West Legal Studies 37 (1999)
11. Li, W., Long, Y.: An analysis on the real right attribute of PE fund partnership shares--challenge launched by modern financial organizition to traditional real right theory and rules. Journal of Gansu Institute of Political Science and Law 111, 15 (2010)
12. Blackett, M.: Ord, partnership, pp. 137–149. R. Butterworths (2002)
13. Xiang, H., Liu, X.: A Research on the stimulus mechanism of PE payment contract. Economy Longitude and Latitude 1, 16 (2009)

The Design and Research of SaaS-Based Financial Reimbursement System

Dacan Li[1], Yuanyuan Gong[1], and Nana Shen[2]

[1] School of Software and Microelectronics, Peking University, Beijing, China
{Pkulee,pkugong}@126.com
[2] School of Management, Hebei University of Technology, Tianjin, China
recallsnly@yahoo.com.cn

Abstract. With the development of computer software and network technology, many large enterprises have applied professional management software to realize enterprise informatization. As the frontier of informatization, enterprise finance becomes starting point. However, in SMEs (small and medium enterprises) of China, the traditional reimbursement model, a very important field of finance, has lasted for decades with little change. Therefore, how to implement network rebuilding for traditional reimbursement model of SMEs becomes focus. However, the emergence of SaaS model can decrease the threshold of finance informatization for SMEs. This paper aims at breaking the traditional reimbursement model to achieve reimbursement business network of SMEs through designing SaaS-based financial reimbursement system. In this paper, the SaaS service pattern is firstly introduced. Then, the framework and functional structure of SaaS -based financial reimbursement system are designed. Finally, the network topology structure is analyzed.

Keywords: Saas, Finance Reimbursement, System Architecture, System Design.

1 Introduction

Faced with the present global digitization and informatization, many enterprises, especially state-owned large enterprises and institutions, have been using modern information technology to transform and reconstruct traditional business process, but it is difficult for SMEs to realize informatization management by using professional management software. As the frontier of informatization, enterprise finance becomes the starting point of enterprise informatization. However, in SMEs(small and medium enterprises) of China, the traditional reimbursement business model, a very important field of finance, has lasted for decades with little change. Therefore, how to implement network rebuilding for traditional reimbursement model of SMEs becomes focus.

Financial reimbursement system, with the functions of online finance, business reimbursement, budget control, internal control process and other functional modules, is an important subsystem of enterprise informatization. Popular software architectures are B/S structure and C/S structure, but many problems will appear

J. Luo (Ed.): Affective Computing and Intelligent Interaction, AISC 137, pp. 861–868.
springerlink.com

when software architecture is designed, for instance, financial officers require that the system should have a good dynamic interaction while the staff beyond the office need to reimburse through the browser; C/S structure, with huge clients, its installation and maintenance cost is high and requirements to clients are critical while B/S structure requires no client installation; the data processing and report performance ability of C/S structure are both very strong while its development cost is high, system upgrade is not so convenient. However, the emergence of SaaS-based new software service model can exactly solve the problems inevitably appearing in C/S and B/S structure, specifically, the traditional software licensing fees are reduced or abolished, besides, application software are deployed in a unified server by vendors, so the expenses of server hardware, network security device and software upgrade and maintenance of end-users are eliminated, consequently, the customers can obtain the software and services needed via the Internet without the additional IT investments except a PC and internet connection, resulting that financial management informatizaon become more and more accessible. Therefore, the emergence of SaaS model brings hope for financial informatization of SMEs.

With the development of Internet technology and the mature of application software, SaaS (Software as a Service) rose at the beginning of twenty-first Century as an innovative software application model. SaaS-based financial reimbursement system is designed to achieve the intelligence, networking of financial management through the network, and then to construct a new financial management system in the information age. So for the majority of SMEs, SaaS is the best way to implement informatization by using advanced technology.

2 Saas Service Model Analysis

2.1 Saas Service Model Overview

SaaS (Software as a Service) is a software application model, that is, software deployed as a hosted service , so it is accessed over the Internet and can be used as long as the rental. Internet users can manage the business activities through the Internet, for the SMEs, which can not only use the services they need, but also eliminate capital investments in the system construction and maintenance, hardware purchase, etc. The emergence of SaaS technology brings many advantages, mainly in the following aspects:

(1) Fast on-line.

(2) Saving a lot of construction funds.

(3)Data security. By the unified management of data center, data can be guaranteed in many safety factors, such as the computer lab facilities, electricity, personnel and so on.

2.2 SaaS Maturity Model

In the SaaS architecture system, generally speaking, the basic requirements of software system are scalability, multi-tenant and customization through configuration. According to different support levels of these three requirements, SaaS application four-level maturity model are proposed, the difference between each lever and the former lies in whether it introduces some or all of the aforementioned three elements. The following describes them:

The first-level maturity model---custom level: In this model, each tenant has a specific running instance, that is, there is a software application system running on the hardware platform of SaaS service provider.

The second-level maturity model---configurable level: Compared with the first level, the difference is that in this level each tenant uses the same code segment while the users in the first level need to be well individually customized.

The third-level maturity model---configurable, efficient multi-user support level: In this level, service providers use the same instance to service for all tenants, so that all tenants share the running instance.

The fourth level maturity model: Compared with the third level, the load balancing strategy is added in this level.

Multi-tenant single-instance application architecture is the real SaaS application architecture, that is, the third level of maturity model, service providers use the same instance to service for all tenants, so that all tenants share the running instance, software system and operating platform, which would greatly reduce the overhead of running instances. The third-level maturity model is shown in Fig.1:

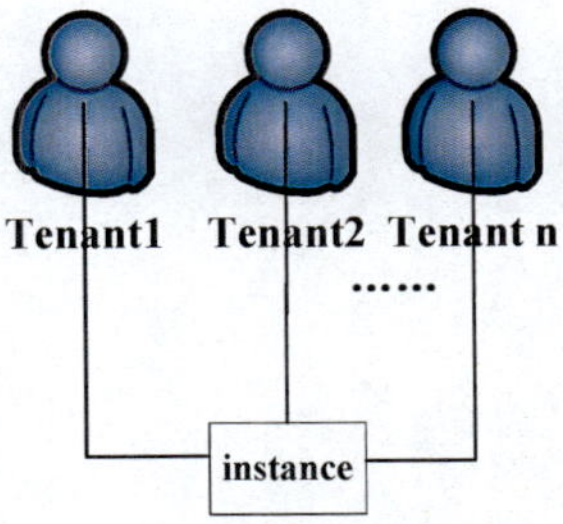

Fig. 1. The Third-level Maturity Model

3 System Design

SaaS-based financial reimbursement system mainly concerns the following concepts: tenants, traditional users, system function.

Tenants: Mainly refer to companies or units, who rent SaaS services and become users of product mean tenants.

Traditional users: After leasing the service, tenants can give the service to enterprises or employees for their own use, we call these users traditional users.

System function: By the function of the atom, financial reimbursement system is divided into many basic, independent, non-overlapping functions.

SaaS-based financial reimbursement system is divided into SaaS basic platform and financial reimbursement platform. The former conducts unified management for the whole system, mainly for software service operators. The latter primarily services for business tenants. The overall system level diagram is shown in Fig.2:

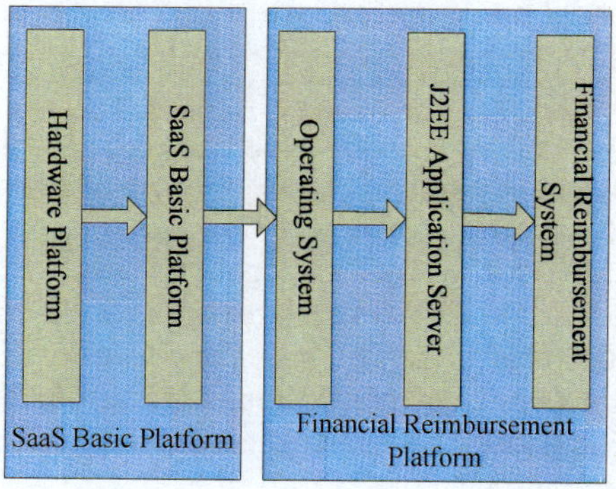

Fig. 2. The Overall System Level Diagram

3.1 SaaS Basic Platform Design

To design a better architecture, many factors need analysis, for instance, structural, maintainability, safety and so on. Because the system is SaaS-based platform financial reimbursement system, safety is significantly important, SaaS basic platform design is shown in Fig.3:

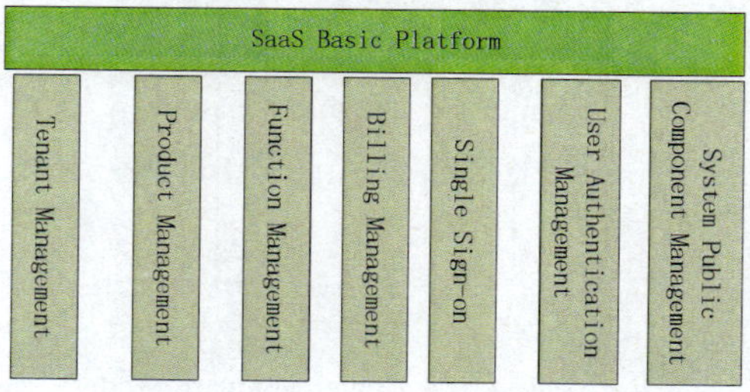

Fig. 3. SaaS Basic Platform

The SaaS basic platform provides services for software leasing personnel, which mainly provides tenant management, product management, functional management, billing management, single sign-on, user authentication management, system public assembly management and so on, including the following contents:

(1) The system administrator can view the information about tenants and traditional users, as well as delete, ban and add information of users.

(2) The system administrator can modify the sub-functions of user system, and manage the sub-functions.

(3) To enable or ban sub-functions and add subsystem.

(4) The cost of tenants and detailed running account can both be viewed.

(5) The platform also provides authentication information, which guarantees the security of users.

(5) The single sign-on of the platform pays attention to the needs of users, making the users not need to log in again when accessing to other subsystems. Besides, the platform scalability is also considered.

3.2 Financial Reimbursement Subsystem Function Structure Design

According to the system analysis, reimbursement system is designed to regulate the reimbursement business processes, improve efficiency, save staff reimbursement time, and achieve staff reimbursement networking. System integrates the modern financial management concepts and provides fast and convenient network solution for financial reimbursement.

In function, the system completes the online finance, business reimbursement, budget control, internal control process and other auxiliary functions to achieve the control of the approval process. Specifically they include the following aspects:

(1) Online Finance: creating online financial portal, achieving financial management informatizaition.

(2) Business Reimbursement: expenses reimbursement, leadership approval, financial audits, electronic payments.

(3) Budgetary Control: budget assignment, budget breakdown, real-time control, real-time analysis.

(4) Internal Control Processes: electronic processes, rights and responsibilities clear, uniform rules, electronic signatures.

(5) Other Functions: approval of commission, business application, certificate generation, payroll inquiries.

Financial reimbursement subsystem functional block diagram is shown in Fig.4:

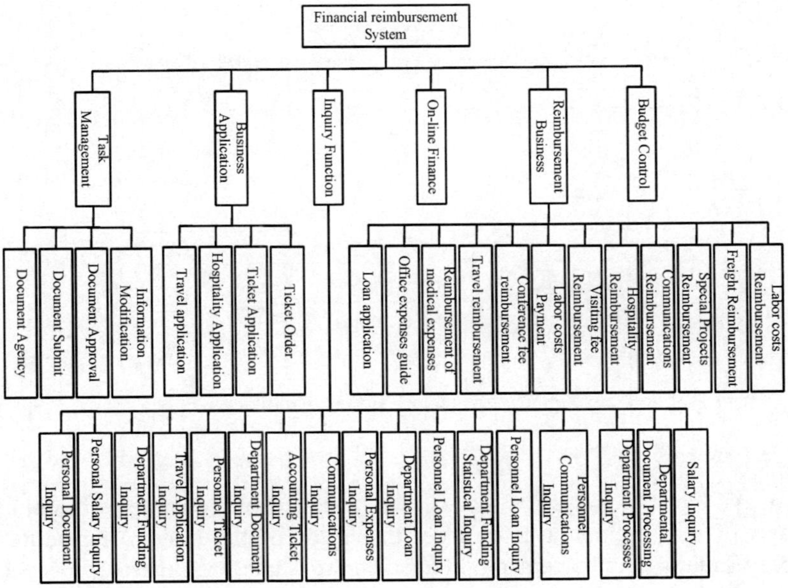

Fig. 4. Financial Reimbursement Subsystem Functional Block diagram

3.3 System Architecture Design

SaaS-based financial reimbursement system architecture provides firstly an open, soft renting service-based and then a complete technology framework of service integration, content aggregation to support a variety of business to get services and information resources required in various financial reimbursement business performance through service providers, including task management, business application, online finance, budget control and other functions. The overall architecture is shown in Fig.5, from bottom to top being four levels: the basic layer, data layer, application layer and presentation layer.

(1)Basic Layer

Basic layer is the physical basis of the entire system and provides environmental support for the operation of the system, including hardware equipment, software equipment, SaaS basic platform.

(2)Data Layer

Data layer is the core of the system, the implementation of data layer can produce various types of information and data, including relational database, encrypted file management, and other unstructured information processing.

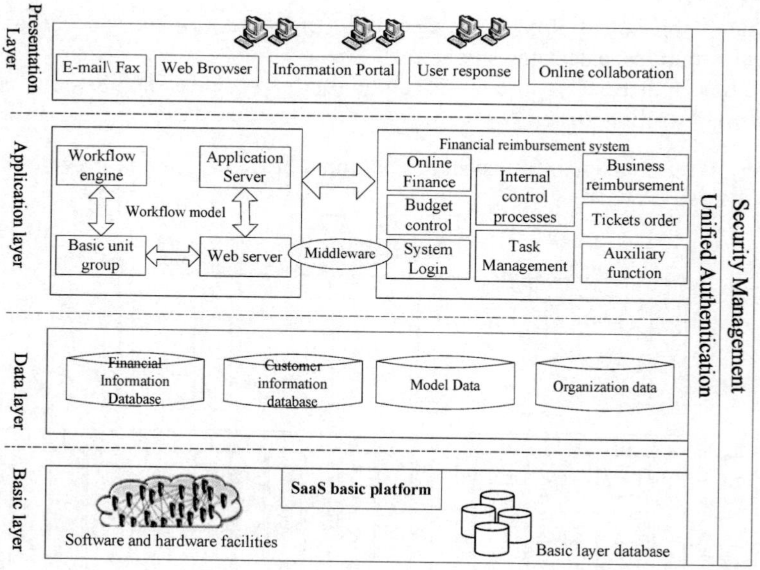

Fig. 5. The Overall System Architecture

(3)Application Layer

Business application system are various management subsystems customized on the basis of the support layer to meet the needs of the system, including online finance, business reimbursement, budget control, internal control processes, task management, etc. It also includes the internal interfaces among systems, external interfaces between business application system and the external information system.

Through these interfaces, the information integration in the departments can be achieved.

(4)Presentation Layer

Through the portal, presentation layer provides a unified login system, as well as personalized page display. Besides it also supports mobile terminal login.

3.4 The System Network Topology Design

System network topology design, that is, the design of hardware environment, includes two parts, one part is hardware environment of SaaS service providers, including hardware platform, software platform, database and application services, primarily responsible for providing services for customers, such as the identification of users, validation and delivery of the license. The other part of the hardware environment is to provide hardware support for the financial reimbursement system, that is, the financial reimbursement system generally require a certain amount of servers with network interconnection devices and clients to operate the financial reimbursement system, the security of network transmission is generally considered, so equipment is added to ensure information security in the network. Network topology is shown in Fig.6:

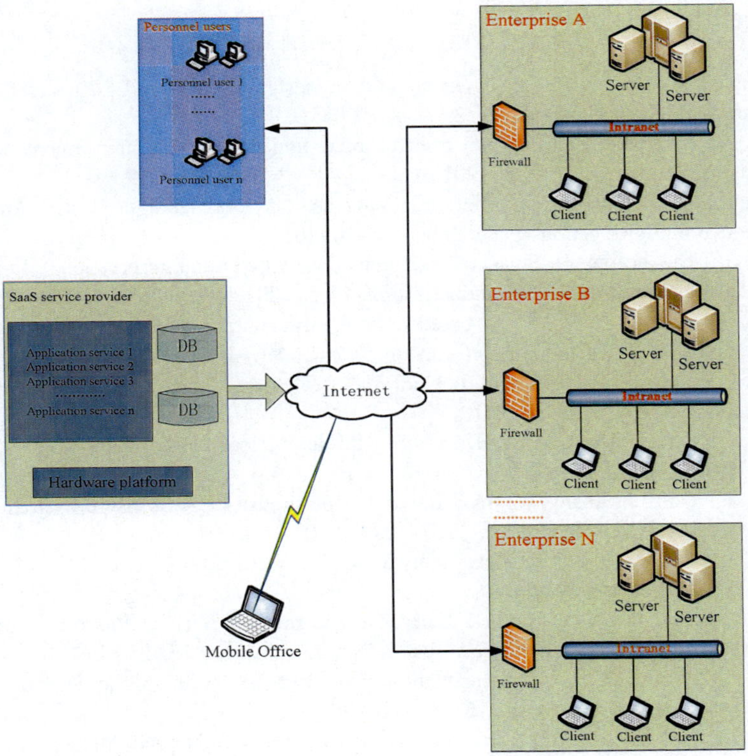

Fig. 6. System Network Topology Structure

4 Conclusion

Financial reimbursement system being the carrier, the system architecture, function modules and network topology supporting systems are all designed in SaaS model through analysis of financial reimbursement process. Overall, the SaaS model-based financial reimbursement application services can connect with its domestic and international accounts and online banking system, so that enterprises will possess more scalability in the development of financial informatization; in addition,, its flexible template and workflow architecture can adapt to different requirements of various departments, and users only need to enter the system to complete the online business, making cross-regional financial reimbursement business can be achieved; finally, the user login is controlled by physical digital authentication in the financial reimbursement system, which makes the security of the system can be fully protected.

With the feature of low maintenance cost, low investment risk, low application threshold and other advantages, SaaS has become a new trend in software industry. However, SaaS still faces many challenges in the development process, such as technical difficulties, no clear business model. Thus, with gradually mature condition, SaaS-based model of financial reimbursement application will be more extensively developed.

References

1. Cheng, Y.: Financial reimbursement system design of C/S and B/S mixed structure. Journal of Ningbo Polytechnic 12, 29–32 (2008)
2. Li, G.: Architecture design and research of online financial reimbursement information systems. Economist, 164–165 (2010)
3. Zheng, Z.: Network financial reimbursement system design. Civil Architectural Engineering Information Technology 3, 83–86 (2011)
4. Li, S.: The design of Financial expenses card and online reservation reimbursement system. Chinese Education Informatization, 49–51 (2010)
5. Ma, X., Wang, R.: Workflow-based online financial reimbursement system technology research and implementation. Information Science, 107–108 (2010)
6. Zhen, R., Lu, L.: SaaS-Based OA Platform Architecture Design and Implementation. The Computer Age (12), 19–20 (2010)
7. Zhao, Y.: SaaS-Based System Data Safety Strategy Research. Software Guide 9, 143–144 (2010)
8. He, Y., Yang, J., Zhang, C.: SaaS-Based Textile Industry OA Software System Research. Information System Engineering, 107–108 (2010)
9. Liu, J.: SaaS-Based System Data Safety Strategy Investigation. Internet Safety Technology and Application, 68–69 (2009)
10. Zhu, Z., Li, P.: A SaaS-based Enterprise Information Service Platform Architecture. Journal of Xian University of Posts and Telecommunications 15, 43–47 (2010)
11. Niu, B., Zhao, W.: A Implementation plan of SaaS-based Public Service Platform. Computer and Modernization, 157–160 (2010)
12. Xu, X.: SaaS-Based SMEs Modernization Study. Journal of Ningbo Polytechnic 14, 86–87 (2010)

Multiple Model Comparative Analysis on Spatial-Temporal Scan Statistics of Agglomeration Economies

Peian Wang and Yongping Bai

College of Geography and Environmental Science, Northwest Normal University, Lanzhou
136848090@qq.com

Abstract. Agglomeration problem has always been one of key problems and hot issues in regional economy research area, and the positioning agglomeration is the most important issue to be solved when agglomeration phenomenon is analyzed. Given that the agglomeration has a high degree of sensitivity, when using spatial autocorrelation means to detect the agglomeration, the selection for the scale is vulnerable to be influenced by the subjective judgments of the researchers, so there is a possible selection bias problem. In contrast, the scanning statistics method put forward by Kulldorff and other scholars has showed its unique advantages. The research explored the agglomeration issue of the employed population in industry field among areas in Zhejiang province from 2000 to 2009. Through theoretical and empirical analyses, differences concerned with the agglomeration issue between spatial autocorrelation and spatial-temporal scanning statistics were compared from the following three aspects: scale selection, scale transformation and spatial-temporal integration. And the research further confirmed that spatial-temporal scanning statistics has significant advantages. It is not only an effective solution to the bias problem of artificial selection, realizes the scaling and transformation, but also more effectively integrates the spatial-temporal analysis advantages such as three-dimensional, dynamic and multiple-scale characteristics.

Keywords: Scaling matrix, spatial autocorrelation, spatial-temporal proximity, scan statistics, agglomeration, Zhejiang.

1 Introduction

In recent years, a complete set of theory system relating to research on agglomeration economics is developed gradually by domestic and oversea scholars, from population agglomeration to industrial agglomeration; agglomeration theory has become the classical theory of city and regional economic development. But so far, the means and tools are able to carry on quantitative analysis is still relatively limited, the main reasons may be that to determine and scale of metric space is difficult to be unified. However, with the in-depth development of econometrics and spatial analysis technology, theory and model of quantitative analysis agglomeration gradually become richer. In addition, to adopt Gini coefficient, Ellison-Glaeser index and

J. Luo (Ed.): Affective Computing and Intelligent Interaction, AISC 137, pp. 869–879.
springerlink.com

spatial concentration ratio and other means to carry on quantitative analysis for agglomeration phenomenon[1], some foreign scholars have developed model and tool that adopt spatial autocorrelation means to carry on research on agglomeration problem, so far, this means have been used widely. The spatial autocorrelation analysis means may initially derive from biometric research [2], now has become one of important analysis means in geography. There are two types of spatial autocorrelation analysis means which includes global spatial autocorrelation and local spatial autocorrelation, among them, global spatial autocorrelation index was put forward by Moran in 1948, which inflects the similarity of spatial adjacency or adjacent area cell attribute value, but in the initial stage of Moran it cannot judge the nature of agglomeration, therefore, later Getis and Ord based on global Moran index put forward global G coefficient, with quantitative means to make judgment for types of spatial degree agglomeration. After that, based on Moran index and Geary coefficient, Ansalin has developed LISA local analysis means of spatial autocorrelation[3]. Above mentioned several means provide more positive tool for quantitative research on agglomeration phenomenon, but as one exploratory spatial data analysis model and means, it itself is still inadequacies, and there is much room for improvement.

At present, in many studies, means of spatial autocorrelation is commonly used to quantitative analysis agglomeration problem, but wide practice has proved that this kind of means have relatively obvious defects. For example, there are disputes in construct of spatial weights matrix, different researchers always have certain difference in selection of weights, and due to selection difference, and the interpretation for the analysis results is different as well. In addition, due to the basis of spatial autocorrelation modeling is entirely dependent on spatial cross-section data and then be produced, and the temporal influencing factor should not be totally considered, however, for research on agglomeration problem, spatial-temporal interaction is a essential problem, so put aside temporal factor, pure research spatial phenomenon, which is still a defect that should not be neglected. With the in-depth development of spatial analysis technology and the joint efforts of researchers, from 1995 to 2006 Kulldorff and other professors from America Harvard University combined the formers' achievement, by means of probability distribution model, put forward and developed a set means of scan statistics, this kind of means not only full brought temporal variable, also better solved scale transformation and other problems in agglomeration research. Thus, this research seeks to theory and empirical analysis, to make a comparison for these two kinds of means, and carry on deep analysis of comparison results.

2 Data Sources and Handling

The study was selected among a number of employees from every cities, towns and over its scale of Zhejiang province of industrial enterprises and total number of people of year-end to carry on related analysis, the used data for analysis mainly from 2001-2010*"Zhejiang province statistical yearbook"*.Due to analysis adopted three different kind of software, each software for type and format of the data has different requirement, so data is need to be handled. First, using ArcGIS9.3to import industrial employees from every cities, towns of Zhejiang province and population data into data sheet, to generate format file in Shp (Geoda9.5-I can be analyzed directly). And then, in order to easy to locate LISA maps in Sam4.0,on the basis of software requirement to carry on coordinate and index transformation for Shp file. When using scan statistics software Clusterseer0.2.3, to carry on line interchange for coordinate data of every cities, towns and population data, and then carry on line interchange for a number of industrial employees from 2000 to 2009, to make time, region coordinate and population data be matched year after year, region after region.

3 Means of Spatial Autocorrelation

Exploratory spatial data analysis with spatial association measures as its core, which is one kind of means to explore spatial structure, distributional pattern and others under different spatial association. Since Ciff (1981), Getis (1992), Ord (1992) and others brought means of spatial statistics into socioeconomic study area, with the development of means of measure statistics and 3S technology, means of ESDA has been widely used and developed. Spatial autocorrelation as the important analysis means of ESDA, its core idea is through construction spatial weight matrix, to carry on global or local spatial autocorrelation, and to analysis whether certain spatial cell and surrounding spatial cell in some certain attribute value has quantitative spatial connection. Moran's, Geary's C, Getis, Join count and others are all commonly used analysis means for spatial autocorrelation, the basic principles of these means are the same, but there is difference in the scope of application and emphasis. By means of analysis from Sunlin Zhang, Guozhi Hong[4], Yanguang Chen[2] and other scholars, it is considered that Moran's I is one kind of relatively typical and wider application means

3.1 Global(Local) Spatial Autocorrelation

The calculating formulas of global spatial autocorrelation Moran's I is developed gradually on the basis of the covariance of statistics related coefficient.

$$I = \frac{n}{\sum_{i=1}^{n}\sum_{j=1}^{n} W_{ij}} \times \frac{\sum_{i=1}^{n}\sum_{j=1}^{n} W_{ij}(x_i - \bar{x})(x_j - \bar{x})}{\sum_{i=1}^{n}(x_i - \bar{x})^2} . \quad (1)$$

W_{ij} is spatial weight matrix, which show changing status of spatial adjacency or adjacent area scale. Normal distribution or random distribution can be used to check statistic significance of Moran's I.

According to above formulas to calculate got Moran's I value range of coefficient between -1 and 1, when I is positive value that represents positive correlation, when I is negative value that represents negative correlation, when I is zero that represents irrelevant, and the bigger the value is, which represents stronger spatial correlation the attribute value is. In addition, Moran scatter diagram can much more visually show the spatial lag factor and data to carry on visualizing analysis. Global spatial autocorrelation supposed that spatial is coessential, that is there is only one kind of trend full of entire region [5], which only from whole to probe whether object of study has agglomeration in study region, it is difficult to carry through accurate positioning for agglomeration scope, therefore, it is necessary to use local spatial autocorrelation to confirm specific agglomeration region and agglomeration scope. Gi(d), LISA[3]and Moran scatter diagram (*Anselin, 1996*) were commonly used local spatial autocorrelation analysis index. The study will use integrate of LISA and Moran scatter diagram ways to carry through analysis.

3.2 Construction of Spatial Weight Matrix

Numerous study confirm that the occurrence of pattern and process of geography object of study, spatial-temporal distribution, intercoupling and other features all are depend on scale of existence[6]. The change of scale has great influence on study results, the different spatial scale shown different spatial information. Due to the difference of object of study and aim, and plus that scale itself has variety meaning [7]; scale matching problem becomes a relatively complex problem. The selection of scale much more depend on viewer themselves, due to the acknowledge and comprehensive difference of viewer, there is certain difference in selecting of study scale. So how to confirm a reasonable spatial scale has long been one of the most difficult and also most disputed problems in spatial statistics study.

The premise to carry through spatial autocorrelation analysis is first to definet"spatial weighted matrix", that is under what kind of spatial scale criterion to analysis the relationship among regions, the selection of spatial weighted matrix by its nature, is the quantization express of spatial scale criterion, from its further nature should be interpreted as "*spatial scale matrix*" is much more reasonable, because for a region, its neighbor may at least one or multiple, for different spatial scale, the number of neighbor is different as well, the matrix here involved in calculation that aims at adjust the scale of adjoin region, to come to next step to adjust related coefficient. In addition, besides spatial factors itself, under different temporal factors, the spatial relationship is continuously get changed as well, so it seems relatively reasonable that the concept of "*spatial scale matrix*"also should be extended for "*spatial-temporal scale matrix*".

At present, there are two kinds of commonly used means of matrix built for spatial autocorrelation analysis. One is built according to different adjoin criterion, the other is built according to different distance criterion.

4 Means of Scan Statistics

Figuratively, the process of scan statistics is just like when we use magnifier to observe an object that adopted dynamic agglomeration process. The principles of scan statistics is as same as it, the model uses roundness scan window, which continuously auto transformation the scan scale, to compare the likelihood ratio that before and after window change, and then uses probability function ratio to carry through examine, to come to the next step to find out agglomeration region of most bigger significance [8].In recent years, with the gradually development and perfection of means of scan statistics, the means of scan statistics have been widely used in medicine,ecology,geography,sociology,demography,criminology,catastrophology and other many areas[9-10]. Particularly some scholars from Swiss, Bulgaria and other countries, by means of pioneering study, has gradually extended the means of scan statistics to socioeconomic area, among them Devis Tuia1[11],Hoje Kang[12],Hanna Mach[13]and other scholars uses means of scan statistics carried through research on economic agglomeration problems, which provides favorable lesson for research on socioeconomic problems. Therefore, this study also exploratory uses this kind of means in agglomeration problems of industrial employees in Zhejiang province.

Modeling. There are three essential features should be considered when using spatial-temporal scan statistics: shape and structure of region need to be scanned; shape and size of scanned window; probability distribution type that on the condition of null assumption. In modeling window can be varied shapes, such as roundness window (Kulldorff and Nagarwalla,1995) [14],ellipse window (Kulldorff etal., 2006) [15]or random polygon window (Patil and Taillie, 2003, Duczmal etal., 2005) [16].It is the premise that selected proper window to accurately carry through spatial scan. It is suitable to use roundness scan window, the specific model[17] as follows:

Let n_z to present actual employees that scanned roundness window in Z. m_z to present total number of people in scanned window Z area, μ (z) is the expected employees in scanned window in Z that on the condition of null assumption. At the same time, let total employees in all area of G is n_G,, while total number of all area is m_G, expected employees of all area is μ (G).

$$\mu(Z)=\frac{n_G}{m_G}\times m_z \ . \tag{2}$$

$$\mu(G)=\sum\mu(Z) \ . \tag{3}$$

To further build logarithm likelihood function value of scan window.

$$L(Z)=\frac{e^{-n_G}}{n_G!}\times\left(\frac{n_z}{\mu(Z)}\right)^{n_z}\left(\frac{n_G-n_z}{\mu(G)-\mu(Z)}\right)^{n_G-n_z}\prod_{\chi_i\in Z}\mu(\chi_i) \ . \tag{4}$$

$$L_0=\frac{e^{-n_G}}{n_G!}\times\left(\frac{n_G}{\mu(G)}\right)^{n_G}\prod_{\chi_i\in Z}\mu(\chi_i) \ . \tag{5}$$

L(z) is the likelihood function value of Z in spatial-temporal window, L_0 is the likelihood function value that get on the condition of null assumption. Therefore, according to formula ③ and ④ we can get formula ⑤.

$$\frac{L(Z)}{L_0}=\frac{\left(\frac{n_z}{\mu(Z)}\right)^{n_z}\left(\frac{n_G-n_z}{\mu(G)-\mu(Z)}\right)^{n_G-n_z}}{\left(\frac{n_G}{\mu(G)}\right)^{n_G}}. \quad (6)$$

For formula ⑤ take the logarithm, we can get LLR (Log Likelikood Ratio) ⑥.

$$\mathrm{LLR}=n_z\log(\frac{n_z}{\mu(Z)})+(n_G-n_z)\log\left(\frac{n_G-n_z}{\mu(G)-\mu(z)}\right)-n_G\log(\frac{n_G}{\mu(G)}). \quad (7)$$

That the biggest LLR in Z of spatial-temporal scan window can be represented.

$$\mathrm{MaxLLR}=\frac{\max(L(Z))}{L_0}=\mathrm{Max}\left(\frac{L(Z)}{L_0}\right). \quad (8)$$

5 Empirical Analysis

5.1 Spatial Autocorrelation Analysis

Taking the case of spatial agglomeration analysis of industrial employees in Zhejiang province, Geoda0.9.5i software and Sam4.0 software were used separately, by means of selecting different spatial adjacency criterion, to carry out spatial autocorrelation analysis. As shown in Table 1, due to different adjacency criterion, which result in the Moran's I value also express tremendous difference, under several different adjacency criterion, what Minimum Spanning Tree criterion got Moran's I value is 0.681, the result is the biggest one; while Delaunay Triangulation criterion got Moran's I value is0.144, the result is the smallest one. In these several adjacency means, the result of Queen and Rook is completely same, which shows under this two kinds of means, scale change has no influence on spatial related coefficient. But the adjacency results of Delaunay,K-Nearest and Gabriel compared to other several kinds of adjacency result has great difference, which shows that adjacency criterion has great influence on spatial agglomeration scope. If combined with Moran scatter diagram, it can be found that due to difference of selection of adjacency criterion, the region of obvious agglomeration is of great difference as well. Therefore, the results got by means of adjacency criterion that lack of essential stability and reliability, when carrying through agglomeration analysis, it is required to carry through careful identify and judge according to data features.

In addition, when using global Moran's I and local LISA index to analysis spatial agglomeration, in the process of building adjacency criterion or selecting of distance scale, the influence of temporal is usually ignored, but in actual problem, the spatial and temporal as the two nature expression of inherent attribution of object of study[18], Neither of the two can be dispensed, so just from different spatial scale to analysis agglomeration is not all-conclusive, it is necessary to consider the time-space interactive of spatial-temporal, when Tobler put forward geography "first law", just considered the spatial relevance of phenomenon, he considered that relevance between closed things and far away things is much more higher, and not clearly considered the influence factor of temporal, so in this sense *"first law"* should be extended to spatial-temporal relevance will be more complete. For this, many scholars have noticed this problem, and definitely put forward related definition, concept and some exploratory quantitative analysis means, for example, academician Xiaowen Li and others put forward "*spatial-temporal proximity*" [19],Taolv, Youhui Cao put forward "*spatial-temporal contiguity*" [20] concept and so on. In addition, some scholars from China and abroad also developed corresponding analysis software used for spatial-temporal analysis, such as Satscan, Stars, Clusterseer, Changeanalys and so on, spatial-temporal analysis technology now is becoming hotspot and key point of spatial statistics study.

Table 1. Comparison for different adjacency criterion spatial autocorrelation results

Means of Adjacency	Adjacency Criterion	Moran's I	P-Value <5%	Permutation
Rook	Up Down Left Right	0.3417	0.003	999
Queen	The Up and Down or so and Diagonal	0.3417	0.002	999
K-Nearest	K-Nearest Criterion	0.2265	0.005	999
Delaunay	Delaunay Triangulation Criterion	0.144	0.004	199
Gabriel	Gabriel Criterion	0.294	0.005	199
Relative	Relative Neighbourhood	0.346	0.005	199
Min Tree	Minimum Spanning Tree Criterion	0.681	0.01	199

5.2 Scan Statistics Analysis

Using scan statistics model to carry through spatial-temporal scan for agglomeration of every cities, towns industrial employees in Zhejiang province, in order to illustrate this, at the same time using Cluserseer to carry through pure spatial scan, pure spatial scan selected the data in the year of 2009,while spatial-temporal selected data from year of 2000 to 2009. From scan results (Table2) we can see that, it is show that 2009 pure spatial scanned results have two possible agglomeration region, the first most likely cluster mainly concentrated in north area of Zhejiang province, which includes Ningbo city, Cixi county, Yuyao city, Fenghua city, Shangyu city, Shaoxing downtown and Shaoxing county and other 24 regions, the LLR is 369.18.The second most likely cluster includes Leqing city, Wenling city and Wenzhou city and other 6 regions, the LLR is 31.86.Obviously,due to pure spatial scan statistics not considered the influence of temporal factor, so the scope of this two agglomeration region are all rather bigger, seems has some nebulous. Instead, the spatial-temporal scan results of 2000 to 2009 shown, after brought temporal variable, through long-playing examine and screening, the first most likely cluster region filtered Haiyan county, Changxing county, Anji county, and Tonglu county and other 9 regions, the efficiency of exploratory improved greatly, makes the key region become more highlight and clear. Meanwhile, compared with pure spatial scan results, the LLR improved obviously 2 times, reach to 753.639. While the second most likely cluster has filtered Wenling city and Ruian city, the LLR improved obviously 2.5times as well. From above mentioned results we can see that, spatial-temporal scan not only can through spatial scan to effectively identify the agglomeration region, also can combine with temporal trends, to carry through auto identification and filter of agglomeration region, so as to obviously improved the accuracy of exploratory. From last figure it can be intuitively seen that, the change status of agglomeration region from pure spatial scan to before and after of spatial-temporal scan.

Table 2. Comparison for spatial-temporal scan statistics and pure spatial scan statistics results

Time	Classification	Obvious Region	Log Likelikood Ratio	P-Value	Conclusion Comparison
Pure Spatial Scan Statistics in 2009	First Cluster	24 Regions	369.18	0.001	LLR is relatively lower, the scope of identification is oversize, is rather nebulous
	Second Cluster	6 Regions	31.86	0.003	LLR is relatively lower, the scope is oversize.
Spatial-Temporal Scan Statistics from 2000 to 2009	First Cluster	15 Regions	753.639	0.001	LLR improved obviously, by means of filter and the scope become smaller.
	Second Cluster	4 Regions	78.91	0.001	LLR improved obviously, by means of filter and the scope become smaller.

Further to carry through comparison for results of spatial autocorrelation and scan statistics. From last figure of *figure 1* we can see that, using spatial autocorrelation Queen adjacency matrix (with representative adjacency means) to carry through ratio analysis, got the higher agglomeration regions totally concentrated in northeast regions of Zhejiang province, mainly includes Hangzhou,Shaoxing,Cixi,and Ningbo and other cities and towns, besides Huzhou, Deqing, Shangyu and other regions, it is rather close to the pure spatial results and spatial-temporal results, but with respect to scan statistics, the spatial autocorrelation exploratory of agglomeration regions without finding Wenzhou, Leqing and other cities and towns where located in southeast of Zhejiang province, because from the actual development condition, the numbers of industrial development scale and actual employees of Wenzhou, Leqing and regions is rather highlight, in whole Zhejiang province and even whole country all have rather important influence, so in this context, the exploratory of spatial autocorrelation is less integrity; keeps on using spatial autocorrelation K-nearest adjacency results to analysis, it is can be found that after adjacency scale changed,Ninghai county also be brought into agglomeration regions, but from the actual industrial development condition in the year of 2009,the value of gross output in Ninghai county raked in 33 place within the Zhejiang province, industrial employees also ranked in 25 place, so it is obvious that K-nearest results watered agglomeration regions, while at the same time, but Shaoxing county was repulsed out of agglomeration regions, it is formed "islet "in whole scope of agglomeration regions, view from the actual industrial development condition of Shaoxing county, it is obviously not accord with reality; at last, keeps on observing the results of Queen second order adjacency, can be found that Hangzhou city also be repulsed out of agglomeration regions, with this to compare with actual situation, error is self-evident. So from the analysis results of spatial autocorrelation, not only selection of adjacency criterion has great influence on analysis results, also stability and reliability of conclusion also lack of enough guarantee. By contrast, the results of spatial-temporal scan statistics are nearer to actual situation.

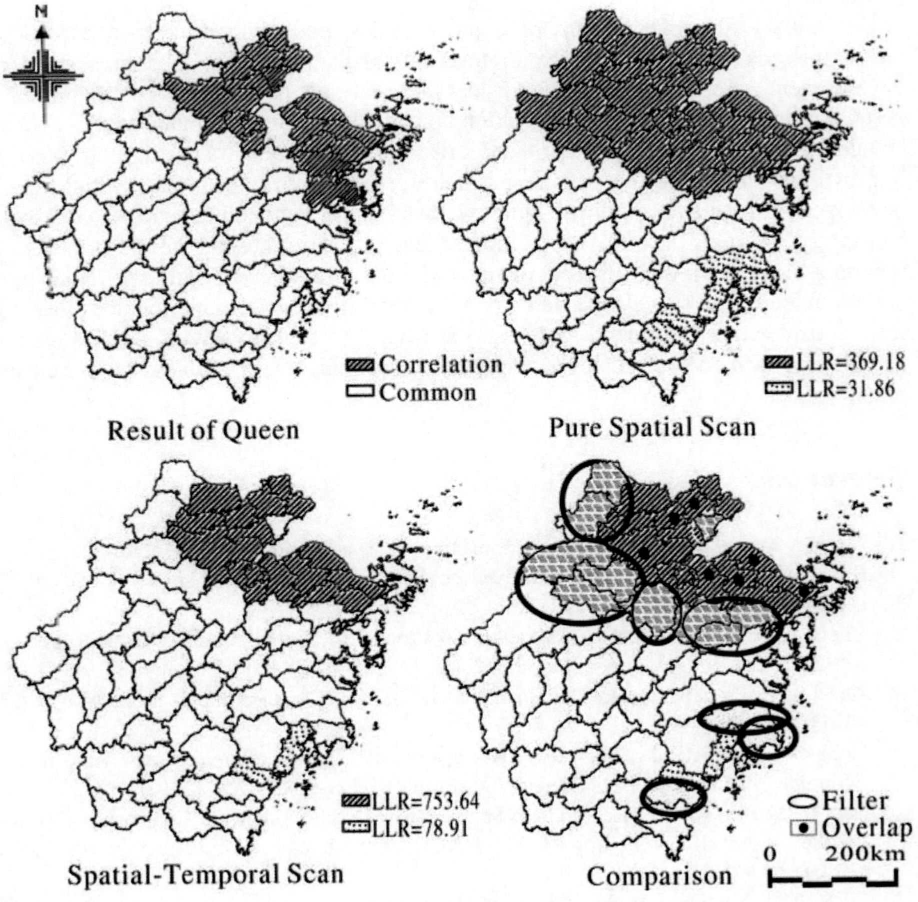

Fig. 1. Comparison of results maps for Spatial autocorrelation and spatial-temporal (spatial) scan statistics

6 Summary and Discussion

By comparing it can be seen that in agglomeration probing aspects, there are three obvious advantages for temporal-spatial scan statistics means :(1)The advantage of scale selection. By comparison with ways of spatial autocorrelation artificial selection adjoining or distance matrix, the aim of scale auto transformation is realized by the means of scan statistics, the instability factor of subjective selection scale is avoided;(2)The advantage of scale transformation. The scale selected by the spatial autocorrelation analysis is static, single, and unable to carry out multi-scale analysis, while the means of scan statistics can entirely realize the transformation from static, single scale to dynamic, three-dimensional multi-scale;(3)The advantage of space-time integration In the process of analyzing agglomeration problems, temporal factor is often ignored by spatial autocorrelation, while spatial-temporal scan statistics is fully integrated with spatial-temporal technology, good analytical results can be achieved.

Of course, although means of scan statistics have above-mentioned various advantages, as one means of exploratory spatial data analysis, it is in a stage of development, so there are still some deficiencies and limitations. For example, a certain unit of space, its shape may often be irregular, so the scanning window shape should also be varied, although Kulldorff and other scholars tried their best to improve the shape of the scanning window, the window from the initial round developed into the current ellipse window, there is still not enough space to cope with constantly changing spatial shape. Since then, scholars have put forward Flexscan scanning means[21], but this kind of means is just limited to a purely spatial scan, is not extended to the space-time domain. Moreover, the scan statistic method does not fully consider non-stationary space-time factors, therefore, follow-up studies should be more focused on such issues to improve the accuracy and usefulness of space-time scan.

References

1. Ge, Y., Yao, S., Pu, Y., et al.: Application of Spatial Autocorrelation for the Spatial Patterns of Urbanization and Localization Economy. Human Geography 83(3), 20–25 (2005)
2. Chen, Y.: Reconstructing the Mathematical Process of Spatial Autocorrelation Based on Moran'Statistics. Geographic Research 28(6), 1450–1462 (2009)
3. Anselin, L.: Local Indicators of Spatial Association-LISA. Geographical Analysis 27(2), 93–115 (1995)
4. Hong, G., Hu, H., Li, X.: Analysis of Regional Growth Convergence with Spatial Econometrics in China. Acta Geographica Sinica 65(12), 1548–1558 (2010)
5. Cao, F., Wu, J., Xu, M.: Research on Spatial Disparity of Economy at County Level Based on the Spatial Econometrics Mode. Areal Researech and Development 29(6), 23–29 (2010)
6. Li, S., Cai, Y.: Some Scaling Issues of Geography. Geographical Research 24(1), 11–18 (2005)
7. Li, X.: Scale and Economic Geography Inquiry. Economic Geography 25(4), 433–436 (2005)

8. Kulldorff: Spatial Scan Statistic. Communications in Statistics. Theory and Methods 26, 1481–1496 (1997)
9. Kim, Y., Kelly, M.: A Bootstrap Based Space-Time Surveillance Model with an Application to Crime Occurrences. Journal of Geographical Systems 10(2), 141–165 (2008)
10. Li, D., Fang, Z., Yu, Y.: Detect Investment Fund Performance Persistence with Scan Statistics. Operations Research and Management Science 15(1), 82–87 (2006)
11. Tuia, D., Kaiser, C., Da Cunha, A., Kanevski, M.: Socio-Economic Cluster Detection with Spatial Scan Statistics. Case Study: Services at Intra-Urban Scale. In: Geocomputation, National University of Ireland (2007)
12. Kang, H.: Detecting Agglomeration Processes Using Space-Time Clustering Analyses. The Annals of Regional Science (45), 291–311 (2010)
13. Maoh, H., Kanaroglou, P.: Geographic Clustering of Firms and Urban Form: a Multivariate Analysis. Journal of Geographical Systems 9(1), 29–52 (2007)
14. Kulldorff, M., Nagarwalla, N.: Spatial Disease Clusters: Detection and Inference. Statistics in Medicine 14, 799–819 (1995)
15. Kulldor, M., Huang, L., Pickle, L., et al.: An Ellipse Spatial Scan Statistic. Statistics in Medicine 25, 3929–3943 (2006)
16. Duczmal, L., Assuncao, R.: A simulated Annealing Strategy for Detection of Arbitrarily Shaped Spatial Clusters. Computational Statistics and Data Analysis 45, 269–286 (2005)
17. Yin, F., Feng, Z., Li, X., et al.: The Early Warning System Based on the Prospective Space-Time Permutation Statistic. Journal of Hygiene Research 36(4), 455–459 (2007)
18. Jiang, X., Zhou, Y.: From Space to Time A Research on Spatio-Temporal Data Model. Journal of Jilin University (Earth Science Edition) 26(3), 480–485 (2006)
19. Li, X., Cao, C., Chang, C.: The First Law of Geography and Spatial-Temporal Proximity. Nature Magazine 29(2), 69–71 (2007)
20. Lv, T., Cao, Y.: Construction of Spatial Autocorrelation Method of Spatial-Temporal Proximity and Its Application. Geographical Research 29(2), 351–360 (2010)
21. Takahashi, K., Tango, T.: A Comparison of SaTScan and FleXScan for Outbreak Detection and Monitoring. Advances in Disease Surveillance 5, 67–72 (2008)

The Theoretical Analysis on the Allocation of Urban Resources Based on Agglomeration Economies

Peian Wang and Yongping Bai

College of Geography and Environmental Science, Northwest Normal University, Lanzhou
136848090@qq.com

Abstract. The abstract should summarize the contents of the paper and should As the hot issue in the economic field, the phenomenon of agglomeration economy has always been paid much attention all the time, however, the researches from the standpoint of the allocation of resources are very limited. Therefore, in combination of the spatial-temporal property of agglomeration economies, the research adopts the quantitative analysis software Geoda 9.5 and Sam 4.0 to make multi-angular spatial autocorrelation analysis on the levels of agglomeration economies in 70 cities and counties of Zhejiang Province from 2001 to 2010. Starting from the types of the urbanization agglomeration economies and the localization agglomeration economies, the paper makes comparison on the aggregation scale and scope of the factor flow of the regional resource allocation respectively; meanwhile, the paper makes inspection on the analysis results from different perspectives through Sam 4.0 Software, and comes to the conclusions as below: The agglomeration economies have remarkable economic effect on the allocation of resources. The urbanization agglomeration economies and localization agglomeration economies have significant differentiation on the allocation of resources, and the resource allocation effect of urbanization agglomeration is significantly higher than that of the localization agglomeration economies.

Keywords: Agglomeration economies, Factor flow, Urbanization economies, Localization economies, Spatial autocorrelation, Allocation of resources.

1 The Research Content and Review

Resources are the foundation on which the human society relies for survival and development. The scarcity of resources obliges the people to seek for the optimal mode of resource allocation; therefore, the allocation of resources has always been one of the hot issues in the research field of economics. The winner of Nobel economics prize, the American economist- Koopmans consideres that the essence of economics lies in the research on the optimal utilization of the scarce resources [1]. The premise of the resource allocation lies in that the resources have the characteristics of mobility, which are specifically reflected in: on one hand, the resources can be transferred under the natural force; on the other hand, the resources can be transferred through human and biological action. The latter is the research emphasis of the resource allocation. The allocation of resources [2] can be regarded as

J. Luo (Ed.): Affective Computing and Intelligent Interaction, AISC 137, pp. 881–890.
springerlink.com

the specific reflection and evolution process of the combination relationship between the resources and other economic factors and among resources in the aspect of time structure, space structure and industrial structure, etc. The basic objective of resource allocation is to realize the optimal allocation of resources and the sustainable utilization, and the power comes from the driving mechanism that the economic entity is in pursuit of maximizing its own interests. According to the types, the allocation of resources can be divided into two types, the allocation of natural resource and social resources. No matter the natural resources or social resources, both variety and quantity are limited. Only by making reasonable distribution on the limited resources can we maximally meet the social demands, and get the optimal social and economic benefits. Therefore, the allocation of resources must follow the basic allocation principle in combination of the economic benefit, ecological benefit and social benefit. The purpose of studying the resource allocation is to explore the intrinsic mechanism of resource allocation through the scientific theory and approach, and seek the way of allocating resources reasonably, so as to realize the maximization of value enhancement when the resources participate in the production process.

At present, the correlation studies on resource allocation at home and abroad are characterized with multi-factor and multi-scale. In summary, they focus on the micro-study oriented with some kind of factors and the macro-study oriented with the regional space. From the micro-scale, lots of studies focus on many aspects including mineral resources, water resources, land resources, and human resources, etc. From the macro-scale, the research basically focuses on the allocation pattern, allocation method and spatial mode of the resources. For instance, in terms of the research on the flow of natural resources, DongYu and other scholars draw lessons from "the thinking field" in physics to explore the essence of mechanics for the resources flow through the theory of resource field, and point out that the resource potential difference is the intrinsic nature of the resource flow, and the resultant force of resource field force and the external force is the direct power of the resources flow [3]; Cheng Sheng kui and other scholars expound the practical significance and scientific significance of the research on the flow of natural resources, and summarize some key problems in the field of the research on the flow of natural resources[4]; Bai Yong Xiu puts forward the theoretical framework of urbanization resource allocation through the innovation-oriented urban construction and the case analysis, and the make systematic research on structural optimization and function optimization of the central cities in the urbanization resource allocation of the western regions[5]. Yan Jun yin and other scholars establish multi-objective standards of the optimal allocation of resources bases on the finiteness and non-renewability of the regional resources, and establish the evaluation index system of the reasonable development scale and development speed of the regional resources, and make detailed empirical analysis on the reasonable development of mineral resources in Hebei Province[6]; Liu Yu hui and others study the rationality of the spatial allocation of the resources and put forward the ideal model of spatial allocation of the resources [7]; Zhang Yao Jun and others analyze the development path of the resource-based cities under the background of globalization, and make in-depth discussion on the sustainable development of the resource-based cities[8]; Dong Suo Cheng, Shi Guang yi and others systematically summarize the research theories of resource allocation at home and abroad in recent years, and reaffirm the significance of strengthening the research on the resources economies and world resources [9], and put forward that the resources researches play a significant role in guaranteeing the national resource security, promoting the sustainable utilization of resources, accelerating the transformation of the economic growth mode and constructing the resource-conserving society.

Throughout the above-mentioned researches, the researches on resource allocation from the urban scale are still limited. However, the city is the forefront of the economic activities. In this forefront, from industry cluster to agglomeration economies, the economic phenomenon with agglomeration feature is becoming more and more distinguishing feature for urban development. However, the agglomeration economy is always accompanied by the agglomeration and diffusion of various resources, and has significant driving effect on the allocation of resources. Therefore, the perspective of agglomeration economies plays a significant role in studying the allocation of resources. The agglomeration economies can be divided into urbanization economics and localization economics. The concept of agglomeration economies follows Marshall's concept concerning the internal economics and external economics. However, in 1937, Hu Fu explicitly put forward the localization economics (*Lin Jin Zhong and other scholars translate it as "the localized economics"*) in his book of Location Theory and Shoe & Leather Industry [10]. Essentially, the agglomeration economy is realized by scale economy and scope economy. Dividing the agglomeration economies into the localization economics (or specialization economics) and urbanization economics is mainly to stress on the specialization and diversification of agglomeration. Localization economics emphasizes the facilitation effect on the production efficiency caused by the industry diversity of regional economy system. If the urbanization economics is dominant in this region, the more the size of urban population has, the completer the industrial categories becomes, and the more efficient the allocation of resources has, then the faster the regional economics increases. Conversely, if the localization economics is dominant in this region, then the homogeneity of the industrial specialization degree, the spatial proximity and the resource allocation will determine the development speed of economic development in this region, and the choice on the types of agglomeration economies is one of the important factors to be considered for the regional economic policy [11]. From the micro scale, the industry cluster plays a direct driving role in the formation of agglomeration economies, and the important component of the agglomeration economies. The industry cluster can be regarded as the spatial agglomeration of a lot of closely-related enterprises and the related supporting institutions and the phenomenon of the powerful and sustainably competitive advantages in a particular field (usually centering on a leading industry) [12]. The fundamental power of the formation of agglomeration economies comes from the pursuit of economic entity for maximizing its own interests and minimizing the cost. All other forms of activities focus on this core, including the price, market, location and other factors, etc. Agglomeration economy realizes the low-cost operation and high-yield development of economy through the flow of various resources, which is the essence of agglomeration allocation of all resources. The agglomeration economy is just generated based on this purpose. The external effect of agglomeration economies shows that the agglomeration can realize the advantages of the complementary resources, industry convergence, and division of labor based on specialization and technical exchange, etc. Therefore, the agglomeration not only can obtain the influence and promotion of specialization in the same industry within the agglomeration area, but also can win the support of diversified development. No matter from the resource acquisition, logistics support, information exchange, or talent cultivation, it will have a very favorable effect. The formation of agglomeration economies is the three-dimensional, professional and diversified "*super-large factory*", it breaks through the adverse pattern of the economic entity which developed independently in the past, and blends into a big economic circle where the opportunity and competition are close to reach. In this "*super-large factory*", all kinds of resources can obtain the most efficient allocation through the free flow in the agglomeration-diffusion way of the factors.

2 Data Sources and Processing

Kindly The measuring index of urbanization agglomeration economy indicates the ratio of the industrial working population in some city(county)to the total population in this city(county). The measuring index of localization agglomeration economy indicates the ratio of the industrial working population in some city(county)to the social working population in this city(county)[13]. The research will choose the relevant data of 70 cities (counties) in Zhejiang Province from 2001 to 2010 and make analysis. The research adopts three different kinds of analysis softwares and each software has different requirements on data type and format. Therefore, we need to make relevant data processing. Firstly, by virtue of ArcGIS9.3, we import the annual data of industrial working population in 70 cities(counties) in Zhejiang Province from 2001 to 2010, the population data of the that year and the working population data of the whole society into the excel sheet, and generate the shp formatted file ((Geoda0.9.5i can make direct analysis). Then, for the convenience of making positioning on LISA maps in the Sam 4.0, we make coordinate and data-transpose processing on shp formatted file according to the software requirements, so as to make the time and area coordinate match with the number of employees year by year and region by region.

3 Means of Spatial Autocorrelation

Exploratory spatial data analysis with spatial association measures as its core, which is one kind of means to explore spatial structure, distributional pattern and others under different spatial association. Since Ciff(1981), Getis(1992), Ord(1992)and others brought means of spatial statistics into socioeconomic study area, with the development of means of measure statistics and 3S technology, means of ESDA has been widely used and developed. Spatial autocorrelation as the important analysis means of ESDA, its core idea is through construction spatial weight matrix, to carry on global or local spatial autocorrelation, and to analysis whether certain spatial cell and surrounding spatial cell in some certain attribute value has quantitative spatial connection. Moran's, Geary's C, Getis, Join count and others are all commonly used analysis means for spatial autocorrelation, the basic principles of these means are the same, but there is difference in the scope of application and emphasis. By means of analysis from Sunlin Zhang, Yanguang Chen[14] , Guozhi Hong[15] and other scholars, it is considered that Moran's I is one kind of relatively typical and wider application means.

3.1 Global (Local) Spatial Autocorrelation

The calculating formulas of global spatial autocorrelation Moran's I is developed gradually on the basis of the covariance of statistics related coefficient.

$$I = \frac{n}{\sum_{i=1}^{n}\sum_{j=1}^{n} W_{ij}} \times \frac{\sum_{i=1}^{n}\sum_{j=1}^{n} W_{ij}(x_i - \bar{x})(x_j - \bar{x})}{\sum_{i=1}^{n}(x_i - \bar{x})^2} . \quad (1)$$

Among them, Wij is spatial weight matrix, which show changing status of spatial adjacency or adjacent area scale. Normal distribution (use N to present) or random distribution (use R to present) can be used to check statistic significance of Moran's I.

(1) On the condition of normality assumption, the expected value E_N (I)and variance VAR_N(I) of Moran's I respectively are:

$$E_N(I) = \frac{-1}{(n-1)} \quad VAR_N(I) = \frac{n^2 S_1 - nS_2 + 2S_0}{S_0(n^2-1)} - E_N(I)^2 . \quad (2)$$

(2) On the condition of random assumption, the expected value E_R(I) and variance VAR_R(I)of Moran's I respectively are:

$$E_R(I) = -\frac{1}{(n-1)}$$

$$VAR_R(I) = \frac{\{n[(n^2-3n+3)S_1 - nS_2 + 3S_0^2]\} - \{k[(n^2-n)S_1 - 2nS_2 + 6S_0^2]\}}{(n-1)(n-2)(n-3)S_0^2} - E_R(I)^2 . \quad (3)$$

According to above formulas to calculate got Moran's I value range of coefficient between -1 and 1, when I is positive value that represents positive correlation, when I is negative value that represents negative correlation, when I is zero that represents irrelevant, and the bigger the value is, which represents stronger spatial correlation the attribute value is. In addition, Moran scatter diagram can much more visually show the spatial lag factor and data to carry on visualizing analysis. Global spatial autocorrelation supposed that spatial is coessential, that is there is only one kind of trend full of entire region [16], which only from whole to probe whether object of study has agglomeration in study region, it is difficult to carry through accurate positioning for agglomeration scope, therefore, it is necessary to use local spatial autocorrelation to confirm specific agglomeration region and agglomeration scope. Gi(d), LISA[17]and Moran scatter diagram (*Anselin, 1996*) were commonly used local spatial autocorrelation analysis index. The study will use integrate of LISA and Moran scatter diagram ways to carry through analysis.

3.2 Construction of Spatial Weight Matrix

Numerous study confirm that the occurrence of pattern and process of geography object of study, spatial-temporal distribution, intercoupling and other features all are depend on scale of existence[18]. The change of scale has great influence on study results, the different spatial scale shown different spatial information. Due to the difference of object of study and aim, and plus that scale itself has variety meaning [19]; scale matching problem becomes a relatively complex problem. The selection of scale much more depend on viewer themselves, due to the acknowledge and comprehensive difference of viewer, there is certain difference in selecting of study scale. So how to confirm a reasonable spatial scale has long been one of the most difficult and also most disputed problems in spatial statistics study.

The premise to carry through spatial autocorrelation analysis is first to definet"spatial weighted matrix", that is under what kind of spatial scale criterion to analysis the relationship among regions, the selection of spatial weighted matrix by its nature, is the quantization express of spatial scale criterion, from its further nature should be interpreted as "*spatial scale matrix*" is much more reasonable, because for a region, its neighbor may at least one or multiple, for different spatial scale, the number of neighbor is different as well, the matrix here involved in calculation that aims at adjust the scale of adjoin region, to come to next step to adjust related coefficient. In addition, besides spatial factors itself, under different temporal factors, the spatial relationship is continuously get changed as well, so it seems relatively reasonable that the concept of "*spatial scale matrix*"also should be extended for "*spatial-temporal scale matrix*".

At present, there are two kinds of commonly used means of matrix built for spatial autocorrelation analysis. One is built according to different adjoin criterion, the other is built according to different distance criterion.

In the spatial autocorrelation analysis, the research will adopt the commonly-used Queen (up, down, left and right and diagonal) adjacent matrix in Geoda software and the multi-scale value of distance transformation in Sam 4.0 software to make correlation analysis.

4 Empirical Analysis

The fundamental approach of resource allocation is realized by the factor flow, which is the dynamic expression of resource allocation, therefore, no matter the urbanization agglomeration economy, or localization agglomeration economy, their effect on the allocation of resources can be reflected by the agglomeration-diffusion effect of the factor flows of the urban core resources. In a broad sense, the factor flows of the urban core resources mainly include the population flow, materials flow, capital flow, technology flow, information flow and energy flow. The factor flows are the main components of the urban resource system, and the forms of resource distribution is realized by the movement of these factors. In order to compare the resource allocation ability of urbanization agglomeration economy and localization agglomeration economy, we choose the volume of passenger traffic, the volume of freight, the balance of loans to financial institutions, the number of patent granting, the telecom income and the industrial power consumption of each city (county) in Zhejiang Province in 2010 as the characteristic index of above-mentioned factor flows. All these indexes can reflect the allocation of different factor flows objectively, and make analysis on their agglomeration scale and agglomeration scope through the spatial autocorrelation method accordingly. Through comparison, it can be seen from Table 1, both urbanization agglomeration economy and localization agglomeration economy have significant effect on the allocation of resources, and the factor flow of each resource, the values of urbanization agglomeration economy and localization agglomeration economy and F value in the table have the significant correlations. Especially, the correlation between the urbanization economy and industrial power consumption index is the most significant. In addition, the agglomeration-diffusion intensity and scope of different economic factors are also different; moreover, both urbanization agglomeration economy and localization agglomeration economy have significant difference in the allocation of factor flows. Meanwhile, in several factor flows, expect for the information flow index, the values of other factor flows and urbanization economy and F value are all significantly higher than that of the localization economy. Therefore, we can prove that the resource allocation ability of urbanization agglomeration economy is significantly higher than the localization agglomeration economy in these two economic types.

Table 1. The correlation analysis on the resource agglomeration scale of urbanization and localization agglomeration economies

The Flows of Resources	Characteristic Index	Urbanization Economies		Localization Economies		P-Value < 5%
		Pearson's r	Pearson's r F	Pearson's r	Pearson's r F	
Population Flow	Volume of Passenger Traffic	0.536	27.872	0.478	20.426	< 0.001
Materials Flow	Volume of Freight	0.456	18.066	0.405	13.557	0.001
Capital Flow	Loans to Financial Institutions	0.418	14.641	0.346	9.389	0.002
Technology Flow	The Number of Patent Granting	0.542	28.713	0.487	21.424	< 0.001
Information Flow	Telecom Business Income	0.522	25.89	0.522	25.787	0.005
Energy Flow	Industrial Power Consumption	0.61	40.989	0.543	28.866	< 0.001

In order to further analyze resource allocation performance of the two agglomeration economies, Sam 4.0 Analysis Software was adopted to compare the agglomeration intensity and range of each element flow from different dimensionalities. During this process, these agglomeration intensities and ranges will also be observed. By analyzing Result Table 2, we may find that, the agglomeration intensities and ranges of each element are quite different, among which, the agglomeration and diffusing effect of industrial electricity consumption of cities and counties in Zhejiang Province is the most remarkable, with Moran's I value reaching to 0.3224. Comparatively, the Moran's I value of financial organization's loan balance is merely 0.0816. Moreover, the spatial autocorrelation of cargo transportation volume is also quite weak, only 0.1201. The agglomeration intensities and ranges of rest elements are quite close. However, when observing resource agglomeration range with a dynamic dimensionality, from multiple spatial dimensionalities, and based on LISA Maps (omit), it can be found that, regions with obvious agglomeration nature are still Ningbo, Cixi, Yuyao and other several cities. These regions are all overlaid areas with obvious effect from urbanization economics and localization economics. Thus, it can be demonstrated that, the complex agglomeration economies formed by mutual influence of the aforementioned two agglomeration economies will show a most remarkable effect on resource allocation.

Table 2. Agglomeration Scale of Resource Element Flow

The Flows of Resources	Characteristic Index	Moran's I	Distance Scale	DistCntr	P< 5%	Regio -ns
Population Flow	Volume of Passenger Traffic	0.2105	8.86-64.08	36.831	0.015	13 Cities
Materials Flow	Volume of Freight	0.1201	140.63-159.72	150.175	0.002	6 Cities
Capital Flow	Loans to Financial Institutions	0.0816	151.75-160.51	156.131	0.004	7 Cities
Technology Flow	The Number of Patent Granting	0.2551	7.96-68.03	37.991	0.04	6 Cities
Information Flow	Telecom Business Income	0.2652	8.86-47.68	28.272	0.01	5 Cities
Energy Flow	Industrial Power Consumption	0.3224	8.86-64.08	36.831	0.001	12 Cities

5 Conclusion and Discussion

Based on above analysis, we may come to following conclusions: ① Agglomeration economies have remarkable economic effect on resource allocation. Both urbanization economics and localization economics have obvious influence on resource allocation; ② Different economic elements have different intensity and range in agglomeration and diffuseness. In the aspect of resource allocation, agglomeration economies have obvious space and time feature, with their allocation complying with the law of "space-time adjacency"; ③ Urbanization economics and localization economics have obvious difference in resource allocation. The effect of urbanization economics in resource allocation is stronger than that of localization economics; ④ Complex economy mode formed by the interaction between urbanization economics and localization economics can fully realized optimized resource allocation. Regions of this type can bring into play the advantage of professional economy, as well as promote the development of diversified economy; ⑤ Agglomeration economies show obvious space and time characteristics in resource allocation, and have significant practical function in regulating and controlling the type and scale of agglomeration economies. This is concerned with reasonable utilization of resources and regional economic development, which will as well cause a great influence on the development direction of the entire region.

References

1. Chen, J., Huang, M.: Analysis on the Recession of Corporeal Economy and the Rising of Acknowledge from the Dimensionality of Resource Economics. World Economic Papers 4, 56–58 (1999)
2. Liu, C., Chen, X., Liu, C., et al.: Review of Foreign River Basin Water Resources Allocation Models. Journal of Hehai University 33(5), 522–525 (2005)
3. Dong, Y., Xie, G.: Pilot Study of Resources Field Theory and Its Application to Resources Flow. Scientia Geographica Sinica 21(5), 407–411 (2001)
4. Cheng, S., Shen, L., Min, Q.: New Perspective of Resources Science Research-Process and Effect of Natural Resources Flow. Resources Science 28(2), 199–200 (2006)
5. Bai, Y.: Research on Resource Allocation System for Western Urbanization. Press of China's Economy, Shan Xi (2011)
6. Yan, J., Zhao, G., Sun, W.: Research on Regional Mineral Resource Allocation Based on Sustainable Development. Ecological Economics 5, 133–139 (2006)
7. Liu, Y., Guo, X., Xu, G.: The Research of Resource Space Allocation of Wuhan-Ezhou-Huanggang Urban Development Zone in Wuhan Metropolis. Urban Planning Forum 185(7), 17–22 (2009)
8. Zhang, Y., Cheng, S., Min, Q.: A Research on Sustainabled Evelopment of Resource Based Cities Underglobalization. Progess in Geography 21(3), 249–258 (2002)
9. Dong, S., Shi, G., Shen, L.: Progress and Its Prospects of Resource Economic and World Resources in China. Journal of Natural Resources 25(9), 1432–1445 (2010)
10. Lin, J.: Urban Agglomeration Economy Theories and Evolution. Economic Review 4, 92–98 (2007)
11. Ge, Y., Yao, S., Pu, Y., et al.: Application of Spatial Autocorrelation for the Spatial Patterns of Urbanization and Localization Economy. Human Geography 83(3), 20–25 (2005)
12. Porter, M.E.: Clusters and the New Economics of Competition. Harvard Business Review 11, 30–35 (1998)
13. Moomaw, R.L.: Agglomeration Economics: Localization or Urbanization. Urban Studies 25, 150–161 (1998)
14. Chen, Y.: Reconstructing the Mathematical Process of Spatial Autocorrelation Based on Moran's Statistics. Geographic Research 28(6), 1450–1462 (2009)
15. Hong, G., Hu, H., Li, X.: Analysis of Regional Growth Convergence with Spatial Econometrics in China. Acta Geographica Sinica 65(12), 1548–1558 (2010)

16. Cao, F., Wu, J., Xu, M.: Research on Spatial Disparity of Economy at County Level Based on the Spatial Econometrics Mode. Areal Researech and Development 29(6), 23–29 (2010)
17. Anselin, L.: Local Indicators of Spatial Association-LISA. Geographical Analysis 27(2), 93–115 (1995)
18. Li, S., Cai, Y.: Some Scaling Issues of Geography. Geographical Research 24(1), 11–18 (2005)
19. Li, X.: Scale and Economic Geography Inquiry. Economic Geography 25(4), 433–436 (2005)

Evaluating the Demand of Investment and Financing in Rural Infrastructure Construction in Jiangsu Province

Yang Guo and Ying Shen*

Department of Tourism, Fudan University, Shanghai 200433, China
ying_shen@fudan.edu.cn, fdgy@fudan.edu.cn

Abstract. Rural infrastructure construction is significant for promoting the coordinated development of urban and rural areas. How to realize the goal and task of rural infrastructure construction in Jiangsu Province is an urgent issue demanding a solution. This thesis starts from introducing the evaluation model and operation mechanism of investment and financing system, and then evaluates the investment in improving the rural infrastructure construction in Jiangsu Province, and discusses the effective methods to accelerate the synthetic infrastructure construction of the rural areas in Jiangsu Province.

Keywords: Rural infrastructure, investment and financing demand, model evaluation.

1 Introduction

China is intent on constructing a new rural system. The key to this project is to formulate new guideline for coordinating development between rural and urban district, to which infrastructure construction is an indispensable step. In order to actualize the overall target to "realize a well-off life province-wide and construct a new Jiangsu out of the old one", during the process of coordination between rural and urban development, efforts should be made to extend infrastructure construction from urban district to rural district, so as to equalize national public service. What we mean by coordinating infrastructure construction between rural and urban district is essentially an extension from urban to rural district and a substantial step towards integration of infrastructure construction. To put it in a more specific way, there are following tasks to do: coordination in water supply, drainage, gas supply, electricity supply, telecommunication, sewage and garbage treatment and flood control etc. The following two steps should be taken to guarantee drinking water security in rural areas. First, drinking water security projects should be carried out in a more efficient way. Second, dredging projects should be carried out among small and mid-volume rivers as well as the lake ecospheres. Public transportation network and logistics network should be built to connect rural areas to urban areas. Sanitation projects should be carried out to improve rural sanitation condition and hence its habitability. What's more, to equalize public service between urban and rural district is also of great importance. Security mechanism should be built and improved with a tilted focus toward rural district,

* Corresponding author.

J. Luo (Ed.): Affective Computing and Intelligent Interaction, AISC 137, pp. 891–901.
springerlink.com

covering education, medical service, cultural service, employment and social welfare development. Public resources should be reallocated between urban and rural district to improve financial security in rural district, which is expected to support public service construction. Efforts should be targeted to equalize public service between rural and urban district.

Jiangsu Province is an important battlefield for urban-rural integration in Yangtze Delta. Overall target and component tasks are all important issues waiting for immediate research. Fiscal and financial channels are two indispensable tracks to successful rural reformation. However, in the past, fiscal and financial channels are working on their own respectively to support rural development. As a result, fiscal support could not bring out the leverage effect it is supposed to exert and rural financing is so stagnant that it can not effectively satisfy basic needs of rural infrastructure construction. In this research, we would try a perspective of investment-financing system model evaluation and operation mechanism, packaging fiscal and financial tracks together, to discuss how to effectively accelerate infrastructure construction in rural areas of Jiangsu Province. Investment-financing framework for rural infrastructure construction is composed of two parts: first of all, government is the primary investor. The optimization and widening up of investment channels should embrace the following tasks: to set up a provincial-level infrastructure fund, to increase on fiscal income and to mobilize principal assets via rural land reform. Second, multi-channel investment including commercial bank loans, corporate investment and governmental investment should be joined together. Besides, government should establish a support system, composed of planning control, industrial policies and fiscal policies, to aid this investment-financing system for rural infrastructure construction.

2 The Ultimate Crux in Rural Infrastructure Construction and Status Quo of Fiscal Investment

There have been marked achievements in rural infrastructure construction as well as rural production and living conditions in Jiangsu Province. However, generally speaking, rural infrastructure lags behind, unable to meet the demand of agricultural development, countryside progress and farmer income increase, indexes of a new rural structure. Aspirations to make a fundamental change to the present situation of the rural district, to lessen urban-rural gap, to coordinate urban-rural development and after all to formulate a new paradigm which integrates urban-rural economic development could only be actualized by enduring efforts.

Issues concerning agriculture, farmer and rural areas are serious impediment against efforts to balance regional development and such impediment is a nation-wide problem. First of all, low-level infrastructure has become a bottleneck of agricultural development and of increase in farmer's income. This problem is especially poignant concerning irrigation infrastructure. Nation-wide, there are more than 900 million acres of farmlands, almost half of the whole nation's farmlands, which are in want of irrigation. The effective utilization coefficient of irrigation water (signified by η, referring to a result from actual effectively utilized irrigation volume, counting out vertical leakage and horizontal loss, on a certain area of irrigated land and during a certain irrigation season, divided by overall irrigation volume at the ditch entrance; this coefficient is composed of two parts: canal-system water utilization coefficient and field water utilization coefficient) turns out only 0.46, 60% of its international counterpart. Irrigation quota is generally high and probability of irrigation is low. The average probability of irrigation of reservoir-based water source in main mid-sized irrigation district turns out to be 56%. More than 300 million acres of farmlands suffer from aridity on an annual base. Irrigation districts want an extra supply of 30 billion m^3 water during years of a medium aridity. Yield loss caused by aridity amounts to tens and billions of kilograms annually. Besides, farming and living conditions are really poor in rural areas. Drinking water supply, transportation and environmental protection jeopardize everyday life in some areas. Measured by drinking water safety indexes such as quality, quantity, accessibility and supply guarantee, 170 million rural population of the nation is endangered by insufficient drinking water safety. Concerning traffic, till the end of 2009, 92.7% rural areas in the nation had their roads cemented, 3.4% lagging behind the desired target of "11th five-year plan". At the same time, farmers in most areas resort to plant stalks, such as straws, and timbers as major fuel. Clean energy such as marsh gas, natural gas and solar energy accounts for only a small percentage. This impedes improvement in rural living condition and construction in rural eco-environment. Finally, social security development in rural district seriously lags behind that in urban district. There in rural districts, middle school and primary school education facilities are deficient. Medical service facilities are deficient. Reproduction control service lags behind demand. Public cultural facilities and public physical education facilities are deficient. All these problems are looking for fundamental solution. Resources of public service facilities in rural areas are relatively deficient and hence low per capita share. Chances for rural residents to access to equal public service, such as quality education, medical service, culture and entertainment, physical training and environmental protection, are much lower than those for urban residents.

Let's hark back to rural infrastructure construction of Jiangsu Province. The provincial government has been increasing investment on "agriculture, farmer and rural areas", by providing medical support to farmers, carrying out integrated agricultural development and reconstructing road etc. Yet fiscal resource only cannot meet all the needs of integrated infrastructure development in the rural district, which, to some degree, impedes agricultural and industrial development and upgrading. At present, economic development in Jiangsu Province displays regional imbalance. Rural districts in northern Jiangsu and mid Jiangsu are underdeveloped, lagging behind the overall economic development target of the province. Public services are allocated in an unbalanced pattern between underdeveloped rural district and well-developed urban district.

Concerning the fiscal capital input in rural infrastructure construction, the agriculture-targeted fiscal capital (approximated by expenditure on agriculture, forestry, water supply and gas supply) displays a rising trend. Yet from its percentage of overall fiscal expenditure, we could see fiscal support for basic facilities construction in rural areas is really limited. At the same time, fiscal input for agriculture, rural areas and farmers from provincial government of Jiangsu is rising in amount. Yet compared with its contribution to GDP, agriculture gets disproportionately negligible fiscal support. As a matter of fact, province-wide, compared with its contribution to GDP, agriculture decreases in fiscal support year by year. Developed countries allocate an agriculture-oriented fiscal expenditure averaged at 10%, which is much higher than that of Jiangsu Province.

Table 1. Fiscal subsidy to agriculture and its percentage of fiscal budget (unit: ¥100 million)

Year	2000	2001	2002	2003	2004	2005
Subsidy to agriculture	37.84	43.30	51.14	65.93	75.68	97.57
Percentage	6.40%	5.93%	5.94%	6.29%	5.77%	5.83%

Table 2. Percentage of Gross Agricultural Product to GDP and Percentage of Agriculture expenditure to Overall Fiscal Expenditure

Year	2000	2001	2002	2003	2004	2005
Percentage of Gross Agricultural Product to GDP	11.97%	11.44%	10.53%	8.89%	8.48%	7.01%
Percentage of Agriculture expenditure to Overall Fiscal Expenditure	5.86%	5.53%	4.87%	4.70%	4.35%	3.95%

Data source: Panel *Put Fiscal-financing role into full play to support new countryside construction in Jiangsu Province*, Nanjing Branch of People's Bank

3 Model Evaluation of Investment-Financing Demand for Rural Infrastructure

B-J model is a prediction methodology based on random timing. This model analyzes the stability of time series. Time-reliant variables cluster together, though a single variable composing the time series display strong uncertainty. The modeled time series is produced by stationary random process with zero-mean. The randomness of process is invariable on the time scale. Mathematics model could approximately describe regularity of series change; all sample points fluctuate in a random pattern.

This paper sets several years' infrastructure investment series into the model and evaluates investment scale of infrastructure. Then reference is made from ratio of rural fixed asset investment to overall fixed asset investment. Demand for rural infrastructure investment is thus calculated. Data for this model comes from overall fixed asset investment of Jiangsu Province from 1981 to 2007.

Table 3. Overall Fixed Asset Investment (unit: ¥100 million)

Year	1981	1982	1983	1984	1985	1986	1987	1988	1989
Investment	60.5	76.21	105.27	130.98	191.93	241.23	317.12	371.87	320.23
Year	1990	1991	1992	1993	1994	1995	1996	1997	1998
Investment	356.3	439.98	711.7	1144.2	1331.13	1680.17	1949.53	2203.09	2535.5
Year	1999	2000	2001	2002	2003	2004	2005	2006	2007
Investment	2742.65	2995.43	3302.96	3849.24	5335.8	6827.59	8739.71	10071.42	12268.07

Data source: Jiangsu Statistics 2008

From stability test of all data in table 3, time curve y_t is recorded in figure 1.

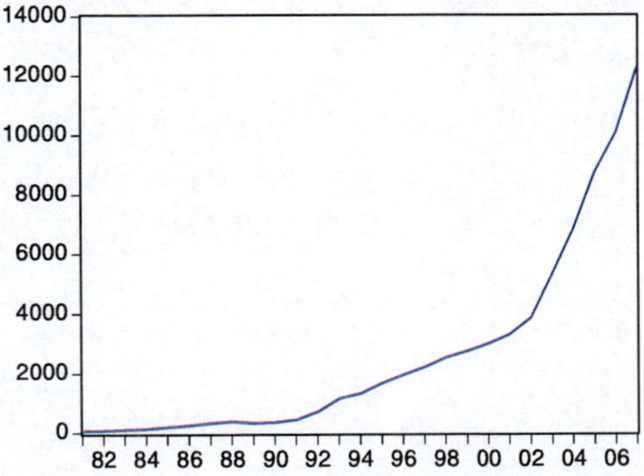

Fig. 1. Overall fixed investment (unit: ¥ 100 million)

Statistics software is run to make correlation analysis on overall fixed investment, producing data of both autocorrelation and partial correlation. From figure 2, we could see autocorrelation displays property of tailing attenuation, while partial correlation comes to a close after stage 1, from which we judge stage 1 is auto-regression.

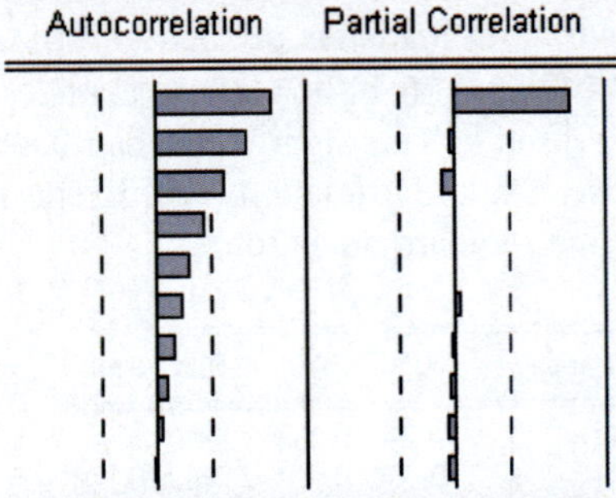

	AC	PAC	Q-Stat	Prob
1	0.797	0.797	19.122	0.000
2	0.625	-0.026	31.368	0.000
3	0.460	-0.083	38.275	0.000
4	0.332	-0.008	42.037	0.000
5	0.235	-0.003	44.003	0.000
6	0.176	0.032	45.159	0.000
7	0.126	-0.018	45.783	0.000
8	0.078	-0.039	46.033	0.000
9	0.028	-0.042	46.067	0.000
10	-0.024	-0.050	46.094	0.000

Fig. 2. Auto-correlation and Partial-correlation of Function y_t

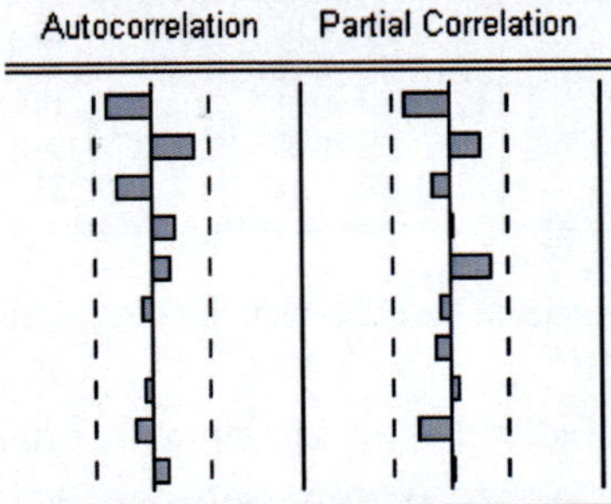

	AC	PAC	Q-Stat	Prob
1	-0.312	-0.312	2.7458	0.098
2	0.291	0.214	5.2278	0.073
3	-0.245	-0.124	7.0692	0.070
4	0.155	0.011	7.8402	0.098
5	0.121	0.281	8.3359	0.139
6	-0.071	-0.069	8.5170	0.203
7	0.003	-0.112	8.5173	0.289
8	-0.049	0.056	8.6125	0.376
9	-0.115	-0.225	9.1712	0.422
10	0.112	0.017	9.7359	0.464

Fig. 3. First Order Difference Test

First order difference test is made on overall fixed asset investment. Auto-correlation coefficient and partial correlation coefficient, resulting from the test, falls in the range of twice the standard deviation. We then postulate that parameters for B-J model are $p=1, q=0$.

Thus we formulate model B-J（1，0）：$y_t = C + C(1)y_{t-1} + u_t$.

Data of 1981-2007 are settled as sample. Data of 2008 is set aside as test for prediction. Parameter estimates as: ：$C = -6.40014, C(1) = 1.213228$。

Variable	Coefficient	Std. Error	t-Statistic	Prob.
C	30.01547	307.7947	0.097518	0.9231
AR(1)	1.213228	0.018887	64.23656	0.0000

R-squared	0.994217	Mean dependent var	2701.512
Adjusted R-squared	0.993976	S.D. dependent var	3335.635
S.E. of regression	258.8850	Akaike info criterion	14.02445
Sum squared resid	1608515.	Schwarz criterion	14.12123
Log likelihood	-180.3178	F-statistic	4126.336
Durbin-Watson stat	1.489331	Prob(F-statistic)	0.000000

Fig. 4. Parameters

B-J（1，0）model is followed down here. Statistics prediction turns out overall fixed asset investment for 2008 as ￥149.331106 billion while actually it was ￥150.655 billion. Percentage error is measured as less significant than 0.8%.Result of residual test has displayed white noise. Auto correlation coefficient and partial correlation coefficient is less than twice the standard deviation.

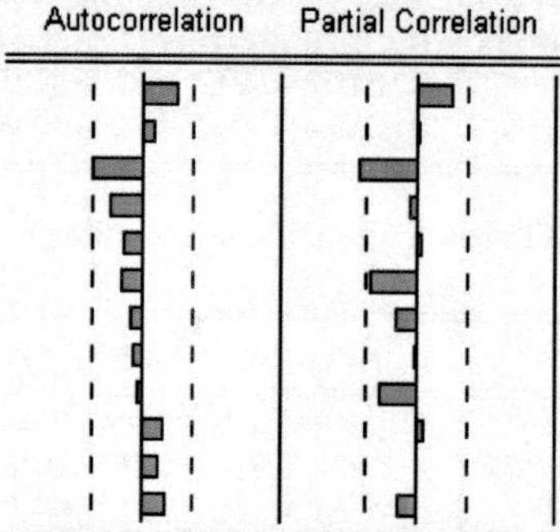

	AC	PAC	Q-Stat	Prob
1	0.254	0.254	1.8839	
2	0.084	0.020	2.0962	0.148
3	-0.382	-0.436	6.7166	0.035
4	-0.242	-0.057	8.6495	0.034
5	-0.132	0.035	9.2502	0.055
6	-0.163	-0.347	10.214	0.069
7	-0.091	-0.149	10.530	0.104
8	-0.072	-0.021	10.738	0.150
9	-0.032	-0.281	10.781	0.214
10	0.164	0.057	12.003	0.213
11	0.126	0.005	12.779	0.236
12	0.166	-0.143	14.220	0.221

Fig. 5. Auto correlation and Partial correlation

Here, from B-J (1,0) model ,we can predict that infrastructure investment turns out to be:

Table 4. Overall fixed investment demand prediction (unit: ¥ 100 million)

Year	2009	2010	2011	2012	2013
Investment	18265.4	22153.7	26871.1	32594.4	39538.0

From the table we can see infrastructure investment for 2009-2013 amounts to ¥ 1.394226 thousand billion. According to Jiangsu Statistics 2008, rural fixed asset investment turned out ¥36.9083 billion, taking up 24.5% of the ¥150.605 billion of overall fixed asset investment. Overall fixed asset investment in Jiangsu turned out ¥ 187.516 billion in 2009, close to statistics prediction. So we made a rough estimate of infrastructure investment for 2009-2013 to be ¥ 1.394226 thousand billion. We postulate that capital sources and their respective percentage of rural infrastructure investment might follow their counterpart in overall fixed asset investment, which is manifested in table 5.

Table 5. Statistics of Overall Fixed Capital Investment Sources (Unit: ¥100 million)

	Sum total	Capital from national budget	Domestic loan	Foreign investment	Self-financing	Others
Capital	15060.5	153.86	1818.06	1394.98	10624.51	2210.27
Percentage (%)	100	0.9	11.2	8.6	65.6	13.6

Data source: Jiangsu Statistics 2009

We postulate that capital sources for and their respective percentage of rural infrastructure construction investment would copy their counterparts concerning overall fixed capital investments. Hence, capital from national budget would turn out 30.743 billion, domestic loan would turn out 382.576 billion, foreign investment would turn out 293.763 billion, self-financing would turn out 2.2408 thousand billion and other sources would turn out 46.4556 billion.

4 The Necessity of Fiscal Support for Rural Infrastructure Construction and its Economic Effect

Rural infrastructure, to a large degree, falls into the realm of public product. A majority of primary government and organization in rural areas are not fiscally strong, which results in serious deficiency of public products and hence low level of income and social welfare for farmers. As a result, external forces, the most important of which is fiscal support for public product, should be introduced to actualize agriculture development. Government-subsidized rural public product gives orientation and displays multiplier effect. Fiscal support for rural public product is an external force facilitating increase in agriculture both from individual farmers and from the community. Fiscal input direction, deducted interest and subsidy orientation could directly reflect policy orientation, serving as a clear signal for corporate and nongovernmental capitals to enter agriculture. What's more, for input beyond fiscal budget, economic incentives should be supplemented to attract capital flow to agriculture. Primary fiscal input could bring about multiplier effect. Multiplier effect, on one hand, would attract financial capital as well as non-governmental capital in the next round of fiscal input, on the other hand targeted project would directly or indirectly stimulate agricultural production and increase farmer's income, which would stimulate farmers' consumption and hence demand and hence desirable economic and social benefits.

More fundamentally, fiscal support for rural infrastructure construction answers to an intrinsic appeal for lessening urban-rural gap and for social harmony. Household income, whether urban or rural, has displayed a marked increase than that of past. At the same time, income gap between urban household and rural household is enlarging, its real gap much more marked than nominal gap. For example in 2009, urban-rural income turned out a ratio of 3.3:1. But as a matter of fact, a majority of farmers' income is not monetary income which could turn into purchase power. While concerning urban household income, a certain amount of invisible income is beyond reach of statistics. As a result, real income ratio might be much larger than 3.3:1. To lessen urban-rural gap and to create rural employment needs, great improvement in infrastructure esp. road, electricity supply and telecommunication should be achieved as prerequisite. An urban-rural binary opposition pattern of economic structure has been in place for a long time. It means product supply prioritizes urban area and urban citizen against their rural counterparts, which results in gigantic urban-rural difference of consumption in education, public medical service, social welfare and infrastructure. Lessening urban-rural difference in supply of public product and equalize consumption right for rural residents in basic education, public medical service, social welfare and infrastructure are prerequisites for rural-urban coordination and hence a well-off nation.

Fiscal support for rural infrastructure construction is a must for economic development. Infrastructure, providing material base for economic growth, keeps pace with economic output in most cases. According to "world development report" issued by the World Bank, when infrastructure stock increases by 1%, so does GDP. During the take-off stage of economy where industrialization is taking place, infrastructure investment plays a pivotal role for economic growth in the target area. Deficient supply of rural infrastructure has impeded increase in agricultural productivity and agricultural competitiveness. Fiscal support for infrastructure construction is a must for enhancing agricultural competitiveness and ensuring wholesome and continual development of national economy. Fiscal input in rural infrastructure could lay a foundation for long-effect mechanism for increase in farmers' income, which would stimulate and drive up rural consumption demand. The ensuing income effect, consumption effect and employment effect could play lasting and effective role to expand domestic demand and to drive up economic growth.

Infrastructure investment is essential to economic growth. According to Rostow, a necessary but insufficient condition to economic take-off is preceding construction from social basic capitals. According to Romer and Lucas, external economies fostered by infrastructure would induce scale economy and growth in investment return. If factors of production are free to flow, they would flow to areas with better infrastructure.

Infrastructure investment is vital to economic growth in rural areas. During the primary stage when factors of production are put in, capital stock of infrastructure is yet to get accumulated and thus infrastructure does not display properties as public product. Under such circumstance, infrastructure's role on economic growth is displayed by multiplier effect. According to multiplier effect theory put forward by Keynes, when effective demand is deficient, increase in investment flow will induce initial increment in national income, and then in a circulative manner will cause increase in consumption and income. It means increment in national economy will turn out to be a multiple to investment increment. When rural infrastructure has been accumulated to be capital stock, it will be shared by farmers in this region with zero or low cost, which will drive up rural economic growth. From long-term perspective, sound infrastructure helps to play down transaction fees, play up transaction efficiency, expand labor division network and finally actualize endogenous economic growth.

The problems facing China in the recent years are: on one hand haphazard investment has resulted in serious surplus in productivity in some industries, seeing productivity greatly exceeding market demand; on the other hand infrastructure in rural areas are seriously lacking. If excessive productivity in those over-invested industries could be guided over to rural infrastructure construction, rural infrastructure construction could be accelerated to improve production and living conditions in rural areas, to promote income increase for farmers and hence to improve their consumption power. Besides, surplus productivity of over-invested industries could be fully consumed, leading to a chain effect to other relevant industries, so as to contribute to the wholesome development of national economy. Rural infrastructure construction is a necessary condition to rural economic growth though it by no means invites over-investments. Rational planning should be carried out to ensure that best utility could be made into full play out of given amount of investment. Over-investment will result in low efficiency of constructed facilities, stagnant performance of invested fund and low return. Investment in rural facility construction is highly regional, marginal and dispersed. Without effective foresighted planning for fiscal investment, rural infrastructure construction could not go in the right way. Occupancy of too much capital in infrastructure means deficient capital in other facility construction projects and other production products, which will cause impediment to rural economic development from the other side of the spectrum.

5 Conclusion and Prospect

We've analyzed economic effect of fiscal support and evaluated investment and financing based on five-year fiscal budgeting. From this, we can see that model evaluation concerning investment on and financing in rural infrastructure construction does provide reference. An investment-financing mechanism bundling fiscal support and financing together, as a solution to capital source for rural infrastructure construction, is a must. Rural infrastructure, falling into the category of public product and quasi public product, has an extensively positive external effect. Local governments should shoulder responsibilities to play up rural infrastructure construction. In a time when China is intent on social and economic integration between rural and urban areas as well as on solving the urban-rural binary contradiction, central government is tilting its allocation of investment budget toward rural construction. A fixed capital investment increment mechanism, which takes into account growth in fiscal income, is in urgent need for reference by the central government. The incremented amount should be targeted at rural construction and more specifically construction of rural infrastructure. Jurisdictional division should be made clear among different tiers of governments for a more efficient and effective rural construction and agricultural investment. Local governments should be more obliged for fiscal investment concerning agriculture, farmer and rural areas, and should couple this obligation with proper public fiscal system. At the same time, fiscal investment should play a role of introduction, conduction and demonstration, guiding farmers to contribute labor to those infrastructures that would benefit them immediately, for an actual improvement in rural production and living conditions. Fiscal, investing, pricing and financing tools should be employed flexibly to induce non-governmental capitals invested to construct a brand new countryside.

From the study of fiscal-financial capital input evaluation model, we concludes that under the premise of marked increase in local fiscal income with a sustainable pattern, input should be made on a rational pace and a sound fiscal investment-financing mechanism should be established. An investment-financing framework and mechanism bundling fiscal and financial capital should be formulated for rural construction, which will guarantee fiscal input in new countryside construction, social security, science and technology, public sanitation, public security, environmental protection and eco-construction. Hence, supply capacity of public product and public service would be reinforced. Actual livelihoods issues in rural areas would be solved. Rural infrastructure would be improved in all ways and developed in a leapfrog pattern.

References

1. Rural Development Institute of China Academy of Social Science, Department of Rural Social and Economic Survey of National Bureau of Statistics China: Analysis and prediction of rural economic situation in China. Social Science Academic Press, Beijing (2010)
2. Fan, S., Zhang, L., Zhang, X.: Economic Growth, Regional Difference and Poverty — Research on Public Investment in Rural China. China Agriculture Press, Beijing (2002)
3. Chen, K., Qiu, R.: Practice of and Reflection on Rural Water Resources Development in Jiangsu Province. Jiangsu Water Resources (1), 43–46 (2009)
4. Ding, X., Zhang, Y.: Theory and Practice of Public Fiscal Coverage of Countryside. Management World (10), 1–7 (2007)
5. Liu, H., Zhu, P.: Research on Rural Household-based Public Product Preference and Public Product Investment Decision. Social Science in Nanjing (9), 147–151 (2008)
6. Li, R.: Quantitative Analysis on Investment Benefit from Rural Public Infrastructure. Journal of Agrotechnic Economics (2), 5–9 (2003)

7 Ma, S.: Computation on Pull Effect of Infrastructure Construction Investment on Economic Growth. Statistical Research (10), 30–33 (2001)
8. Zhang, Z.: Reflection on Introducing Non-governmental Fund to Rural Infrastructure Construction in China. Rural Economy (3), 66–68 (2004)
9. Liu, X., Xin, X., Mao, X.: Impact of Infrastructure Investment on Rural Household Income and Expenditure in Poverty-stricken Rural Areas. China Rural Survey (1), 31–36 (2003)
10. Zhou, H.: Status Quo and Issues of Rural Water Resource Construction in Jiangsu Province. Modern Economic Research (8), 59–62 (2006)

The Research on Cointegration between Different Prices in International Stock Market

Dali Zhao

School of Economics, Huazhong University of Science & Technology
wangmeist@gmail.com

Abstract. According to the daily data from 1990 to 2005, this paper investigate whether there is cointegration relationship between S & P 500 index and Britain's FTSE 100 index, Hong Kong's Hang Seng index, the Nikkei 225 index or not in a empirical way. It showed that there were cointegration between S & P 500 index and Britain's FTSE 100 index, the Nikkei 225 index independently, while there was no cointegration between S & P500 Index and the Hang Seng Index in Hong Kong. This conclusion was in line with international capital markets operation.

Keywords: Index of stock, Unit root test, Cointegration test, ECM.

1 Introduction

As global economic integration, the economies between countries are jointly influenced by the fluctuations in the world economy, and the performance of stock markets reflecting the national economic operation situation will reflect synchronous fluctuations in the long-term with a balanced relationship. On the issue of linkage between the stock market, domestic and foreign scholars have studied a lot. Karolyi and Stu (1995) used the data in ISSM database from 1988 to 1992 to investigate the jointly fluctuations of Japanese company which listed in both New York and Tokyo stock markets, two issues linked to the stock market. He concluded that the jointly relationship between the two markets was close, and when the fluctuations in the market became remarkable, this covariance enhanced. Some scholars studied the emerging East Asia stock index except China, pointing out that the Southeast Asian financial crisis can not contribute significantly to the improvement of these emerging stock markets and the world stock market linkages.

Table 1. Stock index change 1990～2005

	closing 1990	closing prices 1995	Growth percentage
S&P500	353.4	615.93	42.6%
FTSE100	2422.7	3689.3	34.3%
HSI	2836.57	10073.387	71.4%
NIKKEI 225	38915.87	19868.15	-95.8%

Data resource: Wind database

J. Luo (Ed.): Affective Computing and Intelligent Interaction, AISC 137, pp. 903–908.
springerlink.com

From 1990 to 1995 the Standard Poor's 500 index • (This article referred to as the S & P500), Britain's FTSE 100 index (This article referred to as the FTSE100) and the Hong Kong Hang Seng Index (This article referred to as HSI) expressed upward trend, while the Nikkei 225 stock index (This article Referred to as NIKKEI 225) showed a downward trend (see table 1). This paper will answer two questions as follows: was there an inherent connection between these changes in trends, and was there a stable balance relationship between them?

2 Model and Method

2.1 ADF Test

The standard unit root test is Dickey and Fuller the DF test. Because in this process it does not consider the autocorrelation in residuals. Therefore, to solve this problem there is a method of joining the dependent variable lags behind many of the bands to remove residual correlation, which is ADF test. The ADF will be obtained by the τ statistics compared with the critical value, if it exists, the unit root assumptions were rejected, that sequence is stationary, or, there are no reasons to refuse the null assumption that the original sequence exist unit root, a non-stationary sequence.

2.2 Cointegration Test

There is a general assumption that the time series are stationary when employing traditional regression methods, while many an economic phenomena are non-stationary time series, there will be pseudo regression when employing traditional ordinary least squares. For that reason, Engle and Granger (1987) first proposed a new method named Cointegration to deal with non-stationary sequences. The basic idea of this approach is that in two or more non-stationary sequences, it is probable to find a balanced relationship between variables. There are two time series or more, and if their linear combination is a stationary I (0) process, they are cointegration. There are two test approaches to judge whether some I (1) sequences are cointegrated or not: one is from Engle and Granger (1987) based on the return of the residuals of the EG two-step cointegration test; another Johansen and is Juselius (1990) in the VAR system with maximum likelihood estimate more variables to test cointegration relations.

3 Results and explanation

3.1 Unit Root Test

In data processing, in order to eliminate heteroscedasticity and data fluctuations, the author firstly employed natural logarithm of the four time series, to form four new time sequences LS & P500, LFTSE100, LHSI, and LNIKKEI225.

Table 2. ADF test results

Variable	(c, t, p)	ADF value	1% critc value	5% critc value	10% critc value	Results
LS&P500	(c,0,23)	0.28922	-3.434325	-2.863183	-2.567693	no stationary
LFTSE100	(c,0,23)	-0.322533	-3.434325	-2.863183	-2.567693	no stationary
LHSI	(c,0,23)	-1.062895	-3.434325	-2.863183	-2.567693	no stationary
LNIKKEI225	(c,t,23)	-2.671420	-3.963922	-3.412685	-3.128313	no stationary
ΔLS&P500	(c,0,23)	-37.9825	-3.434328	-2.863184	-2.567693	stationary
ΔLFTSE100	(c,0,23)	-37.83161	-3.434328	-2.863184	-2.567693	stationary
ΔLHSI	(c,0,23)	-36.93047	-3.434328	-2.863184	-2.567693	stationary
ΔLNIKKEI225	(c,t,23)	-29.65881	-3.963922	-3.412685	-3.128313	stationary

Note: 1, (c, t, p)is the style of ADF and whether includes intercept and time trend or not and the lag ranks; t, c, p is determined by AIC criterion。 2, Δrepresents first order difference 。 3, critic value is obtained in corresponding level。

Then this paper employed unit root test, the results can be seen in Table 2. Test results showed that the sequences of LS & P500, LFTSE100, LHSI, and LNIKKEI225 are no stationary with 10% significant level in the ADF test value. And after first order difference of the four sequences, the formation of new sequences ΔLS & P500, ΔLFTSE100, ΔLHSI, ΔLNIKKEI225 are all stationary according to less than 1 percent of the criterion in ADF test. According the ADF tests above, we know that the time series LS & P500, LFTSE100, LHSI, LNIKKEI225 bands were all I (1), which we can test co-integration.

3.2 Cointegration and VECM

(1) S&P500 index and Ftse100 index

First regress LFTSE100 with L S&P500 employing OLS:

LS&P500= − 1.786588+0.989587LFTSE100

(-27.31937) (119.8664)

$$R^2 = 0.901889 \quad F = 14367.94$$

Then the unit root test for the residuals of the model above showed that there was no unit root and the test style was (c, t, 23). t value is in parentheses and the following model are the same. The Eviews software outputs are presented in the table 3.

Table 3. The result of ADF test of residuals

ADF Test Statistic	-2.863425	1% Critical Value	-3.434328
		5% Critical Value	-2.863184
		10% Critical Value	-2.567639

The results showed that ADF test is less than 5 percent criterion, we can refused the null assumption in 5% significant level, the residual sequence is stationary. Thus we can determiner that Standard• Poor's 500 index and Britain's FTSE 100 index were cointegrated.

At last we set Error Correction Model, including error correction variable:

$$\Delta LS\&P500 = 0.000276 + 0.296065\Delta LFTSE100 - 0.001573ecm(-1)$$
$$(1.62296) \quad (13.96492) \quad (-0.462207)$$

$$R^2 = 0.111766 \quad F = 98.22015$$

ECM showed that in the short term Standard • Poor's 500 index were influenced by Britain's FTSE 100 index movements. Error correction factor of the system reflects the error of their own departure from the role of a balanced mechanism, the results of the regression of the error correction factor is -0.001573, noting that the error rate reached 0.157 percent, and error correction capability is not strong as well as coefficient was not significant; symbol was the negative which suggested that this amendment is the reverse, in line with that principle.

(2) Standard • Poor's 500 index and Hong Kong's Hang Seng index

Firstly we regress LS&P500 on LFTSE100:

$$LS\&P500 = 3.186352 + 0.330458LHSI$$
$$(94.98369) \quad (85.49683)$$
$$R^2 = 0.823842 \quad F = 7309.707$$

Then the unit root test for the residuals of the model above showed that there was no unit root and the test style was (c, t, 23). The Eviews software outputs are presented in the table 4.

Table 4. The result of ADF test of residuals

ADF Test Statistic	-1.344725	1% Critical Value	-3.434328
		5% Critical Value	-2.863184
		10% Critical Value	-2.567639

The results showed that, even in 10% of the significant level, residual sequence is not stationary, there was no cointegrations between Standard • Poor's 500 Index and the Hang Seng Index in Hong Kong's.

(3) Standard • Poor's 500 index and the Nikkei 225 index

First we make a simple estimation between LS&P500 and LNIKKEI225:

$$LS\&P500 = 11.52874 - 0.550661LNIKKEI225$$
$$(84.14576) \quad (-39.99124)$$

$$R^2 = 0.505739 \quad F = 1599.299$$

Then the unit root test for the residuals of the model above showed that there was no unit root and the test style was (c, t, 23). The Eviews software outputs are presented in the table 5.

Table 5. The result of ADF test of residuals

ADF Test Statistic	-3.183828	1% Critical Value	-3.963926
		5% Critical Value	-3.412687
		10% Critical Value	-3.128314

The results showed that ADF test is less than 5 percent criterion, we can refused the null assumption in 5% significant level, the residual sequence is stationary. Thus we could determiner that Standard • Poor's 500 index and LNIKKEI225 index were cointegrated.

At last we set Error Correction Model, including error correction variable:

$$\Delta LS\&P500=0.000387+0.074331\Delta LNIKKEI225-0.002728ecm(-1)+$$

$$(2.175952) \quad (6.332035) \quad (-1.727272)$$

$$R^2 = 0.02919 \qquad F = 21.59153$$

ECM showed that in the short term Standard • Poor's 500 index were influenced by LNIKKEI225 index movements. Error correction factor of the system reflects the error of their own departure from the role of a balanced mechanism, the results of the regression of the error correction factor is 0.002728, noting that the error rate reached 0.272 percent, and error correction capability is not strong as well as coefficient was weak significant; symbol was the negative which suggested that this amendment is the reverse, in line with that principle.

4 Conclusions

Through the above empirical analysis, we found that from 1990 to 2005 there was a cointegration between Standard • Poor's 500 index and Britain's FTSE 100 index, as well as between Standard • Poor's 500 index and the Nikkei 225 index. While there was no cointegration between Standard • Poor's 500 Index and the Hang Seng Index in Hong Kong. Through the establishment of error correction model between Standard • Poor's 500 index and Britain's FTSE 100 index, in the short term Standard• Poor's 500 index were influenced by Britain's FTSE 100 index fluctuations , but the ability of error correction is not very strong in the Standard • Poor's 500. There was a similar conclusion between Standard• Poor's 500 index and the Nikkei index of 225 stock index.

References

1. Gujarati, D.N.: Basic Econometrics, 4th edn. McGraw-Hill Companies, Inc., New York (2003)
2. Engle, R.F., Granger, C.W.J.: Cointegration and Error Correction: Reprentation, Estimation and Testing. Econometrica 55(2), 251–276 (1987)
3. Huang, B.-N., Yang, C.-W., Hu, J.W.-S.: Causality and Cointegration of Stock Markets Among the United States, Japan and South China Growth Triangle. International Review of Financial Analysis 9(3), 281–297 (2000)

4. Woodridge, J.M.: Introductory Econometrics: A Modern Approach. South-Western College Publishing a division of Thomson Learning (2000)
5. Johansen, S.: Estimation and Hypothesis Testing of Cointegration Vectors in Gaussian Autoregressive Vector Models. Econometrica 59(2), 1551–1580 (1991)
6. Karolyi, G.A., Stulz, R.M.: Why do Markets over together? An Investigation of U.S.-Japan Stock Return using ADRS. The Review of Financial Studies 32(3), 5–33 (1996)
7. Phylaktis, K., Ravazollo, F.: Stock Market Linkages immerging Markets: Implications for International Portfolio Diversification as Business School, Emerging Markets Group, Discussion Paper. 20020301 (2002)

Pricing Capacity-Booking Fees for Semiconductor Fabs with Outsourcing Alternatives

Muh-Cherng Wu and Cheng-Hang Tsai

Department of Industrial Engineering and Management,
National Chiao Tung University, 30070 Hsin-Chu City, Taiwan

Abstract. A semiconductor foundry fab is a factory that manufactures integrated circuits (IC) designed by a large number of customers (called IC design houses). We consider a foundry fab that requires the customers to pay a booking fee for reserving an amount of capacity. And the capacity reserved for the customers can be either provided by the fab itself or through outsourcing capacity from other fabs. The outsourcing activity is implemented by subsequently paying booking fee at a fixed rate to the capacity-providers. This paper develops a method to determine the optimum booking price to charge design houses by the foundry fab in order to maximize the profit of the fab. The proposed capacity-booking paradigm and solution method help resolving the information distortion problem in demand forecasting.

Keywords: Capacity booking, demand forecast, information distortion.

1 Introduction

The supply chain of semiconductor products involves the following four stages: design, manufacturing, packaging, and testing. The investment of semiconductor manufacturing is capitally intensive. Therefore, a business model called *semiconductor foundry* arises. A foundry company is a *make-to-order* production system, which manufactures semiconductor products designed by its customers, typically called IC design houses or *design houses* briefly. To avoid conflict of interest, a foundry fab would not involve any design business; therefore, many design houses are attracted to use its manufacturing capacity. How to effectively allocate the capacity of a foundry fab to its design houses is very important.

To make effective allocation, a foundry fab would ask customers to make demand forecast. In practice, customers attempt to boost their demands in order to catch opportunities of unexpected uprising demand. This in turn may mislead the fab to expand its capacity and consequently results in an oversupplied situation. Much research thus addressed how to resolve the demand boosting issue.

To overcome the forecast distortion problem, Hsu and Wang [1] proposed to actively adjust the forecasts reported by customers according to their historical accuracy. Some studies suggested the design of a bonus incentive system to encourage customers to "tell-the-truth" [2-5]. In their studies, demand forecast

J. Luo (Ed.): Affective Computing and Intelligent Interaction, AISC 137, pp. 909–916.
springerlink.com

(essentially stochastic) is modeled by a random variable. The bonus system is designed to encourage customers to report the mean of the random variable; and a

This research advocates a "*booking-by-down-payment*" mechanism. That is, customers are requested by the fab to pay booking fees (down payment) for reserving capacity, and violation of a booking commitment would lead to lose of booking fee. The pricing decision of capacity-booking fees is a *trade-off decision* for the fab. The higher the booking fee, the higher the penalty of overbooking, and the lower the booking quantity; in turn, the fab may underestimate demand and lose orders. Conversely, the lower the booking fee, the higher the booking quantity, which may lead to the overestimation of future demand and result in capacity underutilization.

In the context of *booking-by-down-payment*, this research also addresses the issues of *capacity outsourcing*. Capacity expansion for semiconductor manufacturing requires a long lead time (at least six months). To promptly increase capacity supply, a fab tends to outsource capacity from other fabs because semiconductor fabs are often closely located to each other. That is, for the fab of interest, we assume that its capacity available to booking can be increased by outsourcing from other fabs. The capacity outsourced by the fab is also through a reservation process, in which the booking price per unit of capacity is a *constant*. Then, the fab has two decisions to be made: (1) determining the capacity-booking price for customers, and (2) determining the booking volume of outsourcing. This research proposes an approach to make the two decisions in order to maximize the profit of the fab.

2 Problem Assumptions

In the research problem, there are three parties: *the fab*, *design houses (customers of the fab)*, and *other fabs*. The customers buy capacity from the fab; the fab if needed outsource part of the demanding capacity from other fabs. Assumptions about the interactions and behaviors of the three parties are described below.

Assumption 1: *business deal of capacity booking*. Capacity-buyers have to pay an amount of booking fee to reserve capacity. The booking fee, charged by per unit of capacity, will be reimbursed if the booking commitment is fulfilled; and will be taken as a penalty charge otherwise. A capacity-buyer typically refers to design houses (i.e., customers), but refers to the fab wherever outsourcing capacity is addressed.

Assumption 2: capacity preparation policy. The fab provides capacity to design houses in two ways, by self-production or by outsourcing. The fab intends to strictly keep its commitments to customers. That is, the sum of outsourced-capacity and self-capacity of the fab should equal the volume committed to design houses.

Assumption 3: capacity selling policy. If the actual demands of the design houses are higher than the booking quantity and less than the self-capacity of the fab, then the fab will still sell the remaining self-capacity to the customer.

Assumption 4: booking prices. The booking price for the outsourced-capacity is assumed to be a *constant*. Yet, the capacity-booking price announced to design houses by the foundry fab is a *decision variable*.

Assumption 5: aggregate demand forecast. This research assumes that customers and the fab have a consensus about the demand forecast of the whole market, which is a random variable.

Assumption 6: rational decision makers. The fab and its customers are all rational decision makers. They aim to maximize their profit in making the decisions of booking price and booking quantity.

This research addresses only the *aggregated demand* of all the customers. To facilitate presentation, all the customers of the fab are aggregated into one, called the *virtual customer*, (simply called *the customer* hereafter in the paper).

3 Decision Models

This research determines the price for booking capacity based on a *response-based* paradigm. That is, a decision of booking price will lead the customer to respond an optimal booking-quantity, which maximizes the customer profit. The booking price together with the responded quantity forms a *feasible portfolio*, which essentially corresponds to a profit level for the fab. Pricing the booking fee is equivalent to identify a portfolio that maximizes the fab profit. The responded quantity in the optimal portfolio can be used to determine the outsourcing-quantity to be booked by the fab. Figure 1 shows the response-based model of the pricing decision.

The response-based paradigm involves two decision models. The first model, describing the decision of the customer, aims to find the functional relationship between booking price and the booking quantity responded by the customer. With the functional relationship, the second model, describing the decision of the fab, aims to determine a booking price that can maximize the profit of the fab.

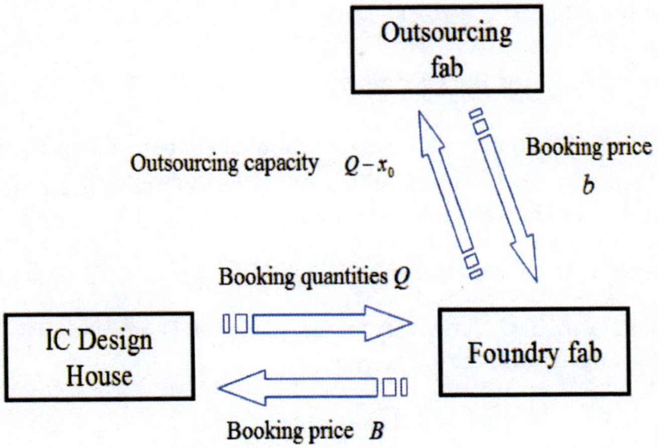

Fig. 1. Response-based decision model

3.1 Notation

Parameters and Random Variable

b : booking price of outsourcing-capacity ($/wafer)
p_M : product selling price of the fab ($/wafer)
v_M : variable manufacturing cost of the fab ($/wafer)
C_M : fixed cost of the fab ($)
p_B : product selling price of the customer ($/wafer)
s_B : stockout loss of the customer ($/wafer)
v_0 : cost of the fab incurred due to outsourcing capacity ($/wafer)
x_0 : the present capacity of the fab in the decision horizon (wafers)
Y : forecast demand of the customer, a random variable
$f(y)$: probability density function of Y
$F(y) = \int_{-\infty}^{y} f(t)dt$: cumulative density function of Y

Decision Variables

B : booking price determined by the fab ($/wafer)
Q : booking quantity responded by the customer (wafer)

Other Variables and Notation

π_M : the profit of the fab ($)
π_B : the profit of the customer ($)
$(x)^+ = \max(0, x)$

3.2 Decision Model for the Customer

The profit of the customer π_B can be formulated below, where the first term represents the profit earned due to selling, the second term indicates the loss due to overbooking, and the third term denotes the stockout loss.

$$\pi_B = (p_B - p_M) \cdot Min(Max(x_0, Q), y) - B \cdot (Q - y)^+ - s_B \cdot (y - Min(Max(x_0, Q), y)) \quad (1)$$

With a random variable y involved in the model, this research assumes that the customer aims to maximize the expected profit

$$E[\pi_B] = \int_0^{\infty} \pi_B \cdot f(y)dy. \quad (2)$$

The expected profit involves two decision variables, B and Q, where B is determined by the fab and Q is by the customer. At a particular B, we can determine an optimal Q

that maximizes $E[\pi_B]$. That is, in the derivation of an optimal Q for the customer, B can be regarded as a parameter. As a result, the optimal Q can be represented by a function $Q = g(B)$. To facilitate obtaining the function $Q = g(B)$, the decision model is addressed in the following two cases.

Case (1): $Q > x_0$

The expected profit $E[\pi_B]$ can be derived as follows.

$$E[\pi_B] = (p_B - p_M)\cdot\left(\int_0^Q y\cdot f(y)dy + \int_Q^\infty Q\cdot f(y)dy\right) - B\cdot\int_0^Q (Q-y)\cdot f(y)dy -$$
$$s_B\cdot\int_Q^\infty (y-Q)\cdot f(y)dy$$

To obtain the optimal Q, we have to ensure $\dfrac{dE[\pi_B]}{dQ} = 0$, which subsequently leads to the following relationship between Q and B.

$$F(Q) = \frac{p_B - p_M + s_B}{B + p_B - p_M + s_B}. \tag{3}$$

Case (2): $Q \le x_0$:

$$E[\pi_B] = (p_B - p_M)\cdot\left(\int_0^{x_0} y\cdot f(y)dy + \int_{x_0}^\infty x_0\cdot f(y)dy\right) - B\cdot\int_0^Q (Q-y)\cdot f(y)dy$$
$$- s_B\cdot\int_{x_0}^\infty (y - x_0)\cdot f(y)dy.$$

Likewise, we have to ensure $\dfrac{dE[\pi_B]}{dQ} = 0$ and in turn obtain the following equation.

$$-B\cdot F(Q) = 0. \tag{4}$$

3.3 Decision Model for the Fab

The profit of the fab can be modeled as follows.

$$\begin{aligned}\pi_M &= p_M\cdot Min(Max(x_0, Q), y) - C_M - v_M\cdot Min(x_0, y)\\ &- v_o\cdot Min((y-x_0)^+, (Q-x_0)^+) + B\cdot(Q-y)^+ - b\cdot((Q-x_0)^+ - (y-x_0)^+)^+\end{aligned} \tag{5}$$

In eq (5), the first term describes the revenue through selling. The second term denotes the fixed manufacturing cost of the fab, and the third term indicates the variable manufacturing cost of the fab. The fourth term denotes the cost incurred due to outsourcing. The fifth term indicates the revenue, due to the overbooking of the customer. The last term indicates the loss caused by the overbooking of the fab in outsourcing capacity.

This research assumes that the fab aims to maximize the expected profit $E[\pi_M] = \int_0^\infty \pi_M \cdot f(y)dy$. The expected profit $E[\pi_M]$ involves two decision variables Q and B, which are constrained by the function $Q = g(B)$ that describes the responded behavior of the customer. With the function $Q = g(B)$ being available, $E[\pi_M]$ can be regarded as involving only one variable B. To simplify the evaluation of the integrals, we address the problem of finding an optimal B in the following two cases.

Case (1): $Q > x_0$

$$
\begin{aligned}
E[\pi_M] &= p_M \cdot \left(\int_0^{g(B)} y \cdot f(y) \cdot dy + \int_{g(B)}^{\infty} g(B) \cdot f(y) \cdot dy\right) - C_M \\
&- v_M \cdot \left(\int_0^{x_0} y \cdot f(y) \cdot dy + \int_{x_0}^{\infty} x_0 \cdot f(y) \cdot dy\right) \\
&- v_o \cdot \left(\int_{x_0}^{g(B)} (y - x_0) \cdot f(y) \cdot dy + \int_{g(B)}^{\infty} (g(B) - x_0) \cdot f(y) \cdot dy\right) + B \cdot \int_0^{g(B)} (g(B) - y) \cdot f(y) \cdot dy \\
&- b \cdot \left(\int_0^{x_0} (g(B) - x_0) \cdot f(y) \cdot dy + \int_{x_0}^{g(B)} (g(B) - y) \cdot f(y) \cdot dy\right).
\end{aligned}
$$

Case (2): $Q \leq x_0$

$$
E[\pi_M] = (p_M - v_M) \cdot \left(\int_0^{x_0} y \cdot f(y) \cdot dy + \int_{x_0}^{\infty} x_0 \cdot f(y) \cdot dy\right) - C_M + B \cdot \int_0^{Q} (Q - y) \cdot f(y) \cdot dy
$$

4 Numerical Examples

Numerical examples of solving the research problem are described below. The scenario of the examples involves three entities: *Fab_A*, *Fab_B* and *Customer_C*. *Fab_A* is a foundry fab that provides capacity to the *Customer_C*, which is a virtual entity created by aggregating all customers of *Fab_A*. *Fab_A* may outsource capacity from *Fab_B*. The demand of *Customer_C* is a trimmed normal distribution.

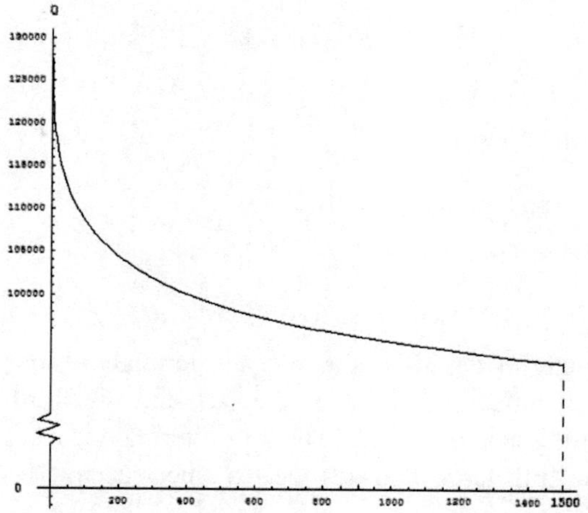

Fig. 2. Relationship between Q and B

Fig. 2 shows the result of the first decision model. The relation between Q and B is a monotonically decreasing function. Fig. 3 shows the profit of *Fab_A* at various levels of B. The profit is maximum at B = 247, which is the optimal decision suggested to *Fab_A*. From Fig.3, the optimal B decision will yield Q =102.9.

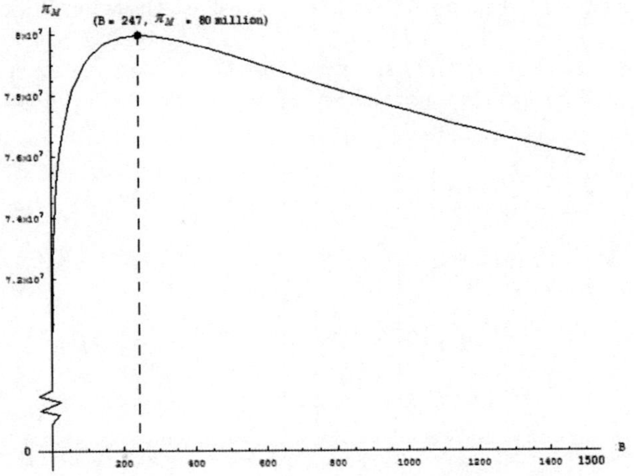

Fig. 3. Relationship between π_M and B

5 Concluding Remarks

The information of demand forecast provided by customers might be intentionally distorted because the customers tend to ensure the availability of capacity when the actual demand turns out higher than their expectation. This research advocates that the customers pay a booking fee for reserving capacity and the capacity booked by the customer is provided by the concerned fab and if needed partially through outsourcing capacity from other fabs. We develop a method for pricing the optimal booking fees in order to maximize the profit of the concerned fab.

This research solves the pricing problem based on a *response-based paradigm* that involves two decision models. The first model is to describe the booking behavior of the customers and consequently yields a function $Q = g(B)$. That is, given a B (booking price), the customers will respond a Q (booking quantity). Based on the $Q = g(B)$ function, the second model designed for describing the behavior of the concerned fab is used to determine an optimal booking price.

References

1. Hsu, H.M., Wang, W.P.: Modeling demand driven with forecasting regulation safety stock setting and capacity analysis. International Journal of Industrial Engineering-Theory, Applications, and Practice 7(3), 195–201 (2000)

2. Mallik, S.: Contracting over Multiple Parameters: Capacity Allocation in Semiconductor Manufacturing. European Journal of Operational Research 182, 174–193 (2007)
3. Karabuk, S., Wu, S.D.: Incentive Schemes for Semiconductor Capacity Allocation: A Game Theoretic Analysis. Production and Operations Management 14, 175–188 (2005)
4. Mallik, S., Harker, P.T.: Coordinating supply chains with competition: capacity allocation in semiconductor manufacturing. European Journal of Operational Research, 330–347 (2004)
5. Cachon, G.P., Lariviere, M.A.: Capacity choice and allocation: strategic behavior and supply chain performance. Management Science 45(8), 1091–1108 (1999)

The Solution of Verdier Equation

Tian Shangguan, Wenjun Lu, and Yirong Ying

College of Economics, Shanghai University, Shanghai, 200444, China

Abstract. This paper researched on the optimal model about the Verdier Equation and found its analytic solution based on the particular conditions, it maybe enables us to solve the according optimization problem.

Keywords: Iteration, Verdier Equation, capital flow.

1 Introduction

Optimal model, as a considerably mature mathematical method, has its widespread applications in natural science such as physics. In the field of applied Economics, it is very common to build optimal models to research the various problems about the opening of capital account. There are some different constraint conditions when building the optimal model to research the opening of capital account, and those constraint conditions are listed below as table 1.

Table 1. Optimal model

Objective function	Constraint condition bound	Research model
Infinite integral form	integral form	consumption model
Utility function form	linear constraint condition	tax decision model
Utility function form	nonlinear constraint condition	optimal tax model
Damage function form	implicit constraint condition	currency crisis model
Form of expectation with Infinite integral	differentiable form	optimal portfolio

Recently, Verdier (2008) considers the following optimal problem with recursive constraints.

$$\max_{\{c_t, z_t, k_t\}} \sum \beta^t \frac{C_t^{1-\sigma}-1}{1-\sigma} \tag{1}$$

$$s.t.\ (1+n)(1+g)Z_{t+1} = (1-\delta)Z_t - C_t + \left[aK_t^\rho + (1-a)Z_t^\rho\right]^{\frac{\eta}{\rho}} \tag{2}$$

$$a\eta\left(\frac{d_t}{v_t}\right)^\rho \frac{y_t}{d_t} = r + \delta . \tag{3}$$

This system can be written as a single log-linear law of motion for Z_t. After taking a first-order Taylor log-linear approximation of Equation (3), we have

J. Luo (Ed.): Affective Computing and Intelligent Interaction, AISC 137, pp. 917–922.
springerlink.com

$$(\rho-1)\hat{d}_t+\left(\frac{\eta}{\rho}-1\right)a\rho\left(\frac{d^*}{v^*}\right)^{\rho}\hat{d}_t+\left(\frac{\eta}{\rho}-1\right)(1-a)\rho\left(\frac{z^*}{v^*}\right)^{\rho}\hat{z}_t=0, \quad (4)$$

where 'hats' refers to a variable's log-deviation from its steady state denoted by a '∗'. This expression can be used to write Equation (2) and (3) into a single log-linear law of motion for z which has the familiar solution:

$$\log z_t=\lambda^t\log z_0+\left(1-\lambda^t\right)\log z^*, \quad (5)$$

where $1-\lambda$ denotes the convergence speed. Since y, k and d can be written as functions of z. Equation (5) implies that the change in net foreign debt takes the form

$$\log d_t-\log d_0=-\left(1-\lambda^t\right)\log d_0+\left(1-\lambda^t\right)\log d^*. \quad (6)$$

What is the function behind this result and what does it imply for the dynamics of debt and capital flows? To understand the intuition, it is useful to suppose $\rho=0$, i.e. the production technology is Cobb-Douglas. For now, this is both innocuous and useful for deriving the qualitative predictions of the model as well as the reduced form to be estimated. The model retains the same qualitative predictions while the intuition is made clearer. If $\rho=0$, then $Y_t=K_t^{\alpha}Z_t^{\eta}(\theta_t L_t)^{1-\alpha-\eta}$.

In that case the combination of the credit constraint $(k_t=d_t)$, the small-open-economy assumption $(R_{kt}=r+\delta)$ and profit maximization imply that Equation (3) becomes

$$d_t=k_t=\frac{\alpha}{r+\delta}y_t. \quad (7)$$

Therefore k/y has a constant path to the steady state. The production function can therefore be written as $y_t=Bz_t^{\varepsilon}$ with $B=\left(\frac{\alpha}{r+\delta}\right)^{\frac{\alpha}{1-\alpha}}$ and $\varepsilon=\frac{\eta}{1-\alpha}$. Given profit maximization and this collapsed production functions, Equation (2) can be rewritten as follows:

$$(1+n)(1+g)Z_{t+1}=(1-\alpha)y_t-c_t+(1-\delta)Z_t. \quad (8)$$

With a constant saving rate, equilibrium requires that domestic savings musts equal investment in domestic capital minus net factor payments on debt, so that

$$(1+n)(1+g)Z_{t+1}=(1-\delta)Z_t+sBZ_t^{\varepsilon}, \quad (9)$$

where $\varepsilon = \dfrac{\eta}{1-\alpha}$, $B = \left(\dfrac{\alpha}{\gamma+\delta}\right)^{\frac{\alpha}{1-\alpha}}$.

Obviously, equation (9) is a first-order nonlinear recursive. However, it generates a small problem: we can only get the general solution of equation (9) because it lacks the initial condition in order to find the particular solution.

To analyze it easily, adding an initial condition here: $Z|_{t=0} = Z_0$.

It is necessary to reduce the constraint condition of optimization problem (1) so as enable us to solve problem (1) easily. Therefore, we introduce notations:

$$a = \frac{1-\delta}{(1+n)(1+g)}, \quad b = \frac{sB}{(1+n)(1+g)}.$$

Then equation (2) can be transformed to

$$\begin{cases} Z_{t+1} = aZ_t + bZ_t^{\varepsilon} \\ Z|_{t=0} = Z_0. \end{cases} \tag{10}$$

2 The Solution of Verdier Equation

We can refer equation (3) as Verdier equation and conclude its resolve solution on some particular conditions.

Equation (3) can be referred as Verdier equation and its analytic solution can be solved out on the condition of some particular forms.

2.1 In the Case $\varepsilon = 2$

It is obviously that Verdier equation degenerates to a geometric series with common ratio of $(a+b)$ when $\varepsilon = 1$, namely $Z_{t+1} = (a+b)Z_t$. This case can be omitted.

If $a = 0$, system of equations (10) can be transformed to

$$Z_{t+1} = Z_0^{2^{t+1}} b^{2^{t+1}-1}. \tag{11}$$

If $a \neq 0$, by using linear transformation $Z_t = mx_t + n$, the system of equations (10) can be modified as

$$m(x_{t+1} + l) = bm^2(x_t + l)^2 + \left[nb^2 + (a-1)n + ml - bm^2l^2\right], \tag{12}$$

where $l = \dfrac{a+2bn}{2bm}$.

Let

$$\begin{cases} nb^2 + (a-1)n + ml + bm^2l^2 = 0 \\ m = bm^2. \end{cases}$$

From the above system of equations, m and n can be determined as follows:

$$\begin{cases} m = \dfrac{1}{b} \\ n = \dfrac{-l(1+l)}{ab-b+b^3}. \end{cases}$$

Then equation (12) can be transformed to

$$(x_{t+1} + l) = (x_t + l)^2.$$

Simulating equation (11), the above recursive form can be solved out as follows:

$$x_t = (x_0 + l)^{2^t} - l.$$

Assert the above equation into Z_t recursive, then getting the following formula

$$Z_t = \frac{1}{b}\left[(x_0 + l)^{2^t} - l\right] - \frac{l(1+l)}{ab-b+b^3}. \tag{13}$$

2.2 In the Case $\varepsilon = -1$

In this situation, equation (10) can be looked as the fractional iteration form

$$Z_{t+1} = aZ_t + \frac{b}{Z_t}. \tag{14}$$

If fixed point exists in the iteration equation (14), it can be denoted as

$$\overline{Z} = \lim_{t \to \infty} Z_t.$$

Known from equation (14), the fixed points $\overline{Z}$ satisfy algebra equation below

$$\lambda = a\lambda + \frac{b}{\lambda}.$$

If $a < 1$, there exist two fixed points in equation (14). We denote them as

$$\overline{Z}_{\pm} = \pm\sqrt{\frac{b}{1-a}}.$$

According to equation (14), it can be easily derived as continued product form

$$Z_{t+1} = Z_0 \prod_{k=1}^{t} \left(a + \frac{b}{Z_k^2} \right). \tag{15}$$

According to the character of the infinite product, conditions of converge or diverge can be concluded from equation (15) easily.

2.3 In the Case $a << 1$

If $a << 1$, it can be easily concluded from (10) as follows:

$$\ln Z_{t+1} \left(1 - a \cdot \frac{Z_t}{Z_{t+1}} \right) = \varepsilon \ln Z_t + \ln b. \tag{16}$$

If $|x| < 1$, taking advantage of the approximation below:

$$\ln(1-x) = -2\ln 2 - \ln x.$$

Equation (16) can be turned to

$$\ln Z_{t+1} - \frac{\varepsilon + 1}{2} \ln Z_t = \ln 2\sqrt{ab}. \tag{17}$$

Let $x_{t+1} = \ln Z_{t+1}$, then we obtain the first-order recursive formula

$$\begin{cases} x_{t+1} - \dfrac{1+\varepsilon}{2} x_t = \ln 2\sqrt{ab} \\ x|_{t=0} = x_0. \end{cases} \tag{18}$$

System of equations (18) is a recursive formula with arithmetic-geometric series, according to the elementary algebra, after a simple linear transformation, its analytic solution can be concluded as follows

$$x_t = \left(x_0 + \frac{2\ln 2\sqrt{ab}}{\varepsilon - 1} \right) \left(\frac{1+\varepsilon}{2} \right)^t - \frac{2\ln 2\sqrt{ab}}{\varepsilon - 1}.$$

Asserting it back into the variable Z_t, and then getting the following formula

$$Z_t = \frac{Z_0^{\left(\frac{1+\varepsilon}{2}\right)^t}}{(4ab)^{\frac{1}{\varepsilon-1}}} \cdot \exp\left\{ \frac{2\ln 2\sqrt{ab}}{\varepsilon - 1} \cdot \left(\frac{\varepsilon + 1}{2} \right)^t \right\}. \tag{19}$$

That is,

$$Z_t = \frac{Z_0^{\left(\frac{1+\varepsilon}{2}\right)^t}}{(4ab)^{\frac{1-\left(\frac{1+\varepsilon}{2}\right)^t}{\varepsilon-1}}}. \tag{20}$$

3 Conclusion

It is important to discuss the convergence or diverge of recursive constraints condition (9) or (10) for solving optimal problem (1).

If $a<1$, then Z_t is convergent, and there exist at least a fixed points in equation (14), it local at $\overline{Z}_+ = \sqrt{\frac{b}{1-a}}$. Next, analyze the convergence character of equation (20). If $Z_0 > 4ab$, and $-3<\varepsilon<1$, then Z_t is convergent to $(4ab)^{\frac{1}{1-\varepsilon}}$. If $Z_0 < 4ab$, and $\varepsilon>1$ or $\varepsilon<-3$, then Z_t is also convergent to $(4ab)^{\frac{1}{1-\varepsilon}}$. Otherwise, If $Z_0 > 4ab$, and $\varepsilon>1$ or $\varepsilon<-3$, then Z_t is divergent.

Acknowledgments. This research was supported by National Natural Science Foundation of China (71171128) and Research Fund of Program Foundation of Education Ministry of China (10YJA790233).

References

1. Genevieve, V.: What drives long-term capital flows? A theoretical and empirical investigation. Journal of International Economics 74, 120–142 (2008)
2. Alfaro, L., Kalemli-Ozcan, S., Volosovych, V.: Why Doesn't Capital Flow from Rich to Poor Countries? An Empirical Investigation. Mimeo 32, 363–394 (2003)
3. Gollin, D.: Getting income shares right. Journal of Political Economy 110, 458–474 (2002)
4. Lane, P.R.: International trade and economic convergence: the credit channel. Oxford Economic Papers 53, 221–240 (2001)
5. Verdier, G.: The Role of Capital Flows in Neoclassical Open-Economy Models with Imperfect Capital Markets. Mimeo 26, 34–46 (2003)

Harmonious Labor-Relations Research from the Perspective of Management

Zhonghao Zhu[1] and Jianhua Fang[2]

[1] Business School Hohai University, Nanjing, China
zzh5566@yeah.net
[2] Business School of Jiujiang University, Jiujiang, Jiangxi, China
fangerli@163.com

Abstract. The Harmonious Labor Relations is a kind of harmonious labor relations the two sides of the benign state, is also a state of balance is to promote social justice and equity, the main body of labor relations through a variety of communication and coordination mechanisms for the two sides of the operation of the construction of a the balance of rights and obligations of the state. In this paper, the social responsibility of management, management ethics, motivation and other perspectives to explore the logic of building a harmonious labor relationship based.

Keywords: social responsibility, management ethics, motivation theory, Harmonious Labor Relations.

1 Introduction

Sixteenth Party Central Committee in the "CPC Central Committee on Strengthening the Party's governing capability of the decision" for the first time that "building a socialist harmonious society" concept, and made a harmonious socialist society defined four principles: a comprehensive implementation of respecting labor, respect knowledge, respect talent and respect creation, enhance the creativity of the whole society; to properly coordinate the interests of all relations, properly handle contradictions among the people; strengthen social construction and management, social management system innovation; sound working mechanism to maintain social stability. Harmonious labor relations as the most direct manifestation of respect for labor, constructed in the current background of a harmonious society, harmonious labor relations research has a strong theoretical and practical significance.

2 To Clarify the Concept

To study Management Science and harmonious labor relations logic, we must labor relations, and harmonious labor relations for a definition of the nature and content. As a traditional concept of labor relations, different countries have different interpretations, also known as industrial relations, labor relations, employment

J. Luo (Ed.): Affective Computing and Intelligent Interaction, AISC 137, pp. 923–928.
springerlink.com

relations, labor relations, labor relations and employee relations so. In our definition of labor relations are not the same, some scholars believe that the management of labor relations and employee groups and employee personal rights and obligations between the arrangements and the benefits of cooperation and conflict caused by the sum of (Yang ren, 2000). More people think that labor relations is the employee and the employer (including all types of enterprises, individual businesses, institutions, etc.) in the realization of the labor process to establish the socio-economic relations. Broadly speaking, living in urban and rural nature of any worker and any employer to engage in labor and formed between the result of social relations are all areas of labor relations. In a narrow sense, the reality of the economic life of labor relations is in accordance with national labor laws and regulations, for acts arising from employment relationship is between the employer and employee rights and obligations.

On the "harmonious labor relations," the nature and content of the academic and there is no uniform definition, it is generally believed that a socialist harmonious society urgently needs the harmonious labor relations, what is the harmonious labor relations? Some scholars believe that: the so-called Labor relations, is in a state of some organizations to communicate, according to harmonious and orderly participation in a proactive, fair, just, harmonious labor relations (Wangxian Sen, 2005). Also suggested that: the essence of harmonious labor relations between the interests of the principal labor relations harmony, is the subject of labor relations rights and obligations of both sides of the balance (Pei Chi, 2005). Some people think that a socialist harmonious labor relations "in the state, society, business, consistent with the fundamental interests of workers, based on the two sides of labor relations have different specific interests, but the relative balance of rights and obligations of labor relations; is able to work The relationship between the interests of the main demands include legal and institutional framework, implementation and security according to law, labor relations; is the ability to market regulation and state intervention through the combination of self-resolve and eliminate conflicts of interest, promote social justice and equitable labor relations "(GAO Ai-di, 2006). Chang Kay believes that the so-called harmonious society, the law is mainly a balance of social rights of all parties, the labor relations, is the capital and the right to work to balance.

We believe that harmonious labor relations as an ideal description, to give its precise definition under, or a very difficult task. Literally understood on the basis of harmonious labor relations is a kind of harmonious labor relations in the benign state of labor relations is the main rights and obligations of both sides balance. Thus, we can give a harmonious working relationship and such a definition: In order to promote social justice and equity, the labor relations of the main mechanisms for bilateral communication and coordination through a variety of building the operation of a balance of rights and obligations.

3 Social Responsibility and Harmonious Labor Relations

Corporate social responsibility is good for the community to pursue long-term goal of an obligation, it goes beyond the requirements of legal and economic obligations. Management theorists on the social responsibility are two different views: the classical perspective and socio-economic point of view. Classical view holds that the only social responsibility of management is to maximize profits. This view of the most outspoken supporters is an economist and Nobel Prize winner. Freedman that the primary responsibility of managers is from the best interests of shareholders to engage in business activities. He also stressed that any case, when given a free hand to organize resources manager for the "social benefits", the operating costs are increasing. These costs can either pass on to consumers through higher prices, or reduce the dividend yield is absorbed by shareholders. Socio-economic point of view that the social responsibility of managers is not only to create profit, including the protection and promotion of social welfare. Company is not only independent entity accountable to their shareholders. They are also socially responsible, community recognized by various laws and regulations of the company's establishment, and by purchasing their products and services to support. In addition, the socio-economic view of supporters that the organization is not just economic institutions.

Enterprises should take how much and what type of social responsibility, which has been in favor of and against the social responsibility debate. We believe that labor relations is the most basic social relations, harmonious labor relations is the foundation of social harmony. Business as a member of society, maintain social stability, maintain social harmony is part of corporate social responsibility, thus building a harmonious labor relations is a way of corporate social responsibility. Harmonious labor relations, long term, not only does not hinder business development and reduce corporate profits, but its concept of operation and management of enterprises, means a higher demand, which is exactly You Improving business productivity, changing way of promoting economic development an important opportunity, it makes business from the past in order to compete with low labor costs as the basic model of development tools to build a harmonious working relationship to improve the innovation model of development as the basic means change, which will help to promote their long-term health development.

4 Management Ethics and Harmonious Labor Relations

What moral and harmonious labor relations management what kind of logical relationship? Robbins, in his management book, "Management" elaborated on the four perspectives on corporate ethics: utilitarianism, rights, concept, concept of justice and social contract theory of integration theory. Utilitarian view of morality is the result or consequence of the development in full accordance with ethical decision-making. Utilitarianism to encourage efficiency and productivity, and meet the goal of maximizing profits, as utilitarian view of these characteristics, utilitarianism will also cause the rights of some stakeholders to be ignored. The second moral is the moral right to view, this point of view more attention to respecting and protecting individual freedoms and privileges, including privacy, freedom of thought, freedom of speech, life and security and the rights provided by law. The third point is theoretical concept of moral justice, which requires managers to implement a fair and impartial and to strengthen the rules, and in the process to comply with all laws and regulations. Finally a kind of morality is the integration of social contract theory, this view should be based on empirical factors (what is) and normative factors (what should be) to develop ethical decision-making, based on two kinds of "contract" of integration: allow companies to address and determine the basic rules of acceptable the general contract, as well as community members deal with an acceptable pattern of behavior among a more specific master contract.

Robbins also said that while most managers will adopt a utilitarian ethic, but because managers are changing their environment, emphasis on individual rights, social justice and community standards of the trend means that managers need to be non-utilitarian standards foundation of ethics. Managers in global organizations to clear the moral principles, in January 1999 the World Economic Forum, UN Secretary-General called on world business leaders "to adopt and participation", "Global Compact", the contract outlines the global business in human rights, labor and the environment in nine basic principles. Which are discussed in harmony with our labor relations have linked the six, they are: Principle 1: In the context of their impact and to maintain respect for internationally recognized human rights; Principle 2: not participate in any disregard and trample on human rights behavior; Principle 3: Maintain freedom of association; recognize the right of collective bargaining; Principle 4: the total elimination of all forms of compulsory labor; principle of elimination of child labor; Principle 6: eliminate any discrimination in employment and occupation.

Therefore, we believe that with the globalization of economic development and the management of the environment changes, we managers morality will change. Consistent with the ethical principles of global organizations is imperative, which established ethical principles, is the pursuit of harmonious labor relations, nor is our current discord in the missing labor relations. According to China's "Labor Law," China's domestic enterprises and other economic organizations, legal rights and interests of employees entitled to the following: (1) equal employment and choice of occupation; (2) the right to obtain payment for labor; and (3) rest and vacation rights; (4) access to labor safety and health protection of the rights; (5) the right to receive vocational training; (6) The right to social insurance and welfare rights; (7) the right to social welfare; (8) drew the right to labor disputes; (9) to participate in the democratic management of trade unions and workers rights ; (10) and other labor rights law. But the employer prejudice the legitimate interests of all workers widespread acts, such as ignoring and trampling on human rights workers and the use of child labor in Shanxi "illegal brick kilns" event; recruitment of sex discrimination when hiring and employment discrimination; malicious withholding of wages, "migrant workers" events; malicious exploitation of workers social insurance, welfare, rest and holidays with events is everywhere.

5 Motivation Theory and Labor Relations

Incentives in Management is always a hot topic. Some scholars believe that the motivational incentive is essentially continuous process, the process refers to the incentive induced by the needs of the engine, and then act to achieve their goals by the cycle. Need is a source of motivation and also the promotion of social productive forces, changes in social relations of production of a powerful driving force. Demand theory is the foundation of all motivation theory. Management theorists on the incentive to do so Robbins definition: a high level of effort the individual hopes to achieve organizational goals and aspirations, the condition that such efforts can meet some of the individual needs of its three key elements: effort, organizational goals and needs. Meanwhile, Robbins also believe that incentives can be seen as a process needs to be met. Since the need is the source of motivation, then motivation theory and the relationship between a harmonious labor relations, we can find from the need for theory.

Management needs of the more influential theories: Maslow on needs theory, Herzberg's motivation - hygiene theory, McClelland's three needs theory. Maslow believed that each person has five levels of needs: physiological needs, security needs, social needs, esteem needs, self-fulfillment needs, and its built in accordance with the pyramid approach. Herzberg's Motivation - Hygiene Theory, also known as two-factor theory. The theory that the impact of human behavior, there are two positive factors, namely, health factors and incentives. Health factors, also known as the maintenance factors, such factors on the behavior of people who only serve to maintain the enthusiasm, maintenance of the status quo. Incentives, also known as satisfaction factors that affect people's working enthusiasm of the major internal factors. McClelland's achievement theory that people need basic survival needs are met before we can produce Achievement, power needs and ownership needs.

From the theoretical analysis of these needs, we can easily find people to inspire their work, to be the guarantor of a lower level needs are met, otherwise, not only can not inspire enthusiasm, but increased dissatisfaction, and dampen the enthusiasm of staff work. We believe that harmonious labor relations are very consistent with incentive process. From the party and the government on "the development of harmonious labor relations," remarks of view, "the development of harmonious labor relations," the concrete manifestation of that is called development process to effectively maintain a good, good, and Bao Zhang achieve good health of the people rights, security rights, labor rights, education, and other subsistence and development, that is to change the original harmony of labor relations, to resolve discord arising under the labor relations problems, and make it grow labor relations rights and obligations of the principal balance of the state of the road. If the current tension between employers and employees more prominent problems exist: Corporate social insurance system is flawed and ignore the labor protection belong to the lower-level hierarchy of needs, from the perspective of motivation theory, and even lower-level workers need all Debu to satisfy the enthusiasm of the staff can not really get excited, the staff is definitely no lack of enthusiasm high efficiency, high efficiency is not no high profits; their safety needs are not protected, then employees need to be met out of the question ownership, no employees a sense of belonging can not be truly low-cost, high efficiency work, from the business point of view, it is impossible to achieve real profit maximization.

6 Conclusion

Enterprises as an integral part of society, with economic development, globalization and changes in regulatory environment, consistent with the ethical principles the global organization is imperative to build a harmonious working relationship between corporate social responsibility is a natural way, is the pursuit of advanced business management ethics the only way the principle. Motivation theory from the point of view, first of all to ensure harmonious relations between workers with lower-level needs are met, which is bound to truly lay the foundation for enterprises to maximize profits. Therefore, in the long run, building a harmonious labor relations, not only does not hinder the development of enterprises, but also help promote the healthy development of enterprises.

References

1. Bureau of the CPC Central Committee Propaganda Department of the theory. Implementation of the "Labor Contract Law" thinking 2008, theory of hot face to face (2008)
2. Ding, M.: Managerial Psychology. China Science and Technology University Press, Hefei (2001)
3. Robbins, S.P., Kurt, M.: Management, 7th edn. China Renmin University Press, Beijing (2003)
4. Liu, T., Luo, Y.: Chinese Harmonious Labor Relations Review. Marxism with the Reality (bimonthly) (6) (2007)

Design on Motivate Mechanisms of Knowledge Workers

Xiaolin Chen

School of Business Administration, Northeastern University, Shenyang, China
xlchen@mail.neu.edu.cn

Abstract. Although scholars and managers have realized that knowledge workers in knowledge-based organizations should be motivated effectively, it seems that some of the motivate mechanisms they proposed have not worked well. The reason may be that those mechanisms are for all knowledge workers under the assuming that their motivation factors are the same. In fact, knowledge workers have different kinds, and every kind of knowledge workers have different needs in their different career stages. Therefore, the motivate mechanism should not be the same. In this paper, different kinds and motivate factors of knowledge workers are analyzed, and then motivate mechanisms are put forwarded for different kinds of knowledge workers.

Keywords: Knowledge worker, motivate mechanism, motivate factor.

1 Introduction

According to Drucker, a knowledge worker is one who relies on knowledge rather than skills to perform his or her job [1]. Teachers, doctors and lawyers are example of knowledge workers who master and exert symbol, concept, knowledge and information to create knowledge, and they have intelligence input, creative ability and authority to fulfill their job [2]. In the knowledge economy era, human capital and knowledge capital have been the source of enterprise keeping competitive advantage. The competition between the enterprises finally depends on the knowledge workers who are knowledge carrier. How to effectively motivate knowledge worker has become the center of gravity of enterprise human resources management, and also become a core proposition of the modern enterprise keeping sustainable development.

Nowadays, scholars and managers have realized that knowledge workers in knowledge-based organizations should be motivated effectively, and then they put forward the corresponding theories and practice methods. As to motivation theories, Stott and Walker refer to Maslow's theory to indicate that motivation for knowledge work comes from his three highest hierarchical levels, including belongingness, esteem and self-actualization [3-4]. Their implication is that knowledge workers do not share knowledge because of money or to improve their relations with their co-workers. Their motivation rather comes from their desire for self-actualization. From other motivation theories, motivation factors also include: the wish to earn wages, to expend mental or physical energy, to contribute to the production of goods or services, the desire for social interaction and social status [5], the wish to survive,

J. Luo (Ed.): Affective Computing and Intelligent Interaction, AISC 137, pp. 929–933.
springerlink.com

enjoy, belong, play, the desire for recognition and respect, the need for achievement, affiliation and power [6]. An important instrument for introducing some order into this smorgasbord was offered by Herzberg's two-factor theory that was based on the distinction between motivation factors and maintenance or hygiene factors. Basing on several empirical studies, Herzberg asserts that salary, working conditions, status and interpersonal relations are hygiene factors. While challenge of work, promotional opportunities, sense of achievement, recognition of job done, and sense of responsibility may act as motivators. In addition, a sixth factor of the desire for operational autonomy, which is defined as the balance between regulatory needs and regulatory capacities, was distinguished by Hindriks [7]. Other studies also show the relevance of this factor as a distinct motivator [8].

According to the above research, regulators have conducted some beneficial attempts in practice. For example, through the administrative intervention in organizations, let the knowledge workers work together, in order to make them trust each other and share their knowledge. Other organizations provide the knowledge workers that willing to share knowledge with rewards. But it seems that some of the motivate mechanisms they proposed have not worked well. Because the activities of sharing knowledge innovation neither can be forced to realize, nor can be totally realized by rewards. The reason is that those mechanisms are for all knowledge workers under the assuming that their motivation factors are the same. In fact, knowledge workers have different kinds, and every kind of knowledge workers have different needs in their different career stages. Therefore, the motivate mechanism should not be the same. In this paper, different kinds and motivate factors of knowledge workers are analyzed, and then motivate mechanisms are put forwarded for different kinds of knowledge workers.

2 Types of Knowledge Workers and Analysis of Motivate Factors

Alexandra Durcikova and Andrea Everard, who according to the knowledge content and its employees in knowledge sharing of participation in different degrees, presented a 2×2 model of employee typology shown in figure 1, which shows four categories, namely seeing eye dogs, foxes, peacocks and ants type [9].

Knowledge workers in the upper left hand cell have knowledge and will share it, and are identified as Seeing Eye dogs. They will be able to explore the unknown area, and share this knowledge with colleagues and team partners. This type of knowledge worker is the mainstay of the organization, they cannot only promote knowledge innovation, also can make the organization form culture atmosphere of keeping learning and sharing knowledge. This kind of knowledge workers often work as leaders within the organization or directors of some the project, they work hard mainly based on organizational development sense of responsibility, the second is to challenge for the new field of achievement, then is the need of personal growth.

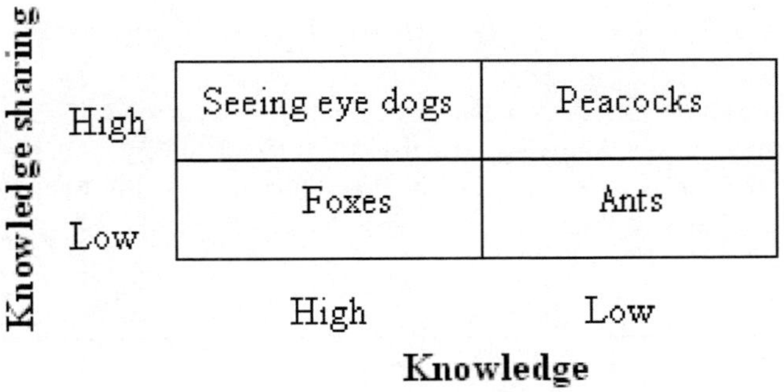

Fig. 1. The knowledge worker typology (KT)

The cell in the lower left hand side of the matrix represents knowledge workers who have knowledge but do not share it with others. They hoard their knowledge while that does not preclude them from seeking knowledge from others. One of the reasons that this kind of knowledge workers do not share their knowledge or contribute to management systems may be because that they believe unique knowledge can be a source of power. Another reason may be that they feel that they are just dropping their knowledge into a black hole. They also have certain authority in the organization, but may not in a leadership position, they work hard is mainly based on personal development needs and obtain sense of achievement to challenge new work, and it still has a certain proportion of sense of responsibility to organizational development.

Knowledge workers who fall in the upper right hand cell have low knowledge yet are willing to share. The professional ability of these employees needs to be enhanced, and they show or exaggerate their knowledge content eager for the respect of profession. Since the knowledge they transmit may not be accurate and that will not lead to an increase in an organization's knowledge base and may even have grave consequences. They work hard firstly for the needs of the individual's development and sense of achievement they acquire in new challenge work, secondly is based on sense of responsibility to organizational development.

Finally, knowledge workers who have low knowledge and are not willing to share it fall into the lower right hand cell of the matrix. According to Alexandra Durcikova and Andrea Everard, these kinds of knowledge workers are mainly those employees whose work are the routine tasks in organizations. Their knowledge content for work is not like other types of knowledge workers, or the motivation to knowledge sharing for promoting personal knowledge ability is not so strong. They work hard for personal development and reward, secondly for getting the sense of accomplishment and sense of responsibility based on organizational development.

From the above analysis, we can see that incentive factors are different for all types of knowledge workers to work hard. The difference is not the incentive factor itself, but in these factors for each type of knowledge workers at different levels. So, for each type of knowledge worker we should make distinct incentive measures.

3 Motivate Mechanisms to Knowledge Workers

Based on the above analysis, this paper puts forward the incentive model to knowledge workers. As shown in figure 2, personal ability and the organization culture have influence to knowledge worker behavior, by his behavior reaches to organizational objective, the personal goal or both, it will have further incentive effect.

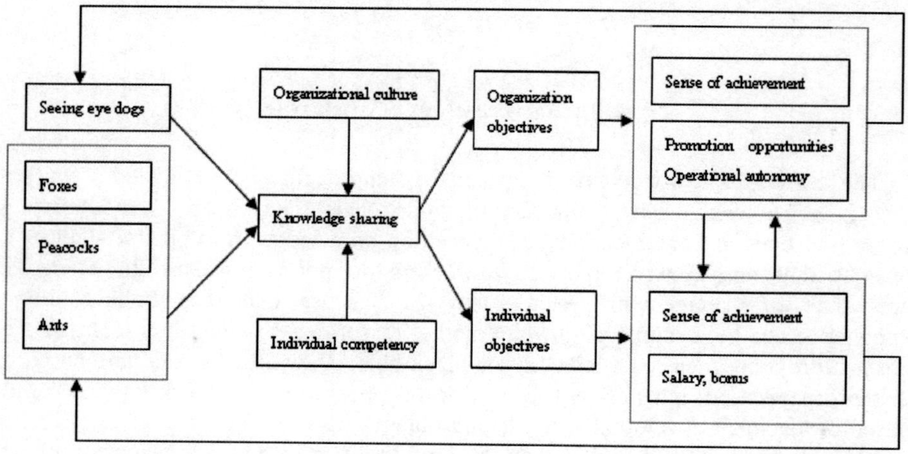

Fig. 2. Motivation model for knowledge workers to share their tacit knowledge in organizations

According to Herzberg's two-factor theory, the hygiene factors only contribute to motivation in a roundabout, primarily negative way and the motivation factors act as motivators. This is particularly true for "seeing eye dogs" which share their knowledge mainly because they expect or hope for recognition and appreciation of their knowledge work, promotional opportunities or because of a sense of responsibility since they are often assigned to be responsible for the development of their organizations. So it is suggested that this kind of knowledge workers be motivated to share their tacit knowledge mainly by motivators such as the sense of achievement, the promotion opportunities and more operational autonomy etc. At the same time, some hygiene factors such as bonuses and salary should be combined sometimes since they may affect the knowledge sharing in some degree too.

As for the other three kinds of knowledge workers, a two-trajectory system is suggested. First, their hope to develop their individual knowledge competences should be supported and the instruments may be relevant to the hygiene factors such as salary, status, interpersonal relations etc. While this is only for a short period of time since the hygiene factors can only reinforce the will of tacit knowledge sharing when they are present, but they will lead to dissatisfaction and, therefore, to a decreased motivation when absent. So second, motivators of spirit rewards such as the sense of achievement, the promotion opportunities and more operational autonomy are suggested to be use in long term.

There are something that should be noticed when organizations use the above motivation mechanisms. First, some reasonable knowledge protection mechanisms such as IP protection should be permitted in organizations and this may promote the "foxes" to share more of their tacit knowledge. Second, organizations should take precautions to prevent "peacocks" from emerging or sharing their knowledge, since the simple equation that knowledge sharing is good for organizations cannot be sustained. Knowledge can be augmented if people truly learn from each other, while it may also prove detrimental to knowledge if inadequate representations of knowledge are transferred between people. Third, organizations should not only motivate the "ants" to share their knowledge but also should help them develop their knowledge competences quickly, that will reinforce the sense of belongingness of this kind of knowledge workers, and this is also one of the objectives of organizations. Obviously, these above measures will make the KT model shown in figure 1 a dynamic one that is good for the knowledge innovation in organizations.

The motivation model is not static, but should be dynamic. On one hand, as for the position of the "seeing eye dogs" entails a significant amount of mental exertion, other knowledge workers are required to come in and renew the batch of ideas. On the other hand, if the organizational culture encourages learning and respects for individual development, with the ability of other types of knowledge workers rising, their behavioral target will be changed from the basic pursuit of the bonus and reward to personal growth. When personal development reaches, if the organization offers for knowledge workers enough development space, they will consciously combine personal and organizational development goals, that is the sense of responsibility to organizational development will gradually strengthened. And when their work can help to realize the goal of organizational development, it will have further incentive effect to personal behaviors. So, in the process of roles mixing and ascending of all types of knowledge workers, the backbone forces of organization will continuously be strengthened, the knowledge innovation ability of organization will continue to be promoted.

References

1. Drucker, P.: The Future that has already Happened. The Futurist, New York (1998)
2. Study Report of Ansheng Consulting Company: Workforce of the 21st Century (1998)
3. Stott, K., Walker, A.: Teams, Teamwork & Teambuilding: The Manager's Complete Guide to Teams in Organizations. Prentice Hall, New York (1995)
4. Tampoe, M.: Motivating Knowledge Workers-the Challenge for the 1990s. In: Myers, P.S. (ed.) Knowledge Management and Organizational Design, pp. 179–190. Butterworth-Heinemann, Boston (1996)
5. Vroom, V.H.: Work and Motivation. Wiley, New York (1964)
6. McClelland, D.C.: Assessing Human Motivation. General Learning Press, Morristown (1971)
7. Herzberg, F.: Work and the Nature of Man. Granada Publishing, London (1968)
8. De Sitter, L.U.: Synergetic Production: Human Resources Mobilization in Production: an Introduction to Structuration. Van Gorcum, Assen (1994)
9. Durcikova, A., Everard, A.: An Employee Typology: A Knowledge Management Perspective. In: Eighth Americas Conference on Information Systems (2002)

The Application of Competence Model in the Decision of Excellent Managers

Yang Yuqing and Zhou Zhihong

Automobile Management Institute, Bengbu, Anhui Province, China

Abstract. To select the excellent managers in enterprises and public institutions, this paper designs the competence model and the selection process of excellent managers. To raise selection quality, a new portfolio of decision model has been developed in formal decision and explains the combining use of G_1 weighting, many Vague target decision based on the grey relation and ideal point method. Finally, a case study is given.

Keywords: competence model, selection process, grey relation, ideal point method, application.

1 Instruction

The selection mechanism of competence model covers shortages of traditional selection device. Making use of competence model helps enterprise and public institution find outstanding leaders and improve the quality of selecting key talents.

2 The Structure of Competence Model

Ways of document searching, behavior event interview, expert group discussion, questionnaire method, help find the core competence factors and construct the competence model framework, which in turn discover that the capability of excellent leaders are mainly focusing on three aspects, namely, leadership and work force, scope of knowledge and working experience as well as responsibility and performance consciousness (as is shown in Table 1).

Table 1. Outstanding Leader Competence Model

Competence Dimensions	Competence Factors
Leadership and Work	Capability of coordination and communication C_{11}
	Capability of resolution and time management C_{12}

J. Luo (Ed.): Affective Computing and Intelligent Interaction, AISC 137, pp. 935–944.
springerlink.com

Table 1. *(continued)*

Force CM_1	Capability of stability maintenance, conflict resolution, guard consciousness C_{13}
	Consciousness of uncompromising integrity, democracy life style, cooperation and collaboration C_{14}
	Consciousness of personal trainingC_{15}
Scope of Knowledge and Working Experience CM_2	Educational background, training status, foreign-related knowledge, general knowledge C_{21}
	Capability of insight, investigating problems, decision making C_{22}
	Capability of learning, innovation and flexibility C_{23}
Responsibility and Performance Consciousness CM_3	Leadership capabilityC_{31}
	Confidence, patience and high stress tolerance, anti-fatigue capability C_{32}

3 The Selection Process Based on Competence Model

Firstly, announce the criteria of selection on public or Internet and organize the enrollment. The juries are composed of leaders and experts in enterprises and public institutions.

Secondly, select participants by using personality test software, following with holding the thesis defense on questions involved with position competition.

Thirdly, by synthesizing the result of personality test and thesis defense, the juries grade each participant based on competence model by 0/1 score system (qualified for 1 score, unqualified for 0). The top 50% of the participants, ranking by total score, go to the next round as candidate. The rest will be enrolled into training talent pool.

At last, for those candidates entering the next round, the juries will judge them by the following methods.

4 The Formal Selection Decision Based on Competence Model

So far, the most commonly used selection model is Analytic Hierarchy Process plus Fuzzy Comprehensive Evaluation. Based on the characteristics of competence model and the effectiveness of selection, a new model of portfolio decision has been developed:

1—G1 Weighting Method: Calculate the weight of the competence factor within each dimension.

2—Vague Fuzzy Decision: Represent candidates' assessment result from both negative and positive sides in view of the behavior description within competence factor.

3—Analysis of Grey Relation: Ensure candidates ranking under each dimension are based on the comparison of grey relation between candidates and ideal leaders.

4—TOPSIS: Calculate the degree of approximation between the ranking under the competence factor and ideal point, then work out the official candidate.

4.1 G1 Weighting Method

It is better to choose combination weighting method as G1 method to confirm weighting coefficient[1]. The first step we need to do is to order the evaluation index, then to assign the value (compare the adjacent index in turn), which give the condition of strong rank preservation. The steps are as followed:

(1)Confirm Order Relation

Choose the most important element as you think in the index set $\{x_1, x_2, \cdots x_n\}$, mark it as x_1^*; then select the most important one from the rest of n-1 index set, mark it as x_2^*; and so on. In this way, it makes the only order relation, that is $x_1^* \succ x_2^* \succ \ldots \succ x_n^*$

(2)Comparison Judgment of Corresponding Importance

Suppose the comparison of the importance between x_{k-1} and x_k is ω_{k-1}/ω_k, of which the ideal judgment is $\omega_{k-1}/\omega_k = r_k$, $k=n, n-1, n-2, \ldots, 3, 2,$ please refer to the following table of r_k 's assigned value:

Table 2. Assigning Reference Form

r_k	Instruction	r_k	Instruction
1.0	Index x_{k-1} and index x_k are equally important	1.6	Index x_{k-1} is especially more important than index x_k
1.2	Index x_{k-1} is a little more important than index x_k	1.8	Index x_{k-1} is extremely more important than index x_k
1.4	Index x_{k-1} is obviously more important than index x_k		

(3)The calculation of weighting coefficient ω_k

If the rational assigned value of r_k matches the relation $r_{k-1} > 1/r_k$, the weighting coefficient ω_n's calculating formula is as follows:

$$\omega_n = (1 + \sum_{k=2}^{n} \prod_{i=k}^{n} r_i)^{-1} \tag{1}$$

4.2 The Vague Multi-target Decision Based on Grey Related Theory

(1) Vague set theory

The basic idea is[2]: Assume U as a domain of discourse, for any elements in the range of x, it can use a true membership function $t_v(x)$ and a false membership function $f_v(x)$ in V of vague set to describe the range of membership function, which means $t_v(x) \leq \mu_v(x) \leq 1 - f_v(x)$, these two ranges become a subinterval [$t_v(x)$,1-$f_v(x)$] of range [0,1], the support, against degree and unknown degree of one of the object in the range are $t_v(x)$、$f_v(x)$ and 1-$t_v(x)$ - $f_v(x)$, which makes Vague set is more powerful to judge than other obscure judging methods when dealing with uncertain information.

(2) Grey Relation Analysis

The basic idea is to judge whether the relation is strong based on the similarity of sequenced curve's similarity. The closer of the curve, the stronger relation of corresponding sequence is, vice versa. Its mathematical definition is as follows:

Assume the reference sequence:

$X_0(k) = \{x_0(1), x_0(2),..., x_0(n)\},\ k = 1,2,..., n$

Compare sequence: $X_i(k) = \{x_i(1), x_i(2),..., x_i(n)\},\ i = 1,2,..., n$

The grey relation coefficient of compare sequence X_i and reference sequence X_0 on the k point:

$$\gamma_i(k) = \frac{\min_i \min_k |x_0(k) - x_i(k)| + \xi \max_i \max_k |x_0(k) - x_i(k)|}{|x_0(k) - x_i(k)| + \xi \max_i \max_k |x_0(k) - x_i(k)|} \tag{2}$$

ξ as the distinguishing coefficient, $\xi \in [0,1]$, normally referencing $\xi \leq 0.5$ as the most appropriate.

The target weighted vector $\omega_i = (\omega_1, \omega_2, ..., \omega_n)$, the calculating formula of weighted vector's grey relation is:

$$R_i(X_0, X_i) = \sum_{k=1}^{n} \omega_i r_i(k) \tag{3}$$

(3) The Vague Multi-target Decision based on Grey Related Theory

Firstly, confirm the ideal plan satisfying the target conditions set and make it as reference order, then calculate the relation between candidate plan and ideal plan from the positive and negative aspects, and come to the ordering result of candidate plan.

Assume the candidate plan set $A = \{A_1, A_2, ..., A_m\}$, target conditions set $C = \{C_1, C_2, ..., C_n\}$, the characteristics of Plan A_i under Condition C is shown with Vague set as follows:

$$A_i = \{(C_1,[t_{i1},1-f_{i1}]),(C_2,[t_{i2},1-f_{i2}]),...(C_n,[t_{in},1-f_{in}])\}$$

Among them, t_{ij} represents that Plan A_i satisfies the level of Target Condition C_j, f_{ij} represents that Plan A_i does not satisfy the level of Target Condition C_j, $t_{ij}, f_{ij} \in [0,1]$, $t_{ij} + f_{ij} \leq 1$. From the positive aspect, confirm the ideal plan of decision as: $A_0^t(a_0^t(1), a_0^t(2), ..., a_0^t(k), ..., a_0^t(n))$; from the negative aspect, confirm the ideal plan of decision as:

$$A_0^f(a_0^f(1), a_0^f(2), ..., a_0^f(k), ..., a_0^f(n))$$

The reference order of ideal plan is as follows:

$$A_0\left\{\lfloor a_0^t(1), (1-a_0^f(1))\rfloor, ..., \lfloor a_0^t(k), (1-a_0^f(k))\rfloor, ..., \lfloor a_0^t(n), (1-a_0^f(n))\rfloor\right\}$$

Referring to formula (2) and (3), comes the relation $R_i \lfloor R_i^t, R_i^f \rfloor$ between candidate plan and ideal plan A_0:

$$\gamma_i^t(k) = \frac{\min_i \min_k \Delta_i^t(k) + \xi \max_i \max_k \Delta_i^t(k)}{\Delta_i^t(k) + \xi \max_i \max_k \Delta_i^t(k)} \tag{4}$$

$$R_i^t = \sum_{k=1}^{n} \omega_i \gamma_i^t \tag{5}$$

Among them $k=1,2,...n, \xi=0.5$, $\Delta_i^t(k) = \left|t_{ik} - A_0^t(k)\right|$. Similarly, comes to the solution of $\gamma_i^f(k)$ and R_i^f.

In comparing multi-plans, order according the value of real relation R^t. If the value R^t equal, compare the value of fake relation R^f. Finally confirm the advantages and disadvantages order of every plan.

4.3 Ideal Point Method

Ideal Point Method is similar to relation analysis principles. It refers to seeking for the plan closest to the ideal point as the final evaluation decision plan, so as to avoid the final decision differences caused by various evaluation and decision results. The process of decision is as follows:

Assume the candidate plan set $A=\{A_i\}(i=1,2,...m)$, attribute set $D=\{D_j\}(j=1,2,...,p)$, the attribute value of the i plan in j attribute is x_{ij}, the decision matrix X is indicated as:

$$X = \{x_{ij}\} = \begin{pmatrix} x_{11} & x_{12} & \cdots & x_{1p} \\ x_{21} & x_{22} & \cdots & x_{2p} \\ \cdots & \cdots & \cdots & \cdots \\ x_{m1} & x_{m2} & \cdots & x_{mp} \end{pmatrix} \quad Y = \{y_{ij}\} = \begin{pmatrix} y_{11} & y_{12} & \cdots & y_{1p} \\ y_{21} & y_{22} & \cdots & y_{2p} \\ \cdots & \cdots & \cdots & \cdots \\ y_{m1} & y_{m2} & \cdots & y_{mp} \end{pmatrix}$$

Assume I_1 and I_2 are extremely-large index set and extremely-small index set respectively, choose standardized function for X to construct a normalized decision matrix Y. The calculating formula of standardized function is:

$$y_{ij} = \frac{x_{ij} - \min_j x_{ij}}{\max_j x_{ij} - \min_j x_{ij}} (X_{ij} \in I_1) \quad (6) \qquad y_{ij} = \frac{\max_j x_{ij} - x_{ij}}{\max_j x_{ij} - \min_j x_{ij}} (X_{ij} \in I_2) \quad (7)$$

According to the normalized decision matrix Y, confirm the approximately ideal decision plan set I* as: $I^* = (y_1^*, y_2^*, ..., y_p^*)$

To solve the corresponding approximation α_i and corresponding nearness T_i between every plan and ideal point:

$$\alpha_i = \sqrt{\sum_{j=1}^{p} (y_{ij} - Y_j^*)^2} \; (i = 1,2,...m) \quad (8) \qquad T_i = 1 - \frac{\sum_{j=1}^{p} y_{ij} Y_j^*}{\sum_{j=1}^{p} (Y_j^*)^2} (i = 1,2,..., m) \quad (9)$$

Choose the ordering principle as: the one with a smaller T_i value is better; if the T_i value is equivalent, the one with a smaller α_i value is better.

5 Demonstration

According to the above selecting process based on competence model, conduct the primary selection and formal selection of excellent managers in enterprises and public institutions. By judging the benchmark competence of overall excellent managers, five candidates are primarily selected into the formal selection. The formal selection method is introduced as follows:

5.1 Formal Selection Index Weight

G_1 method is applied to confirm the corresponding importance of factors in every dimension of excellent managers competence model. Hereby takes the weight of five competence factors in leadership and work force dimension as an example:

(1) Confirmation of Order Relationship

After the confirmation of the evaluating group, the most important index of the five is coordination and communication ability (X_1), followed by resolution and time management (X_2), the third is stability maintenance, conflict resolution, guard consciousness (X_3), the fourth is uncompromising integrity, democracy life style, cooperation and collaboration (X_4), and the final is concern and subordinates training, consciousness of talents training (X_5). Therefore the order relationship can be confirmed as: $x_1 \succ x_2 \succ x_3 \succ x_4 \succ x_5$.

(2) Comparative Judgment of Corresponding Importance

The comparative judgment of the corresponding importance between x_{k-1} and x_k given by the evaluating group is: $r_2 = \frac{\omega_1}{\omega_2} = 1.4$ $r_3 = \frac{\omega_2}{\omega_3} = 1.2$ $r_4 = \frac{\omega_3}{\omega_4} = 1.2$ $r_5 = \frac{\omega_4}{\omega_5} = 1.4$

(3)Weight Calculation

According to formula (1), weight ω_k is calculated as follows: $r_2r_3r_4r_5 = 2.419$, $r_3r_4r_5 = 1.728$, $r_4r_5 = 1.440$, $r_5 = 1.200$,

$$r_2r_3r_4r_5 + r_3r_4r_5 + r_4r_5 + r_5 = 6.787; \quad \omega_5 = (1+6.787)^{-1} \approx 0.128;$$

$\omega_4 = \omega_5 r_5 \approx 0.154$; $\omega_3 = \omega_4 r_4 \approx 0.184$; $\omega_2 = \omega_3 r_3 \approx 0.222$; $\omega_1 = \omega_2 r_2 \approx 0.312$

(4) Synthesis

Similarly, calculate the weight of every competence factor in the Scope of Knowledge and Working Experience Dimension and the Responsibility and Performance Consciousness Dimension. The result is synthesized as follows: (See Table 3)

Considering the comprehensive quality of excellent managers, equal the importance of three dimensions, namely, Leadership and Work Force, Scope of Knowledge and Working Experience, Responsibility and Performance Consciousness.

Table 3. Weight Table of Formal Selecting Index

Competence Dimensions	Competence Factors	Weight
Leadership and Work Force CM_1	Capability of coordination and communication C_{11}	0.312
	Capability of resolution and time management C_{12}	0.222
	Capability of stability maintenance, conflict resolution, guard consciousness C_{13}	0.184
	Consciousness of uncompromising integrity, democracy life style, cooperation and collaboration C_{14}	0.154
	Consciousness of personal trainingC_{15}	0.128
Scope of Knowledge and Working Experience CM_2	Educational background, training status, foreign-related knowledge, general knowledge C_{21}	0.433
	Capability of insight, investigating problems, decision making C_{22}	0.309
	Capability of learning, innovation and flexibility C_{23}	0.258
Responsibility and Performance Consciousness CM_3	Leadership capabilityC_{31}	0.583
	Confidence, patience and high stress tolerance, anti-fatigue capability C_{32}	0.417

5.2 Advantages and Disadvantages Ordering of Every Dimension of Candidates

The evaluating group individually judge five candidates based on Vague set theory. For instance, 10 members evaluate a certain candidate's "Coordination and Communication Ability", implement behaviors descriptive and comparative analysis referring to character test and coordination and communication ability in defense materials. 5 members regard the candidate as completely own this competence, 1 completely not own, 4 basically own, then the evaluating result of the candidate's "coordination and communication ability" is shown as [0.5, 0.9]; other competence factors analogize.

(1) Evaluating Result Formulation base on Vague set

Assume A_1、A_2、A_3、A_4、A5 are five candidates of excellent managers, C_{11}、C_{12}、C_{13}、C_{14}、C_{15} are 5 evaluating goals of Leadership and Work Force dimension, the evaluating result of the candidates is formulated by the following Vague set:

$$A_1 = \{(C_{11},[0.5,0.9]), (C_{12},[0.6,0.8]), (C_{13},[0.3,0.7]), (C_{14},[0.4,0.8]), (C_{15},[0.6,0.7])\}$$

$$A_2 = \{(C_{11},[1,1]), (C_{12},[0.7,0.9]), (C_{13},[0.6,0.8]), (C_{14},[0.5,0.7]), (C_{15},[0.4,0.8])\}$$

$$A_3 = \{(C_{11},[0.3,0.5]), (C_{12},[0.4,0.6]), (C_{13},[0.5,0.8]), (C_{14},[0.4,0.7]), (C_{15},[0.3,0.6])\}$$

$$A_4 = \{(C_{11},[0.5,0.8]), (C_{12},[0.6,0.8]), (C_{13},[0.3,0.8]), (C_{14},[0.4,0.9]), (C_{15},[0.6,0.8])\}$$

$$A_5 = \{(C_{11},[0.3,0.7]), (C_{12},[0,0]), (C_{13},[0.5,0.6]), (C_{14},[0.4,0.8]), (C_{15},[0.3,0.6])\}$$

Confirm the ideal plan: $A_0{}^t = \{1,1,1,1,1\}$, $A_0{}^f = \{0,0,0,0,0\}$

(2) Calculating Relation

$\Delta_1{}^t = \{0.5,0.4,0\ \ .7,0.6,0.4\ \}$ $\Delta_2^t = \{0,0.3,0.4,0.5,0.6\}$ $\Delta_3^t = \{0.7,0.6,0.5,0.6,0.7\}$

$\Delta_4^t = \{0.5,0.4,0.7,0.6,0.4\}$ $\Delta_5^t = \{0.7,1,0.5,0.6,0.7\}$

Then $\min_i \min_k \Delta_i{}^t(k) = 0$, $\max_i \max_k \Delta_i{}^t(k) = 1$; Calculate the real relation of the five goals evaluating result between every candidate and the ideal ones according to formula (4) and (5) (See Table 4):

Table 4. Coefficient Table of Real Relation

	$\gamma_i^t(1)$	$\gamma_i^t(2)$	$\gamma_i^t(3)$	$\gamma_i^t(4)$	$\gamma_i^t(5)$	R_i^t
$A_1(\gamma_1{}^t)$	0.500	0.556	0.417	0.454	0.556	0.497
$A_2(\gamma_2{}^t)$	1.000	0.625	0.556	0.500	0.454	0.688
$A_3(\gamma_3{}^t)$	0.417	0.454	0.500	0.454	0.417	0.446
$A_4(\gamma_4{}^t)$	0.500	0.556	0.417	0.454	0.556	0.497
$A_5(\gamma_5{}^t)$	0.417	0.333	0.500	0.454	0.417	0.419

Similarly, the fake relation of the five goals evaluating result between every candidate and the ideal ones (See Table 5):

Therefore, come to the overall relation between every candidate and the ideal ones: $R_1[0.497\ ,0.724\ ]$, $R_2[0.688\ ,0.815\ ]$, $R_3[0.446\ ,0.577\ ]$, $R_4[0.497\ ,0.732\ ]$, $R_5[0.419\ ,0.552\ ]$

(3) Advantages and Disadvantages Ordering

Advantages and Disadvantages Ordering of every candidate in Leadership and Work Force Dimension: $R_2 \succ R_4 \succ R_1 \succ R_3 \succ R_5$

Table 5. Coefficient Table of Fake Relation

	$\gamma_i^f(1)$	$\gamma_i^f(2)$	$\gamma_i^f(3)$	$\gamma_i^f(4)$	$\gamma_i^f(5)$	R_i^f
$A_1(\gamma_1^f)$	0.833	0.714	0.625	0.714	0.625	0.724
$A_2(\gamma_2^f)$	1.000	0.833	0.714	0.625	0.714	0.815
$A_3(\gamma_3^f)$	0.500	0.556	0.714	0.625	0.556	0.577
$A_4(\gamma_4^f)$	0.714	0.714	0.714	0.833	0.714	0.732
$A_5(\gamma_5^f)$	0.625	0.333	0.556	0.714	0.556	0.552

Similarly, Advantages and Disadvantages Ordering in Scope of Knowledge and Working Experience: $R_3 \succ R_2 \succ R_5 \succ R_1 \succ R_4$

Advantages and Disadvantages Ordering in Responsibility and Performance Consciousness: $R_5 \succ R_1 \succ R_3 \succ R_4 \succ R_2$

5.3 Candidates' Formal Selecting Decision

According to the advantages and disadvantages ordering result of candidates in every dimension, the detailed calculating process is as follows:

(1) Confirm the Decision Matrix

Note the No. 1 as 1 and so on, arrange the ordering and scoring table of every candidate (See Table 6):

Table 6. Ordering and Scoring Table of Every Candidate

Competence Dimensions	Ordering and Scoring of Candidate in Every Dimension				
	A_1	A_2	A_3	A_4	A_5
Leadership and Work Force	3	1	4	2	5
Scope of Knowledge and Working Experience	4	2	1	5	3
Responsibility and Performance Consciousness	2	5	3	4	1

Decision Matrix $X = \begin{pmatrix} 3 & 4 & 2 \\ 1 & 2 & 5 \\ 4 & 1 & 3 \\ 2 & 5 & 4 \\ 5 & 3 & 1 \end{pmatrix}$ Normalized Matrix $Y = \begin{pmatrix} 0.5 & 0.25 & 0.75 \\ 1 & 0.75 & 0 \\ 0.25 & 1 & 0.5 \\ 0.75 & 0 & 0.25 \\ 0 & 0.5 & 1 \end{pmatrix}$

(2) Normalized Decision Matrix

Since the ordering and scoring is an extremely-small index, normalize Decision Matrix X as Y according to Formula (7).

(3) Confirm the Ideal Candidate Plan

According to Normalized Decision Matrix Y, come to the ideal point plan set as I^*=(1, 1, 1)

(4) Comprehensive Analysis of Selection Decision

Calculate the corresponding approximation α_i and corresponding nearness T_i between five candidates and the ideal ones according to formula (8) and (9) (See Table 7):

Table 7. Corresponding Approximation and Corresponding Nearness of Every Candidate

	A1	A2	A3	A4	A5
α_i	0.935	1.031	0.901	1.274	1.118
T_i	0.500	0.417	0.417	0.667	0.500

From the selection ordering principles, comes that Candidate A_3 achieved the best result in selection decision and therefore enacted as excellent candidate.

6 Conclusion

This paper by using selection decision model, improving the selection quality to some extent. Besides, in the process of constructing excellent managers competence model, the sample size should be expanded.

References

1. Guo, Y.: Comprehensive Evaluation Theories and Methods. Science Press (2002)
2. Gau, W.L., Buehrer, D.J.: Vague sets. IEEE Trans. on Systems Man Cybernetic, 23 (1993)

The Employment and Functions of Internet in the Teaching of English for Golf Major*

Zhi Li and Hong Zhu

Foreign Languages Teaching Department
Tianjin University of Commerce, Tianjin, China
Aprilfool@eyou.com, agnes_zhu@163.com

Abstract. This paper aims to discuss the employment and functions of Internet in the teaching of English for golf major, which includes the use of campus LAN for the construction of network platform and the use of the Internet for students' autonomous learning. With the full use of the Internet, golf English teaching will achieve its 3s goal, that is, student-centered, subject-centered society-centered.

Keywords: Golf English teaching, construction of network platform, 3s goals.

1 Introduction

With golf recognized officially as an Olympic event in 2016, a growing number of colleges and universities have established golf majors to meet the need of the development of golf industry. Statistics shows that at least 83 institutions of higher learning have established golf major or golf related courses, among which 31 universities and college, 25 technological academies, and 27 trade schools [1].

Due to the swift development of golf majors in domestic universities, the teaching of English course for golf major has become one of the urgent problems to be solved by English teachers. Yet, researches done concerning the teaching of English for golf major are very limited among which is Dai Jing probe into the practicality of the teaching of English for golf major[2]. A lot left to be done in the field of teaching of English for golf major.

2 Special Features of English for Golf Major

Golf English can be defined as a branch of ESP (English for Specific Purposes), which is used in the context of activities involving the sports of golf. It is the language being or to be used by those who are or will be engaged in the field of golf. In order to master golf English, one must have a good foundation of English language skills, a thorough knowledge of the sports of golf, and solid golf techniques, which are the indispensible factors deciding one's proficiency in golf English.

* This article is supported by Tianjin University of Commerce Education and Teaching Reform Project (504-11JGXMYB49).

J. Luo (Ed.): Affective Computing and Intelligent Interaction, AISC 137, pp. 945–947.
springerlink.com

Among the above mentioned three factors, the last two constitutes the most difficult part in the teaching and learning of English for golf major. The knowledge of the sports of golf covers a wide range of topics, such as golf's present and past, golf rules and etiquettes, golf course, golf equipment, important golf events, major golf organizations, among other things. And the mastery of golf techniques requires a lot of hard training and practicing, in which the imitating of world famous golf players is essential.

In the teaching of English for golf major, Internet is a very valuable and indispensible resource whose function cannot be underestimated. The use of internet in the teaching of English for golf major can be divided into two main parts, which are the use of campus LAN for the construction of network platform and the use of the Internet for students' autonomous learning.

3 Use of Campus LAN

The use of campus LAN (local area network) is for the purpose of the construction of learning platform, which is of vital importance for the teaching and learning of English for golf major. This platform is made up of three parts which are students' functional section, teachers' functional section, and resource management section. Students' section has such components as personal data, course introduction, shared resources, on-line test, on-line communication, and BBS. Teachers' functional section is made up of results enquiry, on-line communication, and on-line scoring. Resource management mainly includes system maintenance, data process, teaching arrangement, results statistics, and on-line statistics of class hours.

Main tasks of network learning platform are to ensure students' autonomous learning under teachers' supervision. Teachers upload audio, video and other type of resources for students to finish on their own in a certain period of time. On one hand, students can practice and strengthen their learning contents and methods that they obtain from classroom. On the other hand, through a great deal of language practice such as listening, speaking, reading, writing, translating, they can improve their language application skills comprehensively. On-line communication of network platform can realize real-time interaction not only between teachers and students but also among students themselves. Besides that, BBS provide a platform for non-real-time online discussions.

4 Use of the Internet

The use of the Internet is for the purpose of autonomous learning, which serves as extension and supplement of classroom learning and playing an important role in enhancing and strengthening classroom learning. The autonomous learning based on internet mainly includes paying attention in real time to annual major international golf competitions, such as Men's majors--the Masters in April, the U.S. Open in June, the British Open in July, and the PGA Championship in August. During the time of the events, a lot of news reports can be found in the official networks of PGA, the Masters or other major news networks, such as CNN. And a wide range of audio and visual real-time materials tracking those events are available in the Internet. In surfing the

Internet and tracking the real-time events, students can familiarize themselves with all kinds of golf terms, which poses as a barrier for the majority of students in their way to mastering golf English. What's more, students can have the chance of learning golf techniques from the world first rank golfers.

5 Conclusions

In a word, Internet provides us a means to keep pace with the world and is a very important tool in the teaching of English for golf major. With the full use of the Internet, golf English teaching will achieve its 3s goal, that is, student-centered, subject-centered society-centered.

References

1. Dai, J., Dai, L.Z.: Vocabulary Teaching Exploration on Golf Professional English. Journal of Hunan International Economics University 11, 150–151 (2011) (in Chinese)
2. Dai, J.: On the Practicality of the Teaching of English for Golf Major. Journal of Suzhou Education Institute, 119–120 (2007) (in Chinese)

The Construction of Practical Teaching Mode of English for Golf Major*

Zhi Li and Hong Zhu

Foreign Languages Teaching Department, Tianjin University of Commerce, Tianjin, China
Aprilfool@eyou.com, agnes_zhu@163.com

Abstract. With the rapid development of golf major in China's colleges and universities, a practical teaching mode of golf English suitable for colleges and universities is what calls for further researches. This essay aims at establishing a practical teaching mode of English for golf major based on "classroom-internet-golf course". The teaching mode consists of three dimensions—teaching in the classroom, autonomous learning on the internet, application on golf courses, and the ultimate goal of which is to achieve the circulation and complementation of teaching, learning and utilizing.

Keywords: English for golf major, teaching mode, practical teaching.

1 Introduction

With rapid development of economic globalization and the popularization of golf in the world, golf industry has become the most promising and burgeoning industry in the 21st century. Golf industry is rapidly increasing its proportion in gross national economic product and golf's role in business communication can no longer be neglected. With golf recognized officially as an Olympic event in 2016, more and more people are attracted to golf courses. A growing number of colleges establish their golf majors to meet the need of the development of golf industry.

Due to swift development of golf majors in universities in China, a vast variety and a large amount of researches have been done in the field related to Golf. Researches done on the teaching Golf English are relatively few compared with other aspects related with golf. Dai Jing Studies the practicality of Golf English teaching[1] and the teaching of vocabulary for golf major[2], Tan Yan focuses on the problems and counter measures in the teaching of Golf English[3]. A practical teaching mode of golf English suitable for colleges and universities is a topic which calls for further researches.

This essay aims at establishing dynamic practical teaching mode of English for golf major based on "classroom-internet-golf course" (as is shown in Fig. 1) so as to achieve the purpose of the dynamic circulation and mutual complementation of teaching, learning and utilizing.

* This article is supported by Tianjin University of Commerce Education and Teaching Reform Project (504-11JGXMYB49).

J. Luo (Ed.): Affective Computing and Intelligent Interaction, AISC 137, pp. 949–951.
springerlink.com

2 Practical Teaching Mode of English Course for Golf Major

This essay will approach golf English practical teaching from three dimensions. The first dimension is teaching in classroom, which include basic language skills as well as English knowledge and abilities related to golf major. The former comprise students'

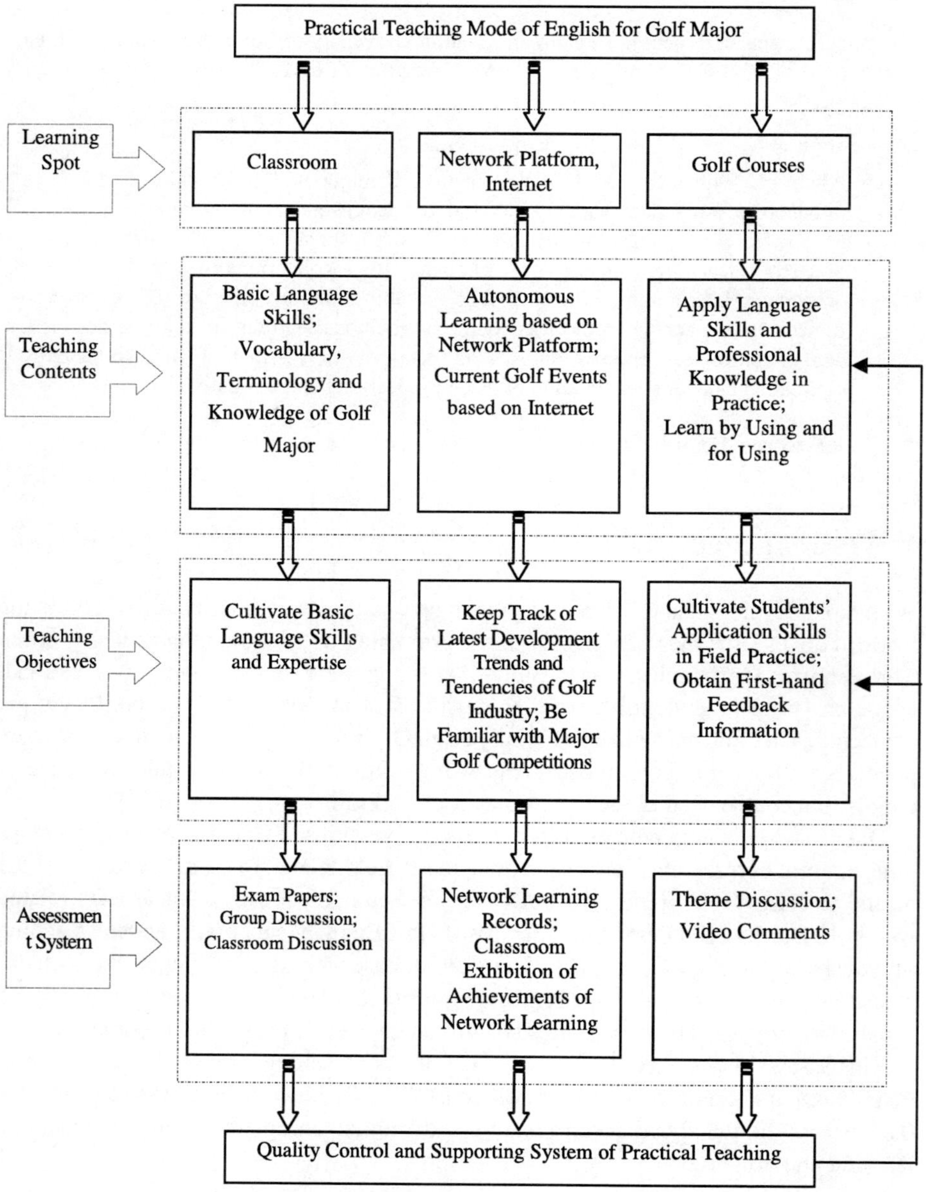

Fig. 1. Practical Teaching Mode of English for Golf Major

listening, speaking, reading, writing and translating skills in English. The latter involve golf history and development, famous golf competitions, well-known golfers, golf etiquettes and rules, decisions on the rules of golf, golf course, golf course design, golf course maintenance, golf techniques, golf facility management and business golf, with a large quantity of terminology and vocabulary for golf major.

The second dimension is autonomous learning based on network, serving as extension and supplement of classroom learning and playing an important role in enhancing and strengthening classroom learning. Autonomous learning based on network can be divided into two sections. The first section refers to autonomous learning under supervision of teachers. Teachers and students can communicate easily through network learning platform which is constructed based on campus LAN (local area network). The second section refers to network autonomous learning based on internet, mainly including paying attention in real time to annual major international golf competitions, such as the Masters in April, the U.S. Open in June, the British Open in July, and the PGA Championship in August.

The third dimension is the application based on golf course. Due to the unique features of golf major, students have plenty of opportunities of field practice in golf course. Sometimes, they can also have chances to assist the organization of golf competitions in the courses. Teachers can give students specific tasks in advance to enable them to practice with missions and complete summary afterwards. Students can also raise their own questions during the practice and bring back to classroom to discuss and solve them collectively. Therefore, students can check the practicability of what they've learnt in classroom, at the same time, report to teachers what they need to obtain and master most in practice so that these contents can then be supplemented in teaching section.

3 Conclusion

This research is only a tentative effort to meet the need of cultivating practical golf talents with a good command of English. By constructing and implementing practical teaching mode of English for golf major, we've succeeded in making teaching, learning and utilizing complementary to one another. We've proved that English teaching activities based on "classroom-internet-golf course" are a feasible ways for the reform of English teaching for golf major and for the improvement of practicability of English teaching.

References

1. Dai, J.: On the Practicality of the Teaching of English for Golf Major. Journal of Suzhou Education Institute, 119–120 (2007) (in Chinese)
2. Dai, J., Dai, L.Z.: Vocabulary Teaching Exploration on Golf Professional English. Journal of Hunan International Economics University 11, 150–151 (2011) (in Chinese)
3. Tan, Y., Dai, Z.M.: A Probe into the Teaching of Golf English for Higher Vocational Colleges. Foreign Languages Teaching 6, 50–51 (2010) (in Chinese)

Design and Implementation of RFID-Based Assistant Platform of Special Equipments Management

Guoli Ji[1,*], Xuefei Peng[1], Wei Lin[1], Meishuang Tang[2], and Xinjian Qi[1]

[1] Department of Automation, Xiamen University, Xiamen, Fujian 361005, China
[2] Modern Educational Technical and Practical Training Center
Xiamen University, Xiamen, Fujian 361005, China
glji@xmu.edu.cn, pxfheb@126.com, linwei_xm@vip.163.com, tangms@xmu.edu.cn, xinjianqi1988@163.com

Abstract. In this paper, a solution of wireless internet interactive platform based on Radio Frequency Identification (RFID) is proposed for the inspection and supervision of electromechanical and pressure bearing special equipments. Using RFID technology and General Packet Radio Service (GPRS) wireless telecommunication technology as well as online application service of Web Service, we establish a multiparty interactive platform based on Dot Net (.net) framework. It can not only achieve some practical functions such as quick equipments identification, basic data maintenance, equipments circulation tracing and exceptions management, but also provide some expansion interfaces to meet the demands of different service demands. Furthermore, it is designed for different kinds of users to supervise special equipments jointly. Compared with current manual way, our solution cuts expenditure cost effectively, improves inspection efficiency and quality, and reduces accidents greatly. It is also a good reference for management of other kinds of equipments.

Keywords: RFID, special equipments, management.

1 Introduction

The management and supervision of special equipments (lifts, pressure vessels, hoisting machineries, etc) is a work which many countries including China have paid great attention to. With the social and economic development in recent years, the special equipments ownership in China has been increased rapidly. However, there are many defects in current supervision of special equipments, such as low efficiency, poor communication channels, inadequate control of illegal equipments, shortage of monitoring, lack of effective emergency plans, low public participation, etc, which are difficult to be resolved in the existing model. Therefore, more efficient managements and technologies are required to manage and supervise special equipments.

The inspection and supervision of special equipments are phased and periodic, rather than real-time, including checking whether corresponding operators are equipped, checking whether the devices are registered or due, the inspection of

* Corresponding author.

J. Luo (Ed.): Affective Computing and Intelligent Interaction, AISC 137, pp. 953–958.
springerlink.com

Welding and appearance, etc. Here, RFID is introduced. Radio Frequency Identification (RFID) is an automatic identification solution that streamlines identification and data acquisition [1], which has been widely applied to a variety of fields [2]. We can find some applications and research of RFID in the management of equipments. Ko,C.H [3] integrates RFID, database, and scheduling theory technology with a web-based system to improve facility and equipment maintenance efficiency. Chae, S [4] applies RFID to the prevention of collision accidents with heavy equipments such as hydraulic excavators and cranes. Atsuko, M [5] tries to construct a safety management system of medical equipments by using the RFID tag and PDA. In these studies and applications, RFID technology has played an important technical support role in solving the difficulties of device management, inspection and repair. In this study, we apply RFID to the inspection and supervision of different kinds of special equipments. Combining the equipments with dynamic information and with the PDA and GPRS wireless communications access, the original problem of poor communication is solved effectively. Moreover, we integrate individual equipments into the same system for more effective management, and eventually build a multiple interactive Web platform using .Net framework development technology and Web Service online application services.

2 Management Program Design

Special equipments are classified into three categories: A. Devices whose application place are fixed, such as boilers, elevators, etc; B. Devices whose application place are not fixed and personal users are purchasers, such as cranes; C. Devices whose application place is not fixed and personal users are not purchasers, such as cylinders. According to these classifications, we design a special equipments management implementation model as shown in Fig. 1, by fully integrating RFID automatic identification technology and PDAs.

Users are available to interact with the platform by computer browsers or handheld devices. After logging in the platform by a computer browser, the user is assigned a role (including general public, operator, administrator, etc) by the system to carry out the corresponding permissions. We design authority with the reference of role-based access control RBAC (Role-Based Access Control) model [6] to manage large quantity of users and corresponding authority. It endows one or more users with a role defined under different authorities, thus makes the configuration more flexible and comprehensive.

The operator is required to login the platform with a PDA when inspecting the equipments. In the unit, RFID tags are pasted on the surface of the devices where tags are safe and can be seen easily. The basic information such as the electronic tag ID, equipment code and type, and inspect information such as the inspect type, inspectors and inspect date, are all stored in the chips. During the operating process, the inspectors scan tags to identify the device first, using handheld devices equipped with RFID reader module(issuing tag is needed when first time). Then the inspectors fill in the electronic inspection records and modify the data with pre-installed modules in handheld devices. Finally, they rewrite the tag and exchange the data with the platform using the handheld device's GPRS wireless network module to keep a real-time update with the server.

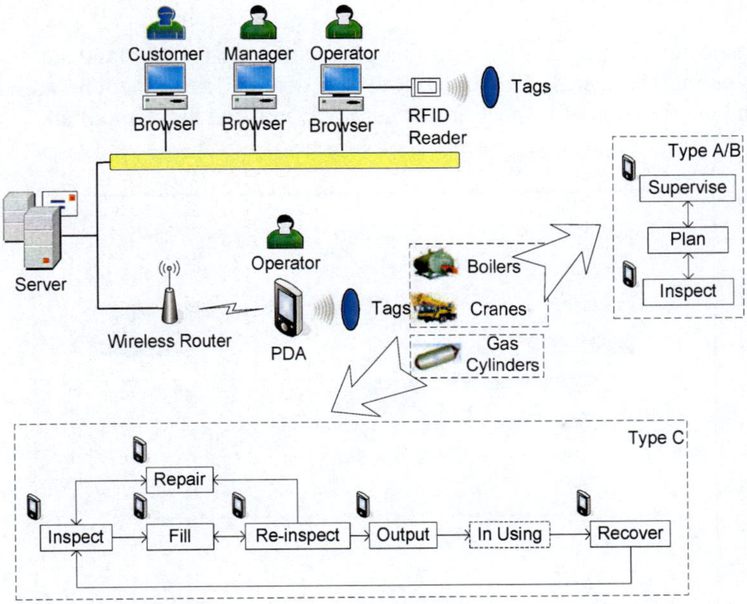

Fig. 1. Special equipments management model

Fig.1 shows the management model of class A (B) and class C. The locations of special equipments of class A (B) are fixed, and the inspectors inspect the devices according to the given plans. As special equipments of class C are used circularly, every link of the cycle needs logs. So we define the link of the process as inspect, fill, re-inspect, output, recover, repair, etc. If a cylinder is damaged after being filled or has problems when inspected, it needs repair and gets into the cycle from the link of inspect. These processes cycle according to the direction shown in Fig.1. Here the fill and re-inspect, inspect and repair are reversible. When the next cycle starts, the user is usually replaced. In both modes, the operators need to update the real-time data of the tags and upload it to the Web Service platform in every link.

Compared with current mode, the new platform has many great advantages. (1) Under B/S mode, the user can view the device information and inspect plans anytime, anywhere; (2) Devices are identified automatically and the data is uploaded in real time using the wireless network to improve the efficiency; (3) The data of the tags is compared with the server automatically to improve the accuracy; (4) The equipment requires a tag before getting registered, thus the user can judge whether it is illegal according to that; (5) Automatic inspect plans for managers are maked to distribute before equipments are due, and an initial mechanism of monitoring is established; (6)

In the public community, the public can register, login using a browser and make comments and suggestions; (7) The user can track the whole flow process of cylinders.

3 Function Design

We have designed three functions in this platform: 1. Inspection and supervision; 2. Limited public service; 3. Public service. The detailed sub-functions are shown in Fig.2, and the function of inspection and supervision is the most important.

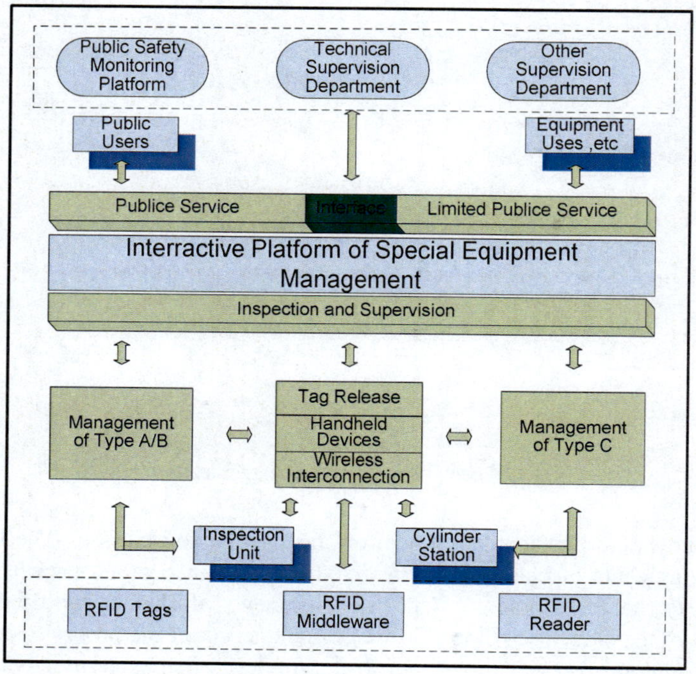

Fig. 2. Function design

3.1 Inspection and Supervision

The application of RFID technology (tags issuing module, handheld devices, etc.) plays a central role in the platform, which the implementation of various types of special equipments management is based on. RFID tag technology, middleware and reader technology provide the technical support for the realization of these functions.

In Management of Class A (B), to meet the needs of routine work, the main design contains the management of equipment data, equipment inspection, equipment supervision and related unit information. The main user is from inspecting unit. In Management of Class C (Class cylinder), according to the characteristics of Class C equipment's liquidity, sharing and gas filling, the main design contains distribution, filling management, inspection management and inventory management. The main users are from the filling station and inspection station.

3.2 Limited Public Service

The main users are from units which are related with special equipments directly. Equipment users could query the information and inspection reports of their equipments, or apply to register new devices. Production units could register the sold equipments, which will ensure the accuracy and data from the beginning. Maintenance units can consult the data and inspection records of the equipments which they are responsible for.

3.3 Public Service

The public service is designed for public users. The users can browse the given information, such as parts of the information of special equipments, the information of supervision and maintenance. Also, they can register an account to join in the discussion of topics related with special equipments.

In addition to these three functions, our platform provides interfaces for the platform of public safety regulation, supervision departments and other platforms. Therefore, the users could filter and make statistics of equipments with low safety factor.

4 Function Realization

The technology framework of this platform adopts multi-layer structure. The users can make operations such as scheduling, statistic, searching, tracking, management and inspection in the display layer. After business logic layer, the data interacts with the database, through the interfaces provided by Web Service which presents SOAP request and receives SOAP response.

Presentation layer provides users with the login and operation pages, including the interface of the browser and handheld device. Front page of the browser is developed with DIV + CSS + JS, and background service with ASP.NET (C#) technologies. Client of PDA is coded in C language.

Business Logic Layer contains components such as data entry validation, verification of the legitimacy of the user, add, modify, and query operations, to ensure the normal operation of platform. These components are integrated in the platform between the various modules.

We define a series of API of Web Service and the corresponding XML format in the web Service layer. The platform packages fields into standard XML format, transforms data with the external API exposed from Web Service, analyzes XML according to definitions, and finally presents to users in presentation layer or operates the database. Also, we use Message Digest Algorithm MD5 to encrypt the main information, generates the fingerprint of each file to verify whether it is tampered maliciously.

Currently the database model and the overall framework of the platform have been designed completely. Cylinder tracking management system and PDA functions have also been independently developed, and we are testing them with inspection unit. Further more, EMC management system and Comment Network for elevator have been developed and tested in Xiamen, China.

5 Summary and Forecast

We proposed an RFID-based platform management of special equipments, and achieved a good result in the practical application. Combining RFID and GPRS wireless communication technology with the B/S development approach, our platform enables the users to fully collect and control the information of events during the inspection and supervision of special equipments, which includes the cylinder's circulation links of filling, out, recycle, inspect, etc. The resource can be shared with other regulatory and inspection department and the public regulation social service. These functions enhanced the data consistency and timeliness, improved the efficiency of the corresponding departments, and ensured the maximum normal operation of equipments and the safety of people and property.

By further integrating the sensor technology, the platform will allow the real-time monitoring. Efforts are also underway to build the Internet of Things for supervision and management of special equipments.

Acknowledgments. This project was funded by funds from the Fundamental Research Funds for the Central Universities of Xiamen University (Nos. 2011121047, 201112G018 and CXB2011035), the Key Research Project of Fujian Province of China (No. 2009H0044) and Xiamen University National 211 3rd Period Project of China (No. 0630-E72000).

References

1. Wang, L.C., Lin, Y.C., Lin, P.H.: Dynamic mobile RFID-based supply chain control and management system in construction. In: Advanced Engineering Informatics, vol. 21, pp. 377–390. Elsevier Sci. Ltd., Oxford (2007)
2. Yin, S.Y.L., Tserng, H.P., Wang, J.C., Tsai, S.C.: Developing a precast production management system using RFID technology. In: Automation in Construction, pp. 677–691. Elsevier Science, Amsterdam (2009)
3. Ko, C.H.: RFID-based building maintenance system. In: Automation in Construction, vol. 18, pp. 275–284. Elsevier Science, Amsterdam (2009)
4. Chae, S., Yoshida, T.: Application of RFID technology to prevention of collision accident with heavy equipment. In: Automation in Construction, vol. 19, pp. 368–374. Elsevier Science, Amsterdam (2010)
5. Atsuko, M., Akiko, S., Terutaka, M., Shoko, T., Kenichi, F., Toshinari, N., Masaki, M., Hiroyuki, H., Hiroshi, I.: Management of Medical Equipment for Prevention of Accidents-13.56MHz RFID System. In: World Congress on Medical Physics and Biomedical Engineering, vol. 14, pp. 474–476. Springer, Heidelberg (2006)
6. Sandhu, R.S., Coyne, E.J.: Role-Based Access Control Models. J. IEEE Computer 29(2), 38–47 (1996)

Game Analysis on Economy Sustainable Development System of New and Traditional Energy

Ning Ai and Herui Cui

Department of Economic and Management,
North China Electric Power University Baoding, 071003 China
chris851ai@yahoo.com.cn, cuiherui1967@126.com

Abstract. At present, China is transferring the resources of traditional energy into the resources of new energy gradually. How many resources the center should transfer is a core issue in a economic period. This paper introduces the sustainable degree concept of energy resources use and finds the trategic selection by built the game model of the central and traditional, the new energy. Then,by analysing the strategic selection of center and the response of old and new energy to the central selection, we can get the best selection of center that accord with countrywide advantage.

Keywords: Traditional energy, new energy, sustainable degree, game.

1 Introduction

With the advancement of human technology, energy has become the material basis of economic development increasingly. China is a developing country, still in the traditional energy stage basically, depends on traditional energy in a short time mainly[1].But the environment pollution and destruction of ecological resources by the traditional energy is very serious, we must strive to develop new energy industries and transfer the resources of traditional energy into new energy[2].However, because of the restriction of the current level of technology, and the needs of economic development, it is not reality to transfer the resources of traditional energy into the new energy simply and largely in a short time[3].So that, transferring the resources of traditional energy into new energy should get a sustainable way[4].

At present, China is transferring the resources of traditional energy into the new energy gradually[5]. The National Energy Commission(NEC) launched a plan of new energy development, adjust configuration of energy fast, develop clean energy vigorously, advance proportion of reproducible energy vigorously and wash out behindhand productivity of traditional energy[6].As the decision-maker, how many resources the center should transfer is a ineluctability problem. There is nothingness in the research of how the center to transfer the resources of traditional energy into new energy[7,8]. Invest in new energy, not only embodies the spirit of global citizen participation, but also obtains good returns[9].Sinoma, S, who use game theory analysis of the tourism sustainable development; Yujie Tao, who use game theory analysis of river water resources sustainable development. This paper will use game.

J. Luo (Ed.): Affective Computing and Intelligent Interaction, AISC 137, pp. 959–967.
springerlink.com

By analysing the strategic selection of center and the response of old and new energy to the central selection, we can get the best selection of center that accord with countrywide advantage.

This paper builds on the next two assumptions: First, both traditional energy and new energy want to get more resources in their development. Second, as the decision-maker, the center must make the decision that accord with countrywide advantage.

For convenience of discussion, this paper simply carves up the national economic system into traditional and new economic subsystem. Then, introduces the sustainable degree concept and builds the strategic selection of the game model. Then, by analysing the strategic selection of center and the response of old and new energy to the central selection, we can get the best selection of center that accord with countrywide advantage.

2 The Sustainable Degree of Energy Resources

2.1 Basic Concept

In order to measure the sustainable degree of energy resources use, we introduce a concept—the sustainable degree of energy resources use(sustainable degree) that can reflect the status of further using of resources. The sign "h" denotes it. In the economic system of energy development, the sign "M" denotes the amount of the most potential resources of this economic system in a economic period. If exhaust all the resources "M" for pursuing the largest energy development, this economic system doesn't achieve the best economic level in the view of the using of resources. If the amount of resources into the economic system is "e", $r = 1 - \frac{e}{M}$ denotes remanent coefficient of potential resources in the economic period, $0 \le r \le 1$ .This paper define the sustainable degree "h" is a increasingly function of "r". Larger the remanent coefficient of resources is, higher the sustainable degree is. On the other side, economic development needs for decision. The decision-maker has a subjective attitude of sustainable degree when he is making decision. The different decision-maker have different attitude for the same economic status. The different decision-maker have different attitude for the same status of expending of resources. We use sign "λ" to denote the gene of attitude, $0 \le \lambda \le 1$.

2.2 Mathematical Definition

Define $h = h(\lambda, r)$ is sustainable degree of this economic system and suppose $h(\lambda, r)$ suffice the next three condition:

$$\frac{dh}{dr} \ge 0, \frac{dh}{d\lambda} \le 0 \tag{1}$$

$$\frac{d^2h}{dr^2} \le 0, \frac{d^2h}{d\lambda^2} \ge 0, \frac{d^2h}{drd\lambda} \ge 0 \tag{2}$$

$$h = h(\lambda, 0) = 0, h(\lambda, 1) = 1, h = h(0, r) = 1 \tag{3}$$

From (1),(3),we can know $0 \le h = h(\lambda, r) \le 1$.

2.3 Economic Analysis

The economic meanings of these assumption are the next:

In(1), $\frac{dh}{dr} \ge 0$ means larger the remanent coefficient of resources is, higher the sustainable degree is. $\frac{dh}{d\lambda} \le 0$ means larger the gene of attitude is, lower the sustainable degree is.

In(2), $\frac{d^2h}{dr^2} \le 0$ accord with the economic general assumption. It means the contribution from the remanent coefficient of resources to the sustainable degree is diminishing marginal. In order to explain the second condition, $\frac{dh}{d\lambda}$ is negative, it denotes the amount of diminishing marginal from the sustainable degree to the gene of attitude. $\frac{d^2h}{d\lambda^2} \ge 0$ means $\frac{dh}{d\lambda}$ is a increasingly function of λ.That is to say, in the face of the same consumption of resources, the amount of diminishing marginal will reduce when the decision-maker whose gene of attitude is higher increases the gene. The third condition $\frac{d^2h}{drd\lambda} \ge 0$ means $\frac{dh}{dr}$ is a increasingly function of λ.That is to say, the contribution from r to the sustainable degree is a increasingly function of λ.It means the more importance the decision-maker attach to the sustainable degree, the more importance the decision-maker attach to the marginal contribution of the remanent coefficient of resources.

In(3), $h = h(\lambda, 0) = 0$ means the sustainable degree is 0 when the resources exhausted. $h = h(\lambda, 1) = 1$ means the sustainable degree is 1 when there is no use of resources. $h = h(0, r) = 1$ means the decision-maker who don't attach any importance to the sustainable degree think that the sustainable degree is 1 like the early days of the economic period in spite of consumption of resources.

From the explanation, the three condition of $h = h(\lambda, r)$ is reasonable in economic view. This paper use this function $h = h(\lambda, r)$ to measure the sustainable degree.

3 The Game Model of Center, Traditional and New Energy Strategic Selection

3.1 Build the Model

Suppose that the center make a decision that the amount of transfer in the resources of traditional energy is E. For simplicity, suppose that this part of resources are all transfer into the new energy development in order to eliminate the contradiction of lack of resources in new energy. Traditional and new energy for its own economic development, decided to invest their own resources for a certain amount. The amount of new energy is e_1, the amount of traditional energy is e_2.So the total amount of consumption of traditional energy resources is $e_2 + E$ and the total amount of consumption of new energy resources is e_1.The total amount of new energy resources using is $e_1 + E$ and The total amount of traditional energy resources using is e_2.

Energy industry can not develop without the use of resources, we focus on analysis of resource utilization play a key role in energy development in the economic system. Ignore other factors or fix other factors unchanged, suppose that the development of industry decided by the resource utilization.

Suppose the function of new development economic increasing is f, the total amount of new energy resources using is $e_1 + E$ and its amount of economic growth is $Q_1 = f(e_1 + E)$.

Considerring the traditional energy, transferring the resources from traditional energy to new energy will advance the traditional energy use the resources better. The allocation of traditional energy resources E will advance the traditional energy development of economic growth. So that we can suppose that the traditional energy development is an increasing function of economic growth. After traditional energy re-enterring the amount of resources is e_2, so that the amount of its economic growth is $Q_2 = g(E, e_2)$.

3.2 Basic Assumptions

We can suppose the two function $f(e)$、 $g(E,e)$ like the next:

$$f'(e) \geq 0 \tag{4}$$

$$f''(e) \leq 0 \tag{5}$$

$$\frac{dg}{de} \geq 0, \frac{dg}{dE} \geq 0 \tag{6}$$

$$\frac{d^2 g}{de^2} \leq 0, \frac{d^2 g}{dE^2} \leq 0 \tag{7}$$

The assumption is that the production function of economics in the rate of change on the elements of the conventional assumptions.

3.3 Mathematical Analysis

The discussion above only concerned the aspect of economic growth. Introduction has point out that the economic decision-makers have to consider sustainable development issues. So then consider the sustainable degree.

Development of new energy should take into account their own economic growth Q_1 and the sustainable degree h_1.

$$h_1 = h(\lambda_1, 1 - \frac{e_1}{M_1}) \underline{\underline{\Delta}} h_1 = h(\lambda_1, 1 - q_1 e_1) \tag{8}$$

M is the amount of resources of new energy. $q_1 \underline{\underline{\Delta}} \frac{1}{M_1}$, $q_1 e_1$ is the coefficient of new energy development resource consumption. λ_1 is a gene of attitude of new energy to sustainable degree. In order to balance the two sides, define the new energy resources utilization level of utility function u_1, $u_1 = u_1(Q_1, h_1)$. $u_2 \underline{\underline{\Delta}} u_2(Q_2, h_2)$, the expression of h_2 is:

$$h_2 = h(\lambda_2, 1 - \frac{e_2 + E}{M_2}) \underline{\underline{\Delta}} h_2 = h(\lambda_2, 1 - q_2(e_2 + E)) \tag{9}$$

M_2 is the amount of resources of traditional energy. $q_2 \underline{\underline{\Delta}} \frac{1}{M_2}$, $q_2(e_2 + E)$ is coefficient of traditional energy development resource consumption. λ_2 is a gene of attitude of new energy to sustainable degree. Similarly, the center needs to balance countrywide economic increasing $Q = Q_1 + Q_2$ and the sustainable degree h.

$$h = h(\lambda, 1 - \frac{e_1 + e_2 + E}{M}) \underline{\underline{\Delta}} h = h(\lambda, 1 - q(e_1 + e_2 + E)) \tag{10}$$

$M = M_1 + M_2$ is the amount of resources of country in this economic period. $q \underline{\underline{\Delta}} \frac{1}{M}$, $q(e_1 + e_2 + E)$ is coefficient of countrywide resource consumption. λ is a gene of attitude of center to sustainable degree. Define the development of new energy resource utilization level of utility function u, $u = u(Q, h)$. u_1、u_2 and u should appease the conventional assumptions of utility function:

$$\frac{du}{dQ} \geq 0, \frac{du}{dh} \geq 0 \tag{11}$$

$$\frac{d^2u}{dQ^2} \leq 0, \frac{d^2u}{dh^2} \leq 0 \tag{12}$$

3.4 Economic Analysis

The game model of center and traditional, new energy strategic selection is:

First of all, the center make decision. Transfer the resources E from traditional energy to new energy. Secondly, traditional and new energy decide to dive the resources e_1、 e_2 into themselves. This game model can be described by the game tree like the picture1.The broken line means the central utility depends on the traditional and new energy.

Solving the game is solving the nash equilibrium (E^*, e_1^*, e_2^*) of center and traditional, new energy. So that, analyse the response of traditional and new energy to the central selection.

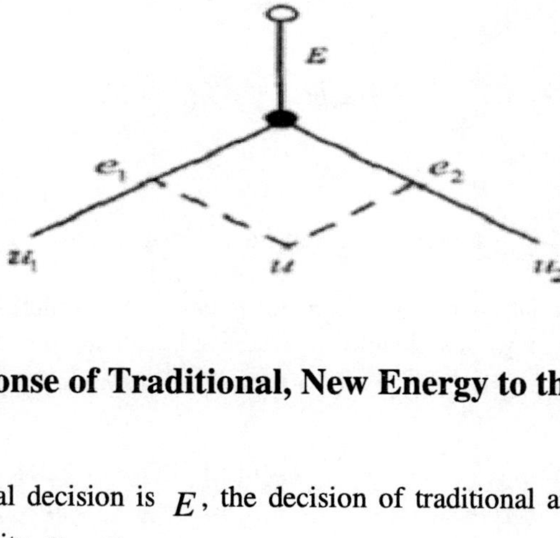

4 The Response of Traditional, New Energy to the Central Selection

Suppose the central decision is E, the decision of traditional and new energy is to maximize their utility u_1、 u_2.

$$\max_{e_1} u_1 = u_1(f(e_1 + E), h(\lambda_1, 1 - q_1 e_1)) \tag{13}$$

Solving $\frac{du_1}{de_1} = 0$ can get the best formula of new energy($h_2^{'}$ means the second die partielle ableitung of h):

$$\frac{du_1}{df} f' - \frac{du_1}{dh} q_1 h_2' = 0 \tag{14}$$

Formula(14)can confirm the best function of new energy $e_1^* = e_1^*(Q_1, E)$.

So that:

$$\max_{e_2} u_2 = u_2(g(E, e_2), h(\lambda_2, 1 - q_2(e_2 + E))) \tag{15}$$

Solving $\frac{du_2}{de_2} = 0$ can get the best formula of traditional energy:

$$\frac{du_2}{dg} g' - \frac{du_2}{dh} q_2 h_2' = 0 \tag{16}$$

Formula(16)can confirm the best function of traditional energy $e_2^* = e_2^*(Q_2, E)$.

5 The Best Selection of the Center That Accord with Countrywide Advantage

5.1 Select Examples

Adopting some idiographic function, analyse the best decision of center.

Firstly, the economic increasing function of traditional and new energy is choosen a linear function. So it is $f(e_1 + E) = A(e_1 + E), g(E, e_2) = B(E)e_2, A \rangle 0, B(E) \rangle 0$. A、B(E) reflect the contribution of economic increasing of the two subsystem. And there is a conventional assumption $B'(E) \geq 0, B''(E) \geq 0$.

Secondly, we can suppose that the center and traditional, new energy have the same gene of attitude of the sustainable degree λ .The sustainable degree function is $h(\lambda, r) = r^\lambda, 0 \leq \lambda, r \leq 1$. $h(\lambda, r)$ can satisfy the condition from(1)to(3)of all about h .All the utility function is

$u_1(Q_1, h_1) = Q_1 h_1, u_2(Q_2, h_2) = Q_2 h_2, u(Q, h) = Qh$.

5.2 Mathematical Analysis

So that, we can get that:

$$(1 - q_1 e_1) - \lambda q_1 (e_1 + E) = 0 \tag{17}$$

$$(1 - q_2(e_2 + E)) - \lambda q_2 e_2 = 0 \tag{18}$$

So that, we can get that:

$$e_1^* = \frac{1}{q_1(1+\lambda)} - \frac{\lambda}{1+\lambda}E \tag{19}$$

$$e_2^* = \frac{1}{q_2(1+\lambda)} - \frac{1}{1+\lambda}E \tag{20}$$

\So that, we can get that:

$$\max_E u = [A(E+e_1) + B(E)e_2^*][1-q(e_1^*+e_2^*+E)]^{\lambda} \tag{21}$$

$e_1^* + e_2^* + E$ is constant has no connection with E.So that, we can get:

$$\max_E v = A(E+e_1^*) + B(E)e_2^* \tag{22}$$

From $\frac{du}{dE} = 0$, we can get that:

$$A(1+\frac{de_1^*}{dE}) + B'(E)e_2^* + B(E)\frac{de_2^*}{dE} = 0 \tag{23}$$

So that, we can get that:

$$A + B'(E)(\frac{1}{q_2} - E) - B(E) = 0 \tag{24}$$

Because of $\frac{1}{q_2} = M_2$, we can get that:

$$A + B'(E)(M_2 - E) - B(E) = 0 \tag{25}$$

This formula can confirm the best decision $E^* = E^*(M_2, A)$.It has property like the next:

$$1 + [B'(E)(M_2 - E) - B'(E)]\frac{dE}{dA} = B'(E)\frac{dE}{dA} \tag{26}$$

$$[2B'(E) - B''(E)(Q_2 - E)]\frac{dE}{dA} = 1 \tag{27}$$

So that, we can know that: $\frac{dE}{dA} \rangle 0$.

6 Conclusions

This paper builds the game model of the central and traditional, the new energy strategic selection. In the game, the three sides should consider economic development and resources sustainable using. The results of analysis indicates that higher the efficiency of the use of resources by new energy is, more the resources diversion from traditional energy to new energy is from the countrywide economic advantage. So that, the new energy development should get a way that efficiency of the use of resources is high. Otherwise, the diversion is disadvantage for countrywide economic development.

Acknowledgments. Thanks for my tutor who gives me the most valuable helps, thanks for my family that gives me important supports and thanks for my classmates' valuable helps.

References

1. Wu, G., Tian, L.: A Strategic Analysis of the Sustainable Energy Developmentin Jiangsu Province. Journal of Jiangsu University of Science and Technology (Social Sciences) (4), 32–35 (2000)
2. Yang, Y.: The Way to the Low-carbon:the Development of New Energy is Urgent. China Ocean Shipping Monthly (2), 24–26 (2010)
3. Fu, Y., Tian, L.: Statistical Verifying Estimation and Application of Logistic Model in the Forecast of Energy Consuming in Jiangsu Province. Journal of Jiangsu University of Science and Technology (Social Sciences) (1), 17–19 (2001)
4. Zhang, G.: Strategic Thinking of Chinese New Energy Industrial Development. Research on Economics and Management (11), 5–9 (2009)
5. Xie, S.: Economic Game Theory. Fudan University Press, Shanghai (2002)
6. Zhang, W.: Game Theory and Information Economics. Shanghai Joint Publishing, Shanghai People's Publishing House, Shanghai (1996)
7. Wang, R., Shang, Q.: Chinese Four Main New Energy Plan to Change. Outlook Weekly (12), 15 (2009)
8. Zhu, S.: Three Questions of New Energy Development. Agricultural Engineering Technology (New Energy Industry) (2), 5 (2010)
9. Liu, Z.: New Energy: 2010 Investment Highlights. Financial Digest (3), 33 (2010)

Sports Right of the College Student and Its Assurance

Zhai Huanan and Zhang Junfeng

Wuhan Institute of Physical Education, Physical Education Department,
Wuhan Hubei 430079

Abstract. On the basis of the theory of constructing college student's sports rights system, take some universities in Wuhan as an example, through the investigation of the present situation of college student's sports right, put forward the feasible advice of maintaining college student's sports right, and provide theoretical reference for healthy and sustainable development of college student's sports.

Keywords: college student, sports right, assurance.

1 Introduction

"The national fitness regulations" promulgated by the State Council in August 2009 highlights that citizen's right of taking part in sports activities listed in all kinds of laws in our country is specifically confirmed as the actual rights. The article 4 in this regulations expressly says, "citizens shall have the right to participate in the national fitness activities. The Local government at various levels shall protect citizens' right to participate in the national fitness activities." Research shows sports right is a kind of human right set by the national laws to ensure citizens' health and improve their body. College student is the most vigorous and most dynamic social group, Cultivation of their sports right consciousness and its actual assurance are the manifestation of ensuring all rights of our citizens and the basic guarantee of developing all kinds of national fitness activities. Cultivation of college students' sports rights and its assurance can greatly maintain students' sports rights and fundamentally promote the sports rights consciousness and the assurance level of the whole society.

Based on the college students' sports in our country and the expert consultation, this paper preliminarily establishes the basic model of college student's sports right framework, through the social investigation of part of Wuhan college students, studies the situation of ensuring college student's sports rights, and make some thinkings of maintaining current college student's sports right. The research uses Delphi method, three-round consultations and interviews of 30 related experts are made, and the expert advice about sports right system are obtained. The special questionnaires about college student's sports right are designed and delivered to college students in 10 colleges in Wuhan such as Central China Normal University and Wuhan University. 650 questionnaires are delivered, 620 are recovered, and the recovery rate is 95.4%. 24 invalid questionnaires are eliminated, 596 valid questionnaires are recovered, and the valid recovery rate is 96.1%. At the same time the reliability and validity inspection are done.

J. Luo (Ed.): Affective Computing and Intelligent Interaction, AISC 137, pp. 969–976.
springerlink.com

2 Establishment of College Student's Sports Right System

"The Constitution of People's Republic of China" sets," the state develops sports industry and mass sports activities to improve people's health."Sports Law of People's Republic of China" in 1995 makes specific regulations about citizens' sports right. But, whether the constitution or the law, no definite statement is made about what sports right citizens have, so far, sports right still belongs to the category of presumption rights, its connotation needs to be clarified and detailed. It is unknown what content college student's sports rights include theoretically. Therefore, it is necessary to establish college student's sports right system. In the actual research, through questionnaire and consultation of 30 experts, centralization and coordination degree of the experts' opinions are analyzed by mathematical method. Through several round communication with the relevant experts, each index gradually tends to agree. So it can be said that maintaining college student's sports right has legal meaning and actual sports meaning, and should be paid attention to by high schools and the whole society. The establishment of college student's sports right system should follow scientific, integral, objective and guiding principle. The basic framework (chart1) of current student's sports right system is obtained after three-round investigation. Each kind of sports right also includes several observing points (or second level index), it should also be "better rough than precise" (limited by the length of the paper, this paper doesn't include second level index).

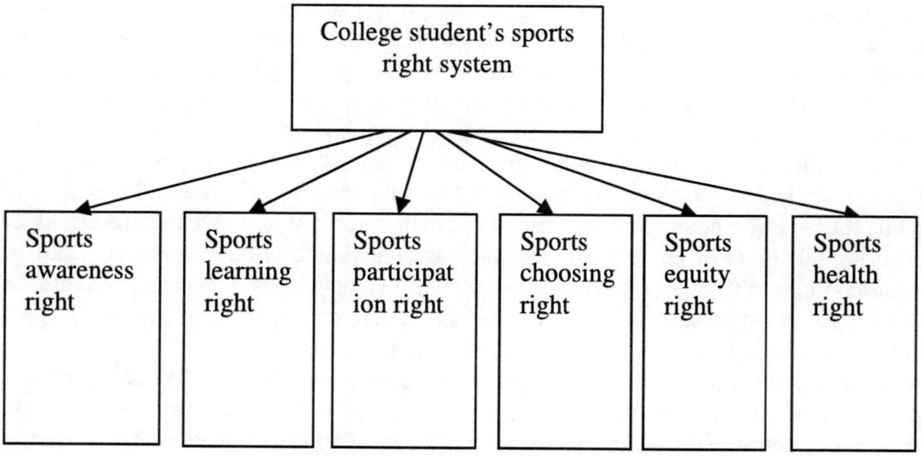

Chart 1. Sports right content system

It can be concluded that college student's sports rights include six basic aspects, that is, sports awareness right, sports learning right, sports participation right, sports choosing right, sports equity right and sports health right. The sports awareness right means college students have the right to know the sports information and improve participation through all kinds of platform and media. The sports awareness right is the basis of all sports rights, it directly affects the formation of college student's sports concept and awareness. The sports learning right is the right to obtain sports knowledge and skill through practice at college. The sports participation right ensures college students to take part in sports activities, college students have the right to take part in various sports training lessons and sports competitions according to sports teaching and competition plan, they also have the right to use various sports facilities. The sports choosing right reflects college students can freely choose sports activities in and out of class. The sports equity right ensures students have the equal chance in sports learning and sports activities. Ensuring college student's sports equity right actually ensures college student's legal human right, it's a more comprehensive social right. The sports health right means students can obtain and maintain physical and mental health through sports activities, and at the same time school can provide healthy and safe service for college students

3 Situation Analysis of College Student's Sports Right Assurance

So, whether college student's sports right is ensured? The self assessment of college students generally reflects their subjective opinion on the satisfaction of sports right requirement and its assurance level. In order to know the situation of college student's sports right assurance, the research picks students in 10 colleges in Wuhan as the investigation objects, including 5 key colleges and 5 ordinary colleges. It generally covers science, agriculture, medicine, teaching and sports and such majors. On the whole, the investigation result generally reflects the basic situation of current college student's sports right assurance.

The investigation result shows as the following table 1.

Table 1. The investigation result of the situation of college student's sports right assurance n=596

Assurance situation / index	ensured		Basically ensured		Not ensured	
	N	%	N	%	N	%
Sports awareness right	132	22.2	120	20.1	344	57.7
Sports learning right	120	20.1	150	25.2	326	54.7
Sports participation right	144	24.2	132	22.1	320	53.7
Sports choosing right	114	19.1	126	21.2	356	59.7
Sports equity right	108	18.1	138	23.2	344	57.7
Sports health right	126	21.1	108	18.1	350	58.7

On the whole, college students are not so satisfied with their assurance of sports right. Their satisfaction with the above six sports rights is merely 18.1%-24.2%, while "basically satisfied" is only about 20%, these two in total do not meet half of the whole number. The number of certain sports right is "not ensured" is 54%-60%. The investigation shows current college students have great desire of sports right. Sports equity and sports health rights are basically ensured and their satisfaction degree is quite high. Sports awareness right, sports participation right, sports learning right and sports choosing right are comparatively weak. Therefore, on the social background of ruling country by law and ruling school by law, college student's sports right assurance should be paid great attention by high school leaders and the entire society.

3.1 College Student's Sports Awareness Right Assurance Situation

It can be seen from the table that 22.2% investigated college students think their "sports awareness right" is ensured, 20.1% of them think it's basically ensured, the other 57.7% think their sports awareness right is not ensured.

Sports awareness right reflects student's requirement and satisfaction degree of school sports information. College students are the subject of high school education activities. As the place of inheriting civilization and spreading knowledge, school has the responsibility and obligations to spread all kinds of sports information to college students, and to propagandize the nation's sports policies and guidelines so that college students can fully make sure their sports rights and obligations. That the investigated college students are not satisfied with sports awareness right shows that colleges and sports teachers don't have enough propaganda of the sports information and reflects school's weakness in the platform of spreading sports information.

3.2 College Student's Sports Learning Right Assurance Situation

It can be seen from the table that 20.1% and 25.25 investigated college students are satisfied and basically satisfied with their sports learning right, while 54.7% are not so satisfied, it means college student's sports learning right assurance needs to be optimized. The further interview shows college students on the whole approve offering sports course in college, but they don't approve the way it's offered. Many college students are not interested in the organized sports lesson but free sports club. Obviously, traditional sports teaching constrains student's freedom, their personality and hobbies. The observation shows besides inefficient sports teaching, college students don't require much about sports lesson, the situation "sports lesson is playing" in many colleges is very severe. Therefore, further reformation of college sports lesson is imperative.

3.3 College Student's Sports Participation Right Assurance Situation

"The Education Law of People's Republic of China" specifies in article 42 that the educatees have the right "to take part in various activities according to the teaching plan and to use teaching facilities, equipment and books data". "The Sports Law of People's Republic of China specifies in article 46, "public sports facilities shall be open to the society, convenient for people to carry out sports activities, implement preferential measures to the students, the elderly and the disabled, and improve the utilization rate of sports facilities." These provide the rights for college students to take part in various kinds of sports activities in and out of school.

At present, the factors restricting college students to participate in sports activities both come from the students themselves and come from objective reasons. 10 year enrollment expansion of colleges satisfy most students' desire for colleges, but because of seriously insufficient school sports facilities, sports conditions are limited. In the 596 investigated college students, 231 are satisfied with school sports fields and facilities, it's 38.8% of the total number. The investigation proves that some colleges are inadequate in the assurance of students' sports participation right. The reason may be the overall size of the school sports facilities is limited, the opening time and the number of people are restrained, some sports fields even charge partial expenses, causing some college students not free to take part in all kinds of sports activities.

3.4 College Student's Sports Choosing Right Assurance Situation

Whether college students should have the option of sports has always been an controversial issue. Chinese traditional education idea has always considered teachers as the main body in education activities, while students as the object, students learn what teachers teach. It is the same with sports education in colleges. Long time since, the organization and management of sports teaching are generally in accordance with the regulated teaching syllabus, standarded teaching model and unified evaluation index of the college students. In the investigated 596 college students, 240 are satisfied or basically satisfied with sports choosing right, accounting for 40.3% of the total number. It means colleges don't pay enough attention to and don't have enough assurance of college student's sports choosing right. "The teaching guidelines of physical education in the national ordinary university" issued by the ministry of education in August 2002 points out, "fully play the main role of students and the leading role of teachers......, under the guidance of teachers, students should be free to choose the teaching content, teachers and teaching time, creating vivid, lively and active learning atmosphere. The above "three freedom" guarantee college students' sports choosing right fundamentally. And, of course, it needs to improve college sports teachers' professional quality and professional level and improve the school's sports activities conditions in order to meet the students' need of choosing sports activities.

3.5 College Student's Sports Equity Right Assurance Situation

College student's sports equity right here refers to college students have equal opportunities in learning at school and participating in sports activities. From the broad sense, ensuring college student's sports equity right means ensuring college student's equal human right, it embodies social justice fundamentally. The realization of sports equity right requires adopting the teaching mode that fit the college students' physical and mental development in the process of teaching, which ensures every college student has equal opportunity in participating in sports. In this premise, different students should be treated differently.

In the investigated college students, most are satisfied with sports equity right, which suggests that the assurance of sports equity right is fine, colleges can carry out relevant state policies, treat students equally, make fully use of sports means, and develop college students' comprehensive quality specifically.

3.6 College Student's Sports Health Right Assurance Situation

The article 5 in the first chapter of "Sports Law of People's Republic of China" says, "the state provides special protection of sports activities for youths, children and young people to improve their physical and mental health." The article 23 in the third chapter says, "schools shall establish a medical examination system for college students. The education, sports and administration departments shall strengthen the monitoring of students' physique." The above two regulations can be concluded to college students' sports health right. Sports health right shall include three aspects: first, college students should obtain physical and mental health through sports activities; Second, the school sports equipments should meet the standards of college students' healthy physical exercise; Third, school and the relevant education departments must set up a perfect monitoring system of college students' physique and regularly investigate the health situation of college students to ensure the students' physical and mental health.

In the investigated 596 college students, 196 think school regularly investigates health, it is only 32.9% of the total number, which means colleges don't have enough survey of students' health. Though our country promulgates "The State Student's Physical Health Standard", but part of the colleges only carry out it in the form. 293 students think the sports facilities and equipments meet the health standard", it's 49.2% of the total number.

On the whole, from the evaluation of sports health right of 596 college students, we can see that at present colleges have done something to ensure college students' sports health right, but still need to improve in paying attention to students' health and health standard of sports facilities.

4 Reflection on College Student's Sports Right Assurance

"The National Fitness Regulations" issued by the state council specifies citizens and college students have the right to engage in sports fitness activities according to the law, it suggests the significance of college students enjoying sports rights from a side. Only when the citizens and college students are fully aware of the existence of sports rights and use them can sports activities truly become the important content of modern people's lifestyle, it can be said that the realization of the citizens' sports health right including college students means our Chinese people are really strong and our overall national power is improved. In view of the practical significance of college students' sports rights assurance, it's necessary to emphasis on the cultivation of college student's sports right awareness and take all kinds of measures to maintain college student's sports right. Based on this, society, school and college students themselves should coordinate to optimize the assurance of college students' sports rights.

4.1 Social Aspect

(1) Perfect the relevant legislation. It has very realistic necessity to establish and perfect sports right policy that fit college student's current situation and mate with the "national fitness regulations" according to the gap between our existing sports rights legislation and college students' demand of rights. And at the same time, the establishment of the state's relevant policy requires changing the traditional concept of emphasizing on the

need of society and country to the concept of promoting the people oriented and concerning people's livelihood. It should be fully realized that in the current society, sports is a kind of important human right cause that meet the needs of individuals, so it should be regarded and used as citizen's basic right.

(2) Enhance the education of law. Because people in our country lack of the knowledge of civil rights, more efforts should be done to promote the knowledge of law so that citizens can fully understand all kinds of rights and obligations, and form the consciousness of law and right that "everyone bears law in mind". Only in this way, can people take legal weapon to protect their legal sports rights through legal means when their rights are violated.

4.2 School Aspect

(1) Pay attention to and maintain college students' sports rights. The school should clarify college students' various sports rights, and organize them to learn and practice to secure and maintain their own legal sports rights. College sports departments should improve the teaching staff and the equipment to truly satisfy the demand of college students' sports rights.

(2) Change the traditional education concept. The school should be fully aware of the importance of maintaining college students' sports rights, and enhance the school sports work with the attitude of being comprehensively responsible to college students' physical health. In the organization of work and school sports activities, we must promote college students to participate from the legal perspective. In addition, the proportion of the sports health fund in the whole education funds should be specified in order to meet the need of school sports competitions and other activities.

(3) Make the sports education as the main channel of maintaining sports rights. Sports education is very significant for college students to learn to sports knowledge and skills, form sporting interest and specialty, enhance the sense of social responsibility and the ability to plan a healthy life. Making the sports education as the main channel of maintaining college students' sports rights can promote the reform of physical education and improve the actual effect of PE curriculum construction.

4.3 College Students Themselves

(1) Correct understanding of the right of participation in sports activities. Contemporary college students should clearly know that practicing and fulfilling the right of participation in sports is not just about individual interests but also a social responsibility and the important embodiment of college students' social adaptation ability. Enjoying the right of participation in sports activities is not only the basic condition of promoting one's own health, but also a preparation of fulfilling their social responsibility. College students themselves should pay attention to and cherish the opportunity of sports class in order to become the real embodiment and practicer of sports rights.

(2) Perform the obligation of participation in sports activities. College students shoulder the responsibility of building a great country. Taking some time to participate in sports in accordance with the state regulations is not only cherishing the due right of improving their physical and mental health, but also fulfilling the due obligations for

the society, making them become the qualified citizens of learning, abiding, using and protecting law. At the same time, college students have the obligations to propagandize the significance of "sports improve health" to the society and lead people to join in the sports fitness activities.

5 Conclusion

(1) Sports right is a kind of human right regulated by the state laws to meet the benefit and effect in obtaining health and strengthening body. Ensuring college student's sports right is the inevitable demand of ensuring the benign social operation under the condition of socialist market economy, it is also an important guarantee of promoting college students' physical and mental health. Establishing college students' sports rights system mainly includes sports awareness right, sports learning right, sports participation right, sports choosing right, sports equity right and sports health right.

(2) The investigation of the college students in 10 colleges in wuhan shows student's sports right awareness and the assurance level are low, the demand for sports activities is urgent. Specifically, in the current college student's sports right system, sports equity right and sports health right are basically ensured and their satisfaction degree is higher. Sports awareness right, sports participation right, sports learning right and sports choosing right are very weak. The research shows that college student's sports right assurance needs to be paid attention to by the school management, and to rely on the corresponding environment.

(3) In today's society, sports is becoming one of the living content and living ways of modern people. It's very urgent to maintain college student's sports right and cultivate its awareness so that sports can trully come into modern people's life. This paper does some research about maintaining college student's sports right and puts forword the thinkings and measures of optimizing college student's sports right from society, school and college students themselves three aspects.

References

1. The State Council of People's Republic of China. National Fitness Regulations. China Legal Publishing House (November 2009)
2. Tang, H.: Broad View of National Fitness Sports. Hubei People's Press (April 2010)
3. Xia, Y.: Origin of Human Right Concept - Historical Philosophy of Right. China University of Political Science and Law (1992)
4. Yu, S.: Sports Right of Citizens. Sports Science (6), 27–30 (1993)
5. Wang, Y.: The Essence Analysis of Sports Right. Zhejiang Sports Science (3), 30–33 (2006)
6. Zhang, H.: Discussion of School Sports Legislation of People's Republic of China. Wuhan Institute of Physical Education Journal (7), 15–18 (2007)

Author Index